CAMBRIDGE STUDIES IN ADVANCED MATHEMATICS 68

LÉVY PROCESSES AND INFINITELY DIVISIBLE DISTRIBUTIONS

Lévy processes are rich mathematical objects and constitute perhaps the most basic class of stochastic processes with a continuous time parameter. This book is intended to provide the reader with comprehensive basic knowledge of Lévy processes, and at the same time serve as an introduction to stochastic processes in general. No specialist knowledge is assumed and proofs are given in detail. Systematic study is made of stable and semi-stable processes, and the author gives special emphasis to the correspondence between Lévy processes and infinitely divisible distributions. All serious students of random phenomena will find that this book has much to offer.

Now in paperback, this corrected edition contains a brand new supplement discussing relevant developments in the area since the book's initial publication.

Ken-iti Sato is Professor Emeritus at Nagoya University, Japan.

CAMBRIDGE STUDIES IN ADVANCED MATHEMATICS

All the titles listed below can be obtained from good booksellers or from Cambridge University Press. For a complete series listing visit: www.cambridge.org/mathematics.

Already published

103 E. Frenkel *Langlands correspondence for loop groups*
104 A. Ambrosetti & A. Malchiodi *Nonlinear analysis and semilinear elliptic problems*
105 T. Tao & V. H. Vu *Additive combinatorics*
106 E. B. Davies *Linear operators and their spectra*
107 K. Kodaira *Complex analysis*
108 T. Ceccherini-Silberstein, F. Scarabotti & F. Tolli *Harmonic analysis on finite groups*
109 H. Geiges *An introduction to contact topology*
110 J. Faraut *Analysis on Lie groups: An introduction*
111 E. Park *Complex topological K-theory*
112 D. W. Stroock *Partial differential equations for probabilists*
113 A. Kirillov, Jr *An introduction to Lie groups and Lie algebras*
114 F. Gesztesy *et al. Soliton equations and their algebro-geometric solutions, II*
115 E. de Faria & W. de Melo *Mathematical tools for one-dimensional dynamics*
116 D. Applebaum *Lévy processes and stochastic calculus (2nd Edition)*
117 T. Szamuely *Galois groups and fundamental groups*
118 G. W. Anderson, A. Guionnet & O. Zeitouni *An introduction to random matrices*
119 C. Perez-Garcia & W. H. Schikhof *Locally convex spaces over non-Archimedean valued fields*
120 P. K. Friz & N. B. Victoir *Multidimensional stochastic processes as rough paths*
121 T. Ceccherini-Silberstein, F. Scarabotti & F. Tolli *Representation theory of the symmetric groups*
122 S. Kalikow & R. McCutcheon *An outline of ergodic theory*
123 G. F. Lawler & V. Limic *Random walk: A modern introduction*
124 K. Lux & H. Pahlings *Representations of groups*
125 K. S. Kedlaya *p-adic differential equations*
126 R. Beals & R. Wong *Special functions*
127 E. de Faria & W. de Melo *Mathematical aspects of quantum field theory*
128 A. Terras *Zeta functions of graphs*
129 D. Goldfeld & J. Hundley *Automorphic representations and L-functions for the general linear group, I*
130 D. Goldfeld & J. Hundley *Automorphic representations and L-functions for the general linear group, II*
131 D. A. Craven *The theory of fusion systems*
132 J. Väänänen *Models and games*
133 G. Malle & D. Testerman *Linear algebraic groups and finite groups of Lie type*
134 P. Li *Geometric analysis*
135 F. Maggi *Sets of finite perimeter and geometric variational problems*
136 M. Brodmann & R. Y. Sharp *Local cohomology (2nd Edition)*
137 C. Muscalu & W. Schlag *Classical and multilinear harmonic analysis, I*
138 C. Muscalu & W. Schlag *Classical and multilinear harmonic analysis, II*
139 B. Helffer *Spectral theory and its applications*
140 R. Pemantle & M. C. Wilson *Analytic combinatorics in several variables*
141 B. Branner & N. Fagella *Quasiconformal surgery in holomorphic dynamics*
142 R. M. Dudley *Uniform central limit theorems (2nd Edition)*

Lévy Processes and Infinitely Divisible Distributions

Revised Edition

KEN-ITI SATO
Nagoya University, Japan

CAMBRIDGE
UNIVERSITY PRESS

University Printing House, Cambridge CB2 8BS, United Kingdom

Cambridge University Press is part of the University of Cambridge.

It furthers the University's mission by disseminating knowledge in the pursuit of education, learning and research at the highest international levels of excellence.

www.cambridge.org
Information on this title: www.cambridge.org/9781107656499

Originally published in Japanese as *Kahou Katei* by Kinokuniya, Tokyo.

First published in English by Cambridge University Press, 1999

First published in English 1999
Reprinted 2005
Corrected paperback edition 2013

A catalogue record for this publication is available from the British Library

ISBN 978-1-107-65649-9 Paperback

Contents

Preface to the revised edition

After the publication of this book in 1999 progress continued in the theory of Lévy processes and infinitely divisible distributions. Of the various directions let me mention two.

1. Fluctuation theory of Lévy processes on the line has been studied in many papers. It is a development of Wiener–Hopf factorizations. The publication of two books, Kyprianou [**303**] and Doney [**103**], provided fresh impetus. Lévy processes without positive jumps were deeply analyzed in this connection such as in Kuznetsov, Kyprianou, and Rivero [**301**]. Following Lamperti [**306**], the relation between selfsimilar (in an extended sense) Markov processes on the positive half line and exponential functionals of Lévy processes on the real line was studied in Bertoin and Yor [**30, 31**] and others. In the study of stable processes Kyprianou, Pardo, and Watson [**304**] combined this line of research and Wiener–Hopf factorizations.

2. A comprehensive treatment of infinitely divisible distributions on the line appeared in the monograph [**503**] by Steutel and van Harn, which discussed a lot of subjects not treated in this book. The analysis of the law of $\int_0^t \mathrm{e}^{B_s+as}\mathrm{d}s$ (exponential functional of Brownian motion with drift) was explored by Yor and others, for example in Matsumoto and Yor [**341**]. Further, the law of $\int_0^\infty \mathrm{e}^{-X_s}\mathrm{d}Y_s$ for a two-dimensional Lévy process $\{(X_t, Y_t)\}$ began to attract attention such as in Lindner and Sato [**321, 322**]. Tail behaviors related to subexponentiality are another subject in one dimension. In higher dimensions Watanabe's results [**563**] on densities of stable distributions opened a new horizon.

Subordinators (increasing Lévy processes) and their applications were described in Bertoin's lecture [**27**]. We can find a unique approach to them in Schilling, Song, and Vondraček [**472**]. On stochastic differential equations based on Lévy processes, Kunita's paper [**299**] and Applebaum's book [**6**] should be mentioned.

In this new printing a Supplement of 30 pages is attached at the end. The purpose of the Supplement is twofold. First, among a great many areas of progress it covers some subjects that I am familiar with (Sections 59, 62, 63, 64, and a part of 57). Second, it includes some materials closely connected with the original ten chapters (Sections 56, 58, 60, 61 and a part of 57). Changes to the text of the first printing are only few, which

I considered necessary. Any addition or deletion was avoided, with the exception of a few inserted lines in Remarks 15.12 and 37.13 and Definition 51.9. Naturally all numberings except reference numbers remain the same. Thus I chose to preserve the contents of the first printing, refraining from partial improvement. Newly added references are restricted to those cited in the Supplement and in this Preface. They are marked with asterisks in the list of references. For readers of the first printing a list of corrections and changes in the ten chapters is posted on my website (http://ksato.jp/).

There is no writing on the history of the study of Lévy processes and infinitely divisible distributions. But Notes at the end of each chapter of this book and my article [**451**] point out some epoch-making works.

I would like to thank Alex Lindner, Makoto Maejima, René Schilling, and Toshiro Watanabe for valuable comments on the published book and on this printing in preparation. Bibliographical remarks by Alex and René were very helpful. Encouragement from the late Hiroshi Tanaka is my cherished memory.

Ken-iti Sato
Nagoya, 2013

Preface to the first printing

Stochastic processes are mathematical models of random phenomena in time evolution. Lévy processes are stochastic processes whose increments in nonoverlapping time intervals are independent and whose increments are stationary in time. Further we assume a weak continuity called stochastic continuity. They constitute a fundamental class of stochastic processes. Brownian motion, Poisson processes, and stable processes are typical Lévy processes. After Paul Lévy's characterization in the 1930s of all processes in this class, many researches have revealed properties of their distributions and behaviors of their sample functions. However, Lévy processes are rich mathematical objects, still furnishing attractive problems of their own. On the other hand, important classes of stochastic processes are obtained as generalizations of the class of Lévy processes. One of them is the class of Markov processes; another is the class of semimartingales. The study of Lévy processes serves as the foundation for the study of stochastic processes.

Dropping the stationarity requirement of increments for Lévy processes, we get the class of additive processes. The distributions of Lévy and additive processes at any time are infinitely divisible, that is, they have the nth roots in the convolution sense for any n. When a time is fixed, the class of Lévy processes is in one-to-one correspondence with the class of infinitely divisible distributions. Additive processes are described by systems of infinitely divisible distributions.

This book is intended to provide comprehensive basic knowledge of Lévy processes, additive processes, and infinitely divisible distributions with detailed proofs and, at the same time, to serve as an introduction to stochastic processes. As we deal with the simplest stochastic processes, we do not assume any knowledge of stochastic processes with a continuous parameter. Prerequisites for this book are of the level of the textbook of Billingsley [**34**] or that of Chung [**80**].

Making an additional assumption of selfsimilarity or some extensions of it on Lévy or additive processes, we get certain important processes. Such are stable processes, semi-stable processes, and selfsimilar additive processes. We give them systematic study. Correspondingly, stable, semi-stable, and selfdecomposable distributions are treated. On the other hand, the class of Lévy processes contains processes quite different from selfsimilar, and intriguing time evolution in distributional properties appears.

There are ten chapters in this book. They can be divided into three parts. Chapters 1 and 2 constitute the basic part. Essential examples and a major tool for the analysis are given in Chapter 1. The tool is to consider Fourier transforms of probability measures, called characteristic functions. Then, in Chapter 2, characterization of all infinitely divisible distributions is given. They give description of all Lévy processes and also of all additive processes. Chapters 3, 4, and 5 are the second part. They develop fundamental results on which subsequent chapters rely. Chapter 3 introduces selfsimilarity and other structures. Chapter 4 deals with decomposition of sample functions into jumps and continuous motions. Chapter 5 is on distributional properties. The third part ranges from Chapter 6 to Chapter 10. They are nearly independent of each other and treat major topics on Lévy processes such as subordination and density transformation, recurrence and transience, potential theory, Wiener–Hopf factorizations, and unimodality and multimodality.

We do not touch extensions of Lévy processes and infinitely divisible distributions connected with Lie groups, hypergroups, and generalized convolutions. There are many applications of Lévy processes to stochastic integrals, branching processes, and measure-valued processes, but they are not included in this book. Risk theory, queueing theory, and stochastic finance are active fields where Lévy processes often appear.

The original version of this book is *Kahou katei* written in Japanese, published by Kinokuniya at the end of 1990. The book is enlarged and material is rewritten. Many recent advances are included and a new chapter on potential theory is added. Exercises are now given to each chapter and their solutions are at the end of the volume.

For many years I have been happy in collaborating with Makoto Yamazato and Toshiro Watanabe. I was encouraged by Takeyuki Hida and Hiroshi Kunita to write the original Japanese book and the present book. Frank Knight and Toshiro Watanabe read through the manuscript and gave me numerous suggestions for correction of errors and improvement of presentation. Kazuyuki Inoue, Mamoru Kanda, Makoto Maejima, Yumiko Sato, Masaaki Tsuchiya, and Makoto Yamazato pointed out many inaccuracies to be eliminated. Part of the book was presented in lectures at the University of Zurich [**446**] as arranged by Masao Nagasawa. The preparation of this book was made in AMSLaTeX; Shinta Sato assisted me with the computer. My heartfelt thanks go to all of them.

Ken-iti Sato
Nagoya, 1999

Remarks on notation

$\mathbb{N}$, $\mathbb{Z}$, $\mathbb{Q}$, $\mathbb{R}$, and $\mathbb{C}$ are, respectively, the collections of all positive integers, all integers, all rational numbers, all real numbers, and all complex numbers.

$\mathbb{Z}_+$, $\mathbb{Q}_+$, and $\mathbb{R}_+$ are the collections of nonnegative elements of $\mathbb{Z}$, $\mathbb{Q}$, and $\mathbb{R}$, respectively.

For $x \in \mathbb{R}$, positive means $x > 0$; negative means $x < 0$. For a sequence $\{x_n\}$, increasing means $x_n \leq x_{n+1}$ for all n; decreasing means $x_n \geq x_{n+1}$ for all n. Similarly, for a real function f, increasing means $f(s) \leq f(t)$ for $s < t$, and decreasing means $f(s) \geq f(t)$ for $s < t$. When the equality is not allowed, we say strictly increasing or strictly decreasing.

$\mathbb{R}^d$ is the d-dimensional Euclidean space. Its elements $x = (x_j)_{j=1,\ldots,d}$, $y = (y_j)_{j=1,\ldots,d}$ are column vectors with d real components. The inner product is $\langle x, y\rangle = \sum_{j=1}^d x_j y_j$; the norm is $|x| = (\sum_{j=1}^d {x_j}^2)^{1/2}$. The word d-variate is used in the same meaning as d-dimensional.

For sets A and B, $A \subset B$ means that all elements of A belong to B. For A, $B \subset \mathbb{R}^d$, $z \in \mathbb{R}^d$, and $c \in \mathbb{R}$, $A+z = \{x+z \colon x \in A\}$, $A-z = \{x-z \colon x \in A\}$, $A+B = \{x+y \colon x \in A,\, y \in B\}$, $A-B = \{x-y \colon x \in A,\, y \in B\}$, $cA = \{cx \colon x \in A\}$, $-A = \{-x \colon x \in A\}$, $A \setminus B = \{x \colon x \in A \text{ and } x \notin B\}$, $A^c = \mathbb{R}^d \setminus A$, and $\mathrm{dis}(z, A) = \inf_{x \in A} |z-x|$. $\overline{A}$ is the closure of A.

$\mathcal{B}(\mathbb{R}^d)$ is the Borel σ-algebra of $\mathbb{R}^d$. For any $B \in \mathcal{B}(\mathbb{R}^d)$, $\mathcal{B}(B)$ is the σ-algebra of Borel sets included in B. $\mathcal{B}(B)$ is also written as $\mathcal{B}_B$.

$\mathrm{Leb}(B)$ is the Lebesgue measure of a set B. $\mathrm{Leb}(\mathrm{d}x)$ is written $\mathrm{d}x$.

$\int g(x,y)\mathrm{d}_x F(x,y)$ is the Stieltjes integral with respect to x for fixed y.

The symbol δ_a represents the probability measure concentrated at a.

$[\mu]_B$ is the restriction of a measure μ to a set B.

The expression $\mu_1 * \mu_2$ represents the convolution of finite measures μ_1 and μ_2; $\mu^n = \mu^{n*}$ is the n-fold convolution of μ. When $n = 0$, μ^n is understood to be δ_0.

Sometimes $\mu(B)$ is written as μB. Thus $\mu(a,b] = \mu((a,b])$.

A non-zero measure means a measure not identically zero.

$1_B(x)$ is the indicator function of a set B, that is, $1_B(x) = 1$ for $x \in B$ and 0 for $x \notin B$.

$a \wedge b = \min\{a, b\}$, $a \vee b = \max\{a, b\}$.

The expression $\operatorname{sgn} x$ represents the sign function; $\operatorname{sgn} x = 1, 0, -1$ according as $x > 0, = 0, < 0$, respectively.

$P[A]$ is the probability of an event A. Sometimes $P[A]$ is written as PA.

$E[X]$ is the expectation of a random variable X. $E[X; A] = E[X 1_A]$. Sometimes $E[X]$ is written as EX.

$\operatorname{Var} X$ is the variance of a real random variable X.

$X \stackrel{\mathrm{d}}{=} Y$ means that X and Y are identically distributed. See p. 3 for the meaning of $\{X_t\} \stackrel{\mathrm{d}}{=} \{Y_t\}$.

P_X is the distribution of X.

The abbreviation a. s. denotes almost surely, that is, with probability 1. The abbreviation a. e. denotes almost everywhere, or almost every, with respect to the Lebesgue measure. Similarly, μ-a. e. denotes almost everywhere, or almost every, with respect to a measure μ.

$D([0,\infty),\mathbb{R}^d)$ is the collection of all functions $\xi(t)$ from $[0,\infty)$ to $\mathbb{R}^d$ such that $\xi(t)$ is right-continuous, $\xi(t+) = \lim_{h\downarrow 0} \xi(t+h) = \xi(t)$ for $t \geq 0$, and $\xi(t)$ has left limits $\xi(t-) = \lim_{h\downarrow 0} \xi(t-h) \in \mathbb{R}^d$ for $t > 0$.

I is the identity matrix. A' is the transpose of a matrix A. For an $n \times m$ real matrix A, $\|A\|$ is the operator norm of A as a linear transformation from $\mathbb{R}^m$ to $\mathbb{R}^n$, that is, $\|A\| = \sup_{|x|\leq 1} |Ax|$. (However, the prime is sometimes used not in this way. For example, together with a stochastic process $\{X_t\}$ taking values in $\mathbb{R}^d$, we use $\{X'_t\}$ for another stochastic process taking values in $\mathbb{R}^d$; X'_t is not the transpose of X_t.)

Sometimes a subscript is written larger in parentheses, such as $X_t(\omega) = X(t,\omega)$, $X_t = X(t)$, $S_n = S(n)$, $T_x = T(x)$, $x_n = x(n)$, and $t_k = t(k)$.

The integral of a vector-valued function or the expectation of a random variable on $\mathbb{R}^d$ is a vector with componentwise integrals or expectations.

$\#A$ is the number of elements of a set A.

The expression $f(t) \sim g(t)$ means that $f(t)/g(t)$ tends to 1.

The symbol $\square$ denotes the end of a proof.

CHAPTER 1

Basic examples

1. Definition of Lévy processes

In this chapter we give basic definitions concerning probability spaces, stochastic processes, and Lévy processes, and describe the main properties of characteristic functions, which we use in this book systematically. Then we introduce Poisson processes, compound Poisson processes, and Brownian motions. They are Lévy processes of fundamental importance.

DEFINITION 1.1. A *probability space* $(\Omega, \mathcal{F}, P)$ is a triplet of a set Ω, a family $\mathcal{F}$ of subsets of Ω, and a mapping P from $\mathcal{F}$ into $\mathbb{R}$ satisfying the following conditions.

(1) $\Omega \in \mathcal{F}$, $\emptyset \in \mathcal{F}$ ($\emptyset$ is the empty set).
(2) If $A_n \in \mathcal{F}$ for $n = 1, 2, \ldots$, then $\bigcup_{n=1}^{\infty} A_n$ and $\bigcap_{n=1}^{\infty} A_n$ are in $\mathcal{F}$.
(3) If $A \in \mathcal{F}$, then $A^c \in \mathcal{F}$. (A^c is $\Omega \setminus A$, the complement of A.)
(4) $0 \le P[A] \le 1$, $P[\Omega] = 1$, and $P[\emptyset] = 0$.
(5) If $A_n \in \mathcal{F}$ for $n = 1, 2, \ldots$ and they are disjoint (that is, $A_n \cap A_m = \emptyset$ for $n \ne m$), then $P[\bigcup_{n=1}^{\infty} A_n] = \sum_{n=1}^{\infty} P[A_n]$.

In the terminology of measure theory, a probability space is a measure space with total measure 1. In general, if $\mathcal{F}$ is a family of subsets of Ω satisfying (1), (2), and (3), then $\mathcal{F}$ is called a *σ-algebra* on Ω. The pair $(\Omega, \mathcal{F})$ is called a *measurable space.* A mapping P with the properties (4) and (5) is called a *probability measure.* For a probability space $(\Omega, \mathcal{F}, P)$, any set A in $\mathcal{F}$ is called an *event,* and $P[A]$ is called the *probability* of the event A.

The collection of all Borel sets on $\mathbb{R}^d$, denoted by $\mathcal{B}(\mathbb{R}^d)$, is called the Borel σ-algebra. It is the σ-algebra generated by the open sets in $\mathbb{R}^d$ (that is, the smallest σ-algebra that contains all open sets in $\mathbb{R}^d$). A real-valued function $f(x)$ on $\mathbb{R}^d$ is called measurable, if it is $\mathcal{B}(\mathbb{R}^d)$-measurable.

DEFINITION 1.2. Let $(\Omega, \mathcal{F}, P)$ be a probability space. A mapping X from Ω into $\mathbb{R}^d$ is an $\mathbb{R}^d$-valued *random variable* (or random variable on $\mathbb{R}^d$) if it is $\mathcal{F}$-measurable, that is, $\{\omega\colon X(\omega) \in B\}$ is in $\mathcal{F}$ for each $B \in \mathcal{B}(\mathbb{R}^d)$.

We write $P[\{\omega\colon X(\omega) \in B\}]$ as $P[X \in B]$. As a mapping of B, this is a probability measure on $\mathcal{B}(\mathbb{R}^d)$, which we denote by $P_X(B)$ and call the *distribution* (or *law*) *of* X. In general, probability measures on $\mathcal{B}(\mathbb{R}^d)$ are called distributions on $\mathbb{R}^d$.

If two random variables X, Y on $\mathbb{R}^d$ (not necessarily defined on a common probability space) have an identical distribution, namely $P_X = P_Y$, we write

$$X \stackrel{\mathrm{d}}{=} Y. \tag{1.1}$$

If X is a real-valued (i.e. $\mathbb{R}$-valued) random variable and if the integral $\int_\Omega X(\omega)P(\mathrm{d}\omega)$ exists, then it is called the *expectation* of X and denoted by $E[X]$, or EX. If X is a random variable on $\mathbb{R}^d$, and $f(x)$ is a bounded measurable function on $\mathbb{R}^d$, then

$$E[f(X)] = \int_{\mathbb{R}^d} f(x)P_X(\mathrm{d}x). \tag{1.2}$$

A random variable X is said to have a property A *almost surely* (abbreviated as a. s.), or with probability 1, if there is $\Omega_0 \in \mathcal{F}$ with $P[\Omega_0] = 1$ such that $X(\omega)$ has the property A for every $\omega \in \Omega_0$. It is not required that the set $\{\omega\colon X(\omega)$ has property $A\}$ belongs to $\mathcal{F}$, but sometimes we write, by abuse, $P[\,X(\omega)$ has property $A\,] = 1$.

Usually we fix a probability space and random variables are supposed to be defined on it. An important concept concerning a family of random variables is independence.

DEFINITION 1.3. Let X_j be an $\mathbb{R}^{d_j}$-valued random variable for $j = 1, \ldots, n$. The family $\{X_1, \ldots, X_n\}$ is *independent* if, for every $B_j \in \mathcal{B}(\mathbb{R}^{d_j})$, $j = 1, \ldots, n$,

$$P[\,X_1 \in B_1, \ldots, X_n \in B_n\,] = P[\,X_1 \in B_1\,] \ldots P[\,X_n \in B_n\,].$$

Often we say that $X_1, \ldots, X_n$ are independent instead of saying that the family $\{X_1, \ldots, X_n\}$ is independent. An infinite family of random variables is independent, if every finite subfamily of it is independent.

DEFINITION 1.4. A family $\{X_t\colon t \geq 0\}$ of random variables on $\mathbb{R}^d$ with parameter $t \in [0, \infty)$ defined on a common probability space is called a *stochastic process*. It is written as $\{X_t\}$. As is explained in Remarks on notation, X_t and $X_t(\omega)$ are sometimes written as $X(t)$ and $X(t, \omega)$. For any fixed $0 \leq t_1 < t_2 < \cdots < t_n$,

$$P[\,X(t_1) \in B_1, \ldots, X(t_n) \in B_n\,]$$

determines a probability measure on $\mathcal{B}((\mathbb{R}^d)^n)$. The family of the probability measures over all choices of n and $t_1, \ldots, t_n$ is called the *system of finite-dimensional distributions* of $\{X_t\}$. A stochastic process $\{Y_t\}$ is called a *modification* of a stochastic process $\{X_t\}$, if

$$P[\,X_t = Y_t\,] = 1 \quad \text{for } t \in [0, \infty). \tag{1.3}$$

Two stochastic processes $\{X_t\}$ and $\{Y_t\}$ (not necessarily defined on a common probability space) are *identical in law*, written as

$$\{X_t\} \stackrel{\mathrm{d}}{=} \{Y_t\}, \tag{1.4}$$

if the systems of their finite-dimensional distributions are identical. Considered as a function of t, $X(t,\omega)$ is called a *sample function*, or *sample path*, of $\{X_t\}$. Sometimes we use the word stochastic process also for a family having an interval different from $[0,\infty)$ as its set of indices, for example, $\{X_t\colon t \in [s,\infty)\}$.

DEFINITION 1.5. A stochastic process $\{X_t\}$ on $\mathbb{R}^d$ is *stochastically continuous* or *continuous in probability* if, for every $t \geq 0$ and $\varepsilon > 0$,

$$\lim_{s\to t} P[\,|X_s - X_t| > \varepsilon\,] = 0. \tag{1.5}$$

Stochastic processes are mathematical models of time evolution of random phenomena. So the index t is usually taken for time. Thus we freely use the word *time* for t. The most basic stochastic process modeled for continuous random motions is the Brownian motion and that for jumping random motions is the Poisson process. These two belong to a class called Lévy processes. Lévy processes are, speaking only of essential points, stochastic processes with stationary independent increments. How important this class is and what rich structures it has will be gradually revealed in this book. First we give its definition.

DEFINITION 1.6. A stochastic process $\{X_t\colon t \geq 0\}$ on $\mathbb{R}^d$ is a *Lévy process* if the following conditions are satisfied.

(1) For any choice of $n \geq 1$ and $0 \leq t_0 < t_1 < \cdots < t_n$, random variables $X_{t_0}, X_{t_1} - X_{t_0}, X_{t_2} - X_{t_1}, \ldots, X_{t_n} - X_{t_{n-1}}$ are independent (independent increments property).
(2) $X_0 = 0$ a. s.
(3) The distribution of $X_{s+t} - X_s$ does not depend on s (temporal homogeneity or stationary increments property).
(4) It is stochastically continuous.
(5) There is $\Omega_0 \in \mathcal{F}$ with $P[\Omega_0] = 1$ such that, for every $\omega \in \Omega_0$, $X_t(\omega)$ is right-continuous in $t \geq 0$ and has left limits in $t > 0$.

A Lévy process on $\mathbb{R}^d$ is called a d-dimensional Lévy process. Dropping the condition (5), we call any process satisfying (1)–(4) a *Lévy process in law*. We define an *additive process* as a stochastic process satisfying the conditions (1), (2), (4), and (5). An *additive process in law* is a stochastic process satisfying (1), (2), and (4).

The conditions (1) and (3) together are expressed as the stationary independent increments property. Under the conditions (2) and (3), the condition (4) can be replaced by

$$\lim_{t\downarrow 0} P[\,|X_t| > \varepsilon\,] = 0 \quad \text{for } \varepsilon > 0. \tag{1.6}$$

We will see in Chapter 2 that any Lévy process in law has a modification which is a Lévy process. Similarly any additive process in law has a modification which is an additive process. Thus the condition (5) is not essential.

Lévy defined additive processes without assuming the conditions (4) and (5). But such processes are reducible to the additive processes defined above. See Notes at the end of Chapter 2.

EXAMPLE 1.7. Let $\{X_t\}$ be a Lévy process on $\mathbb{R}^d$ and $h(t)$ be a strictly increasing continuous function from $[0,\infty)$ into $[0,\infty)$ satisfying $h(0) = 0$. Then $\{X_{h(t)}\}$ is an additive process on $\mathbb{R}^d$. If $h(t) = ct$ with $c > 0$, then $\{X_{h(t)}\}$ has temporal homogeneity and it is a Lévy process.

A theorem of Kolmogorov guarantees the existence of a stochastic process with a given system of finite-dimensional distributions. Let $\Omega = (\mathbb{R}^d)^{[0,\infty)}$, the collection of all functions $\omega = (\omega(t))_{t\in[0,\infty)}$ from $[0,\infty)$ into $\mathbb{R}^d$. Define X_t by $X_t(\omega) = \omega(t)$. A set

$$C = \{\omega\colon X(t_1,\omega) \in B_1, \ldots, X(t_n,\omega) \in B_n\} \tag{1.7}$$

for $0 \le t_1 < \cdots < t_n$ and $B_1, \ldots, B_n \in \mathcal{B}(\mathbb{R}^d)$ is called a cylinder set. Consider the σ-algebra $\mathcal{F}$ generated by the cylinder sets, called the Kolmogorov σ-algebra.

THEOREM 1.8 (Kolmogorov's extension theorem). *Suppose that, for any choice of n and $0 \le t_1 < \cdots < t_n$, a distribution $\mu_{t_1,\ldots,t_n}$ is given and that, if $B_1, \ldots, B_n \in \mathcal{B}(\mathbb{R}^d)$ and $B_k = \mathbb{R}^d$, then*

$$\begin{aligned} &\mu_{t_1,\ldots,t_n}(B_1 \times \cdots \times B_n) \\ &\quad = \mu_{t_1,\ldots,t_{k-1},t_{k+1},\ldots,t_n}(B_1 \times \cdots \times B_{k-1} \times B_{k+1} \times \cdots \times B_n). \end{aligned} \tag{1.8}$$

Then, there exists a unique probability measure P on $\mathcal{F}$ that has $\{\mu_{t_1,\ldots,t_n}\}$ as its system of finite-dimensional distributions.

This theorem is in Kolmogorov [**293**]. Proofs are found also in Breiman [**66**] and Billingsley [**34**].

Construction of the direct product of probability spaces is often needed.

THEOREM 1.9. *Let $(\Omega_n, \mathcal{F}_n, P_n)$ be probability spaces for $n = 1, 2, \ldots$. Let $\Omega = \Omega_1 \times \Omega_2 \times \cdots$ and let $\mathcal{F}$ be the σ-algebra generated by the collection of sets*

$$C = \{\omega = (\omega_1, \omega_2, \ldots)\colon \omega_k \in A_k \text{ for } k = 1, \ldots, n\}, \tag{1.9}$$

over all n and all $A_k \in \mathcal{F}_k$ for $k = 1, \ldots, n$. Then there exists a unique probability measure P on $\mathcal{F}$ such that

$$P[C] = P_1[A_1] \ldots P_n[A_n]$$

for each C of (1.9).

Proof is found in Halmos [**180**] and Fristedt and Gray [**151**]. If $\Omega_n = \mathbb{R}^d$ and $\mathcal{F}_n = \mathcal{B}(\mathbb{R}^d)$ for each n, then Theorem 1.9 is a special case of a discrete time version of Theorem 1.8.

We give the definition of a random walk. It is a basic object in probability theory. A Lévy process is a continuous time analogue of a random walk.

DEFINITION 1.10. Let $\{Z_n : n = 1, 2, \ldots\}$ be a sequence of independent and identically distributed $\mathbb{R}^d$-valued random variables. Let $S_0 = 0$, $S_n = \sum_{j=1}^n Z_j$ for $n = 1, 2, \ldots$. Then $\{S_n : n = 0, 1, \ldots\}$ is a *random walk* on $\mathbb{R}^d$, or a d-dimensional random walk.

For any distribution μ on $\mathbb{R}^d$, there exists a random walk such that Z_n has distribution μ. This follows from Theorem 1.9.

Two families $\{X_t\}$, $\{Y_s\}$ of random variables are said to be independent if, for any choice of $t_1, \ldots, t_n$ and $s_1, \ldots, s_m$, the two multi-dimensional random variables $(X_{t_j})_{j=1,\ldots,n}$ and $(Y_{s_k})_{k=1,\ldots,m}$ are independent. A sequence of events $\{A_n : n = 1, 2, \ldots\}$ is said to be independent, if the sequence of random variables $\{1_{A_n}(\omega) : n = 1, 2, \ldots\}$ is independent. For a sequence of events $\{A_n\}$, the upper limit event and the lower limit event are defined by

$$\limsup_{n\to\infty} A_n = \bigcap_{n=1}^{\infty} \bigcup_{k=n}^{\infty} A_k \quad \text{and} \quad \liminf_{n\to\infty} A_n = \bigcup_{n=1}^{\infty} \bigcap_{k=n}^{\infty} A_k,$$

respectively.

PROPOSITION 1.11 (Borel–Cantelli lemma). (i) *If $\sum_{n=1}^{\infty} P[A_n] < \infty$, then $P[\limsup_{n\to\infty} A_n] = 0$.*

(ii) *If $\{A_n : n = 1, 2, \ldots\}$ is independent and $\sum_{n=1}^{\infty} P[A_n] = \infty$, then we have $P[\limsup_{n\to\infty} A_n] = 1$.*

A sequence of $\mathbb{R}^d$-valued random variables $\{X_n : n = 1, 2, \ldots\}$ is said to *converge stochastically*, or *converge in probability*, to X if, for each $\varepsilon > 0$, $\lim_{n\to\infty} P[\,|X_n - X| > \varepsilon\,] = 0$. This is denoted by

$$X_n \to X \quad \text{in prob.}$$

If $\{X_n\}$ converges stochastically to X and X', then $X = X'$ a. s. A sequence $\{X_n\}$ is said to *converge almost surely* to X, denoted by

$$X_n \to X \quad \text{a. s.},$$

if $P[\lim_{n\to\infty} X_n(\omega) = X(\omega)\,] = 1$.

PROPOSITION 1.12. (i) *If $X_n \to X$ a. s., then $X_n \to X$ in prob.*

(ii) *If $X_n \to X$ in prob., then a subsequence of $\{X_n\}$ converges a. s. to X.*

It follows from (i) that, if $\{X_t\}$ is a Lévy process, then

$$X_t = X_{t-} \quad \text{a. s.} \quad \text{for any fixed } t > 0, \tag{1.10}$$

where X_{t-} denotes the left limit at t. For $t_n \uparrow t$ implies $X_{t_n} \to X_{t-}$ a. s. and $X_{t_n} \to X_t$ in prob. Among the five conditions in the definition of a Lévy process the condition (4) is implied by (2), (3), and (5). In fact, for any $t_n \downarrow 0$, X_{t_n} converges to 0 a. s. and hence in prob., which implies (1.6).

PROPOSITION 1.13 (Inheritance of independence). *Suppose that, for each $j = 1, \ldots, k$, $X_{j,n} \to X_j$ in prob. as $n \to \infty$. If the family $\{X_{j,n} \colon j = 1, \ldots, k\}$ is independent for each n, then the family $\{X_j \colon j = 1, \ldots, k\}$ is independent.*

Proofs of Propositions 1.11–1.13 are found in [**34**], [**80**] and others.

The concept of independence is extended to σ-algebras (though we will not use this extension often). Let $(\Omega, \mathcal{F}, P)$ be a probability space. Sub-σ-algebras $\mathcal{F}_1, \mathcal{F}_2, \ldots$ of $\mathcal{F}$ are said to be independent if, for any $A_n \in \mathcal{F}_n$, $n = 1, 2, \ldots$, $\{A_n\}$ is independent. Given a family of random variables $\{X_t \colon t \in T\}$, where T is an arbitrary set, we say that a sub-σ-algebra $\mathcal{G}$ is the σ-algebra generated by $\{X_t \colon t \in T\}$ and write $\mathcal{G} = \sigma(X_t \colon t \in T)$ if

(1) X_t is $\mathcal{G}$-measurable for each t,
(2) $\mathcal{G}$ is the smallest σ-algebra that satisfies (1).

In general, for a family $\mathcal{A}$ of subsets of Ω, the smallest σ-algebra that contains $\mathcal{A}$ is called the σ-algebra generated by $\mathcal{A}$ and denoted by $\sigma(\mathcal{A})$. A random variable X and σ-algebra $\mathcal{F}_1$ are said to be independent if $\sigma(X)$ and $\mathcal{F}_1$ are independent.

THEOREM 1.14 (Kolmogorov's 0–1 law). *Let $\{\mathcal{F}_n \colon n = 1, 2, \ldots\}$ be an independent family of sub-σ-algebras of $\mathcal{F}$. If an event A belongs to the σ-algebra $\sigma(\bigcup_{n=m}^{\infty} \mathcal{F}_n)$ for each m, then $P[A]$ is 0 or 1.*

Proofs are found in [**34**], [**80**] and others. The following fact (sometimes called Dynkin's lemma, see [**81**], [**121**]) will be used.

PROPOSITION 1.15. *Let $\mathcal{A}$ be a collection of subsets of Ω such that*

(1) $A \in \mathcal{A}$ and $B \in \mathcal{A}$ imply $A \cap B \in \mathcal{A}$.

Let $\mathcal{C} \supset \mathcal{A}$ and suppose the following.

(2) If $A_n \in \mathcal{C}$, $n = 1, 2, \ldots$, and $\{A_n\}$ is increasing, then $\bigcup_{n=1}^{\infty} A_n \in \mathcal{C}$.
(3) If $A \in \mathcal{C}$, $B \in \mathcal{C}$, and $A \supset B$, then $A \setminus B \in \mathcal{C}$.
(4) $\Omega \in \mathcal{C}$.

Then $\mathcal{C} \supset \sigma(\mathcal{A})$.

The proof of the following proposition on evaluation of some expectations shows the strength of Proposition 1.15.

PROPOSITION 1.16. *Let X and Y be independent random variables on $\mathbb{R}^{d_1}$ and $\mathbb{R}^{d_2}$, respectively. If $f(x,y)$ is a bounded measurable function on $\mathbb{R}^{d_1} \times \mathbb{R}^{d_2}$, then $g(y) = E[f(X,y)]$ is bounded and measurable and $E[f(X,Y)] = E[g(Y)]$.*

Proof. Let $\mathcal{C}$ be the collection of sets $A \in \mathcal{B}(\mathbb{R}^{d_1} \times \mathbb{R}^{d_2})$ such that $f = 1_A(x,y)$ satisfies the conclusion above. Here 1_A is the indicator function of the set A (see Remarks on notation). Let $\mathcal{A}$ be the collection of sets $A = A_1 \times A_2$ with $A_1 \in \mathcal{B}(\mathbb{R}^{d_1})$ and $A_2 \in \mathcal{B}(\mathbb{R}^{d_2})$. It follows from the definition of independence that $\mathcal{A} \subset \mathcal{C}$. Since $\mathcal{A}$ and $\mathcal{C}$ satisfy (1)–(4) of Proposition 1.15 with $\Omega = \mathbb{R}^{d_1} \times \mathbb{R}^{d_2}$, we have $\mathcal{C} = \mathcal{B}(\mathbb{R}^{d_1} \times \mathbb{R}^{d_2})$. For general f use approximation by linear combinations of functions of the form $1_A(x,y)$. $\square$

2. Characteristic functions

The primary tool in the analysis of distributions of Lévy processes is characteristic functions, or Fourier transforms, of distributions. We will give definitions, properties, and examples of characteristic functions.

DEFINITION 2.1. The *characteristic function* $\widehat{\mu}(z)$ of a probability measure μ on $\mathbb{R}^d$ is

$$\widehat{\mu}(z) = \int_{\mathbb{R}^d} e^{i\langle z,x\rangle} \mu(dx), \quad z \in \mathbb{R}^d. \tag{2.1}$$

The characteristic function of the distribution P_X of a random variable X on $\mathbb{R}^d$ is denoted by $\widehat{P}_X(z)$. That is

$$\widehat{P}_X(z) = \int_{\mathbb{R}^d} e^{i\langle z,x\rangle} P_X(dx) = E[e^{i\langle z,X\rangle}].$$

DEFINITION 2.2. A sequence of probability measures μ_n, $n = 1, 2, \ldots,$ *converges* to a probability measure μ, written as

$$\mu_n \to \mu \quad \text{as } n \to \infty,$$

if, for every bounded continuous function f,

$$\int_{\mathbb{R}^d} f(x)\mu_n(dx) \to \int_{\mathbb{R}^d} f(x)\mu(dx) \quad \text{as } n \to \infty.$$

When μ and μ_n are finite measures, the convergence $\mu_n \to \mu$ is defined in the same way. When $\{\mu_t\}$ are probability measures with a real parameter, we say that

$$\mu_s \to \mu_t \quad \text{as } s \to t,$$

if

$$\int_{\mathbb{R}^d} f(x)\mu_s(\mathrm{d}x) \to \int_{\mathbb{R}^d} f(x)\mu_t(\mathrm{d}x) \quad \text{as } s \to t$$

for every bounded continuous function f. This is equivalent to saying that $\mu_{s_n} \to \mu_t$ for every sequence s_n that tends to t.

We say that B is a μ-continuity set if the boundary of B has μ-measure 0. The convergence $\mu_n \to \mu$ is equivalent to the condition that $\mu_n(B) \to \mu(B)$ for every μ-continuity set $B \in \mathcal{B}(\mathbb{R}^d)$.

A sequence of random variables $\{X_n\}$ on $\mathbb{R}^d$ converges in probability to X if and only if the distribution of $X_n - X$ converges to δ_0 (distribution concentrated at 0). The next fact is frequently used.

PROPOSITION 2.3. *If $X_n \to X$ in probability, then the distribution of X_n converges to the distribution of X.*

DEFINITION 2.4. The *convolution* μ of two distributions μ_1 and μ_2 on $\mathbb{R}^d$, denoted by $\mu = \mu_1 * \mu_2$, is a distribution defined by

$$\mu(B) = \iint_{\mathbb{R}^d \times \mathbb{R}^d} 1_B(x+y)\mu_1(\mathrm{d}x)\mu_2(\mathrm{d}y), \qquad B \in \mathcal{B}(\mathbb{R}^d). \tag{2.2}$$

The convolution of two finite measures on $\mathbb{R}^d$ is defined by the same formula.

The convolution operation is commutative and associative. If X_1 and X_2 are independent random variables on $\mathbb{R}^d$ with distributions μ_1 and μ_2, respectively, then $X_1 + X_2$ has distribution $\mu_1 * \mu_2$.

The following are the principal properties of characteristic functions. In (v) we will use the following terminology: $\widetilde{\mu}$ is the *dual* of μ and $\mu^\sharp$ is the *symmetrization (of a probability measure)* of μ if $\widetilde{\mu}(B) = \mu(-B)$, $-B = \{-x : x \in B\}$, and $\mu^\sharp = \mu * \widetilde{\mu}$. When $d = 1$, another name of the dual of μ is the *reflection* of μ. If μ is identical with its dual, it is called *symmetric*.

PROPOSITION 2.5. *Let μ, μ_1, μ_2, μ_n be distributions on $\mathbb{R}^d$.*

(i)(Bochner's theorem) *We have that $\widehat{\mu}(0) = 1$ and $|\widehat{\mu}(z)| \le 1$, and $\widehat{\mu}(z)$ is uniformly continuous and nonnegative-definite in the sense that, for each $n = 1, 2, \ldots,$*

$$\sum_{j=1}^{n}\sum_{k=1}^{n} \widehat{\mu}(z_j - z_k)\xi_j\bar{\xi}_k \ge 0 \text{ for } z_1, \ldots, z_n \in \mathbb{R}^d,\ \xi_1, \ldots, \xi_n \in \mathbb{C}. \tag{2.3}$$

Conversely, if a complex-valued function $\varphi(z)$ on $\mathbb{R}^d$ with $\varphi(0) = 1$ is continuous at $z = 0$ and nonnegative-definite, then $\varphi(z)$ is the characteristic function of a distribution on $\mathbb{R}^d$.

(ii) *If $\widehat{\mu}_1(z) = \widehat{\mu}_2(z)$ for $z \in \mathbb{R}^d$, then $\mu_1 = \mu_2$.*

(iii) *If* $\mu = \mu_1 * \mu_2$, *then* $\widehat{\mu}(z) = \widehat{\mu}_1(z)\widehat{\mu}_2(z)$. *If* X_1 *and* X_2 *are independent random variables on* $\mathbb{R}^d$, *then*

$$\widehat{P}_{X_1+X_2}(z) = \widehat{P}_{X_1}(z)\widehat{P}_{X_2}(z).$$

(iv) *Let* $X = (X_j)_{j=1,\ldots,n}$ *be an* $\mathbb{R}^{nd}$*-valued random variable, where* $X_1, \ldots, X_n$ *are* $\mathbb{R}^d$*-valued random variables. Then* $X_1, \ldots, X_n$ *are independent if and only if*

$$\widehat{P}_X(z) = \widehat{P}_{X_1}(z_1) \ldots \widehat{P}_{X_n}(z_n) \quad \textit{for } z = (z_j)_{j=1,\ldots,n},\ z_j \in \mathbb{R}^d.$$

(v) *Suppose that* $\widetilde{\mu}$ *is the dual of* μ *and* $\mu^\sharp$ *is the symmetrization of* μ. *Then* $\widehat{\widetilde{\mu}}(z) = \widehat{\mu}(-z) = \overline{\widehat{\mu}(z)}$ *and* $\widehat{\mu^\sharp}(z) = |\widehat{\mu}(z)|^2$.

(vi) *If* $\mu_n \to \mu$, *then* $\widehat{\mu}_n(z) \to \widehat{\mu}(z)$ *uniformly on any compact set.*

(vii) *If* $\widehat{\mu}_n(z) \to \widehat{\mu}(z)$ *for every* z, *then* $\mu_n \to \mu$.

(viii) *If* $\widehat{\mu}_n(z)$ *converges to a function* $\varphi(z)$ *for every* z *and* $\varphi(z)$ *is continuous at* $z = 0$, *then* $\varphi(z)$ *is the characteristic function of some distribution.*

(ix) *Let* n *be a positive integer. If* μ *has a finite absolute moment of order* n, *that is,* $\int |x|^n \mu(\mathrm{d}x) < \infty$, *then* $\widehat{\mu}(z)$ *is a function of class* C^n *and, for any nonnegative integers* $n_1, \ldots, n_d$ *satisfying* $n_1 + \cdots + n_d \leq n$,

$$\int x_1^{n_1} \ldots x_d^{n_d} \mu(\mathrm{d}x) = \left[\left(\frac{1}{\mathrm{i}} \frac{\partial}{\partial z_1} \right)^{n_1} \cdots \left(\frac{1}{\mathrm{i}} \frac{\partial}{\partial z_d} \right)^{n_d} \widehat{\mu}(z) \right]_{z=0}.$$

(x) *Let* n *be a positive even integer. If* $\widehat{\mu}(z)$ *is of class* C^n *in a neighborhood of the origin, then* μ *has finite absolute moment of order* n.

(xi) *Let* $-\infty < a_j < b_j < \infty$ *for* $j = 1, \ldots, d$ *and* $B = [a_1, b_1] \times \cdots \times [a_d, b_d]$. *If* B *is a* μ*-continuity set, then*

$$\mu(B) = \lim_{c \to \infty} (2\pi)^{-d} \int_{[-c,c]^d} \widehat{\mu}(z)\mathrm{d}z \int_B \mathrm{e}^{-\mathrm{i}\langle x,z \rangle} \mathrm{d}x.$$

(xii) *If* $\int |\widehat{\mu}(z)|\mathrm{d}z < \infty$, *then* μ *is absolutely continuous with respect to the Lebesgue measure, has a bounded continuous density* $g(x)$, *and*

$$g(x) = (2\pi)^{-d} \int_{\mathbb{R}^d} \mathrm{e}^{-\mathrm{i}\langle x,z \rangle} \widehat{\mu}(z)\mathrm{d}z.$$

The assertion (xi) contains the inversion formula, which strengthens the one-to-one property (ii).

In the one-dimensional case the properties above are proved in Billingsley [**34**], Breiman [**66**], Chung [**80**], Fristedt and Gray [**151**], and many other books. In general dimensions see Dudley [**110**], pp. 233–240, 255, Cuppens [**91**], pp. 16, 37, 41, 53, 54, and also Linnik and Ostrovskii [**325**], pp. 169–173.

When μ is a distribution on $[0,\infty)$, the *Laplace transform* of μ is defined by

$$L_\mu(u) = \int_{[0,\infty)} \mathrm{e}^{-ux}\mu(\mathrm{d}x) \quad \text{for } u \geq 0. \tag{2.4}$$

PROPOSITION 2.6. *Let μ, μ_1, and μ_2 be distributions on $[0,\infty)$.*
(i) If $L_{\mu_1}(u) = L_{\mu_2}(u)$ for $u \geq 0$, then $\mu_1 = \mu_2$.
(ii) If $\mu = \mu_1 * \mu_2$, then $L_\mu(u) = L_{\mu_1}(u)L_{\mu_2}(u)$.

Proof. (i) For any complex w with $\operatorname{Re} w \leq 0$ we can define $\Phi_j(w) = \int \mathrm{e}^{wx}\mu_j(\mathrm{d}x)$, $j = 1, 2$. These are analytic on $\{w\colon \operatorname{Re} w < 0\}$. For the integral $\int_{[0,n]} \mathrm{e}^{wx}\mu_j(\mathrm{d}x)$ is analytic since we can differentiate under the integral sign, and this sequence is uniformly bounded and convergent to $\Phi_j(w)$ pointwise as $n \to \infty$. If $w = -u < 0$, then $\Phi_1(w) = \Phi_2(w)$. Hence $\Phi_1(w) = \Phi_2(w)$ on $\{w\colon \operatorname{Re} w < 0\}$ by the unique determination theorem for analytic functions. Since $\Phi_j(w)$ is continuous on $\{w\colon \operatorname{Re} w \leq 0\}$ and $\Phi_j(w) = \widehat{\mu}_j(z)$ for $w = \mathrm{i}z$, we have $\widehat{\mu}_1(z) = \widehat{\mu}_2(z)$, which implies $\mu_1 = \mu_2$.
(ii) If $\mu = \mu_1 * \mu_2$, then the definition (2.2) and Fubini's theorem give $L_\mu(u) = L_{\mu_1}(u)L_{\mu_2}(u)$. □

Feller [**139**] contains a direct proof of (i) and a formula to express μ in terms of $L_\mu(u)$.

EXAMPLE 2.7. ($d = 1$) Let $c > 0$. The *Poisson distribution* with mean c is defined by

$$\mu\{k\} = \mathrm{e}^{-c}c^k/k! \quad \text{for } k = 0, 1, 2, \ldots,$$

while $\mu(B) = 0$ for any B containing no nonnegative integer. We have

$$\widehat{\mu}(z) = \exp(c(\mathrm{e}^{\mathrm{i}z} - 1)), \qquad z \in \mathbb{R}, \tag{2.5}$$

$$L_\mu(u) = \exp(c(\mathrm{e}^{-u} - 1)), \qquad u \geq 0. \tag{2.6}$$

EXAMPLE 2.8. ($d = 1$) The *nondegenerate Gaussian distribution* on $\mathbb{R}$ with mean γ and variance a is defined by

$$\mu(B) = (2\pi a)^{-1/2} \int_B \mathrm{e}^{-(x-\gamma)^2/(2a)}\mathrm{d}x,$$

where $a > 0$ and $\gamma \in \mathbb{R}$. We have

$$\widehat{\mu}(z) = \exp(-\tfrac{1}{2}az^2 + \mathrm{i}\gamma z), \qquad z \in \mathbb{R}. \tag{2.7}$$

EXAMPLE 2.9. Let $A = (A_{jk})$ be a $d \times d$ positive-definite symmetric matrix and A^{-1} be its inverse. Then Ax is in $\mathbb{R}^d$ for $x \in \mathbb{R}^d$ (recall that x is a column vector). Denote the determinant of A by $\det A$. Let

$\gamma = (\gamma_j)_{j=1,\dots,d} \in \mathbb{R}^d$. Define the *nondegenerate Gaussian distribution* μ on $\mathbb{R}^d$with mean vector γ and covariance matrix A by

$$\mu(B) = (2\pi)^{-d/2}(\det A)^{-1/2} \int_B \mathrm{e}^{-\langle x-\gamma, A^{-1}(x-\gamma)\rangle/2} \mathrm{d}x, \quad B \in \mathcal{B}(\mathbb{R}^d).$$

(The general definition of nondegeneracy will be given in Definition 24.16.) Its characteristic function is

$$\widehat{\mu}(z) = \exp(-\tfrac{1}{2}\langle z, Az\rangle + \mathrm{i}\langle \gamma, z\rangle), \qquad z \in \mathbb{R}^d. \tag{2.8}$$

By mean vector and covariance matrix we mean that

$$\gamma_j = \int x_j \mu(\mathrm{d}x), \qquad A_{jk} = \int (x_j - \gamma_j)(x_k - \gamma_k)\mu(\mathrm{d}x).$$

EXAMPLE 2.10. Extending the preceding example, we call μ a *Gaussian distribution* on $\mathbb{R}^d$ if

$$\widehat{\mu}(z) = \exp(-\tfrac{1}{2}\langle z, Az\rangle + \mathrm{i}\langle \gamma, z\rangle), \qquad z \in \mathbb{R}^d,$$

where γ is in $\mathbb{R}^d$ and A is a nonnegative-definite symmetric matrix. To see the existence of μ for any given γ and A, let X be a random variable with nondegenerate Gaussian distribution with mean 0 and covariance matrix I (I is the identity matrix), and let $Y = BX + \gamma$ with a symmetric matrix B. Then Y has characteristic function

$$E[\mathrm{e}^{\mathrm{i}\langle z,Y\rangle}] = E[\mathrm{e}^{\mathrm{i}\langle z,BX+\gamma\rangle}] = E[\mathrm{e}^{\mathrm{i}\langle Bz,X\rangle}]\mathrm{e}^{\mathrm{i}\langle z,\gamma\rangle} = \mathrm{e}^{-\langle z,B^2 z\rangle/2+\mathrm{i}\langle z,\gamma\rangle}.$$

Choosing an orthogonal matrix C such that $D = C'AC$ is diagonal, let D_1 be a diagonal matrix with ${D_1}^2 = D$. Then $A = B^2$ with $B = CD_1C'$. It is proved by using Proposition 2.5(ix) and (x) (or using the Y above) that μ has mean vector γ and covariance matrix A. If $d = 1$, then A is a nonnegative real number.

EXAMPLE 2.11. ($d = 1$) The *Cauchy distribution* μ with parameters $\gamma \in \mathbb{R}$ and $c > 0$ is given by

$$\mu(B) = \pi^{-1}c \int_B ((x-\gamma)^2 + c^2)^{-1}\mathrm{d}x, \qquad B \in \mathcal{B}(\mathbb{R}).$$

We have

$$\widehat{\mu}(z) = \mathrm{e}^{-c|z|+\mathrm{i}\gamma z}, \qquad z \in \mathbb{R}. \tag{2.9}$$

EXAMPLE 2.12. The *d-dimensional Cauchy distribution* with parameters $\gamma \in \mathbb{R}^d$ and $c > 0$ is defined by

$$\mu(B) = \pi^{-(d+1)/2}\Gamma((d+1)/2)c \int_B (|x-\gamma|^2 + c^2)^{-(d+1)/2}\mathrm{d}x$$

for $B \in \mathcal{B}(\mathbb{R}^d)$. Let us show that the characteristic function is

$$\widehat{\mu}(z) = \mathrm{e}^{-c|z|+\mathrm{i}\langle\gamma,z\rangle}, \qquad z \in \mathbb{R}^d. \tag{2.10}$$

After the change of variables $x' = c^{-1}(x-\gamma)$, we may assume that $\gamma = 0$ and $c = 1$. We prove by induction that

$$c_d \int_{\mathbb{R}^d} \mathrm{e}^{\mathrm{i}\langle z,x\rangle}(|x|^2+1)^{-(d+1)/2}\mathrm{d}x = \mathrm{e}^{-|z|},$$

where $c_d = \pi^{-(d+1)/2}\Gamma((d+1)/2)$. The case $d=1$ is given in the preceding example. Assume that the equality is true for $d-1$. Let $\varphi(z)$ be the left-hand side of the equality. Since $\varphi(z)$ is invariant under orthogonal transformations, it is a function only of $|z|$. Therefore we may assume that the dth component z_d is 0. Then, for $z' = (z_j)_{j=1,\dots,d-1}$,

$$\begin{aligned}\varphi(z) &= c_d \int_{\mathbb{R}^{d-1}} \mathrm{e}^{\mathrm{i}\langle y,z'\rangle}\mathrm{d}y \int_{-\infty}^{\infty} (|y|^2+u^2+1)^{-(d+1)/2}\mathrm{d}u \\ &= c_d \int_{\mathbb{R}^{d-1}} \mathrm{e}^{\mathrm{i}\langle y,z'\rangle}(|y|^2+1)^{-d/2}\mathrm{d}y \int_{-\infty}^{\infty} (u^2+1)^{-(d+1)/2}\mathrm{d}u.\end{aligned}$$

Through $u = \tan\theta$ we get

$$\int_{-\infty}^{\infty} (u^2+1)^{-(d+1)/2}\mathrm{d}u = 2\int_0^{\pi/2} \cos^{d-1}\theta\mathrm{d}\theta = \pi^{1/2}\Gamma(d/2)/\Gamma((d+1)/2).$$

Hence the equality for d is proved. Simultaneously we get $\mu(\mathbb{R}^d) = 1$.

The d-dimensional Gaussian distribution with a diagonal covariance matrix A is the direct product of d Gaussian distributions on $\mathbb{R}$. But none of the d-dimensional Cauchy distributions is the direct product of one-dimensional ones.

EXAMPLE 2.13. ($d=1$) Let

$$\mu(B) = c(2\pi)^{-1/2}\int_{B\cap(0,\infty)} \mathrm{e}^{-c^2/(2x)}x^{-3/2}\mathrm{d}x, \qquad B \in \mathcal{B}(\mathbb{R})$$

with $c > 0$. This is the case of *one-sided strictly stable distributions* of index $1/2$, which we shall study in Chapter 3. We can check $\mu(\mathbb{R}) = 1$ by $c^2/x = y^2$ as

$$c(2\pi)^{-1/2}\int_0^{\infty} \mathrm{e}^{-c^2/(2x)}x^{-3/2}\mathrm{d}x = 2(2\pi)^{-1/2}\int_0^{\infty} \mathrm{e}^{-y^2/2}\mathrm{d}y = 1.$$

Let us find its Laplace transform

$$L_\mu(u) = c(2\pi)^{-1/2}\int_0^{\infty} \mathrm{e}^{-ux-c^2/(2x)}x^{-3/2}\mathrm{d}x.$$

Differentiation in $u > 0$ and the change of variables $ux = c^2/(2y)$ lead to

$$\begin{aligned}L'_\mu(u) &= -c(2\pi)^{-1/2}\int_0^{\infty} \mathrm{e}^{-ux-c^2/(2x)}x^{-1/2}\mathrm{d}x \\ &= -c^2(4\pi u)^{-1/2}\int_0^{\infty} \mathrm{e}^{-uy-c^2/(2y)}y^{-3/2}\mathrm{d}y = -c(2u)^{-1/2}L_\mu(u).\end{aligned}$$

Noting that $L_\mu(u)$ is continuous on $\{u \geq 0\}$ and $L_\mu(0) = 1$, we see that

$$L_\mu(u) = \exp(-c(2u)^{1/2}), \qquad u \geq 0. \tag{2.11}$$

The characteristic function is

$$\widehat{\mu}(z) = \exp(-c|z|^{1/2}(1 - \mathrm{i}\,\mathrm{sgn}\, z)). \tag{2.12}$$

In fact, let $\Phi(w) = \int \mathrm{e}^{wx}\mu(\mathrm{d}x)$ for complex w with $\mathrm{Re}\, w \leq 0$. As is shown in the proof of Proposition 2.6, $\Phi(w)$ is analytic on $\{\mathrm{Re}\, w < 0\}$ and continuous on $\{\mathrm{Re}\, w \leq 0\}$. For $w = -u$ we have $\Phi(-u) = \exp(-c(2u)^{1/2})$ by (2.11). Hence we have

$$\Phi(w) = \exp[-c(-2w)^{1/2}]$$

by choosing an appropriate branch of the square root function. That is,

$$\Phi(w) = \exp[-c|2w|^{1/2}\mathrm{e}^{(\mathrm{i}/2)\arg(-2w)}],$$

with $\arg(-2w)$ being continuous on $\{\mathrm{Re}\, w \leq 0, w \neq 0\}$ and vanishing for negative real w. Hence $\arg(-2w) = -(\pi/2)\,\mathrm{sgn}\, z$ for $w = \mathrm{i}z$ with non-zero real z. This implies (2.12).

EXAMPLE 2.14. ($d = 1$) The *exponential distribution* with parameter $\alpha > 0$ is defined by

$$\mu(B) = \alpha \int_{B\cap(0,\infty)} \mathrm{e}^{-\alpha x}\mathrm{d}x.$$

We have

$$L_\mu(u) = \alpha/(\alpha + u), \qquad u \geq 0, \tag{2.13}$$

$$\widehat{\mu}(z) = \alpha/(\alpha - \mathrm{i}z), \qquad z \in \mathbb{R}. \tag{2.14}$$

The mean of μ is $1/\alpha$.

EXAMPLE 2.15. ($d = 1$) For $c > 0$ and $\alpha > 0$,

$$\mu(B) = (\alpha^c/\Gamma(c)) \int_{B\cap(0,\infty)} x^{c-1}\mathrm{e}^{-\alpha x}\mathrm{d}x$$

is the Γ-*distribution* with parameters c, α. Sometimes c is called shape parameter and α scale parameter. If $c = 1$, then μ is exponential. We get

$$L_\mu(u) = (1 + \alpha^{-1}u)^{-c}, \qquad u \geq 0, \tag{2.15}$$

$$\widehat{\mu}(z) = (1 - \mathrm{i}\alpha^{-1}z)^{-c} = \exp[-c\log(1 - \mathrm{i}\alpha^{-1}z)], \qquad z \in \mathbb{R}, \tag{2.16}$$

where log is the principal value (that is, the imaginary part is in $(-\pi, \pi]$). The mean of μ is c/α. When $c = n/2$ with $n \in \mathbb{N}$ and $\alpha = 1/2$, statisticians call μ the χ^2-distribution with n degrees of freedom.

EXAMPLE 2.16. ($d = 1$) The *geometric distribution* with parameter p, $0 < p < 1$:

$$\mu\{k\} = pq^k, \qquad k \in \mathbb{Z}_+,$$
$$L_\mu(u) = p(1 - q\mathrm{e}^{-u})^{-1}, \qquad u \geq 0,$$
$$\widehat{\mu}(z) = p(1 - q\mathrm{e}^{\mathrm{i}z})^{-1}, \qquad z \in \mathbb{R},$$

where $q = 1 - p$. The *negative binomial distribution* with parameters $c > 0$ and p, $0 < p < 1$ is its generalization:

$$\begin{aligned} \mu\{k\} &= \binom{-c}{k} p^c(-q)^k \\ &= (k!)^{-1}(-c)(-c-1)\ldots(-c-k+1)p^c(-q)^k, \qquad k \in \mathbb{Z}_+, \\ L_\mu(u) &= p^c(1 - q\mathrm{e}^{-u})^{-c}, \qquad u \geq 0, \\ \widehat{\mu}(z) &= p^c(1 - q\mathrm{e}^{\mathrm{i}z})^{-c}, \qquad z \in \mathbb{R}. \end{aligned}$$

Notice that the parameter c is not restricted to positive integers. For $w = q\mathrm{e}^{\mathrm{i}z}$, $(1 - w)^{-c}$ stands for $\mathrm{e}^{-c\log(1-w)}$ (log is the principal value).

EXAMPLE 2.17. ($d = 1$) Let n be a positive integer, $0 < p < 1$, and $q = 1 - p$. The *binomial distribution* with parameters n, p is

$$\mu\{k\} = \binom{n}{k} p^k q^{n-k}, \qquad k = 0, 1, \ldots, n.$$

We have

$$L_\mu(u) = (p\mathrm{e}^{-u} + q)^n, \qquad u \geq 0,$$
$$\widehat{\mu}(z) = (p\mathrm{e}^{\mathrm{i}z} + q)^n, \qquad z \in \mathbb{R}.$$

EXAMPLE 2.18. ($d = 1$) The *uniform distribution* on $[-a, a]$ for $a > 0$ is

$$\mu(B) = (2a)^{-1} \int_{B\cap[-a,a]} \mathrm{d}x,$$
$$\widehat{\mu}(z) = (\sin az)/(az).$$

with the understanding that $(\sin az)/(az) = 1$ for $z = 0$.

EXAMPLE 2.19. The distribution concentrated at a single point $\gamma \in \mathbb{R}^d$ is the *δ-distribution* at γ and denoted by δ_γ. Its characteristic function is $\mathrm{e}^{\mathrm{i}\langle\gamma,z\rangle}$.

3. Poisson processes

We define and construct Poisson processes.

DEFINITION 3.1. A stochastic process $\{X_t \colon t \geq 0\}$ on $\mathbb{R}$ is a *Poisson process* with parameter $c > 0$ if it is a Lévy process and, for $t > 0$, X_t has Poisson distribution with mean ct.

THEOREM 3.2 (Construction). *Let $\{W_n : n = 0, 1, \ldots\}$ be a random walk on $\mathbb{R}$, defined on a probability space $(\Omega, \mathcal{F}, P)$, such that $T_n = W_n - W_{n-1}$ has exponential distribution with parameter $c > 0$. Define X_t by*

$$X_t(\omega) = n \text{ if and only if } W_n(\omega) \le t < W_{n+1}(\omega). \tag{3.1}$$

Then, $\{X_t\}$ is a Poisson process with parameter c.

A characteristic property of a random variable T with exponential distribution is that

$$P[\,T > s + t \,|\, T > s\,] = P[\,T > t\,], \qquad s \ge 0,\, t \ge 0, \tag{3.2}$$

called lack of memory. Here, for an event A with positive probability, $P[B|A]$ is the conditional probability of B given A, that is,

$$P[B|A] = P[B \cap A]/P[A]. \tag{3.3}$$

The property (3.2) follows easily from the definition of exponential distribution. Conversely, if a nonnegative random variable has the property of lack of memory, then its distribution is either exponential or δ_0. When we consider a model of arrival of customers at a service station and assume that the length of interval of successive arrivals has lack of memory, W_n is the waiting time until the arrival of the nth customer, X_t is the number of customers who arrived before the time t, and $\{X_t\}$ is a Poisson process.

Proof of Theorem 3.2. The random walk $\{W_n\}$ increases to ∞ almost surely, since

$$P[\,W_n \le t\,] \le P[\,T_1 \le t, \ldots, T_n \le t\,] = (P[\,T_1 \le t])^n \to 0, \qquad n \to \infty.$$

So we can define $\{X_t\}$ by (3.1). Obviously $X_0 = 0$ a.s. Example 2.15 says that W_n has Γ-distribution with parameters n, c. We have

$$P[\,X_t = n\,] = \mathrm{e}^{-ct}(n!)^{-1}(ct)^n, \qquad t > 0,\, n \ge 0. \tag{3.4}$$

In fact,

$$P[\,X_t = n\,] = P[\,W_n \le t < W_n + T_{n+1}\,] = P[\,(W_n, T_{n+1}) \in B\,],$$

where $B = \{(x, y) : 0 \le x \le t < x + y\}$, and, using independence of W_n and T_{n+1}, we have

$$\begin{aligned} P[\,X_t = n\,] &= c^{n+1}((n-1)!)^{-1} \iint_B x^{n-1}\mathrm{e}^{-cx}\mathrm{e}^{-cy}\mathrm{d}x\mathrm{d}y \\ &= c^{n+1}((n-1)!)^{-1} \int_0^t x^{n-1}\mathrm{e}^{-cx}\mathrm{d}x \int_{t-x}^{\infty} \mathrm{e}^{-cy}\mathrm{d}y = \mathrm{e}^{-ct}(n!)^{-1}(ct)^n. \end{aligned}$$

Next we show that

$$P[\,W_{n+1} > t + s \,|\, X_t = n\,] = \mathrm{e}^{-cs}, \qquad t > 0,\, s \ge 0,\, n \ge 0. \tag{3.5}$$

Calculation of the same sort as used for (3.4) leads to

$$P[X_t = n, W_{n+1} > t+s] = P[W_n \le t, W_n + T_{n+1} > t+s]$$
$$= c^{n+1}((n-1)!)^{-1} \int_0^t x^{n-1} e^{-cx} dx \int_{t+s-x}^{\infty} e^{-cy} dy = e^{-c(t+s)}(n!)^{-1}(ct)^n.$$

This and (3.4) yield (3.5) by the definition of the conditional probability.

Let $n \ge 0$ and $m \ge 1$. Let us consider the conditional distribution of $(W_{n+1} - t, T_{n+2}, \dots, T_{n+m})$ given $X_t = n$. It is equal to the distribution of $(T_1, T_2, \dots, T_m)$. To show this, let $P[W_n \le t < W_{n+1}] = a$ and observe that, for any $s_1, \dots, s_m \ge 0$,

$$P[W_{n+1} - t > s_1, T_{n+2} > s_2, \dots, T_{n+m} > s_m \mid X_t = n]$$
$$= P[W_n \le t, W_{n+1} - t > s_1, T_{n+2} > s_2, \dots, T_{n+m} > s_m]/a$$
$$= P[W_n \le t, W_{n+1} - t > s_1]P[T_{n+2} > s_2, \dots, T_{n+m} > s_m]/a$$
$$= P[W_{n+1} - t > s_1 \mid X_t = n]P[T_{n+2} > s_2, \dots, T_{n+m} > s_m]$$
$$= P[T_1 > s_1]P[T_2 > s_2, \dots, T_m > s_m]$$
$$= P[T_1 > s_1, T_2 > s_2, \dots, T_m > s_m].$$

Here we have used (3.5).

Now it follows that

$$P[X_{t+s} - X_t = m] = P[X_s = m], \qquad t > 0,\ s > 0. \tag{3.6}$$

In fact,

$$P[X_t = n, X_{t+s} - X_t = m] = P[X_t = n, X_{t+s} = n+m]$$
$$= P[X_t = n]P[W_{n+m} \le t+s < W_{n+m+1} \mid X_t = n]$$
$$= P[X_t = n]P[W_m \le s < W_{m+1}] = P[X_t = n]P[X_s = m],$$

where we have written $W_{n+m} \le t+s < W_{n+m+1}$ as

$$(W_{n+1} - t) + T_{n+2} + \cdots + T_{n+m} \le s < (W_{n+1} - t) + T_{n+2} + \cdots + T_{n+m+1}.$$

Addition over n gives (3.6).

The same argument shows that, for $0 \le t_0 < t_1 < \cdots < t_k$,

$$P[X_{t_0} = n_0, X_{t_1} - X_{t_0} = n_1, \dots, X_{t_k} - X_{t_{k-1}} = n_k]$$
$$= P[X_{t_0} = n_0, X_{t_1} = n_0 + n_1, \dots, X_{t_k} = n_0 + \cdots + n_k]$$
$$= P[X_{t_0} = n_0]P[X_{t_1 - t_0} = n_1, \dots, X_{t_k - t_0} = n_1 + \cdots + n_k].$$

Repeating this, we get the independent increments property, as

$$P[X_{t_0} = n_0, X_{t_1} - X_{t_0} = n_1, \dots, X_{t_k} - X_{t_{k-1}} = n_k]$$
$$= P[X_{t_0} = n_0]P[X_{t_1 - t_0} = n_1] \dots P[X_{t_k - t_{k-1}} = n_k]$$
$$= P[X_{t_0} = n_0]P[X_{t_1} - X_{t_0} = n_1] \dots P[X_{t_k} - X_{t_{k-1}} = n_k].$$

Sample functions of $\{X_t\}$ are right-continuous step functions with jumps of height 1. So (5) of Definition 1.6 is satisfied. This implies (4) also, and all the conditions are now checked. □

Let us describe, for the process $\{X_t\}$ constructed above, the conditional distribution of the positions of $W_1, \ldots, W_n$ in $[0, t]$ given $X_t = n$. For any interval I, the number of jumps of $X_t(\omega)$, $t \in I$, is denoted by $J(I) = J(I)(\omega)$.

PROPOSITION 3.3. *For $0 < s < t$ and $n \geq 1$, the conditional distribution of X_s given that $X_t = n$ is binomial with parameters n, s/t. For $0 = t_0 < t_1 < \cdots < t_k = t$ and $I_j = (t_{j-1}, t_j]$, the conditional distribution of $(J(I_1), \ldots, J(I_k))$ given that $X_t = n$ is multinomial with parameters n, $(t_1 - t_0)/t, \ldots, (t_k - t_{k-1})/t$, that is,*

$$
\begin{aligned}
(3.7) \qquad & P[\, J(I_1) = n_1, \ldots, J(I_k) = n_k \mid X_t = n \,] \\
&= \frac{n!}{(n_1!) \ldots (n_k!)} \left(\frac{t_1 - t_0}{t} \right)^{n_1} \cdots \left(\frac{t_k - t_{k-1}}{t} \right)^{n_k}
\end{aligned}
$$

for any nonnegative integers $n_1, \ldots, n_k$ with $n_1 + \cdots + n_k = n$.

Proof. Using the independent increments property and the Poisson distribution, we have

$$
\begin{aligned}
&\text{left-hand side of (3.7)} \\
&= P[\, X_{t_1} = n_1,\, X_{t_2} = n_1 + n_2, \ldots, X_{t_k} = n_1 + \cdots + n_k \,]/P[\, X_t = n \,] \\
&= \mathrm{e}^{-ct_1} \frac{(ct_1)^{n_1}}{n_1!} \ldots \mathrm{e}^{-c(t_k - t_{k-1})} \frac{(c(t_k - t_{k-1}))^{n_k}}{n_k!} \left(\mathrm{e}^{-ct} \frac{(ct)^n}{n!} \right)^{-1},
\end{aligned}
$$

which proves (3.7). □

PROPOSITION 3.4. *Let $n \geq 1$ and $t > 0$. The conditional distribution of $W_1, \ldots, W_n$ given that $X_t = n$ coincides with the distribution of the order statistics $V_1 \leq V_2 \leq \cdots \leq V_n$ obtained from n samples $Z_1, \ldots, Z_n$ from the population with uniform distribution on $[0, t]$.*

Proof. The random variables $Z_1, \ldots, Z_n$ are independent, each with uniform distribution on $[0, t]$. Arrangement of $Z_1(\omega), \ldots, Z_n(\omega)$ in increasing order gives $V_1(\omega) \leq \cdots \leq V_n(\omega)$. The assertion to be proved is that

$$
(3.8) \qquad P[\, W_1 \leq t_1, \ldots, W_n \leq t_n \mid X_t = n \,] = P[\, V_1 \leq t_1, \ldots, V_n \leq t_n \,]
$$

for any $t_1, \ldots, t_n \in [0, t]$. It suffices to show it for $0 \leq t_1 \leq \cdots \leq t_n \leq t$. Further, we may assume that $0 < t_1 < \cdots < t_n < t$. Let $t_0 = 0$, $t_{n+1} = t$, and $I_j = (t_{j-1}, t_j]$ for $j = 1, \ldots, n+1$. Let $J'(I_j)$ be the number of $V_1, \ldots, V_n$ that fall into I_j. Then, for any nonnegative integers $m_1, \ldots, m_{n+1}$ with

$m_1 + \cdots + m_{n+1} = n$, we have

$$P[\,J'(I_1) = m_1, \ldots, J'(I_{n+1}) = m_{n+1}\,]$$
$$= \frac{n!}{(m_1!)\ldots(m_{n+1}!)}\left(\frac{t_1 - t_0}{t}\right)^{m_1}\cdots\left(\frac{t_{n+1} - t_n}{t}\right)^{m_{n+1}},$$

which equals

$$P[\,J(I_1) = m_1, \ldots, J(I_{n+1}) = m_{n+1} \,|\, X_t = n\,]$$

by the preceding proposition. This shows (3.8) for $0 < t_1 < \cdots < t_n < t$. □

4. Compound Poisson processes

We study compound Poisson distributions and processes, which generalize Poisson distributions and processes.

DEFINITION 4.1. A distribution μ on $\mathbb{R}^d$ is *compound Poisson* if, for some $c > 0$ and for some distribution σ on $\mathbb{R}^d$ with $\sigma\{0\} = 0$,

$$\widehat{\mu}(z) = \exp(c(\widehat{\sigma}(z) - 1)), \qquad z \in \mathbb{R}^d. \tag{4.1}$$

The Poisson distribution is a special case where $d = 1$ and $\sigma = \delta_1$.

DEFINITION 4.2. Let $c > 0$ and let σ be a distribution on $\mathbb{R}^d$ with $\sigma(\{0\}) = 0$. A stochastic process $\{X_t : t \geq 0\}$ on $\mathbb{R}^d$ is a *compound Poisson process* associated with c and σ if it is a Lévy process and, for $t > 0$, X_t has a compound Poisson distribution

$$E[\mathrm{e}^{\mathrm{i}\langle z, X(t)\rangle}] = \exp(tc(\widehat{\sigma}(z) - 1)), \qquad z \in \mathbb{R}^d. \tag{4.2}$$

The c and σ are uniquely determined by $\{X_t\}$. Here we use $\sigma\{0\} = 0$; see Exercise 6.15.

THEOREM 4.3 (Construction). *Let $\{N_t : t \geq 0\}$ be a Poisson process with parameter $c > 0$ and $\{S_n : n = 0, 1, \ldots\}$ be a random walk on $\mathbb{R}^d$ defined on a common probability space $(\Omega, \mathcal{F}, P)$. Assume that $\{N_t\}$ and $\{S_n\}$ are independent and $P[\,S_1 = 0\,] = 0$. Define*

$$X_t(\omega) = S_{N_t(\omega)}(\omega). \tag{4.3}$$

Then $\{X_t\}$ is a compound Poisson process satisfying (4.2), where σ is the distribution of S_1.

Proof. Let $B, B_0, \ldots \in \mathcal{B}(\mathbb{R}^d)$. We have, from (4.3),

$$\begin{aligned} P[\,X_t \in B\,] &= P[\,S_{N_t} \in B\,] \\ &= \sum_{n=0}^{\infty} P[\,N_t = n, S_n \in B\,] = \sum_{n=0}^{\infty} P[\,N_t = n\,]P[\,S_n \in B\,]. \end{aligned} \tag{4.4}$$

For $0 \leq t_0 < t_1$ we have

$$P[\,X_{t_0} \in B_0,\, X_{t_1} - X_{t_0} \in B_1\,]$$

$$
\begin{aligned}
&= P[\, S_{N(t_0)} \in B_0,\, S_{N(t_1)} - S_{N(t_0)} \in B_1 \,] \\
&= \sum_{n_0, n_1} P[\, N_{t_0} = n_0,\, N_{t_1} - N_{t_0} = n_1,\, S_{n_0} \in B_0,\, S_{n_0+n_1} - S_{n_0} \in B_1 \,] \\
&= \sum P[\, N_{t_0} = n_0 \,] P[\, N_{t_1 - t_0} = n_1 \,] P[\, S_{n_0} \in B_0 \,] P[\, S_{n_1} \in B_1 \,] \\
&= P[\, X_{t_0} \in B_0 \,] P[\, X_{t_1 - t_0} \in B_1 \,].
\end{aligned}
$$

Here we used the stationary independent increments property of the Poisson process and the random walk. Letting $B_0 = \mathbb{R}^d$, we get $X_{t_1} - X_{t_0} \stackrel{\mathrm{d}}{=} X_{t_1 - t_0}$.

Similarly, for $0 \leq t_0 < t_1 < \cdots < t_k$,

$$
\begin{aligned}
&P[\, X_{t_0} \in B_0,\, X_{t_1} - X_{t_0} \in B_1, \ldots, X_{t_k} - X_{t_{k-1}} \in B_k \,] \\
&\quad = \sum_{n_0, \ldots, n_k} P[\, N_{t_0} = n_0,\, N_{t_1} - N_{t_0} = n_1, \ldots, N_{t_k} - N_{t_{k-1}} = n_k,\, S_{n_0} \in B_0, \\
&\qquad\qquad S_{n_0+n_1} - S_{n_0} \in B_1, \ldots, S_{n_0+\cdots+n_k} - S_{n_0+\cdots+n_{k-1}} \in B_k \,] \\
&\quad = \sum P[\, N_{t_0} = n_0 \,] P[\, N_{t_1 - t_0} = n_1 \,] \ldots P[\, N_{t_k - t_{k-1}} = n_k \,] \\
&\qquad \times P[\, S_{n_0} \in B_0 \,] P[\, S_{n_1} \in B_1 \,] \ldots P[\, S_{n_k} \in B_k \,] \\
&\quad = P[\, X_{t_0} \in B_0 \,] P[\, X_{t_1} - X_{t_0} \in B_1 \,] \ldots P[\, X_{t_k} - X_{t_{k-1}} \in B_k \,].
\end{aligned}
$$

Hence (1) and (3) of Definition 1.6 of a Lévy process are shown. The properties (2) and (5) are obvious; (4) follows from (5). As to the characteristic function,

$$
\begin{aligned}
E[\mathrm{e}^{\mathrm{i}\langle z, X(t) \rangle}] &= \sum_{n=0}^{\infty} P[\, N_t = n \,] E[\mathrm{e}^{\mathrm{i}\langle z, S(n) \rangle}] \\
&= \sum_{n=0}^{\infty} \mathrm{e}^{-ct} (n!)^{-1} (ct)^n \widehat{\sigma}(z)^n = \exp(ct(\widehat{\sigma}(z) - 1)),
\end{aligned}
$$

as asserted. □

Imagine the expenditure of an insurance company. If the number of accidents in t days is a Poisson process and payments to them are independent and identically distributed with distribution σ, then the total payment in t days by the company is a compound Poisson process X_t for $d = 1$. Thus compound Poisson processes are important in risk theory.

In the process $\{X_t\}$ constructed above, $N_t(\omega)$ represents the number of jumps of $X_s(\omega)$, $s \in (0, t]$. The nth jumping time $W_n(\omega)$ for $X_t(\omega)$ coincides with W_n in the preceding section. The amount $Z_n(\omega)$ of the nth jump of $X_t(\omega)$ equals $S_n(\omega) - S_{n-1}(\omega)$.

Remark 4.4. Let $\mathcal{B}_0$ be the collection of Borel sets B in $(0, \infty) \times (\mathbb{R}^d \setminus \{0\})$ such that $B \subset (0, t] \times (\mathbb{R}^d \setminus \{0\})$ for some t. For $B \in \mathcal{B}_0$ denote by $J(B, \omega)$

the number of js satisfying $(W_j(\omega), Z_j(\omega)) \in B$. Let ρ be the product measure $c\,\mathrm{Leb}(\mathrm{d}t)\sigma(\mathrm{d}x)$ on $(0,\infty) \times (\mathbb{R}^d \setminus \{0\})$. Then we can prove the following.

(i) Let $B \in \mathcal{B}_0$. If $\rho(B) = 0$, then $J(B) = 0$ a.s. If $\rho(B) > 0$, then $J(B)$ has Poisson distribution with mean $\rho(B)$.

(ii) If $B_1, \ldots, B_k \in \mathcal{B}_0$ are disjoint, then $J(B_1), \ldots, J(B_k)$ are independent.

(iii) If $B_1, \ldots, B_k \in \mathcal{B}_0$ are disjoint and $\bigcup_{j=1}^k B_j = (0,t] \times (\mathbb{R}^d \setminus \{0\})$, then the conditional distribution of $(J(B_1), \ldots, J(B_k))$ given that $N_t = n$ is multinomial with parameters $n, (ct)^{-1}\rho(B_1), \ldots, (ct)^{-1}\rho(B_k)$.

The proof will be given in Chapter 4 in a more general form.

PROPOSITION 4.5. *Let $c > 0$ and let σ be a distribution on $(0,\infty)$. Let $\{X_t\}$ be the compound Poisson process on $\mathbb{R}$ associated with c and σ. Then X_t is increasing in t almost surely, and*

$$E[\mathrm{e}^{-uX_t}] = \exp\left[tc\int_{(0,\infty)} (\mathrm{e}^{-ux} - 1)\sigma(\mathrm{d}x)\right] \quad \textit{for } u \geq 0. \tag{4.5}$$

Proof. Let $\{X_t^0\}$ be the process constructed in Theorem 4.3 from c and σ. We have $X_t \stackrel{\mathrm{d}}{=} X_t^0$ for each t and their characteristic functions equal the right-hand side of (4.2). Thus $P[X_t \geq 0] = P[X_t^0 \geq 0] = 1$. Hence, for any $s < t$, $P[X_t \geq X_s] = P[X_t - X_s \geq 0] = P[X_{t-s} \geq 0] = 1$. It follows that

$$P[\,X_s \leq X_t \text{ for any } s, t \in \mathbb{Q} \text{ satisfying } 0 \leq s < t\,] = 1.$$

Hence X_t is increasing in t almost surely by the right-continuity. By using Proposition 2.6(ii) we see that $E[\mathrm{e}^{-uX_t^0}]$ equals the right-hand side of (4.5) in the same way as in the last part of the proof of Theorem 4.3. Then note that $E[\mathrm{e}^{-uX_t}] = E[\mathrm{e}^{-uX_t^0}]$. □

EXAMPLE 4.6. Let $d = 1$, $0 < p < 1$, $q = 1 - p$, $c = -\log p$, and let σ be concentrated on positive integers and

$$\sigma\{k\} = c^{-1}k^{-1}q^k, \qquad k = 1, 2, \ldots.$$

Then the distribution at time t of the corresponding compound Poisson process $\{X_t\}$ is negative binomial with parameters t, p (Example 2.16). In particular, for $t = 1$, it is geometric with parameter p. In fact,

$$L_\sigma(u) = c^{-1}\sum_{k=1}^{\infty} \mathrm{e}^{-ku}k^{-1}q^k = -c^{-1}\log(1 - q\mathrm{e}^{-u}), \quad u \geq 0,$$

$$E[\mathrm{e}^{-uX_t}] = \exp[-t\log(1 - q\mathrm{e}^{-u}) - tc] = p^t(1 - q\mathrm{e}^{-u})^{-t}.$$

The Bessel function $J_\nu(z)$ with order $\nu \in \mathbb{R}$ is defined by

$$J_\nu(z) = \left(\frac{z}{2}\right)^\nu \sum_{k=0}^{\infty} \frac{(-1)^k (z/2)^{2k}}{k!\,\Gamma(\nu + k + 1)} \quad \text{for } \nu \neq -1, -2, \ldots, \tag{4.6}$$

$$J_{-n}(z) = \lim\nolimits_{\nu \to -n} J_\nu(z) = (-1)^n J_n(z) \quad \text{for } n = 1, 2, \ldots, \tag{4.7}$$

where z is a variable in $\mathbb{C}\setminus(-\mathbb{R}_+)$ ($-\mathbb{R}_+$ is the set of nonpositive reals) and $(z/2)^\nu$ is the branch analytic on $\mathbb{C}\setminus(-\mathbb{R}_+)$ and positive on $\mathbb{R}_+\setminus\{0\}$. The modified Bessel functions $I_\nu(z)$ and $K_\nu(z)$ are defined by

$$I_\nu(z) = \mathrm{e}^{-\mathrm{i}\nu\pi/2} J_\nu(\mathrm{i}z) \quad \text{for } \nu \in \mathbb{R}, \tag{4.8}$$

$$K_\nu(z) = \frac{\pi}{2\sin\nu\pi}(I_{-\nu}(z) - I_\nu(z)) \quad \text{for } \nu \notin \mathbb{Z}, \tag{4.9}$$

$$\begin{aligned} K_n(z) &= K_{-n}(z) = \lim_{\nu\to n} K_\nu(z) \\ &= \frac{(-1)^n}{2}\left[\frac{\partial I_{-\nu}}{\partial\nu} - \frac{\partial I_\nu}{\partial\nu}\right]_{\nu=n} \quad \text{for } n = 1, 2, \ldots. \end{aligned} \tag{4.10}$$

Thus we have

$$I_\nu(z) = (z/2)^\nu \sum_{k=0}^{\infty} \frac{(z/2)^{2k}}{k!\,\Gamma(\nu+k+1)} \quad \text{for } \nu \neq -1, -2, \ldots, \tag{4.11}$$

where $z \in \mathbb{C}\setminus \mathrm{i}\mathbb{R}_+$ and $(z/2)^\nu$ is the branch analytic on $\mathbb{C}\setminus \mathrm{i}\mathbb{R}_+$ and positive on $\mathbb{R}_+\setminus\{0\}$. From the definition we have

$$I_{-n}(z) = I_n(z) \quad \text{for } n \in \mathbb{Z}, \tag{4.12}$$

$$K_{-\nu}(z) = K_\nu(z) \quad \text{for } \nu \in \mathbb{R}. \tag{4.13}$$

EXAMPLE 4.7. Let $d = 1$ and $\sigma = p\delta_1 + q\delta_{-1}$ with $0 < p < 1$ and $q = 1 - p$. Let $\{X_t\}$ be the compound Poisson process with this σ and $c = 1$. Then, for $t > 0$,

$$P[\,X_t = k\,] = \mathrm{e}^{-t}(p/q)^{k/2} I_k(2(pq)^{1/2}t), \qquad k \in \mathbb{Z}. \tag{4.14}$$

In the case $p = q = 1/2$, we have $E[\mathrm{e}^{\mathrm{i}zX_t}] = \mathrm{e}^{t(\cos z - 1)}$ and

$$P[\,X_t = k\,] = \mathrm{e}^{-t} I_k(t), \qquad k \in \mathbb{Z}. \tag{4.15}$$

Let us show (4.14). For integers $n \geq k \geq 0$ with $n - k = 2j$ being even, $S_n = k$ occurs if and only if among the n jumps $(n+k)/2$ are positive and $(n-k)/2$ are negative. Thus

$$P[\,S_n = k\,] = \binom{n}{(n+k)/2} p^{(n+k)/2} q^{(n-k)/2} = \binom{k+2j}{k+j} p^{k+j} q^j.$$

Hence,

$$P[\,X_t = k\,] = \sum_{j=0}^{\infty} \frac{\mathrm{e}^{-t} t^{k+2j}}{(k+2j)!} \binom{k+2j}{k+j} p^{k+j} q^j = \mathrm{e}^{-t}\left(\frac{p}{q}\right)^{k/2} \sum_{j=0}^{\infty} \frac{((pq)^{1/2}t)^{k+2j}}{j!\,(k+j)!}.$$

This is (4.14) for $k \geq 0$ by (4.11). By the same formula with p and q interchanged, we get the expression for $P[\,X_t = -k\,]$.

5. Brownian motion

It was by N. Wiener [**573**] in 1923 that Brownian motion in the mathematical sense was defined and shown to exist. Since then it has been deeply investigated by P. Lévy, S. Kakutani, K. Itô, and many others. Its fine structure is still being studied in these days.

DEFINITION 5.1. A stochastic process $\{X_t\colon t \geq 0\}$ on $\mathbb{R}^d$ defined on a probability space $(\Omega, \mathcal{F}, P)$ is a *Brownian motion*, or a *Wiener process*, if it is a Lévy process and if,

(1) for $t > 0$, X_t has a Gaussian distribution with mean 0 and covariance matrix tI (I is the identity matrix),
(2) there is $\Omega_0 \in \mathcal{F}$ with $P[\Omega_0] = 1$ such that, for every $\omega \in \Omega_0$, $X_t(\omega)$ is continuous in t.

A Brownian motion on $\mathbb{R}^d$ is sometimes called d-dimensional Brownian motion. Several methods are known for proving the existence of a Brownian motion (see Billingsley [**34**], Breiman [**66**], Hida [**200**], Knight [**289**]). We can prove that any Lévy process satisfying (1) in Definition 5.1 automatically fulfills the condition (2), using that, by definition, a Lévy process has right-continuous sample functions with left limits almost surely. A proof will be given in Chapter 2, together with the existence of general Lévy processes. In this section we set aside the problem of the existence of a Brownian motion and exhibit some of its properties.

PROPOSITION 5.2. *Let $\{X(t)\}$ be a stochastic process on $\mathbb{R}^d$ and let $X_1(t), \ldots, X_d(t)$ be the components of $X(t)$. Then the following are equivalent.*

(1) *$\{X(t)\}$ is a d-dimensional Brownian motion.*
(2) *$\{X_j(t)\}$ is a one-dimensional Brownian motion for each j and $\{X_1(t)\}, \ldots, \{X_d(t)\}$ are independent.*

Proof. Assume (2). Let $0 \leq t_0 < \cdots < t_n$. Since the family

$$\{X_j(t_l) - X_j(t_{l-1})\colon l = 1, \ldots, n,\ j = 1, \ldots, d\} \tag{5.1}$$

is independent, the family

$$\{X(t_l) - X(t_{l-1})\colon l = 1, \ldots, n\} \tag{5.2}$$

is independent. As it is easy to check the other conditions in Definition 1.6, $\{X(t)\}$ is a Lévy process. For $0 \leq s < t$

$$E[\exp(\mathrm{i}z(X_j(t) - X_j(s)))] = \exp(-2^{-1}(t-s)z^2), \qquad z \in \mathbb{R},$$

and hence, by the independence of the components,

$$E[\exp(\mathrm{i}\langle z, X(t) - X(s)\rangle)] = \exp(-2^{-1}(t-s)|z|^2), \qquad z \in \mathbb{R}^d. \tag{5.3}$$

Almost sure continuity of $X(t)$ follows from that of the components. Therefore $\{X(t)\}$ is a d-dimensional Brownian motion.

Conversely assume (1). Since the family (5.2) is independent for $0 \leq t_0 < \cdots < t_n$, $\{X_j(t)\}$ has independent increments for each j. The other conditions are obviously satisfied and $\{X_j(t)\}$ is a one-dimensional Brownian motion for each j. Let us show their independence. As (5.3) means

$$E\left[\exp\left(\mathrm{i}\sum_{j=1}^{d} z_j(X_j(t)-X_j(s))\right)\right] = \prod_{j=1}^{d}\exp(-\tfrac{1}{2}(t-s)z_j{}^2)$$

for $0 \le s < t$, the family $\{X_j(t) - X_j(s)\colon j = 1,\ldots,d\}$ is independent by Proposition 2.5(iv). Hence, for $0 \le t_0 < \cdots < t_n$, the independence of the family (5.2) implies that of the family (5.1). Write $Y_j = (X_j(t_l) - X_j(t_{l-1})\colon l = 1,\ldots,n)$ and $Z_j = (X_j(t_l)\colon l = 1,\ldots,n)$. It follows from the independence of (5.1) that $\{Y_j\colon j = 1,\ldots,d\}$ is independent. Thus $\{Z_j\colon j = 1,\ldots,d\}$ is independent. Therefore, for any choice of $t_{1,1},\ldots,t_{1,m(1)},\ldots,t_{d,1},\ldots,t_{d,m(d)} \in [0,\infty)$, define $W_j = (X_j(t_{j,l})\colon l = 1,\ldots,m(j))$ and find that $\{W_j\colon j = 1,\ldots,d\}$ is independent (it is enough to consider the whole of $t_{1,1},\ldots,t_{d,m(d)}$ as $t_1,\ldots,t_n$). This shows that $\{X_1(t)\},\ldots,\{X_d(t)\}$ are independent. □

PROPOSITION 5.3. *Let $\{X(t)\}$ be a d-dimensional Brownian motion. For any choice of $t_1,\ldots,t_n \in [0,\infty)$, $(X(t_l)\colon l = 1,\ldots,n)$ has Gaussian distribution on $\mathbb{R}^{nd}$ with mean 0 and the covariance matrix is determined by*

$$E[X_j(t_l)X_k(t_m)] = \delta_{jk}(t_l \wedge t_m). \tag{5.4}$$

Here δ_{jk} is 1 or 0 according as $j = k$ or $j \ne k$.

Proof. In general, if $(Y_l\colon l = 1,\ldots,n)$ has Gaussian distribution on $\mathbb{R}^n$ with mean 0 and if $Z_m = \sum_{l=1}^{n} c_{ml}Y_l$, $m = 1,\ldots,n'$, with real numbers c_{ml}, then $(Z_m\colon m = 1,\ldots,n')$ is Gaussian distributed on $\mathbb{R}^{n'}$ with mean 0. Let $0 = t_0 \le t_1 \le \cdots \le t_n$. Since $(X_j(t_l) - X_j(t_{l-1})\colon l = 1,\ldots,n,\ j = 1,\ldots,d)$ is Gaussian distributed on $\mathbb{R}^{nd}$ with mean 0, we see that $(X_j(t_l)\colon l = 1,\ldots,n,\ j = 1,\ldots,d)$ is also Gaussian distributed on $\mathbb{R}^{nd}$ with mean 0. To see (5.4) for $j \ne k$, use the independence of the components. For $j = k$ (5.4) follows from

$$E[X_j(s)X_j(t)] = E[X_j(s\wedge t)^2] = s \wedge t,$$

completing the proof. □

We note invariance of the Brownian motion under some transformations. The property (ii) is the selfsimilarity with exponent 1/2, which will be studied in Chapter 3.

THEOREM 5.4. *Let $\{X(t)\}$ be a d-dimensional Brownian motion.*

(i) *$\{-X(t)\}$ is a d-dimensional Brownian motion.*

(ii) *For each $c > 0$, $\{c^{-1/2}X(ct)\}$ is a d-dimensional Brownian motion.*

(iii) *Define $Y(t) = tX(t^{-1})$ for $t > 0$ and $Y(0) = 0$. Then $\{Y(t)\}$ is a d-dimensional Brownian motion.*

Proof. The assertion (i) follows from the symmetry of Gaussian distributions with mean 0. To see (ii), notice that, for $0 \le s < t$,

$$E[\exp(\mathrm{i}\langle z, c^{-1/2}X(ct) - c^{-1/2}X(cs)\rangle)]$$
$$= \exp(-\tfrac{1}{2}(ct - cs)|c^{-1/2}z|^2) = \exp(-\tfrac{1}{2}(t-s)|z|^2).$$

The other conditions are easy to check. Let us show (iii). For any choice of $t_1, \ldots, t_n$, $(Y(t_l) \colon l = 1, \ldots, n)$ is Gaussian distributed with mean 0 as in the preceding proposition. Further, for any positive s, t, we have

$$E[Y_j(s)Y_k(t)] = stE[X_j(s^{-1})X_k(t^{-1})] = \delta_{jk}(s \wedge t).$$

It follows that $\{Y(t)\}$ is identical in law with a Brownian motion. Let Ω_0 be the set for $\{X(t)\}$ in Definition 5.1. For any $\omega \in \Omega_0$, $Y(t, \omega)$ is continuous in $t > 0$ by the definition of $Y(t)$. We claim that $Y(t) \to 0$ as $t \to 0$ a.s. Define

$$\Omega_1 = \bigcap_{n=1}^{\infty} \bigcup_{m=1}^{\infty} \bigcap_{t \in \mathbb{Q} \cap (0, 1/m)} \{|X(t)| \le 1/n\}$$

and define Ω_1' in the same way with $Y(t)$ in place of $X(t)$. Then $\Omega_0 \cap \Omega_1' = \Omega_0 \cap \{\lim_{t \downarrow 0} Y(t) = 0\}$ and $\Omega_0 \cap \Omega_1 = \Omega_0 \cap \{\lim_{t \downarrow 0} X(t) = 0\} = \Omega_0 \cap \{X(0) = 0\}$. We have $P[\Omega_1'] = P[\Omega_1]$ by the identity in law of $\{Y(t)\}$ and $\{X(t)\}$. Thus $P[\Omega_0 \cap \Omega_1'] = P[\Omega_0 \cap \Omega_1] = 1$. □

We give some properties of sample functions of a Brownian motion. In the rest of this section let $\{X(t)\}$ be a Brownian motion. The dimension is 1 except in Theorem 5.8.

THEOREM 5.5 (Behavior for large t). ($d = 1$) *Fix a sequence $t_n \uparrow \infty$. Then*

$$\limsup_{n \to \infty} X(t_n) = \infty \qquad a.\,s., \tag{5.5}$$

$$\liminf_{n \to \infty} X(t_n) = -\infty \qquad a.\,s. \tag{5.6}$$

Proof. Since $X(t_n) \stackrel{\mathrm{d}}{=} t_n^{1/2} X(1)$, we have

$$P[\,X(t_n) > K\,] = P[\,X(1) > t_n{}^{-1/2}K\,] \to 1/2, \qquad n \to \infty,$$

for any K. By Fatou's lemma

$$P[\,X(t_n) > K \text{ for infinitely many } n\,]$$
$$= E\Big[\limsup_{n \to \infty} 1_{\{X(t_n) > K\}}\Big] \ge \limsup_{n \to \infty} E[1_{\{X(t_n) > K\}}] = 1/2.$$

Hence $P[\,\limsup_{n \to \infty} X(t_n) > K\,] \ge 1/2$. Therefore

$$P\Big[\limsup_{n \to \infty} X(t_n) = \infty\Big] \ge 1/2.$$

Let $t_0 = 0$ and let $Z_n = X(t_n) - X(t_{n-1})$. Then $\{Z_n\}$ is independent and $X(t_n) = Z_1 + \cdots + Z_n$. We have

$$\left\{\limsup_{n\to\infty} X(t_n) = \infty\right\} = \left\{\limsup_{n\to\infty}(X(t_n) - X(t_m)) = \infty\right\}$$
$$\in \sigma(Z_{m+1}, Z_{m+2}, \dots).$$

for each m. So Kolmogorov's 0–1 law (Theorem 1.14) says that this event has probability 0 or 1. Since the probability is not less than 1/2, it must be 1. By the symmetry implied by Theorem 5.4(i), (5.6) is automatic from (5.5). □

THEOREM 5.6 (Behavior for small t). ($d = 1$) *Let*

$$T_0(\omega) = \inf\{t > 0\colon X_t(\omega) > 0\},$$
$$T_0'(\omega) = \inf\{t > 0\colon X_t(\omega) < 0\}$$

for $\omega \in \Omega$. *Then*

$$T_0 = 0 \qquad a.\ s., \tag{5.7}$$

$$T_0' = 0 \qquad a.\ s. \tag{5.8}$$

Proof. Let $t_n \downarrow 0$. Use $Y(t) = tX(t^{-1})$. It follows from Theorems 5.4(iii) and 5.5 that

$$P[\,X(t_n) > 0 \text{ for infinitely many } n\,]$$
$$= P[\,Y(t_n) > 0 \text{ for infinitely many } n\,]$$
$$= P[\,X({t_n}^{-1}) > 0 \text{ for infinitely many } n\,] = 1.$$

This shows (5.7). The symmetry leads (5.7) to (5.8). □

THEOREM 5.7 (Non-monotonicity). ($d = 1$) *Almost surely there is no interval in which* $X(t, \omega)$ *is monotone.*

Proof. Let $[a, b] \subset [0, \infty)$. Using the set Ω_0 in (2) of Definition 5.1, let

$$A^{[a,b]} = \{\omega \in \Omega_0\colon X(t, \omega) \text{ is increasing in } t \in [a, b]\}.$$

We claim that $A^{[a,b]}$ is an event with probability 0. For $t_{n,k} = a + k(b-a)/n$, let

$$A_{n,k} = \{\omega \in \Omega_0\colon X(t_{n,k-1}, \omega) \le X(t_{n,k}, \omega)\}.$$

Then

$$A^{[a,b]} = \bigcap_{n=1}^{\infty} \bigcap_{k=1}^{n} A_{n,k} \in \mathcal{F}.$$

Since

$$P[A_{n,k}] = P[\,X(t_{n,k}) - X(t_{n,k-1}) \ge 0\,] = \tfrac{1}{2}$$

and $\{A_{n,k}\colon k = 1, \dots, n\}$ is independent, we have $P\left[\bigcap_{k=1}^{n} A_{n,k}\right] = 2^{-n}$. Hence $P[A^{[a,b]}] = 0$. The set of $\omega \in \Omega_0$ such that $X(t, \omega)$ is increasing in some interval is the union of $A^{[a,b]}$ with $a, b \in \mathbb{Q} \cap [0, \infty)$, $a < b$, and hence, has probability 0. Similarly the set of $\omega \in \Omega_0$ such that $X(t, \omega)$ is decreasing in some interval is of probability 0. □

THEOREM 5.8. *Let $\{X(t)\}$ be a d-dimensional Brownian motion. Fix $t > 0$ and let*

$$\Delta_n: \qquad 0 = t_{n,0} < t_{n,1} < \cdots < t_{n,N(n)} = t \tag{5.9}$$

be a sequence of partitions of $[0, t]$ such that

$$\operatorname{mesh}(\Delta_n) = \max_{1 \le k \le N(n)} (t_{n,k} - t_{n,k-1}) \to 0 \text{ as } n \to \infty. \tag{5.10}$$

Let

$$Y_n = \sum_{k=1}^{N(n)} |X(t_{n,k}) - X(t_{n,k-1})|^2. \tag{5.11}$$

Then

$$E[|Y_n - td|^2] \to 0 \quad \text{as } n \to \infty.$$

Proof. Each component $\{X_j(t)\}$ of $\{X(t)\}$ is a one-dimensional Brownian motion and

$$|X(t_{n,k}) - X(t_{n,k-1})|^2 = \sum_{j=1}^{d} |X_j(t_{n,k}) - X_j(t_{n,k-1})|^2.$$

Hence it suffices to prove the assertion for $d = 1$. Assuming $d = 1$, we have

$$E[Y_n] = \sum_{k=1}^{N(n)} E[|X(t_{n,k}) - X(t_{n,k-1})|^2] = \sum_{k=1}^{N(n)} (t_{n,k} - t_{n,k-1}) = t.$$

Thus $E[|Y_n - t|^2]$ is the variance of Y_n. Use the fact that a Gaussian random variable on $\mathbb{R}$ with mean 0 and variance t has fourth moment $3t^2$. Then

$$\begin{aligned}
\operatorname{Var} Y_n &= \sum_{k=1}^{N(n)} \operatorname{Var}(|X(t_{n,k}) - X(t_{n,k-1})|^2) \\
&= \sum_{k=1}^{N(n)} \left(E[|X(t_{n,k}) - X(t_{n,k-1})|^4] - (E[|X(t_{n,k}) - X(t_{n,k-1})|^2])^2\right) \\
&= 2 \sum_{k=1}^{N(n)} (t_{n,k} - t_{n,k-1})^2 \le 2t \operatorname{mesh}(\Delta_n) \to 0,
\end{aligned}$$

which shows the assertion. □

Existence of a nowhere differentiable continuous function is known. Such a function was first constructed by K. Weierstrass in the 1870s as an infinite sum of trigonometric functions. Sample functions of a Brownian motion have this property.

THEOREM 5.9 (Nowhere differentiability). ($d = 1$) *Almost surely $X_t(\omega)$ is nowhere differentiable in t.*

Proof (Dvoretzky, Erdős, and Kakutani [**117**]). For positive integers K, N, and n define

$$A_n(K, N) = \{\omega\colon \text{ there is } s \in [0, N - N/n) \text{ such that } |t - s| \le 2N/n \\ \text{implies } |X_t(\omega) - X_s(\omega)| \le K|t - s|\}.$$

We have

$$\{\omega\colon X_t(\omega) \text{ is differentiable at some } t \in [0, N)\} \subset \bigcup_{K=1}^{\infty} \bigcup_{n=1}^{\infty} A_n(K, N).$$

Here, by differentiable at $t = 0$ we mean differentiable at $t = 0$ on the right. Now we fix K and N and write $A_n(K, N)$ as A_n. Let

$$Y_{n,k} = \max_{l=0,1,2} |X_{(k+l)N/n} - X_{(k+l-1)N/n}|, \qquad k = 0, \ldots, n - 2,$$

where, for $k = 0$, the maximum is taken over $l = 1, 2$. Define

$$B_n = \bigcup_{k=0}^{n-2} \{Y_{n,k} \le 4KN/n\}.$$

Then $A_n \subset B_n$ and

$$P[B_n] \le {a_n}^2 + (n - 2){a_n}^3 \quad \text{with} \quad a_n = P[\,|X_{N/n}| \le 4KN/n\,].$$

Since $X_{N/n} \stackrel{\mathrm{d}}{=} (N/n)^{1/2} X_1$,

$$a_n = P[\,|X_1| \le 4K(N/n)^{1/2}\,] \le 8K(N/n)^{1/2}(2\pi)^{-1/2}.$$

It follows that $P[B_n] \to 0$ as $n \to \infty$, and hence $P[\liminf B_n] = 0$. As A_n increases with n, we have

$$\bigcup_{n=1}^{\infty} A_n = \liminf_{n\to\infty} A_n \subset \liminf_{n\to\infty} B_n,$$

which completes the proof. □

REMARK 5.10. Theorem 5.8 implies that, almost surely, Brownian sample functions are not of bounded variation on any interval $[0, t]$. The non-monotonicity follows also from this fact, also from Theorem 5.6 and the Markov property (Proposition 10.7), and also from Theorem 5.9. Infinite variation on any interval is also a consequence of Theorem 5.9, as functions of bounded variation are differentiable almost everywhere. It will be proved again in Theorem 21.9. If, in Theorem 5.8, we make an additional assumption that Δ_{n+1} is a refinement of Δ_n for every n, then Y_n converges to td almost surely. This is Lévy's result. Proofs are found in Doob [**106**], p. 395, and Knight [**289**], p. 23. This fact is usually expressed by saying that the quadratic variation on $[0,t]$ of Brownian sample functions is td. However, recall that the sequence $\{\Delta_n\}$ here does not depend on ω. Lévy [**318**], p. 190, asserts that, almost surely, there is a sequence of partitions $\Delta_n = \Delta_n(\omega)$ in (5.9) depending on ω, satisfying (5.10), such that $Y_n(\omega)$ of (5.11) tends to infinity. The proof is found in Freedman [**147**], p. 48. See Remark 49.14 in Chapter 9.

A sample function $X(t,\omega)$ on $\mathbb{R}$ is said to have an *increase time* $t_0 > 0$ if there is $\varepsilon > 0$ such that $X(t_0 - s, \omega) \leq X(t_0, \omega) \leq X(t_0 + s, \omega)$ for all $s \in [0, \varepsilon)$. Theorem 5.7 can be strengthened so that, almost surely, Brownian sample functions have no increase time. This fact was discovered by Dvoretzky, Erdős, and Kakutani [**117**]. See Knight [**289**], Adelman [**4**], and Burdzy [**71**] for proofs.

6. Exercises 1

E 6.1. Prove that the Poisson distribution has characteristic function (2.5) and Laplace transform (2.6).

E 6.2. Prove that the nondegenerate Gaussian distribution on $\mathbb{R}^d$ of Example 2.9 has characteristic function (2.8).

E 6.3. Prove that the Cauchy distribution on $\mathbb{R}$ of Example 2.11 has characteristic function (2.9).

E 6.4. Prove that the Γ-distribution of Example 2.15 has Laplace transform (2.15) and characteristic function (2.16).

E 6.5. Show that the characteristic function $\widehat{\mu}(z)$ of a distribution μ on $\mathbb{R}$ defined by

$$\mu(\mathrm{d}x) = c 1_{\{|x|>2\}}(x)(x^2 \log|x|)^{-1}\mathrm{d}x$$

with a positive constant c is differentiable at $z = 0$, but that μ has infinite absolute moment of order 1.

E 6.6. Let X and X_n, $n = 1, 2, \ldots$, be random variables on $\mathbb{R}^d$. Prove that $X_n \to X$ in prob. if and only if every subsequence of $\{X_n\}$ contains a further subsequence that converges to X almost surely.

E 6.7. Let S be the space of $\mathbb{R}^d$-valued random variables defined on a probability space $(\Omega, \mathcal{F}, P)$, where two random variables equal a.s. are identified. Define

$$r(X,Y) = E\left[\frac{|X-Y|}{1+|X-Y|}\right]$$

for $X, Y \in S$. Show that r is a metric on S and that $r(X_n, X) \to 0$ is equivalent to the condition that $X_n \to X$ in prob.

E 6.8. Let X_n, $n = 1, 2, \ldots$, be random variables on $\mathbb{R}^d$ and let $\gamma \in \mathbb{R}^d$. Show that $X_n \to \gamma$ in prob. if and only if the distribution of X_n converges to δ_γ.

E 6.9. Let $\mathfrak{P}$ be the class of all probability measures on $\mathbb{R}^d$. Show that we can define a metric $r(\mu_1, \mu_2)$ on $\mathfrak{P}$ so that the convergence in this metric is equivalent to the convergence of probability measures in Definition 2.2.

E 6.10. Show that, if X is Gaussian distributed on $\mathbb{R}$ with mean 0 and variance 1, then $1/X^2$ has a one-sided strictly stable distribution of index 1/2 with $c = 1$ as in Example 2.13.

E 6.11. Prove that, for any probability measure μ on $\mathbb{R}^d$,

$$\mathrm{Re}\,(1 - \widehat{\mu}(2z)) \leq 4\,\mathrm{Re}\,(1 - \widehat{\mu}(z)) \quad \text{for } z \in \mathbb{R}^d.$$

E 6.12. Show that, for any probability measure μ on $\mathbb{R}$,

$$\textstyle\int_{-\infty}^{\infty} |x|\mu(\mathrm{d}x) = \pi^{-1} \int_{-\infty}^{\infty} (\mathrm{Re}\,(1 - \widehat{\mu}(z))/z^2)\mathrm{d}z.$$

E 6.13. If X is a random variable on $\mathbb{R}^d$ with $E|X| < \infty$, then the vector EX in $\mathbb{R}^d$ is defined componentwise. Show that $|EX| \leq E|X|$.

E 6.14. Let X and Y be independent random variables on $\mathbb{R}^d$ with $E|X| < \infty$, $E|Y| < \infty$, and $EY = 0$. Show that $E|X + Y| \geq E|X|$.

E 6.15. Let $\{X_t\}$ be a compound Poisson process on $\mathbb{R}^d$. Show that c and σ with which $\{X_t\}$ is associated (Definition 4.2) are uniquely determined by $\{X_t\}$.

E 6.16. Let $\{X_t\}$ be a compound Poisson process on $\mathbb{R}^d$ associated with c and σ. Let B be a Borel set in $\mathbb{R}^d$ such that $0 \notin B$, and let $T_B(\omega) = \inf\{t > 0\colon X_t(\omega) \in B\}$, the first time that $X_t(\omega)$ hits B. We understand that the infimum of the empty set is ∞. Show that, for any $q \geq 0$ and Borel sets $C \subset \mathbb{R}^d \setminus B$ and $D \subset B$,

$$E[\mathrm{e}^{-qT_B};\, 0 < T_B < \infty,\, X_{T_B-} \in C,\, X_{T_B} \in D] = E\big[\textstyle\int_0^{T_B} \mathrm{e}^{-qt} 1_C(X_t) c\sigma(D - X_t)\mathrm{d}t\big],$$

where $X_{T_B-} = \lim_{\varepsilon \downarrow 0} X_{T_B - \varepsilon}$.

E 6.17. Let $\{Z_n\colon n = 1, 2, \ldots\}$ be independent and identically distributed random variables on $\mathbb{R}$. Show the following.

(i) For any sequence $\{a_n\}$ increasing to ∞, $Z_n/a_n \to 0$ in prob.

(ii) For some sequence $\{a_n\}$ increasing to ∞, $Z_n/a_n \to 0$ a. s.

(iii) If $P[|Z_1| > a] > 0$ for every $a > 0$, then there is a sequence $\{a_n\}$ increasing to ∞ such that the probability that Z_n/a_n tends to 0 is 0.

E 6.18. Let $\{X_n\}$ be a sequence of independent random variables on $\mathbb{R}$ such that $X_n \to 0$ in prob. Let $\{a_n\}$ be a sequence increasing to ∞. Then, can we say that $X_n/a_n \to 0$ a. s.?

Notes

Basic original references on Lévy processes and additive processes are Lévy's two books, [**317**] and [**318**]. Skorohod's [**491**] and [**493**] are the first and the second edition of a book, but each of them is rich in contents with its own merit. Chapter 4 of Gihman and Skorohod [**165**] is similar to a part of them. Although published only in Japanese, Itô's books [**222, 227**] should be mentioned as rigorous introductions to Lévy and additive processes. Bertoin's recent book [**26**] is an excellent monograph on Lévy processes with emphasis on path properties. Nice introductions to stochastic processes and their applications are Billingsley [**34**], Resnick [**413**], and Karlin and Taylor [**260, 261**]. Freedman [**147**] contains elementary treatment of Brownian motion; Theorem 5.7 and further related properties of sample functions are described there. Detailed exposition of Brownian motion is found in Lévy [**318**], Itô and McKean [**228**], Hida [**200**], Knight [**289**], Durrett [**113**], Karatzas and Shreve [**258**], and Revuz and Yor [**415**]. Exclusively treating Poisson processes is Kingman [**286**].

The name *Lévy process* is now used in many books and research papers. The name *additive process* for a process with independent increments is not widely employed at present, but it is used by Lévy [**318**] (*processus additif*) and Itô [**224, 225**]. It is in a broader sense without assuming stochastic continuity and $X_0 = 0$; see Notes at the end of Chapter 2. Other names are *differential process* by Doob [**104**] and Itô and McKean [**228**], and *decomposable process* by Loève [**327**].

Example 4.7 follows Feller [**139**].

CHAPTER 2

Characterization and existence of Lévy and additive processes

7. Infinitely divisible distributions and Lévy processes in law

In this chapter we define infinitely divisible distributions, determine their characteristic functions, show that they correspond to Lévy processes in law, and then prove that any Lévy process in law has a modification which is a Lévy process. So the collection of all infinitely divisible distributions is in one-to-one correspondence with the collection of all Lévy processes, when two processes identical in law are considered as the same. We also characterize additive processes in law and show that every additive process in law has a modification which is an additive process. Our method is based on transition functions of Markov processes.

Denote by μ^{n*} or μ^n the n-fold convolution of a probability measure μ with itself, that is,

$$\mu^n = \mu^{n*} = \underbrace{\mu * \cdots * \mu}_{n}.$$

DEFINITION 7.1. A probability measure μ on $\mathbb{R}^d$ is *infinitely divisible* if, for any positive integer n, there is a probability measure μ_n on $\mathbb{R}^d$ such that $\mu = {\mu_n}^n$.

Since the convolution is expressed by the product in characteristic functions, μ is infinitely divisible if and only if, for each n, an nth root of the characteristic function $\widehat{\mu}(z)$ can be chosen in such a way that it is the characteristic function of some probability measure.

EXAMPLES 7.2. Gaussian, Cauchy, and δ-distributions on $\mathbb{R}^d$ are infinitely divisible. Poisson, geometric, negative binomial, exponential, and Γ-distributions on $\mathbb{R}$ are infinitely divisible. So are the one-sided strictly stable distribution of index 1/2 on $\mathbb{R}$ of Example 2.13 and compound Poisson distributions on $\mathbb{R}^d$. These facts are seen from the form of their characteristic functions in Section 2. That is, the nth roots of these distributions are obtained by taking the parameters appropriately. On the other hand, uniform and binomial distributions are not infinitely divisible. In fact, no probability measure (other than δ) with bounded support is infinitely divisible, as will be shown in Section 24. Another proof that uniform distributions are not infinitely divisible is given by Lemma 7.5, because their characteristic functions have zeros.

EXAMPLE 7.3. If $\{X_t\}$ is a Lévy process on $\mathbb{R}^d$, then, for every t, the distribution of X_t is infinitely divisible. To see this, let $t_k = kt/n$. Let $\mu = P_{X_t}$ and $\mu_n = P_{X(t_k)-X(t_{k-1})}$, which is independent of k by temporal homogeneity. Then $\mu = \mu_n{}^n$, since

$$X_t = (X_{t_1} - X_{t_0}) + \cdots + (X_{t_n} - X_{t_{n-1}}),$$

the sum of n independent identically distributed random variables. This is the beginning of the intimate relation between Lévy processes and infinitely divisible distributions.

We begin with a simple lemma.

LEMMA 7.4. *If μ_1 and μ_2 are infinitely divisible, then $\mu_1 * \mu_2$ is infinitely divisible.*

Proof. For each n, $\mu_1 = \mu_{1,n}{}^n$ and $\mu_2 = \mu_{2,n}{}^n$ with some $\mu_{1,n}$ and $\mu_{2,n}$. Hence $\mu_1 * \mu_2 = (\mu_{1,n} * \mu_{2,n})^n$. □

By applying the lemma above to Gaussian distributions (Example 2.10) and compound Poisson distributions (Definition 4.1), we see that μ is infinitely divisible if

$$\widehat{\mu}(z) = \exp\left[-\tfrac{1}{2}\langle z, Az\rangle + i\langle \gamma, z\rangle + \int_{\mathbb{R}^d} (e^{i\langle z,x\rangle} - 1)\, \nu(dx)\right]$$

with A symmetric nonnegative-definite, $\gamma \in \mathbb{R}^d$, and ν a finite measure. We will show, in Section 8, that the characteristic function of a general infinitely divisible distribution has a form which is a generalization of the above. The generalization consists in allowing ν to be an infinite measure satisfying certain conditions.

Now we will show in several lemmas that, for any infinitely divisible distribution μ, the nth root is uniquely defined and further, for any $t \geq 0$, the tth power of μ is definable.

LEMMA 7.5. *If μ is infinitely divisible, then $\widehat{\mu}(z)$ has no zero, that is, $\widehat{\mu}(z) \neq 0$ for any $z \in \mathbb{R}^d$.*

Proof. For each n there is μ_n such that $\widehat{\mu}(z) = \widehat{\mu}_n(z)^n$. By Proposition 2.5(v) $|\widehat{\mu}_n(z)|^2 = |\widehat{\mu}(z)|^{2/n}$ is a characteristic function. Define $\varphi(z)$ by

$$\varphi(z) = \lim_{n\to\infty} |\widehat{\mu}_n(z)|^2 = \begin{cases} 1 & \text{if } \widehat{\mu}(z) \neq 0, \\ 0 & \text{if } \widehat{\mu}(z) = 0. \end{cases}$$

Since $\widehat{\mu}(0) = 1$ and $\widehat{\mu}(z)$ is continuous, $\varphi(z) = 1$ in a neighborhood of 0. It follows from Proposition 2.5(viii) that $\varphi(z)$ is a characteristic function.

Hence $\varphi(z)$ is continuous on $\mathbb{R}^d$. Hence $\varphi(z) = 1$ for all $z \in \mathbb{R}^d$, which shows that $\widehat{\mu}(z) \neq 0$ everywhere. □

The converse of the lemma above is not true. For example, a binomial distribution with parameters n, p has characteristic function without zero if $p \neq 1/2$, but it is not infinitely divisible.

The next lemma is some complex analysis.

LEMMA 7.6. *Suppose that $\varphi(z)$ is a continuous function from $\mathbb{R}^d$ into $\mathbb{C}$ such that $\varphi(0) = 1$ and $\varphi(z) \neq 0$ for any z. Then, there is a unique continuous function $f(z)$ from $\mathbb{R}^d$ into $\mathbb{C}$ such that $f(0) = 0$ and $\mathrm{e}^{f(z)} = \varphi(z)$. For any positive integer n there is a unique continuous function $g_n(z)$ from $\mathbb{R}^d$ into $\mathbb{C}$ such that $g_n(0) = 1$ and $g_n(z)^n = \varphi(z)$. They are related as $g_n(z) = \mathrm{e}^{f(z)/n}$.*

We write $f(z) = \log \varphi(z)$ and $g_n(z) = \varphi(z)^{1/n}$ and call them the *distinguished logarithm* and the *distinguished nth root* of φ, respectively. Note that $f(z)$ is *not* a composite function of $\varphi(z)$ and a fixed branch of the logarithmic function. That is, $\varphi(z_1) = \varphi(z_2)$ does not imply $f(z_1) = f(z_2)$. More generally, we define, for $t \geq 0$, $\varphi(z)^t = \mathrm{e}^{tf(z)}$, and call it the *distinguished t th power* of φ. We apply this to characteristic functions. Suppose that $\widehat{\mu}(z) \neq 0$ for all z. Then $\widehat{\mu}(z)^t$ is defined for every $t \geq 0$, but it is not always a characteristic function as the remark after the proof of Lemma 7.5 shows. If $\widehat{\mu}(z)^t$ is the characteristic function of a probability measure, then this probability measure is denoted by μ^{t*} or μ^t.

Proof of Lemma 7.6. For $z \in \mathbb{R}^d$ let C_z be the directed line segment from 0 to z:

$$C_z\colon \quad w(t) = tz, \quad 0 \leq t \leq 1.$$

Then, $\varphi(w(t))$, $0 \leq t \leq 1$, draws a curve in $\mathbb{C}\setminus\{0\}$. Let $\log^{(\mathbb{C})} w$, $w \in \mathbb{C}\setminus\{0\}$, be the complex logarithmic function (multi-valued):

$$\log^{(\mathbb{C})} w = \log |w| + \mathrm{i} \arg w,$$

where $\arg w$ is determined up to addition of $2n\pi$, $n \in \mathbb{Z}$. Let $h_z(t)$, $0 \leq t \leq 1$, be the unique branch of $\log^{(\mathbb{C})} \varphi(tz)$ such that $h_z(0) = 0$ and $h_z(t)$ is continuous in t. Define $f(z) = h_z(1)$. Then,

$$\mathrm{e}^{f(z)} = \mathrm{e}^{h_z(1)} = \varphi(z), \quad f(0) = h_0(1) = 0.$$

We claim that $f(z)$ is continuous. Fix z_0. Let $\Gamma_{z_0 z}$ be the closed curve in $\mathbb{R}^d$ defined by

$$z(t) = \begin{cases} tz_0 & \text{for } 0 \leq t \leq 1, \\ (t-1)z + (2-t)z_0 & \text{for } 1 \leq t \leq 2, \\ (3-t)z & \text{for } 2 \leq t \leq 3. \end{cases}$$

It draws a triangle $0, z_0, z, 0$. Let $\theta_{z_0 z}(t)$ be the branch of $\arg \varphi(z(t))$ which is continuous in $t \in [0,3]$ and satisfies $\theta_{z_0 z}(0) = 0$. The set $\{\varphi(tz_0) \colon t \in [0,1]\}$ is away from 0 with a positive distance and $\max_{t \in [0,1]} |\varphi(tz) - \varphi(tz_0)|$ is small if z is close to z_0. It follows that there is a neighborhood U of z_0 such that, for all $z \in U$, the rotation number around the origin of the closed curve $\varphi(z(t))$, $0 \le t \le 3$, is zero. Therefore $\theta_{z_0 z}(3) = 0$ for $z \in U$. Thus $\operatorname{Im} f(z) = \theta_{z_0 z}(2)$ for $z \in U$. Hence $\operatorname{Im} f(z)$ is close to $\operatorname{Im} f(z_0)$ if z is sufficiently close to z_0. Hence, for every $\varepsilon > 0$, there is a neighborhood V of z_0 such that, for $z \in V$, $|f(z) - f(z_0)| < \varepsilon$.

Uniqueness. Suppose that $\widetilde{f}(z)$ is continuous, $\widetilde{f}(0) = 0$, and $e^{\widetilde{f}(z)} = \varphi(z)$. Then $h_z(t) = \widetilde{f}(tz)$ from the uniqueness of $h_z(t)$, and hence $\widetilde{f}(z) = h_z(1) = f(z)$.

The nth root. The complex nth root function of w is $|w|^{1/n} e^{i(1/n)\arg w}$ (multi-valued). Starting from this we can hold the same discussion as above to see the existence and uniqueness of $g_n(z)$. On the other hand, $e^{f(z)/n}$ satisfies the desired conditions. Hence $g_n(z) = e^{f(z)/n}$. □

Lemmas 7.5 and 7.6 imply that, if μ is infinitely divisible, then, for each positive integer n, a distribution μ_n satisfying $\mu = \mu_n{}^n$ is unique and $\widehat{\mu}_n(z) = \widehat{\mu}(z)^{1/n}$, that is, $\mu_n = \mu^{1/n}$. However, it is known that, in general, $\nu_1 * \nu_1 = \nu_2 * \nu_2$ for two probability measures ν_1, ν_2 does not imply $\nu_1 = \nu_2$ (Feller [**139**], p. 506).

LEMMA 7.7. *Suppose that $\varphi(z)$ and $\varphi_n(z)$, $n = 1, 2, \ldots$, are continuous functions from $\mathbb{R}^d$ into $\mathbb{C}$ such that $\varphi(0) = \varphi_n(0) = 1$ and $\varphi(z) \neq 0$ and $\varphi_n(z) \neq 0$ for any z. If $\varphi_n(z) \to \varphi(z)$ uniformly on any compact set, then $\log \varphi_n(z) \to \log \varphi(z)$ uniformly on any compact set.*

Proof. Look at the construction of the distinguished logarithm in Lemma 7.6. □

LEMMA 7.8. *If $\{\mu_k\}$ is a sequence of infinitely divisible distributions and $\mu_k \to \mu$, then μ is infinitely divisible.*

Proof. We claim that $\widehat{\mu}(z) \neq 0$. Since $\widehat{\mu}_k(z) \to \widehat{\mu}(z)$, we have $|\widehat{\mu}_k(z)|^{2/n} \to |\widehat{\mu}(z)|^{2/n}$ for $n = 1, 2, \ldots$ as $k \to \infty$. By Proposition 2.5(v) $|\widehat{\mu}_k(z)|^{2/n}$ is a characteristic function. As $|\widehat{\mu}(z)|^{2/n}$ is continuous, it is a characteristic function by Proposition 2.5(viii). We have $|\widehat{\mu}(z)|^2 = (|\widehat{\mu}(z)|^{2/n})^n$. Hence $|\widehat{\mu}(z)|^2$ is the characteristic function of an infinitely divisible distribution. Hence $\widehat{\mu}(z) \neq 0$ by Lemma 7.5 as claimed. Recall that the convergence $\widehat{\mu}_k(z) \to \widehat{\mu}(z)$ is uniform on any compact set by Proposition 2.5(vi). It follows from Lemma 7.7 that $\log \widehat{\mu}_k(z) \to \log \widehat{\mu}(z)$. Therefore $\widehat{\mu}_k(z)^{1/n} \to \widehat{\mu}(z)^{1/n}$ as $k \to \infty$ for any n. Since $\widehat{\mu}(z)^{1/n}$ is continuous, $\widehat{\mu}(z)^{1/n}$ is the characteristic function of a probability measure again by Proposition 2.5(viii). Hence μ is infinitely divisible. □

LEMMA 7.9. *If μ is infinitely divisible, then, for every $t \in [0, \infty)$, μ^t is definable and infinitely divisible.*

Proof. We have a distribution $\mu^{1/n}$ for any positive integer n. It is infinitely divisible, since $\widehat{\mu}(z)^{1/n} = (\widehat{\mu}(z)^{1/(nk)})^k$ for any k. Hence, for any positive integers m and n, $\mu^{m/n}$ is also infinitely divisible by Lemma 7.4. For any irrational number $t > 0$, choose rational numbers r_n approaching t. Then $\widehat{\mu}(z)^{r_n} \to \widehat{\mu}(z)^t$ and $\widehat{\mu}(z)^t$ is continuous. Hence $\widehat{\mu}(z)^t$ is a characteristic function by Proposition 2.5(viii). The corresponding distribution is infinitely divisible by Lemma 7.8. □

Obviously μ^0 equals δ_0. Now we will show the correspondence between infinitely divisible distributions and Lévy processes in law.

THEOREM 7.10. (i) *If $\{X_t \colon t \geq 0\}$ is a Lévy process in law on $\mathbb{R}^d$, then, for any $t \geq 0$, P_{X_t} is infinitely divisible and, letting $P_{X_1} = \mu$, we have $P_{X_t} = \mu^t$.*

(ii) *Conversely, if μ is an infinitely divisible distribution on $\mathbb{R}^d$, then there is a Lévy process in law $\{X_t \colon t \geq 0\}$ such that $P_{X_1} = \mu$.*

(iii) *If $\{X_t\}$ and $\{X'_t\}$ are Lévy processes in law on $\mathbb{R}^d$ such that $P_{X_1} = P_{X'_1}$, then $\{X_t\}$ and $\{X'_t\}$ are identical in law.*

In the theorem above, μ is said to be the infinitely divisible distribution corresponding to the Lévy process in law $\{X_t\}$; conversely, $\{X_t\}$ is said to be the Lévy process in law corresponding to the infinitely divisible distribution μ.

Proof of theorem. (i) Let $\{X_t\}$ be a Lévy process in law. The infinite divisibility of P_{X_t} is the same as the case of a Lévy process in Example 7.3. Let $\mu = P_{X_1}$. Since $\mu = (P_{X_{1/n}})^n$, we have $P_{X_{1/n}} = \mu^{1/n}$. Hence $P_{X_{m/n}} = \mu^{m/n}$. If $t > 0$ is irrational, choose rational numbers r_n such that $r_n \to t$. We have $X_{r_n} \to X_t$ in probability, hence $P_{X(r_n)} \to P_{X_t}$. Hence $P_{X_t} = \mu^t$. Here we use stochastic continuity of $\{X_t\}$ and Proposition 2.3.

(ii) Let μ be infinitely divisible. Then μ^t is a distribution with characteristic function $\mathrm{e}^{t \log \widehat{\mu}(z)}$. Hence

$$\mu^s * \mu^t = \mu^{s+t}, \tag{7.1}$$

$$\mu^0 = \delta_0, \tag{7.2}$$

$$\mu^t \to \delta_0 \quad \text{as } t \downarrow 0. \tag{7.3}$$

Let us construct the corresponding Lévy process in law. Consider Ω, $\mathcal{F}$, and $X_t(\omega) = \omega(t)$ in Kolmogorov's extension theorem 1.8. For any $n \geq 0$ and any $0 \leq t_0 < t_1 < \cdots < t_n$, define

$$\mu_{t_0,\ldots,t_n}(B_0 \times \cdots \times B_n) \tag{7.4}$$

$$= \int \cdots \int \mu^{t_0}(\mathrm{d}y_0) 1_{B_0}(y_0) \mu^{t_1 - t_0}(\mathrm{d}y_1) 1_{B_1}(y_0 + y_1)$$
$$\times \ldots \mu^{t_n - t_{n-1}}(\mathrm{d}y_n) 1_{B_n}(y_0 + \cdots + y_n).$$

Then $\mu_{t_0,\ldots,t_n}$ is extended to a probability measure on $\mathcal{B}((\mathbb{R}^d)^{n+1})$ and the family $\{\mu_{t_0,\ldots,t_n}\}$ satisfies the consistency condition by (7.1). Hence, by Kolmogorov's extension theorem, we get a unique measure P on $\mathcal{F}$ such that

$$P[\, X_{t_0} \in B_0, \ldots, X_{t_n} \in B_n \,] = \mu_{t_0,\ldots,t_n}(B_0 \times \cdots \times B_n). \tag{7.5}$$

In particular, X_t has distribution μ^t. Let us show that $\{X_t : t \geq 0\}$ is a Lévy process in law. If $0 \leq t_0 < \cdots < t_n$, then we have, from (7.4) and (7.5),

$$\begin{aligned} E[f(X_{t_0}, \ldots, X_{t_n})] &= \int \cdots \int f(y_0, y_0 + y_1, \ldots, y_0 + \cdots + y_n) \mu^{t_0}(\mathrm{d}y_0) \\ &\qquad \times \mu^{t_1 - t_0}(\mathrm{d}y_1) \ldots \mu^{t_n - t_{n-1}}(\mathrm{d}y_n) \end{aligned} \tag{7.6}$$

for any bounded measurable function f. Let $z_1, \ldots, z_n \in \mathbb{R}^d$ and

$$f(x_0, \ldots, x_n) = \exp\Big(\mathrm{i} \sum_{j=1}^{n} \langle z_j, x_j - x_{j-1} \rangle \Big).$$

Then

$$\begin{aligned} E\Big[\exp\Big(\mathrm{i} \sum_{j=1}^{n} \langle z_j, X_{t_j} - X_{t_{j-1}} \rangle \Big) \Big] &= \int \cdots \int \exp\Big(\mathrm{i} \sum_{j=1}^{n} \langle z_j, y_j \rangle \Big) \mu^{t_1 - t_0}(dy_1) \ldots \mu^{t_n - t_{n-1}}(\mathrm{d}y_n) \\ &= \prod_{j=1}^{n} \int \exp(\mathrm{i} \langle z_j, y_j \rangle) \mu^{t_j - t_{j-1}}(\mathrm{d}y_j). \end{aligned}$$

It follows that

$$E[\exp(\mathrm{i}\langle z_j, X_{t_j} - X_{t_{j-1}} \rangle)] = \int \exp(\mathrm{i}\langle z_j, y_j \rangle) \mu^{t_j - t_{j-1}}(\mathrm{d}y_j),$$

which shows that $X_{t_j} - X_{t_{j-1}}$ has distribution $\mu^{t_j - t_{j-1}}$ and that

$$E\Big[\exp\Big(\mathrm{i} \sum_{j=1}^{n} \langle z_j, X_{t_j} - X_{t_{j-1}} \rangle \Big) \Big] = \prod_{j=1}^{n} E[\exp(\mathrm{i}\langle z_j, X_{t_j} - X_{t_{j-1}} \rangle)]. \tag{7.7}$$

By Proposition 2.5(iv) this says that $\{X_t\}$ has independent increments. The convergence (7.3) says that $X_t \to 0$ in prob. as $t \downarrow 0$ (Exercise 6.8). Hence $P[\,|X_s - X_t| > \varepsilon\,] = P[\,|X_{|s-t|}| > \varepsilon\,] \to 0$ as $s \to t$. Hence we have (1), (2),(3), and (4) of Definition 1.6. That is, $\{X_t\}$ is a Lévy process in law.

(iii) Let $\{X_t\}$ and $\{X'_t\}$ be Lévy processes in law and $X_1 \stackrel{\mathrm{d}}{=} X'_1$. Then, by (i), $X_t \stackrel{\mathrm{d}}{=} X'_t$. It follows that $X_{s+t} - X_s \stackrel{\mathrm{d}}{=} X'_{s+t} - X'_s$ for any t and s. Hence

$$(X_{t_0}, X_{t_1} - X_{t_0}, \ldots, X_{t_n} - X_{t_{n-1}}) \stackrel{\mathrm{d}}{=} (X'_{t_0}, X'_{t_1} - X'_{t_0}, \ldots, X'_{t_n} - X'_{t_{n-1}})$$

for $0 \le t_0 < \cdots < t_n$ by independence. Since $(X_{t_0}, \ldots, X_{t_n})$ is a function of $(X_{t_0}, X_{t_1} - X_{t_0}, \ldots, X_{t_n} - X_{t_{n-1}})$, we get

$$(X_{t_0}, \ldots, X_{t_n}) \stackrel{\mathrm{d}}{=} (X'_{t_0}, \ldots, X'_{t_n}),$$

completing the proof. □

REMARK 7.11. Even if $\{X_t\}$ has stationary independent increments and starts at the origin, the assertion (i) of Theorem 7.10 is not true unless $\{X_t\}$ is stochastically continuous. In this case, the distribution of X_t is infinitely divisible but is not always equal to μ^t. For example, let $f(t)$ be a function such that $f(t) + f(s) = f(t+s)$ for all nonnegative t and s but that $f(t)$ is not a constant multiple of t, and let $X_t = f(t)$. Such a function is given by G. Hamel [**181**].

8. Representation of infinitely divisible distributions

The following theorem gives a representation of characteristic functions of all infinitely divisible distributions. It is called the *Lévy–Khintchine representation* or the *Lévy–Khintchine formula* . ssIt was obtained on $\mathbb{R}$ around 1930 by de Finetti and Kolmogorov in special cases, and then by Lévy in the general case. It was immediately extended to $\mathbb{R}^d$. An essentially simpler proof on $\mathbb{R}$ was given by Khintchine. This theorem is fundamental to the whole theory. Let $D = \{x \colon |x| \le 1\}$, the closed unit ball.

THEOREM 8.1. (i) *If μ is an infinitely divisible distribution on $\mathbb{R}^d$, then*

$$\widehat{\mu}(z) = \exp\Big[-\tfrac{1}{2}\langle z, Az\rangle + \mathrm{i}\langle \gamma, z\rangle + \int_{\mathbb{R}^d} (\mathrm{e}^{\mathrm{i}\langle z,x\rangle} - 1 - \mathrm{i}\langle z, x\rangle 1_D(x))\nu(\mathrm{d}x)\Big], \quad z \in \mathbb{R}^d, \tag{8.1}$$

where A is a symmetric nonnegative-definite $d \times d$ matrix, ν is a measure on $\mathbb{R}^d$ satisfying

$$\nu(\{0\}) = 0 \quad \textit{and} \quad \int_{\mathbb{R}^d} (|x|^2 \wedge 1)\nu(\mathrm{d}x) < \infty, \tag{8.2}$$

and $\gamma \in \mathbb{R}^d$.

(ii) *The representation of* $\widehat{\mu}(z)$ *in (i) by* A, ν, *and* γ *is unique.*

(iii) *Conversely, if* A *is a symmetric nonnegative-definite* $d \times d$ *matrix,* ν *is a measure satisfying (8.2), and* $\gamma \in \mathbb{R}^d$, *then there exists an infinitely divisible distribution* μ *whose characteristic function is given by (8.1).*

DEFINITION 8.2. We call (A, ν, γ) in Theorem 8.1 the *generating triplet* of μ. The A and the ν are called, respectively, the *Gaussian covariance matrix* and the *Lévy measure* of μ. When $A = 0$, μ is called *purely non-Gaussian.*

COROLLARY 8.3. *If* μ *has the generating triplet* (A, ν, γ), *then* μ^t *has the generating triplet* $(tA, t\nu, t\gamma)$.

REMARK 8.4. The integrand of the integral in the right-hand side of (8.1) is integrable with respect to ν, because it is bounded outside of any neighborhood of 0 and

$$e^{i\langle z,x\rangle} - 1 - i\langle z, x\rangle 1_D(x) = O(|x|^2) \quad \text{as} \quad |x| \to 0$$

for fixed z. There are many other ways of getting an integrable integrand. Let $c(x)$ be a bounded measurable function from $\mathbb{R}^d$ to $\mathbb{R}$ satisfying

$$c(x) = 1 + o(|x|) \quad \text{as } |x| \to 0, \tag{8.3}$$

$$c(x) = O(1/|x|) \quad \text{as } |x| \to \infty. \tag{8.4}$$

Then (8.1) is rewritten as

$$\widehat{\mu}(z) = \exp\Big[-\tfrac{1}{2}\langle z, Az\rangle + i\langle \gamma_c, z\rangle + \int_{\mathbb{R}^d} (e^{i\langle z,x\rangle} - 1 - i\langle z, x\rangle c(x))\nu(dx)\Big] \tag{8.5}$$

with $\gamma_c \in \mathbb{R}^d$ defined by

$$\gamma_c = \gamma + \int_{\mathbb{R}^d} x(c(x) - 1_D(x))\nu(dx). \tag{8.6}$$

(Here it is enough to assume $c(x) = 1 + O(|x|)$, $|x| \to 0$, instead of (8.3), but we will use (8.3) in Theorem 8.7.) The following are examples of $c(x)$ sometimes used:

$$c(x) = 1_{\{|x|\le\varepsilon\}}(x) \quad \text{with } \varepsilon > 0,$$

$$c(x) = 1/(1 + |x|^2),$$

$$c(x) = 1_{\{|x|\le 1\}}(x) + 1_{\{1<|x|\le 2\}}(x)(2 - |x|),$$

and, in the case $d = 1$,

$$c(x) = 1_{[-1,1]}(x) + 1_{(1,\infty)}(x)/x - 1_{(-\infty,-1)}(x)/x,$$

$$c(x) = (\sin x)/x.$$

We denote the triplet in (8.5) by $(A, \nu, \gamma_c)_c$. It is also called a generating triplet and (8.5) is also called a Lévy–Khintchine representation. But, when we mention the generating triplet (A, ν, γ) without subscript to the parentheses, it means the one in (8.1). More generally, if $c(x)$ is a measurable function and if, for every z, $e^{i\langle z,x\rangle} - 1 - i\langle z, x\rangle c(x)$ is integrable with respect to a given Lévy measure ν, then we have (8.5) with γ_c of (8.6). The triplet is again called a generating triplet, written as $(A, \nu, \gamma_c)_c$. Thus, if ν satisfies the additional condition $\int_{|x|\le 1} |x|\nu(dx) < \infty$, then using the zero function as c, we get

$$\text{(8.7)} \qquad \widehat{\mu}(z) = \exp\left[-\tfrac{1}{2}\langle z, Az\rangle + i\langle \gamma_0, z\rangle + \int_{\mathbb{R}^d} (e^{i\langle z,x\rangle} - 1)\nu(dx)\right]$$

with $\gamma_0 \in \mathbb{R}^d$. This is the representation by triplet $(A, \nu, \gamma_0)_0$. The constant γ_0 here is called the *drift* of μ. If ν satisfies $\int_{|x|>1} |x|\nu(dx) < \infty$, then, letting $c(x)$ be a constant function 1, we have the representation by triplet $(A, \nu, \gamma_1)_1$:

$$\text{(8.8)} \quad \widehat{\mu}(z) = \exp\left[-\tfrac{1}{2}\langle z, Az\rangle + i\langle \gamma_1, z\rangle + \int_{\mathbb{R}^d} (e^{i\langle z,x\rangle} - 1 - i\langle z, x\rangle)\nu(dx)\right].$$

We call the constant γ_1 the *center* of μ. It will be shown in Chapter 5 that finiteness of $\int_{|x|>1} |x|\nu(dx)$ is equivalent to finiteness of $\int_{\mathbb{R}^d} |x|\mu(dx)$ and that $\gamma_1 = \int_{\mathbb{R}^d} x\,\mu(dx)$, the mean of μ (Example 25.12). Thus the center and the mean are identical. We note that in the triplets A and ν are invariant no matter what function $c(x)$ we choose.

EXAMPLES 8.5. Consider infinitely divisible distributions as in Examples 7.2. The Lévy measure ν is zero when and only when μ is Gaussian. If μ is compound Poisson, then $A = 0$, $\nu = c\sigma$, and $\gamma_0 = 0$. If $d = 1$ and μ is Poisson, then $A = 0$, $\nu = c\delta_1$, and $\gamma_0 = 0$. If $\mu = \delta_a$, then $A = 0$, $\nu = 0$, and $\gamma = \gamma_0 = a$. If $d = 1$ and μ is negative binomial as in Example 2.16, then $A = 0$, $\gamma_0 = 0$, and ν is concentrated on the positive integers with $\nu(\{k\}) = ck^{-1}(1-p)^k$ (see Example 4.6). The generating triplets of the exponential distributions and the Γ-distributions will be calculated in Example 8.10. That of the one-sided strictly stable distribution with index 1/2 will be in Example 8.11. The Cauchy distribution will be treated in Chapter 3.

There are two ways of approaching Theorem 8.1. One is a probabilistic way, which analyzes the structure of jumps of sample functions of a general Lévy process and thus obtains the representation (8.1) as its distribution at a fixed time. Lévy [**312**] showed the theorem in 1934 in this way. The analysis of sample functions was made deeper by Itô [**220**] and given the form of the Lévy–Itô decomposition. The second is an analytic method initiated by Khintchine [**277**] in 1937, which directly derives the representation. We use the analytic method. The Lévy–Itô decomposition of sample functions will be proved later by using the Lévy–Khintchine representation.

LEMMA 8.6. *For any $u \in \mathbb{R}$ and $n \in \mathbb{N}$*

$$\mathrm{e}^{\mathrm{i}u} = \sum_{k=0}^{n-1} \frac{(\mathrm{i}u)^k}{k!} + \theta \frac{|u|^n}{n!}$$

with some $\theta \in \mathbb{C}$ satisfying $|\theta| \le 1$.

Proof. Immediate from the identity

$$\mathrm{e}^{\mathrm{i}u} = \sum_{k=0}^{n-1} \frac{(\mathrm{i}u)^k}{k!} + \frac{\mathrm{i}^n}{(n-1)!} \int_0^u (u-v)^{n-1} \mathrm{e}^{\mathrm{i}v} \mathrm{d}v$$

for $u \in \mathbb{R}$. □

First we prove parts (ii) and (iii) of the theorem.

Proof of Theorem 8.1(ii). Suppose that $\widehat{\mu}(z)$ is expressed by (8.1) with A, ν, and γ. First note that

$$(8.9) \qquad |\mathrm{e}^{\mathrm{i}\langle z,x\rangle} - 1 - \mathrm{i}\langle z,x\rangle 1_D(x)| \le \tfrac{1}{2}|z|^2|x|^2 1_{\{|x|\le 1\}}(x) + 2 \cdot 1_{\{|x|>1\}}(x)$$

by Lemma 8.6. It follows that, by Lebesgue's convergence theorem, the expression in the square brackets in (8.1) is continuous in z. Thus

$$\log \widehat{\mu}(sz) = -\tfrac{1}{2}s^2\langle z, Az\rangle + \mathrm{i}s\langle \gamma, z\rangle + \int_{\mathbb{R}^d} (\mathrm{e}^{\mathrm{i}\langle sz,x\rangle} - 1 - \mathrm{i}\langle sz, x\rangle 1_D(x))\nu(\mathrm{d}x)$$

for $s \in \mathbb{R}$. Using (8.9) and Lebesgue's theorem again, we see that

$$s^{-2}\log \widehat{\mu}(sz) \to -\tfrac{1}{2}\langle z, Az\rangle \quad \text{as } s \to \infty.$$

Hence A is uniquely determined by μ. Let $\psi(z) = \log \widehat{\mu}(z) + \frac{1}{2}\langle z, Az\rangle$ and $C = [-1,1]^d$. Then we can prove

$$(8.10) \qquad \int_C (\psi(z) - \psi(z+w))\mathrm{d}w = 2^d \int_{\mathbb{R}^d} \mathrm{e}^{\mathrm{i}\langle z,x\rangle}\left(1 - \prod_{j=1}^d \frac{\sin x_j}{x_j}\right)\nu(\mathrm{d}x),$$

where $\frac{\sin x_j}{x_j}$ is understood to be 1 when $x_j = 0$. In fact, since

$$\psi(z) - \psi(z+w) = \int_{\mathbb{R}^d} (\mathrm{e}^{\mathrm{i}\langle z,x\rangle} - \mathrm{e}^{\mathrm{i}\langle z+w,x\rangle} + \mathrm{i}\langle w, x\rangle 1_D(x))\nu(\mathrm{d}x) - \mathrm{i}\langle \gamma, w\rangle$$

and

$$\begin{aligned} |\mathrm{e}^{\mathrm{i}\langle z,x\rangle} - \mathrm{e}^{\mathrm{i}\langle z+w,x\rangle} + \mathrm{i}\langle w,x\rangle| &\le |1 - \mathrm{e}^{\mathrm{i}\langle w,x\rangle} + \mathrm{i}\langle w,x\rangle| + |\langle w,x\rangle|\,|1 - \mathrm{e}^{\mathrm{i}\langle z,x\rangle}| \\ &\le \tfrac{1}{2}|w|^2|x|^2 + |w|\,|x|^2|z|, \end{aligned}$$

we can use Fubini's theorem and get

$$\int_C (\psi(z) - \psi(z+w))\mathrm{d}w = \int_{\mathbb{R}^d} \mathrm{e}^{\mathrm{i}\langle z,x\rangle}\nu(\mathrm{d}x) \int_C (1 - \mathrm{e}^{\mathrm{i}\langle w,x\rangle})\mathrm{d}w,$$

which shows (8.10). Now let

$$\rho(dx) = 2^d\left(1 - \prod_{j=1}^{d} \frac{\sin x_j}{x_j}\right)\nu(dx).$$

Then ρ is a finite measure, since

$$\prod_{j=1}^{d} \frac{\sin x_j}{x_j} = 1 - \tfrac{1}{6}|x|^2 + O(|x|^4) \quad \text{as } |x| \to 0.$$

The right-hand side of (8.10) is the Fourier transform of ρ. Hence, by Proposition 2.5(ii), ρ is uniquely determined by ψ, that is, by μ. Since $\nu(\{0\}) = 0$, ν is uniquely determined by μ. Now γ is determined from μ, A, and ν by (8.1). □

Proof of Theorem 8.1(iii). Given A, ν, and γ, let $\varphi(z)$ be the right-hand side of (8.1). Let

$$\varphi_n(z) = \exp\left[-\tfrac{1}{2}\langle z, Az\rangle + \mathrm{i}\langle\gamma, z\rangle + \int_{|x|>1/n} (\mathrm{e}^{\mathrm{i}\langle z,x\rangle} - 1 - \mathrm{i}\langle z, x\rangle 1_D(x))\nu(\mathrm{d}x)\right].$$

Since the measure ν restricted to $\{|x| > 1/n\}$ is finite, this $\varphi_n(z)$ is the convolution of a Gaussian and a compound Poisson distribution and, hence, is the characteristic function of an infinitely divisible distribution (Example 7.2 and Lemma 7.4). As $n \to \infty$, it converges to $\varphi(z)$. On the other hand, $\varphi(z)$ is continuous, as is noticed at the beginning of the proof of part (ii). Hence $\varphi(z)$ is the characteristic function of an infinitely divisible distribution by Proposition 2.5(viii) and Lemma 7.8. □

The result below incorporates an essential part of the proof of part (i). Let us write $f \in C_\sharp$ if f is a bounded continuous function from $\mathbb{R}^d$ to $\mathbb{R}$ vanishing on a neighborhood of 0.

THEOREM 8.7. *Let $c(x)$ be a bounded continuous function from $\mathbb{R}^d$ to $\mathbb{R}$ satisfying (8.3) and (8.4). Suppose that μ_n $(n = 1, 2, \dots)$ are infinitely divisible distributions on $\mathbb{R}^d$ and that each $\widehat{\mu}_n(z)$ has the Lévy–Khintchine representation by generating triplet $(A_n, \nu_n, \beta_n)_c$. Let μ be a probability measure on $\mathbb{R}^d$. Then $\mu_n \to \mu$ if and only if μ is infinitely divisible and $\widehat{\mu}(z)$ has the Lévy–Khintchine representation by the generating triplet $(A, \nu, \beta)_c$ with A, ν, β satisfying the following three conditions.*

(1) *If $f \in C_\sharp$, then*

$$\lim_{n\to\infty} \int_{\mathbb{R}^d} f(x)\nu_n(\mathrm{d}x) = \int_{\mathbb{R}^d} f(x)\nu(\mathrm{d}x).$$

(2) *Define symmetric nonnegative-definite matrices* $A_{n,\varepsilon}$ *by*

$$\langle z, A_{n,\varepsilon} z\rangle = \langle z, A_n z\rangle + \int_{|x|\le\varepsilon} \langle z, x\rangle^2 \nu_n(\mathrm{d}x).$$

Then

$$\lim_{\varepsilon\downarrow 0} \limsup_{n\to\infty} |\langle z, A_{n,\varepsilon} z\rangle - \langle z, Az\rangle| = 0 \quad \textit{for } z \in \mathbb{R}^d.$$

(3) $\beta_n \to \beta$.

If we use (8.1) for the Lévy–Khintchine representation, Theorem 8.7 cannot be proved. This is because of the discontinuity of $1_D(x)$.

Proof of theorem. Assume that $\mu_n \to \mu$. Then, μ is infinitely divisible (Lemma 7.8) and $\widehat{\mu}(z) \neq 0$ (Lemma 7.5). It follows from Lemma 7.7 and Proposition 2.5(vi) that

$$\log \widehat{\mu}_n(z) \to \log \widehat{\mu}(z) \quad \text{uniformly on any compact set.} \tag{8.11}$$

Define $\rho_n(dx) = (|x|^2 \wedge 1)\nu_n(dx)$. We claim that $\{\rho_n\}$ is tight in the sense that

$$\sup_n \rho_n(\mathbb{R}^d) < \infty, \tag{8.12}$$

$$\lim_{l\to\infty} \sup_n \int_{|x|>l} \rho_n(\mathrm{d}x) = 0. \tag{8.13}$$

Suppose, for a moment, that (8.12) and (8.13) are proved. Then by the selection theorem there is a subsequence $\{\rho_{n_k}\}$ which converges to some finite measure ρ. The selection theorem in [**91**], p. 29 applies. Alternatively, we can use, after normalizing, a similar theorem in [**34**], [**110**], [**139**] on families of probability measures, because we can assume $\inf_n \rho_n(\mathbb{R}^d) > 0$ as the conclusion is evident in the case $\rho_n(\mathbb{R}^d) \to 0$. Define ν by $\nu(\{0\}) = 0$ and $\nu(\mathrm{d}x) = (|x|^2 \wedge 1)^{-1}\rho(\mathrm{d}x)$ on $\{|x| > 0\}$. The measure ρ may have a point mass at 0, but it is ignored in defining ν. Let

$$g(z,x) = \mathrm{e}^{\mathrm{i}\langle z,x\rangle} - 1 - \mathrm{i}\langle z,x\rangle c(x), \tag{8.14}$$

which is bounded and continuous in x for fixed z. We have

$$\begin{aligned} \log \widehat{\mu}_n(z) &= -\tfrac{1}{2}\langle z, A_n z\rangle + \mathrm{i}\langle \beta_n, z\rangle + \int g(z,x)\nu_n(\mathrm{d}x) \\ &= -\tfrac{1}{2}\langle z, A_{n,\varepsilon} z\rangle + \mathrm{i}\langle \beta_n, z\rangle + I_{n,\varepsilon} + J_{n,\varepsilon}, \end{aligned} \tag{8.15}$$

where

$$I_{n,\varepsilon} = \int_{|x|\le\varepsilon} (g(z,x) + \tfrac{1}{2}\langle z,x\rangle^2)(|x|^2\wedge 1)^{-1}\rho_n(\mathrm{d}x),$$

$$J_{n,\varepsilon} = \int_{|x|>\varepsilon} g(z,x)(|x|^2\wedge 1)^{-1}\rho_n(\mathrm{d}x).$$

Let E be the set of $\varepsilon > 0$ for which $\int_{|x|=\varepsilon} \rho(\mathrm{d}x) = 0$. Then

$$\lim_{k\to\infty} J_{n_k,\varepsilon} = \int_{|x|>\varepsilon} g(z,x)(|x|^2 \wedge 1)^{-1}\rho(\mathrm{d}x) \quad \text{for } \varepsilon \in E.$$

Hence

$$\lim_{E\ni\varepsilon\downarrow 0}\lim_{k\to\infty} J_{n_k,\varepsilon} = \int_{\mathbb{R}^d} g(z,x)\nu(\mathrm{d}x). \tag{8.16}$$

Furthermore we get

$$\lim_{\varepsilon\downarrow 0}\sup_n |I_{n,\varepsilon}| = 0 \tag{8.17}$$

from (8.12) since $(g(z,x)+\frac{1}{2}\langle z,x\rangle^2)(|x|^2\wedge 1)^{-1}$ tends to 0 by (8.3) as $x \to 0$. Considering the real part and the imaginary part of (8.15) separately and using (8.11), (8.16), and (8.17), we get

$$\lim_{E\ni\varepsilon\downarrow 0}\limsup_{k\to\infty}\langle z, A_{n_k,\varepsilon}z\rangle = \lim_{E\ni\varepsilon\downarrow 0}\liminf_{k\to\infty}\langle z, A_{n_k,\varepsilon}z\rangle, \tag{8.18}$$

$$\limsup_{k\to\infty}\langle \beta_{n_k}, z\rangle = \liminf_{k\to\infty}\langle \beta_{n_k}, z\rangle, \tag{8.19}$$

and both sides in (8.18) and (8.19) are finite. By (8.19) there is β such that $\beta_{n_k} \to \beta$. Since each side of (8.18) is a nonnegative quadratic form of z, it is equal to $\langle z, Az\rangle$ with some symmetric, nonnegative-definite A. In (8.18) we can drop the restriction of ε to the set E, because $\langle z, A_{n,\varepsilon}z\rangle$ is monotone in ε. It follows that $\widehat{\mu}(z)$ has representation (8.5) with these A, ν, and β (in place of γ_c) and that (1), (2), and (3) hold with $n\to\infty$ via the subsequence $\{\mu_{n_k}\}$. The A, ν, and β in the triplet $(A,\nu,\beta)_c$ are unique, because we already proved part (ii) of Theorem 8.1. As we can begin our discussion with any subsequence of $\{\mu_n\}$, this uniqueness ensures that (1) and (3) hold for the whole sequence $\{\mu_n\}$. Now, looking back over our argument, we see that

$$\lim_{\varepsilon\downarrow 0}\limsup_{n\to\infty}\langle z, A_{n,\varepsilon}z\rangle = \lim_{\varepsilon\downarrow 0}\liminf_{n\to\infty}\langle z, A_{n,\varepsilon}z\rangle = \langle z, Az\rangle. \tag{8.20}$$

This is equivalent to (2). This finishes proof of the 'only if' part, provided that (8.12) and (8.13) are true.

Proof of (8.12) and (8.13) is as follows. Let $[-h,h]^d = C(h)$. We have

$$\begin{aligned} &-\int_{C(h)} \log\widehat{\mu}_n(z)\mathrm{d}z \\ &\quad= \tfrac{1}{2}\int_{C(h)}\langle z, A_n z\rangle \mathrm{d}z - \int_{\mathbb{R}^d}\nu_n(\mathrm{d}x)\int_{C(h)} g(z,x)\mathrm{d}z \\ &\quad\geq (2h)^d \int_{\mathbb{R}^d}\left(1-\prod_{j=1}^d \frac{\sin hx_j}{hx_j}\right)\nu_n(\mathrm{d}x). \end{aligned} \tag{8.21}$$

The leftmost member of (8.21) tends to $-\int_{C(h)} \log \widehat{\mu}(z)\mathrm{d}z$ as $n \to \infty$. Letting $h = 1$, and noting

$$\inf_x \left(1 - \prod_{j=1}^{d} \frac{\sin x_j}{x_j}\right)(|x|^2 \wedge 1)^{-1} > 0,$$

we see that (8.12) is true. Since

$$\lim_{h \downarrow 0} \frac{-1}{(2h)^d} \int_{C(h)} \log \widehat{\mu}(z)\mathrm{d}z = 0,$$

it is shown that, for any $\varepsilon > 0$, there are n_0 and h_0 such that

$$\int_{\mathbb{R}^d} \left(1 - \prod_{j=1}^{d} \frac{\sin h_0 x_j}{h_0 x_j}\right) \nu_n(\mathrm{d}x) < \varepsilon \quad \text{for } n \geq n_0.$$

If $|x| > 2\sqrt{d}/h_0$, then $|x_{j_0}| > 2/h_0$ for some j_0 and

$$1 - \prod_{j=1}^{d} \frac{\sin h_0 x_j}{h_0 x_j} \geq 1 - \left|\frac{\sin h_0 x_{j_0}}{h_0 x_{j_0}}\right| \geq 1 - \frac{1}{h_0 |x_{j_0}|} > \frac{1}{2}.$$

Hence

$$\tfrac{1}{2} \int_{|x| > 2\sqrt{d}/h_0} \nu_n(\mathrm{d}x) < \varepsilon \quad \text{for } n \geq n_0.$$

Hence

$$\tfrac{1}{2} \int_{|x| > 2\sqrt{d}/h_0} \rho_n(\mathrm{d}x) < \varepsilon \quad \text{for } n \geq n_0.$$

This proves (8.13).

Let us prove the 'if' part. Define $\rho_n(\mathrm{d}x) = (|x|^2 \wedge 1)\nu_n(\mathrm{d}x)$ as above, and $\rho(\mathrm{d}x) = (|x|^2 \wedge 1)\nu(\mathrm{d}x)$. Let the set E be as above. Then we get (8.16) from condition (1). Since conditions (1) and (2) imply uniform boundedness (8.12) of $\{\rho_n\}$, we get (8.17) also. Hence, using (2) and (3), we have

$$\lim_{n \to \infty} \log \widehat{\mu}_n(z) = -\tfrac{1}{2}\langle z, Az\rangle + \mathrm{i}\langle \beta, z\rangle + \int g(z, x)\nu(\mathrm{d}x).$$

The right-hand side is equal to $\log \widehat{\mu}(z)$. Therefore $\mu_n \to \mu$. □

Now, using the 'only if' part of Theorem 8.7, completion of the proof of Theorem 8.1 is easy.

Proof of Theorem 8.1(i). We are given an infinitely divisible probability measure μ. Choose $t_n \downarrow 0$ arbitrarily. Define μ_n by

$$\widehat{\mu}_n(z) = \exp[t_n^{-1}(\widehat{\mu}(z)^{t_n} - 1)] = \exp\left[t_n^{-1} \int_{\mathbb{R}^d \setminus \{0\}} (\mathrm{e}^{\mathrm{i}\langle z, x\rangle} - 1)\mu^{t_n}(\mathrm{d}x)\right].$$

The distribution μ_n is compound Poisson. Note that

$$\widehat{\mu}_n(z) = \exp[t_n^{-1}(\mathrm{e}^{t_n \log \widehat{\mu}(z)} - 1)] = \exp[t_n^{-1}(t_n \log \widehat{\mu}(z) + O(t_n^2))]$$

for each z as $n \to \infty$. Hence $\widehat{\mu}_n(z) \to \mathrm{e}^{\log \widehat{\mu}(z)} = \widehat{\mu}(z)$. Since μ_n has the representation (8.5) in Theorem 8.7, we can apply the theorem and conclude that $\widehat{\mu}(z)$ has the Lévy–Khintchine representation with triplet $(A, \nu, \beta)_c$. This representation can be written in the form (8.1). □

COROLLARY 8.8. *Every infinitely divisible distribution is the limit of a sequence of compound Poisson distributions.*

Proof. See the proof of Theorem 8.1(i). □

COROLLARY 8.9. *Let $t_n \downarrow 0$. If ν is the Lévy measure of an infinitely divisible distribution μ, then, for any $f \in C_\sharp$,*

$$t_n^{-1} \int_{\mathbb{R}^d} f(x) \mu^{t_n}(\mathrm{d}x) \to \int_{\mathbb{R}^d} f(x) \nu(\mathrm{d}x).$$

Proof. The distribution μ_n in the proof of Theorem 8.1(i) has Lévy measure $[t_n^{-1} \mu^{t_n}]_{\mathbb{R}^d \setminus \{0\}}$. Condition (1) in Theorem 8.7 gives the result. □

EXAMPLE 8.10 (Γ-distribution). Let μ be a Γ-distribution with parameters c, α as in Example 2.15. Let us show that

$$\widehat{\mu}(z) = \exp\left[c \int_0^\infty (\mathrm{e}^{\mathrm{i}zx} - 1) \frac{\mathrm{e}^{-\alpha x}}{x} \mathrm{d}x\right].$$

This is representation (8.7) with $A = 0$, $\nu(dx) = c1_{(0,\infty)}(x) x^{-1} \mathrm{e}^{-\alpha x} dx$, and $\gamma_0 = 0$. This μ is not a compound Poisson distribution, because ν has total mass ∞. By (2.15) and by

$$\begin{aligned}\log(1 + \alpha^{-1} u) &= \int_0^u \frac{\mathrm{d}y}{\alpha + y} = \int_0^u \mathrm{d}y \int_0^\infty \mathrm{e}^{-\alpha x - yx} \mathrm{d}x \\ &= \int_0^\infty \mathrm{e}^{-\alpha x} \mathrm{d}x \int_0^u \mathrm{e}^{-yx} \mathrm{d}y,\end{aligned}$$

the Laplace transform is expressed as

$$L_\mu(u) = \exp\left[c \int_0^\infty (\mathrm{e}^{-ux} - 1) \frac{\mathrm{e}^{-\alpha x}}{x} \mathrm{d}x\right].$$

Extending this equality to the left half plane $\{w \in \mathbb{C} \colon \mathrm{Re}\, w \le 0\}$ by analyticity inside and continuity to the boundary, we get

$$\int_{\mathbb{R}} \mathrm{e}^{wx} \mu(\mathrm{d}x) = \exp\left[c \int_0^\infty (\mathrm{e}^{wx} - 1) \frac{\mathrm{e}^{-\alpha x}}{x} \mathrm{d}x\right].$$

Let $w = \mathrm{i}z$, $z \in \mathbb{R}$, to get $\widehat{\mu}(z)$.

EXAMPLE 8.11 (One-sided strictly $\frac{1}{2}$-stable distribution). The Laplace transform of the μ of Example 2.13 is

$$L_\mu(u) = \mathrm{e}^{-c\sqrt{2u}} = \exp\left[c(2\pi)^{-1/2} \int_0^\infty (\mathrm{e}^{-ux} - 1) x^{-3/2} \mathrm{d}x\right], \quad u \ge 0.$$

The last equality is obtained as follows:

$$\begin{aligned}\int_0^\infty (\mathrm{e}^{-ux}-1)x^{-3/2}\mathrm{d}x &= -\int_0^\infty x^{-3/2}\mathrm{d}x\int_0^x u\mathrm{e}^{-uy}\mathrm{d}y\\ &= -u\int_0^\infty \mathrm{e}^{-uy}\mathrm{d}y\int_y^\infty x^{-3/2}\mathrm{d}x\\ &= -2u\int_0^\infty \mathrm{e}^{-uy}y^{-1/2}\mathrm{d}y = -2u^{1/2}\Gamma(\tfrac{1}{2}) = -2\sqrt{\pi u}.\end{aligned}$$

Extending the expression to the left half plane, we get, on the imaginary axis,

$$\widehat{\mu}(z) = \exp\left[c(2\pi)^{-1/2}\int_0^\infty (\mathrm{e}^{\mathrm{i}zx}-1)x^{-3/2}\mathrm{d}x\right], \qquad z\in\mathbb{R}.$$

This is the form (8.7) with $A=0$, $\nu(\mathrm{d}x) = 1_{(0,\infty)}(x)c(2\pi)^{-1/2}x^{-3/2}\mathrm{d}x$, and $\gamma_0 = 0$. Again μ is not a compound Poisson distribution.

REMARK 8.12. All infinitely divisible distributions in Example 7.2 are such that their infinite divisibility is obvious if we look at explicit forms of their characteristic functions. That is, their nth roots in the convolution sense are obtained by taking their parameters appropriately. But there are many other infinitely divisible distributions whose infinite divisibility is more difficult to prove. We list some such distributions on $\mathbb{R}$ with the papers where their infinite divisibility is proved. Here c's are normalizing constants. We have chosen scaling and translation appropriately.
Student's t-distribution (Grosswald [**174**], Ismail [**217**])

$$\mu(\mathrm{d}x) = c(1+x^2)^{-(\alpha+1)/2}\mathrm{d}x, \qquad \alpha\in(0,\infty);$$

Pareto distribution (Steutel [**499**], Thorin [**532**])

$$\mu(\mathrm{d}x) = c\,1_{(0,\infty)}(x)(1+x)^{-\alpha-1}\mathrm{d}x, \qquad \alpha\in(0,\infty);$$

F-distribution (Ismail and Kelker [**218**])

$$\mu(\mathrm{d}x) = c\,1_{(0,\infty)}(x)x^{\beta-1}(1+x)^{-\alpha-\beta}\mathrm{d}x, \qquad \alpha,\beta\in(0,\infty);$$

Gumbel distribution (extreme value distribution of type 1 in [**241**]) (Steutel [**500**])

$$\mu(-\infty,x] = \mathrm{e}^{-\mathrm{e}^{-x}};$$

Weibull distribution (extreme value distribution of type 3 in [**241**]) with parameter $0<\alpha\le 1$ (Steutel [**499**])

$$\mu(-\infty,x] = \begin{cases} 0, & x\le 0,\\ 1-\mathrm{e}^{-x^\alpha}, & x>0\end{cases}$$

(Weibull with $\alpha > 1$ is not infinitely divisible, see Exercise 29.10); log-normal distribution (the distribution of X when $\log X$ is Gaussian distributed) (Thorin [**533**])

$$\mu(\mathrm{d}x) = c\,1_{(0,\infty)}(x)x^{-1}\mathrm{e}^{-\alpha(\log x)^2}\mathrm{d}x, \qquad \alpha \in (0,\infty);$$

logistic distribution (Steutel [**501**])

$$\mu(-\infty, x] = (1+\mathrm{e}^{-x})^{-1} \quad \text{for } x \in \mathbb{R};$$

half-Cauchy distribution (Bondesson [**54**])

$$\mu(\mathrm{d}x) = 2\pi^{-1}1_{(0,\infty)}(x)(1+x^2)^{-1}\mathrm{d}x.$$

See Exercise 55.1 for Pareto and Weibull with $0 < \alpha \le 1$. Other such examples are mixtures of Γ-distributions of parameter c when c is fixed in $(0,2]$ (Remark 51.13). There are many infinitely divisible distributions on $\mathbb{R}$ with densities expressible by Bessel functions (Feller [**139**], Hartman [**182**], Pitman and Yor [**377**], Ismail and Kelker [**218**], Yor [**596**], see also Example 30.11, Exercises 34.1, 34.2, and 34.15). On $\mathbb{R}^d$ the following are known to be infinitely divisible (Takano [**513, 514**]):

$$\mu(\mathrm{d}x) = c\,\mathrm{e}^{-|x|}\mathrm{d}x; \tag{8.22}$$

$$\mu(\mathrm{d}x) = c(1+|x|^2)^{-\alpha-(d/2)}\mathrm{d}x, \qquad \alpha \in (0,\infty). \tag{8.23}$$

9. Additive processes in law

Infinitely divisible distributions have close connections not only with Lévy processes but also with additive processes.

THEOREM 9.1. *If $\{X_t\colon t \ge 0\}$ is an additive process in law on $\mathbb{R}^d$, then, for every t, the distribution of X_t is infinitely divisible.*

We prove this theorem from the next result. This is one of the fundamental limit theorems on sums of independent random variables, conjectured by Kolmogorov and proved by Khintchine.

DEFINITION 9.2. A double sequence of random variables $\{Z_{nk}\colon k = 1, 2, \ldots, r_n;\ n = 1, 2, \ldots\}$ on $\mathbb{R}^d$ is called a *null array* if, for each fixed n, $Z_{n1}, Z_{n2}, \ldots, Z_{nr_n}$ are independent and if, for any $\varepsilon > 0$,

$$\lim_{n\to\infty} \max_{1\le k\le r_n} P[\,|Z_{nk}| > \varepsilon\,] = 0. \tag{9.1}$$

The sums $S_n = \sum_{k=1}^{r_n} Z_{nk}$, $n = 1, 2, \ldots$, are called the *row sums*.

THEOREM 9.3 (Khintchine [**278**]). *Let $\{Z_{nk}\}$ be a null array on $\mathbb{R}^d$ with row sums S_n. If, for some $b_n \in \mathbb{R}^d$, $n = 1, 2, \ldots$, the distribution of $S_n - b_n$ converges to a distribution μ, then μ is infinitely divisible.*

First we give a centering of random variables on $\mathbb{R}^d$. Define a mapping τ from $\mathbb{R}^d$ to $\mathbb{R}^d$ by

$$\tau_j(x) = \tau_j(x_1, \dots, x_d) = \begin{cases} -1, & x_j \le -1, \\ x_j, & -1 < x_j \le 1, \\ 1, & 1 < x_j, \end{cases} \tag{9.2}$$

where $\tau(x) = (\tau_j(x))_{j=1,\dots,d}$ in components.

LEMMA 9.4. *For any random variable X on $\mathbb{R}^d$, there exists $a \in \mathbb{R}^d$ satisfying*

$$E[\tau(X - a)] = 0. \tag{9.3}$$

Proof. The function $E[\tau_j(X - a)]$ of $a = (a_k)_{k=1,\dots,d}$ depends only on a_j. As a function of a_j, it is continuous and decreasing. It tends to 1 and -1 as a_j tends to $-\infty$ and ∞, respectively. Hence it vanishes at some a_j. This gives the jth coordinate of the desired point a. □

LEMMA 9.5. *Let X be a random variable on $\mathbb{R}^d$ with distribution μ satisfying $E[\tau(X)] = 0$. Then,*

$$|\widehat{\mu}(z) - 1| \le (2 + \sqrt{d}|z| + \tfrac{1}{2}|z|^2)E[|\tau(X)|^2], \qquad z \in \mathbb{R}^d. \tag{9.4}$$

Proof. Notice that $\widehat{\mu}(z) - 1 = E[e^{i\langle z,X\rangle} - 1 - i\langle z, \tau(X)\rangle]$. If $|x_p| \le 1$ for all p, then

$$|e^{i\langle z,x\rangle} - 1 - i\langle z, \tau(x)\rangle| \le \tfrac{1}{2}\langle z, \tau(x)\rangle^2 \le \tfrac{1}{2}|z|^2|\tau(x)|^2$$

by Lemma 8.6. If $|x_p| > 1$ for some p, then

$$|e^{i\langle z,x\rangle} - 1 - i\langle z, \tau(x)\rangle| \le 2 + \sqrt{d}|z| \le (2 + \sqrt{d}|z|)|\tau(x)|^2.$$

Hence (9.4) follows. □

Proof of Theorem 9.3. Choose $a_{nk} \in \mathbb{R}^d$ satisfying $E[\tau(Z_{nk} - a_{nk})] = 0$, using Lemma 9.4. Denote the distribution of $Z_{nk} - a_{nk}$ by μ_{nk}, and let

$$\gamma_n = \sum_{k=1}^{r_n} a_{nk} - b_n.$$

We claim that

$$\exp\left[i\langle \gamma_n, z\rangle + \sum_{k=1}^{r_n}(\widehat{\mu}_{nk}(z) - 1)\right] \to \widehat{\mu}(z), \qquad n \to \infty. \tag{9.5}$$

This will prove the infinite divisibility of μ by Lemma 7.8, since the left-hand side of (9.5) is the charactersitic function of a translated compound Poisson distribution. Such a sequence of infinitely divisible distributions is sometimes called the associated sequence. The property (9.1), combined with the estimate

$$|\widehat{P}_{Z_{nk}}(z) - 1| = \left|\int (\mathrm{e}^{\mathrm{i}\langle z,x\rangle} - 1) P_{Z_{nk}}(\mathrm{d}x)\right| \leq \varepsilon|z| + 2P[\,|Z_{nk}| > \varepsilon\,],$$

shows that, for any $M > 0$,

$$\sup_{|z|\leq M} \max_{1\leq k\leq r_n} |\widehat{P}_{Z_{nk}}(z) - 1| \to 0, \qquad n \to \infty. \tag{9.6}$$

We have

$$\max_{1\leq k\leq r_n} |a_{nk}| \to 0, \qquad n \to \infty. \tag{9.7}$$

In fact, if $a = (a_l)_{1\leq l\leq d}$, $a_j = -\varepsilon$, and $0 < \varepsilon < 1$, then, for sufficiently large n,

$$\begin{aligned} E[\tau_j(Z_{nk} - a)] &\geq (\varepsilon/2)P[\,|(Z_{nk})_j| < \varepsilon/2\,] - P[\,(Z_{nk})_j \leq -\varepsilon/2\,] \\ &\geq \varepsilon/4 - \varepsilon/8 > 0 \end{aligned}$$

by (9.1). Hence the proof of Lemma 9.4 shows that $\min_{1\leq k\leq r_n}(a_{nk})_j > -\varepsilon$ for large n. Similarly, if $a_j = \varepsilon$ and $0 < \varepsilon < 1$, then, for large n,

$$\begin{aligned} E[\tau_j(Z_{nk} - a)] &\leq -(\varepsilon/2)P[\,|(Z_{nk})_j| < \varepsilon/2\,] + P[\,(Z_{nk})_j \geq \varepsilon/2\,] \\ &\leq -\varepsilon/4 + \varepsilon/8 < 0 \end{aligned}$$

and hence $\max_{1\leq k\leq r_n}(a_{nk})_j < \varepsilon$. This shows (9.7). It folows from (9.1) and (9.7) that, for any $\varepsilon > 0$,

$$\max_{1\leq k\leq r_n} P[\,|Z_{nk} - a_{nk}| > \varepsilon\,] \to 0, \qquad n \to \infty. \tag{9.8}$$

Further, for any $M > 0$,

$$\sup_{|z|\leq M} \max_{1\leq k\leq r_n} |\widehat{\mu}_{nk}(z) - 1| \to 0, \qquad n \to \infty, \tag{9.9}$$

just as (9.6) follows from (9.1). If n is so large that the left-hand side of (9.9) is less than $1/2$, then $\log \widehat{\mu}_{nk}(z)$ is defined continuously on $\{|z| \leq M\}$ as the principal branch of the logarithm of $\widehat{\mu}_{nk}(z)$. Thus

$$\log \widehat{\mu}_{nk}(z) = \log(1 + \theta_{nk}(z)) = \theta_{nk}(z)(1 + \rho_{nk}(z)),$$

where

$$\theta_{nk}(z) = \widehat{\mu}_{nk}(z) - 1, \qquad \rho_{nk}(z) = \sum_{l=1}^{\infty}(-1)^l(1+l)^{-1}\theta_{nk}(z)^l.$$

We have

$$|\rho_{nk}(z)| \leq |\theta_{nk}(z)|/(1 - |\theta_{nk}(z)|)$$

and, by (9.9),

$$\sup_{|z|\leq M} \max_{1\leq k\leq r_n} |\rho_{nk}(z)| \to 0, \qquad n \to \infty.$$

Let

$$v_n = \sum_{k=1}^{r_n} E[\,|\tau(Z_{nk} - a_{nk})|^2\,].$$

Apply Lemma 9.5. Then

$$\sum_{k=1}^{r_n} |\theta_{nk}(z)| \le (2 + \sqrt{d}|z| + \tfrac{1}{2}|z|^2)v_n \quad \text{for } |z| \le M.$$

Hence

$$\sum_{k=1}^{r_n} \log \widehat{\mu}_{nk}(z) = \sum_{k=1}^{r_n} (\widehat{\mu}_{nk}(z) - 1) + o(v_n), \qquad n \to \infty, \tag{9.10}$$

uniformly on $\{|z| \le M\}$. Next we show that

$$v_n = O(1), \qquad n \to \infty. \tag{9.11}$$

Choose $h > 0$ so small that $|\widehat{\mu}(z) - 1| < 1/2$ on the set $C_h = \{z = (z_j)_{1\le j\le d} \colon |z_j| \le h \text{ for } 1 \le j \le d\}$. Choose, for a moment, M so that $\{|z| \le M\} \supset C_h$. Since

$$e^{i\langle \gamma_n, z\rangle} \prod_{k=1}^{r_n} \widehat{\mu}_{nk}(z) \to \widehat{\mu}(z) \tag{9.12}$$

uniformly on any compact set by our assumption, we see that, uniformly on C_h,

$$i\langle \gamma_n, z\rangle + \sum_{k=1}^{r_n} \log \widehat{\mu}_{nk}(z) \to \log \widehat{\mu}(z), \qquad n \to \infty.$$

Use (9.10), integrate over C_h, and multiply by $-(2h)^{-d}$. Then

$$\sum_{k=1}^{r_n} \int_{\mathbb{R}^d} \left(1 - \prod_{j=1}^{d} \frac{\sin hx_j}{hx_j}\right) \mu_{nk}(dx) + o(v_n) + o(1) = c_1,$$

where

$$c_1 = -(2h)^{-d} \int_{C_h} \log \widehat{\mu}(z) dz.$$

Hence

$$c_1 \ge c_2 \sum_{k=1}^{r_n} \int_{\mathbb{R}^d} |\tau(x)|^2 \mu_{nk}(dx) + o(v_n) + o(1) = c_2 v_n + o(v_n) + o(1),$$

where

$$c_2 = \inf_{x\in\mathbb{R}^d} \left(1 - \prod_{j=1}^{d} \frac{\sin hx_j}{hx_j}\right) \frac{1}{|\tau(x)|^2} > 0.$$

This shows (9.11). Now (9.10), (9.11), and (9.12) yield, for any $M > 0$,

$$\widehat{\mu}(z) = \exp\left[\mathrm{i}\langle\gamma_n, z\rangle + \sum_{k=1}^{r_n}(\widehat{\mu}_{nk}(z) - 1) + o(1)\right] + o(1), \qquad n \to \infty,$$

uniformly on $\{|z| \le M\}$. This completes the proof of (9.5). □

In order to prove Theorem 9.1, we give a lemma.

LEMMA 9.6. *A stochastically continuous process $\{X_t : t \ge 0\}$ is uniformly stochastically continuous on any finite interval $[0, t_0]$, that is, for every $\varepsilon > 0$ and $\eta > 0$, there is $\delta > 0$ such that, if s and t are in $[0, t_0]$ and satisfy $|s - t| < \delta$, then $P[\,|X_s - X_t| > \varepsilon\,] < \eta$.*

Proof. Fix ε and η. For any t there is $\delta_t > 0$ such that $P[\,|X_s - X_t| > \varepsilon/2\,] < \eta/2$ for $|s - t| < \delta_t$. Let $I_t = (t - \delta_t/2, t + \delta_t/2)$. Then $\{I_t : t \in [0, t_0]\}$ covers the interval $[0, t_0]$. Hence there is a finite subcovering $\{I_{t_j} : j = 1, \ldots, n\}$ of $[0, t_0]$. Let δ be the minimum of $\delta_{t_j}/2$, $j = 1, \ldots, n$. If $|s - t| < \delta$ and $s, t \in [0, t_0]$, then $t \in I_{t_j}$ for some j, hence $|s - t_j| < \delta_{t_j}$ and

$$P[\,|X_s - X_t| > \varepsilon\,] \le P[\,|X_s - X_{t_j}| > \varepsilon/2\,] + P[\,|X_t - X_{t_j}| > \varepsilon/2\,] < \eta. \quad \square$$

Proof of Theorem 9.1. Fix $t > 0$ and let $t_{nk} = kt/n$ for $n = 1, 2, \ldots$ and $k = 0, 1, \ldots, n$. Let $r_n = n$ and $Z_{nk} = X(t_{nk}) - X(t_{n,k-1})$ for $k = 1, \ldots, n$. Then $\{Z_{nk}\}$ is a null array by Lemma 9.6. The row sum S_n equals X_t. Hence we can apply Theorem 9.3 with $\mu = P_{X_t}$ and $b_n = 0$. □

THEOREM 9.7. (i) *Let $\{X_t : t \ge 0\}$ be an additive process in law on $\mathbb{R}^d$ and, for $0 \le s \le t < \infty$, let $\mu_{s,t}$ be the distribution of $X_t - X_s$. Then $\mu_{s,t}$ is infinitely divisible and*

(9.13) $\qquad \mu_{s,t} * \mu_{t,u} = \mu_{s,u}$ *for* $0 \le s \le t \le u < \infty$,

(9.14) $\qquad \mu_{s,s} = \delta_0$ *for* $0 \le s < \infty$,

(9.15) $\qquad \mu_{s,t} \to \delta_0$ *as* $s \uparrow t$,

(9.16) $\qquad \mu_{s,t} \to \delta_0$ *as* $t \downarrow s$.

(ii) *Conversely, if $\{\mu_{s,t} : 0 \le s \le t < \infty\}$ is a system of probability measures on $\mathbb{R}^d$ satisfying (9.13)–(9.16), then there is an additive process in law $\{X_t : t \ge 0\}$ such that, for $0 \le s \le t < \infty$, $X_t - X_s$ has the distribution $\mu_{s,t}$.*

(iii) *If $\{X_t\}$ and $\{X'_t\}$ are additive processes in law on $\mathbb{R}^d$ such that $X_t \overset{\mathrm{d}}{=} X'_t$ for any $t \ge 0$, then $\{X_t\}$ and $\{X'_t\}$ are identical in law.*

Proof. (i) Fix $s \ge 0$. Then $\{X_{s+t} - X_s,\ t \ge 0\}$ is an additive process in law. Hence $\mu_{s,t}$ is infinitely divisible by Theorem 9.1. The property (9.13) comes from $(X_t - X_s) + (X_u - X_t) = X_u - X_s$ and from the independence of $X_t - X_s$ and $X_u - X_t$; (9.14) expresses $X_s - X_s = 0$; (9.15) and (9.16) express the stochastic continuity.

(ii) We can make a proof in the same way as that of Theorem 7.10(ii). Replace μ^{t-s} by $\mu_{s,t}$ in the argument there. The properties (7.1)–(7.3) are replaced by (9.13)–(9.16).

(iii) Suppose that $\{X_t\}$ and $\{X'_t\}$ are additive processes in law satisfying $X_t \stackrel{\mathrm{d}}{=} X'_t$ for $t \geq 0$. Let $\mu_{s,t}$ and $\mu'_{s,t}$ be the distributions of $X_t - X_s$ and $X'_t - X'_s$, respectively, for $0 \leq s \leq t < \infty$. Then $\mu_{0,t} = \mu'_{0,t}$ since $X_0 = X'_0 = 0$. Since $\mu_{0,t}$ is infinitely divisible, $\widehat{\mu}_{0,t}(z) \neq 0$ by Lemma 7.5. Hence it follows from $\mu_{0,s} * \mu_{s,t} = \mu_{0,t}$ and $\mu'_{0,s} * \mu'_{s,t} = \mu'_{0,t}$ that $\mu_{s,t} = \mu'_{s,t}$. The rest of the proof is the same as that of Theorem 7.10(iii). □

THEOREM 9.8. (i) *Suppose that* $\{X_t \colon t \geq 0\}$ *is an additive process in law on* $\mathbb{R}^d$. *Let* $(A_t, \nu_t, \gamma(t))$ *be the generating triplet of the infinitely divisible distribution* $\mu_t = P_{X_t}$ *for* $t \geq 0$. *Then, the following conditions are satisfied.*

(1) $A_0 = 0$, $\nu_0 = 0$, $\gamma(0) = 0$.

(2) *If* $0 \leq s \leq t < \infty$, *then* $\langle z, A_s z\rangle \leq \langle z, A_t z\rangle$ *for* $z \in \mathbb{R}^d$ *and* $\nu_s(B) \leq \nu_t(B)$ *for* $B \in \mathcal{B}(\mathbb{R}^d)$.

(3) *As* $s \to t$ *in* $[0, \infty)$, $\langle z, A_s z\rangle \to \langle z, A_t z\rangle$ *for* $z \in \mathbb{R}^d$, $\nu_s(B) \to \nu_t(B)$ *for* $B \in \mathcal{B}(\mathbb{R}^d)$ *with* $B \subset \{x \colon |x| > \varepsilon\}$, $\varepsilon > 0$, *and* $\gamma(s) \to \gamma(t)$.

(ii) *Let* $\{\mu_t \colon t \geq 0\}$ *be a system of infinitely divisible probability measures on* $\mathbb{R}^d$ *with generating triplets* $(A_t, \nu_t, \gamma(t))$ *satisfying the conditions (1)–(3). Then there exists, uniquely up to identity in law, an additive process in law* $\{X_t \colon t \geq 0\}$ *on* $\mathbb{R}^d$ *such that* $P_{X_t} = \mu_t$ *for* $t \geq 0$.

Proof. (i) The process $\{X_t\}$ determines, by Theorem 9.7, the system $\{\mu_{s,t} \colon 0 \leq s \leq t < \infty\}$. Then, $\mu_t = \mu_{0,t}$. The property (1) is obvious. Since $\mu_{0,t} = \mu_{0,s} * \mu_{s,t}$ for $s \leq t$, the infinite divisibility of $\mu_{s,t}$ gives the property (2). As $s \to t$, μ_s tends to μ_t by the stochastic continuity of $\{X_t\}$. Hence, for any sequence s_n that tends to t,

$$\lim_{\varepsilon \downarrow 0} \limsup_{n \to \infty} \left| \langle z, A_{s_n} z\rangle + \int_{|x|<\varepsilon} \langle z, x\rangle^2 \nu_{s_n}(\mathrm{d}x) - \langle z, A_t z\rangle \right| = 0 \quad \text{for any } z,$$

by Theorem 8.7. Fix t_0 such that s_n and t are in $[0, t_0]$. Then,

$$0 \leq \int_{|x|<\varepsilon} \langle z, x\rangle^2 \nu_{s_n}(\mathrm{d}x) \leq \int_{|x|<\varepsilon} \langle z, x\rangle^2 \nu_{t_0}(\mathrm{d}x) \to 0 \quad \text{as } \varepsilon \downarrow 0.$$

Hence we see that $\lim_{n\to\infty} |\langle z, A_{s_n} z\rangle - \langle z, A_t z\rangle| = 0$. It follows from $\nu_s \leq \nu_{t_0}$ for $s \in [0, t_0]$ that $\nu_s(\mathrm{d}x) = g_s(x)\nu_{t_0}(\mathrm{d}x)$ with some measurable function $g_s(x)$ satisfying $0 \leq g_s(x) \leq 1$. If $s < s' \leq t_0$, then $g_s(x) \leq g_{s'}(x)$, ν_{t_0}-a. e. If s_n tends monotonically to t, then $g_{s_n}(x)$ tends to $g_t(x)$, ν_{t_0}-a. e., and hence $\nu_{s_n}(B) \to \nu_t(B)$ for any Borel set B in $\{x \colon |x| > \varepsilon\}$ with $\varepsilon > 0$. It follows that, as $s \to t$, $\nu_s(B) \to \nu_t(B)$ for every such B and the integral

$$\int_{\mathbb{R}^d} (\mathrm{e}^{\mathrm{i}\langle z,x\rangle} - 1 - \mathrm{i}\langle z,x\rangle 1_{\{|x|\le 1\}}(x))\nu_s(\mathrm{d}x)$$

tends to the integral with respect to ν_t. Recalling the Lévy–Khintchine representation and using $\log \widehat{\mu}_s(z) \to \log \widehat{\mu}_t(z)$, we see that $\gamma(s) \to \gamma(t)$.

(ii) We use Theorem 9.7. Using (2), define, for $0 \le s \le t < \infty$, $\mu_{s,t}$ to be the infinitely divisible distribution with generating triplet $(A_t - A_s, \nu_t - \nu_s, \gamma(t) - \gamma(s))$. Then $\mu_{0,t} = \mu_t$ by (1). In order to prove our assertion it is enough to show that $\{\mu_{s,t}\}$ satisfies (9.13)–(9.16). Among them (9.13) and (9.14) are clear. It follows from (2) and (3) that $\log \widehat{\mu}_s(z) \to \log \widehat{\mu}_t(z)$ as $s \to t$. Hence (9.15) and (9.16) follow. □

REMARK 9.9. Let $\{X_t\}$ be an additive process in law having (A_t, ν_t, γ_t) as the generating triplet of the distribution of X_t. We call $\{(A_t, \nu_t, \gamma_t) \colon t \ge 0\}$ the system of generating triplets of $\{X_t\}$. By Theorem 9.7(iii) it determines $\{X_t\}$ up to identity in law.

There is a unique measure $\widetilde{\nu}$ on $[0,\infty) \times \mathbb{R}^d$ such that

$$\widetilde{\nu}([0,t] \times B) = \nu_t(B) \quad \text{for } t \ge 0 \text{ and } B \in \mathcal{B}(\mathbb{R}^d). \tag{9.17}$$

The measure $\widetilde{\nu}$ satisfies

$$\widetilde{\nu}(\{t\} \times \mathbb{R}^d) = 0 \quad \text{for } t \ge 0, \tag{9.18}$$

$$\int_{[0,t]\times\mathbb{R}^d} (1 \wedge |x|^2)\widetilde{\nu}(\mathrm{d}(s,x)) < \infty \quad \text{for } t \ge 0. \tag{9.19}$$

Conversely, if a measure $\widetilde{\nu}$ on $[0,\infty) \times \mathbb{R}^d$ satisfies (9.18) and (9.19), then the measures ν_t, $t \ge 0$, defined by (9.17) satisfy the conditions for ν_t described in (1), (2), and (3) of Theorem 9.8(i).

In fact, (9.17) defines $\widetilde{\nu}$ from ν_t, $t \ge 0$, uniquely on the algebra generated by the sets of the form $[0,t] \times B$ with $t \ge 0$ and $B \in \mathcal{B}(\mathbb{R}^d)$, and this $\widetilde{\nu}$ is extended uniquely to the σ-algebra generated by this algebra by a standard argument in measure theory. This σ-algebra is the Borel σ-algebra of $[0,\infty) \times \mathbb{R}^d$. The property (9.18) comes from (1) and (3) of Theorem 9.8(i), and (9.19) is equivalent to $\int (1 \wedge |x|^2)\nu_t(\mathrm{d}x) < \infty$. The converse is easy to check.

In Chapter 4 we will consider $\widetilde{\nu}$ as a measure on $(0,\infty) \times (\mathbb{R}^d \setminus \{0\})$. This is possible because $\{(s,x) \colon s = 0 \text{ or } x = 0\}$ has $\widetilde{\nu}$-measure 0. If the corresponding process $\{X_t\}$ is a Lévy process in law, then $\widetilde{\nu}$ is the product of the Lebesgue measure on $[0,\infty)$ and the Lévy measure ν of $\{X_t\}$.

10. Transition functions and the Markov property

An important property of Lévy and additive processes in law is the Markov property. In this section we define Markov processes by using transition functions, and then characterize Lévy and additive processes in law as Markov processes with spatially homogeneous transition functions.

DEFINITION 10.1. A mapping $P_{s,t}(x, B)$ of $x \in \mathbb{R}^d$ and $B \in \mathcal{B}(\mathbb{R}^d)$ with $0 \le s \le t < \infty$ is called a *transition function* on $\mathbb{R}^d$ if

(1) it is a probability measure as a mapping of B for any fixed x;
(2) it is measurable in x for any fixed B;
(3) $P_{s,s}(x, B) = \delta_x(B)$ for $s \ge 0$;
(4) it satisfies

$$\int_{\mathbb{R}^d} P_{s,t}(x, \mathrm{d}y)\, P_{t,u}(y, B) = P_{s,u}(x, B) \quad \text{for } 0 \le s \le t \le u. \tag{10.1}$$

If, in addition,

(5) $P_{s+h,t+h}(x, B)$ does not depend on h,

then it is called a *temporally homogeneous transition function* and it is given by $P_t(x, B)$ such that

$$P_t(x, B) = P_{s,s+t}(x, B) \quad \text{for } s \ge 0.$$

The property (4) is called the *Chapman–Kolmogorov identity*. In the case of a temporally homogeneous transition function it is written as

$$\int_{\mathbb{R}^d} P_s(x, \mathrm{d}y)\, P_t(y, B) = P_{s+t}(x, B) \quad \text{for } s \ge 0, t \ge 0. \tag{10.2}$$

The properties (1) and (2) imply that $\int P_{s,t}(x, \mathrm{d}y) f(y)$ is measurable in x for any bounded measurable function $f(x)$, for any f is expressible as the limit of finite linear combinations of indicator functions of Borel sets.

If a transition function $P_{s,t}(x, B)$ on $\mathbb{R}^d$ is given, then, for any $a \in \mathbb{R}^d$, the following stochastic process $\{Y_t \colon t \ge 0\}$ can be constructed by using Kolmogorov's extension theorem. Let $\Omega^0 = (\mathbb{R}^d)^{[0,\infty)}$, the collection of all functions ω from $[0, \infty)$ into $\mathbb{R}^d$, and $Y_t(\omega) = \omega(t)$ for $t \ge 0$. Let $\mathcal{F}^0$ be the σ-algebra generated by Y_t, $t \ge 0$. Define, for any $0 \le t_0 < \cdots < t_n$ and $B_0, \ldots, B_n$,

$$\begin{aligned} \mu^{0,a}_{t_0,\ldots,t_n}&(B_0 \times \cdots \times B_n) \\ &= \int P_{0,t_0}(a, \mathrm{d}x_0) 1_{B_0}(x_0) \int P_{t_0,t_1}(x_0, \mathrm{d}x_1) 1_{B_1}(x_1) \\ &\quad \times \int P_{t_1,t_2}(x_1, \mathrm{d}x_2) 1_{B_2}(x_2) \ldots \int P_{t_{n-1},t_n}(x_{n-1}, \mathrm{d}x_n) 1_{B_n}(x_n). \end{aligned} \tag{10.3}$$

This $\mu^{0,a}_{t_0,\ldots,t_n}$ is uniquely extended to a probability measure on $(\mathbb{R}^d)^{n+1}$ and the family $\{\mu^{0,a}_{t_0,\ldots,t_n}\}$ satisfies the consistency condition by virtue of (4).

Then Theorem 1.8 applies. Thus a unique probability measure $P^{0,a}$ on $\mathcal{F}^0$ extending this family exists. Likewise, given $s \geq 0$ and $a \in \mathbb{R}^d$, we consider the restriction $P_{t,u}(x,B)$ with $s \leq t \leq u < \infty$, $\Omega^s = (\mathbb{R}^d)^{[s,\infty)}$, and $Y_t(\omega) = \omega(t)$, $t \geq s$. We define $\mu^{s,a}_{t_0,\ldots,t_n}$ for $s \leq t_0 < t_1 < \cdots < t_n$ similarly. Thus we get the probability measure $P^{s,a}$ on the σ-algebra $\mathcal{F}^s$ generated by Y_t, $t \geq s$. In the case of a temporally homogeneous transition function $P_t(x,B)$, (10.3) is replaced by

$$\begin{aligned}(10.4)\quad \mu^a_{t_0,\ldots,t_n}&(B_0 \times \cdots \times B_n)\\ &= \int P_{t_0}(a, dx_0)1_{B_0}(x_0) \int P_{t_1-t_0}(x_0, dx_1)1_{B_1}(x_1)\\ &\quad \times \int P_{t_2-t_1}(x_1, dx_2)1_{B_2}(x_2) \ldots \int P_{t_n-t_{n-1}}(x_{n-1}, dx_n)1_{B_n}(x_n)\end{aligned}$$

and the probability measure $P^{0,a}$ is denoted by P^a.

DEFINITION 10.2. A stochastic process $\{X_t \colon t \geq 0\}$ defined on a probability space $(\Omega, \mathcal{F}, P)$ is called a *Markov process with transition function* $\{P_{s,t}(x,B)\}$ *and starting point* a, if it is identical in law with the process $\{Y_t \colon t \geq 0\}$ defined above on $(\Omega^0, \mathcal{F}^0, P^{0,a})$. The process $\{Y_t\}$ is the *path space representation* of the process $\{X_t\}$. If, in addition, the transition function is temporally homogeneous, then $\{X_t\}$ is called a *temporally homogeneous Markov process*. In the same way, a stochastic process $\{X_t \colon t \geq s\}$ defined on some $(\Omega, \mathcal{F}, P)$ is called a *Markov process having transition function* $\{P_{t,u}(x,B)\}$ *and starting from* a *at time* s, if it is identical in law with the process $\{Y_t \colon t \geq s\}$ defined on $(\Omega^s, \mathcal{F}^s, P^{s,a})$ as above.

DEFINITION 10.3. A transition function $P_{s,t}(x,B)$ on $\mathbb{R}^d$ is said to be *spatially homogeneous* (or *translation invariant*) if

$$P_{s,t}(x,B) = P_{s,t}(0, B-x) \tag{10.5}$$

for any s, t, x, and B, where $B - x = \{y - x \colon y \in B\}$.

Let us characterize additive processes in law as Markov processes with spatially homogeneous transition functions.

THEOREM 10.4. (i) *Let* $\{X_t\}$ *be an additive process in law on* $\mathbb{R}^d$. *Define* $P_{s,t}(x,B)$ *by*

$$P_{s,t}(x,B) = P[\,X_t - X_s \in B - x\,] \quad \text{for } 0 \leq s \leq t. \tag{10.6}$$

Then $P_{s,t}(x,B)$ *is a spatially homogeneous transition function and* $\{X_t\}$ *is a Markov process with this transition function and starting point* 0.

(ii) *Conversely, if* $\{X_t\}$ *is a stochastically continuous Markov process on* $\mathbb{R}^d$ *with spatially homogeneous transition function and starting point* 0, *then* $\{X_t\}$ *is an additive process in law.*

Proof. (i) Obviously $P_{s,t}(x, B)$ satisfies (1), (3), and (10.5). The definition (10.6) says that

$$P_{s,t}(x, B) = \mu_{s,t}(B - x) = \int \mu_{s,t}(\mathrm{d}y) 1_B(x + y),$$

where $\mu_{s,t}$ is the distribution of $X_t - X_s$. Hence (2) is also clear. For any bounded measurable function f, we have

$$\int P_{s,t}(x, \mathrm{d}y) f(y) = \int \mu_{s,t}(\mathrm{d}y) f(x + y). \tag{10.7}$$

Therefore, for $s \leq t \leq u$,

$$\begin{aligned}\int P_{s,t}(x, \mathrm{d}y) P_{t,u}(y, B) &= \int \mu_{s,t}(\mathrm{d}y) P_{t,u}(x + y, B) \\ &= \int \mu_{s,t}(\mathrm{d}y) \mu_{t,u}(B - x - y) = (\mu_{s,t} * \mu_{t,u})(B - x) \\ &= \mu_{s,u}(B - x) = P_{s,u}(x, B),\end{aligned}$$

which is the Chapman–Kolmogorov identity. Hence $P_{s,t}(x, B)$ is a spatially homogeneous transition function. Construct the process $\{Y_t\}$ as above with $a = 0$ from this transition function. We see that it coincides with the construction in the proof of Theorem 9.7. Hence $\{Y_t\}$ is the additive process in law corresponding to the system $\{\mu_{s,t}\}$. Hence $\{X_t\} \stackrel{\mathrm{d}}{=} \{Y_t\}$ by Theorem 9.7. Therefore $\{X_t\}$ is a Markov process with $P_{s,t}(x, B)$ as a transition function and starting at 0.

(ii) Suppose that $P_{s,t}(x, B)$ is a spatially homogeneous transition function and that $\{X_t\}$ is a stochastically continuous Markov process with this transition function and starting point 0. Define $\mu_{s,t}$ by

$$\mu_{s,t}(B) = P_{s,t}(0, B).$$

Then

$$P_{s,t}(x, B) = P_{s,t}(0, B - x) = \mu_{s,t}(B - x) = \int \mu_{s,t}(\mathrm{d}y) 1_B(x + y).$$

Hence we have (10.7). For $0 \leq t_0 < \cdots < t_n$, $P[X_{t_0} \in B_0, \ldots, X_{t_n} \in B_n]$ equals the right-hand side of (10.3) with $a = 0$. Using (10.7), we rewrite successively the integrals in $x_n, \ldots, x_0$. Thus we get

$$\begin{aligned}&P[X_{t_0} \in B_0, \ldots, X_{t_n} \in B_n] \\ &\quad = \int \cdots \int \mu_{0,t_0}(\mathrm{d}x_0) 1_{B_0}(x_0) \mu_{t_0,t_1}(\mathrm{d}x_1) 1_{B_1}(x_0 + x_1) \\ &\qquad\qquad \times \ldots \mu_{t_{n-1},t_n}(\mathrm{d}x_n) 1_{B_n}(x_0 + \cdots + x_n).\end{aligned}$$

Hence, for every bounded measurable function f, $E[f(X_{t_0}, \ldots, X_{t_n})]$ equals the right-hand side of (7.6) with $\mu^{t_0}, \mu^{t_1 - t_0}, \ldots, \mu^{t_n - t_{n-1}}$ replaced by μ_{0,t_0},

$\mu_{t_0,t_1}, \ldots, \mu_{t_{n-1},t_n}$. Therefore, by the same proof as that of Theorem 7.10, $\{X_t\}$ has independent increments and $X_t - X_s$, $0 \leq s \leq t$, has distribution $\mu_{s,t}$. Hence $\{X_t\}$ is an additive process in law. □

In the temporally homogeneous case, the preceding theorem gives a characterization of Lévy processes in law as temporally homogeneous Markov processes with spatially homogeneous transition functions.

THEOREM 10.5. (i) *Let μ be an infinitely divisible distribution on $\mathbb{R}^d$ and let $\{X_t\}$ be the Lévy process in law corresponding to μ. Define $P_t(x, B)$ by*

$$P_t(x, B) = \mu^t(B - x). \tag{10.8}$$

Then $P_t(x, B)$ is a temporally and spatially homogeneous transition function and $\{X_t\}$ is a Markov process with this transition function and starting point 0.

(ii) *Conversely, any stochastically continuous, temporally homogeneous Markov process on $\mathbb{R}^d$ with spatially homogeneous transition function and starting point* 0 *is a Lévy process in law.*

Proof is similar.

We will give the basic property of Markov processes. The expectation under the probability measure $P^{s,a}$ on Ω^s is denoted by $E^{s,a}$.

PROPOSITION 10.6. *Consider $\{Y_t : t \geq 0\}$, the path space representation of a Markov process with a transition function $P_{s,t}(x, B)$. Let $0 \leq t_0 < \cdots < t_n$ and let $f(x_0, \ldots, x_n)$ be a bounded measurable function. Then $E^{0,a}[f(Y_{t_0}, \ldots, Y_{t_n})]$ is measurable in a and*

$$\begin{aligned} E^{0,a}[f(Y_{t_0}, \ldots, Y_{t_n})] &= \int P_{0,t_0}(a, \mathrm{d}x_0) \int P_{t_0,t_1}(x_0, \mathrm{d}x_1) \int P_{t_1,t_2}(x_1, \mathrm{d}x_2) \\ &\quad \times \cdots \int P_{t_{n-1},t_n}(x_{n-1}, \mathrm{d}x_n)\, f(x_0, x_1, \ldots, x_n). \end{aligned} \tag{10.9}$$

Moreover, for any $0 \leq s_0 < \cdots < s_m \leq s$ and for any bounded measurable function $g(x_0, \ldots, x_m)$, we have

$$\begin{aligned} E^{0,a}[g(Y_{s_0}, \ldots, Y_{s_m})\, f(Y_{s+t_0}, \ldots, Y_{s+t_n})] \\ = E^{0,a}[g(Y_{s_0}, \ldots, Y_{s_m})\, E^{s,Y_s}[f(Y_{s+t_0}, \ldots, Y_{s+t_n})]]. \end{aligned} \tag{10.10}$$

The meaning of the right-hand side of (10.10) needs explanation:

$$E^{s,Y_s}[f(Y_{s+t_0}, \ldots, Y_{s+t_n})]$$

signifies $h(Y_s)$ with $h(x) = E^{s,x}[f(Y_{s+t_0}, \ldots, Y_{s+t_n})]$, so that the right-hand side of (10.10) means $E^{0,a}[g(Y_{s_0}, \ldots, Y_{s_m})h(Y_s)]$. The property (10.10) is referred to as the *Markov property*.

Proof of Proposition 10.6. Measurability of $E^{0,a}[f(Y_{t_0})]$ in a and (10.9) for $n=0$ are immediate from the definition. Suppose that $n \geq 1$. We claim that, if $k(x_0, \dots, x_n)$ is bounded and measurable, then, for $0 \leq s \leq t$,

$$\int P_{s,t}(x_{n-1}, \mathrm{d}x_n) k(x_0, \dots, x_n) \tag{10.11}$$

is bounded and measurable in $(x_0, \dots, x_{n-1})$. In fact, the boundedness is obvious and the measurability is proved first for

$$k(x_0, \dots, x_n) = 1_{B_0}(x_0) \dots 1_{B_n}(x_n)$$

with Borel sets $B_0, \dots, B_n$, second for $k(x_0, \dots, x_n) = 1_B(x_0, \dots, x_n)$ with a Borel set B by using Proposition 1.15, and then for a general k by approximation. By the bounded measurability of (10.11), the right-hand side of (10.9) is well-defined. As to the equality in (10.9), it is nothing but (10.3) if $f(x_0, \dots, x_n) = 1_{B_0}(x_0) \dots 1_{B_n}(x_n)$, and extension to a general f is similar. The measurability with respect to a of the left-hand side of (10.9) comes from the equality (10.9).

Let us prove (10.10). Since the measurability of the $h(x)$ defined above is proved similarly, the right-hand side of (10.10) is well-defined. Let $g(x_0, \dots, x_m) = 1_{C_0}(x_0) \dots 1_{C_m}(x_m)$ and $f(x_0, \dots, x_n) = 1_{B_0}(x_0) \dots 1_{B_n}(x_n)$. Then the left-hand side of (10.10) equals

$$P^{0,a}[\, Y_{s_0} \in C_0, \dots, Y_{s_m} \in C_m, Y_s \in \mathbb{R}^d, Y_{s+t_0} \in B_0, \dots, Y_{s+t_n} \in B_n \,].$$

Rewrite this probability by the transition function as in (10.3), integrate $n+1$ times, and use (10.9) twice. Then we get (10.10) for our g and f. Extension to general g and f is similar. □

Lévy processes have the following property, which is stronger than the Markov property.

PROPOSITION 10.7. *Let $\{X_t : t \geq 0\}$ be a Lévy process on $\mathbb{R}^d$. Then, for any $s \geq 0$, $\{X_{s+t} - X_s : t \geq 0\}$ is a Lévy process identical in law with $\{X_t : t \geq 0\}$; $\{X_{s+t} - X_s : t \geq 0\}$ and $\{X_t : 0 \leq t \leq s\}$ are independent.*

Proof. Fix s and let $Z_t = X_{s+t} - X_s$. Since $Z_0 = 0$ and $Z_{t_2} - Z_{t_1} = X_{s+t_2} - X_{s+t_1}$, the definition of the Lévy process for $\{X_t\}$ implies that $\{Z_t\}$ is a Lévy process. The rest of the assertion is immediate. □

REMARK 10.8. Sometimes it is useful to consider a random starting point. Given a transition function $P_{s,t}(x, B)$, let $\{Y_t\}$ and $(\Omega^0, \mathcal{F}^0, P^{0,a})$ be as in Definition 10.2. Using Proposition 1.15, we can prove from Proposition 10.6 that $P^{0,a}[A]$ is measurable in a for any $A \in \mathcal{F}^0$. For a probability measure ρ on $\mathbb{R}^d$, define

$$P^{0,\rho}[A] = \textstyle\int_{\mathbb{R}^d} \rho(\mathrm{d}a) P^{0,a}[A].$$

A stochastic process $\{X_t : t \geq 0\}$ defined on a probability space $(\Omega, \mathcal{F}, P)$ is called a Markov process with transition function $P_{s,t}(x, B)$ and *initial distribution* ρ, if it is identical in law with the process $\{Y_t\}$ on the probability space $(\Omega^0, \mathcal{F}^0, P^{0,\rho})$.

11. Existence of Lévy and additive processes

We have shown that, for any infinitely divisible probability measure μ, there exists, uniquely up to identity in law, a Lévy process in law with distribution μ at time 1. Now we will show that any Lévy process in law has a Lévy process modification, which establishes the correspondence between the infinitely divisible probability measures and the Lévy processes. More generally, we will deal with additive processes in law. For this purpose we will give a sufficient condition for a Markov process to have sample paths right-continuous with left limits. This is a result of Dynkin [**120**] and Kinney [**287**]. A sufficient condition for sample path continuity is also given, which proves the existence of the Brownian motion.

Denote the ε-neighborhood of x by $D_\varepsilon(x) = \{y\colon |y-x| < \varepsilon\}$, and its complement by $D_\varepsilon(x)^c$. Suppose that we are given a transition function $P_{s,t}(x, B)$ on $\mathbb{R}^d$. Let

$$\alpha_{\varepsilon,T}(u) = \sup\{P_{s,t}(x, D_\varepsilon(x)^c)\colon x \in \mathbb{R}^d \text{ and } s,t \in [0,T] \text{ with } 0 \le t-s \le u\}. \tag{11.1}$$

THEOREM 11.1. *Let $\{X_t\colon t \ge 0\}$ be a Markov process on $\mathbb{R}^d$ defined on $(\Omega, \mathcal{F}, P)$ with transition function $\{P_{s,t}(x,B)\}$ and a fixed starting point. If*

$$\lim_{u\downarrow 0} \alpha_{\varepsilon,T}(u) = 0 \quad \textit{for any } \varepsilon > 0 \textit{ and } T > 0, \tag{11.2}$$

then there is a Markov process $\{X'_t\colon t \ge 0\}$ defined on the probability space $(\Omega, \mathcal{F}, P)$ such that

$$P[\,X_t = X'_t\,] = 1 \quad \textit{for } t \ge 0, \tag{11.3}$$

and $X'_t(\omega)$ is right-continuous with left limits as a function of t for every ω. This $\{X'_t\}$ automatically satisfies

$$P[\,X'_t = X'_{t-}\,] = 1 \quad \textit{for } t > 0. \tag{11.4}$$

If, moreover, the transition function satisfies

$$\lim_{u\downarrow 0} \frac{1}{u}\alpha_{\varepsilon,T}(u) = 0 \quad \textit{for any } \varepsilon > 0 \textit{ and } T > 0, \tag{11.5}$$

then there is $\Omega_1 \in \mathcal{F}$ with $P[\Omega_1] = 1$ such that, for every $\omega \in \Omega_1$, $X'_t(\omega)$ is continuous as a function of t.

Let $M \subset [0,\infty)$ and $\varepsilon > 0$. We say that $X_t(\omega)$, with ω fixed, has ε-oscillation n times in M, if there are $t_0, t_1, \ldots, t_n$ in M such that $t_0 < t_1 < \cdots < t_n$ and $|X_{t_j}(\omega) - X_{t_{j-1}}(\omega)| > \varepsilon$ for $j = 1, \ldots, n$. We say that $X_t(\omega)$ has ε-oscillation infinitely often in M, if, for every n, $X_t(\omega)$ has ε-oscillation n times in M. Let

$$\Omega_2 = \{\,\omega\colon \lim_{s\in\mathbb{Q}, s\downarrow t} X_s(\omega) \text{ exists in } \mathbb{R}^d \text{ for every } t \ge 0 \text{ and}$$

$$\lim_{s\in\mathbb{Q}, s\uparrow t} X_s(\omega) \text{ exists in } \mathbb{R}^d \text{ for every } t>0\},$$

$$A_{N,k} = \{\omega\colon X_t(\omega) \text{ does not have } \tfrac{1}{k}\text{-oscillation infinitely often in } [0,N]\cap\mathbb{Q}\},$$

$$\Omega_2' = \bigcap_{N=1}^{\infty}\bigcap_{k=1}^{\infty} A_{N,k}.$$

Then $\Omega_2' \in \mathcal{F}$, since $\mathbb{Q}$ is countable.

LEMMA 11.2. $\Omega_2' \subset \Omega_2$.

Proof. Let $\omega \in \Omega_2'$. If $t_n \in \mathbb{Q}$ is strictly decreasing to t, then, for every k, there is n_0 such that

$$|X_{t_n}(\omega) - X_{t_{n_0}}(\omega)| \le \frac{1}{k} \quad \text{for } n \ge n_0.$$

Hence $\lim_{n\to\infty} X_{t_n}(\omega)$ exists in $\mathbb{R}^d$. It follows that $\lim_{s\in\mathbb{Q},s\downarrow t} X_s(\omega)$ exists in $\mathbb{R}^d$. Similarly for $\lim_{s\in\mathbb{Q},s\uparrow t} X_s(\omega)$. □

Actually $\Omega_2 = \Omega_2'$, but this fact is not needed.

LEMMA 11.3. *If $\{X_t\}$ is stochastically continuous and $P[\Omega_2'] = 1$, then there is $\{X_t'\}$ satisfying (11.3) such that $X_t'(\omega)$ is right-continuous with left limits for every ω.*

Proof. Use Lemma 11.2. For $\omega \in \Omega_2'$, define $X_t'(\omega) = \lim_{s\in\mathbb{Q},s\downarrow t} X_s(\omega)$. For $\omega \notin \Omega_2'$, define $X_t'(\omega) = 0$. It follows from this definition that $X_t'(\omega)$ is right-continuous with left limits. If $s_n \in \mathbb{Q}$ and $s_n \downarrow t$, then $X_{s_n} \to X_t$ in probability, while $X_{s_n} \to X_t'$ a. s. by $P[\Omega_2'] = 1$. Hence $P[X_t = X_t'] = 1$ for $t \ge 0$. □

We define, for $M \subset [0,\infty)$,

$$B(p,\varepsilon,M) = \{\omega\colon X_t(\omega) \text{ has } \varepsilon\text{-oscillation } p \text{ times in } M\}.$$

LEMMA 11.4. *Let p be a positive integer. Let $0 \le s_1 < \cdots < s_m \le u \le t_1 < \cdots < t_n \le v \le T$ and $M = \{t_1,\ldots,t_n\}$. If $\{X_t\colon t\ge 0\}$ is a Markov process with the transition function $P_{s,t}(x,B)$ and a fixed starting point, then*

$$E[Z(\omega)1_{B(p,4\varepsilon,M)}(\omega)] \le E[Z](2\alpha_{\varepsilon,T}(v-u))^p \tag{11.6}$$

for every $Z = g(X_{s_1},\ldots,X_{s_m})$ with a nonnegative measurable g.

It is important that the right-hand side of (11.6) does not depend on n.

Proof of lemma. It is enough to prove the lemma for the path space representation. So we assume that $\{X_t\}$ is itself that representation. We use induction in p. First, let $p = 1$. Let

$$C_k = \{\,|X_{t_j} - X_u| \le 2\varepsilon \text{ for } j = 1,\ldots,k-1 \text{ and } |X_{t_k} - X_u| > 2\varepsilon\,\},$$

$$D_k = \{ |X_v - X_{t_k}| > \varepsilon \}.$$

Then $C_1, \ldots, C_n$ are disjoint and

$$\begin{aligned} B(1, 4\varepsilon, M) &\subset \bigcup_{k=1}^{n} \{ |X_{t_k} - X_u| > 2\varepsilon \} = \bigcup_{k=1}^{n} C_k \\ &\subset \{ |X_v - X_u| \geq \varepsilon \} \cup \bigcup_{k=1}^{n} (C_k \cap D_k). \end{aligned}$$

Hence

$$\begin{aligned} E^{0,a}[Z 1_{B(1,4\varepsilon,M)}] &\leq E^{0,a}[Z 1_{\{|X_v - X_u| \geq \varepsilon\}}] + \sum_{k=1}^{n} E^{0,a}[Z 1_{C_k} 1_{D_k}] \\ &\leq E^{0,a}[Z] \alpha_{\varepsilon,T}(v-u) + \sum_{k=1}^{n} E^{0,a}[Z 1_{C_k}] \alpha_{\varepsilon,T}(v-u) \end{aligned}$$

by the Markov property, where a is the starting point of $\{X_t\}$. Thus

$$E^{0,a}[Z 1_{B(1,4\varepsilon,M)}] \leq 2E^{0,a}[Z] \alpha_{\varepsilon,T}(v-u).$$

This is (11.6) for $p = 1$.

Now suppose that the statement is true for $p-1$ in place of p. Let

$$\begin{aligned} F_k = \{ & X_t \text{ has } 4\varepsilon\text{-oscillation } p-1 \text{ times in } \{t_1, \ldots, t_k\}, \\ & \text{but does not have } 4\varepsilon\text{-oscillation } p-1 \text{ times in } \{t_1, \ldots, t_{k-1}\}\}, \\ G_k = \{ & X_t \text{ has } 4\varepsilon\text{-oscillation once in } \{t_k, t_{k+1}, \ldots, t_n\}\}. \end{aligned}$$

Then $F_1, \ldots, F_n$ are disjoint and

$$B(p-1, 4\varepsilon, M) = \bigcup_{k=1}^{n} F_k.$$

Furthermore,

$$B(p, 4\varepsilon, M) \subset \bigcup_{k=1}^{n} (F_k \cap G_k),$$

In fact, if $\omega \in B(p, 4\varepsilon, M)$, then $X_t(\omega)$ has 4ε-oscillation p times in some $\{t_{n_0}, t_{n_1}, \ldots, t_{n_p}\}$ with $n_0 < n_1 < \cdots < n_p$ and hence there is $k \leq n_{p-1}$ such that $\omega \in F_k$. At the same time $\omega \in G_k$ because $|X(t_{n_{p-1}}) - X(t_{n_p})| > 4\varepsilon$. Consequently

$$E^{0,a}[Z 1_{B(p,4\varepsilon,M)}] \leq \sum_{k=1}^{n} E^{0,a}[Z 1_{F_k} 1_{G_k}] \leq \sum_{k=1}^{n} E^{0,a}[Z 1_{F_k}] 2\alpha_{\varepsilon,T}(v-u)$$

by the Markov property and by the case $p = 1$. It follows that

$$\begin{aligned} E^{0,a}[Z 1_{B(p,4\varepsilon,M)}] &\leq E^{0,a}[Z 1_{B(p-1,4\varepsilon,M)}] 2\alpha_{\varepsilon,T}(v-u) \\ &\leq E^{0,a}[Z] (2\alpha_{\varepsilon,T}(v-u))^p. \end{aligned}$$

Thus the statement is true for p. $\square$

Proof of Theorem 11.1. We assume (11.2). This guarantees stochastic continuity of $\{X_t\}$. In order to show existence of a right-continuous modification $\{X'_t\}$ with left limits, it is enough to prove that $P[\Omega'_2] = 1$, as Lemma 11.3 says. Hence it is enough to prove that $P[A_{N,k}{}^c] = 0$ for any fixed N and k. Using (11.2), choose l such that $2\alpha_{1/(4k),N}(N/l) < 1$. Now

$$\begin{aligned} P[A_{N,k}{}^c] &= P\big[\, X_t \text{ has } \tfrac{1}{k}\text{-oscillation infinitely often in } [0,N]\cap\mathbb{Q}\,\big] \\ &\le \sum_{j=1}^{l} P\big[\, X_t \text{ has } \tfrac{1}{k}\text{-oscillation infinitely often in } \big[\tfrac{j-1}{l}N, \tfrac{j}{l}N\big]\cap\mathbb{Q}\,\big] \\ &= \sum_{j=1}^{l} \lim_{p\to\infty} P\big[B\big(p, \tfrac{1}{k}, \big[\tfrac{j-1}{l}N, \tfrac{j}{l}N\big]\cap\mathbb{Q}\big)\big]. \end{aligned}$$

Enumerate the elements of $[\frac{j-1}{l}N, \frac{j}{l}N]\cap\mathbb{Q}$ as $t_1, t_2, \ldots$. Since

$$P\big[B\big(p, \tfrac{1}{k}, \{t_1, \ldots, t_n\}\big)\big] \le \big(2\alpha_{1/(4k),N}\big(\tfrac{N}{l}\big)\big)^p$$

by Lemma 11.4, we get, letting $n \to \infty$,

$$P\big[B\big(p, \tfrac{1}{k}, \big[\tfrac{j-1}{l}N, \tfrac{j}{l}N\big]\cap\mathbb{Q}\big)\big] \le \big(2\alpha_{1/(4k),N}\big(\tfrac{N}{l}\big)\big)^p.$$

The right-hand side goes to 0 as $p \to \infty$. Hence $P[A_{N,k}{}^c] = 0$, which was to be proved. As the stochastic continuity of $\{X_t\}$ implies that of $\{X'_t\}$, we obtain the assertion (11.4).

Now assume (11.5). In order to see continuity of paths of $\{X'_t\}$, that is, to see the existence of Ω_1 having the desired property, it is enough to show that, for each N, there is $H_N \in \mathcal{F}$ such that $P[H_N] = 1$ and

$$H_N \subset \{\, X'_{t-} = X'_t \text{ for any } t \in (0, N]\,\}.$$

Fix N, and let

$$\begin{aligned} R_{l,\varepsilon}(\omega) &= \#\{\, j \in \{1, 2, \ldots, l\} \colon |X'_{jN/l}(\omega) - X'_{(j-1)N/l}(\omega)| > \varepsilon\,\}, \\ R_\varepsilon(\omega) &= \#\{\, t \in (0, N] \colon |X'_t(\omega) - X'_{t-}(\omega)| > \varepsilon\,\}. \end{aligned}$$

Then, $R_{l,\varepsilon}$ is $\mathcal{F}$-measurable and

$$R_\varepsilon(\omega) \le \liminf_{l\to\infty} R_{l,\varepsilon}(\omega).$$

Since

$$\begin{aligned} E[R_{l,\varepsilon}] &= E\left[\sum_{j=1}^{l} 1_{\{|X'_{jN/l} - X'_{(j-1)N/l}| > \varepsilon\}}\right] \\ &= \sum_{j=1}^{l} P[\,|X'_{jN/l} - X'_{(j-1)N/l}| > \varepsilon\,] \le l\,\alpha_{\varepsilon,N}\big(\tfrac{N}{l}\big), \end{aligned}$$

we have $\lim_{l\to\infty} E[R_{l,\varepsilon}] = 0$ by (11.5). Hence $E[\liminf_{l\to\infty} R_{l,\varepsilon}] = 0$ by Fatou's lemma. Now let

$$H_N = \bigcap_{k=1}^{\infty} \left\{ \liminf_{l\to\infty} R_{l,1/k} = 0 \right\}.$$

Then H_N is the desired event, because $H_N \subset \{ R_\varepsilon = 0 \text{ for any } \varepsilon > 0 \}$. □

Now we can show the existence of Lévy processes and additive processes.

THEOREM 11.5. *Let $\{X_t\}$ be a Lévy or additive process in law on $\mathbb{R}^d$. Then it has a modification which is, respectively, a Lévy or additive process.*

Proof. It is enough to consider an additive process in law $\{X_t\}$. By Theorem 10.4, it is a Markov process with spatially homogeneous transition function (10.6). Therefore

$$P_{s,t}(x, D_\varepsilon(x)^c) = P_{s,t}(0, D_\varepsilon(0)^c) = P[\,|X_t - X_s| \geq \varepsilon\,].$$

Hence $\alpha_{\varepsilon,T}(u) \to 0$ as $u \to 0$, by virtue of the uniform stochastic continuity of Lemma 9.6. Now Theorem 11.1 applies. □

We can prove Theorem 11.5 without using Theorem 11.1, if we like. Indeed, given an additive process in law $\{X_t\}$, we will construct in Section 20 another probability space and an additive process $\{Y_t\}$ defined there satisfying $\{Y_t\} \overset{\mathrm{d}}{=} \{X_t\}$. On the way we will need the fact that if $\{X_t\}$ is Gaussian at any t, then $\{Y_t\}$ can be chosen to have continuous sample functions, but we can prove this directly from properties of Gaussian distributions. Using $\{Y_t\}$ thus constructed, we can show the existence of an additive process modification of $\{X_t\}$.

COROLLARY 11.6. *For every infinitely divisible distribution μ on $\mathbb{R}^d$, there is a Lévy process $\{X_t\}$ such that $P_{X_1} = \mu$. It is unique up to identity in law.*

The Lévy process $\{X_t\}$ in Corollary 11.6 is called *the Lévy process corresponding to μ*. Thus, to each infinitely divisible distribution in Example 7.2, there corresponds a Lévy process. Poisson and compound Poisson processes in Sections 3 and 4 respectively correspond to Poisson and compound Poisson distributions. The Lévy process on $\mathbb{R}^d$ corresponding to a Cauchy distribution is called a *Cauchy process.* The Lévy process on $\mathbb{R}$ corresponding to an exponential distribution is called a Γ*-process,* since it has Γ-distribution at any t.

THEOREM 11.7. *If $\{X_t\}$ is an additive process on $\mathbb{R}^d$ with a Gaussian distribution at each t, then $\{X_t\}$ has continuous paths a. s., that is, there is $\Omega_1 \in \mathcal{F}$ such that $P[\Omega_1] = 1$ and, for every $\omega \in \Omega_1$, $X_t(\omega)$ is continuous in t.*

The Lévy–Itô decomposition in Chapter 4 will show that no other additive processes have continuous paths a. s.

Proof of theorem. Step 1. Consider the case where $d = 1$ and X_t has characteristic function $\mathrm{e}^{-tz^2/2}$. Then $\{X_t\}$ satisfies (11.5) and we can use the second half of Theorem 11.1. In fact, for $\varepsilon > 0$,

$$P[|X_s| > \varepsilon] = \frac{2}{\sqrt{2\pi s}} \int_\varepsilon^\infty \mathrm{e}^{-x^2/(2s)} \mathrm{d}x = \frac{2}{\sqrt{2\pi}} \int_{\varepsilon/\sqrt{s}}^\infty \mathrm{e}^{-x^2/2} \mathrm{d}x$$
$$\leq \frac{2\sqrt{s}}{\varepsilon\sqrt{2\pi}} \mathrm{e}^{-\varepsilon^2/(2s)},$$

since

$$\int_c^\infty \mathrm{e}^{-x^2/2}\, \mathrm{d}x \leq \int_c^\infty \mathrm{e}^{-x^2/2}\left(1 + \frac{1}{x^2}\right) \mathrm{d}x = \frac{1}{c}\mathrm{e}^{-c^2/2} \text{ for } c > 0, \tag{11.7}$$

and it follows that

$$\frac{1}{u}\alpha_{\varepsilon,T}(u) = \frac{1}{u} \sup_{s \leq u\wedge T} P[|X_s| > \varepsilon] \leq \sup_{s \leq u\wedge T} \frac{1}{s} P[|X_s| > \varepsilon] \to 0 \text{ as } u \downarrow 0.$$

Thus the one-dimensional Brownian motion $\{X_t^0\}$ exists.

Step 2. Consider the general case of $d = 1$. Let $\{X_t\}$ be an additive process on $\mathbb{R}$ with generating triplets $(A_t, 0, \gamma_t)$, where A_t is continuous and increasing, $A_0 = 0$, γ_t is continuous, and $\gamma_0 = 0$. Using the $\{X_t^0\}$ above, let $Y_t = X_{A_t}^0 + \gamma_t$. Then $\{Y_t\}$ is an additive process with continuous paths a. s. and $E[\mathrm{e}^{\mathrm{i}zY_t}] = \mathrm{e}^{-A_t z^2/2 + \mathrm{i}\gamma_t z}$. Hence $\{Y_t\} \overset{\mathrm{d}}{=} \{X_t\}$. We claim that $\{X_t\}$ also has continuous paths a. s. Fix N and define

$$R_{l,\varepsilon} = \#\{j \in \{1, 2, \ldots, l\} \colon |X_{jN/l} - X_{(j-1)N/l}| > \varepsilon\},$$
$$R'_{l,\varepsilon} = \#\{j \in \{1, 2, \ldots, l\} \colon |Y_{jN/l} - Y_{(j-1)N/l}| > \varepsilon\}.$$

The continuity of $\{Y_t\}$ implies $\limsup_{l\to\infty} R'_{l,\varepsilon} = 0$ a. s. for any $\varepsilon > 0$. It follows from $\{X_t\} \overset{\mathrm{d}}{=} \{Y_t\}$ that

$$\{R_{l,\varepsilon} \colon l = 1, 2, \ldots\} \overset{\mathrm{d}}{=} \{R'_{l,\varepsilon} \colon l = 1, 2, \ldots\}.$$

Therefore $\limsup_{l\to\infty} R_{l,\varepsilon} = 0$ a. s. for $\varepsilon > 0$. The final part of the proof of Theorem 11.1 shows that $\{X_t\}$ has continuous paths a. s.

Step 3. The case $d \geq 2$. Let $\{X_t\}$ be an additive process on $\mathbb{R}^d$ with $(A_t, 0, \gamma_t)$. Express $X_t = (X_j(t))_{j=1,\ldots,d}$ componentwise. Then, for each j, $\{X_j(t)\}$ is an additive process on $\mathbb{R}$ with generating triplets $(A_j(t), 0, \gamma_j(t))$, where $A_j(t)$ is the (j, j)-entry of the matrix A_t and $\gamma_j(t)$ is the jth component of γ_t. Hence $\{X_j(t)\}$ has continuous paths a. s. by Step 2. Since the continuity of X_t is equivalent to the continuity of the components, $\{X_t\}$ has continuous paths a. s. □

COROLLARY 11.8. *The Brownian motion on $\mathbb{R}^d$ exists.*

The correspondence of a Lévy process $\{X_t\}$ and an infinitely divisible distribution μ through $P_{X_1} = \mu$ has been established. The generating triplet (A, ν, γ) of μ is called the *generating triplet* of the Lévy process $\{X_t\}$. Sometimes we call $\{X_t\}$ the Lévy process *generated by* (A, ν, γ). The A and ν are called, respectively, the *Gaussian covariance matrix* and the *Lévy measure* of $\{X_t\}$. If $\int_{|x|\le 1} |x|\, \nu(\mathrm{d}x) < \infty$, then the γ_0 in the triplet $(A, \nu, \gamma_0)_0$ (see (8.7)) is called the *drift* of $\{X_t\}$. If $\int_{|x|>1} |x|\, \nu(\mathrm{d}x) < \infty$, then the γ_1 in the triplet $(A, \nu, \gamma_1)_1$ (see (8.8)) is called the *center* of $\{X_t\}$. If $A = 0$, then $\{X_t\}$ is called *purely non-Gaussian*. It is important in the theory of Lévy processes to grasp various probabilistic properties in relation to conditions on their generating triplets, drifts, or centers. The basic classification of Lévy processes is the following.

DEFINITION 11.9. Let $\{X_t\}$ be a Lévy process on $\mathbb{R}^d$ with generating triplet (A, ν, γ). It is said to be of
type A if $A = 0$ and $\nu(\mathbb{R}^d) < \infty$;
type B if $A = 0$, $\nu(\mathbb{R}^d) = \infty$, and $\int_{|x|\le 1}|x|\, \nu(\mathrm{d}x) < \infty$;
type C if $A \ne 0$ or $\int_{|x|\le 1} |x|\, \nu(\mathrm{d}x) = \infty$.

We will use linear transformation of Lévy processes.

PROPOSITION 11.10. *Let $\{X_t\}$ be a Lévy process on $\mathbb{R}^d$ with generating triplet (A, ν, γ) and let U be an $n \times d$ matrix. Then $\{UX_t\}$ is a Lévy process on $\mathbb{R}^n$ with generating triplet (A_U, ν_U, γ_U) given by*

$$A_U = UAU', \tag{11.8}$$

$$\nu_U = [\nu U^{-1}]_{\mathbb{R}^n \setminus \{0\}}, \tag{11.9}$$

$$\gamma_U = U\gamma + \int Ux(1_E(Ux) - 1_D(x))\, \nu(\mathrm{d}x), \tag{11.10}$$

where $(\nu U^{-1})(B) = \nu(\{\, x \colon Ux \in B\,\})$, $D = \{\, x \in \mathbb{R}^d \colon |x| \le 1\,\}$, $E = \{\, y \in \mathbb{R}^n \colon |y| \le 1\,\}$ and $[\nu U^{-1}]_{\mathbb{R}^n \setminus \{0\}}$ is the restriction of the measure νU^{-1} to $\mathbb{R}^n \setminus \{0\}$.

As a special case, projections of Lévy processes are again Lévy processes.

Proof of proposition. It is easy to see that $\{UX_t\}$ is a Lévy process. To calculate its generating triplet,

$$\begin{aligned} E[\mathrm{e}^{\mathrm{i}\langle z, UX_1\rangle}] &= E[\mathrm{e}^{\mathrm{i}\langle U'z, X_1\rangle}] \\ &= \exp\Big[-\tfrac{1}{2}\langle U'z, AU'z\rangle + \mathrm{i}\langle \gamma, U'z\rangle \\ &\qquad + \int (\mathrm{e}^{\mathrm{i}\langle U'z, x\rangle} - 1 - \mathrm{i}\langle U'z, x\rangle 1_D(x))\nu(\mathrm{d}x)\Big] \end{aligned}$$

$$= \exp\Big[-\tfrac{1}{2}\langle z, UAU'z\rangle + \mathrm{i}\langle \gamma_U, z\rangle$$
$$+ \int (\mathrm{e}^{\mathrm{i}\langle z,y\rangle} - 1 - \mathrm{i}\langle z,y\rangle 1_E(y))\nu U^{-1}(\mathrm{d}y)\Big],$$

where γ_U is defined by (11.10). Noting that

$$\int (|y|^2 \wedge 1)\nu U^{-1}(\mathrm{d}y) = \int (|Ux|^2 \wedge 1)\nu(\mathrm{d}x)$$
$$\leq (\|U\|^2 \vee 1)\int (|x|^2 \wedge 1)\nu(\mathrm{d}x) < \infty,$$

we get (11.8) and (11.9). □

12. Exercises 2

E 12.1. Let μ be a probability measure on $\mathbb{R}^d$. Show that, if, for some sequence $n_k \uparrow \infty$ of positive integers, there are probablity measures μ_{n_k} such that $\mu = {\mu_{n_k}}^{n_k}$, then μ is infinitely divisible.

E 12.2 (Extension of the uniqueness part of Theorem 8.1). For $k = 1, 2$ let A_k be $d \times d$ symmetric matrices, $\gamma_k \in \mathbb{R}^d$, and ν_k be signed measures on $\mathbb{R}^d$ satisfying $\nu_k(\{0\}) = 0$ and $\int (|x|^2 \wedge 1)|\nu_k|(\mathrm{d}x) < \infty$. Let

$$\varphi_k(z) = \exp\big[-\tfrac{1}{2}\langle z, A_k z\rangle + \mathrm{i}\langle \gamma_k, z\rangle + \textstyle\int (\mathrm{e}^{\mathrm{i}\langle z,x\rangle} - 1 - \mathrm{i}\langle z,x\rangle 1_D(x))\nu_k(\mathrm{d}x)\big],$$

where $D = \{x\colon |x| \leq 1\}$. Show that, if $\varphi_1(z) = \varphi_2(z)$ for all $z \in \mathbb{R}^d$, then $A_1 = A_2$, $\gamma_1 = \gamma_2$, and $\nu_1 = \nu_2$.

E 12.3. Let μ be a probability measure on $\mathbb{R}^d$. Suppose that (8.1) holds with a symmetric $d \times d$ matrix A, $\gamma \in \mathbb{R}^d$, and a signed measure ν on $\mathbb{R}^d$ satisfying $\nu(\{0\}) = 0$ and $\int (|x|^2 \wedge 1)|\nu|(\mathrm{d}x) < \infty$. Show that, if A is not nonnegative-definite or if $\nu(B) < 0$ for some Borel set B, then μ is not infinitely divisible. (For application of this result, see an example of Gnedenko and Kolmogorov [**167**], p. 81, and the solution of E 12.4.)

E 12.4. Show that there is a random variable $X = (X_j)_{1 \leq j \leq d}$ on $\mathbb{R}^d$ such that the distribution of X is not infinitely divisible but every linear combination $a_1 X_1 + \cdots + a_d X_d$ of the components has an infinitely divisible distribution.

E 12.5. A set M of probability measures on $\mathbb{R}^d$ is called *precompact* if any sequence μ_n, $n = 1, 2, \ldots$, in M has a subsequence μ_{n_k}, $k = 1, 2, \ldots$, that converges to a probability measure on $\mathbb{R}^d$. Show that a set M of infinitely divisible probability measures on $\mathbb{R}^d$ is precompact if and only if the generating triplets $(A_\mu, \nu_\mu, \gamma_\mu)$ of $\mu \in M$ satisfy the following conditions:

(1) $\sup_{\mu \in M} \|A_\mu\| < \infty$;
(2) $\sup_{\mu \in M} \int_{\mathbb{R}^d} (|x|^2 \wedge 1)\nu_\mu(\mathrm{d}x) < \infty$;
(3) $\lim_{l \to \infty} \sup_{\mu \in M} \int_{|x|>l} \nu_\mu(\mathrm{d}x) = 0$;
(4) $\sup_{\mu \in M} |\gamma_\mu| < \infty$.

E 12.6. Let μ_1 and μ_2 be probability measures on $[0,\infty)$. Show that if ${\mu_1}^n = {\mu_2}^n$ for some $n \in \mathbb{N}$, then $\mu_1 = \mu_2$. (This fact has the following extension. If both μ_1 and μ_2 have the infima of their supports (see Section 24 for the definition) at 0 and if, for some $n \in \mathbb{N}$, the restrictions of ${\mu_1}^n$ and ${\mu_2}^n$ to an interval $[0,a)$ are identical, then the restrictions of μ_1 and μ_2 to this interval $[0,a)$ are identical. See Rossberg, Jesiak, and Siegel [**424**], p. 85.)

E 12.7. Give another proof of Lemma 9.6 by using Exercise 6.7.

E 12.8. Let $X = (X_1, \ldots, X_{n+m})$ be an $\mathbb{R}^{n+m}$-valued random variable with infinitely divisible distribution with generating triplet (A, ν, γ). Let $X^{(1)} = (X_1, \ldots, X_n)$ and $X^{(2)} = (X_{n+1}, \ldots, X_{n+m})$. Write $A = (A_{jk})_{j,k=1,\ldots,n+m}$, and

$$E^{(1)} = \{x \in \mathbb{R}^{n+m} \colon x_j = 0 \text{ for } j = n+1, \ldots, n+m\},$$
$$E^{(2)} = \{x \in \mathbb{R}^{n+m} \colon x_j = 0 \text{ for } j = 1, \ldots, n\}.$$

Show that $X^{(1)}$ and $X^{(2)}$ are independent if and only if $A_{jk} = 0$ for $1 \le j \le n$ and $n+1 \le k \le n+m$ and ν is supported on $E^{(1)} \cup E^{(2)}$.

E 12.9. Let X, $X^{(1)}$, and $X^{(2)}$ be as in E 12.8. Show that $X^{(1)}$ and $X^{(2)}$ are independent if and only if, for every choice of j and k with $1 \le j \le n$ and $n+1 \le k \le n+m$, X_j and X_k are independent.

E 12.10. Let $d \ge 2$. Let $X = (X_1, \ldots, X_d)$ be an $\mathbb{R}^d$-valued random variable with infinitely divisible distribution. Let (A, ν, γ) be its generating triplet . Show the following.

(i) $X_1, \ldots, X_d$ are independent if and only if A is diagonal and ν is supported on the union of the coordinate axes.

(ii) If $X_1, \ldots, X_d$ are pairwise independent, that is, if X_j and X_k are independent for every choice of j and k with $j \ne k$, then $X_1, \ldots, X_d$ are independent.

E 12.11. The definition of symmetrization (Section 2) has a drawback, that the symmetrization of a symmetric distribution is not itself. Prove the following.

(i) A probability measure μ on $\mathbb{R}^d$ is symmetric if and only if $\widehat{\mu}(z)$ is real. Symmetry of μ does not imply $\widehat{\mu}(z) \ge 0$. If μ is symmetric and infinitely divisible, then $\widehat{\mu}(z) > 0$.

(ii) For any probability measure μ on $\mathbb{R}^d$, $\operatorname{Re}\widehat{\mu}(z)$ is the characteristic function of a symmetric probability measure μ_0 and $\mu_0(B) = \frac{1}{2}(\mu(B) + \mu(-B))$ for $B \in \mathcal{B}(\mathbb{R}^d)$.

(iii) If μ is infinitely divisible, μ_0 is not necessarily infinitely divisible.

(iv) If μ is a probability measure on $\mathbb{R}^d$, $|\widehat{\mu}(z)|$ is not necessarily the characteristic function of a probability measure.

E 12.12. For each $n = 1, 2, \ldots$, let Z_{nk}, $k = 1, \ldots, r_n$, be independent random variables on $\mathbb{R}^d$. Prove that, if

$$\max_{1 \le k \le r_n} |\widehat{P}_{Z_{nk}}(z) - 1| \to 0 \quad \text{as } n \to \infty$$

for each $z \in \mathbb{R}^d$, then $\{Z_{nk}\}$ is a null array.

E 12.13. Let $\{X_t\}$ be the Γ-process with $EX_1 = 1$ and $\{Y_t\}$ be the Poisson process with $EY_1 = 1$. Show that, for $t \in \mathbb{N}$ and $x \in (0, \infty)$, $P[\,X_t > x\,] = P[\,Y_x \leq t-1\,]$.

E 12.14. Show that (11.5) is not a necessary condition for an additive process to have continuous paths a. s.

E 12.15. Let $\{X_t\}$ be a stochastic process on $\mathbb{R}$ and $\mu = P_{X_1}$. If μ is infinitely divisible and $P_{X_t} = \mu^t$, is the process $\{X_t\}$ a Lévy process in law?

Notes

Lévy [**312, 317, 318**] determined the structure of additive processes, and thus the structure of infinitely divisible distributions at the same time. Not only this, he has given an idea of how to determine a general process with independent increments without assuming stochastic continuity. It is, essentially, an additive process in our sense with a nonrandom function added and also with independent jumps at a countable number of fixed times added. For precise formulation and proof see the books of Doob [**106**], Loève [**327**], and Itô [**223, 225**].

The theory of infinitely divisible distributions is based on the Lévy–Khintchine representation. The theory was established by the works of Khintchine [**278**], [**279**] and Lévy [**317**]. Further development is seen in the books of Gnedenko and Kolmogorov [**167**], Linnik [**324**], Linnik and Ostrovskii [**325**], and Rossberg, Jesiak, and Siegel [**424**].

As mentioned in Section 8, there are two ways of approaching the Lévy–Khintchine representation. Itô [**220, 222, 225**] and Skorohod [**491, 493**] take the probabilistic way. Khintchine [**279**], Gnedenko and Kolmogorov [**167**], Loève [**327**], Feller [**139**], and Breiman [**66**] use the analytic method. Ito [**225**] also explains this method. They consider the one-dimensional case. Our treatment in the d-dimensional case partly follows Cuppens [**91**] and Parthasarathy [**367**]. The author learned the expression in (2) of Theorem 8.7 from Masaaki Tsuchiya. The usual expression is that of (8.20). The proof of Theorem 9.3 follows Cuppens [**91**].

Doob [**104**] gave the proof that additive processes have modifications right-continuous with left limits. In many books this fact is proved by the martingale theory, which was fully developed by Doob after 1940. However, in this book, we do not use martingales (except in the proof of Theorem 46.4). We have proved, following [**164**], the Dynkin–Kinney theorem on regularity of paths of Markov processes, and applied it to additive processes. Sometimes the Dynkin–Kinney theorem is also proved by the martingale theory (see Chung [**81**]).

The terminology of types A, B, and C of Lévy processes follows Pecherskii and Rogozin [**368**]. Linear transformation of infinitely divisible probability measures (Proposition 11.10) is extensively used by Maruyama [**340**]. The results in Exercises 12.8, 12.9, and 12.10 are given in Sato [**432, 434**]; a related paper is Dwass and Teicher [**119**].

When $\{X_t\}$ is the Brownian motion on $\mathbb{R}$, the process $\{Y_t\}$ defined by $Y_t = e^{aX_t+b}$ with $a \neq 0$ and b real is called the geometric Brownian motion. If $b = 0$, then the distribution of Y_t is log-normal with $\alpha = (2ta^2)^{-1}$ in Remark 8.12.

CHAPTER 3

Stable processes and their extensions

13. Selfsimilar and semi-selfsimilar processes and their exponents

If $\{X_t\colon t \geq 0\}$ is the Brownian motion on $\mathbb{R}^d$, then, for any $a > 0$, the process $\{X_{at}\colon t \geq 0\}$ is identical in law with the process $\{a^{1/2}X_t\colon t \geq 0\}$ (Theorem 5.4). This means that any change of time scale for the Brownian motion has the same effect as some change of spatial scale. This property is called selfsimilarity of a stochastic process. There are many selfsimilar Lévy processes other than the Brownian motion; they constitute an important class called strictly stable processes. Stable processes are a slight generalization; they are Lévy processes for which change of time scale has the same effect as change of spatial scale and addition of a linear motion. Stable distributions were introduced in the 1920s by Lévy and stable processes have been extensively studied since the 1930s. In this chapter we give representations of stable and strictly stable processes. Further we determine selfsimilar additive processes. Extension of the notion of selfsimilarity to semi-selfsimilarity is also studied. Semi-stable, selfdecomposable, and semi-selfdecomposable distributions appear.

In this section, we will define stable, strictly stable, semi-stable, and strictly semi-stable distributions first, and then define processes with these names as Lévy processes corresponding to these distributions. On the other hand we will define selfsimilarity, broad-sense selfsimilarity, semi-selfsimilarity, and broad-sense semi-selfsimilarity for general stochastic processes, and prove the existence of the exponent – a quantity that expresses the relationship between the change of time scale and that of spatial scale. These notions applied to Lévy processes give equivalent definitions of the stable processes and the like.

DEFINITION 13.1. Let μ be an infinitely divisible probability measure on $\mathbb{R}^d$. It is called *stable* if, for any $a > 0$, there are $b > 0$ and $c \in \mathbb{R}^d$ such that

$$\widehat{\mu}(z)^a = \widehat{\mu}(bz)e^{i\langle c,z\rangle}. \tag{13.1}$$

It is called *strictly stable* if, for any $a > 0$, there is $b > 0$ such that

$$\widehat{\mu}(z)^a = \widehat{\mu}(bz). \tag{13.2}$$

It is called *semi-stable* if, for some $a > 0$ with $a \neq 1$, there are $b > 0$ and $c \in \mathbb{R}^d$ satisfying (13.1). It is called *strictly semi-stable* if, for some $a > 0$ with $a \neq 1$, there is $b > 0$ satisfying (13.2).

DEFINITION 13.2. Let $\{X_t \colon t \geq 0\}$ be a Lévy process on $\mathbb{R}^d$. It is called a stable, strictly stable, semi-stable, or strictly semi-stable process if the distribution of X_t at $t = 1$ is, respectively, stable, strictly stable, semi-stable, or strictly semi-stable.

EXAMPLE 13.3. If μ is Gaussian on $\mathbb{R}^d$, then

$$\widehat{\mu}(z) = \mathrm{e}^{-\langle z, Az\rangle/2 + \mathrm{i}\langle \gamma, z\rangle}$$

and μ is stable, as it satisfies (13.1) with $b = a^{1/2}$ and $c = (-a^{1/2} + a)\gamma$. Thus, if μ is Gaussian with mean 0, then it is strictly stable. If μ is Cauchy on $\mathbb{R}^d$ with parameter $c > 0$ and $\gamma \in \mathbb{R}^d$ (Example 2.12), then

$$\widehat{\mu}(z) = \mathrm{e}^{-c|z| + \mathrm{i}\langle \gamma, z\rangle}$$

and μ is strictly stable, satisfying (13.2) with $b = a$. If

$$\nu = \sum_{n=-\infty}^{\infty} b^{-n\alpha} \delta_{b^n x_0} \tag{13.3}$$

with $b > 1$, $0 < \alpha < 2$, and $x_0 \in \mathbb{R}^d \setminus \{0\}$, then $\int (|x|^2 \wedge 1)\nu(\mathrm{d}x) < \infty$ and the infinitely divisible distribution μ with generating triplet $(0, \nu, 0)$ is semi-stable with $a = b^\alpha$ in (13.1), but it is not stable, as is seen from

$$\widehat{\mu}(z) = \exp\left[\sum_{n=-\infty}^{\infty} b^{-n\alpha}\left(\mathrm{e}^{\mathrm{i}\langle z, b^n x_0\rangle} - 1 - \mathrm{i}\langle z, b^n x_0\rangle 1_D(b^n x_0)\right)\right].$$

DEFINITION 13.4. Let $\{X_t \colon t \geq 0\}$ be a stochastic process on $\mathbb{R}^d$. It is called *selfsimilar* if, for any $a > 0$, there is $b > 0$ such that

$$\{X_{at} \colon t \geq 0\} \stackrel{\mathrm{d}}{=} \{bX_t \colon t \geq 0\}. \tag{13.4}$$

It is called *broad-sense selfsimilar* if, for any $a > 0$, there are $b > 0$ and a function $c(t)$ from $[0, \infty)$ to $\mathbb{R}^d$ such that

$$\{X_{at} \colon t \geq 0\} \stackrel{\mathrm{d}}{=} \{bX_t + c(t) \colon t \geq 0\}. \tag{13.5}$$

It is called *semi-selfsimilar* if, for some $a > 0$ with $a \neq 1$, there is $b > 0$ satisfying (13.4). It is called *broad-sense semi-selfsimilar* if, for some $a > 0$ with $a \neq 1$, there are $b > 0$ and a function $c(t)$ satisfying (13.5).

The connection of these notions to stable processes and the like is as follows.

PROPOSITION 13.5. *Let $\{X_t : t \geq 0\}$ be a Lévy process on $\mathbb{R}^d$. Then it is selfsimilar, broad-sense selfsimilar, semi-selfsimilar, or broad-sense semi-selfsimilar if and only if it is, respectively, strictly stable, stable, strictly semi-stable, or semi-stable.*

Proof. Let $\mu = P_{X_1}$. Suppose that $\{X_t\}$ is semi-stable. By the definition there is a positive $a \neq 1$ for which (13.1) holds with some b and c. The Lévy processes $\{X_{at}\}$ and $\{bX_t + tc\}$ correspond to the distributions with characteristic functions $\widehat{\mu}(z)^a$ and $\widehat{\mu}(bz)\mathrm{e}^{\mathrm{i}\langle c,z\rangle}$, respectively. Hence, by Theorem 7.10(iii),

$$\{X_{at}\} \stackrel{\mathrm{d}}{=} \{bX_t + tc\}, \tag{13.6}$$

and hence $\{X_t\}$ is broad-sense semi-selfsimilar. Conversely, if $\{X_t\}$ is broad-sense semi-selfsimilar, then it follows from (13.5) that $P_{X_a} = P_{bX_1 + c(1)}$, that is, $\widehat{\mu}(z)^a = \widehat{\mu}(bz)\mathrm{e}^{\mathrm{i}\langle c(1),z\rangle}$, and $\{X_t\}$ is semi-stable. (At the same time it is shown that $c(t) = tc(1)$.) The other assertions are proved similarly. □

Before turning to the existence of the exponent, we give some definitions and lemmas.

DEFINITION 13.6. A probability measure μ on $\mathbb{R}^d$ is *trivial* if it is a δ-distribution; otherwise it is *non-trivial*. A random variable X on $\mathbb{R}^d$ is *constant* or *trivial* if its distribution P_X is trivial. X is *non-constant* or *non-trivial* if P_X is non-trivial. X is *non-zero* if $P_X \neq \delta_0$. A stochastic process $\{X_t : t \geq 0\}$ on $\mathbb{R}^d$ is a *trivial* process or a *deterministic* process if, for every $t \geq 0$, X_t is trivial. It is a *non-trivial* process if it is not a trivial process. It is a *zero* process if, for every $t \geq 0$, $P_{X_t} = \delta_0$. It is a *non-zero* process if it is not a zero process.

LEMMA 13.7. (i) *Let X be a non-zero random variable on $\mathbb{R}^d$. Suppose that $b_1, b_2 \in (0,\infty)$ satisfy $b_1 X \stackrel{\mathrm{d}}{=} b_2 X$. Then $b_1 = b_2$.*

(ii) *Let X be a non-constant random variable on $\mathbb{R}^d$. Suppose that $b_1, b_2 \in (0,\infty)$ and $c_1, c_2 \in \mathbb{R}^d$ satisfy $b_1 X + c_1 \stackrel{\mathrm{d}}{=} b_2 X + c_2$. Then $b_1 = b_2$ and $c_1 = c_2$.*

Proof. (i) Suppose that $b_1 \neq b_2$. Then $X \stackrel{\mathrm{d}}{=} bX$ with some $b \in (0,1)$. Hence $X \stackrel{\mathrm{d}}{=} b^n X$ for $n = 1, 2, \ldots$ and, letting $n \to \infty$, we have $X \stackrel{\mathrm{d}}{=} 0$ a.s. This proves (i).

(ii) We have $X \stackrel{\mathrm{d}}{=} {b_2}^{-1}(b_1 X + c_1 - c_2)$. So we assume that $X \stackrel{\mathrm{d}}{=} bX + c$ with $b > 0$ and $c \in \mathbb{R}^d$ and claim that $b = 1$ and $c = 0$. Let X_1 and X_2 be independent random variables, each of which has the same distribution as X. Then

$$X_1 - X_2 \stackrel{\mathrm{d}}{=} (bX_1 + c) - (bX_2 + c) = b(X_1 - X_2).$$

The random variable $X_1 - X_2$ is non-zero, because X is non-constant. Hence $b = 1$ by (i). Therefore $X \stackrel{\mathrm{d}}{=} X + nc$ for $n = 1, 2, \ldots$. It follows that $c = 0$. $\square$

LEMMA 13.8. *Let $\{X_t \colon t \geq 0\}$ be a non-trivial stochastic process on $\mathbb{R}^d$. If it satisfies (13.5), then b and $c(t)$ are uniquely determined by a.*

Proof. Suppose that

$$\{X(at)\} \stackrel{\mathrm{d}}{=} \{b_1 X(t) + c_1(t)\} \stackrel{\mathrm{d}}{=} \{b_2 X(t) + c_2(t)\}.$$

If $X(t)$ is non-constant, then we have $b_1 = b_2$ and $c_1(t) = c_2(t)$ for this t by Lemma 13.7. By non-triviality such a t exists. Hence $b_1 = b_2$. Now $c_1(t) = c_2(t)$ follows even if $X(t)$ is constant at t, because $b_1 X(t) + c_1(t) \stackrel{\mathrm{d}}{=} b_1 X(t) + c_2(t)$. $\square$

We use the following lemma.

LEMMA 13.9. *If μ is a probability measure on $\mathbb{R}^d$ satisfying $|\widehat{\mu}(z)| = 1$ on a neighborhood of $z = 0$, then μ is trivial.*

Proof. If we consider the components of a random variable with distribution μ, the assertion reduces to the one-dimensional case. So we assume that $d = 1$. For each $z \neq 0$ in a neighborhood of 0, there is $\theta \in \mathbb{R}$ such that $\widehat{\mu}(z) = \mathrm{e}^{\mathrm{i}\theta}$. Thus

$$1 = \mathrm{e}^{-\mathrm{i}\theta}\widehat{\mu}(z) = \int \mathrm{e}^{\mathrm{i}(zx - \theta)} \mu(\mathrm{d}x) = \int \cos(zx - \theta)\mu(\mathrm{d}x).$$

Hence μ is concentrated on the points $x = z^{-1}(2n\pi + \theta)$, $n \in \mathbb{Z}$. If μ is non-trivial, then any two points x_1, x_2 with positive μ-measure satisfy $|x_1 - x_2| \geq 2\pi |z|^{-1}$. This is absurd, as $|z|$ can be chosen arbitrarily small. $\square$

LEMMA 13.10. *Let Z and W be non-constant random variables on $\mathbb{R}^d$. Let Z_n be random variables on $\mathbb{R}^d$, $b_n > 0$, and $c_n \in \mathbb{R}^d$. If $P_{Z_n} \to P_Z$ and $P_{b_n Z_n + c_n} \to P_W$ as $n \to \infty$, then $b_n \to b$ and $c_n \to c$ with some $b \in (0, \infty)$ and $c \in \mathbb{R}^d$ as $n \to \infty$, and $bZ + c \stackrel{\mathrm{d}}{=} W$.*

Proof. Write $P_{Z_n} = \mu_n$, $P_Z = \mu$, and $P_W = \rho$. Then $\widehat{\mu}_n(z) \to \widehat{\mu}(z)$ and $\widehat{\mu}_n(b_n z)\mathrm{e}^{\mathrm{i}\langle c_n, z\rangle} \to \widehat{\rho}(z)$ uniformly on any compact set. Let b_∞ be a limit point of $\{b_n\}$ in $[0, \infty]$. If $b_\infty = 0$, then, letting $n \to \infty$ via the subsequence n_k satisfying $b_{n_k} \to b_\infty$, we get $|\widehat{\mu}_{n_k}(b_{n_k} z)| \to |\widehat{\mu}(0)| = 1$, which shows that $|\widehat{\rho}(z)| = 1$ and W is constant by Lemma 13.9, contradicting the assumption. If $b_\infty = \infty$, then $|\widehat{\mu}_{n_k}(z)| = |\widehat{\mu}_{n_k}(b_{n_k} b_{n_k}^{-1} z)| \to |\widehat{\rho}(0)| = 1$ and hence $|\widehat{\mu}(z)| = 1$, contradicting the assumption again. It follows that $0 < b_\infty < \infty$. There is $\varepsilon > 0$ such that $\widehat{\mu}(b_\infty z) \neq 0$ for $|z| \leq \varepsilon$. It follows that

$$\mathrm{e}^{\mathrm{i}\langle c_{n(k)}, z\rangle} \to \widehat{\rho}(z)/\widehat{\mu}(b_\infty z)$$

uniformly in z with $|z| \le \varepsilon$ as $k \to \infty$. Hence

$$\mathrm{i}\langle c_{n_k}, z\rangle = \log \mathrm{e}^{\mathrm{i}\langle c_{n(k)}, z\rangle} \to \log(\widehat{\rho}(z)/\widehat{\mu}(b_\infty z))$$

for $|z| \le \varepsilon$, as in Lemma 7.7, with the branch of the logarithm taken continuous in z and equal to 0 at $z = 0$. Hence, c_{n_k} tends to some $c_\infty \in \mathbb{R}^d$. Now we have $W \stackrel{\mathrm{d}}{=} b_\infty Z + c_\infty$. By Lemma 13.7 this shows that the original sequences b_n and c_n tend to b_∞ and c_∞, respectively. □

THEOREM 13.11. *Let $\{X_t : t \ge 0\}$ be a broad-sense semi-selfsimilar, stochastically continuous, non-trivial process on $\mathbb{R}^d$ with $X_0 = \text{const}$ a.s. Denote by Γ the set of all $a > 0$ such that there are $b > 0$ and $c(t)$ satisfying (13.5). Then:*

(i) *There is $H > 0$ such that, for every $a \in \Gamma$, $b = a^H$.*

(ii) *The set $\Gamma \cap (1, \infty)$ is non-empty. Let a_0 be the infimum of this set. If $a_0 > 1$, then $\Gamma = \{a_0{}^n : n \in \mathbb{Z}\}$ and $\{X_t\}$ is not broad-sense selfsimilar. If $a_0 = 1$, then $\Gamma = (0, \infty)$ and $\{X_t\}$ is broad-sense selfsimilar.*

Proof. By Lemma 13.8, b and $c(t)$ in (13.5) are uniquely determined by a. We write $b = b(a)$ and $c(t) = c(t, a)$. The set Γ has the following properties.

(1) $1 \in \Gamma$ and $b(1) = 1$.
(2) If $a \in \Gamma$, then $a^{-1} \in \Gamma$ and $b(a^{-1}) = b(a)^{-1}$.
(3) $\Gamma \cap (1, \infty)$ is non-empty.
(4) If a and a' are in Γ, then $aa' \in \Gamma$ and $b(aa') = b(a)b(a')$.
(5) If $a_n \in \Gamma$ $(n = 1, 2, \ldots)$ and $a_n \to a$ with $0 < a < \infty$, then $a \in \Gamma$ and $b(a_n) \to b(a)$.

The property (1) is obvious. If (13.5) holds, then

$$\{X(a^{-1}t)\} \stackrel{\mathrm{d}}{=} \{b^{-1}X(t) - b^{-1}c(a^{-1}t)\}.$$

Thus (2) holds. Since Γ contains an element other than 1, the property (3) follows from (2). If a and a' are in Γ, then

$$\begin{aligned} \{X(aa't)\} &\stackrel{\mathrm{d}}{=} \{b(a)X(a't) + c(a't, a)\} \\ &\stackrel{\mathrm{d}}{=} \{b(a)b(a')X(t) + c(a't, a) + b(a)c(t, a')\}, \end{aligned} \tag{13.7}$$

which shows (4). To prove (5), write $b_n = b(a_n)$ and $c_n(t) = c(t, a_n)$. Then $X_{a_n t} \stackrel{\mathrm{d}}{=} b_n X_t + c_n(t)$ and $X_{a_n t} \to X_{at}$ in prob. If t is such that X_{at} is non-constant, then X_t is non-constant and, by Lemma 13.10, $b_n \to b$ and $c_n(t) \to c(t)$ for some $b \in (0, \infty)$ and $c(t) \in \mathbb{R}^d$. Since such a t exists by the non-triviality of $\{X_t\}$, we have $b_n \to b$. Now, for any t, the last part of the proof of Lemma 13.10 shows that $c_n(t)$ tends to some $c(t)$. Hence $\{X_{at}\} \stackrel{\mathrm{d}}{=} \{bX_t + c(t)\}$, which shows (5).

We denote by $\log \Gamma$ the set of $\log a$ with $a \in \Gamma$. Then, by (1)–(5), $\log \Gamma$ is a closed additive subgroup of $\mathbb{R}$ and $(\log \Gamma) \cap (0, \infty) \neq \emptyset$. Denote the infimum of $(\log \Gamma) \cap (0, \infty)$ by r_0. Suppose that $r_0 > 0$. Then we have $r_0 \in \log \Gamma$ and $r_0 \mathbb{Z} = \{r_0 n \colon n \in \mathbb{Z}\} \subset \log \Gamma$. If there is $r \in (\log \Gamma) \setminus (r_0 \mathbb{Z})$, then $nr_0 < r < (n+1)r_0$ with some $n \in \mathbb{Z}$, and hence $r - nr_0 \in \log \Gamma$ and $0 < r - nr_0 < r_0$, a contradiction. This shows that, if $r_0 > 0$, then $\log \Gamma = r_0 \mathbb{Z}$. If $r_0 = 0$ and there is r in $\mathbb{R} \setminus (\log \Gamma)$, then we have $(r - \varepsilon, r + \varepsilon) \subset \mathbb{R} \setminus (\log \Gamma)$ with some $\varepsilon > 0$ by the closedness of $\log \Gamma$, and, choosing $s \in \log \Gamma$ satisfying $0 < s < 2\varepsilon$, we get $r - \varepsilon < ns < r + \varepsilon$ with some $n \in \mathbb{Z}$ and $ns \in \log \Gamma$, which is absurd. This shows that, if $r_0 = 0$, then $\log \Gamma = \mathbb{R}$. Letting $a_0 = \mathrm{e}^{r_0}$, we see that the assertion (ii) of the theorem is proved.

We claim the following.

(6) If $a > 1$ and $a \in \Gamma$, then $b(a) > 1$.

In fact, suppose that $a > 1$, $a \in \Gamma$, and $b(a) \leq 1$. Fix t. Then

$$\widehat{\mu}_{a^n t}(z) = \widehat{\mu}_t(b(a)^n z) \mathrm{e}^{\mathrm{i}\langle z, c(t, a^n) \rangle} \quad \text{for } n \in \mathbb{Z}, z \in \mathbb{R}^d.$$

Hence

$$|\widehat{\mu}_{a^n t}(b(a)^{-n} z)| = |\widehat{\mu}_t(z)| \quad \text{for } n \in \mathbb{Z}, z \in \mathbb{R}^d.$$

Since $X(0)$ is constant, we have $|\widehat{\mu}_{a^n t}(w)| \to 1$ uniformly in w in any compact set as $n \to -\infty$. Since $|b(a)^{-n} z| \leq |z|$ for $n \leq 0$, we have

$$0 \leq 1 - |\widehat{\mu}_{a^n t}(b(a)^{-n} z)| \leq \sup_{|w| \leq |z|} (1 - |\widehat{\mu}_{a^n t}(w)|) \to 0$$

as $n \to -\infty$. It follows that $|\widehat{\mu}_t(z)| = 1$ and hence X_t is constant by Lemma 13.9. Since t is arbitrary, this contradicts the non-triviality. This proves (6). Now we prove the assertion (i). Suppose that $a_0 > 1$. Let $H = (\log b(a_0))/(\log a_0)$. Then $H > 0$ by (6). Any a in Γ is written as $a = {a_0}^n$ with $n \in \mathbb{Z}$. Hence $b(a) = b(a_0)^n = {a_0}^{Hn} = a^H$. In the case $a_0 = 1$, we have $\Gamma = (0, \infty)$ and the properties (4) and (5) yield the existence of $H \in \mathbb{R}$ satisfying $b(a) = a^H$. Also in this case, the property (6) shows that $H > 0$. This proves (i). □

DEFINITION 13.12. The H in Theorem 13.11 is called the *exponent* of the non-trivial broad-sense semi-selfsimilar process. It is uniquely determined by the process. If a is in $\Gamma \cap (1, \infty)$, then a and a^H are called, respectively, an *epoch* and a *span* of the process. Instead of broad-sense semi-selfsimilar with exponent H, we sometimes say *broad-sense H-semi-selfsimilar*. The semi-selfsimilarity implies the broad-sense semi-selfsimilarity. Thus we say that $\{X_t\}$ is semi-selfsimilar with exponent H or *H-semi-selfsimilar*. if it has exponent H as a broad-sense semi-selfsimilar process and if it is semi-selfsimilar. Similarly we use the words *broad-sense H-selfsimilar* and *H-selfsimilar*.

REMARK 13.13. Let $\{X_t\}$ be a semi-selfsimilar, stochastically continuous, non-zero process on $\mathbb{R}^d$ with $X_0 = 0$ a. s. Let $\Gamma = \{a > 0\colon$ there is $b > 0$ satisfying (13.4)$\}$. Then we can prove the statements (i) and (ii) of Theorem 13.11 with broad-sense semi-selfsimilar replaced by semi-selfsimilar. The H thus determined equals the exponent in Definition 13.12, if $\{X_t\}$ is non-trivial. If $\{X_t\}$ is a non-zero, trivial, semi-selfsimilar, stochastically continuous process with $X_0 = 0$ a. s. and if a is an epoch, then $X_t = t^H g(\log t)$ a. s., where g is a continuous periodic function with period $\log a$. Also in this case, H is called the exponent.

Broad-sense selfsimilar processes are related to selfsimilar processes in the following way.

PROPOSITION 13.14. *If $\{X_t\}$ is broad-sense selfsimilar, stochastically continuous, non-trivial, and $X_0 =$ const a.s., then there is a continuous function $k(t)$ from $[0,\infty)$ to $\mathbb{R}^d$ such that $\{X_t - k(t)\}$ is selfsimilar and $X_0 - k(0) = 0$ a. s.*

Proof. By Theorem 13.11 $\{X_t\}$ has the exponent $H > 0$. For any $a > 0$ there is unique $c_a(t)$ such that $\{X_{at}\} \stackrel{\mathrm{d}}{=} \{a^H X_t + c_a(t)\}$. If $a \to a_0 > 0$, then $c_a(t)$ is convergent, because X_{at} and $a^H X_t$ are convergent in probability. It follows that $c_a(t) \to c_{a_0}(t)$ as $a \to a_0 > 0$. As $a \downarrow 0$, $c_a(t)$ tends to X_0. So let $c_0(t) = X_0$. We have

$$c_{aa'}(t) = c_a(a't) + a^H c_{a'}(t) \quad \text{for } a > 0,\ a' > 0,$$

as in (13.7). Now let $k(t) = c_t(1)$. Then $k(t)$ is continuous on $[0,\infty)$ and

$$\begin{aligned}\{X_{at} - k(at)\} &\stackrel{\mathrm{d}}{=} \{a^H X_t + c_a(t) - c_{at}(1)\}\\ &= \{a^H X_t - a^H c_t(1)\} = \{a^H (X_t - k(t))\},\end{aligned}$$

which is the desired property. □

Proposition 13.14 can be extended. The statement remains valid if we replace *selfsimilar* by *semi-selfsimilar* (Maejima and Sato [**333**]).

THEOREM 13.15. *Let $\{X_t\colon t \geq 0\}$ be a non-trivial semi-stable process on $\mathbb{R}^d$ with exponent H as a broad-sense semi-selfsimilar process. Then $H \geq 1/2$.*

Proof. By Proposition 13.5 $\{X_t\}$ is broad-sense semi-selfsimilar and, by Theorem 13.11, the exponent $H > 0$ exists. We have (13.6) with $b = a^H$, where $a > 1$ is an epoch and $b > 1$ is a span. Let the generating triplet of $\{X_t\}$ be (A, ν, γ). Then that of $\{X_{at}\}$ is $(aA, a\nu, a\gamma)$. We define, for any $r > 0$, a transformation T_r of measures ρ on $\mathbb{R}^d$ by

$$(T_r\rho)(B) = \rho(r^{-1}B) \quad \text{for } B \in \mathcal{B}(\mathbb{R}^d). \tag{13.8}$$

Then, by using Proposition 11.10, we see that the generating triplet of $\{a^H X_t+tc\}$ is $(a^{2H}A, T_b\nu, \gamma(a))$ with some $\gamma(a) \in \mathbb{R}^d$ and $b = a^H$. Therefore, by the uniqueness, $aA = a^{2H}A$ and $a\nu = T_b\nu$. By the non-triviality, we have $A \neq 0$ or $\nu \neq 0$. If $A \neq 0$, then $H = 1/2$. Suppose that $\nu \neq 0$. It follows from $a\nu = T_b\nu$ that $a^{-1}\nu = T_{b^{-1}}\nu$. Iteration gives

$$a^n\nu = T_{b^n}\nu \quad \text{for } n \in \mathbb{Z}. \tag{13.9}$$

Let

$$S_n(b) = \{x \in \mathbb{R}^d \colon b^n < |x| \leq b^{n+1}\} \quad \text{for } n \in \mathbb{Z}. \tag{13.10}$$

Then $S_n(b) = b^n S_0(b)$ and $\nu(S_n(b)) = (T_{b^{-n}}\nu)(S_0(b)) = a^{-n}\nu(S_0(b))$. The set $\{x \colon 0 < |x| \leq 1\}$ is partitioned into $S_{-n-1}(b)$, $n \in \mathbb{Z}_+$, and the set $\{x \colon |x| > 1\}$ is partitioned into $S_n(b)$, $n \in \mathbb{Z}_+$. It follows that $\nu(S_0(b)) \neq 0$. Since (13.9) is equivalent to $a^n \int 1_B(x)\nu(\mathrm{d}x) = \int 1_B(b^n x)\nu(\mathrm{d}x)$, we have

$$a^n \int f(x)\nu(\mathrm{d}x) = \int f(b^n x)\nu(\mathrm{d}x)$$

for any nonnegative measurable function f. Thus

$$\begin{aligned}\int_{S_n(b)} |x|^2\nu(\mathrm{d}x) &= \int 1_{S_0(b)}(b^{-n}x)|x|^2\nu(\mathrm{d}x) \\ &= a^{-n}\int 1_{S_0(b)}(x)|b^n x|^2\nu(\mathrm{d}x) = a^{-n(1-2H)}\int_{S_0(b)} |x|^2\nu(\mathrm{d}x).\end{aligned}$$

Since $\int_{|x|\leq 1} |x|^2\nu(\mathrm{d}x) < \infty$, we get $\sum_{n\leq -1} a^{-n(1-2H)} < \infty$, that is, $H > 1/2$. □

DEFINITION 13.16. In the theorem above, $\alpha = 1/H$ is called the *index* of the non-trivial semi-stable process $\{X_t\}$ or of the distribution $\mu = P_{X_1}$. A span $b > 1$ of the process is also called a *span* of μ. Sometimes we say *α-semi-stable* instead of semi-stable with index α. The definition of the index applies to stable processes and distributions, as they are special cases of semi-stable. We say *α-stable* for stable with index α. By Theorem 13.15 the index α satisfies $0 < \alpha \leq 2$.

REMARK 13.17. Theorem 13.15 does not apply to trivial semi-stable processes. But, if $\{X_t\}$ is a non-zero, trivial, strictly semi-stable process, then there is $c \neq 0$ such that $X_t = tc$ a.s., and the exponent $H = 1$ and the index $\alpha = 1/H = 1$ of $\{X_t\}$ are defined. In this case $\{X_t\}$, or the distribution $\mu = P_{X_1} = \delta_c$, is strictly 1-stable.

From now on, when we talk about an α-semi-stable (or α-stable) distribution or process with $\alpha \neq 1$, the non-triviality is implicitly assumed. When we consider a 1-semi-stable (or 1-stable) distribution or process, we are assuming that the distribution is not δ_0 or the process is non-zero.

EXAMPLE 13.18. The indices of the distributions in Example 13.3 are as follows. The non-trivial Gaussian distribution is 2-stable. The Brownian motion is a strictly 2-stable process. The Cauchy distribution is strictly 1-stable. Thus the Cauchy process is a strictly 1-stable process. The infinitely divisible distribution with Lévy measure (13.3) is α-semi-stable.

14. Representations of stable and semi-stable distributions

Let us determine the characteristic functions of stable, strictly stable, semi-stable, and strictly semi-stable distributions on $\mathbb{R}^d$.

THEOREM 14.1. *Let μ be infinitely divisible and non-trivial on $\mathbb{R}^d$. Then the following three statements are equivalent: μ is Gaussian; μ is 2-stable; μ is 2-semi-stable.*

Proof. Gaussian implies 2-stable as we saw in Example 13.18. By the definition, 2-stable implies 2-semi-stable. Suppose that μ is 2-semi-stable. The proof of Theorem 13.15 shows that, if $\nu \neq 0$, then $H > 1/2$, that is, $\alpha < 2$. Hence $\nu = 0$, which means that μ is Gaussian. □

THEOREM 14.2. *Let μ be infinitely divisible on $\mathbb{R}^d$ and suppose that $\mu \neq \delta_0$. Then the following three statements are equivalent: μ is Gaussian with mean 0; μ is strictly 2-stable; μ is strictly 2-semi-stable.*

Proof. Gaussian with mean 0 implies strictly 2-stable, if it is not δ_0. Strictly 2-stable implies strictly 2-semi-stable. Suppose that μ is strictly 2-semi-stable, that is, $\widehat{\mu}(z)^a = \widehat{\mu}(a^{1/2}z)$ with some positive $a \neq 1$. Then μ is Gaussian by Theorem 14.1, that is, $\widehat{\mu}(z) = \mathrm{e}^{-\langle z, Az\rangle/2 + \mathrm{i}\langle \gamma, z\rangle}$. It follows that $a\gamma = a^{1/2}\gamma$, and hence $\gamma = 0$. □

Let $S = \{x \in \mathbb{R}^d \colon |x| = 1\}$. It is the unit sphere (if $d \geq 2$) or $\{1, -1\}$ (if $d = 1$). For $b > 1$, let $S_n(b)$ be the set in (13.10). The transformation T_r of measures is defined for $r > 0$ as in (13.8). Note that, if ρ is a measure concentrated on $S_n(b)$, then $T_{b^m}\rho$ is concentrated on $S_{n+m}(b)$. If μ is the distribution of a random variable X, then $T_r\mu$ is the distribution of rX. The restriction of a measure ρ to a Borel set E is denoted by $[\rho]_E$.

THEOREM 14.3. *Let μ be infinitely divisible and non-trivial on $\mathbb{R}^d$ with generating triplet (A, ν, γ). Let $0 < \alpha < 2$.*

(i) *Let $b > 1$. Then the following three statements are equivalent:*

(1) *μ is α-semi-stable with b as a span;*

(2) *$A = 0$ and*

$$\nu = b^{-\alpha} T_b \nu; \tag{14.1}$$

(3) *$A = 0$ and, for each integer n, the measure ν on $S_n(b)$ is determined by the measure ν on $S_0(b)$ by*

$$[\nu]_{S_n(b)} = b^{-n\alpha} T_{b^n}([\nu]_{S_0(b)}). \tag{14.2}$$

(ii) *The following statements are equivalent:*

(1) μ *is* α*-stable;*

(2) $A = 0$ *and*

$$\nu = b^{-\alpha} T_b \nu \quad \textit{for every } b > 0; \tag{14.3}$$

(3) $A = 0$ *and there is a finite measure* λ *on* S *such that*

$$\nu(B) = \int_S \lambda(\mathrm{d}\xi) \int_0^\infty 1_B(r\xi) \frac{\mathrm{d}r}{r^{1+\alpha}} \quad \textit{for } B \in \mathcal{B}(\mathbb{R}^d). \tag{14.4}$$

Proof. (i) Suppose that (1) holds. Consider the corresponding semi-stable process. Since $0 < \alpha < 2$, the proof of Theorem 13.15 shows that $A = 0$. Thus $\nu \neq 0$, since μ is non-trivial. The Lévy measure ν satisfies $a\nu = T_b \nu$ with $a = b^\alpha$, as in the proof of Theorem 13.15. That is, (2) holds. The condition (2) implies (3), since we have (13.9), that is,

$$b^{n\alpha}\nu(B) = (T_{b^n}\nu)(B) = \nu(b^{-n}B) \quad \text{for } B \in \mathcal{B}(\mathbb{R}^d),\ n \in \mathbb{Z},$$

which gives (14.2) for $B \subset S_n(b)$. Now suppose that (3) holds. For any $B \in \mathcal{B}(\mathbb{R}^d)$, let $B_n = B \cap S_n(b)$. Then, by (14.2),

$$\begin{aligned}\nu(B) &= \sum_{n\in\mathbb{Z}} [\nu]_{S_n(b)}(B_n) = \sum_{n\in\mathbb{Z}} b^{-n\alpha}[\nu]_{S_0(b)}(b^{-n}B_n)\\ &= \sum_{n\in\mathbb{Z}} b^{-n\alpha}\nu(b^{-n}B \cap S_0(b)).\end{aligned}$$

Therefore

$$\begin{aligned}(T_b\nu)(B) = \nu(b^{-1}B) &= \sum_{n\in\mathbb{Z}} b^{-n\alpha}\nu(b^{-n-1}B \cap S_0(b))\\ &= b^\alpha \sum_{n\in\mathbb{Z}} b^{-n\alpha}\nu(b^{-n}B \cap S_0(b)).\end{aligned}$$

Hence we get (2). Let us see that (2) implies (1). Consider a random variable X whose distribution is μ. Then $\widehat{\mu}(bz)$ is the characteristic function of the distribution of bX. By Proposition 11.10, the distribution of bX has generating triplet $(0, T_b\nu, \gamma(b))$ with some $\gamma(b) \in \mathbb{R}^d$. On the other hand, $\widehat{\mu}(z)^a$ is the characteristic function of the distribution with generating triplet $(0, a\nu, a\gamma)$. Hence $\widehat{\mu}(z)^a = \widehat{\mu}(bz)\mathrm{e}^{\mathrm{i}\langle c,z\rangle}$ with $a = b^\alpha$ and some c, that is, μ is α-semi-stable having b as a span.

(ii) Suppose that (1) holds. Then μ is α-semi-stable and any $b > 1$ can be its span. Hence, by (i), (14.1) holds for any $b > 1$. Since the property $\nu(B) = b^{-\alpha}\nu(b^{-1}B)$ for all $B \in \mathcal{B}(\mathbb{R}^d)$ implies $\nu(bB) = b^{-\alpha}\nu(B)$ for all $B \in \mathcal{B}(\mathbb{R}^d)$, (14.1) remains true with b replaced by b^{-1}. Hence we have (2).

Assume the condition (2). Let us write, for $E \subset (0,\infty)$ and $C \subset S$,

$$EC = \{x \in \mathbb{R}^d \setminus \{0\} \colon |x| \in E \text{ and } |x|^{-1}x \in C\}. \tag{14.5}$$

Define a finite measure λ on S by

$$\lambda(C) = \alpha\nu((1,\infty)C) \quad \text{for } C \in \mathcal{B}(S). \tag{14.6}$$

Define $\nu'(B)$ by the right-hand side of (14.4). Then ν' is a measure on $\mathbb{R}^d$ with $\nu'(\{0\}) = 0$ and, for $b > 0$ and $C \in \mathcal{B}(S)$,

$$\begin{aligned}\nu'((b,\infty)C) &= \lambda(C)\int_b^\infty \frac{\mathrm{d}r}{r^{1+\alpha}} = \alpha^{-1}b^{-\alpha}\lambda(C) = b^{-\alpha}\nu((1,\infty)C)\\ &= \nu(b(1,\infty)C) = \nu((b,\infty)C)\end{aligned}$$

by (14.3). It follows that $\nu'(B) = \nu(B)$ for all $B \in \mathcal{B}(\mathbb{R}^d \setminus \{0\})$. (Here we have used Proposition 1.15. Fix $\varepsilon > 0$ and consider the set $\{x\colon |x| > \varepsilon\}$. Let $\mathcal{A}_\varepsilon$ be the collection of all sets of the form $(b,\infty)C$ with $b > \varepsilon$ and $C \in \mathcal{B}(S)$. Use of Proposition 1.15 leads to the conclusion that $\nu' = \nu$ on $\sigma(\mathcal{A}_\varepsilon)$, where $\sigma(\mathcal{A}_\varepsilon)$ is the collection of all Borel sets in $\{x\colon |x| > \varepsilon\}$. It follows that $\nu' = \nu$ on $\mathcal{B}(\mathbb{R}^d \setminus \{0\})$.) Thus we have (3).

If we assume (3), then we get (2), since we obtain (14.3) from (14.4). Assume the condition (2). By (i), μ is α-semi-stable and any $b > 1$ can be chosen as a span. That is, for any $b > 1$, there is $c \in \mathbb{R}^d$ such that $\widehat{\mu}(z)^{b^\alpha} = \widehat{\mu}(bz)\mathrm{e}^{\mathrm{i}\langle c,z\rangle}$. Since z is variable, it follows that

$$\widehat{\mu}(z)^{b^{-\alpha}} = \widehat{\mu}(b^{-1}z)\mathrm{e}^{\mathrm{i}\langle -b^{-\alpha-1}c,z\rangle}$$

for any $b > 1$. Hence μ is α-stable. □

REMARK 14.4. In Theorem 14.3(ii), the measure λ on S is uniquely determined by μ, because (14.4) implies (14.6). We call any positive constant multiple of λ a *spherical part* of the Lévy measure ν. For any non-zero finite measure λ on S and for any $0 < \alpha < 2$, we can find an α-stable distribution μ with Lévy measure ν defined by (14.4). In fact, it follows from (14.4) that

$$\int_{\mathbb{R}^d} f(x)\nu(\mathrm{d}x) = \int_S \lambda(\mathrm{d}\xi)\int_0^\infty f(r\xi)\frac{\mathrm{d}r}{r^{1+\alpha}} \tag{14.7}$$

for any nonnegative measurable function f. Hence $\int(|x|^2 \wedge 1)\nu(\mathrm{d}x)$ is finite, so that there is an infinitely divisible distribution μ with Lévy measure ν and Theorem 14.3(ii) applies.

We can consider $r^{-1-\alpha}\,\mathrm{d}r$ as a radial part of the Lévy measure of the α-stable process. Notice that, as α decreases, $r^{-1-\alpha}$ gets smaller for $0 < r < 1$ and bigger for $1 < r < \infty$. Roughly speaking, an α-stable process moves mainly by big jumps if α is close to 0, and mainly by small jumps if α is close to 2. This tendency is clearly visible in the computer simulation of paths for $d = 1$ in Janicki and Weron [**238**]. Rigorous analysis of the behavior of paths will begin in the next chapter.

It follows from Theorem 14.3(i) that, for any $0 < \alpha < 2$, any finite measure ρ on $S_0(b)$ can be extended to the Lévy measure of an α-semi-stable distribution. Indeed, define ν by $[\nu]_{S_n(b)} = b^{-n\alpha}T_{b^n}\rho$ for $n \in \mathbb{Z}$. Then

$\nu(S_n(b)) = b^{-n\alpha}\rho(S_0(b))$ and

$$\int_{S_n(b)} |x|^2\nu(\mathrm{d}x) = b^{n(2-\alpha)}\int_{S_0(b)} |x|^2\rho(\mathrm{d}x)$$

for $n \in \mathbb{Z}$. Hence $\int(|x|^2 \wedge 1)\nu(\mathrm{d}x) < \infty$ and, for any γ, the triplet $(0,\nu,\gamma)$ generates an infinitely divisible distribution, which is α-semi-stable. Thus the Lévy measure of an α-semi-stable distribution for $0 < \alpha < 2$ can be any of discrete, continuous singular, absolutely continuous, and their mixtures. On the other hand, if $d = 1$, then $S = \{1, -1\}$ and any non-trivial α-stable distribution with $0 < \alpha < 2$ has absolutely continuous Lévy measure

$$\nu(\mathrm{d}x) = \begin{cases} c_1 x^{-1-\alpha}\mathrm{d}x & \text{on } (0,\infty), \\ c_2|x|^{-1-\alpha}\mathrm{d}x & \text{on } (-\infty,0) \end{cases}$$

with $c_1 \geq 0$, $c_2 \geq 0$, $c_1 + c_2 > 0$ by Theorem 14.3(ii).

If an α-semi-stable distribution μ on $\mathbb{R}$, $0 < \alpha < 2$, has an absolutely continuous Lévy measure ν, then

$$\nu(\mathrm{d}x) = \begin{cases} g_1(\log x)x^{-1-\alpha}\mathrm{d}x & \text{on } (0,\infty), \\ g_2(\log|x|)|x|^{-1-\alpha}\mathrm{d}x & \text{on } (-\infty,0), \end{cases}$$

and g_1 and g_2 are nonnegative measurable functions satisfying

$$g_j(\log x) = g_j(\log x + \log b) \quad \text{a.e.} \quad \text{for } j = 1, 2,$$

where b is a span. This is proved from (14.1).

PROPOSITION 14.5. *Let μ be non-trivial and α-semi-stable on $\mathbb{R}^d$ with $0 < \alpha < 2$. Let ν be its Lévy measure. Then, $\int_{|x|\leq 1}|x|\nu(\mathrm{d}x)$ is finite if and only if $\alpha < 1$. The integral $\int_{|x|>1}|x|\nu(\mathrm{d}x)$ is finite if and only if $\alpha > 1$. The total mass of ν is always infinite.*

Proof. Let b be a span of μ. We have $\nu(S_n(b)) = b^{-n\alpha}\nu(S_0(b))$ and

$$\int_{S_n(b)} |x|\nu(\mathrm{d}x) = b^{n(1-\alpha)}\int_{S_0(b)} |x|\nu(\mathrm{d}x)$$

for $n \in \mathbb{Z}$ as in the proof of Theorem 13.15. Note that $0 < \nu(S_0(b)) < \infty$. Now the assertions are clear, because $b > 1$. □

REMARK 14.6. By Proposition 14.5 and (14.7) we have the following representation of a non-trivial α-stable distribution μ with $0 < \alpha < 2$. If $0 < \alpha < 1$, then μ has drift γ_0 and

$$\widehat{\mu}(z) = \exp\left[\int_S \lambda(\mathrm{d}\xi)\int_0^\infty (e^{i\langle z,r\xi\rangle} - 1)\frac{\mathrm{d}r}{r^{1+\alpha}} + i\langle\gamma_0, z\rangle\right]. \tag{14.8}$$

If $1 < \alpha < 2$, then μ has center γ_1 and

$$(14.9)\qquad \widehat{\mu}(z) = \exp\left[\int_S \lambda(\mathrm{d}\xi)\int_0^\infty (\mathrm{e}^{\mathrm{i}\langle z,r\xi\rangle} - 1 - \mathrm{i}\langle z,r\xi\rangle)\frac{\mathrm{d}r}{r^{1+\alpha}} + \mathrm{i}\langle\gamma_1,z\rangle\right].$$

These are special cases of (8.7) and (8.8). We will show later in Example 25.12 that the center γ_1 is equal to the mean. If $\alpha = 1$, then

$$(14.10)\qquad \widehat{\mu}(z) = \exp\left[\int_S \lambda(\mathrm{d}\xi)\int_0^\infty (\mathrm{e}^{\mathrm{i}\langle z,r\xi\rangle} - 1 - \mathrm{i}\langle z,r\xi\rangle 1_{(0,1]}(r))\frac{\mathrm{d}r}{r^2} + \mathrm{i}\langle\gamma,z\rangle\right].$$

THEOREM 14.7. *Let μ be infinitely divisible on $\mathbb{R}^d$ with generating triplet (A,ν,γ) and suppose that $\mu \neq \delta_0$.*

(i) *Let $0 < \alpha < 1$. Then μ is strictly α-semi-stable having a span b if and only if μ is α-semi-stable having a span b and the drift $\gamma_0 = 0$.*

(ii) *Let $\alpha = 1$. Then μ is strictly 1-semi-stable having a span b if and only if either μ is 1-semi-stable having a span b, $\nu \neq 0$, and*

$$(14.11)\qquad \int_{1<|x|\leq b} x\nu(\mathrm{d}x) = 0,$$

or $A = 0$, $\nu = 0$, and $\gamma \neq 0$.

(iii) *Let $1 < \alpha < 2$. Then μ is strictly α-semi-stable having a span b if and only if μ is α-semi-stable having a span b and the center $\gamma_1 = 0$.*

(iv) *Let $0 < \alpha < 1$. Then μ is strictly α-stable if and only if μ is α-stable and the drift $\gamma_0 = 0$.*

(v) *Let $\alpha = 1$. Then μ is strictly 1-stable if and only if either μ is 1-stable, $\nu \neq 0$, and the measure λ in Theorem 14.3(ii) satisfies*

$$(14.12)\qquad \int_S \xi\lambda(\mathrm{d}\xi) = 0,$$

or $A = 0$, $\nu = 0$, and $\gamma \neq 0$.

(vi) *Let $1 < \alpha < 2$. Then μ is strictly α-stable if and only if μ is α-stable and the center $\gamma_1 = 0$.*

Proof. (i) $0 < \alpha < 1$. Let μ be α-semi-stable having a span b. Then μ is non-trivial and, by (14.1) and Proposition 14.5,

$$\begin{aligned}\widehat{\mu}(bz) &= \exp\left[\int (\mathrm{e}^{\mathrm{i}\langle z,bx\rangle} - 1)\nu(\mathrm{d}x) + \mathrm{i}b\langle\gamma_0,z\rangle\right]\\ &= \exp\left[b^\alpha\int (\mathrm{e}^{\mathrm{i}\langle z,x\rangle} - 1)\nu(\mathrm{d}x) + \mathrm{i}b\langle\gamma_0,z\rangle\right].\end{aligned}$$

Therefore, $\widehat{\mu}(bz) = \widehat{\mu}(z)^{b^\alpha}$ if and only if $\gamma_0 = 0$.

(ii) $\alpha = 1$. Suppose that μ is non-trivial and 1-semi-stable having a span b. Let

$$c = \int (bx1_D(x) - bx1_D(bx))\nu(\mathrm{d}x).$$

Using (14.1), we get

$$\begin{aligned}\widehat{\mu}(bz) &= \exp\left[\int (\mathrm{e}^{\mathrm{i}\langle z,bx\rangle} - 1 - \mathrm{i}\langle z, bx\rangle 1_D(x))\nu(\mathrm{d}x) + \mathrm{i}b\langle\gamma, z\rangle\right] \\ &= \exp\left[\int (\mathrm{e}^{\mathrm{i}\langle z,bx\rangle} - 1 - \mathrm{i}\langle z, bx\rangle 1_D(bx))\nu(\mathrm{d}x) + \mathrm{i}b\langle\gamma, z\rangle - \mathrm{i}\langle c, z\rangle\right] \\ &= \widehat{\mu}(z)^b \mathrm{e}^{-\mathrm{i}\langle c,z\rangle}.\end{aligned}$$

Since

$$c = b\int_{b^{-1}<|x|\le 1} x\nu(\mathrm{d}x) = b\int_{1<|x|\le b} x\nu(\mathrm{d}x)$$

again by (14.1), μ is strictly 1-semi-stable if and only if (14.11) holds.

(iii) $1 < \alpha < 2$. Using Proposition 14.5 and the expression (8.8) involving γ_1 with $A = 0$, the proof is the same as for (i).

(iv) $0 < \alpha < 1$. Proof is similar to that of (i).

(v) $\alpha = 1$. Assume that μ is 1-stable and non-trivial. If μ is strictly 1-stable, then μ is strictly 1-semi-stable with an arbitrary $b > 1$ being a span, which implies (14.12), and conversely. Note that

$$\int_{1<|x|\le b} x\nu(\mathrm{d}x) = \int_S \lambda(\mathrm{d}\xi)\int_1^b r\xi\frac{\mathrm{d}r}{r^2} = (\log b)\int_S \xi\lambda(\mathrm{d}\xi).$$

(vi) $1 < \alpha < 2$. Proof is similar to that of (iii). □

The theorem just given clarifies the relation between stable and strictly stable processes.

THEOREM 14.8. (i) *If* $\{X_t\}$ *is* α*-stable with* $0 < \alpha < 1$ *or* $1 < \alpha \le 2$*, then, for some* $k \in \mathbb{R}^d$*,* $\{X_t - tk\}$ *is strictly* α*-stable.*

(ii) *If* $\{X_t\}$ *is* 1*-stable but not strictly* 1*-stable, then, for any choice of a function* $k(t)$*,* $\{X_t - k(t)\}$ *is not strictly* 1*-stable.*

(iii) *The statements (i) and (ii) remain valid if we replace* α*-stable and* 1*-stable by* α*-semi-stable and* 1*-semi-stable, respectively.*

Proof. We use Theorem 14.7 for $\mu = P_{X_1}$.

(i) If $0 < \alpha < 1$, then choose $k = \gamma_0$. If $1 < \alpha < 2$, then choose $k = \gamma_1$. If $\alpha = 2$, choose $k = \gamma =$ mean of μ. In each case, $\{X_t - tk\}$ is proved to be strictly α-stable.

(ii) Let $\{X_t\}$ be 1-stable and not strictly 1-stable. Then $\{X_t\}$ is non-trivial. If $k(t)$ is not linear in t, then $\{X_t - k(t)\}$ is not a Lévy process. Consider a Lévy process $\{X_t - tk\}$ with $k \in \mathbb{R}^d$. It has a common Lévy measure ν with $\{X_t\}$. The measure λ determined by ν does not satisfy (14.12), since $\{X_t\}$ is not strictly 1-stable.

(iii) The proof is the same. $\square$

Later we will use the following consequence.

PROPOSITION 14.9. *Let $0 < \alpha < 2$. Let μ be α-semi-stable having a span b on $\mathbb{R}^d$. Then*

$$\log \widehat{\mu}(z) = \begin{cases} -|z|^\alpha(\eta_1(z) + \mathrm{i}\eta_2(z)) + \mathrm{i}\langle\gamma_0, z\rangle & \text{for } 0 < \alpha < 1, \\ -|z|\,(\eta_1(z) + \mathrm{i}\eta_2(z)) + \mathrm{i}\langle\gamma, z\rangle & \text{for } \alpha = 1, \\ -|z|^\alpha(\eta_1(z) + \mathrm{i}\eta_2(z)) + \mathrm{i}\langle\gamma_1, z\rangle & \text{for } 1 < \alpha < 2, \end{cases} \tag{14.13}$$

where γ_0 is the drift, γ is that in (8.1), and γ_1 is the center, and where $\eta_1(z)$ is a nonnegative function continuous on $\mathbb{R}^d \setminus \{0\}$ satisfying $\eta_1(bz) = \eta_1(z)$ and $\eta_2(z)$ is a real function continuous on $\mathbb{R}^d \setminus \{0\}$ satisfying

$$\eta_2(bz) = \begin{cases} \eta_2(z) & \text{for } 0 < \alpha < 1 \text{ or } 1 < \alpha < 2, \\ \eta_2(z) + \left\langle \frac{z}{|z|}, \theta \right\rangle & \text{for } \alpha = 1 \end{cases} \tag{14.14}$$

with $\theta = \int_{1<|x|\le b} x\nu(\mathrm{d}x)$.

Proof. We have (14.13) with

$$|z|^\alpha \eta_1(z) = \textstyle\int (1 - \cos\langle z, x\rangle)\nu(\mathrm{d}x),$$

$$|z|^\alpha \eta_2(z) = \begin{cases} -\int \sin\langle z, x\rangle \nu(\mathrm{d}x) & \text{for } 0 < \alpha < 1, \\ -\int (\sin\langle z, x\rangle - \langle z, x\rangle 1_{\{|x|\le 1\}}(x))\nu(\mathrm{d}x) & \text{for } \alpha = 1, \\ -\int (\sin\langle z, x\rangle - \langle z, x\rangle)\nu(\mathrm{d}x) & \text{for } 1 < \alpha < 2. \end{cases}$$

Hence η_1 and η_2 are continuous on $\mathbb{R}^d \setminus \{0\}$ and $\eta_1(z) \ge 0$. Since the property (14.1) gives

$$|bz|^\alpha \eta_1(bz) = \textstyle\int (1 - \cos\langle z, x\rangle)(T_b\nu)(\mathrm{d}x) = b^\alpha \int (1 - \cos\langle z, x\rangle)\nu(\mathrm{d}x),$$

we have $\eta_1(bz) = \eta_1(z)$ on $\mathbb{R}^d \setminus \{0\}$. Similarly, $\eta_2(bz) = \eta_2(z)$ on $\mathbb{R}^d \setminus \{0\}$ if $\alpha \ne 1$. If $\alpha = 1$, then

$$|bz|\eta_2(bz) = -\textstyle\int (\sin\langle z, x\rangle - \langle z, x\rangle 1_{\{|x|\le 1\}})(T_b\nu)(\mathrm{d}x) + b\langle z, \theta\rangle = b|z|\eta_2(z) + b\langle z, \theta\rangle$$

with $\theta = \int_{1/b<|x|\le 1} x\nu(\mathrm{d}x) = \int_{1<|x|\le b} x\nu(\mathrm{d}x)$. $\square$

Stable distributions have another representation of their characteristic functions.

THEOREM 14.10. *Let $0 < \alpha < 2$. If μ is α-stable and non-trivial on $\mathbb{R}^d$, then there are a finite non-zero measure λ_1 on S and τ in $\mathbb{R}^d$ such that*

$$\widehat{\mu}(z) = \exp\left[-\int_S |\langle z, \xi\rangle|^\alpha \left(1 - \mathrm{i}\tan\frac{\pi\alpha}{2}\,\mathrm{sgn}\langle z, \xi\rangle\right)\lambda_1(\mathrm{d}\xi) + \mathrm{i}\langle\tau, z\rangle\right] \quad \textit{for } \alpha \ne 1, \tag{14.15}$$

$$\widehat{\mu}(z) = \exp\left[-\int_S \left(|\langle z,\xi\rangle| + \mathrm{i}\frac{2}{\pi}\langle z,\xi\rangle \log|\langle z,\xi\rangle|\right)\lambda_1(\mathrm{d}\xi) + \mathrm{i}\langle\tau, z\rangle\right] \quad \textit{for } \alpha = 1. \tag{14.16}$$

The measure λ_1 and the vector τ are uniquely determined by μ. Conversely, for any finite non-zero measure λ_1 on S and any $\tau \in \mathbb{R}^d$, the right-hand side of (14.15) or (14.16) is the characteristic function of a non-trivial α-stable distribution. If $0 < \alpha < 1$ or $1 < \alpha < 2$, then a necessary and sufficient condition for a non-trivial, α-stable distribution μ to be strictly α-stable is that $\tau = 0$. A necessary and sufficient condition for a non-trivial 1-stable distribution μ to be strictly 1-stable is that

$$\int_S \xi\lambda_1(\mathrm{d}\xi) = 0. \tag{14.17}$$

The Γ-function $\Gamma(s)$ is extended from $(0,\infty)$ to any $s \in \mathbb{R}$ with $s \neq 0, -1, -2, \ldots$ by $\Gamma(s+1) = s\Gamma(s)$.

LEMMA 14.11.

$$\int_0^\infty (\mathrm{e}^{\mathrm{i}r} - 1)r^{-1-\alpha}\,\mathrm{d}r = \Gamma(-\alpha)\mathrm{e}^{-\mathrm{i}\pi\alpha/2} \quad \textit{for } 0 < \alpha < 1, \tag{14.18}$$

$$\int_0^\infty (\mathrm{e}^{\mathrm{i}r} - 1 - \mathrm{i}r)r^{-1-\alpha}\,\mathrm{d}r = \Gamma(-\alpha)\mathrm{e}^{-\mathrm{i}\pi\alpha/2} \quad \textit{for } 1 < \alpha < 2, \tag{14.19}$$

$$\int_0^\infty (\mathrm{e}^{\mathrm{i}zr} - 1 - \mathrm{i}zr1_{(0,1]}(r))r^{-2}\,\mathrm{d}r = -\frac{\pi z}{2} - \mathrm{i}z\log z + \mathrm{i}cz \tag{14.20}$$

for $z > 0$ with

$$c = \int_1^\infty r^{-2}\sin r\,\mathrm{d}r + \int_0^1 r^{-2}(\sin r - r)\,\mathrm{d}r. \tag{14.21}$$

Proof. Let $0 < \alpha < 1$. As in Example 8.11 we have

$$\textstyle\int_0^\infty (\mathrm{e}^{-ur} - 1)r^{-1-\alpha}\mathrm{d}r = -\alpha^{-1}\Gamma(1-\alpha)u^\alpha = \Gamma(-\alpha)u^\alpha.$$

It follows that

$$\textstyle\int_0^\infty (\mathrm{e}^{wr} - 1)r^{-1-\alpha}\mathrm{d}r = \Gamma(-\alpha)(-w)^\alpha$$

for any complex number $w \neq 0$ with $\operatorname{Re} w \leq 0$. Indeed, both sides are regular on $\{w\colon \operatorname{Re} w < 0\}$, continuous on $\{w\colon \operatorname{Re} w \leq 0, w \neq 0\}$, and identical for any negative real w. The branch in the right-hand side is chosen as $(-w)^\alpha = |w|^\alpha \mathrm{e}^{\mathrm{i}\alpha\arg(-w)}$ with $\arg(-w) \in (-\pi, \pi]$. Now, letting $w = \mathrm{i}$, we get (14.18).

If $1 < \alpha < 2$, then integration by parts and (14.18) give (14.19) as follows:

$$\begin{aligned}\textstyle\int_0^\infty (\mathrm{e}^{\mathrm{i}r} - 1 - \mathrm{i}r)r^{-1-\alpha}\,\mathrm{d}r &= -\alpha^{-1}\textstyle\int_0^\infty (\mathrm{e}^{\mathrm{i}r} - 1 - \mathrm{i}r)\mathrm{d}(r^{-\alpha})\\ &= \alpha^{-1}\textstyle\int_0^\infty r^{-\alpha}\mathrm{i}(\mathrm{e}^{\mathrm{i}r} - 1)\mathrm{d}r = \mathrm{i}\alpha^{-1}\Gamma(1-\alpha)\mathrm{e}^{-\mathrm{i}\pi(\alpha-1)/2} = \Gamma(-\alpha)\mathrm{e}^{-\mathrm{i}\pi\alpha/2}.\end{aligned}$$

Using $\int_0^\infty r^{-2}(1-\cos r)\mathrm{d}r = \pi/2$ (see solution to Exercise 6.12), the left-hand side of (14.20) is as follows:

$$\begin{aligned}
&\textstyle\int_0^\infty r^{-2}(\cos zr - 1)\mathrm{d}r + \mathrm{i}\int_0^1 r^{-2}(\sin zr - zr)\mathrm{d}r + \mathrm{i}\int_1^\infty r^{-2}\sin zr\,\mathrm{d}r \\
&\quad = z\textstyle\int_0^\infty r^{-2}(\cos r - 1)\mathrm{d}r + \mathrm{i}\int_0^{1/z} r^{-2}(\sin zr - zr)\mathrm{d}r \\
&\quad\quad + \mathrm{i}\textstyle\int_{1/z}^\infty r^{-2}\sin zr\,\mathrm{d}r - \mathrm{i}\int_{1/z}^1 r^{-2} zr\mathrm{d}r \\
&\quad = -\frac{\pi z}{2} + \mathrm{i}z\Big(\textstyle\int_0^1 r^{-2}(\sin r - r)\mathrm{d}r + \int_1^\infty r^{-2}\sin r\,\mathrm{d}r - \log z\Big).
\end{aligned}$$

This is the equality (14.20). □

Proof of Theorem 14.10. We write the inner product simply as $\langle z,\xi\rangle = z\xi$. If $0<\alpha<1$, then note that (14.18) and its complex conjugate

$$\int_0^\infty (\mathrm{e}^{-\mathrm{i}r} - 1)r^{-1-\alpha}\,\mathrm{d}r = \Gamma(-\alpha)\mathrm{e}^{\mathrm{i}\pi\alpha/2}$$

yield

$$\begin{aligned}
\int_0^\infty (\mathrm{e}^{\mathrm{i}rz\xi} - 1)r^{-1-\alpha}\,\mathrm{d}r &= |z\xi|^\alpha\Gamma(-\alpha)\exp\Big[-\mathrm{i}\frac{\pi\alpha}{2}\operatorname{sgn}(z\xi)\Big] \\
&= \Gamma(-\alpha)\Big(\cos\frac{\pi\alpha}{2}\Big)|z\xi|^\alpha\Big(1 - \mathrm{i}\tan\frac{\pi\alpha}{2}\operatorname{sgn}(z\xi)\Big),
\end{aligned}$$

and rewrite (14.8). We get (14.15) if we define λ_1 as a constant multiple of λ. If $1<\alpha<2$, then, using (14.9) and (14.19) in place of (14.8) and (14.18), we get (14.15) similarly. If $\alpha=1$, then (14.20) and its complex conjugate give

$$\int_0^\infty (\mathrm{e}^{\mathrm{i}rz\xi} - 1 - \mathrm{i}rz\xi 1_{(0,1]}(r))r^{-2}\,dr = -\frac{\pi}{2}|z\xi| - \mathrm{i}z\xi\log|z\xi| + \mathrm{i}cz\xi$$

(with the convention that $z\xi\log|z\xi| = 0$ if $z\xi = 0$) and hence we get (14.16) from (14.10), defining λ_1 as a constant multiple of λ and choosing τ appropriately. The uniqueness of λ_1 and τ and the converse assertion come from Remark 14.4. If $0<\alpha<1$, then τ equals the drift γ_0. If $1<\alpha<2$, then τ is the center γ_1. Hence, in these cases, the condition for strict stability is that $\tau=0$, as can be seen from Theorem 14.7. In the case $\alpha=1$, the condition (14.12) for strict stability is equivalent to (14.17). □

DEFINITION 14.12. A measure ρ on $\mathbb{R}^d$ is *symmetric* if $\rho(B)=\rho(-B)$ for $B\in\mathcal{B}(\mathbb{R}^d)$. It is *rotation invariant* if $\rho(B)=\rho(U^{-1}B)$ for every orthogonal matrix U, where $U^{-1}B = \{U^{-1}x\colon x\in B\}$. (If $d=1$, then rotation invariance is tantamount to symmetry.) A stochastic process $\{X_t\}$ on $\mathbb{R}^d$ is symmetric or rotation invariant, respectively, if $\{X_t\} \overset{\mathrm{d}}{=} \{-X_t\}$ or if $\{X_t\} \overset{\mathrm{d}}{=} \{UX_t\}$ for every orthogonal matrix U.

THEOREM 14.13. *Let $0 < \alpha < 2$. If μ is symmetric and α-stable on $\mathbb{R}^d$, then*

$$\widehat{\mu}(z) = \exp\left[-\int_S |\langle z, \xi\rangle|^\alpha \lambda_1(\mathrm{d}\xi)\right] \tag{14.22}$$

with a symmetric finite non-zero measure λ_1 on S. The measure λ_1 is uniquely determined by μ. For any symmetric finite non-zero measure λ_1 on S, there is a symmetric α-stable distribution μ satisfying (14.22).

Proof. Let $\widetilde{\mu}$ be the dual of μ, that is, $\widehat{\widetilde{\mu}}(z) = \widehat{\mu}(-z)$. If μ is symmetric α-stable, then $\mu = \widetilde{\mu}$ and Theorem 14.10 tells us that $\lambda_1 = T_{-1}\lambda_1$ and $\tau = -\tau$. Hence (14.15) and (14.16) reduce to (14.22) in this case. The uniqueness of λ_1 and the existence of μ for any given λ_1 are included in Theorem 14.10. □.

THEOREM 14.14. *A non-trivial probability measure μ on $\mathbb{R}^d$ is rotation invariant and α-stable with $0 < \alpha \le 2$ if and only if*

$$\widehat{\mu}(z) = \mathrm{e}^{-c|z|^\alpha} \tag{14.23}$$

with $c > 0$. An α-stable distribution μ with $0 < \alpha < 2$ is rotation invariant if and only if the measure λ_1 in Theorem 14.10 is uniform on S and $\tau = 0$. Rotation invariant 1-stable distributions on $\mathbb{R}^d$ are Cauchy.

Proof. Suppose that μ is rotation invariant and α-stable with $0 < \alpha < 2$. It is symmetric, and hence (14.22) holds, that is,

$$\widehat{\mu}(z) = \exp\left[-|z|^\alpha \int_S |\langle |z|^{-1}z, \xi\rangle|^\alpha \lambda_1(\mathrm{d}\xi)\right].$$

Since $\widehat{\mu}(z) = \widehat{\mu}(Uz)$ for any orthogonal U, $\int |\langle \zeta, \xi\rangle|^\alpha \lambda_1(\mathrm{d}\xi)$ is constant in $\zeta \in S$. Hence (14.23). It also follows that $\nu = \nu U^{-1}$ by Proposition 11.10 and that λ_1 is uniform on S, since λ_1 is the spherical part of ν. The converse is obvious.

If μ is rotation invariant and 2-stable, then Theorem 14.1 combined with $\widehat{\mu}(z) = \widehat{\mu}(Uz)$ for any U leads to (14.23) with $\alpha = 2$.

The probability measure μ satisfying (14.23) with $\alpha = 1$ is Cauchy with $\gamma = 0$ as in Example 2.12. □

Let us discuss the one-dimensional case, $d = 1$. First we rewrite Theorem 14.10.

THEOREM 14.15. *Let $d = 1$ and $0 < \alpha < 2$. If μ is non-trivial and α-stable, then*

$$\widehat{\mu}(z) = \exp\left[-c|z|^\alpha\left(1 - \mathrm{i}\beta \tan\frac{\pi\alpha}{2}\operatorname{sgn} z\right) + \mathrm{i}\tau z\right] \quad \textit{for } \alpha \neq 1, \tag{14.24}$$

$$\widehat{\mu}(z) = \exp\left[-c|z|\left(1 + \mathrm{i}\beta\frac{2}{\pi}(\operatorname{sgn} z)\log|z|\right) + \mathrm{i}\tau z\right] \quad \textit{for } \alpha = 1 \tag{14.25}$$

with $c > 0$, $\beta \in [-1, 1]$, *and* $\tau \in \mathbb{R}$. *Here* c, β, *and* τ *are uniquely determined by* μ. *Conversely, for every* $c > 0$, $\beta \in [-1, 1]$, *and* $\tau \in \mathbb{R}$, *there is a non-trivial* α*-stable distribution* μ *satisfying (14.24) or (14.25). A necessary and sufficient condition for a non-trivial* α*-stable distribution* μ *to be strictly* α*-stable is that* $\tau = 0$ *or that* $\beta = 0$, *according as* $\alpha \neq 1$ *or* $\alpha = 1$.

Proof. We have $S = \{1, -1\}$. Let $\lambda_1\{1\} = c_1$ and $\lambda_1\{-1\} = c_2$ for the measure λ_1 in Theorem 14.10 and set $c = c_1 + c_2$ and $\beta = (c_1 - c_2)/(c_1 + c_2)$. Then (14.15) and (14.16) become (14.24) and (14.25). Thus the assertions are obtained from Theorem 14.10. □

DEFINITION 14.16. Let $d = 1$ and $0 < \alpha < 2$. A non-trivial α-stable process $\{X_t\}$ with $\mu = P_{X_1}$ satisfying (14.24) or (14.25) is called a *stable process on* $\mathbb{R}$ *with parameters* (α, β, τ, c). The parameter τ is identical with the drift γ_0 if $0 < \alpha < 1$, and with the center γ_1 if $1 < \alpha < 2$. The parameter β represents non-symmetry of the Lévy measure ν. That is, ν is symmetric if and only if $\beta = 0$; concentrated on $(0, \infty)$ if and only if $\beta = 1$; concentrated on $(-\infty, 0)$ if and only if $\beta = -1$. (Many different definitions of the parameters of α-stable processes on $\mathbb{R}$ are found in books and papers. The reader should be careful.)

It is shown in the above that a 1-stable distribution on $\mathbb{R}$ is strictly 1-stable if and only if its Lévy measure is symmetric. But, on $\mathbb{R}^d$ with $d \geq 2$, there are strictly 1-stable distributions with non-symmetric Lévy measures (see Theorem 14.7).

EXAMPLE 14.17. The Cauchy process on $\mathbb{R}$ that corresponds to μ of Example 2.11 is strictly 1-stable, having parameters $(1, 0, \tau, c)$. The process corresponding to the distribution of Example 2.13 is strictly $\frac{1}{2}$-stable with parameters $(\frac{1}{2}, 1, 0, c)$. Using this, it is easy to write down the distribution densities in the cases $(\frac{1}{2}, 1, \tau, c)$ and $(\frac{1}{2}, -1, \tau, c)$. We do not know any other non-Gaussian stable processes on $\mathbb{R}$ whose distribution densities are expressible by elementary functions.

REMARK 14.18. If μ is non-trivial and stable on $\mathbb{R}$, then it has a continuous density by Proposition 2.5(xii), since $|\widehat{\mu}(z)| = \mathrm{e}^{-c|z|^\alpha}$ with $c > 0$. Let $\{X_t\}$ be a stable process on $\mathbb{R}$ with parameters (α, β, τ, c), $0 < \alpha < 2$. Let $X_t^0 = X_t - t\tau$. Let $p(t, x)$ and $p^0(t, x)$ be the continuous densities of the distributions of X_t and X_t^0, respectively, for $t > 0$. Then,

$$p(t,x) = t^{-1/\alpha} p(1, t^{-1/\alpha} x + (1 - t^{(\alpha-1)/\alpha})\tau) \quad \text{for } \alpha \neq 1, \tag{14.26}$$

$$p(t,x) = t^{-1} p(1, t^{-1} x - 2\pi^{-1} c\beta \log t) \quad \text{for } \alpha = 1, \tag{14.27}$$

and it follows that

$$p(t,x) = t^{-1/\alpha} p^0(1, t^{-1/\alpha}(x - \tau t)) \quad \text{for } \alpha \neq 1, \tag{14.28}$$

$$p(t,x) = t^{-1} p^0(1, t^{-1}(x - \tau t) - 2\pi^{-1} c\beta \log t) \quad \text{for } \alpha = 1. \tag{14.29}$$

In the case where $\alpha < 1$ and $\beta = 1$, $p^0(t,x) > 0$ if and only if $x > 0$. In the case where $\alpha < 1$ and $\beta = -1$, $p^0(t,x) > 0$ if and only if $x < 0$. Except in these cases, $p^0(t,x)$ is positive on $\mathbb{R}$. These positivity results will be seen from Section 24 in the almost everywhere sense and from the unimodality result in Section 53 in the everywhere sense. See also Remark 28.8. The behavior of $p(t,x)$ as $t \to \infty$ is important in limit theorems for stable processes. It is obtained from the behavior of $p^0(1,x)$ as $x \to \pm\infty$ or $x \to 0$. For instance, if $0 < \alpha < 1$, then (14.28) shows that $p(t,x)t^{1/\alpha} \to p^0(1,0)$ as $t \to \infty$. The asymptotic expansions of $p^0(1,x)$ are obtained by Linnik [**323**], Skorohod [**489**], and others. We give, without proofs, the results (with misprints corrected and with some formal changes) in Zolotarev [**608**]. We can fix the parameter c without loss of generality. Assume that c equals $\cos(\frac{\pi\beta\alpha}{2})$, $\frac{\pi}{2}$, or $\cos(\pi\beta\frac{2-\alpha}{2})$ for $\alpha < 1$, $= 1$, or > 1, respectively. Let $\alpha' = 1/\alpha$. Let $\rho = (1+\beta)/2$ or $= (1-\beta\frac{2-\alpha}{\alpha})/2$, according as $\alpha < 1$ or > 1.

The following (i)–(iii) are representations of $p^0(1,x)$ by convergent power series.

(i) If $\alpha > 1$, then

$$p^0(1,x) = \frac{1}{\pi}\sum_{n=1}^{\infty}(-1)^{n-1}\frac{\Gamma(n\alpha'+1)}{n!}(\sin \pi n\rho)x^{n-1} \quad \text{for } x \in \mathbb{R}. \tag{14.30}$$

(ii) If $\alpha < 1$, then

$$p^0(1,x) = \frac{1}{\pi}\sum_{n=1}^{\infty}(-1)^{n-1}\frac{\Gamma(n\alpha+1)}{n!}(\sin \pi n\rho\alpha)x^{-n\alpha-1} \quad \text{for } x > 0. \tag{14.31}$$

(iii) If $\alpha = 1$ and $\beta > 0$, then

$$p^0(1,x) = \frac{1}{\pi}\sum_{n=1}^{\infty}(-1)^{n-1}nb_n x^{n-1} \quad \text{for } x \in \mathbb{R}, \tag{14.32}$$

where

$$b_n = \frac{1}{n!}\int_0^{\infty}\exp(-\beta u\log u)u^{n-1}\sin\Big(\frac{\pi}{2}(1+\beta)u\Big)du.$$

The following (iv)–(vii) hold for any positive integer N.

(iv) When $\alpha < 1$, $\beta \neq 1$, $x \in \mathbb{R}$, and $x \to 0$,

$$p^0(1,x) = \frac{1}{\pi}\sum_{n=1}^{N}(-1)^{n-1}\frac{\Gamma(n\alpha'+1)}{n!}(\sin \pi n\rho)x^{n-1} + O(x^N). \tag{14.33}$$

(v) When $\alpha > 1$, $\beta \neq -1$, and $x \to \infty$,

$$p^0(1,x) = \frac{1}{\pi}\sum_{n=1}^{N}(-1)^{n-1}\frac{\Gamma(n\alpha+1)}{n!}(\sin \pi n\rho\alpha)x^{-n\alpha-1} + O(x^{-(N+1)\alpha-1}). \tag{14.34}$$

(vi) Either when $\alpha < 1$, $\beta = 1$, $x > 0, x \to 0$ or when $\alpha \geq 1$, $\beta = -1$, $x \to \infty$,

$$p^0(1,x) = \frac{K}{\sqrt{2\pi\alpha}}\xi^{(2-\alpha)/(2\alpha)}e^{-\xi}\Big[1 + \sum_{n=1}^{N}Q_n(\alpha_*)(\alpha_*\xi)^{-n} + O(\xi^{-N-1})\Big], \tag{14.35}$$

where $\xi = |1-\alpha|(x/\alpha)^{\alpha/(\alpha-1)}$ (if $\alpha \neq 1$), $\xi = e^{x-1}$ (if $\alpha = 1$), $K = |1-\alpha|^{-1/\alpha}$ (if $\alpha \neq 1$), $K = 1$ (if $\alpha = 1$), $\alpha_* = \alpha \wedge (1/\alpha)$, $Q_1(\alpha_*) = -\frac{1}{12}(2 + \alpha_* + 2\alpha_*^2)$, and, in general, $Q_n(\alpha_*)$ is a polynomial of degree $2n$ in the variable α_*.

(vii) When $\alpha = 1$, $\beta \neq -1$, and $x \to \infty$,

$$p^0(1,x) = \frac{1}{\pi}\sum_{n=1}^{N} \frac{P_n(\log x)}{n!} x^{-n-1} + O(x^{-N-2}(\log x)^N), \tag{14.36}$$

where $P_1(\log x) = \frac{\pi}{2}(1+\beta)$ and, in general, $P_n(\log x) = \sum_{l=0}^{n-1} r_{ln}(\log x)^l$ with

$$r_{ln} = \sum_{m=l}^{n-1} \binom{n}{m}\binom{m}{l}(-1)^{m-l}\Gamma^{(m-l)}(n+1)\beta^m\big(\tfrac{\pi}{2}(1+\beta)\big)^{n-m} \sin\tfrac{\pi(n-m)}{2}.$$

Here $\Gamma^{(m-l)}$ denotes the $(m-l)$th derivative of the Γ-function.

The simplest special cases are that, as $x \to \infty$,

$$p^0(1,x) \sim \begin{cases} \frac{1}{\pi}\Gamma(\alpha+1)(\sin \pi\rho\alpha)x^{-\alpha-1} & \text{if } \alpha \neq 1,\ \beta \neq -1, \\ \frac{1+\beta}{2}x^{-2} & \text{if } \alpha = 1,\ \beta \neq -1, \\ \frac{K}{\sqrt{2\pi\alpha}}\xi^{(2-\alpha)/(2\alpha)}\mathrm{e}^{-\xi} & \text{if } \alpha \geq 1,\ \beta = -1. \end{cases} \tag{14.37}$$

Notice that $\sin \pi\rho\alpha > 0$ if $\alpha \neq 1$ and $\beta \neq -1$.

The following representation in Zolotarev [**608**] of strictly stable distributions is sometimes useful. It does not need special treatment of the cases $\alpha = 1$ and 2.

THEOREM 14.19. *Let $0 < \alpha \leq 2$. If μ is a strictly α-stable distribution on $\mathbb{R}$, then*

$$\widehat{\mu}(z) = \exp(-c_1|z|^\alpha \mathrm{e}^{-\mathrm{i}(\pi/2)\theta\alpha \operatorname{sgn} z}), \tag{14.38}$$

where $c_1 > 0$ and $\theta \in \mathbb{R}$ with $|\theta| \leq (\frac{2-\alpha}{\alpha}) \wedge 1$. The parameters c_1 and θ are uniquely determined by μ. Conversely, for any c_1 and θ, there is a strictly α-stable distribution μ satisfying (14.38).

REMARK 14.20. Theorem 14.19 includes the case of δ-distributions other than δ_0. See Remark 13.17. In the case $\alpha \neq 2$, the relationship of the two representations of a strictly α-stable distribution in Theorems 14.15 and 14.19 is as follows:

$$c_1 = \begin{cases} c(1+\beta^2(\tan\frac{\pi\alpha}{2})^2)^{1/2}, & \alpha \neq 1, \\ (c^2+\tau^2)^{1/2}, & \alpha = 1, \end{cases}$$

$$\theta = \begin{cases} \frac{2}{\pi\alpha}\arctan(\beta\tan\frac{\pi\alpha}{2}), & \alpha < 1, \\ -\frac{2}{\pi\alpha}\arctan(\beta\tan\frac{\pi(2-\alpha)}{2}), & \alpha > 1, \\ \frac{2}{\pi}\arctan\frac{\tau}{c}, & \alpha = 1, c \neq 0, \\ \operatorname{sgn}\tau, & \alpha = 1, c = 0, \end{cases}$$

where arctan is the value in $(-\frac{\pi}{2}, \frac{\pi}{2})$. When β increases from -1 to 1, the parameter θ increases from -1 to 1 if $\alpha < 1$, and decreases from $\frac{2-\alpha}{\alpha}$ to $-\frac{2-\alpha}{\alpha}$ if $1 < \alpha < 2$. In the case $\alpha = 2$, θ must be 0. Even if μ is trivial and $\mu \neq \delta_0$, the expressions above for c_1 and θ are valid with the understanding that $\widehat{\mu}(z) = \exp(\mathrm{i}\tau z)$ and $c = 0$.

Proof of Theorem 14.19. Suppose that μ is strictly α-stable with $\alpha \neq 2$. Define c_1 and θ by the formulas in the remark above, and write

$$-c|z|^\alpha(1 - \mathrm{i}\beta \tan \tfrac{\pi\alpha}{2} \operatorname{sgn} z) = -c_1|z|^\alpha(\tfrac{c}{c_1} - \mathrm{i}\beta\tfrac{c}{c_1} \tan \tfrac{\pi\alpha}{2} \operatorname{sgn} z)$$

for $\alpha \neq 1$ and

$$-c|z| + \mathrm{i}\tau z = -c_1|z|(\tfrac{c}{c_1} - \mathrm{i}\tau\tfrac{1}{c_1} \operatorname{sgn} z)$$

for $\alpha = 1$. Then, using trigonometric manipulation, we get (14.38). Conversely, (14.38) leads to (14.24) or (14.25), and the parameters correspond in a unique way. The case $\alpha = 2$ is obvious. □

REMARK 14.21. Let us denote the parameters in (14.38) by $(\alpha, \theta, c_1)_Z$ and the continuous density of μ by $p(x, (\alpha, \theta, c_1)_Z)$, assuming non-triviality. The relations

$$p(x, (\alpha, \theta, c_1)_Z) = {c_1}^{-1/\alpha} p({c_1}^{-1/\alpha}x, (\alpha, \theta, 1)_Z), \tag{14.39}$$

$$p(x, (\alpha, \theta, c_1)_Z) = p(-x, (\alpha, -\theta, c_1)_Z) \tag{14.40}$$

are evident. Let $1 \leq \alpha \leq 2$. Zolotarev [**601**] finds a duality between strictly stable distributions on $\mathbb{R}$ with indices α and $\alpha' = 1/\alpha$:

$$p(x, (\alpha, \theta, 1)_Z) = x^{-1-\alpha} p(x^{-\alpha}, (\alpha', \theta', 1)_Z) \quad \text{for } x > 0, \tag{14.41}$$

where $1 + \theta' = \alpha(1 + \theta)$. Feller [**139**], p. 583, points out that this duality follows from (14.30) and (14.31). As θ increases from $-\frac{2-\alpha}{\alpha}$ to $\frac{2-\alpha}{\alpha}$, θ' increases from $2\alpha - 3$ to 1. When $\theta = 0$ (symmetric), $\theta' = \alpha - 1$. Combining (14.40) and (14.41), we get

$$p(x, (\alpha, \theta, 1)_Z) = |x|^{-1-\alpha} p(-|x|^{-\alpha}, (\alpha', \theta'', 1)_Z) \quad \text{for } x < 0, \tag{14.42}$$

where $1 - \theta'' = \alpha(1 - \theta)$. If $\alpha = 2$ and $\theta = 0$, then (14.41) and (14.42) are the relation of Gaussian to one-sided $\frac{1}{2}$-stable (cf. Exercise 6.10). The positive tail of the density with $1 < \alpha < 2$ and $\beta = -1$ is expressed by the behavior near 0 of the density with $\alpha' = 1/\alpha$ and $\beta' = 1$, because $\theta = \frac{2-\alpha}{\alpha}$ and $\theta' = 1$.

15. Selfdecomposable and semi-selfdecomposable distributions

An extension of stable distributions in a direction different from semi-stable is made by selfdecomposable distributions. Further, semi-self-decomposable distributions generalize both semi-stable and selfdecomposable distributions.

DEFINITION 15.1. Let μ be a probability measure on $\mathbb{R}^d$. It is called *selfdecomposable*, or *of class L*, if, for any $b > 1$, there is a probability measure ρ_b on $\mathbb{R}^d$ such that

$$\widehat{\mu}(z) = \widehat{\mu}(b^{-1}z)\widehat{\rho}_b(z). \tag{15.1}$$

It is called *semi-selfdecomposable* if there are some $b > 1$ and some infinitely divisible probability measure ρ_b satisfying (15.1). If μ is semi-selfdecomposable, then b in the definition is called a *span* of μ. (Remarks concerning the definition will be given in Proposition 15.5 and Exercise 18.14.)

EXAMPLE 15.2. Any stable distribution on $\mathbb{R}^d$ is selfdecomposable. Any semi-stable distribution on $\mathbb{R}^d$ with b as a span is semi-selfdecomposable with b as a span.

To prove this, let μ be non-trivial and α-stable, as trivial distributions are evidently selfdecomposable. For any $a > 0$, there is c such that $\widehat{\mu}(z)^a = \widehat{\mu}(a^{1/\alpha}z)e^{i\langle c,z\rangle}$. Given $b > 1$, let $a = b^\alpha$ and notice that

$$\widehat{\mu}(b^{-1}z)\widehat{\mu}(b^{-1}z)^{a-1} = \widehat{\mu}(b^{-1}z)^a = \widehat{\mu}(z)e^{i\langle c,b^{-1}z\rangle}.$$

It follows that $\widehat{\mu}$ satisfies (15.1) with $\widehat{\rho}_b(z) = \widehat{\mu}(b^{-1}z)^{a-1}e^{-i\langle b^{-1}c,z\rangle}$. Hence μ is selfdecomposable. Proof for semi-stable distributions is similar.

The class of selfdecomposable distributions is comprehended as a class of limit distributions described below.

THEOREM 15.3. (i) *Let $\{Z_n\colon n = 1, 2, \dots\}$ be independent random variables on $\mathbb{R}^d$ and $S_n = \sum_{k=1}^n Z_k$. Let μ be a probability measure on $\mathbb{R}^d$. Suppose that there are $b_n > 0$ and $c_n \in \mathbb{R}^d$ for $n = 1, 2, \dots$ such that*

$$P_{b_n S_n + c_n} \to \mu \quad \textit{as } n \to \infty \tag{15.2}$$

and that

$$\{b_n Z_k\colon k = 1, \dots, n;\ n = 1, 2, \dots\} \textit{ is a null array.} \tag{15.3}$$

Then, μ is selfdecomposable.

(ii) *For any selfdecomposable distribution μ on $\mathbb{R}^d$ we can find $\{Z_n\}$ independent, $b_n > 0$, and $c_n \in \mathbb{R}^d$ satisfying (15.2) and (15.3).*

An analogous characterization of semi-selfdecomposable distributions as limit distributions of a certain kind of subsequences of $\{S_n\}$ is possible (Maejima and Naito [**332**]).

LEMMA 15.4. *Suppose that μ is non-trivial. If $\{Z_n\}$ independent, $b_n > 0$, and $c_n \in \mathbb{R}^d$ satisfy (15.2) and (15.3), then $b_n \to 0$ and $b_{n+1}/b_n \to 1$ as $n \to \infty$.*

Proof. The condition (15.3) says that, for any $\varepsilon > 0$,

$$\max_{1\le k\le n} P[\, b_n|Z_k| > \varepsilon\,] \to 0.$$

Suppose that some subsequence $\{b_{n_l}\}$ of $\{b_n\}$ tends to a non-zero b. Then it follows that, for any k, $P[\,|Z_k| > b^{-1}\varepsilon\,] = 0$. Hence $Z_k = 0$ a.s. Therefore μ is trivial, contrary to the assumption. Hence $b_n \to 0$. Let $W_n = b_n S_n + c_n$ and $W_n' = b_{n+1}S_n + c_{n+1}$. We have $P_{W_n} \to \mu$. Since $W_{n+1} = W_n' + b_{n+1}Z_{n+1}$ and $b_{n+1}Z_{n+1} \to 0$ in prob., we have $\widehat{P}_{W_n'}(z) \to \widehat{\mu}(z)$ for any z. Since

$$W_n' = b_{n+1}b_n^{-1}W_n + c_n' \quad \text{with } c_n' = c_{n+1} - b_{n+1}b_n^{-1}c_n,$$

application of Lemma 13.10 gives that $b_{n+1}b_n^{-1} \to b$ and $c'_n \to c$ for some $b > 0$ and $c \in \mathbb{R}^d$ and that $\widehat{\mu}(z) = \widehat{\mu}(bz)e^{i\langle c,z\rangle}$. Now, by Lemma 13.7, $b = 1$ and $c = 0$. □

Proof of Theorem 15.3. (i) Assume that μ is non-trivial. By Lemma 15.4, $b_n \to 0$ and $b_{n+1}/b_n \to 1$. For any $b > 1$ we can find sequences $\{n_l\}$ and $\{m_l\}$ of positive integers going to infinity such that

$$m_l < n_l \quad \text{and} \quad b_{m_l}b_{n_l}^{-1} \to b. \tag{15.4}$$

In fact, let l_0 be so big that $1/l_0 < \log b$. Choose m_l for $l \geq l_0$ in such a way that $|-\log b_{n+1} + \log b_n| < 1/l$ for all $n \geq m_l$. Then, noting that $-\log b_n \to \infty$, we can choose $n_l > m_l$ such that $|-\log b_{n_l} + \log b_{m_l} - \log b| < 1/l$. Then (15.4) is satisfied. Let

$$W_n = b_n S_n + c_n,$$

$$U_l = b_{n_l} \sum_{k=1}^{m_l} Z_k + b_{n_l} b_{m_l}^{-1} c_{m_l},$$

$$V_l = b_{n_l} \sum_{k=m_l+1}^{n_l} Z_k + c_{n_l} - b_{n_l} b_{m_l}^{-1} c_{m_l}.$$

Then $W_{n_l} = U_l + V_l$ and

$$\widehat{P}_{W_{n_l}}(z) = \widehat{P}_{U_l}(z)\widehat{P}_{V_l}(z) \tag{15.5}$$

by the independence. Since $U_l = b_{n_l} b_{m_l}^{-1} W_{m_l}$,

$$\begin{aligned} |\widehat{P}_{U_l}(z) - \widehat{\mu}(b_{n_l} b_{m_l}^{-1} z)| &= |\widehat{P}_{W_{m_l}}(b_{n_l} b_{m_l}^{-1} z) - \widehat{\mu}(b_{n_l} b_{m_l}^{-1} z)| \\ &\leq \sup_{|w| \leq |z|} |\widehat{P}_{W_{m_l}}(w) - \widehat{\mu}(w)| \to 0 \quad \text{as } l \to \infty \end{aligned}$$

by (15.2). Hence $\widehat{P}_{U_l}(z) \to \widehat{\mu}(b^{-1}z)$. By Theorem 9.3 the conditions (15.2) and (15.3) make μ infinitely divisible. Thus $\widehat{\mu}(z)$ does not have zeros by Lemma 7.5 and we get from (15.5)

$$\widehat{P}_{V_l}(z) \to \widehat{\mu}(z)/\widehat{\mu}(b^{-1}z) \quad \text{as } l \to \infty.$$

Since the limit is continuous, it is the characteristic function of a probability measure ρ_b by Proposition 2.5(viii). Now we have (15.1).

(ii) Let μ be selfdecomposable on $\mathbb{R}^d$. Then $\widehat{\mu}(z)$ has no zero. Indeed, suppose that it has a zero. Then there is $z_0 \in \mathbb{R}^d$ such that $\widehat{\mu}(z_0) = 0$ and $\widehat{\mu}(z) \neq 0$ for $|z| < |z_0|$. Hence $\widehat{\rho}_b(z_0) = 0$ for any $b > 1$ from (15.1) and, therefore, by the inequality in Exercise 6.11,

$$1 = \operatorname{Re}(1 - \widehat{\rho}_b(z_0)) \leq 4 \operatorname{Re}(1 - \widehat{\rho}_b(2^{-1}z_0)) = 4 \operatorname{Re}\left(1 - \frac{\widehat{\mu}(2^{-1}z_0)}{\widehat{\mu}(2^{-1}b^{-1}z_0)}\right),$$

which is absurd, because the last member tends to 0 as $b \downarrow 1$. This proves that $\widehat{\mu}(z)$ has no zero. Let $Z_1, Z_2, \ldots$ be independent random variables on $\mathbb{R}^d$ such that

$$E[\mathrm{e}^{\mathrm{i}\langle z, Z_n\rangle}] = \widehat{\rho}_{(n+1)/n}((n+1)z) = \frac{\widehat{\mu}((n+1)z)}{\widehat{\mu}(nz)}.$$

Then

$$E[\mathrm{e}^{\mathrm{i}\langle z, n^{-1}S_n\rangle}] = \prod_{k=1}^{n} E[\mathrm{e}^{\mathrm{i}\langle n^{-1}z, Z_k\rangle}] = \prod_{k=1}^{n} \frac{\widehat{\mu}(\frac{k+1}{n}z)}{\widehat{\mu}(\frac{k}{n}z)} = \frac{\widehat{\mu}(\frac{n+1}{n}z)}{\widehat{\mu}(\frac{1}{n}z)} \to \widehat{\mu}(z)$$

as $n \to \infty$. We have

$$\max_{1\le k\le n} |E[\mathrm{e}^{\mathrm{i}\langle z, n^{-1}Z_k\rangle}] - 1| = \max_{1\le k\le n} \left| \frac{\widehat{\mu}(\frac{k+1}{n}z)}{\widehat{\mu}(\frac{k}{n}z)} - 1 \right| \to 0 \quad \text{as } n \to \infty,$$

since $\widehat{\mu}(z)$ is continuous. Therefore $\{n^{-1}Z_k \colon k = 1, \ldots, n;\ n = 1, 2, \ldots\}$ is a null array by Exercise 12.12. Hence (15.2) and (15.3) hold with $b_n = n^{-1}$ and $c_n = 0$. □

PROPOSITION 15.5. *If μ is selfdecomposable, then μ is infinitely divisible and, for any $b > 1$, ρ_b in (15.1) is uniquely determined and infinitely divisible. If μ is semi-selfdecomposable, then μ is infinitely divisible and ρ_b is uniquely determined.*

Proof. Let μ be selfdecomposable. Then Theorem 9.3 and the proof of Theorem 15.3 show the infinite divisibility of μ and ρ_b. The uniqueness of ρ_b comes from the fact that $\widehat{\mu}(z)$ has no zero by Lemma 7.5.

Let μ be semi-selfdecomposable with span b. Then, replacing z in (15.1) by $b^{-1}z$ to get $\widehat{\mu}(b^{-1}z)$, we obtain

$$\widehat{\mu}(z) = \widehat{\mu}(b^{-2}z)\widehat{\rho}_b(b^{-1}z)\widehat{\rho}_b(z).$$

Repeating this procedure and using Lemma 7.4, we see that, for any n, there is an infinitely divisible distribution ρ_{b^n} such that

$$\widehat{\mu}(z) = \widehat{\mu}(b^{-n}z)\widehat{\rho}_{b^n}(z).$$

Since $\widehat{\mu}(b^{-n}z) \to 1$ as $n \to \infty$, $\widehat{\rho}_{b^n}(z) \to \widehat{\mu}(z)$ for any z. Hence μ is infinitely divisible by Lemma 7.8. Hence $\widehat{\mu}(z)$ has no zero and ρ_b is unique. □

It follows from the proposition above that, for any $b > 1$, any selfdecomposable distribution is semi-selfdecomposable with b as a span.

We recall that to any infinitely divisible distribution there corresponds a Lévy process.

DEFINITION 15.6. The Lévy processes corresponding to selfdecomposable and semi-selfdecomposable distributions are called, respectively, selfdecomposable and semi-selfdecomposable processes.

A limit theorem similar to Theorem 15.3 gives the following characterization of stable distributions.

THEOREM 15.7. *A probability measure μ on $\mathbb{R}^d$ is stable if and only if there are a random walk $\{S_n\}$, $b_n > 0$, and $c_n \in \mathbb{R}^d$ such that $P_{b_nS_n+c_n} \to \mu$ as $n \to \infty$.*

Proof. Suppose that $P_{b_nS_n+c_n} \to \mu$. A trivial distribution is stable. Assume that μ is non-trivial. For any $k \in \mathbb{N}$, consider

$$b_nS_{kn} + kc_n = \sum_{j=0}^{k-1}\{b_n(S_{(j+1)n} - S_{jn}) + c_n\}.$$

The distribution of the right-hand side tends to μ^k as $n \to \infty$, while the distribution of $b_{kn}S_{kn} + c_{kn}$ tends to μ. Application of Lemma 13.10 tells us that there are $b > 0$ and $c \in \mathbb{R}^d$ such that $\widehat{\mu}(z)^k = \widehat{\mu}(bz)e^{i\langle c,z\rangle}$. Hence μ is stable by Exercise 18.4.

Conversely, let μ be stable. Choose a random walk $S_n = \sum_{k=1}^n Z_k$ such that $P_{Z_k} = \mu$. Then the stability implies that $b_nS_n + c_n$ has distribution μ if $b_n > 0$ and c_n are suitably chosen. □

Let us give representation of semi-selfdecomposable distributions. We use T_r and $S_n(b)$ defined in (13.8) and (13.10).

THEOREM 15.8. *Fix $b > 1$. Let μ be an infinitely divisible distribution on $\mathbb{R}^d$ with generating triplet (A, ν, γ). Then μ is semi-selfdecomposable with b as a span if and only if*

$$T_b\nu \geq \nu. \tag{15.6}$$

In other words, μ is semi-selfdecomposable with b as a span if and only if, for any $n \in \mathbb{Z}$,

$$T_b([\nu]_{S_n(b)}) \geq [\nu]_{S_{n+1}(b)}. \tag{15.7}$$

Semi-selfdecomposability does not impose any restriction on A and γ.

Proof of theorem. Let μ be semi-selfdecomposable with b as a span. Then, μ and ρ_b are infinitely divisible by Proposition 15.5 and Definition 15.1, respectively. The generating triplet $(A^{(b)}, \nu^{(b)}, \gamma^{(b)})$ of the distribution with characteristic function $\widehat{\mu}(b^{-1}z)$ is given by

$$A^{(b)} = b^{-2}A,\ \nu^{(b)} = T_{b^{-1}}\nu,\ \text{and}\ \gamma^{(b)} = b^{-1}\gamma + b^{-1}\int_{1<|x|\leq b} x\nu(dx)$$

by Proposition 11.10. Therefore $\nu \geq T_{b^{-1}}\nu$, which is equivalent to (15.6).

Conversely, suppose that (15.6) holds. Define $A^{(b)}$, $\nu^{(b)}$, and $\gamma^{(b)}$ by the formulas above. The matrix $A - A^{(b)}$ is nonnegative-definite, since $b > 1$. Hence there is an infinitely divisible distribution ρ_b generated by $(A - A^{(b)}, \nu - \nu^{(b)}, \gamma - \gamma^{(b)})$. Then ρ_b satisfies (15.1) and μ is semi-selfdecomposable. □

EXAMPLES 15.9. (i) Let $x_1, \ldots, x_m \in S_0(b)$. Suppose that

$$\nu = \sum_{n=-\infty}^{\infty} \sum_{l=1}^{m} k_{l,n} \delta_{b^n x_l}$$

and that $\{k_{l,n}\}$ satisfies $k_{l,n} \geq k_{l,n+1} \geq 0$ for all l and n, $\sum_{n \geq 0} k_{l,n} + \sum_{n<0} b^{2n} k_{l,n} < \infty$ for all l. Then ν is the Lévy measure of a semi-selfdecomposable distribution with b as a span.

(ii) Let $\nu = g(x)dx$, where g is nonnegative and measurable, $\int(|x|^2 \wedge 1)g(x)\mathrm{d}x < \infty$, and $g_n(x) = 1_{S_n(b)}(x)g(x)$, $n \in \mathbb{Z}$, satisfy $g_n(x) \geq b^d g_{n+1}(bx)$. Then ν is the Lévy measure of a semi-selfdecomposable distribution with b as a span.

In both cases it is easy to check (15.6) or (15.7).

Now we discuss selfdecomposable distributions. Let $S = \{\xi \in \mathbb{R}^d \colon |\xi| = 1\}$, the unit sphere (if $d \geq 2$) or the two-point set $\{1, -1\}$ (if $d = 1$).

THEOREM 15.10. *Let μ be an infinitely divisible distribution on $\mathbb{R}^d$ with generating triplet (A, ν, γ). Then, μ is selfdecomposable if and only if*

$$\nu(B) = \int_S \lambda(\mathrm{d}\xi) \int_0^{\infty} 1_B(r\xi) k_\xi(r) \frac{\mathrm{d}r}{r} \tag{15.8}$$

with a finite measure λ on S and a nonnegative function $k_\xi(r)$ measurable in $\xi \in S$ and decreasing in $r > 0$.

Since $\int_0^{\infty} 1_B(r\xi)k_\xi(r)\frac{\mathrm{d}r}{r} = \int_0^{\infty} 1_B(r\xi)k_\xi(r+)\frac{\mathrm{d}r}{r}$, it is measurable in ξ. Selfdecomposability imposes no restriction on A and γ.

COROLLARY 15.11. *A probability measure μ on $\mathbb{R}$ is selfdecomposable if and only if*

$$\widehat{\mu}(z) = \exp\left[-\tfrac{1}{2}Az^2 + \mathrm{i}\gamma z + \int_{-\infty}^{\infty} (\mathrm{e}^{\mathrm{i}zx} - 1 - \mathrm{i}zx1_{[-1,1]}(x)) \frac{k(x)}{|x|} \mathrm{d}x\right], \tag{15.9}$$

where $A \geq 0$, $\gamma \in \mathbb{R}$, $k(x) \geq 0$, $\int_{-\infty}^{\infty}(|x|^2 \wedge 1)\frac{k(x)}{|x|}\mathrm{d}x < \infty$, and $k(x)$ is increasing on $(-\infty, 0)$ and decreasing on $(0, \infty)$.

Proof of Theorem 15.10. Let μ be selfdecomposable. Then, for any $b > 1$, it is semi-selfdecomposable with span b. Hence (15.6) for every $b > 1$. For $C \in \mathcal{B}(S)$ and $r > 0$ let

$$N(r, C) = \nu((r, \infty)C),$$

where the meaning of $(r, \infty)C$ is as in (14.5). Then $N(\mathrm{e}^{-s}, C)$ is convex in s, because $h(s) = N(\mathrm{e}^{-s}, C)$ satisfies

$$h(s+u) - h(s) = \nu((\mathrm{e}^{-s-u}, \mathrm{e}^{-s}]C) \geq \nu((b\mathrm{e}^{-s-u}, b\mathrm{e}^{-s}]C)$$

$$= h(s + u - \log b) - h(s - \log b)$$

for $b > 1$ and $u > 0$ by (15.6). Define

$$\lambda(C) = \int_{(0,\infty)C} (|x|^2 \wedge 1)\nu(\mathrm{d}x) = -\int_0^\infty (r^2 \wedge 1)\mathrm{d}N(r, C).$$

Then, λ is a finite measure on S and, for each $r > 0$, $N(r, C)$ is a measure in C absolutely continuous with respect to λ. For each $s \in \mathbb{R}$ there exists, by the Radon–Nikodým theorem, a nonnegative measurable function $H_\xi(s)$ of $\xi \in S$ such that

$$N(\mathrm{e}^{-s}, C) = \int_C H_\xi(s)\lambda(\mathrm{d}\xi) \quad \text{for } C \in \mathcal{B}(S). \tag{15.10}$$

If $s_1 < s_2$, then

$$H_\xi(s_1) \le H_\xi(s_2) \tag{15.11}$$

for λ-almost every ξ. If $s_1 < s_2$ and $0 < \alpha < 1$, then

$$\alpha H_\xi(s_1) + (1 - \alpha)H_\xi(s_2) \ge H_\xi(\alpha s_1 + (1 - \alpha)s_2) \tag{15.12}$$

for λ-almost every ξ by the convexity of $N(\mathrm{e}^{-s}, C)$ in s. Thus there is $C_1 \in \mathcal{B}(S)$ with $\lambda(S \setminus C_1) = 0$ such that (15.11) and (15.12) hold for all $\xi \in C_1$ and for all rational s_1, s_2, and α satisfying $s_1 < s_2$ and $0 < \alpha < 1$. Define, for $\xi \in C_1$ and $s \in \mathbb{R}$,

$$H_\xi^\sharp(s) = \sup_{s' \in (-\infty, s) \cap \mathbb{Q}} H_\xi(s').$$

Then $H_\xi^\sharp(s)$ is increasing and convex in s and measurable in ξ. We have (15.10) with C and $H_\xi(s)$ replaced by $C \cap C_1$ and $H_\xi^\sharp(s)$, respectively. Thus there is $C_2 \in \mathcal{B}(S)$ such that $C_2 \subset C_1$, $\lambda(S \setminus C_2) = 0$, and $\lim_{s \to -\infty} H_\xi^\sharp(s) = 0$ for $\xi \in C_2$. Now, it follows from the convexity that, for any $\xi \in C_2$, there is an increasing function $h_\xi(u)$ such that

$$H_\xi^\sharp(s) = \int_{-\infty}^s h_\xi(u)\mathrm{d}u.$$

We can choose $h_\xi(u)$ to be left-continuous in u. Then $h_\xi(u)$ is measurable in ξ, since $h_\xi(u) = \lim_{n \to \infty} n(H_\xi^\sharp(u) - H_\xi^\sharp(u - n^{-1}))$. It follows that

$$\begin{aligned} N(r, C) &= \int_{C \cap C_2} H_\xi^\sharp(-\log r)\lambda(\mathrm{d}\xi) = \int_{C \cap C_2} \lambda(\mathrm{d}\xi) \int_{-\infty}^{-\log r} h_\xi(u)\mathrm{d}u \\ &= \int_{C \cap C_2} \lambda(\mathrm{d}\xi) \int_r^\infty h_\xi(-\log v)\frac{\mathrm{d}v}{v}. \end{aligned}$$

Define $k_\xi(r) = h_\xi(-\log r)$ for $\xi \in C_2$. For $\xi \in S \setminus C_2$, we define $k_\xi(r)$ arbitrarily. Then we get (15.8) for $B = (r, \infty)C$. It follows that, for any

$\varepsilon > 0$, (15.8) holds for any Borel set B in $\{|x| > \varepsilon\}$ by Proposition 1.15. Hence (15.8) holds for any $B \in \mathcal{B}(\mathbb{R}^d)$.

Conversely suppose that (15.8) holds with some λ and $k_\xi(r)$. Then we see that $\nu((\mathrm{e}^{-s}, \infty)C)$ is convex in s for every $C \in \mathcal{B}(S)$. It follows that $\nu((r_1, r_2]C) \geq \nu((br_1, br_2]C)$ for any $b > 1$, $0 < r_1 < r_2$, and $C \in \mathcal{B}(S)$. Approximation shows that $\nu(B) \geq \nu(bB)$ for any $b > 1$ and $B \in \mathcal{B}(\mathbb{R}^d)$. Hence μ is semi-selfdecomposable with span b for any $b > 1$, that is, μ is selfdecomposable. □

REMARK 15.12. (i) The λ and $k_\xi(r)$ in the theorem above satisfy

$$\int_S \lambda(\mathrm{d}\xi) \int_0^\infty (r^2 \wedge 1) k_\xi(r) \tfrac{\mathrm{d}r}{r} < \infty. \tag{15.13}$$

Conversely, for any λ and $k_\xi(r)$ satisfying the conditions in the theorem and (15.13), the measure ν defined by (15.8) is the Lévy measure of a selfdecomposable distribution. Here we use Theorem 8.1.

(ii) The measure λ in the theorem can be chosen to satisfy

$$\lambda(C_0) = 0 \quad \text{for } C_0 = \{\xi \in S \colon k_\xi(r) = 0 \text{ for all } r > 0\}, \tag{15.14}$$

since $\nu((0, \infty)C_0) = 0$. The representation of the Lévy measure ν of a selfdecomposable distribution μ has uniqueness in the following sense. If λ, $k_\xi(r)$ and $\lambda^\sharp$, $k_\xi^\sharp(r)$ are both representations of ν in the theorem satisfying the condition (15.14), then we can find a measurable function $c(\xi)$ with $0 < c(\xi) < \infty$ such that

$$\lambda^\sharp(\mathrm{d}\xi) = c(\xi)\lambda(\mathrm{d}\xi) \tag{15.15}$$

and

$$k_\xi^\sharp(r)\mathrm{d}r = c(\xi)^{-1} k_\xi(r)\mathrm{d}r \quad \text{for } \lambda\text{-almost every } \xi. \tag{15.16}$$

In fact, let

$$a(\xi) = \int_0^\infty (r^2 \wedge 1) k_\xi(r) \tfrac{\mathrm{d}r}{r}$$

for ξ such that the right-hand side is positive and finite. Let $a(\xi) = 1$ for other ξ. Define $a^\sharp(\xi)$ from $k_\xi^\sharp(r)$ similarly. Then

$$\int_C a(\xi)\lambda(\mathrm{d}\xi) = \int_{\mathbb{R}^d} 1_C\left(\tfrac{x}{|x|}\right)(|x|^2 \wedge 1)\nu(\mathrm{d}x) = \int_C a^\sharp(\xi)\lambda^\sharp(\mathrm{d}\xi)$$

for any $C \in \mathcal{B}(S)$. Hence $\lambda^\sharp(\mathrm{d}\xi) = c(\xi)\lambda(\mathrm{d}\xi)$ with $c(\xi) = a(\xi)/a^\sharp(\xi)$. Now

$$\int_S \lambda(\mathrm{d}\xi) \int_0^\infty 1_B(r\xi) k_\xi(r) \tfrac{\mathrm{d}r}{r} = \int_S c(\xi)\lambda(\mathrm{d}\xi) \int_0^\infty 1_B(r\xi) k_\xi^\sharp(r) \tfrac{\mathrm{d}r}{r}$$

for any B, and hence $k_\xi(r) = c(\xi)k_\xi^\sharp(r)$ for $(\lambda \times \mathrm{d}r)$-almost every (ξ, r), which implies (15.16). See Lemma 59.3 for another uniqueness condition.

(iii) If $\nu \neq 0$, then the λ and $k_\xi(r)$ in the theorem can be chosen so that $\lambda(S) = 1$, $\int_0^\infty (r^2 \wedge 1) k_\xi(r) \frac{\mathrm{d}r}{r}$ is finite and independent of ξ, and $k_\xi(r)$ is right-continuous in $r > 0$. If two representations λ, $k_\xi(r)$ and $\lambda^\sharp$, $k_\xi^\sharp(r)$ are both chosen in this way, then $\lambda = \lambda^\sharp$ and $k_\xi(\cdot) = k_\xi^\sharp(\cdot)$ for λ-almost every ξ. This is seen from (ii). Recall that a decreasing function has a countable number of jumps at most and hence coincides with its right-continuous modification except at a countable number of points.

EXAMPLE 15.13. Example 8.10 shows that the characteristic function of a Γ-distribution has a form (15.9) with $A = 0$ and $k(x) = 1_{(0,\infty)}(x)ce^{-\alpha x}$. Hence Γ-distributions are selfdecomposable. In particular, exponential distributions are selfdecomposable. Pareto, F-, log-normal, and logistic distributions in Remark 8.12 are selfdecomposable. These facts are shown in the papers cited there. Student's t and half-Cauchy are also selfdecomposable. The former is shown by Halgreen [**178**] and Shanbhag and Sreehari [**476**]; the latter by Diédhiou [**98**]. In the multi-dimensional case Takano [**514**] shows that the distribution (8.23) is selfdecomposable.

EXAMPLE 15.14. A distribution μ on $\mathbb{R}$ with density

$$g(x) = \frac{\alpha\beta}{\alpha+\beta}(e^{-\alpha x}1_{[0,\infty)}(x) + e^{\beta x}1_{(-\infty,0)}(x)) \tag{15.17}$$

with $\alpha > 0$ and $\beta > 0$ is called a *two-sided exponential distribution.* We have by Examples 2.14 and 8.10

$$\begin{aligned}\widehat{\mu}(z) &= \frac{\beta}{\alpha+\beta}\frac{\alpha}{\alpha-iz} + \frac{\alpha}{\alpha+\beta}\frac{\beta}{\beta+iz} = \frac{\alpha\beta}{(\alpha-iz)(\beta+iz)} \\ &= \exp\left[\int_{-\infty}^{\infty}(e^{izx}-1)\frac{k(x)}{|x|}dx\right]\end{aligned}$$

with

$$k(x) = e^{-\alpha x}1_{(0,\infty)}(x) + e^{\beta x}1_{(-\infty,0)}(x).$$

Hence μ is selfdecomposable. The formula above shows that μ is obtained by convolution from an exponential distribution with parameter α and the dual of an exponential distribution with parameter β. It is also easy to check it directly. A distribution having density $\frac{\alpha}{2}e^{-\alpha|x-\gamma|}$ with $\alpha > 0$ and $\gamma \in \mathbb{R}$ is called a *Laplace distribution.* It is the two-sided exponential distribution with $\alpha = \beta$ with a drift γ added.

EXAMPLE 15.15. Let us study a distribution μ on $\mathbb{R}$ with density

$$g(x) = 1/(\pi\cosh x) = 2/(\pi(e^x + e^{-x})), \tag{15.18}$$

following Feller [**139**]. It is the distribution of the so-called stochastic area of the two-dimensional Brownian motion (Lévy [**315**]). We claim that

$$\widehat{\mu}(z) = 1/\cosh(\pi z/2). \tag{15.19}$$

We have, for $x > 0$,

$$\begin{aligned}1/\cosh x = 2e^{-x}/(1+e^{-2x}) = 2\textstyle\sum_{n=0}^{\infty}(-1)^n e^{-(2n+1)x}, \\ |1/\cosh x - 2\textstyle\sum_{n=0}^{N}(-1)^n e^{-(2n+1)x}| \le 2e^{-(2N+3)x}.\end{aligned}$$

Hence

$$\widehat{\mu}(z) = 2\textstyle\int_0^\infty(\cos zx)g(x)dx = (4/\pi)\sum_{n=0}^{\infty}(-1)^n\int_0^\infty(\cos zx)e^{-(2n+1)x}dx$$

$$= (4/\pi)\textstyle\sum_{n=0}^{\infty}(-1)^n(2n+1)/(z^2+(2n+1)^2).$$

The function $1/\cos w$ has the well-known partial fraction expansion

$$1/\cos w = 2\pi\textstyle\sum_{n=0}^{\infty}(-1)^n(n+\frac{1}{2})/((n+\frac{1}{2})^2\pi^2 - w^2)$$

for $w \in \mathbb{C}$, $w \neq \pm(n+\frac{1}{2})\pi$. A partial fraction expansion of $1/\cosh z = 1/\cos \mathrm{i}z$ follows. Hence we obtain (15.19). It follows from (15.19) that

$$(\log \widehat{\mu}(z))'' = -(\pi^2/4)(\cosh(\pi z/2))^{-2}. \tag{15.20}$$

Next we show that

$$(4/\pi^2)\textstyle\int_{-\infty}^{\infty} \mathrm{e}^{\mathrm{i}zx} x/(\mathrm{e}^x - \mathrm{e}^{-x})\mathrm{d}x = (\cosh(\pi z/2))^{-2}. \tag{15.21}$$

In fact, $\mu * \mu$ has density

$$\begin{aligned}\textstyle\int_{-\infty}^{\infty} g(x-y)g(y)\mathrm{d}y &= (4/\pi^2)\mathrm{e}^{-x}\textstyle\int_{-\infty}^{\infty}(1+\mathrm{e}^{2y-2x})^{-1}(1+\mathrm{e}^{-2y})^{-1}\mathrm{d}y \\ &= (2/\pi^2)\mathrm{e}^{-x}\textstyle\int_0^{\infty}(u+\mathrm{e}^{-2x})^{-1}(u+1)^{-1}\mathrm{d}u = (4/\pi^2)x/(\mathrm{e}^x - \mathrm{e}^{-x}),\end{aligned}$$

where $x/(\mathrm{e}^x - \mathrm{e}^{-x})$ is extended to $x = 0$ continuously, and this identity proves (15.21) from (15.19). Now we can prove that

$$\widehat{\mu}(z) = \exp\big[\textstyle\int_{-\infty}^{\infty}(\mathrm{e}^{\mathrm{i}zx} - 1 - \mathrm{i}zx)/(x(\mathrm{e}^x - \mathrm{e}^{-x}))\mathrm{d}x\big]. \tag{15.22}$$

In fact, if we denote the integral in the right-hand side by $h(z)$, then

$$h''(z) = -\textstyle\int_{-\infty}^{\infty}\mathrm{e}^{\mathrm{i}zx}x/(\mathrm{e}^x - \mathrm{e}^{-x})\mathrm{d}x,$$

and hence $(\log \widehat{\mu}(z))'' = g''(z)$ by (15.20) and (15.21). It follows that $\log \widehat{\mu}(z) = g(z) + az + b$ with some a and b, but $a = b = 0$ since, at $z = 0$, $\log \widehat{\mu}(z)$ and $g(z)$ have a common value and a common derivative. Thus (15.22) is shown. It says that μ is selfdecomposable with $k(x) = 1/|2\sinh x|$. At the same time we have shown that the distribution with density $2x/(\pi^2 \sinh x)$ is selfdecomposable with $k(x) = 1/|\sinh x|$.

16. Selfsimilar and semi-selfsimilar additive processes

Selfdecomposable and semi-selfdecomposable distributions studied in the previous section appear in selfsimilar and semi-selfsimilar additive processes in a natural way.

THEOREM 16.1. (i) *If $\{X_t\colon t \geq 0\}$ is a broad-sense selfsimilar additive process on $\mathbb{R}^d$, then, for every $t \geq 0$, the distribution of X_t is selfdecomposable.*

(ii) *If μ is a non-trivial selfdecomposable distribution on $\mathbb{R}^d$, then, for any $H > 0$, there exists, uniquely in law, a non-trivial H-selfsimilar additive process $\{X_t\colon t \geq 0\}$ such that $P_{X_1} = \mu$.*

Proof. (i) Trivial distributions are selfdecomposable. Suppose that $\{X_t\}$ is a non-trivial, broad-sense selfsimilar additive process. Then, it has an exponent $H > 0$ by Theorem 13.11. Hence, for every $a > 0$,

$$\{X_{at}\} \stackrel{\mathrm{d}}{=} \{a^H X_t + c_a(t)\}$$

with some $c_a(t)$. If $s < t$, then, choosing $a = s/t$, we get

$$X_s \stackrel{\mathrm{d}}{=} (s/t)^H X_t + c_{s/t}(t).$$

Let μ_t and $\mu_{s,t}$ be the distributions of X_t and $X_t - X_s$, respectively. Using the additivity, we get

$$\widehat{\mu}_t(z) = \widehat{\mu}_s(z)\widehat{\mu}_{s,t}(z) = \widehat{\mu}_t((s/t)^H z)\mathrm{e}^{\mathrm{i}\langle c_{s/t}(t),z\rangle}\widehat{\mu}_{s,t}(z)$$

for $s < t$. Given $t > 0$ and $b > 1$, choose s so that $(s/t)^H = b^{-1}$. Then $\widehat{\mu}_t(z) = \widehat{\mu}_t(b^{-1}z)\widehat{\rho}_b(z)$ with some ρ_b. Hence μ_t is selfdecomposable.

(ii) Suppose that we are given a non-trivial selfdecomposable distribution μ and $H > 0$. Then μ is infinitely divisible. For any $b > 1$, there is a unique ρ_b such that $\widehat{\mu}(z) = \widehat{\mu}(b^{-1}z)\widehat{\rho}_b(z)$. It follows that ρ_b is continuous in b. Define, for $t > 0$ and $0 < s < t$, μ_t and $\mu_{s,t}$ by $\widehat{\mu}_t(z) = \widehat{\mu}(t^H z)$ and $\widehat{\mu}_{s,t}(z) = \widehat{\rho}_{(t/s)^H}(t^H z)$. Then

$$\widehat{\mu}_t(z) = \widehat{\mu}((s/t)^H t^H z)\widehat{\rho}_{(t/s)^H}(t^H z) = \widehat{\mu}_s(z)\widehat{\mu}_{s,t}(z).$$

Further define $\mu_0 = \delta_0$, $\mu_{0,t} = \mu_t$, and $\mu_{t,t} = \delta_0$. Then $\mu_t = \mu_s * \mu_{s,t}$ for $0 \le s \le t$ and μ_t is continuous in $t \ge 0$. It follows that $\mu_{s,t} \to \delta_0$ as $s \uparrow t$ or $t \downarrow s$ and that $\mu_{s,t} * \mu_{t,u} = \mu_{s,u}$ for $s \le t \le u$. Therefore, by Theorem 9.7, there is, uniquely in law, an additive process $\{X_t\}$ such that $P_{X_1} = \mu$. For any $a > 0$, we have $X_{at} \stackrel{\mathrm{d}}{=} a^H X_t$. Since both $\{X_{at}\}$ and $\{a^H X_t\}$ are additive, it follows from Theorem 9.7 that $\{X_{at}\} \stackrel{\mathrm{d}}{=} \{a^H X_t\}$. Thus $\{X_t\}$ is H-selfsimilar. Since X_t is unique in law, $\{X_t\}$ is unique in law. □

REMARK 16.2. Fix $H > 0$. Let μ be selfdecomposable and non-trivial. Then we obtain the H-selfsimilar additive process $\{X_t\}$ in Theorem 16.1(ii) on the one hand, and the Lévy process $\{Y_t\}$ corresponding to μ on the other. Each of $\{X_t\}$ and $\{Y_t\}$ is unique in law. Both of them have the distribution μ at $t = 1$. They are identical in law if and only if $H \ge 1/2$ and μ is strictly stable with index $1/H$. This follows from Proposition 13.5, Theorem 13.15, and Definition 13.16.

EXAMPLES 16.3. Let us give the H-selfsimilar additive process $\{X_t\}$ and the Lévy process $\{Y_t\}$ in the remark above for some special μ.

(i) Let μ be α-stable on $\mathbb{R}^d$, $\alpha \in (0,1) \cup (1,2]$, and $\mu = \mu_0 * \delta_\gamma$, where μ_0 is strictly α-stable and $\gamma \in \mathbb{R}^d$. Let $\{X_t^0\}$ be the strictly α-stable process with distribution μ_0 at $t = 1$. Let $H = 1/\alpha$. Then $\{X_t\} \stackrel{\mathrm{d}}{=} \{X_t^0 + t^{1/\alpha}\gamma\}$, since the right-hand side is additive and

$$\{X_{at}^0 + (at)^{1/\alpha}\gamma\} \stackrel{\mathrm{d}}{=} \{a^{1/\alpha}X_t^0 + (at)^{1/\alpha}\gamma\} = \{a^{1/\alpha}(X_t^0 + t^{1/\alpha}\gamma)\}.$$

On the other hand, it is obvious that $\{Y_t\} \stackrel{\mathrm{d}}{=} \{X_t^0 + t\gamma\}$. Therefore $\{X_t\} \stackrel{\mathrm{d}}{=} \{Y_t + (t^{1/\alpha} - t)\gamma\}$.

(ii) Let μ be 1-stable on $\mathbb{R}^d$ and let $H = 1$. Then $Y_t \stackrel{\mathrm{d}}{=} tY_1 + (t \log t)c$ with $c = \frac{2}{\pi} \int_S \xi \lambda_1(\mathrm{d}\xi)$ by Exercise 18.6. On the other hand $X_t \stackrel{\mathrm{d}}{=} tX_1$, since $X_{at} \stackrel{\mathrm{d}}{=} aX_t$ for all t and a. Hence $X_t \stackrel{\mathrm{d}}{=} tY_1 \stackrel{\mathrm{d}}{=} Y_t - (t \log t)c$. It follows that $\{X_t\} \stackrel{\mathrm{d}}{=} \{Y_t - (t \log t)c\}$.

(iii) Let μ be strictly α-stable on $\mathbb{R}^d$. Then $\{X_t\} \stackrel{\mathrm{d}}{=} \{Y_{t^{\alpha H}}\}$. In fact, $\{Y_{t^{\alpha H}}\}$ is an additive process satisfying $\{Y_{(at)^{\alpha H}}\} \stackrel{\mathrm{d}}{=} \{a^H Y_{t^{\alpha H}}\}$. This relation is generalized in Proposition 16.5.

(iv) If $d = 1$ and μ is an exponential distribution, then X_t has an exponential distribution for any $t > 0$, while Y_t has a Γ-distribution for $t > 0$. Properties of $\{X_t\}$ and $\{Y_t\}$ have a qualitatitive difference.

EXAMPLE 16.4. Let $\{X_t\}$ be the Brownian motion on $\mathbb{R}^d$ with dimension $d \geq 3$ on a probability space $(\Omega, \mathcal{F}, P)$. Define the last exit time from the ball $\{x \colon |x| \leq r\}$ by

$$L_r(\omega) = \sup\{t \geq 0 \colon |X_t| \leq r\}.$$

This L_r is measurable if we enlarge the σ-algebra $\mathcal{F}$ appropriately. We will prove in Chapter 7 that $L_r(\omega)$ is finite a. s. Using the $\frac{1}{2}$-selfsimilarity of the Brownian motion, we can prove that the process $\{L_r \colon r \geq 0\}$ is 2-selfsimilar. Indeed, for $a > 0$,

$$\begin{aligned} L_{ar} &= \sup\{t \colon |X_t| \leq ar\} = \sup\{t \colon |a^{-1} X_t| \leq r\} \\ &\stackrel{\mathrm{d}}{=} \sup\{t \colon |X(a^{-2}t)| \leq r\} = \sup\{a^2 t \colon |X_t| \leq r\} = a^2 L_r \end{aligned}$$

and, in the same way, $\{L_{ar} \colon r \geq 0\} \stackrel{\mathrm{d}}{=} \{a^2 L_r \colon r \geq 0\}$. Getoor [**160**] proves that this process is an additive process on $[0, \infty)$ and that, for $r > 0$,

$$P[L_r \in B] = 2^{-(d-2)/2} (\Gamma(\tfrac{d-2}{2}))^{-1} r^{d-2} \int_B s^{-d/2} \mathrm{e}^{-r^2/(2s)} \mathrm{d}s \tag{16.1}$$

for any Borel set B in $[0, \infty)$. Hence, by Theorem 16.1, the distribution (16.1) is selfdecomposable. If $d = 3$, then the process is a $\frac{1}{2}$-stable increasing Lévy process, which was proved earlier by Pitman [**376**]. If $d \geq 4$, then the process is not a Lévy process.

PROPOSITION 16.5. *Let $\eta > 0$. If $\{X_t\}$ is a non-trivial H-selfsimilar additive process on $\mathbb{R}^d$, then $\{X_{t^\eta}\}$ is an ηH-selfsimilar additive process. If $\{X_t\}$ is a non-trivial H-semi-selfsimilar additive process on $\mathbb{R}^d$ having a as an epoch, then $\{X_{t^\eta}\}$ is an ηH-semi-selfsimilar additive process having $a^{1/\eta}$ as an epoch.*

Proof. Preservation of the additivity is easy to check. If $\{X_t\}$ is non-trivial and H-selfsimilar, then, for $X'_t = X_{t^\eta}$,

$$\{X'_{a^{1/\eta}t}\} = \{X_{at^\eta}\} \stackrel{\mathrm{d}}{=} \{a^H X_{t^\eta}\} = \{(a^{1/\eta})^{\eta H} X'_t\} \quad \text{for } a > 0.$$

The H-semi-selfsimilar case is also proved in this way. □

THEOREM 16.6. *If $\{X_t\}$ is a non-trivial broad-sense semi-selfsimilar additive process on $\mathbb{R}^d$ having b as a span, then, for every $t \geq 0$, the distribution of X_t is semi-selfdecomposable having b as a span.*

Proof. Discussion almost the same as the proof of (i) of Theorem 16.1 works. We have only to restrict a to an epoch or the reciprocal of an epoch. □

An analogue of part (ii) of Theorem 16.1 in the semi-selfsimilar case is given by the following two theorems.

THEOREM 16.7. *Let $a > 1$, $H > 0$, and $b = a^H$. Suppose that a system $\{\mu_t \colon t \in [1,a)\}$ of non-trivial probability measures on $\mathbb{R}^d$ is given and satisfies the following conditions:*

(1) $\widehat{\mu}_t(z) \neq 0$;

(2) *for any s, t with $1 \leq s \leq t < a$, there exists a probability measure $\mu_{s,t}$ satisfying*

$$\mu_t = \mu_s * \mu_{s,t}; \tag{16.2}$$

(3) *μ_t is continuous in t;*

(4) *$\widehat{\mu}_t(z) \to \widehat{\mu}_1(bz)$ for $z \in \mathbb{R}^d$ as $t \uparrow a$.*

Then there exists, uniquely in law, a non-trivial H-semi-selfsimilar additive process $\{X_t\}$ having a as an epoch and b as a span and satisfying $P_{X_t} = \mu_t$ for $t \in [1,a)$.

Note that, by Theorem 16.6, the distributions μ_t are proved to be semi-selfdecomposable with b as a span.

Proof of theorem. If $a^n \leq t < a^{n+1}$ for some integer n, then we define μ_t by

$$\widehat{\mu}_t(z) = \widehat{\mu}_{a^{-n}t}(b^n z). \tag{16.3}$$

Then μ_t is defined for all $t > 0$. In particular, $\widehat{\mu}_a(z) = \widehat{\mu}_1(bz)$ and $\widehat{\mu}_{a^n}(z) = \widehat{\mu}_1(b^n z) = \widehat{\mu}_a(b^{n-1}z)$. As $t \uparrow a^{n+1}$, we have

$$\widehat{\mu}_t(z) = \widehat{\mu}_{a^{-n}t}(b^n z) \to \widehat{\mu}_1(b^{n+1}z) = \widehat{\mu}_{a^{n+1}}(z)$$

by (4). Combined with (3), this shows that μ_t is continuous in $t \in (0,\infty)$. We claim that

$$\mu_t \to \delta_0 \quad \text{as } t \downarrow 0. \tag{16.4}$$

If (16.4) is not true, then there are a sequence $t_k \downarrow 0$, $z_0 \in \mathbb{R}^d$, and $\varepsilon > 0$ such that $|\widehat{\mu}_{t_k}(z_0) - 1| > \varepsilon$. Choose $n_k \in \mathbb{Z}$ such that $a^{n_k} \leq t_k < a^{n_k+1}$. Then $n_k \downarrow -\infty$ and $\widehat{\mu}_{t_k}(z) = \widehat{\mu}_{s_k}(b^{n_k}z)$ with $s_k = a^{-n_k}t_k$. Choosing a subsequence if necessary, we can assume that s_k tends to some $s \in [1,a]$. Using (3) and (4), we see that $\widehat{\mu}_{s_k}(b^{n_k}z_0) \to \widehat{\mu}_s(0) = 1$, which is a contradiction. Hence (16.4) holds. Define $\mu_0 = \delta_0$. Next we claim that if $0 \leq s \leq t$, then there is a unique $\mu_{s,t}$ satisfying (16.2). The uniqueness follows from (1). If $1 \le s \le t < a$, then the existence of $\mu_{s,t}$ is assumed by (2). If $1 \le s < t = a$, then

$$\widehat{\mu}_{s,t'}(z) = \frac{\widehat{\mu}_{t'}(z)}{\widehat{\mu}_s(z)} \to \frac{\widehat{\mu}_a(z)}{\widehat{\mu}_s(z)} \quad \text{as } t' \uparrow a$$

and the continuity of $\widehat{\mu}_a(z)/\widehat{\mu}_s(z)$ gives the existence of $\mu_{s,a}$ such that $\widehat{\mu}_{s,a}(z) = \widehat{\mu}_a(z)/\widehat{\mu}_s(z)$ by Proposition 2.5(viii). We define $\mu_{a,a} = \delta_0$. Thus we have $\mu_{s,t}$ of (16.2) for $1 \le s \le t \le a$. If $a^n \le s \le t \le a^{n+1}$ with $n \in \mathbb{Z}$, then it follows from

$$\mu_{a^{-n}t} = \mu_{a^{-n}s} * \mu_{a^{-n}s,a^{-n}t}$$

that

$$\widehat{\mu}_t(z) = \widehat{\mu}_{a^{-n}t}(b^n z) = \widehat{\mu}_{a^{-n}s}(b^n z)\widehat{\mu}_{a^{-n}s,a^{-n}t}(b^n z) = \widehat{\mu}_s(z)\widehat{\mu}_{a^{-n}s,a^{-n}t}(b^n z).$$

Hence $\mu_{s,t}$ exists for $a^n \le s \le t \le a^{n+1}$. If $a^m \le s \le a^{m+1} \le a^n \le t \le a^{n+1}$, then $\mu_{s,t}$ is given by

$$\mu_{s,t} = \mu_{s,a^{m+1}} * \mu_{a^{m+1},a^{m+2}} * \cdots * \mu_{a^{n-1},a^n} * \mu_{a^n,t}.$$

Define $\mu_{0,t} = \mu_t$. Now the existence of $\mu_{s,t}$ satisfying (16.2) is shown for all $0 \le s \le t$. It is obvious that

$$\mu_{s,t} * \mu_{t,u} = \mu_{s,u} \quad \text{for } 0 \le s \le t \le u.$$

Now, by Theorem 9.7, we can construct, uniquely in law, an additive process $\{X_t\}$ such that $P_{X_t} = \mu_t$. This process satisfies $X_{at} \stackrel{\mathrm{d}}{=} bX_t$ since, for $a^n \le t < a^{n+1}$, $\widehat{\mu}_{at}(z) = \widehat{\mu}_{a^{-n}t}(b^{n+1}z) = \widehat{\mu}_t(bz)$. Thus $\{X_{at}\} \stackrel{\mathrm{d}}{=} \{bX_t\}$, that is, $\{X_t\}$ is H-semi-selfsimilar with a as an epoch and b as a span. The uniqueness in law of $\{X_t\}$ follows from the uniqueness of P_{X_t}, which is connected with $\{\mu_t \colon t \in [1, a)\}$ by the semi-selfsimilarity. □

THEOREM 16.8. *If μ is a non-trivial semi-selfdecomposable distribution on $\mathbb{R}^d$ with b as a span, then, for any $H > 0$, there is a non-trivial H-semi-selfsimilar additive process $\{X_t\}$ with b as a span such that X_1 has distribution μ.*

Proof. Given a non-trivial semi-selfdecomposable distribution μ with b as a span and given a positive real H, let $a = b^{1/H}$. Then $a > 1$. Let $q(t)$ be an increasing continuous function on $[1, a]$ such that $q(1) = 0$ and $q(a) = 1$. Recalling that μ is infinitely divisible by Proposition 15.5, define μ_t for $1 \le t < a$ by

$$\widehat{\mu}_t(z) = \widehat{\mu}(z)^{1-q(t)}\widehat{\mu}(bz)^{q(t)}. \tag{16.5}$$

Then $\mu_1 = \mu$. The proof is done if the system $\{\mu_t\}$ is shown to satisfy the conditions in Theorem 16.7. Among them (1), (3), and (4) are clear from (16.5). To show (2), compare the two formulas

$$\widehat{\mu}_t(z) = \widehat{\mu}(z)^{1-q(t)}\widehat{\mu}(bz)^{q(s)}\widehat{\mu}(bz)^{q(t)-q(s)},$$

$$\widehat{\mu}_s(z) = \widehat{\mu}(z)^{1-q(t)}\widehat{\mu}(z)^{q(t)-q(s)}\widehat{\mu}(bz)^{q(s)},$$

and note that there is an infinitely divisible distribution ρ such that $\widehat{\mu}(z) = \widehat{\mu}(b^{-1}z)\widehat{\rho}(z)$. Then we see that

$$\widehat{\mu}_t(z) = \widehat{\mu}_s(z)\widehat{\rho}(bz)^{q(t)-q(s)}.$$

The second factor on the right gives $\widehat{\mu}_{s,t}(z)$ satisfying (16.2). The proof is complete. □

EXAMPLE 16.9. Let $b > 1$ and $\nu = \sum_{n=-\infty}^{\infty} k_n \delta_{b^n x_0}$, where $1 < |x_0| \le b$, $k_n \ge k_{n+1} \ge 0$, and $\sum_{n\ge 0} k_n + \sum_{n<0} b^n k_n < \infty$. Let μ be the distribution on $\mathbb{R}^d$ with

$$\widehat{\mu}(z) = \exp \int (e^{i\langle z,x\rangle} - 1)\nu(dx) = \exp \sum_{n=-\infty}^{\infty} k_n (e^{i\langle z, b^n x_0\rangle} - 1).$$

Then μ is semi-selfdecomposable with b as a span (Examples 13.3 and 15.9(i)). Since $\widehat{\mu}(bz)$ equals the expression above for $\widehat{\mu}(z)$ with k_n replaced by k_{n-1}, the distribution μ_t constructed by (16.5) for $1 \le t < a$ is such that

$$\widehat{\mu}_t(z) = \exp \sum_{n=-\infty}^{\infty} ((1 - q(t))k_n + q(t)k_{n-1})(e^{i\langle z, b^n x_0\rangle} - 1).$$

REMARK 16.10. Our proof shows that the process $\{X_t\}$ in Theorem 16.8 is not determined uniquely in law by μ, b, and H, because there is freedom of choice of $q(t)$. Furthermore, for μ, b, and H given, the systems $\{\mu_t : t \in [1, a)\}$ such that (16.5) holds do not exhaust the systems satisfying the conditions (1)–(4) of Theorem 16.7. For the μ in Example 16.9, this is seen from the fact that the following choice of μ_t, $t \in [1, a)$, does work:

$$\widehat{\mu}_t(z) = \exp \sum_{n=-\infty}^{\infty} ((1 - q_n(t))k_n + q_n(t)k_{n-1})(e^{i\langle z, b^n x_0\rangle} - 1),$$

where, for each n, $q_n(t)$ is increasing and continuous on $[1, a]$, $q_n(1) = 0$ and $q_n(a) = 1$.

17. Another view of selfdecomposable distributions

Selfdecomposable distributions can be viewed as limit distributions of a class of Markov processes called processes of Ornstein–Uhlenbeck type. This interpretation gives a new representation of their Lévy measures.

Given a Lévy process $\{Z_t : t \ge 0\}$ on $\mathbb{R}^d$, a real constant c, and a starting point $x \in \mathbb{R}^d$, we want to introduce a new stochastic process $\{X_t : t \ge 0\}$ which is, almost surely, right-continuous with left limits and satisfies, almost surely,

$$X_t(\omega) = x + Z_t(\omega) - c\int_0^t X_s(\omega)ds \quad \text{for } t \ge 0. \tag{17.1}$$

Considered as an equation for $X_t(\omega)$, (17.1) has at most one solution. If both $X_t(\omega)$ and $X'_t(\omega)$ are solutions, then, letting $f(t) = X_t(\omega) - X'_t(\omega)$ for a fixed ω, we see that $f(t)$ is bounded on any finite interval and

$$f(t) = -c\int_0^t f(s)ds.$$

By induction we get

$$f(t) = \frac{(-c)^n}{(n-1)!} \int_0^t (t-s)^{n-1} f(s) \mathrm{d}s, \quad n = 1, 2, \ldots.$$

Since $\sum_{n=1}^{\infty} \frac{|c|^n}{(n-1)!}(t-s)^{n-1}$ is finite, $\frac{(-c)^n}{(n-1)!}(t-s)^{n-1}$ tends to 0 as $n \to \infty$, boundedly in $s \in [0,t]$. Hence $f(t) = 0$. As to the existence of the solution $\{X_t(\omega)\}$ of (17.1), suppose that $\{Z_t\}$ is, almost surely, of bounded variation in t on any finite interval. Then we can give the solution by

$$X_t(\omega) = \mathrm{e}^{-ct} x + \int_0^t \mathrm{e}^{-c(t-s)} \mathrm{d}Z_s(\omega), \tag{17.2}$$

where the last integral is the Stieltjes integral in s. In fact, (17.2) implies

$$\begin{aligned} c\int_0^t X_s(\omega)\mathrm{d}s &= c\int_0^t \mathrm{e}^{-cs} x \mathrm{d}s + c\int_0^t ds \int_0^s \mathrm{e}^{-c(s-u)} \mathrm{d}Z_u(\omega) \\ &= x(1-\mathrm{e}^{-ct}) + \int_0^t (1-\mathrm{e}^{-c(t-u)}) \mathrm{d}Z_u(\omega) \\ &= x + Z_t(\omega) - X_t(\omega), \end{aligned}$$

that is, (17.1). If $g(u)$ is a real step function on $[s,t]$ with $0 \le s < t < \infty$,

$$g(u) = \sum_{j=1}^{n} a_j 1_{(u_{j-1}, u_j]}(u) \quad \text{with } s = u_0 < u_1 < \cdots < u_n = t,$$

then we have

$$E\left[\exp\left(\mathrm{i}\left\langle z, \int_s^t g(u) \mathrm{d}Z_u(\omega)\right\rangle\right)\right] = \exp\left[\int_s^t \psi(g(u)z)\mathrm{d}u\right] \tag{17.3}$$

for $z \in \mathbb{R}^d$, where $\psi(z) = \log \widehat{P}_{Z_1}(z)$. In fact,

$$\begin{aligned} E\left[\exp\left(\mathrm{i}\left\langle z, \int_s^t g(u)\mathrm{d}Z_u(\omega)\right\rangle\right)\right] &= E\left[\exp\left(\mathrm{i}\left\langle z, \sum_{j=1}^n a_j (Z_{u_j} - Z_{u_{j-1}})\right\rangle\right)\right] \\ = \prod_{j=1}^n E\mathrm{e}^{\mathrm{i}\langle a_j z, Z_{u_j} - Z_{u_{j-1}}\rangle} &= \prod_{j=1}^n \mathrm{e}^{(u_j - u_{j-1})\psi(a_j z)} \\ = \exp\left[\sum_{j=1}^n \psi(a_j z)(u_j - u_{j-1})\right] &= \exp\left[\int_s^t \psi(g(u)z)\mathrm{d}u\right]. \end{aligned}$$

By approximation, (17.3) is true for any real continuous function $g(u)$ on $[s,t]$. Thus we have

$$E[\mathrm{e}^{\mathrm{i}\langle z, X_t\rangle}] = \exp\left[\mathrm{i}\mathrm{e}^{-ct}\langle x, z\rangle + \int_0^t \psi(\mathrm{e}^{-c(t-s)} z)\mathrm{d}s\right]. \tag{17.4}$$

Recall that we have assumed that, almost surely, the sample function of $\{Z_t\}$ is of bounded variation on any finite time interval. In Chapter 4 we will see that this assumption is satisfied if and only if $\{Z_t\}$ is of type A or B (Theorem 21.9). We will not take the route to defining (17.2) for $\{Z_t\}$ of type C by introducing a stochastic integral (it is called the Wiener integral when $\{Z_t\}$ is the Brownian motion). Rather we will define a transition function by using the expression (17.4) and construct a Markov process from the transition function.

LEMMA 17.1. *Let $\{Z_t\}$ be a Lévy process on $\mathbb{R}^d$ generated by (G,ρ,β). Let $c\in\mathbb{R}$. Then there is a temporally homogeneous transition function $P_t(x,B)$ on $\mathbb{R}^d$ such that*

$$\int_{\mathbb{R}^d} \mathrm{e}^{\mathrm{i}\langle z,y\rangle} P_t(x,\mathrm{d}y) = \exp\left[\mathrm{i}\mathrm{e}^{-ct}\langle x,z\rangle + \int_0^t \psi(\mathrm{e}^{-cs}z)\mathrm{d}s\right],\quad z\in\mathbb{R}^d, \tag{17.5}$$

where $\psi(z)=\log\widehat{P}_{Z_1}(z)$. For each t and x, $P_t(x,\cdot)$ is infinitely divisible with generating triplet $(A_t,\nu_t,\gamma_{t,x})$ given by

$$A_t = \int_0^t \mathrm{e}^{-2cs}\mathrm{d}s\, G, \tag{17.6}$$

$$\nu_t(B) = \int_{\mathbb{R}^d} \rho(\mathrm{d}y)\int_0^t 1_B(\mathrm{e}^{-cs}y)\mathrm{d}s,\quad B\in\mathcal{B}(\mathbb{R}^d), \tag{17.7}$$

$$\begin{aligned}\gamma_{t,x} &= \mathrm{e}^{-ct}x + \int_0^t \mathrm{e}^{-cs}\mathrm{d}s\,\beta \\ &\quad + \int_{\mathbb{R}^d}\rho(\mathrm{d}y)\int_0^t \mathrm{e}^{-cs}y(1_D(\mathrm{e}^{-cs}y)-1_D(y))\mathrm{d}s,\end{aligned} \tag{17.8}$$

where $D=\{x\colon |x|\le 1\}$.

Proof. We have

$$\psi(z) = -\frac{1}{2}\langle z,Gz\rangle + \int_{\mathbb{R}^d}(\mathrm{e}^{\mathrm{i}\langle z,y\rangle}-1-\mathrm{i}\langle z,y\rangle 1_D(y))\rho(\mathrm{d}y) + \mathrm{i}\langle\beta,z\rangle.$$

Thus $\psi(\mathrm{e}^{-cs}z)$ is continuous in $(s,z)\in[0,\infty)\times\mathbb{R}^d$. Let $s_{n,j}=j2^{-n}t$. Then

$$\int_0^t \psi(\mathrm{e}^{-cs}z)\mathrm{d}s = \lim_{n\to\infty}\frac{t}{2^n}\sum_{j=1}^{2^n}\psi(\mathrm{e}^{-cs_{n,j}}z).$$

We have seen that

$$E[\mathrm{e}^{\mathrm{i}\langle z,Y\rangle}] = \exp\left[\mathrm{i}\mathrm{e}^{-ct}\langle x,z\rangle + \frac{t}{2^n}\sum_{j=1}^{2^n}\psi(\mathrm{e}^{-cs_{n,j}}z)\right],$$

whenever

$$Y = \mathrm{e}^{-ct}x + \sum_{j=1}^{2^n} \mathrm{e}^{-cs_{n,j}}(Z_{s_{n,j}} - Z_{s_{n,j-1}}).$$

The distribution of Y is infinitely divisible by Lemma 7.4. As $\int_0^t \psi(\mathrm{e}^{-cs}z)\mathrm{d}s$ is continuous in z, we see that the right-hand side of (17.5) is the characteristic function of a probability measure by Proposition 2.5(viii) and this measure is infinitely divisible by Lemma 7.8. Thus an infinitely divisible probability measure $P_t(x, \cdot)$ is definable by (17.5). (It is more meaningful in connection with (17.2) to consider

$$Y' = \mathrm{e}^{-ct}x + \sum_{j=1}^{2^n} \mathrm{e}^{-c(t-s_{n,j})}(Z_{s_{n,j}} - Z_{s_{n,j-1}})$$

instead of Y, but Y and Y' have the same distribution.)

Let us show (17.6)–(17.8). Denote the right-hand side of (17.7) by $\nu_t'(B)$. Then

$$\int_{|x|\le 1} |x|^2 \nu_t'(\mathrm{d}x) \le \mathrm{e}^{2|c|t} t \int_{|y|\le \mathrm{e}^{|c|t}} |y|^2 \rho(\mathrm{d}y) < \infty,$$

$$\int_{|x|>1} \nu_t'(\mathrm{d}x) \le t \int_{|y|\ge \mathrm{e}^{-|c|t}} \rho(\mathrm{d}y) < \infty.$$

It follows that

$$\int_0^t \psi(\mathrm{e}^{-cs}z)\mathrm{d}s = -\frac{1}{2}\int_0^t \mathrm{e}^{-2cs}\mathrm{d}s\langle z, Gz\rangle + \mathrm{i}\int_0^t \mathrm{e}^{-cs}\mathrm{d}s\langle \beta, z\rangle + I,$$

$$I = \int_{\mathbb{R}^d} (\mathrm{e}^{\mathrm{i}\langle z,x\rangle} - 1 - \mathrm{i}\langle z, x\rangle 1_D(x))\nu_t'(\mathrm{d}x)$$

$$+ \mathrm{i}\int_0^t \mathrm{d}s \int_{\mathbb{R}^d} \langle z, \mathrm{e}^{-cs}y\rangle(1_D(\mathrm{e}^{-cs}y) - 1_D(y))\rho(\mathrm{d}y).$$

Hence, by (17.5), the generating triplet $(A_t, \nu_t, \gamma_{t,x})$ of $P_t(x, \cdot)$ is described by (17.6)–(17.8).

It follows from (17.5) that, for any bounded measurable function $f(y)$ on $\mathbb{R}^d$,

$$\int_{\mathbb{R}^d} f(y)P_t(x, \mathrm{d}y) = \int_{\mathbb{R}^d} f(\mathrm{e}^{-ct}x + y)P_t(0, \mathrm{d}y). \tag{17.9}$$

Hence $P_t(x, B)$ is measurable in x for any fixed t and B. Obviously $P_0(x, \cdot) = \delta_x$. Further $P_t(x, B)$ satisfies the Chapman–Kolmogorov identity (10.2), because

$$\iint P_t(x, \mathrm{d}y)P_s(y, \mathrm{d}w)\mathrm{e}^{\mathrm{i}\langle z,w\rangle}$$

$$
\begin{aligned}
&= \int P_t(x, \mathrm{d}y) \exp\left[\mathrm{i}\langle y, \mathrm{e}^{-cs} z\rangle + \int_0^s \psi(\mathrm{e}^{-cr} z)\mathrm{d}r\right] \\
&= \exp\left[\mathrm{i}\langle x, \mathrm{e}^{-ct-cs} z\rangle + \int_0^t \psi(\mathrm{e}^{-cr-cs} z)\mathrm{d}r + \int_0^s \psi(\mathrm{e}^{-cr} z)\mathrm{d}r\right] \\
&= \int P_{t+s}(x, \mathrm{d}w)\mathrm{e}^{\mathrm{i}\langle z, w\rangle}.
\end{aligned}
$$

Therefore $P_t(x, B)$ is a temporally homogeneous transition function. □

DEFINITION 17.2. When $c > 0$ and $\{Z_t\}$ is the Brownian motion on $\mathbb{R}^d$, the temporally homogeneous Markov process having the transition function $\{P_t(x, B)\}$ of Lemma 17.1 is called the *Ornstein–Uhlenbeck process* on $\mathbb{R}^d$. When $c > 0$ and $\{Z_t\}$ is a Lévy process on $\mathbb{R}^d$ generated by (G, ρ, β), the temporally homogeneous Markov process with the transition function $\{P_t(x, B)\}$ is called the *process of Ornstein–Uhlenbeck type generated by* (G, ρ, β, c). This process is sometimes called an *Ornstein–Uhlenbeck process driven by a Lévy process.*

REMARK 17.3. If $\{X_t : t \geq 0\}$ is a process of Ornstein–Uhlenbeck type, then it has a modification that has right-continuous sample functions with left limits. To prove this, use the one-point compactification $\mathfrak{X}$ of $\mathbb{R}^d$ and a metric compatible with the topology of $\mathfrak{X}$ and extend Theorem 11.1 to this setting. Then, we can prove that $\{X_t\}$ has a modification not only in $\mathfrak{X}$ but also in $\mathbb{R}^d$. See Dynkin [**121**], Theorem 3.7, Ethier and Kurtz [**133**], or Chung [**81**]. The Ornstein–Uhlenbeck process has a modification having continuous sample functions.

DEFINITION 17.4. A probability measure μ on $\mathbb{R}^d$ is the *limit distribution* of a temporally homogeneous Markov process on $\mathbb{R}^d$ with a transition function $\{P_t(x, B)\}$ if

$$
P_t(x, \cdot) \to \mu \quad \text{as } t \to \infty \tag{17.10}
$$

for any $x \in \mathbb{R}^d$.

THEOREM 17.5. *Fix* $c > 0$.
(i) *If* ρ *satisfies*

$$
\int_{|x|>2} \log|x| \rho(\mathrm{d}x) < \infty, \tag{17.11}
$$

the process of Ornstein–Uhlenbeck type on $\mathbb{R}^d$ *generated by* (G, ρ, β, c) *has a limit distribution* μ *with*

$$
\widehat{\mu}(z) = \exp\left[\int_0^\infty \psi(\mathrm{e}^{-cs} z)\mathrm{d}s\right]. \tag{17.12}
$$

The distribution μ *is selfdecomposable and the generating triplet* (A, ν, γ) *of* μ *is given by*

$$A = \frac{1}{2c}G, \tag{17.13}$$

$$\nu(B) = \frac{1}{c}\int_{\mathbb{R}^d}\rho(\mathrm{d}y)\int_0^\infty 1_B(\mathrm{e}^{-s}y)\mathrm{d}s, \quad B \in \mathcal{B}(\mathbb{R}^d), \tag{17.14}$$

$$\gamma = \frac{1}{c}\beta + \frac{1}{c}\int_{|y|>1}\frac{y}{|y|}\rho(\mathrm{d}y). \tag{17.15}$$

(ii) *For any selfdecomposable distribution* μ *on* $\mathbb{R}^d$, *there exists a unique triplet* (G, ρ, β) *satisfying (17.11) such that* μ *is the limit distribution of the process of Ornstein–Uhlenbeck type generated by* (G, ρ, β, c). *Using* λ *and* $k_\xi(r)$ *in the expression (15.8) for the Lévy measure* ν *of* μ *in Theorem 15.10, we have*

$$\rho(B) = -c\int_S \lambda(\mathrm{d}\xi)\int_0^\infty 1_B(r\xi)\mathrm{d}k_\xi(r), \tag{17.16}$$

where S *is the unit sphere (if* $d \geq 2$*) or* $\{1, -1\}$ *(if* $d = 1$*), and the integral with respect to* $\mathrm{d}k_\xi(r)$ *is the Stieltjes integral in* r.

Sometimes we say that ρ has finite log-moment if (17.11) is satisfied. Processes of Ornstein–Uhlenbeck type which do not satisfy (17.11) are studied in Theorem 17.11 below.

Proof of (i). Let $P_t(x, B)$ be the transition function of Lemma 17.1. As $t \to \infty$,

$$A_t \to \frac{1}{2c}G,$$

$$\begin{aligned}\int_{|x|\leq 1}|x|^2\nu_t(\mathrm{d}x) &= \int_{\mathbb{R}^d}\rho(\mathrm{d}y)\int_0^t |\mathrm{e}^{-cs}y|^2 1_D(\mathrm{e}^{-cs}y)\mathrm{d}s \\ &\to \int_{\mathbb{R}^d}|y|^2\rho(\mathrm{d}y)\int_0^\infty \mathrm{e}^{-2cs}1_{\{|y|\leq \mathrm{e}^{cs}\}}(y)\mathrm{d}s \\ &= \frac{1}{2c}\int_{\mathbb{R}^d}(|y|^2\wedge 1)\rho(\mathrm{d}y),\end{aligned}$$

$$\begin{aligned}\int_{|x|>1}\nu_t(\mathrm{d}x) &= \int_{\mathbb{R}^d}\rho(\mathrm{d}y)\int_0^t 1_{D^c}(\mathrm{e}^{-cs}y)\mathrm{d}s \\ &\to \int_{\mathbb{R}^d}\rho(\mathrm{d}y)\int_0^\infty 1_{\{|y|>\mathrm{e}^{cs}\}}(y)\mathrm{d}s = \frac{1}{c}\int_{|y|>1}\log|y|\rho(\mathrm{d}y),\end{aligned}$$

$$\gamma_{t,x} \to \frac{1}{c}\beta + \int_{\mathbb{R}^d}\rho(\mathrm{d}y)\int_0^\infty \mathrm{e}^{-cs}y 1_{\{1<|y|\leq \mathrm{e}^{cs}\}}(y)\mathrm{d}s$$

$$= \frac{1}{c}\beta + \frac{1}{c}\int_{|y|>1} \frac{y}{|y|}\rho(\mathrm{d}y).$$

The limits are finite by the assumption (17.11). Since

$$|g(z,x)| \le \tfrac{1}{2}|z|^2|x|^2 1_D(x) + 2\cdot 1_{D^c}(x)$$

for $g(z,x) = \mathrm{e}^{\mathrm{i}\langle z,x\rangle} - 1 - \mathrm{i}\langle z,x\rangle 1_D(x)$, the convergences above show that

$$\int_0^\infty \sup_{|z|\le a} |\psi(\mathrm{e}^{-cs}z)|\mathrm{d}s < \infty$$

for every $a > 0$. Hence $\int_0^\infty \psi(\mathrm{e}^{-cs}z)\mathrm{d}s$ is finite and continuous in z. Hence there exists a probability measure μ satisfying (17.12). Now (17.10) follows from (17.5) and (17.12). Thus μ is a limit distribution and infinitely divisible. The calculus above shows that

$$\int_{\mathbb{R}^d} P_t(x,\mathrm{d}y)\mathrm{e}^{\mathrm{i}\langle z,y\rangle} \to \exp\left[-\tfrac{1}{2}\langle z,Az\rangle + \int_{\mathbb{R}^d}\rho(\mathrm{d}y)\int_0^\infty g(z,\mathrm{e}^{-cs}y)\mathrm{d}s + \mathrm{i}\langle\gamma,z\rangle\right]$$

with A and γ satisfying (17.13) and (17.15). Hence the generating triplet of μ is as asserted. Given $b > 1$, we have

$$\widehat{\mu}(b^{-1}z) = \exp\left[\int_t^\infty \psi(\mathrm{e}^{-cs}z)\mathrm{d}s\right],$$

where $t = c^{-1}\log b$. Hence

$$\frac{\widehat{\mu}(z)}{\widehat{\mu}(b^{-1}z)} = \exp\left[\int_0^t \psi(\mathrm{e}^{-cs}z)\mathrm{d}s\right].$$

The right-hand side is the characteristic function of $P_t(0,\cdot)$. Hence μ is selfdecomposable. □

We prepare a lemma to prove part (ii).

LEMMA 17.6. *Let $k(r)$ and $l(r)$ be nonnegative right-continuous functions on $(0,\infty)$ such that $k(r)$ is decreasing and $k(\infty) = 0$, and $l(r)$ is increasing and $l(0+) = 0$. Then*

$$\int_{0+}^\infty k(r)\mathrm{d}l(r) = -\int_{0+}^\infty l(r-)\mathrm{d}k(r), \tag{17.17}$$

admitting $\infty = \infty$. If one side (hence both sides) of (17.17) is finite, then

$$\lim_{r\downarrow 0} k(r)l(r) = \lim_{r\uparrow\infty} k(r)l(r) = 0. \tag{17.18}$$

Proof. The Lebesgue–Stieltjes integral here is identical with the Lebesgue integral with respect to the induced signed measure. So, by Fubini's theorem, we have

$$\int_{0+}^\infty k(r)\mathrm{d}l(r) = -\int_{(0,\infty)}\mathrm{d}l(r)\int_{(r,\infty)}\mathrm{d}k(u)$$

$$= -\int_{(0,\infty)} \mathrm{d}k(u) \int_{(0,u)} \mathrm{d}l(r) = -\int_{0+}^{\infty} l(u-)\mathrm{d}k(u),$$

admitting $\infty = \infty$. If $\int_{0+}^{\infty} k(r)\mathrm{d}l(r) < \infty$, then

$$0 \leq k(r)l(r) \leq \int_{0+}^{r} k(u)\mathrm{d}l(u) \to 0 \quad \text{as } r \downarrow 0,$$

$$0 \leq k(r)l(r) \leq -\int_{r}^{\infty} l(u-)\mathrm{d}k(u) \to 0 \quad \text{as } r \uparrow \infty,$$

which show (17.18). □

Proof of Theorem 17.5(ii). Given a selfdecomposable distribution μ on $\mathbb{R}^d$, we express its Lévy measure ν as

$$\nu(B) = \int_S \lambda(\mathrm{d}\xi) \int_0^{\infty} 1_B(r\xi)k_\xi(r)\frac{\mathrm{d}r}{r}, \quad B \in \mathcal{B}(\mathbb{R}^d), \tag{17.19}$$

where λ is a probability measure on S and $k_\xi(r)$ is nonnegative, measurable in ξ and decreasing in r, and

$$\int_S \lambda(\mathrm{d}\xi) \int_0^{\infty} (r^2 \wedge 1)k_\xi(r)\frac{\mathrm{d}r}{r} < \infty. \tag{17.20}$$

Define a measure $\rho(B)$ by the right-hand side of (17.16). We claim that

$$\int_{|x|\leq 2} |x|^2\rho(\mathrm{d}x) + \int_{|x|>2} \log|x|\rho(\mathrm{d}x) < \infty. \tag{17.21}$$

Let

$$l(u) = \int_0^u (r^2 \wedge 1)\frac{\mathrm{d}r}{r} = \int_0^{\infty} ((\mathrm{e}^{-2t}u^2) \wedge 1)\mathrm{d}t = \begin{cases} \frac{1}{2}u^2, & 0 \leq u \leq 1, \\ \frac{1}{2} + \log u, & u > 1. \end{cases}$$

Then

$$\int_{\mathbb{R}^d} l(|x|)\rho(\mathrm{d}x) = -c\int_S \lambda(\mathrm{d}\xi) \int_0^{\infty} l(r)\mathrm{d}k_\xi(r),$$

which is, by Lemma 17.6, c times the outer integral in (17.20). Therefore we have (17.21). If $B \in \mathcal{B}(\mathbb{R}^d)$ satisfies $B \subset \{x\colon |x| > \varepsilon\}$ for some $\varepsilon > 0$, then, applying Lemma 17.6 to (17.19), we get

$$\begin{aligned}\nu(B) &= -\int_S \lambda(\mathrm{d}\xi) \int_0^{\infty} \mathrm{d}k_\xi(r) \int_0^r 1_B(u\xi)\frac{\mathrm{d}u}{u} \\ &= -\int_S \lambda(\mathrm{d}\xi) \int_0^{\infty} \mathrm{d}k_\xi(r) \int_0^{\infty} 1_B(\mathrm{e}^{-t}r\xi)\mathrm{d}t \\ &= \frac{1}{c}\int_{\mathbb{R}^d} \rho(\mathrm{d}y) \int_0^{\infty} 1_B(\mathrm{e}^{-t}y)\mathrm{d}t.\end{aligned}$$

That is, (17.14) holds. Next, define G and β by (17.13) and (17.15). Then, the process of Ornstein–Uhlenbeck type generated by (G, ρ, β, c) has μ as the limit distribution.

Uniqueness. Suppose that two processes of Ornstein–Uhlenbeck type with a common c have the limit distribution μ. Let $\{Z_t\}$ and $\{Z_t^\sharp\}$ be the associated Lévy processes with $\psi(z) = \log \widehat{P}_{Z_1}(z)$ and $\psi^\sharp(z) = \log \widehat{P}_{Z_1^\sharp}(z)$. Then we have (17.12) and the same identity with $\psi^\sharp$ in place of ψ. It follows that

$$\exp\left[\int_0^t \psi(\mathrm{e}^{-cs}z)\mathrm{d}s\right] = \exp\left[\int_0^t \psi^\sharp(\mathrm{e}^{-cs}z)\mathrm{d}s\right]$$

for every $t > 0$ and $z \in \mathbb{R}^d$ by the argument at the end of the proof of (i). Differentiating the above at $t = 0$, we get $\psi(z) = \psi^\sharp(z)$. □

COROLLARY 17.7. *Fix $c > 0$. If ν is the Lévy measure of a selfdecomposable distribution on $\mathbb{R}^d$, then ν is expressed by (17.14) with a unique measure ρ on $\mathbb{R}^d$ satisfying (17.11) and*

$$\rho\{0\} = 0 \quad \textit{and} \quad \int_{|x|\leq 2} |x|^2 \rho(\mathrm{d}x) < \infty. \tag{17.22}$$

Conversely, for every measure ρ on $\mathbb{R}^d$ satisfying (17.11) and (17.22), the measure ν determined by (17.14) is the Lévy measure of a selfdecomposable distribution.

Proof. Obvious from Theorem 17.5. □

DEFINITION 17.8. A probability measure μ on $\mathbb{R}^d$ is an *invariant distribution* of the temporally homogeneous Markov process on $\mathbb{R}^d$ with transition function $\{P_t(x, B)\}$ if

$$\int_{\mathbb{R}^d} \mu(\mathrm{d}x) P_t(x, B) = \mu(B) \quad \text{for } t > 0 \text{ and } B \in \mathcal{B}(\mathbb{R}^d)$$

or, equivalently, if the Markov process $\{X_t\}$ with this transition function and the initial distribution μ satisfies $P[\, X_t \in B\,] = \mu(B)$ for any t and B.

COROLLARY 17.9. *A process of Ornstein–Uhlenbeck type satisfying (17.11) has a unique invariant distribution, and this distribution is selfdecomposable.*

Proof. Let μ be the limit selfdecomposable distribution of $\{P_t(x, B)\}$ in Theorem 17.5. For any bounded continuous function f we have

$$\int P_s(x, \mathrm{d}y) \int P_t(y, \mathrm{d}z) f(z) = \int P_{s+t}(x, \mathrm{d}z) f(z).$$

Notice that $\int P_t(y, \mathrm{d}z)f(z)$ is continuous in y by (17.9). Letting $s \to \infty$, we get

$$\int \mu(\mathrm{d}y) \int P_t(y, \mathrm{d}z)f(z) = \int \mu(\mathrm{d}z)f(z). \tag{17.23}$$

Hence μ is an invariant distribution. Conversely, if μ_1 is an invariant distribution, then (17.23) holds with μ_1 in place of μ. Letting $t \to \infty$, we get $\int \mu(\mathrm{d}z)f(z) = \int \mu_1(\mathrm{d}z)f(z)$, that is, $\mu = \mu_1$. □

EXAMPLE 17.10. Let $d = 1$ and let $\{Z_t\}$ be a Lévy process with $E[\mathrm{e}^{\mathrm{i}zZ_t}] = \exp\left[\int_0^\infty (\mathrm{e}^{\mathrm{i}zx} - 1)\, \rho(\mathrm{d}x)\right]$, assuming $\int_0^\infty (x \wedge 1)\, \rho(\mathrm{d}x) < \infty$. Consider the process $\{X_t\}$ of Ornstein–Uhlenbeck type determined by $\{Z_t\}$ and $c > 0$. The process $\{Z_t\}$ is increasing in t (Theorem 21.5 or Corollary 24.8). The process $\{X_t\}$ is a model for water storage in a dam with random inflow combined with outflow in the speed proportional to the water level, as (17.1) shows. There exists an invariant measure μ if and only if $\int_{(2,\infty)} \log x\, \rho(\mathrm{d}x) < \infty$; μ is given by

$$\widehat{\mu}(z) = \exp\left[\frac{1}{c}\int_0^\infty (\mathrm{e}^{\mathrm{i}zx} - 1)\rho(x, \infty)\frac{\mathrm{d}x}{x}\right].$$

Every selfdecomposable distribution with support $[0, \infty)$ is obtained in this way.

Let us consider processes of Ornstein–Uhlenbeck type which do not satisfy the condition (17.11).

THEOREM 17.11. *Consider a process of Ornstein–Uhlenbeck type on $\mathbb{R}^d$ generated by (G, ρ, β, c) such that*

$$\int_{|x|>2} \log |x|\, \rho(\mathrm{d}x) = \infty. \tag{17.24}$$

Then, for any x, its transition function $P_t(x, \cdot)$ does not converge to a probability measure as $t \to \infty$. An invariant distribution does not exist.

Proof. Suppose that, for some $x_0 \in \mathbb{R}^d$, $P_t(x_0, \cdot)$ tends to a probabilty measure μ as $t \to \infty$. Since $P_t(x_0, \cdot)$ is infinitely divisible (Lemma 17.1), μ is infinitely divisible (Lemma 7.8). Let ν be the Lévy measure of μ. Then, by Theorem 8.7,

$$\int f(x)\nu_t(\mathrm{d}x) \to \int f(x)\nu(\mathrm{d}x), \quad t \to \infty,$$

for any bounded continuous f vanishing on a neighborhood of 0. Here ν_t is the Lévy measure of $P_t(x_0, \cdot)$. It follows from (17.7) that

$$\int_{|x|>1}\nu_t(\mathrm{d}x) = \int_{|y|>1}\rho(\mathrm{d}y)(t \wedge (c^{-1}\log |y|)),$$

which tends to ∞ by the assumption (17.24). This is absurd. Hence $P_t(x, \cdot)$ does not tend to a probability measure as $t \to \infty$.

Suppose that there is an invariant distribution μ_1. Then

$$\iint \mu_1(\mathrm{d}x)P_t(x, \mathrm{d}y)\mathrm{e}^{\mathrm{i}\langle z,y\rangle} = \int \mu_1(\mathrm{d}y)\mathrm{e}^{\mathrm{i}\langle z,y\rangle}.$$

The right-hand side is $\widehat{\mu}_1(z)$. The left-hand side equals

$$\widehat{\mu}_1(e^{-ct}z)\exp\left[\int_0^t \psi(e^{-cs}z)ds\right]$$

by (17.5). It follows that

$$\lim_{t\to\infty}\exp\left[\int_0^t \psi(e^{-cs}z)ds\right]=\widehat{\mu}_1(z),$$

that is, $P_t(0,\cdot)\to\mu_1$ as $t\to\infty$. This contradicts the fact just proved. □

REMARK 17.12. Let $C(x,a)$ be a cube in $\mathbb{R}^d$ with center $x=(x_j)_{1\le j\le d}$:

$$C(x,a)=[x_1-a,x_1+a]\times\cdots\times[x_d-a,x_d+a].$$

In general, if μ is an infinitely divisible distribution on $\mathbb{R}^d$ with Lévy measure ν, then, for every $x\in\mathbb{R}^d$ and $a>0$,

$$\mu(C(x,a))\le K_d\Big(\int_{|y|>a/\pi}\nu(dy)\Big)^{-1/2}, \tag{17.25}$$

where K_d is a constant which depends only on d. In the case $d=1$, this is Le Cam's estimate (see [**197**]) and proved by the argument in the proof of Lemma 48.3. For general d, this is proved in Sato and Yamazato [**469**]. Therefore

$$P_t(x,C(y,a))\le K_d\Big(\int_{|z|>a/\pi}\nu_t(dz)\Big)^{-1/2} \tag{17.26}$$

for any process of Ornstein–Uhlenbeck type. Under the condition (17.24), the right-hand side of (17.26) tends to 0 as is shown in the proof of Theorem 17.11, and we get

$$\lim_{t\to\infty}\sup_x\sup_y P_t(x,C(y,a))=0 \tag{17.27}$$

for any $a>0$.

18. Exercises 3

E 18.1. Let μ be infinitely divisible on $\mathbb{R}^d$ with generating triplet (A,ν,γ). Prove the following.

(i) μ is symmetric if and only if ν is symmetric and $\gamma=0$.

(ii) The symmetrization $\mu^\sharp$ of μ is infinitely divisible with generating triplet $(2A,2\nu_0,0)$, where $\nu_0(B)=\frac{1}{2}(\nu(B)+\nu(-B))$ for $B\in\mathcal{B}(\mathbb{R}^d)$.

E 18.2. Show that if μ is a probability measure on $\mathbb{R}^d$ with $d\ge 2$, then the following are equivalent:

(1) μ is rotation invariant;
(2) $\widehat{\mu}(z)$ is a function only of $|z|$;
(3) $\widehat{\mu}(z)$ is real and is a function only of $|z|$;
(4) $\mu(B)=\mu(U^{-1}B)$, $B\in\mathcal{B}(\mathbb{R}^d)$, for every orthogonal matrix U with determinant 1.

Note that, if μ is a probability measure on $\mathbb{R}$, then (1), (2), and (3) are equivalent.

E 18.3. Prove that an infinitely divisible distribution μ on $\mathbb{R}^d$ with generating triplet (A,ν,γ) is rotation invariant if and only if $A=aI$ with $a\ge 0$ and I the identity, ν is rotation invariant, and $\gamma=0$.

E 18.4. Let μ be a probability measure on $\mathbb{R}^d$. Do not assume that μ is infinitely divisible. Prove the following.

(i) If, for all $n \in \mathbb{N}$, there are $b_n > 0$ and $c_n \in \mathbb{R}^d$ such that $\widehat{\mu}(z)^n$ equals $\widehat{\mu}(b_n z)e^{i\langle c_n, z\rangle}$, then μ is stable.

(ii) If there are an integer $n \geq 2$, $b > 0$, and $c \in \mathbb{R}^d$ such that $\widehat{\mu}(z)^n = \widehat{\mu}(bz)e^{i\langle c,z\rangle}$, then μ is semi-stable.

E 18.5. Show the following characterization of stable distributions: a probability measure μ on $\mathbb{R}^d$ is stable if and only if, for any $a_1 > 0$ and $a_2 > 0$, we can find $b > 0$ and $c \in \mathbb{R}^d$ satisfying

$$a_1 X + a_2 Y \stackrel{\mathrm{d}}{=} bX + c,$$

where X and Y are independent random variables each with distribution μ. If μ is non-trivial and α-stable, then $b = (a_1{}^\alpha + a_2{}^\alpha)^{1/\alpha}$.

E 18.6. Let $\{X_t\}$ be a non-trivial α-stable process on $\mathbb{R}^d$, $0 < \alpha \leq 2$. Show the following.

(i) For any $a > 0$, $\{X_{at}\} \stackrel{\mathrm{d}}{=} \{a^{1/\alpha}X_t + tc_a\}$ with

$$c_a = \begin{cases} (a - a^{1/\alpha})\tau & \text{if } \alpha \neq 1, \\ a(\log a)\frac{2}{\pi}\int_S \xi\lambda_1(\mathrm{d}\xi) & \text{if } \alpha = 1. \end{cases}$$

Here τ and λ_1 are those of Theorem 14.10 if $\alpha \neq 2$, and τ is the mean of X_1 if $\alpha = 2$. If $d = 1$ and $\alpha = 1$, then $\int_S \xi\lambda_1(\mathrm{d}\xi) = c\beta$ in the parameters in (14.25).

(ii) For any $t > 0$,

$$X_t \stackrel{\mathrm{d}}{=} \begin{cases} t^{1/\alpha}X_1 + (t - t^{1/\alpha})\tau & \text{if } \alpha \neq 1, \\ tX_1 + t(\log t)\frac{2}{\pi}\int_S \xi\lambda_1(\mathrm{d}\xi) & \text{if } \alpha = 1. \end{cases}$$

E 18.7. Show that the function $k(t)$ in Proposition 13.14 is not unique.

E 18.8. The spherical part of the Lévy measure of an α-stable distribution in $\mathbb{R}^d$ is expressed by λ and λ_1 in Theorems 14.3 and 14.10, respectively. For $0 < \alpha < 2$ let c_α be the positive constant such that $\lambda_1 = c_\alpha \lambda$. Prove that

$$c_\alpha = \pi^{1/2}2^{-\alpha}\frac{\Gamma((2-\alpha)/2)}{\alpha\Gamma((1+\alpha)/2)},$$

and see that c_α is continuous in $\alpha \in (0, 2)$.

E 18.9. Let μ be a rotation invariant α-stable distribution with $0 < \alpha < 2$ such that the measure λ in Theorem 14.3(ii) is the uniform probability measure on S. Show that the constant c in (14.23) is then equal to

$$c_0 = 2^{-\alpha}\frac{\Gamma(d/2)\Gamma((2-\alpha)/2)}{\alpha\Gamma((\alpha+d)/2)}.$$

E 18.10. Let $\{X_t\}$ be a strictly α-stable process on $\mathbb{R}$ with $\alpha \neq 1, 2$. Using the parameter θ in Theorem 14.19, show that

$$P[X_t > 0] = \tfrac{1}{2}(1 + \theta).$$

E 18.11. Suppose that μ is α-semi-stable on $\mathbb{R}^d$ with spans b and b'. Show that, if it is strictly α-semi-stable with b as a span, then it is strictly α-semi-stable with b' as a span.

E 18.12. Let μ be infinitely divisible on $\mathbb{R}^d$ and satisfy, for some $a > 1$, $\alpha \in (0,2]$, and $c \in \mathbb{R}^d$, $\widehat{\mu}(z)^a = \widehat{\mu}(-a^{1/\alpha}z)e^{i\langle c,z\rangle}$.

(i) Show that μ is α-semi-stable.

(ii) Suppose that $\alpha \neq 1$. Show that μ is strictly α-semi-stable if and only if $c = 0$.

(iii) Show that if $\alpha = 1$, then μ is strictly 1-semi-stable.

(iv) For $d = 1$ and $\alpha < 2$, characterize the Lévy measure of μ.

E 18.13. Show that a rotation invariant infinitely divisible distribution μ on $\mathbb{R}^d$ is selfdecomposable if and only if its Lévy measure ν is of the form $\nu(B) = \int_B |x|^{-d}k(|x|)dx$ with a nonnegative decreasing function $k(\cdot)$.

E 18.14. Show that there is a probability measure μ such that $\widehat{\mu}(z) = \widehat{\mu}(b^{-1}z)\widehat{\rho}(z)$ for some $b > 1$ with a probability measure ρ which is not infinitely divisible. (Continued in E 29.13.)

E 18.15. Show that $c = 1 - \gamma$ for the constant c in (14.21). Here γ is Euler's constant, $\gamma = 0.5772\ldots$.

E 18.16. Let $0 < \alpha \leq 2$. Let $\{Z_t\}$ be a strictly α-stable process on $\mathbb{R}^d$, and let $\{X_t\}$ be the process of Ornstein–Uhlenbeck type associated with $\{Z_t\}$ and $c > 0$. Prove that $\{X_t\}$ has a strictly α-stable limit distribution. If $\{X_t\}$ has starting point 0, then the distribution of X_t is also strictly α-stable.

E 18.17 (Breiman [**67**]). Let $0 < \alpha \leq 2$ and $c = 1/\alpha$. Let $\{Z_t\}$ be a strictly α-stable process on $\mathbb{R}^d$. Let $\mu = P_{Z(1)}$. Define $X_t = e^{-t/\alpha}Z(e^t)$. Prove that $\{X_t\colon -\infty < t < \infty\}$ has the following properties.

(i) For any t_0, $\{X_{t_0+t}\colon t \geq 0\}$ is the process of Ornstein–Uhlenbeck type associated with $\{Z_t\}$ and c.

(ii) For any t, X_t has distribution μ.

E 18.18. Let $\{X_t\}$ be a non-trivial strictly α-stable process on $\mathbb{R}$ with $0 < \alpha < 2$. Define $Y_t = t^{2/\alpha}X_{1/t}$ for $t > 0$ and $Y_0 = 0$. Show that $Y_t \stackrel{d}{=} X_t$ for any fixed t but that $\{Y_t\}$ is not a Lévy process. Compare this with Theorem 5.4(iii).

E 18.19 (Linnik and Ostrovski [**325**]). Consider a probability density

$$f(x) = c_0 \exp(bx - ce^{ax}), \qquad x \in \mathbb{R},$$

where a, b, and c are positive constants and c_0 is a normalizing constant. Show that $f(x)dx$ is a purely non-Gaussian selfdecomposable distribution with Lévy measure $\nu(dx) = |x|^{-1}e^{bx}(1-e^{ax})^{-1}1_{(-\infty,0)}(x)\,dx$. What is the corresponding measure ρ in Theorem 17.5(ii)? (If X has Γ-distribution with parameters $a^{-1}b$, c, then $a^{-1}\log X$ has distribution $f(x)dx$.)

Notes

A historical monograph that includes study of stable distributions is Lévy [**311**] in 1925. This is before he developed the theory of infinitely divisible distributions. The two books, Lévy [**317**] in 1937 and Khintchine [**279**] in 1938, characterize stable distributions as a class of infinitely divisible distributions. These are mainly in one dimension, but Lévy [**317**] studies also the case of $\mathbb{R}^d$. Thereafter many books on probability theory contain chapters on stable distributions. A monograph for the multi-dimensional case is Linde [**319**]. The primary importance of stable distributions lies in their role in a limit theorem for $b_n \sum_{j=1}^n Z_j + c_n$ for independent identically distributed random variables $\{Z_n\}$ (Theorem 15.7). The class of the distributions of Z_n for a limit stable distribution μ is called the *domain of attraction* of μ. Its description was given by Khintchine, Feller, Lévy, Gnedenko, Dœblin, and others in the 1930s [**167, 212, 426**]. It is intimately connected with the theory of regularly varying functions [**38, 139**]. The middle of the century saw the development of limit theorems for distributions of random variables into those for distributions on the spaces of sample functions of stochastic processes. Many old limit theorems are taken as convergence of the distributions of functionals of stochastic processes. See Billingsley [**33**].

Many papers study various aspects of stable processes. We will treat some of them in later chapters. Reviews on stable processes are found in [**26, 150, 393, 519, 520, 528, 529**]. Some people use the name of stable processes for processes with stable finite-dimensional distributions. Analysis of such processes is the subject of the book of Samorodnitsky and Taqqu [**427**]. A collection of properties of stable distributions on $\mathbb{R}$ is Zolotarev [**608**].

Selfsimilar processes were introduced by Lamperti [**305**] under the name of semi-stable processes. Special ones without independent increments (now called fractional Brownian motions) were discovered earlier by Kolmogorov [**292**]. Many selfsimilar processes and related limit theorems are known. See a review [**331**] of Maejima. The exponents are denoted by H after the name of H. E. Hurst, a scientist working on the flow of the Nile. Semi-selfsimilar processes are introduced by Maejima and Sato [**333**]. Theorem 13.11 on existence of exponents is proved there, but the special cases of stable and semi-stable processes have been known since Lévy.

Semi-stable distributions were studied already by Lévy [**311, 317**]. They are treated in the book [**248**] of Kagan, Linnik, and Rao. Connections with limit theorems for subsequences of $\{b_n \sum_{j=1}^n Z_j + c_n\}$ with $\{Z_n\}$ independent and identically distributed are indicated by Shimizu [**484**], Pillai [**374**], and Kruglov [**297**]. Proposition 14.9 is by Choi [**76**].

Without the name, selfdecomposable distributions are introduced and characterized in Lévy's book [**317**] as a class of limit distributions (Theorem 15.3). Khintchine's book [**279**] seems to be the first to use the name "class L". Loève's book [**327**] uses the name selfdecomposable. Semi-selfdecomposable distributions are introduced by Maejima and Naito [**332**]. Theorem 15.8 on their Lévy measures is by them. Theorem 15.10 is by Lévy in one dimension and by Wolfe [**580**] and Sato [**431**] in many dimensions.

Theorem 16.1 on selfsimilar additive processes is by Sato [**440**]. Theorems 16.6–16.8 are its extension to semi-selfsimilar additive processes made by Maejima and Sato [**333**]. Comparison of the selfsimilar additive process and the Lévy process associated with a common selfdecomposable distribution is proposed by [**440**]. Study of path behaviors of increasing selfsimilar additive processes is made by Watanabe [**558**].

The representation (17.14) of the Lévy measures of selfdecomposable distributions on $\mathbb{R}^d$ was found by Urbanik [**542**]. Lemma 17.1, Theorems 17.5 and 17.11, and Corollary 17.9 are by Sato and Yamazato [**468, 469**]. But, earlier, Wolfe [**582**] noticed essentially the same results in case $d = 1$. A special case given in Example 17.10 was found by Çinlar and Pinsky [**85**]. Jurek and Vervaat [**246**] and Jurek [**243**] also give similar results without considering processes of Ornstein–Uhlenbeck type. The correspondence between $\{Z_t\}$ and μ in Theorem 17.5 is continuous, as formulated in [**469**].

Let $d \geq 2$. If an infinitely divisible distribution μ on $\mathbb{R}^d$ has the property that, for any $a > 0$, there are an invertible $d \times d$ matrix B and $c \in \mathbb{R}^d$ such that $\widehat{\mu}(z)^a = \widehat{\mu}(B'z)e^{i\langle c,z\rangle}$, then μ is called *operator-stable.* If B is expressed as $B = a^Q = \sum_{n=0}^{\infty}(n!)^{-1}(\log a)^n Q^n$, then Q is called an *exponent.* The notion is introduced by Sharpe [**477**] to describe limit distributions of operator normalization $B_n \sum_{j=1}^{n} Z_j + c_n$ for $\{Z_n\}$ independent and identically distributed, where B_n are invertible $d \times d$ matrices. See Jurek and Mason [**245**]. There is no uniqueness for the exponent Q. The corresponding Lévy processes are operator-stable processes. Processes with stable components are special cases of operator-stable processes (see Remark 49.16). Characterization of strictly operator-stable distributions is given by Sato [**438**]. Jajte [**236**] and Łuczak [**328**] consider *operator-semi-stable* distributions. A fundamental work for generalization of selfdecomposable to *operator-selfdecomposable* is Urbanik [**543**]. When Q is a $d \times d$ matrix all eigenvalues of which have positive real parts, μ is called *Q-selfdecomposable* if, for every $b > 1$, there is a probability measure ρ_b such that $\widehat{\mu}(z) = \widehat{\mu}(b^{-Q'}z)\widehat{\rho}_b(z)$. Extending the definition of processes of Ornstein–Uhlenbeck type, the results on selfdecomposable distributions in Sections 15 and 17 are generalized to Q-selfdecomposable distributions. See [**468, 469**].

Between the class of (semi-)selfdecomposable distributions and the class of (semi-)stable distributions there is a chain of subclasses called L_m, $m = 1, 2, \ldots, \infty$. See Urbanik [**544**], Sato [**431**], Maejima and Naito [**332**], and, for operator extensions, Sato and Yamazato [**470**].

CHAPTER 4

The Lévy–Itô decomposition of sample functions

19. Formulation of the Lévy–Itô decomposition

The Lévy–Itô decomposition expresses sample functions of an additive process as a sum of two independent parts – a continuous part and a part expressible as a compensated sum of independent jumps. In general we cannot express the latter part as the sum of jumps, since the sum of all jumps up to time t may be divergent. This is a delicate point. By a compensated sum we mean summation, similar to Cauchy's principal value, of random quantities with means simultaneously subtracted. The decomposition was conceived by Lévy [**312, 317**], and formulated and proved by Itô [**220**] using many pages. The Lévy–Khintchine representation of infinitely divisible distributions was given by Lévy with the decomposition of sample functions in background understanding. Itô made Lévy's background understanding explicit. He derived the Lévy–Khintchine representation from the decomposition of sample functions. We will, however, adopt the reverse course. That is, we will make full use of the Lévy–Khintchine representation in proving the Lévy–Itô decomposition. Loève [**327**] also uses the representation in his proof of the Lévy–Itô decomposition.

We begin with the definition of Poisson random measures. Let

$$\overline{\mathbb{Z}}_+ = \mathbb{Z}_+ \cup \{+\infty\} = \{0, 1, 2, \dots\} \cup \{+\infty\}.$$

We use the following convention on Poisson distributions. Let X be a $\overline{\mathbb{Z}}_+$-valued random variable. We say that X has Poisson distribution with mean 0 if $X = 0$ a.s.; X has Poisson distribution with mean $+\infty$ if $X = +\infty$ a.s.

DEFINITION 19.1. Let $(\Theta, \mathcal{B}, \rho)$ be a σ-finite measure space. A family of $\overline{\mathbb{Z}}_+$-valued random variables $\{N(B) \colon B \in \mathcal{B}\}$ is called a *Poisson random measure* on Θ with *intensity measure* ρ, if the following hold:

(1) for every B, $N(B)$ has Poisson distribution with mean $\rho(B)$;
(2) if $B_1, \dots, B_n$ are disjoint, then $N(B_1), \dots, N(B_n)$ are independent;
(3) for every ω, $N(\cdot, \omega)$ is a measure on Θ.

We write

$$D_{a,b} = D(a, b] \;\, = \{\, x \in \mathbb{R}^d \colon a < |x| \le b \,\} \quad \text{for } 0 \le a < b < \infty,$$
$$D_{a,\infty} = D(a, \infty) = \{\, x \in \mathbb{R}^d \colon a < |x| < \infty \,\} \quad \text{for } 0 \le a < \infty.$$

Thus, $D_{0,\infty} = \mathbb{R}^d \setminus \{0\}$. Further

$$H = (0,\infty) \times (\mathbb{R}^d \setminus \{0\}) = (0,\infty) \times D_{0,\infty}.$$

A point h in H is denoted by $h = (s,x)$ with $s \in (0,\infty)$ and $x \in D_{0,\infty}$. The Borel σ-algebra of H is denoted by $\mathcal{B}(H)$. The integral of $f(h)$ with respect to a measure ρ on H is written as

$$\int_H f(h)\rho(\mathrm{d}h) = \int_{(0,\infty)\times D(0,\infty)} f(s,x)\rho(\mathrm{d}(s,x)).$$

Now we formulate the main theorems of this chapter. Theorem 19.2 deals with a general additive process; Theorem 19.3 with an additive process satisfying $\int_{|x|\le 1} |x|\, \nu_t(\mathrm{d}x) < \infty$ for every t.

THEOREM 19.2. *Let $\{X_t\colon t \ge 0\}$ be an additive process on $\mathbb{R}^d$ defined on a probability space $(\Omega, \mathcal{F}, P)$ with system of generating triplets $\{(A_t, \nu_t, \gamma(t))\}$ and define the measure $\widetilde{\nu}$ on H by $\widetilde{\nu}((0,t] \times B) = \nu_t(B)$ for $B \in \mathcal{B}(\mathbb{R}^d)$. Using Ω_0 from Definition 1.6 of an additive process, define, for $B \in \mathcal{B}(H)$,*

$$J(B,\omega) = \begin{cases} \#\{\, s\colon (s, X_s(\omega) - X_{s-}(\omega)) \in B\,\} & \text{for } \omega \in \Omega_0, \\ 0 & \text{for } \omega \notin \Omega_0. \end{cases} \tag{19.1}$$

Then the following hold.

(i) *$\{J(B)\colon B \in \mathcal{B}(H)\}$ is a Poisson random measure on H with intensity measure $\widetilde{\nu}$.*

(ii) *There is $\Omega_1 \in \mathcal{F}$ with $P[\Omega_1] = 1$ such that, for any $\omega \in \Omega_1$,*

$$X_t^1(\omega) = \lim_{\varepsilon \downarrow 0} \int_{(0,t]\times D(\varepsilon,1]} \{x\, J(\mathrm{d}(s,x),\omega) - x\,\widetilde{\nu}(\mathrm{d}(s,x))\} + \int_{(0,t]\times D(1,\infty)} x\, J(\mathrm{d}(s,x),\omega) \tag{19.2}$$

is defined for all $t \in [0,\infty)$ and the convergence is uniform in t on any bounded interval. The process $\{X_t^1\}$ is an additive process on $\mathbb{R}^d$ with $\{(0, \nu_t, 0)\}$ as the system of generating triplets.

(iii) *Define*

$$X_t^2(\omega) = X_t(\omega) - X_t^1(\omega) \qquad \text{for } \omega \in \Omega_1. \tag{19.3}$$

There is $\Omega_2 \in \mathcal{F}$ with $P[\Omega_2] = 1$ such that, for any $\omega \in \Omega_2$, $X_t^2(\omega)$ is continuous in t. The process $\{X_t^2\}$ is an additive process on $\mathbb{R}^d$ with $\{(A_t, 0, \gamma(t))\}$ as the system of generating triplets.

(iv) *The two processes $\{X_t^1\}$ and $\{X_t^2\}$ are independent.*

When $t = 0$, the set $(0,t]$ is understood to be empty and the integrals in (19.2) are zero. We will prove that $\int_{(0,t]\times D(\varepsilon,1]} \{x\, J(\mathrm{d}(s,x),\omega) - x\,\widetilde{\nu}(\mathrm{d}(s,x))\}$ has mean 0. Its limit as $\varepsilon \downarrow 0$ is called the *compensated sum of jumps.*

Without the subtraction, the sum of jumps $\int_{(0,t]\times D(\varepsilon,1]} x\, J(\mathrm{d}(s,x),\omega)$ may not converge as $\varepsilon \downarrow 0$.

THEOREM 19.3. *Suppose that the additive process* $\{X_t\}$ *in Theorem 19.2 satisfies* $\int_{|x|\le 1} |x|\, \nu_t(\mathrm{d}x) < \infty$ *for all* $t > 0$. *Let* $\gamma_0(t)$ *be the drift of* X_t. *Then, there is* $\Omega_3 \in \mathcal{F}$ *with* $P[\Omega_3] = 1$ *such that, for any* $\omega \in \Omega_3$,

$$X_t^3(\omega) = \int_{(0,t]\times D(0,\infty)} x\, J(\mathrm{d}(s,x),\omega) \tag{19.4}$$

is defined for all $t \ge 0$. *The process* $\{X_t^3\}$ *is an additive process on* $\mathbb{R}^d$ *such that*

$$E[\mathrm{e}^{\mathrm{i}\langle z, X_t^3\rangle}] = \exp\left[\int_{\mathbb{R}^d} (\mathrm{e}^{\mathrm{i}\langle z,x\rangle} - 1)\, \nu_t(\mathrm{d}x)\right]. \tag{19.5}$$

Define

$$X_t^4(\omega) = X_t(\omega) - X_t^3(\omega) \quad \text{for } \omega \in \Omega_3. \tag{19.6}$$

Then, for any $\omega \in \Omega_2 \cap \Omega_3$, $X_t^4(\omega)$ *is continuous in* t *and* $\{X_t^4\}$ *is an additive process on* $\mathbb{R}^d$ *such that*

$$E[\mathrm{e}^{\mathrm{i}\langle z, X_t^4\rangle}] = \exp[-\tfrac{1}{2}\langle z, A_t z\rangle + \mathrm{i}\langle \gamma_0(t), z\rangle]. \tag{19.7}$$

The two processes $\{X_t^3\}$ *and* $\{X_t^4\}$ *are independent.*

Theorems 19.2 and 19.3 are called the *Lévy–Itô decomposition*. When $\{X_t\}$ satisfies $\int_{|x|\le 1} |x|\, \nu_t(\mathrm{d}x) < \infty$ for every t, we call $\{X_t^3\}$ and $\{X_t^4\}$ in Theorem 19.3 the *jump part* and the *continuous part* of $\{X_t\}$, respectively. Also in Theorem 19.2 the processes $\{X_t^1\}$ and $\{X_t^2\}$ are called the jump part and the continuous part of $\{X_t\}$, respectively. But we have to keep two things in mind in this case: first, the definition of $\{X_t^1\}$ and $\{X_t^2\}$ depends on the choice of the representation explained in Remark 8.4 and thus they have intrinsic meaning only up to addition of deterministic processes; second, the sum of jumps of $\{X_t^1\}$ until time t may not converge.

Our strategy for proof of the Lévy–Itô decomposition is as follows. First we construct Poisson random measures. Second, given any additive process $\{X_t\}$, we use its system of generating triplets $\{(A_t, \nu_t, \gamma(t))\}$ and construct an additive process $\{Y_t\}$ such that $\{Y_t\} \stackrel{\mathrm{d}}{=} \{X_t\}$ and $\{Y_t\}$ has the Lévy–Itô decomposition. Third, using the facts that $\{X_t\}$ and $\{Y_t\}$ induce an identical probability measure on the space $\mathbf{D} = D([0,\infty), \mathbb{R}^d)$ of right-continuous paths with left limits with the σ-algebra $\mathcal{F}_{\mathbf{D}}$ generated by the Borel cylinder sets and that all relevant quantities are $\mathcal{F}_{\mathbf{D}}$-measurable, we can prove that $\{X_t\}$ also has the Lévy–Itô decomposition. Thus, we can avoid direct analysis of sample function behavior of a given additive process in Itô's argument.

We need a general existence theorem for Poisson random measures.

PROPOSITION 19.4. *For any given σ-finite measure space $(\Theta, \mathcal{B}, \rho)$, there exists, on some probability space $(\Omega^0, \mathcal{F}^0, P^0)$, a Poisson random measure $\{N(B)\colon B \in \mathcal{B}\}$ on Θ with intensity measure ρ.*

Proof. Step 1. Assume that $\rho(\Theta) < \infty$. If $\rho = 0$, then choose $N(B)$ identically zero. Assume that $\rho(\Theta) > 0$. On some probability space $(\Omega^0, \mathcal{F}^0, P^0)$ we can construct, by Theorem 1.9, a sequence $\{Z_n\colon n = 1, 2, \dots\}$ of independent identically distributed random variables on Θ each with distribution $\rho(\Theta)^{-1}\rho$ and a Poisson random variable Y with mean $\rho(\Theta)$ such that Y and $\{Z_n\}$ are independent. Define $N(B) = 0$ if $Y = 0$, and $N(B) = \sum_{j=1}^{Y} 1_B(Z_j)$ if $Y \geq 1$. Then clearly $N(B)$ satisfies (3) of Definition 19.1. Let $k \geq 2$. Let $B_1, \dots, B_k \in \mathcal{B}$ be disjoint with $\bigcup_{j=1}^{k} B_j = \Theta$, and $n_1, \dots, n_k \in \mathbb{Z}_+$. Let $n_1 + \dots + n_k = n$. We have

$$\begin{aligned} &P[\,N(B_1) = n_1, \dots, N(B_k) = n_k\,] \\ &\quad = P[\,N(B_1) = n_1, \dots, N(B_k) = n_k \mid N(\Theta) = n\,]P[\,N(\Theta) = n\,] \\ &\quad = P\left[\sum_{j=1}^{n} 1_{B_1}(Z_j) = n_1, \dots, \sum_{j=1}^{n} 1_{B_k}(Z_j) = n_k\right] P[\,Y = n\,]. \end{aligned}$$

Here we meet a multinomial distribution and a Poisson distribution. Thus we have

$$\begin{aligned} &P[\,N(B_1) = n_1, \dots, N(B_k) = n_k\,] \\ &\quad = \frac{n!}{(n_1!)\dots(n_k!)} \left(\frac{\rho(B_1)}{\rho(\Theta)}\right)^{n_1} \cdots \left(\frac{\rho(B_k)}{\rho(\Theta)}\right)^{n_k} \mathrm{e}^{-\rho(\Theta)} \frac{\rho(\Theta)^n}{n!} \\ &\quad = \prod_{j=1}^{k} \mathrm{e}^{-\rho(B_j)} \frac{\rho(B_j)^{n_j}}{n_j!}. \end{aligned}$$

Summing over $n_1, \dots, n_k$ except n_j, we get

$$P[\,N(B_j) = n_j\,] = \mathrm{e}^{-\rho(B_j)} \frac{\rho(B_j)^{n_j}}{n_j!}.$$

Therefore N satisfies (1) and (2).

Step 2. Assume that $\rho(\Theta) = \infty$. By the σ-finiteness, there are disjoint sets $\Theta_1, \Theta_2, \dots \in \mathcal{B}$ with $\bigcup_{k=1}^{\infty} \Theta_k = \Theta$ and $\rho(\Theta_k) < \infty$ for each k. Define ρ_k by $\rho_k(B) = \rho(B \cap \Theta_k)$. By using Step 1 and constructing a product probability space by Theorem 1.9, we get independent Poisson random measures $\{N_k(B)\colon B \in \mathcal{B}\}$, $k = 1, 2, \dots$, with intensity measures ρ_k, defined on a probability space $(\Omega^0, \mathcal{F}^0, P^0)$. Let

$$N(B) = \sum_{k=1}^{\infty} N_k(B) \quad \text{for } B \in \mathcal{B}.$$

We claim that $\{N(B)\}$ is a Poisson random measure with intensity measure ρ. Since

$$E[N(B)] = \sum_{k=1}^{\infty} E[N_k(B)] = \sum_{k=1}^{\infty} \rho_k(B) = \rho(B),$$

and since a sum of independent Poisson random variables is again Poisson, $N(B)$ is Poisson distributed if $\rho(B) < \infty$. If $\rho(B) = \infty$, then

$$\sum_{k=1}^{\infty} P[\, N_k(B) \geq 1\,] = \sum_{k=1}^{\infty}(1 - \mathrm{e}^{-\rho_k(B)}) \geq \sum_{k=1}^{\infty}(2^{-1}\rho_k(B) \wedge a) = \infty$$

with some constant $a > 0$ (choose a' as $1 - \mathrm{e}^{-r} \geq \frac{r}{2}$ for $0 \leq r \leq a'$ and note that $1 - \mathrm{e}^{-r} \geq 1 - \mathrm{e}^{-a'}$ for $r \geq a'$). Hence, by the Borel–Cantelli lemma,

$$P[\, N_k(B) \geq 1 \text{ for infinitely many } k\,] = 1 \quad \text{if } \rho(B) = \infty.$$

Hence $N(B) = \infty$ a. s. if $\rho(B) = \infty$. This proves (1). Proofs of the properties (2) and (3) are straightforward. □

The following proposition gives basic properties of Poisson random measures.

PROPOSITION 19.5. *Let $(\Theta, \mathcal{B}, \rho)$ be a measure space with $\rho(\Theta) < \infty$ and $\{N(B) \colon B \in \mathcal{B}\}$ be a Poisson random measure with intensity measure ρ. Let φ be a measurable function from Θ to $\mathbb{R}^d$ and define*

$$Y(\omega) = \int_{\Theta} \varphi(\theta) N(\mathrm{d}\theta, \omega). \tag{19.8}$$

Then the following are true.

(i) *Y is a random variable on $\mathbb{R}^d$ with compound Poisson distribution satisfying*

$$\begin{aligned} E[\mathrm{e}^{\mathrm{i}\langle z,Y\rangle}] &= \exp\left[\int_{\Theta} (\mathrm{e}^{\mathrm{i}\langle z,\varphi(\theta)\rangle} - 1)\rho(\mathrm{d}\theta)\right] \\ &= \exp\left[\int_{\mathbb{R}^d} (\mathrm{e}^{\mathrm{i}\langle z,x\rangle} - 1)(\rho\varphi^{-1})(\mathrm{d}x)\right] \end{aligned} \tag{19.9}$$

for $z \in \mathbb{R}^d$.

(ii) *If $\int_{\Theta} |\varphi(\theta)|^2 \rho(\mathrm{d}\theta) < \infty$, then $E[|Y|^2] < \infty$ and*

$$E[Y] = \int_{\Theta} \varphi(\theta)\rho(\mathrm{d}\theta), \tag{19.10}$$

$$E[|Y - E[Y]|^2] = \int_{\Theta} |\varphi(\theta)|^2 \rho(\mathrm{d}\theta). \tag{19.11}$$

(iii) *Suppose that $B_1, \ldots, B_m$ are disjoint sets in $\mathcal{B}$ and let*

$$Y_k(\omega) = \int_{B_k} \varphi(\theta) N(\mathrm{d}\theta, \omega) \quad \textit{for } k = 1, \ldots, m.$$

Then, $Y_1, \ldots, Y_m$ are independent.

Proof. (i) The $Y(\omega)$ in (19.8) is finite, because $N(\cdot, \omega)$ is supported on a finite number of points. Let, for $p = (p_1, \ldots, p_d) \in \mathbb{Z}^d$, C_p^n be the set of points $x = (x_1, \ldots, x_d) \in \mathbb{R}^d$ such that $2^{-n}(p_j - 1) < x_j \leq 2^{-n} p_j$ for $j = 1, \ldots, d$. Then, $\{C_p^n : p \in \mathbb{Z}^d\}$ covers $\mathbb{R}^d$. Choose a point x_p^n in each C_p^n and define a function φ_n from Θ to $\mathbb{R}^d$ by $\varphi_n(\theta) = x_p^n$ for $\theta \in \varphi^{-1}(C_p^n)$. Then $|\varphi(\theta) - \varphi_n(\theta)| \leq 2^{-n} d^{1/2}$. Let $Y_n(\omega) = \int_\Theta \varphi_n(\theta) N(\mathrm{d}\theta, \omega)$. Then

$$|Y_n(\omega) - Y(\omega)| \leq 2^{-n} d^{1/2} N(\Theta, \omega) \to 0 \quad \text{as } n \to \infty.$$

Since

$$Y_n(\omega) = \sum_{p \in \mathbb{Z}^d} x_p^n N(\varphi^{-1}(C_p^n), \omega),$$

$Y_n(\omega)$ is a random variable on $\mathbb{R}^d$ and

$$E[\mathrm{e}^{\mathrm{i}\langle z, Y_n \rangle}] = \prod_{p \in \mathbb{Z}^d} E[\mathrm{e}^{\mathrm{i}\langle z, x_p^n N(\varphi^{-1}(C_p^n)) \rangle}] = \prod_{p \in \mathbb{Z}^d} \exp[(\mathrm{e}^{\mathrm{i}\langle z, x_p^n \rangle} - 1)\rho(\varphi^{-1}(C_p^n))]$$

$$= \exp\left[\int_\Theta (\mathrm{e}^{\mathrm{i}\langle z, \varphi_n(\theta) \rangle} - 1)\rho(\mathrm{d}\theta)\right]$$

by the defining properties of Poisson random measures. Therefore, Y is a random variable on $\mathbb{R}^d$ and satisfies (19.9), which shows that the distribution of Y is compound Poisson.

(ii) Let $z_{(j)}$, $\varphi_{(j)}(\theta)$, and $Y_{(j)}$ be the jth coordinates of z, $\varphi(\theta)$, and Y, respectively. It follows from the assumption $\int |\varphi(\theta)|^2 \rho(\mathrm{d}\theta) < \infty$ that

$$\frac{1}{\mathrm{i}} \frac{\partial}{\partial z_{(j)}} \int (\mathrm{e}^{\mathrm{i}\langle z, \varphi(\theta) \rangle} - 1)\rho(\mathrm{d}\theta) = \int \varphi_{(j)}(\theta) \mathrm{e}^{\mathrm{i}\langle z, \varphi(\theta) \rangle} \rho(\mathrm{d}\theta),$$

$$\left(\frac{1}{\mathrm{i}} \frac{\partial}{\partial z_{(j)}}\right)^2 \int (\mathrm{e}^{\mathrm{i}\langle z, \varphi(\theta) \rangle} - 1)\rho(\mathrm{d}\theta) = \int \varphi_{(j)}(\theta)^2 \mathrm{e}^{\mathrm{i}\langle z, \varphi(\theta) \rangle} \rho(\mathrm{d}\theta)$$

for $j = 1, \ldots, d$. Differentiating $E[\mathrm{e}^{\mathrm{i}\langle z, Y \rangle}]$ in (19.9) twice with respect to $z_{(j)}$ and letting $z = 0$, we obtain

$$E[Y_{(j)}] = \int \varphi_{(j)}(\theta) \rho(\mathrm{d}\theta),$$

$$E[Y_{(j)}^2] = \left(\int \varphi_{(j)}(\theta) \rho(d\theta)\right)^2 + \int \varphi_{(j)}(\theta)^2 \rho(\mathrm{d}\theta),$$

by the general result on characteristic functions in Proposition 2.5(ix) and (x). Hence we have (19.10) and (19.11).

(iii) Using $\varphi_n(\theta)$ above, let

$$Y_{n,k}(\omega) = \int_{B_k} \varphi_n(\theta) N(\mathrm{d}\theta, \omega) = \sum_{p \in \mathbb{Z}^d} x_p^n N(B_k \cap C_p^n)$$

for $k = 1, \ldots, m$. Since $N(B_k \cap C_p^n)$ with $1 \le k \le m$ and $p \in \mathbb{Z}^d$ are independent, $Y_{n,1}, \ldots, Y_{n,m}$ are independent. Since $Y_{n,k}(\omega) \to Y_k(\omega)$ as $n \to \infty$, it follows that $Y_1, \ldots, Y_m$ are independent by Proposition 1.13. □

20. Proof of the Lévy–Itô decomposition

Let us prove Theorems 19.2 and 19.3. Assume that we are given an additive process $\{X_t : t \ge 0\}$ on $\mathbb{R}^d$ defined on a probability space $(\Omega, \mathcal{F}, P)$. Let $(A_t, \nu_t, \gamma(t))$ be the generating triplet of X_t. By Remark 9.9 there is a unique σ-finite measure $\widetilde{\nu}$ on $H = (0,\infty) \times D_{0,\infty}$ such that $\widetilde{\nu}((0,t] \times B) = \nu_t(B)$ for $t > 0$ and $B \in \mathcal{B}(D_{0,\infty})$. By Proposition 19.4 there exists, on another probability space $(\Omega^0, \mathcal{F}^0, P^0)$, a Poisson random measure $\{N(B) : B \in \mathcal{B}(H)\}$ on H with intensity measure $\widetilde{\nu}$.

LEMMA 20.1. *There is $\Omega_1^0 \in \mathcal{F}^0$ with $P^0[\Omega_1^0] = 1$ such that, for any $\omega \in \Omega_1^0$, the following hold:*

(1) *for any $\varepsilon \in (0,\infty)$ and $t \in (0,\infty)$, the measure $[N(\cdot,\omega)]_{(0,t] \times D_{\varepsilon,\infty}}$ is supported on a finite number of points, each of which has $N(\cdot,\omega)$-measure 1;*

(2) *for any $s \in (0,\infty)$, $N(\{s\} \times D_{0,\infty}, \omega) = 0$ or 1.*

Proof. For each $\omega \in \Omega^0$, $N(\cdot,\omega)$ is a $\overline{\mathbb{Z}}_+$-valued measure. Write $H_{t,\varepsilon} = (0,t] \times D_{\varepsilon,\infty}$. Since $E^0[N(H_{t,\varepsilon})] = \widetilde{\nu}(H_{t,\varepsilon}) < \infty$, we have $P^0[\,N(H_{t,\varepsilon}) < \infty\,] = 1$. If $N(H_{t,\varepsilon}, \omega) = n < \infty$, then the support of $[N(\cdot,\omega)]_{H_{t,\varepsilon}}$ consists of at most n points. Let $t_k \uparrow \infty$ and $\varepsilon_k \downarrow 0$. Define $\Omega' = \bigcap_{k=1}^{\infty} \{N(H_{t_k,\varepsilon_k}) < \infty\}$. Then $P[\Omega'] = 1$ and, for any $\omega \in \Omega'$, $t > 0$, and $\varepsilon > 0$, $[N(\cdot,\omega)]_{H_{t,\varepsilon}}$ is supported on a finite number of points.

Let $\{N^*(B)\}$ be the Poisson random measure constructed in the proof of Proposition 19.4 with

$$\Theta_k = ((0,t_k] \times D_{\varepsilon_k,\infty}) \setminus ((0,t_{k-1}] \times D_{\varepsilon_{k-1},\infty}) \quad \text{for } k \ge 2$$

and $\Theta_1 = (0,t_1] \times D_{\varepsilon_1,\infty}$. Fix Θ_k and let $\{Z_n\}$ be the independent random variables in Step 1 of the proof. Denote the first component in $(0,t_k]$ of Z_n by $Z_{n,1}$. If $n \ne m$, then

$$P^0[\,Z_{n,1} = Z_{m,1}\,] = \int_{(0,t_k]} P^0[\,Z_{n,1} = s\,]\, P^0[\,Z_{m,1} \in \mathrm{d}s\,] = 0$$

by Proposition 1.16 and by $P^0[\,Z_{n,1} = s\,] \le \widetilde{\nu}(\{s\} \times D_{\varepsilon_k,\infty})/\widetilde{\nu}(\Theta_k) = 0$ as in (9.18). Since

$$\begin{aligned}
&\{\,\exists s > 0 \text{ such that } N^*(\Theta_k \cap (\{s\} \times D_{0,\infty})) \ge 2\,\} \\
&\qquad \subset \{\,\exists n, \exists m, \exists s \text{ such that } n \ne m \text{ and } Z_n, Z_m \in \{s\} \times D_{0,\infty}\,\} \\
&\qquad \subset \{\,\exists n, \exists m \text{ such that } n \ne m \text{ and } Z_{n,1} = Z_{m,1}\,\}
\end{aligned}$$

and the last event has probability 0, it follows that $N^*(\Theta_k \cap (\{s\} \times D_{0,\infty})) \le 1$ for all $s > 0$ a. s. With $\widetilde{\Theta}_k = \bigcup_{l=1}^k \Theta_l$ we can similarly prove that $N^*(\widetilde{\Theta}_k \cap (\{s\} \times D_{0,\infty})) \le 1$ for all $s > 0$ a. s. Letting $k \to \infty$, we see that $N^*(\{s\} \times D_{0,\infty}) \le 1$ for all $s > 0$ a. s.

Coming back to the general Poisson random measure $\{N(B)\}$ with intensity measure $\widetilde{\nu}$, let, for any fixed $\varepsilon > 0$, $Y_t = N(H_{t,\varepsilon})$ for $t > 0$ and $Y_0 = 0$. Then $\{Y_t \colon t \ge 0\}$ is an additive process. The stochastic continuity comes from (9.18). Paths of $\{Y_t\}$ are right-continuous step functions with jump sizes being positive integers. Let $U_k = \inf\{t \colon Y_t \ge k\}$. To show $N(\{s\} \times D_{0,\infty}) \le 1$ for all $s > 0$ a. s. is equivalent to showing

$$P^0[\,U_1 < U_2 < \cdots < U_k < \infty\,] = P^0[\,U_k < \infty\,] \quad \text{for each } k.$$

But these probabilities are identical with those for $\{Y_t^*\}$ similarly constructed from $\{N^*(B)\}$. Hence (1) and (2) are proved. □

We prepare some general lemmas. For $0 < t < \infty$ let $D([0,t], \mathbb{R}^d)$ be the set of functions $\xi(s)$ from $[0,t]$ to $\mathbb{R}^d$ right-continuous for $s \in [0,t)$ with left limits for $s \in (0,t]$. We write $\|\xi\|_t = \sup_{s\in[0,t]} |\xi(s)|$. Then, $\|\xi_1 + \xi_2\|_t \le \|\xi_1\|_t + \|\xi_2\|_t$.

LEMMA 20.2. *Fix $t \in (0,\infty)$. Let $\{Z_j(s) \colon s \in [0,t]\}$, $j = 1,2,\ldots$, be independent stochastic processes and $S_0(s) = 0$, $S_n(s) = \sum_{j=1}^n Z_j(s)$ for $n = 1,2,\ldots$. Suppose that, for each j, sample functions of $Z_j(s)$ belong to $D([0,t], \mathbb{R}^d)$ a. s. Then, for any $\varepsilon > 0$ and n,*

$$P\Big[\max_{1\le j\le n} \|S_j\|_t > 3\varepsilon\Big] \le 3 \max_{1\le j\le n} P[\,\|S_j\|_t > \varepsilon\,]. \tag{20.1}$$

Proof. We write $\|\xi\|$ for $\|\xi\|_t$ in the proof. Let $M_0 = 0$ and $M_k = \max_{1\le j\le k} \|S_j\|$ for $k \ge 1$. Let $a > 0$, $b > 0$, and $A_k = \{M_{k-1} \le a+b < \|S_k\|\}$ for $k \ge 1$. Then $A_1, \ldots, A_n$ are disjoint events and $\{M_n > a+b\} = \bigcup_{k=1}^n A_k$. Thus

$$P[\|S_n\| > a] \ge \sum_{k=1}^n P[A_k \cap \{\|S_n\| > a\}] \ge \sum_{k=1}^n P[A_k \cap \{\|S_n - S_k\| \le b\}]$$

$$= \sum_{k=1}^n P[A_k]\, P[\|S_n - S_k\| \le b] \ge P[M_n > a+b] \min_{1\le k\le n} P[\|S_n - S_k\| \le b].$$

Now, choose $a = \varepsilon$ and $b = 2\varepsilon$. Then

$$P[\|S_n\| > \varepsilon] \ge P[M_n > 3\varepsilon]\left(1 - \max_{1\le k\le n} P[\|S_n - S_k\| > 2\varepsilon]\right)$$

$$\ge P[M_n > 3\varepsilon]\left(1 - 2\max_{1\le k\le n} P[\|S_k\| > \varepsilon]\right).$$

Assume that $\max_{1\le k\le n} P[\|S_k\| > \varepsilon] < \frac{1}{3}$. Then

$$\tfrac{1}{3}P[M_n > 3\varepsilon] \le P[\|S_n\| > \varepsilon].$$

Hence we get (20.1) in this case. In the case $\max_{1\le k\le n} P[\|S_k\| > \varepsilon] \ge \frac{1}{3}$, (20.1) is trivial. □

REMARK 20.3. In the proof of the preceding lemma, we get

$$P[M_n > a+b] \le \frac{P[\|S_n\| > a]}{P[M_n \le b/2]}. \tag{20.2}$$

To see this, we have only to note that $P[M_n \le b/2] \le P[\|S_n - S_k\| \le b]$ for $1 \le k \le n$. We will use the inequality (20.2) in Section 25.

LEMMA 20.4. *Consider* $\{S_n(t)\}$, $n = 0, 1, 2, \ldots$, *in Lemma 20.2. If*

$$\lim_{n,m\to\infty} P[\,\|S_n - S_m\|_t > \varepsilon\,] = 0 \quad \textit{for any } \varepsilon > 0, \tag{20.3}$$

then there is a stochastic process $\{\,S(s)\colon s \in [0,t]\}$ *such that sample functions of* $S(s)$ *belong to* $D([0,t],\mathbb{R}^d)$ *a. s. and*

$$\lim_{n\to\infty} \|S_n - S\|_t = 0 \quad \textit{a. s.} \tag{20.4}$$

Proof. Write $\|\xi\|$ for $\|\xi\|_t$ again. Apply Lemma 20.2 to $Z_{n+1}, Z_{n+2}, \ldots$. Then, for $m > n$,

$$P\Big[\max_{n\le j\le m} \|S_j - S_n\| > 3\varepsilon\Big] \le 3 \max_{n\le j\le m} P[\,\|S_j - S_n\| > \varepsilon\,].$$

Noting that

$$P\left[\max_{\substack{n\le j\le m\\ n\le k\le m}} \|S_j - S_k\| > 6\varepsilon\right] \le P\Big[\max_{n\le j\le m} \|S_j - S_n\| > 3\varepsilon\Big],$$

we get

$$P\left[\sup_{\substack{j\ge n\\ k\ge n}} \|S_j - S_k\| > 6\varepsilon\right] \le 3\sup_{j\ge n} P[\,\|S_j - S_n\| > \varepsilon\,].$$

Since the right-hand side tends to 0 as $n \to \infty$ by (20.3), the probability in the left-hand side tends to 0. Hence we have

$$\lim_{n\to\infty} \sup_{\substack{j\ge n\\ k\ge n}} \|S_j - S_k\| = 0 \quad \text{a. s.}$$

As the function space $D([0,t],\mathbb{R}^d)$ is closed under uniform convergence, we get the process $\{S(s)\}$ with the desired properties. □

LEMMA 20.5. *Let $Z_1, \ldots, Z_n$ be independent random variables on $\mathbb{R}^d$ such that $E[|Z_j|^2] < \infty$ and $EZ_j = 0$ for each j. Let $S_j = Z_1 + \cdots + Z_j$. Then*

$$P\Big[\max_{1\le j\le n}|S_j| > \varepsilon\Big] \le \frac{3^3}{\varepsilon^2}E[|S_n|^2] \quad \textit{for } \varepsilon > 0. \tag{20.5}$$

Proof. Letting $Z_j(s) = Z_j$ for all $s \in [0,t]$ and using Lemma 20.2, we get

$$P\Big[\max_{1\le j\le n}|S_j| > 3\varepsilon\Big] \le 3\max_{1\le j\le n} P[\,|S_j| > \varepsilon\,]. \tag{20.6}$$

Denote the pth component by the subscript p. Since we have

$$\begin{aligned} P[\,|S_j| > \varepsilon\,] &\le \frac{1}{\varepsilon^2}E[|S_j|^2] \le \frac{1}{\varepsilon^2}\sum_{p=1}^{d} E[(S_j)_p^2 + (S_n - S_j)_p^2] \\ &= \frac{1}{\varepsilon^2}\sum_{p=1}^{d} E[((S_j)_p + (S_n - S_j)_p)^2] = \frac{1}{\varepsilon^2}E[|S_n|^2] \end{aligned}$$

by using $EZ_j = 0$, (20.6) implies that

$$P\Big[\max_{1\le j\le n}|S_j| > 3\varepsilon\Big] \le \frac{3}{\varepsilon^2}E[|S_n|^2].$$

Replacing 3ε by ε, we get (20.5). □

The inequality (20.5) can be strengthened by replacing 3^3 with 1 in the right-hand side. This is called Kolmogorov's inequality. The proof given in the case $d = 1$ in [**34**], [**80**], [**139**], [**327**] is easily modified for the d-dimensional case. But (20.5) is enough for our purposes.

We come back to the Poisson random measure $\{N(B)\}$ on H with intensity measure $\widetilde{\nu}$.

LEMMA 20.6. *For any sequence $\{\varepsilon_n\}$ with $0 < \varepsilon_n < 1$ and $\varepsilon_n \downarrow 0$, there is $\Omega_2^0 \in \mathcal{F}^0$ with $P^0[\Omega_2^0] = 1$ such that, for any $\omega \in \Omega_2^0$,*

$$\int_{(0,t]\times D(\varepsilon_n,1]} \{x\,N(\mathrm{d}(s,x),\omega) - x\,\widetilde{\nu}(\mathrm{d}(s,x))\} \tag{20.7}$$

converges to an element of $D([0,\infty),\mathbb{R}^d)$ uniformly on any bounded time interval as $n \to \infty$.

Proof. Let $\varepsilon_0 = 1$ and, for $n = 1, 2, \ldots$,

$$\begin{aligned} Z_n(t) &= \int_{(0,t]\times D(\varepsilon_n,\varepsilon_{n-1}]} \{x\,N(\mathrm{d}(s,x)) - x\,\widetilde{\nu}(\mathrm{d}(s,x))\}, \\ S_n(t) &= Z_1(t) + \cdots + Z_n(t). \end{aligned}$$

Then $S_n(t)$ equals the integral in (20.7). By Proposition 19.5, $E[S_n(t)] = 0$ and

$$E[\,|S_m(t) - S_n(t)|^2] = \int_{D(\varepsilon_m,\varepsilon_n]} |x|^2\, \nu_t(\mathrm{d}x) \quad \text{for } m > n.$$

The sample function of $Z_n(t)$ is right-continuous with left limits a. s. Fix t and let $r_0, r_1, r_2, \ldots$ be an enumeration of $([0,t) \cap \mathbb{Q}) \cup \{t\}$ with $r_0 = 0$, $r_1 = t$. Then

$$(20.8) \qquad P\left[\sup_{s\in[0,t]} |S_m(s) - S_n(s)| > \varepsilon\right] = \lim_{q\to\infty} P\left[\max_{0\le j\le q} |S_m(r_j) - S_n(r_j)| > \varepsilon\right].$$

For fixed q, let $0 = s_0 < s_1 < \cdots < s_q = t$ be the ordering of $\{r_0, r_1, \ldots, r_q\}$. Then

$$S_m(t) - S_n(t) = \sum_{j=1}^{q} \int_{(s_{j-1},s_j]\times D(\varepsilon_m,\varepsilon_n]} \{x\, N(\mathrm{d}(s,x)) - x\,\widetilde{\nu}(\mathrm{d}(s,x))\}$$

and the right-hand side is a sum of independent random variables by Proposition 19.5. Use Lemma 20.5. Then we see that the right-hand side of (20.8) is bounded by $3^3\varepsilon^{-2}\int_{D(\varepsilon_m,\varepsilon_n]} |x|^2\, \nu_t(\mathrm{d}x)$, which tends to 0 as $m, n \to \infty$. Hence we can apply Lemma 20.4 and conclude that $\{S_n(t)\}$ converges to a limit $\{S(t)\}$ uniformly on any bounded time interval a. s. The limit is an element of $D([0,\infty), \mathbb{R}^d)$. □

LEMMA 20.7. *Let*

$$(20.9) \qquad S_\varepsilon(t,\omega) = \int_{(0,t]\times D(\varepsilon,1]} \{x\, N(\mathrm{d}(s,x),\omega) - x\,\widetilde{\nu}(\mathrm{d}(s,x))\}.$$

Then there is $\Omega_3^0 \in \mathcal{F}^0$ with $P^0[\Omega_3^0] = 1$ such that, for any $\omega \in \Omega_3^0$, $S_\varepsilon(t,\omega)$ converges uniformly on any bounded time interval as $\varepsilon \downarrow 0$. Define

$$(20.10) \qquad Y_t^1(\omega) = \lim_{\varepsilon\downarrow 0} S_\varepsilon(t,\omega) + \int_{(0,t]\times D(1,\infty)} x\, N(\mathrm{d}(s,x),\omega)$$

for $\omega \in \Omega_3^0$. Then $\{Y_t^1\}$ is an additive process with generating triplets $(0, \nu_t, 0)$.

Proof. Define, for $\xi \in D([0,\infty), \mathbb{R}^d)$,

$$\|\xi\| = \sum_{n=1}^{\infty} 2^{-n}\Big(1 \wedge \sup_{t\in[0,n]} |\xi(t)|\Big).$$

Then $\|\xi_n - \xi\| \to 0$ is equivalent to uniform convergence of $\xi_n(t)$ to $\xi(t)$ on every bounded interval. Note that

$$\limsup_{\varepsilon,\varepsilon'\downarrow 0} \|S_\varepsilon(\cdot,\omega) - S_{\varepsilon'}(\cdot,\omega)\| = \lim_{n\to\infty} \sup_{\varepsilon,\varepsilon'\in(0,1/n)} \|S_\varepsilon(\cdot,\omega) - S_{\varepsilon'}(\cdot,\omega)\|.$$

Further, $S_\varepsilon(\cdot,\omega)$ is approximated in $\|\cdot\|$-distance by $S_{\varepsilon'}(\cdot,\omega)$, $\varepsilon' \in \mathbb{Q}$, since the support of $[N(\cdot,\omega)]_{(0,t]\times D(a,1]}$ consists of a finite number of points for each $a > 0$. Hence,

$$\sup_{\varepsilon,\varepsilon'\in(0,1/n)} \|S_\varepsilon(\cdot,\omega) - S_{\varepsilon'}(\cdot,\omega)\| = \sup_{\varepsilon,\varepsilon'\in\mathbb{Q}\cap(0,1/n)} \|S_\varepsilon(\cdot,\omega) - S_{\varepsilon'}(\cdot,\omega)\|.$$

Choose, for any n, a finite number of points $\varepsilon_j^{(n)}, \varepsilon_j'^{(n)} \in \mathbb{Q}\cap(0,1/n)$ with $j = 1,\ldots,k_n$ in such a way that

$$P\left[\sup_{\varepsilon,\varepsilon'\in(0,1/n)} \|S_\varepsilon(\cdot,\omega) - S_{\varepsilon'}(\cdot,\omega)\| - \max_j \left\|S_{\varepsilon_j^{(n)}}(\cdot,\omega) - S_{\varepsilon_j'^{(n)}}(\cdot,\omega)\right\| > \frac{1}{n}\right]$$

is less than $1/n$. Let $\varepsilon_1, \varepsilon_2, \ldots$ be the rearrangement of all $\varepsilon_j^{(n)}$ and $\varepsilon_j'^{(n)}$ with $j = 1,\ldots,k_n$ and $n = 1,2,\ldots$ in decreasing order. Then, as $n \to \infty$,

$$\sup_{\varepsilon,\varepsilon'\in(0,1/n)} \|S_\varepsilon(\cdot,\omega) - S_{\varepsilon'}(\cdot,\omega)\| - \sup_{j,k\in J(n)} \|S_{\varepsilon_j}(\cdot,\omega) - S_{\varepsilon_k}(\cdot,\omega)\| \to 0 \text{ in prob.},$$

where $\sup_{j,k\in J(n)}$ means the supremum taken over j and k such that ε_j and ε_k are in $(0,1/n)$. The convergence in probability implies convergence a. s. via a subsequence. It follows that

$$\limsup_{\varepsilon,\varepsilon'\downarrow 0} \|S_\varepsilon(\cdot,\omega) - S_{\varepsilon'}(\cdot,\omega)\| = \limsup_{j,k\to\infty} \|S_{\varepsilon_j}(\cdot,\omega) - S_{\varepsilon_k}(\cdot,\omega)\| \quad \text{a. s.}$$

Now use Lemma 20.6. We see that there is $\Omega_3^0 \in \mathcal{F}^0$ with $P^0[\Omega_3^0] = 1$ such that

$$\limsup_{\varepsilon,\varepsilon'\downarrow 0} \|S_\varepsilon(\cdot,\omega) - S_{\varepsilon'}(\cdot,\omega)\| = 0$$

for $\omega \in \Omega_3^0$. Hence $S_\varepsilon(\cdot,\omega)$ converges to a function in $D([0,\infty),\mathbb{R}^d)$ uniformly on any bounded interval as $\varepsilon \downarrow 0$. We define $\{Y_t^1\}$ by (20.10). It follows from Propositions 1.13 and 19.5 that $\{Y_t^1\}$ has independent increments and that

$$\begin{aligned}
E[\mathrm{e}^{\mathrm{i}\langle z,Y_t^1\rangle}] &= \lim_{\varepsilon\downarrow 0} E\exp\left[\mathrm{i}\left\langle z, S_\varepsilon(t) + \int_{(0,t]\times D(1,\infty)} x\, N(\mathrm{d}(s,x))\right\rangle\right] \\
&= \lim_{\varepsilon\downarrow 0} \exp\left[\int_{D(\varepsilon,\infty)} (\mathrm{e}^{\mathrm{i}\langle z,x\rangle} - 1 - \mathrm{i}\langle z,x\rangle 1_{D(\varepsilon,1]}(x))\, \nu_t(\mathrm{d}x)\right] \\
&= \exp\left[\int_{D(0,\infty)} (\mathrm{e}^{\mathrm{i}\langle z,x\rangle} - 1 - \mathrm{i}\langle z,x\rangle 1_{D(0,1]}(x))\, \nu_t(\mathrm{d}x)\right].
\end{aligned}$$

Recall that ν_t is continuous in t in a strong sense mentioned in Remark 9.9. Now $\{Y_t^1\}$ is an additive process with generating triplets $(0, \nu_t, 0)$. □

Let $\{Y_t^2 : t \geq 0\}$ be an additive process on $\mathbb{R}^d$ having continuous paths with generating triplets $(A_t, 0, \gamma_t)$. Its existence is guaranteed by Theorems 9.8, 11.5, and 11.7. Enlarging the probability space $(\Omega^0, \mathcal{F}^0, P^0)$ if necessary, we construct $\{Y_t^2\}$ on Ω^0 in such a way that $\{Y_t^1\}$ and $\{Y_t^2\}$ are independent. Define

$$Y_t = Y_t^1 + Y_t^2. \tag{20.11}$$

Then $\{Y_t\}$ is an additive process with generating triplets (A_t, ν_t, γ_t).

LEMMA 20.8. *There is $\Omega_4^0 \in \mathcal{F}^0$ with $P^0[\Omega_4^0] = 1$ such that, for any $\omega \in \Omega_4^0$ and $B \in \mathcal{B}(H)$,*

$$N(B, \omega) = \#\{\, s\colon (s, Y_s - Y_{s-}) \in B \,\}. \tag{20.12}$$

Proof. Since Y_t^2 is continuous, $Y_s - Y_{s-} = Y_s^1 - Y_{s-}^1$. Let

$$V_\varepsilon(t) = \int_{(0,t]\times D(\varepsilon,\infty)} \{x\, N(\mathrm{d}(s,x)) - 1_{D(\varepsilon,1]}(x) x\, \widetilde{\nu}(\mathrm{d}(s,x))\}. \tag{20.13}$$

Let $\Omega_4^0 \in \mathcal{F}^0$ with $P^0[\Omega_4^0] = 1$ be such that, for $\omega \in \Omega_4^0$, (1) and (2) of Lemma 20.1 hold and $V_\varepsilon(t,\omega)$ tends to $Y_t^1(\omega)$ uniformly on any finite time interval as $\varepsilon \downarrow 0$. Let $\omega \in \Omega_4^0$. We then have

$$Y_s^1(\omega) - Y_{s-}^1(\omega) = \lim_{\varepsilon\downarrow 0}(V_\varepsilon(s,\omega) - V_\varepsilon(s-,\omega)).$$

If $N(\{(s,x)\},\omega) = 1$, then $N(\{s\}\times D_{0,\infty},\omega) = 1$ and $V_\varepsilon(s,\omega) - V_\varepsilon(s-,\omega) = x$ for small ε, and hence $Y_s^1(\omega) - Y_{s-}^1(\omega) = x$. On the other hand, if $N(\{s\}\times D_{0,\infty},\omega) = 0$, then $Y_s^1(\omega) - Y_{s-}^1(\omega) = 0$. This shows (20.12). □

Write $x_t(\xi) = x(t,\xi) = \xi(t)$ for $\xi \in \mathbf{D} = D([0,\infty),\mathbb{R}^d)$. The σ-algebra $\mathcal{F}_{\mathbf{D}}$ is generated by $\{\, x_t,\ t \geq 0\,\}$. Given $\xi \in \mathbf{D}$, jumping times of ξ are countable, but they are not always enumerable in increasing order. We enumerate them in the following way. For each $n = 1, 2, \ldots$, the number of jumps of ξ such that $\xi(t) - \xi(t-) \in D(\frac{1}{n}, \frac{1}{n-1}]$ (replaced by $D(1,\infty)$ if $n = 1$) is finite in any bounded time interval (because otherwise $\xi(t)$ either does not have right limit or does not have left limit at some t). Let these jumping times be $0 < t_{n,1}(\xi) < t_{n,2}(\xi) < \cdots$. If $\#\{\, t\colon \xi(t) - \xi(t-) \in D(\frac{1}{n}, \frac{1}{n-1}]\,\} = k < \infty$, then we let $t_{n,k+1}(\xi) = t_{n,k+2}(\xi) = \cdots = +\infty$.

LEMMA 20.9. *For any n and j, $t_{n,j}(\xi)$ is $\mathcal{F}_{\mathbf{D}}$-measurable.*

Proof. Let $\mathbb{Q}_t = ((0,t)\cap\mathbb{Q})\cup\{t\}$. We have $t_{1,1}(\xi) \leq t$ if and only if there exists $l \in \mathbb{N}$ such that, for any $m \in \mathbb{N}$, there are $r, s \in \mathbb{Q}_t$ such that $r < s < r + \frac{1}{m}$ and $|\xi(s) - \xi(r)| > 1 + \frac{1}{l}$. Hence $t_{1,1}(\xi)$ is $\mathcal{F}_{\mathbf{D}}$-measurable. We have $t_{1,2}(\xi) \leq t$ if and only if there exists $l \in \mathbb{N}$ such that, for any $m \in \mathbb{N}$, there are $r, s \in \mathbb{Q}_t$ satisfying $t_{1,1}(\xi) < r < s < r + \frac{1}{m}$ and $|\xi(s) - \xi(r)| > 1 + \frac{1}{l}$. Hence $t_{1,2}(\xi)$ is $\mathcal{F}_{\mathbf{D}}$-measurable. Similarly we can show $\mathcal{F}_{\mathbf{D}}$-measurability of other $t_{n,j}(\xi)$. □

Proof of Theorem 19.2. We are given an additive process $\{X_t\}$ on $\mathbb{R}^d$ on a probability space $(\Omega, \mathcal{F}, P)$, with generating triplets (A_t, ν_t, γ_t). Using this system of triplets, we have constructed an additive process $\{Y_t\}$ on $(\Omega^0, \mathcal{F}^0, P^0)$, which is identical in law with $\{X_t\}$. From our construction $\{Y_t\}$ has the Lévy–Itô decomposition. We shall prove that $\{X_t\}$ also has the same decomposition. There are $\Omega_0 \in \mathcal{F}$ and $\Omega_5^0 \in \mathcal{F}^0$ with probability one such that sample functions of $X_t(\omega)$ for $\omega \in \Omega_0$ and of $Y_t(\omega)$ for $\omega \in \Omega_5^0$ belong to $\mathbf{D}$. Define mappings $\psi\colon \Omega \to \mathbf{D}$ and $\psi^0\colon \Omega^0 \to \mathbf{D}$ by

$$x_t(\psi(\omega)) = \begin{cases} X_t(\omega) & \text{for } \omega \in \Omega_0, \\ 0 & \text{for } \omega \notin \Omega_0, \end{cases} \tag{20.14}$$

$$x_t(\psi^0(\omega)) = \begin{cases} Y_t(\omega) & \text{for } \omega \in \Omega_5^0, \\ 0 & \text{for } \omega \notin \Omega_5^0. \end{cases} \tag{20.15}$$

By the equality in law of $\{X_t\}$ and $\{Y_t\}$ we have

$$P[\psi^{-1}(G)] = P^0[(\psi^0)^{-1}(G)], \tag{20.16}$$

if G is a cylinder set in $\mathbf{D}$. It follows that, for every $G \in \mathcal{F}_{\mathbf{D}}$, we have $\psi^{-1}(G) \in \mathcal{F}$ and $(\psi^0)^{-1}(G) \in \mathcal{F}^0$, and (20.16) holds. Let us define $P^{\mathbf{D}}[G]$ by the value of (20.16). Then, under $P^{\mathbf{D}}$, $\{x_t\}$ is an additive process identical in law with $\{X_t\}$ and $\{Y_t\}$.

For $\xi \in \mathbf{D}$ and $B \in \mathcal{B}(H)$ define

$$j(B,\xi) = \#\{\, s \in (0,\infty)\colon (s, x_s(\xi) - x_{s-}(\xi)) \in B \,\}.$$

Since the jumping times of ξ are exhausted by $t_{k,j}(\xi)$, $k, j = 1, 2, \ldots$, we have

$$j(B,\xi) = \sum_{k=1}^{\infty} \sum_{j=1}^{\infty} 1_{G(k,j)}(\xi)$$

with

$$G(k,j) = \{\, \xi\colon t_{k,j}(\xi) < \infty \text{ and } x(t_{k,j}(\xi), \xi) - x(t_{k,j}(\xi)-, \xi) \in B \,\}.$$

Since $x(t,\xi)$ is $(\mathcal{B}_{[0,\infty)} \times \mathcal{F}_{\mathbf{D}})$-measurable in (t,ξ) (use the right-continuity in t), we see from Lemma 20.9 that $x(t_{k,j}(\xi),\xi)$ and $x(t_{k,j}(\xi)-,\xi)$ are $\mathcal{F}_{\mathbf{D}}$-measurable in ξ. Hence $G(k,j) \in \mathcal{F}_{\mathbf{D}}$. Consequently $j(B,\xi)$ is $\mathcal{F}_{\mathbf{D}}$-measurable in ξ. Define $J(B,\omega)$, $\omega \in \Omega$, as in the statement of Theorem 19.2. Then

$$J(B,\omega) = j(B, \psi(\omega)) \quad \text{for } \omega \in \Omega_0$$

and, by Lemma 20.8,

$$N(B,\omega) = j(B, \psi^0(\omega)) \quad \text{for } \omega \in \Omega_4^0 \cap \Omega_5^0.$$

Therefore $\{J(B)\}$, $\{N(B)\}$, and $\{j(B)\}$ are identical in law. Hence $\{J(B)\}$ is a Poisson random measure with intensity measure $\widetilde{\nu}$. This proves (i) of Theorem 19.2.

Next, define for $\xi \in \mathbf{D}$

$$u_\varepsilon(t,\xi) = \sum_{k=1}^{\infty}\sum_{j=1}^{\infty}[x_{t_{k,j}(\xi)}(\xi) - x_{t_{k,j}(\xi)-}(\xi)]1_{G(k,j,t,\varepsilon)}(\xi) - \int_{D(\varepsilon,1]} x\,\nu_t(\mathrm{d}x),$$

$$G(k,j,t,\varepsilon) = \{\,\xi\colon t_{k,j}(\xi) \le t \text{ and } x_{t_{k,j}(\xi)}(\xi) - x_{t_{k,j}(\xi)-}(\xi) > \varepsilon\,\}.$$

Only a finite number of summands are non-zero. Define

$$U_\varepsilon(t,\omega) = \begin{cases} u_\varepsilon(t,\psi(\omega)) & \text{for } \omega \in \Omega_0, \\ 0 & \text{for } \omega \notin \Omega_0. \end{cases}$$

Note that $u_\varepsilon(t,\xi)$ is $\mathcal{F}_{\mathbf{D}}$-measurable in ξ. We have

$$U_\varepsilon(t,\omega) = \int_{(0,t]\times D(\varepsilon,\infty)} \{x\,J(\mathrm{d}(s,x),\omega) - 1_{D(\varepsilon,1]}(x)x\,\widetilde{\nu}(\mathrm{d}(s,x))\}$$

for $\omega \in \Omega_0$ and, using $V_\varepsilon(t)$ of (20.13),

$$V_\varepsilon(t,\omega) = u_\varepsilon(t,\psi^0(\omega)) \quad \text{for } \omega \in \Omega_4^0 \cap \Omega_5^0.$$

Let

$$\mathbf{D}_0 = \{\xi\colon u_\varepsilon(t,\xi) \text{ converges uniformly on any bounded interval as } \varepsilon \downarrow 0\}.$$

Using the notation in the proof of Lemma 20.7, we have

$$\mathbf{D}_0 = \left\{\xi\colon \limsup_{\varepsilon,\varepsilon'\downarrow 0} \|u_\varepsilon(\cdot,\xi) - u_{\varepsilon'}(\cdot,\xi)\| = 0\right\}.$$

The limsup is equal to the limit as $n \to \infty$ of the supremum over $\varepsilon, \varepsilon' \in \mathbb{Q} \cap (0,1/n)$. Hence $\mathbf{D}_0 \in \mathcal{F}_{\mathbf{D}}$ and

$$\begin{aligned} &P[\,U_\varepsilon(\xi) \text{ converges uniformly on any bounded interval as } \varepsilon \downarrow 0\,] = P^{\mathbf{D}}[\mathbf{D}_0] \\ &\quad = P^0[\,V_\varepsilon(\xi) \text{ converges uniformly on any bounded interval as } \varepsilon \downarrow 0\,] = 1 \end{aligned}$$

by Lemma 20.7. Hence there is $\Omega_1 \in \mathcal{F}$ with $P[\Omega_1] = 1$ having the property stated in (ii) and $X_t^1(\omega)$ is defined by (19.2) for $\omega \in \Omega_1$. Let $X_t^1(\omega) = 0$ for $\omega \notin \Omega_1$. Let

$$x_t^1(\xi) = \begin{cases} \lim_{\varepsilon\downarrow 0} u_\varepsilon(t,\xi) & \text{for } \xi \in \mathbf{D}_0, \\ 0 & \text{for } \xi \notin \mathbf{D}_0. \end{cases}$$

Since

$$X_t^1(\omega) = x_t^1(\psi(\omega)) \quad \text{for } \omega \in \Omega_0 \cap \Omega_1, \tag{20.17}$$

$$Y_t^1(\omega) = x_t^1(\psi^0(\omega)) \quad \text{for } \omega \in \Omega_3^0 \cap \Omega_4^0 \cap \Omega_5^0, \tag{20.18}$$

we see that $\{X_t^1\}$ and $\{x_t^1\}$ are additive processes identical in law with $\{Y_t^1\}$. This proves the assertion in (ii).

In the last step define

$$x_t^2(\xi) = x_t(\xi) - x_t^1(\xi) \qquad \text{for } \xi \in \mathbf{D}, \tag{20.19}$$

$$X_t^2(\omega) = X_t(\omega) - X_t^1(\omega) \quad \text{for } \omega \in \Omega. \tag{20.20}$$

Then

$$X_t^2(\omega) = x_t^2(\psi(\omega)) \quad \text{for } \omega \in \Omega_0 \cap \Omega_1. \tag{20.21}$$

By (20.11), (20.15), and (20.18), we have

$$Y_t^2(\omega) = x_t^2(\psi^0(\omega)) \quad \text{for } \omega \in \Omega_3^0 \cap \Omega_4^0 \cap \Omega_5^0. \tag{20.22}$$

For $\xi \in \mathbf{D}_0$, $x_t^2(\xi)$ is continuous in t. In fact, if t is a discontinuity point of ξ, then $u_\varepsilon(t,\xi) - u_\varepsilon(t-,\xi) = x_t(\xi) - x_{t-}(\xi)$ for any small ε, and $x_t^1(\xi) - x_{t-}^1(\xi) = x_t(\xi) - x_{t-}(\xi)$, from which $x_t^2(\xi) = x_{t-}^2(\xi)$ follows. If t is a continuity point of ξ, then $u_\varepsilon(t,\xi) = u_\varepsilon(t-,\xi)$ for all ε, and $x_t^1(\xi) = x_{t-}^1(\xi)$, which shows $x_t^2(\xi) = x_{t-}^2(\xi)$ again. Therefore $X_t^2(\omega)$ is continuous in t for $\omega \in \Omega_0 \cap \Omega_1$. It follows from (20.16), (20.17), (20.18), (20.21), and (20.22) that the three processes $\{(X_t^1, X_t^2)\colon t \geq 0\}$, $\{(Y_t^1, Y_t^2)\colon t \geq 0\}$, and $\{(x_t^1, x_t^2)\colon t \geq 0\}$ are identical in law. Thus, $\{X_t^1\}$ and $\{X_t^2\}$ are independent, since $\{Y_t^1\}$ and $\{Y_t^2\}$ are independent; $\{X_t^2\}$ is an additive process identical in law with $\{Y_t^2\}$. The proof of (iii) and (iv) is complete. □

Proof of Theorem 19.3. We assume $\int_{|x|\leq 1} |x|\, \nu_t(\mathrm{d}x) < \infty$. For a Borel set C satisfying $C \subset D_{\varepsilon,\infty}$ with some $\varepsilon > 0$, let $Y'(C) = \int_{(0,t]\times C} |x|\, J(\mathrm{d}(s,x))$. By Propositions 4.5 and 19.5, $Y'(C)$ has a compound Poisson distribution on $[0,\infty)$ and

$$E[\mathrm{e}^{-uY'(C)}] = \exp\left[\int_C (\mathrm{e}^{-u|x|} - 1)\, \nu_t(\mathrm{d}x)\right] \quad \text{for } u > 0.$$

Choosing $C = D_{\varepsilon,\infty}$ and letting $\varepsilon \downarrow 0$, we get

$$\begin{aligned} E\left[\exp\left(-u \int_{(0,t]\times D(0,\infty)} |x|\, J(\mathrm{d}(s,x))\right)\right] \\ = \exp\left[\int_{D(0,\infty)} (\mathrm{e}^{-u|x|} - 1)\, \nu_t(\mathrm{d}x)\right] \quad \text{for } u > 0. \end{aligned} \tag{20.23}$$

The right-hand side goes to 1 as $u \downarrow 0$. Hence

$$\int_{(0,t]\times D(0,\infty)} |x|\, J(\mathrm{d}(s,x)) < \infty \quad \text{a.s.} \tag{20.24}$$

Hence X_t^3 is definable by (19.4) and finite a. s. It follows from (19.2) that

$$X_t^3(\omega) = X_t^1(\omega) + \int_{D(0,1]} x\, \nu_t(\mathrm{d}x).$$

The process $\{X_t^4\}$ defined by (19.6) satisfies

$$X_t^4(\omega) = X_t^2(\omega) - \int_{D(0,1]} x\, \nu_t(\mathrm{d}x).$$

Hence all assertions in Theorem 19.3 are obtained from Theorem 19.2. □

Assuming the existence of the additive process with continuous sample functions for a given system $\{(A_t, 0, \gamma(t))\}$, the proof of Theorem 19.2 above gives a new proof of the existence of an additive process on $\mathbb{R}^d$ with a given system of generating triplets $\{(A_t, \nu_t, \gamma(t))\}$.

21. Applications to sample function properties

From the Lévy–Itô decomposition we can deduce many sample function properties of additive processes. Let us consider continuity, jumping times, increasingness, and variation. For simplicity we discuss only Lévy processes. Thus Theorems 19.2 and 19.3 hold with $\widetilde{\nu}(\mathrm{d}(s,x)) = \mathrm{d}s\nu(\mathrm{d}x)$ (Remark 9.9). In this section let $\{X_t\}$ be a Lévy process on $\mathbb{R}^d$ defined on $(\Omega, \mathcal{F}, P)$ with generating triplet (A, ν, γ). If $\int_{|x|\le 1} |x|\nu(\mathrm{d}x) < \infty$, then the drift of $\{X_t\}$ is denoted by γ_0. We use the event Ω_0 in Definition 1.6 and the jumping number $J(B, \omega)$ defined in (19.1). Let

$$J_\varepsilon(t,\omega) = \int_{(0,t]\times D(\varepsilon,\infty)} J(\mathrm{d}(s,x),\omega),$$

$$X_\varepsilon(t,\omega) = \int_{(0,t]\times D(\varepsilon,\infty)} xJ(\mathrm{d}(s,x),\omega).$$

Let

$$J(t,\omega) = \int_{(0,t]\times D(0,\infty)} J(\mathrm{d}(s,x),\omega) = \lim_{\varepsilon\downarrow 0} \int_{(0,t]\times D(\varepsilon,\infty)} J(\mathrm{d}(s,x),\omega),$$

allowing ∞. $J(t,\omega)$ is the number of jumps of $X_s(\omega)$ in the time interval $(0,t]$.

THEOREM 21.1 (Continuity). *Sample functions of $\{X_t\}$ are continuous a. s. if and only if $\nu = 0$.*

Proof. By Theorem 19.2, the number of jumping times $s \in (0,t]$ satisfying $X_s - X_{s-} \in D_{\varepsilon,\infty}$ has mean $t\int_{|x|>\varepsilon} \nu(\mathrm{d}x)$. Hence the number of jumps is 0 a. s. if and only if $\nu = 0$. □

An $\mathbb{R}^d$-valued function $f(t)$ is *piecewise constant* if there exist $0 = t_0 < t_1 < \cdots < t_n = \infty$ or $0 = t_0 < t_1 < \cdots$, $\lim_{j\to\infty} t_j = \infty$, such that $f(t)$ is constant on each interval $[t_{j-1}, t_j)$. Let us use the classification into types A, B, and C in Definition 11.9.

THEOREM 21.2 (Piecewise constancy). *Sample functions of $\{X_t\}$ are piecewise constant a. s. if and only if it is a compound Poisson or a zero process, that is, if and only if it is of type A with $\gamma_0 = 0$.*

Proof. The 'only if' part. The jumping number $J(t)$ is finite a. s. for each t. By Theorem 19.2(i), $J(t)$ has Poisson distribution with mean $t\nu(\mathbb{R}^d)$.

Hence $\nu(\mathbb{R}^d) < \infty$. Recall that Poisson with mean ∞ would imply that $J(t) = \infty$ a. s. Therefore $X_t = \int_{(0,t]\times D(0,\infty)} xJ(\mathrm{d}(s,x)) = X_t^3$, and we have

$$Ee^{i\langle z,X(t)\rangle} = \exp\left[t\int_{\mathbb{R}^d}(\mathrm{e}^{\mathrm{i}\langle z,x\rangle}-1)\nu(\mathrm{d}x)\right]$$

by Theorem 19.3.

The 'if' part. By Theorem 19.2(i), $E[J(t)] = t\nu(\mathbb{R}^d) < \infty$. Hence $J(t) < \infty$ a. s. Since $A = 0$ and $\gamma_0 = 0$, $X_t = \int_{(0,t]\times D(0,\infty)} xJ(\mathrm{d}(s,x))$ a. s. by Theorem 19.3. Hence $\{X_t\}$ is piecewise constant a. s. □

THEOREM 21.3 (Jumping times). *If $\nu(\mathbb{R}^d) = \infty$, then, almost surely, jumping times are countable and dense in $[0,\infty)$. If $0 < \nu(\mathbb{R}^d) < \infty$, then, almost surely, jumping times are infinitely many and countable in increasing order, and the first jumping time $T(\omega)$ has exponential distribution with mean $1/\nu(\mathbb{R}^d)$.*

Proof. Countability of jumps is a consequence of right-continuity with left limits, as we have seen in the previous section. For $\varepsilon > 0$ and $\omega \in \Omega_0$, let $T_\varepsilon(\omega)$ be the first time that $X_t(\omega)$ jumps with size $> \varepsilon$ (i. e. the first t such that $X_t(\omega) - X_{t-}(\omega) \in D_{\varepsilon,\infty}$). Let $T_\varepsilon(\omega) = \infty$ if $X_t(\omega)$ does not have a jump with size $> \varepsilon$. Since $T_\varepsilon(\omega) \le t$ is equivalent to $\int_{(0,t]\times D(\varepsilon,\infty)} J(\mathrm{d}(s,x),\omega) \ge 1$,

$$P[\,T_\varepsilon \le t\,] = 1 - \exp\left[-t\int_{D(\varepsilon,\infty)}\nu(\mathrm{d}x)\right]$$

by Theorem 19.2(i). Hence, if $\int_{D(\varepsilon,\infty)}\nu(\mathrm{d}x) = c > 0$, then T_ε has exponential distribution with mean $1/c$.

Suppose $\nu(\mathbb{R}^d) = \infty$. Then $\lim_{\varepsilon\downarrow 0} P[T_\varepsilon \le t] = 1$ for any $t > 0$, and hence $\lim_{\varepsilon\downarrow 0} T_\varepsilon = 0$ a. s. Hence there is $H_0 \in \mathcal{F}$ with $P[H_0] = 1$ such that, for any $\omega \in H_0$, the time 0 is a limiting point of jumping times of $X_t(\omega)$. Next use Proposition 10.7. We see that, for any $s > 0$, there is $H_s \in \mathcal{F}$ with $P[H_s] = 1$ such that, for any $\omega \in H_s$, the set of jumping times has s as a limiting point on the right. Consider $H = \bigcap_{s\in\mathbb{Q}_+} H_s$. Jumping times are dense in $[0,\infty)$ for any $\omega \in H$.

Suppose $0 < \nu(\mathbb{R}^d) < \infty$. By Theorem 19.2(i), $J(t)$ has Poisson distribution with mean $t\nu(\mathbb{R}^d)$, and $J(t) < \infty$ a. s. Hence the jumping times are enumerable in increasing order. The first jumping time T has exponential distribution with mean $1/\nu(\mathbb{R}^d)$, because $P[T \le t] = P[J(t) \ge 1] = 1 - \mathrm{e}^{-t\nu(\mathbb{R}^d)}$. It follows that $T(\omega) < \infty$ a. s. Let $T^{(s)}$ be the first jumping time after s. Using Proposition 10.7, we see that $T^{(s)} < \infty$ a. s. Hence there are infinitely many jumps, a. s. □

In the case $0 < \nu(\mathbb{R}^d) < \infty$, we can actually say more: if we denote the nth jumping time by $U_n(\omega)$ and $U_0(\omega) = 0$, then $\{U_n - U_{n-1} \colon n \in \mathbb{N}\}$ constitutes independent identically distributed random variables, each exponentially distributed with mean $1/\nu(\mathbb{R}^d)$, and $\lim_{n\to\infty} U_n(\omega) = \infty$ a. s. To see this, note that $\{J(t)\}$ is a Poisson process with parameter $\nu(\mathbb{R}^d)$.

DEFINITION 21.4. A Lévy process $\{X_t\}$ on $\mathbb{R}$ is said to be *increasing* if $X_t(\omega)$ is increasing as a function of t, a. s. An increasing Lévy process is often called a *subordinator,* in connection with Bochner's subordination to be discussed in Chapter 6.

THEOREM 21.5 (Increasingness). *Let $d = 1$. A Lévy process $\{X_t\}$ on $\mathbb{R}$ is increasing if and only if $A = 0$, $\int_{(-\infty,0)} \nu(\mathrm{d}x) = 0$, $\int_{(0,1]} x\,\nu(\mathrm{d}x) < \infty$, and $\gamma_0 \geq 0$.*

Proof. The 'if' part. It follows from $\int_{(-\infty,0)} \nu(\mathrm{d}x) = 0$ and Theorem 19.2(i) that $J((0,t]\times(-\infty,0)) = 0$ a. s., that is, $\{X_t\}$ does not have negative jumps. Hence, by Theorem 19.3,

$$X_t = \int_{(0,t]\times(0,\infty)} x\,J(\mathrm{d}(s,x)) + t\gamma_0 \quad \text{a. s.},$$

because $X_t^4 = t\gamma_0$. This shows that $\{X_t\}$ is increasing.

The 'only if' part. Since $\{X_t\}$ has no negative jumps, we have, by Theorem 19.2(i), $\nu((-\infty,0)) = 0$. Since an increasing function remains increasing after a finite number of its jumps are deleted, we have $X(t) - X_\varepsilon(t) \geq 0$. Hence

$$\widetilde{X}(t) = \lim_{\varepsilon\downarrow 0} X_\varepsilon(t) = \int_{(0,t]\times(0,\infty)} x\,J(\mathrm{d}(s,x))$$

exists and is bounded above by $X(t)$. By Propositions 4.5 and 19.5 we have

$$\begin{aligned} E[\mathrm{e}^{-uX_\varepsilon(t)}] &= \exp\left[t\int_{(\varepsilon,\infty)} (\mathrm{e}^{-ux} - 1)\,\nu(\mathrm{d}x)\right] \\ &= \exp\left[t\int_{(\varepsilon,\infty)} (\mathrm{e}^{-ux} - 1 + ux1_{(0,1]}(x))\,\nu(\mathrm{d}x) - tu\int_{(\varepsilon,1]} x\,\nu(\mathrm{d}x)\right] \end{aligned}$$

for $u > 0$. As $\varepsilon \downarrow 0$, $E[\mathrm{e}^{-uX_\varepsilon(t)}]$ tends to $E[\mathrm{e}^{-u\widetilde{X}(t)}]$, which is positive, and

$$\int_{(\varepsilon,\infty)} (\mathrm{e}^{-ux} - 1 + ux1_{(0,1]}(x))\,\nu(\mathrm{d}x)$$

tends to the integral over $(0,\infty)$, which is finite. Hence $\int_{(0,1]} x\,\nu(\mathrm{d}x) < \infty$. Now we can use Theorem 19.3. We have $X_t = X_t^3 + X_t^4$, $X_t^3 = \widetilde{X}(t)$, and $\{X_t^4\}$ has generating triplet $(A, 0, \gamma_0)$. But $X_t^4 = X_t - \widetilde{X}(t) \geq 0$ and hence $A = 0$ and $\gamma_0 \geq 0$. □

A consequence of Theorem 21.5 should be contemplated. A Lévy process on $\mathbb{R}$ generated by (A,ν,γ) with $A = 0$, $\nu((-\infty,0)) = 0$, and $\int_{(0,1]} x\,\nu(\mathrm{d}x) = \infty$ has positive jumps only, does not have a Brownian-like part, but it is fluctuating, not increasing, no matter how large γ is. Moreover, it is not increasing in any time interval (by Theorem 21.9(ii) below combined with the Markov property). An explanation is that such a process can exist only with infinitely strong drift in the negative direction, which cancels the divergence of the sum of jumps; but it causes a random continuous motion in the negative direction.

REMARK 21.6. When $\{X_t\}$ is a subordinator, the Laplace transform of its distribution is more convenient than the characteristic function. The general form is as follows:

$$E[\mathrm{e}^{-uX_t}] = \exp\left[t\left(\int_{(0,\infty)} (\mathrm{e}^{-ux} - 1)\nu(\mathrm{d}x) - \gamma_0 u\right)\right] \quad \text{for } u \geq 0. \tag{21.1}$$

This is shown in the proof of Theorem 21.5.

EXAMPLE 21.7. Let $\{X_t\}$ be a stable process on $\mathbb{R}$ with parameters (α, β, τ, c) as in Definition 14.16. It is a subordinator if and only if $0 < \alpha < 1$, $\beta = 1$, and $\tau \geq 0$. This is called a *stable subordinator.* Use Theorem 21.5 and Remark 14.4 to check this. This example continues in Example 24.12.

Let us seek the condition for the sample functions to be of finite variation on any finite interval. For any $\xi \in D([0,\infty), \mathbb{R}^d)$ and $0 \leq t_1 < t_2 < \infty$, define $v((t_1, t_2], \xi)$, the *variation* of ξ on $(t_1, t_2]$, by

$$v((t_1, t_2], \xi) = \sup_{\Delta} \sum_{j=1}^{n} |\xi(s_j) - \xi(s_{j-1})|, \tag{21.2}$$

where Δ is a partition $t_1 = s_0 < s_1 < \cdots < s_n = t_2$ of $(t_1, t_2]$ and the supremum is taken over all partitions of $(t_1, t_2]$. Write

$$v_t(\xi) = \begin{cases} 0 & \text{for } t = 0, \\ v((0, t], \xi) & \text{for } t > 0, \end{cases}$$

which we call the *variation function* of ξ.

LEMMA 21.8. (i) *In the definition (21.2) of $v((t_1, t_2], \xi)$ it is enough to take the supremum over all partitions Δ in which the points of partition $\{s_0, \ldots, s_n\}$ belong to $\mathbb{Q} \cup \{t_1, t_2\}$.*

(ii) *For $0 \leq t_1 < t_2 < t_3 < \infty$,*

$$v((t_1, t_2], \xi) + v((t_2, t_3], \xi) = v((t_1, t_3], \xi). \tag{21.3}$$

(iii) *Suppose that*

$$\sum_{s \in (0,t]} |\xi(s) - \xi(s-)| < \infty \quad \textit{for any } t \in (0, \infty), \tag{21.4}$$

where the sum is taken over all s such that $\xi(s) - \xi(s-) \neq 0$. Then the function ξ_1 defined by

$$\xi_1(t) = \begin{cases} 0 & \text{for } t = 0, \\ \sum_{s \in (0,t]} (\xi(s) - \xi(s-)) & \text{for } 0 < t < \infty \end{cases} \tag{21.5}$$

belongs to $D([0, \infty), \mathbb{R}^d)$ and

$$v_t(\xi_1) = \sum_{s \in (0,t]} |\xi(s) - \xi(s-)| = \sum_{s \in (0,t]} |\xi_1(s) - \xi_1(s-)| \leq v_t(\xi). \tag{21.6}$$

(iv) *Suppose that ξ has finite variation on $(0, t]$ for any $t \in (0, \infty)$. Then $v_t(\xi)$ is right-continuous and increasing on $[0, \infty)$ and (21.4) holds. The function $\xi_2(t)$ defined by*

$$\xi(t) = \xi_1(t) + \xi_2(t) \tag{21.7}$$

is continuous and satisfies

$$v_t(\xi) = v_t(\xi_1) + v_t(\xi_2), \tag{21.8}$$

and $v_t(\xi_2)$ is continuous.

When ξ has finite variation on $(0, t]$ for any $t \in (0, \infty)$, we call ξ_1 and ξ_2, respectively, the *jump part* and the *continuous part* of ξ.

Proof of lemma. The assertion (i) follows from the right-continuity of ξ; (ii) is true because, in the definition of $v((t_1, t_3], \xi)$, it is enough only to consider partitions in which t_2 is a point of partition.

(iii) The jumping times of ξ in $(0, \infty)$ are countable. We enumerate them as $t_1, t_2, \ldots$. Write

$$\eta_n(t) = \begin{cases} 0 & \text{for } t = 0, \\ \sum_{j=1}^n (\xi(t_j) - \xi(t_j-)) 1_{(0,t]}(t_j) & \text{for } 0 < t < \infty. \end{cases}$$

As $n \to \infty$, $\eta_n(t)$ tends to $\xi_1(t)$ uniformly on any bounded interval of t. Hence ξ_1 inherits the right-continuity with left limits. If $t = t_k$, a discontinuity point of ξ, then, for $n \geq k$, $\eta_n(t) - \eta_n(t-) = \xi(t) - \xi(t-)$, which implies $\xi_1(t) - \xi_1(t-) = \xi(t) - \xi(t-)$. If $t > 0$ is a continuity point of ξ, then $\eta_n(t) - \eta_n(t-) = 0$ for every n, and ξ_1 is continuous at t. Hence

$$v_t(\xi_1) \geq \textstyle\sum_{s \in (0,t]} |\xi_1(s) - \xi_1(s-)| = \sum_{s \in (0,t]} |\xi(s) - \xi(s-)|. \tag{21.9}$$

On the other hand,

$$|\xi_1(u) - \xi_1(t)| = \left| \textstyle\sum_{s \in (t,u]} (\xi(s) - \xi(s-)) \right| \leq \sum_{s \in (t,u]} |\xi(s) - \xi(s-)|$$

for $t < u$, which shows that the inequality in (21.9) is an equality. The last inequality in (21.6) is obvious.

(iv) Increasingness of $v_t(\xi)$ is obvious. Let $\varepsilon > 0$. For any partition $t = s_0 < s_1 < \cdots < s_n = u$ of $(t, u]$ with s_1 close enough to t, we have

$$\textstyle\sum_{j=1}^n |\xi(s_j) - \xi(s_{j-1})| < \varepsilon + \sum_{j=2}^n |\xi(s_j) - \xi(s_{j-1})| \leq \varepsilon + v((s_1, u], \xi)$$

$$= \varepsilon + v_u(\xi) - v_{s_1}(\xi) \le \varepsilon + v_u(\xi) - v_{t+}(\xi).$$

Hence $v_u(\xi) - v_t(\xi) \le \varepsilon + v_u(\xi) - v_{t+}(\xi)$. Hence $v_{t+}(\xi) \le v_t(\xi)$. The reverse inequality is evident. Thus $v_t(\xi)$ is right-continuous. Use $t_1, t_2, \ldots$ from the proof of (iii). Since

$$\textstyle\sum_{j=1}^n |\xi(t_j) - \xi(t_j-)| 1_{(0,t]}(t_j) \le v_t(\xi)$$

for every n, ξ satisfies (21.4). Use $\eta_n(t)$ again and let $\zeta_n(t) = \xi(t) - \eta_n(t)$. Together with $\eta_n(t)$, $\zeta_n(t)$ converges uniformly on any bounded interval of t. The limit function is $\xi_2(t)$ in (21.7). If $t = t_k$ for some k, then $\zeta_n(t) - \zeta_n(t-) = 0$ for $n \ge k$, since $\eta_n(t) - \eta_n(t-) = \xi(t) - \xi(t-)$. If $t > 0$ is a continuity point of ξ, then $\zeta_n(t) - \zeta_n(t-) = 0$ for every n. Therefore $\xi_2(t)$ is continuous at every t.

Now let us prove (21.8). It follows from $\xi = \xi_1 + \xi_2$ that $v_t(\xi) \le v_t(\xi_1) + v_t(\xi_2)$. In order to show the reverse inequality, we claim that

$$v_t(\xi) \ge \textstyle\sum_{j=1}^n |\xi(t_j) - \xi(t_j-)| 1_{(0,t]}(t_j) + v_t(\zeta_n) \tag{21.10}$$

for any n. Let $0 = s_0 < \cdots < s_l = t$ be a partition of $(0, t]$ such that $\{s_1, \ldots, s_l\}$ contains all points in the set $\{t_1, \ldots, t_n\} \cap (0, t]$. Let $N \subset \{1, \ldots, l\}$ be the set defined by $\{s_k \colon k \in N\} = \{t_1, \ldots, t_n\} \cap (0, t]$, and let $M = \{1, \ldots, l\} \setminus N$. Choose $\varepsilon > 0$ satisfying $s_k - s_{k-1} > \varepsilon > 0$ for all k. Then

$$\begin{aligned} v_t(\xi) &\ge \textstyle\sum_{k \in N} (|\xi(s_k) - \xi(s_k - \varepsilon)| + |\xi(s_k - \varepsilon) - \xi(s_{k-1})|) \\ &\quad + \textstyle\sum_{k \in M} |\xi(s_k) - \xi(s_{k-1})| \\ &= \textstyle\sum_{k \in N} (|\xi(s_k) - \xi(s_k - \varepsilon)| + |\zeta_n(s_k - \varepsilon) - \zeta_n(s_{k-1})|) \\ &\quad + \textstyle\sum_{k \in M} |\zeta_n(s_k) - \zeta_n(s_{k-1})|. \end{aligned}$$

Letting $\varepsilon \downarrow 0$, we get

$$v_t(\xi) \ge \textstyle\sum_{j=1}^n |\xi(t_j) - \xi(t_j-)| 1_{(0,t]}(t_j) + \sum_{k=1}^l |\zeta_n(s_k) - \zeta_n(s_{k-1})|. \tag{21.11}$$

This proves (21.10). Now every partition $0 = s_0 < \cdots < s_l = t$ of $(0, t]$ satisfies (21.11), because (21.11) is weaker than (21.10). Fixing the partition, let $n \to \infty$ and use (21.5) and the convergence $\zeta_n \to \xi_2$. Then take the supremum over partitions. We obtain $v_t(\xi) \ge v_t(\xi_1) + v_t(\xi_2)$. This proves (21.8).

It remains to prove that $v_t(\xi_2)$ is continuous. The right-continuity is similarly proved to that of $v_t(\xi)$. For any $\varepsilon > 0$, choose a partition $0 = s_0 < \cdots < s_n = t$ such that s_{n-1} is close enough to t. We find

$$\textstyle\sum_{j=1}^n |\xi_2(s_j) - \xi_2(s_{j-1})| < v((0, s_{n-1}], \xi_2) + \varepsilon \le v_{t-}(\xi_2) + \varepsilon,$$

since ξ_2 is continuous. Hence $v_t(\xi_2) \le v_{t-}(\xi_2) + \varepsilon$. Thus $v_t(\xi_2) \le v_{t-}(\xi_2)$. We get $v_t(\xi_2) = v_{t-}(\xi_2)$, as the reverse inequality is evident. □

THEOREM 21.9 (Variation). (i) *Suppose that $\{X_t\}$ is of type A or B. Then, almost surely, the sample function $X_t(\omega)$ has finite variation on $(0, t]$ for any $t \in (0, \infty)$; the variation function $V_t(\omega)$ of $X_t(\omega)$ is a subordinator with*

$$E[\mathrm{e}^{-uV_t}] = \exp\left[t \left(\int_{\mathbb{R}^d} (\mathrm{e}^{-u|x|} - 1)\, \nu(\mathrm{d}x) - u|\gamma_0| \right) \right], \qquad u \ge 0; \tag{21.12}$$

the continuous part of $X_t(\omega)$ is $t\gamma_0$, a. s.

(ii) *If $\{X_t\}$ is of type C, then, almost surely, the sample function $X_t(\omega)$ has infinite variation on $(0,t]$ for any $t \in (0,\infty)$.*

The assertion (i) implies that $\{V_t\}$ has drift $|\gamma_0|$ and Lévy measure ν_0 defined by $\nu_0(B) = \int_{\mathbb{R}^d} 1_B(|x|)\,\nu(\mathrm{d}x)$ for $B \in \mathcal{B}(\mathbb{R})$.

Proof of theorem. Let

$$U_t = \int_{(0,t]\times D(0,\infty)} |x|\, J(\mathrm{d}(s,x)).$$

To prove (i), suppose that $\{X_t\}$ is of type A or B. Recall the proof of Theorem 19.3. We have shown that $U_t < \infty$ a. s., that

$$E[\mathrm{e}^{-uU_t}] = \exp\left[t\int_{D(0,\infty)} (\mathrm{e}^{-u|x|} - 1)\,\nu(\mathrm{d}x)\right] \quad \text{for } u > 0,$$

and that

$$X_t = \int_{(0,t]\times D(0,\infty)} x\, J(\mathrm{d}(s,x)) + t\gamma_0.$$

Since

$$U_t = \sum_{s\in(0,t]} |X_s - X_{s-}| \quad \text{and} \quad X_t = \sum_{s\in(0,t]} (X_s - X_{s-}) + t\gamma_0,$$

Lemma 21.8(iii) says that X_t has finite variation on $(0,t]$ for any $t \in (0,\infty)$ and the variation function V_t of X_t is expressed as

$$V_t = U_t + t|\gamma_0|.$$

For $0 \le s < t$, let $\mathcal{F}_t^s$ be the σ-algebra generated by $\{X_{t_2} - X_{t_1} : t_1, t_2 \in [s,t]\}$. Then $V_t - V_s$ is $\mathcal{F}_t^s$-measurable by Lemma 21.8(i). Hence $\{V_t\}$ has independent increments and

$$E[\mathrm{e}^{-u(V_t - V_s)}] = \exp\left[(t-s)\left(\int_{D(0,\infty)} (\mathrm{e}^{-u|x|} - 1)\,\nu(\mathrm{d}x) - u|\gamma_0|\right)\right].$$

As a function of t, V_t is right-continuous with left limits. We have $V_0 = 0$. Therefore $\{V_t\}$ is a Lévy process and all assertions in (i) are shown.

Let us prove (ii) for $\{X_t\}$ of type C. Let

$$U_\varepsilon(t) = \int_{(0,t]\times D(\varepsilon,\infty)} |x|\, J(\mathrm{d}(s,x)).$$

By Proposition 19.5 we have

$$E[\mathrm{e}^{-uU_\varepsilon(t)}] = \exp\left[t\int_{D(\varepsilon,\infty)} (\mathrm{e}^{-u|x|} - 1)\,\nu(\mathrm{d}x)\right], \quad u > 0.$$

Hence

$$E[\mathrm{e}^{-uU_\varepsilon(t)}] = \exp\left[t\int_{D(\varepsilon,\infty)} (\mathrm{e}^{-u|x|} - 1 + u|x|1_{D(0,1]}(x))\,\nu(\mathrm{d}x)\right.$$
$$\left. - tu\int_{D(\varepsilon,1]} |x|\nu(\mathrm{d}x)\right].$$

Suppose that $\int_{|x|\leq 1}|x|\,\nu(\mathrm{d}x) = \infty$. Then the right-hand side tends to 0 as $\varepsilon \downarrow 0$. Hence $U(t) = \infty$ a.s. for any $t > 0$. Considering $X_t(\omega)1_{[0,t_0]}(t)$ with t_0 fixed and recalling Lemma 21.8(iv), we see that X_t has infinite variation on $(0,t]$ for any $t > 0$, a.s.

Next consider the case $A \neq 0$ and $\nu = 0$. We claim that $\{X_t\}$ has infinite variation on $(0,t]$ for any $t > 0$, a.s. It is enough to prove this in the case $\gamma = 0$. Let $X_j(t)$, $j = 1,\ldots,d$, be the components of X_t. Since $A \neq 0$, there is j such that $\{X_j(t)\}$ is a positive constant multiple of the one-dimensional Brownian motion. Fix t and consider a sequence of partitions $\Delta_n\colon 0 = s_{n,0} < s_{n,1} < \cdots < s_{n,N(n)} = t$ of $(0,t]$ such that $\mathrm{mesh}(\Delta_n) \to 0$ as $n \to \infty$. Let

$$Y_n(\omega) = \sum_{k=1}^{N(n)} |X_j(s_{n,k},\omega) - X_j(s_{n,k-1},\omega)|^2.$$

If $X_j(s,\omega)$ has finite variation on $(0,t]$, then $Y_n(\omega) \to 0$ as $n \to \infty$ by the uniform continuity of $X_j(s,\omega)$. But, by Theorem 5.8, Y_n tends to a positive constant in the L^2-sense as $n \to \infty$. Hence there is a subsequence $\{\Delta_{n(l)}\}$ such that $Y_{n(l)}$ tends a.s. to that positive constant. Hence Y_n tends to 0 only with probability 0. Thus $\{X_j(t)\}$ has infinite variation on $(0,t]$ for any $t > 0$, a.s. The variation of X_t is bigger than or equal to that of $X_j(t)$.

Finally, consider the case $A \neq 0$ and $\int_{|x|\leq 1}|x|\,\nu(\mathrm{d}x) < \infty$. By virtue of Theorem 19.3, $X_t = X_t^3 + X_t^4$, where X_t^3 is the sum of jumps and X_t^4 is a continuous Lévy process with generating triplet $(A,0,\gamma_0)$. We know that X_t^4 has infinite variation on $(0,t]$ for any $t > 0$, a.s. Again apply Lemma 21.8(iv) to $X_t(\omega)1_{[0,t_0]}(t)$ with t_0 fixed. If X_t has finite variation on $(0,t_0]$, then so does X_t^4, leading to a contradiction. □

22. Exercises 4

E 22.1. Let $\{X_t\}$ be a Lévy process on $\mathbb{R}$ with Lévy measure ν satisfying $\nu((0,\infty)) > 0$. Let Y_t be the largest jump of X_s, $0 \leq s \leq t$, that is, $Y_t = \max_{s\in(0,t]}(X_s - X_{s-})$. Show that $P[\,Y_t \geq a\,] = 1 - \mathrm{e}^{-t\nu([a,\infty))}$ for $a > 0$.

E 22.2 (Khintchine [**278**], rediscovered by Ferguson and Klass [**141**]). Let μ be infinitely divisible on $\mathbb{R}$ with $\widehat{\mu}(z) = \exp\int_{(0,\infty)}(\mathrm{e}^{\mathrm{i}zx} - 1)\nu(\mathrm{d}x)$, $\int_{(0,1]} x\nu(\mathrm{d}x) < \infty$. Define $h(s) = \inf\{x > 0\colon \nu((x,\infty)) \leq s\}$, the right-continuous inverse function of $s = \nu((x,\infty))$. Let $\{N_t\}$ be a Poisson process with parameter 1 and let U_n be the nth jumping time of N_t. Show that μ is the distribution of $\sum_{n=1}^{\infty} h(U_n)$. In particular, if μ is a strictly α-stable distribution supported on $[0, \infty)$, $0 < \alpha < 1$, then μ is the distribution of $c \sum_{n=1}^{\infty} U_n^{-1/\alpha}$, where c is a positive constant.

E 22.3 (LePage [**309**]). Let $0 < \alpha < 2$. Let $\{N_t\}$ and $\{U_n\}$ be as in E 22.2 and let $\{Y_n\}$ be independent, identically distributed symmetric random variables with $E[|Y_n|^\alpha] < \infty$. Assume that $\{N_t\}$ and $\{Y_n\}$ are independent. Show that $X = \sum_{n=1}^{\infty} Y_n U_n^{-1/\alpha}$ exists a. s. and the distribution of X is symmetric α-stable.

E 22.4 (Rosinski [**421**]). This is a generalization of compound Poisson processes. Let $\{N_t\}$ and $\{U_n\}$ be as in E 22.2 and $\{Y_n\}$ be independent identically distributed random variables on $\mathbb{R}^k$. We assume that $\{N_t\}$ and $\{Y_n\}$ are independent. Let $h(s, y)$ be a measurable function from $(0, \infty) \times \mathbb{R}^k$ to $\mathbb{R}$. Define $S_0 = 0$, $S_n = \sum_{j=1}^{n} h(U_j, Y_j)$ for $n = 1, 2, \ldots$, and $X_t = S_{N(t)}$. Prove that $\{X_t\}$ is an additive process with

$$E[\mathrm{e}^{\mathrm{i}zX_t}] = \exp \textstyle\int_0^t \mathrm{d}s \int_{\mathbb{R}^k} (\mathrm{e}^{\mathrm{i}zh(s,y)} - 1)\lambda(\mathrm{d}y),$$

where λ is the distribution of Y_n.

E 22.5. Prove the following for a Lévy process $\{X_t\}$ on $\mathbb{R}^d$ defined on a complete probability space $(\Omega, \mathcal{F}, P)$. The probability that sample functions are continuous is 0 or 1. The probability that sample functions are piecewise constant is 0 or 1. In the case $d = 1$, the probability that sample functions are increasing is 0 or 1.

E 22.6. Show that none of the three assertions in E 22.5 is true for a general additive process.

E 22.7. Let $\{X_t\}$ be a Lévy process on $\mathbb{R}^d$. Suppose that there is $\Omega_1 \in \mathcal{F}$ with $P[\Omega_1] > 0$ with the following property. For any $\omega \in \Omega_1$, there exists $t > 0$ such that $X_s(\omega)$, $s \in [0, t]$, is piecewise constant. Prove that $\{X_t\}$ is a compound Poisson process or a zero process.

E 22.8. Let $\{X_t\}$ be a non-trivial semi-stable process on $\mathbb{R}^d$ with index α, $0 < \alpha \leq 2$. Show that, almost surely, the sample functions of $\{X_t\}$ have the following properties. They are continuous if $\alpha = 2$. Their jumping times are dense in $(0, \infty)$ if $\alpha < 2$. They are of finite variation on any bounded interval if $0 < \alpha < 1$. They have infinite variation on any time interval if $1 \leq \alpha \leq 2$.

E 22.9. Let $0 < \alpha < 1$. Prove that, if $\{X_t\}$ is α-stable or α-semi-stable on $\mathbb{R}^d$, then the variation function V_t of X_t is, respectively, an α-stable or α-semi-stable subordinator.

E 22.10. Let $\{Z_t\}$ be a Lévy process on $\mathbb{R}^2$. Denote the first and the second component of Z_t by X_t and Y_t, respectively. Show that if the one-dimensional Lévy processes $\{X_t\}$ and $\{Y_t\}$ are, respectively, Gaussian and purely non-Gaussian, then they are independent. Nice applications of this result are given by Kasahara [**263**].

E 22.11. A subset K of $\mathbb{R}^d$ is a *cone* if it is convex and closed and contains at least two points and if, for any $x \in K$ and $a \geq 0$, ax is in K. A cone K is a *proper cone* if it does not contain a straight line that goes through 0. An $\mathbb{R}^d$-valued function $f(t)$ is said to be K-increasing if $f(t) - f(s) \in K$ whenever $s < t$. Let K be a proper cone. Prove that sample functions of a Lévy process $\{X_t\}$ on $\mathbb{R}^d$ are K-increasing a. s. if and only if $A = 0$, $\nu(\mathbb{R}^d \setminus K) = 0$, $\int_{|x|\le 1} |x|\nu(\mathrm{d}x) < \infty$, and $\gamma_0 \in K$.

Notes

Poisson random measures, or Poisson point processes, are constructed by Kingman [**285**]. Their importance in the theory of Markov processes is observed by Itô [**226**]. Although we take a different approach, our proof of Theorems 19.2 and 19.3 uses techniques of Itô's original proof [**220, 222**]. Lemma 20.2 follows Kwapień and Woyczyński [**302**], p. 15. The proof of Lemma 20.7 is based on Doob's technique in his book [**106**], p. 55. Theorem 21.9 on sample function variation is a completion of the result in Skorohod's book [**493**].

Kunita and Watanabe [**300**] have built up the theory of square integrable martingales, based on a generalization of Itô's formula for stochastic integrals. They prove the Lévy–Itô decomposition as an application of their theory in semimartingales. Theory of convergence of stochastic processes to Lévy processes is developed in the framework of semimartingales and Poisson point processes. See Jacod and Shiryaev [**230**] and Kasahara and Watanabe [**265**].

Khintchine [**278**] gives a series representation similar to E 22.2 for a random variable on $\mathbb{R}$ with a general purely non-Gaussian infinitely divisible distribution. This and extensions of E 22.2, E 22.3, and E 22.4 by LePage [**310**] and Rosinski [**421**] have many applications to stochastic processes with infinitely divisible finite-dimensional distributions.

CHAPTER 5

Distributional properties of Lévy processes

23. Time dependent distributional properties

Let $\{X_t: t \geq 0\}$ be a Lévy process on $\mathbb{R}^d$, and let μ be the distribution of X_1. By the definition, the increments of $\{X_t\}$ are stationary. But, as t goes on, the distribution μ^t of X_t may exhibit time evolution of a qualitative nature. If $\{X_t\}$ is a stable process with index $\alpha \in (0, 2]$, then, for any $s > 0$ and $t > 0$, there are $c > 0$ and $b \in \mathbb{R}^d$ such that, for every Borel set B, $\mu^t(B) = \mu^s(cB + b)$ (in fact, $c = (s/t)^{1/\alpha}$, see Theorem 13.15), that is, μ^t and μ^s coincide under an affine transformation of the state space $\mathbb{R}^d$. Thus stable processes do not have any time evolution of qualitative nature in their distributions. However, general Lévy processes are far more complicated than stable processes.

For example, let $\{X_t\}$ be a Lévy process on $\mathbb{R}$ such that

$$E[\mathrm{e}^{\mathrm{i}zX_t}] = \exp\left[t \int_0^\infty (\mathrm{e}^{\mathrm{i}zx} - 1)\mathrm{e}^{-x}\mathrm{d}x\right]. \tag{23.1}$$

Then μ^t is unimodal (defined later in this section) with mode 0 for $t \leq 2$, but is not unimodal for $t > 2$, as is shown in Example 23.4 below. Thus, unimodality of μ^t possibly depends on the time t. On the other hand, if $\{X_t\}$ is a Lévy process on $\mathbb{R}^d$ with $E|X_{t_0}| < \infty$ for some $t_0 > 0$, then $E|X_t| < \infty$ for all $t > 0$ (Corollary 25.8). Thus, finiteness of the mean of μ^t never depends on the time t. Hence there are two sorts of properties.

DEFINITION 23.1. Consider a property $\mathfrak{P}$ relating to a distribution on $\mathbb{R}^d$. We say that $\mathfrak{P}$ is a *time dependent distributional property in the class of Lévy processes*, if there is a Lévy process $\{X_t\}$ on $\mathbb{R}^d$ such that, for some t_1 and t_2 in $(0, \infty)$, $P_{X(t_1)}$ does have the property $\mathfrak{P}$ and $P_{X(t_2)}$ does not. For any property $\mathfrak{P}$ which is not a time dependent distributional property in the class of Lévy processes, the following dichotomy holds: if $\{X_t\}$ is a Lévy process on $\mathbb{R}^d$, then either P_{X_t} has the property $\mathfrak{P}$ for every $t > 0$ or there is no $t > 0$ for which P_{X_t} has the property $\mathfrak{P}$.

We will see in Section 27 the existence of a subordinator $\{X_t\}$ such that, for some $t_0 > 0$, P_{X_t} is continuous and singular for $t < t_0$ and absolutely continuous for $t > t_0$. Time dependence of the distributions in this case is drastic.

In this chapter we study properties of the distributions μ^t of Lévy processes $\{X_t\}$ on $\mathbb{R}^d$. In other words, we study properties of infinitely divisible distributions on $\mathbb{R}^d$. But we are concerned whether the properties are time dependent or not. In this respect most of the properties we deal with in this chapter and Chapter 10 are classified as follows.

(a) Let $d = 1$. The following are time dependent distributional properties in the class of Lévy processes:

(1) μ is unimodal (Example 23.4);
(2) μ is unimodal with mode 0 (Example 23.3);
(3) ($n \geq 2$ fixed) μ is n-modal (Remark 54.10);
(4) μ is strongly unimodal (Definition 52.2, Example 52.7);
(5) μ is symmetric and unimodal (Remark 54.4);
(6) μ is continuous and singular (Theorem 27.23, Remarks 27.22 and 27.24);
(7) μ is absolutely continuous (Theorem 27.23, Remarks 27.22 and 27.24);
(8) (n fixed) μ has density of class C^n (Example 23.3, Remark 28.7);
(9) (α fixed) $\dim_{\mathrm{R}} \mu = \alpha$ (Notes at the end of this chapter);
(10) ($\alpha > 0$ fixed) $\int_{|x|>1} |x|^{-\alpha} \mathrm{e}^{|x|} \mu(\mathrm{d}x) < \infty$ (Remark 25.9).

(b) Let $d \geq 1$. None of the following is a time dependent distributional property in the class of Lévy processes:

(1) μ is continuous, that is, does not have a point mass (Theorem 27.4);
(2) μ is discrete, that is, concentrated on a countable set (Corollary 27.5);
(3) μ is discrete and has finite entropy (Exercise 29.24);
(4) ($\alpha > 0$ fixed) μ has finite moment of order α, that is, $\int |x|^\alpha \mu(\mathrm{d}x) < \infty$ (Corollary 25.8);
(5) ($\alpha_0 > 0$ fixed) $\int \mathrm{e}^{\alpha|x|\log|x|} \mu(\mathrm{d}x) < \infty$ for $\alpha \in (0, \alpha_0)$ (Remark 26.2);
(6) ($\alpha > 0$, $0 < \beta \leq 1$, and $\gamma \geq 0$ fixed) $\int |x|^\gamma \mathrm{e}^{\alpha|x|^\beta} \mu(\mathrm{d}x) < \infty$ (Corollary 25.8);
(7) ($\alpha > 0$ fixed) $\int_{|x|>1} (\log|x|)^\alpha \mu(\mathrm{d}x) < \infty$ (Corollary 25.8);
(8) ($c \in \mathbb{R}^d$ fixed) $\int \mathrm{e}^{\langle c,x \rangle} \mu(\mathrm{d}x) < \infty$ (Theorem 25.17);
(9) μ is symmetric (Exercise 18.1);
(10) ($d = 1$) μ is concentrated on $[0, \infty)$ (Theorem 24.11);
(11) ($d = 1$) μ is concentrated on $[0, \infty)$ and subexponential (Definition 25.13, Remark 25.14).

For each of the properties (1)–(11) in (b), we shall give a necessary and sufficient condition in terms of the generating triplet (A, ν, γ). But the connection of a time dependent distributional property with the generating triplet is more delicate. Although it is desirable to analyze time evolution of the property for a given Lévy process, usually it is hard to accomplish. Unimodality and n-modality will be studied in Chapter 10.

The definition of unimodality is as follows.

DEFINITION 23.2. A measure ρ on $\mathbb{R}$ is called *unimodal with mode* a if ρ is finite outside of any neighborhood of a and if $\rho(-\infty, x]$ and $\rho(x, \infty)$ are convex on $(-\infty, a)$ and (a, ∞), respectively. That is, ρ is unimodal with mode a if

$$\rho = c\delta_a + f(x)\mathrm{d}x,$$

where $0 \le c \le \infty$ and $f(x)$ is increasing on $(-\infty, a)$, decreasing on (a, ∞), and $\int_{|x-a|>\varepsilon} f(x)\mathrm{d}x < \infty$ for $\varepsilon > 0$. A measure ρ on $\mathbb{R}$ is *unimodal* if, for some a, it is unimodal with mode a.

EXAMPLE 23.3. Let $\{X_t\}$ be a Γ-process. It is a Lévy process on $\mathbb{R}$ with

$$\mu^t = \frac{\alpha^t}{\Gamma(t)} 1_{(0,\infty)}(x) x^{t-1} \mathrm{e}^{-\alpha x} \mathrm{d}x \tag{23.2}$$

for $t > 0$, where $\alpha > 0$. Let $f_t(x)$ be the density of μ^t. If $0 < t \le 1$, then $f_t(x)$ is strictly decreasing on $(0, \infty)$. If $t > 1$, then $f_t(x)$ is strictly increasing on $(0, \frac{t-1}{\alpha})$ and strictly decreasing on $(\frac{t-1}{\alpha}, \infty)$. Hence μ^t is unimodal with mode 0 for $0 \le t \le 1$, and unimodal with mode $\frac{t-1}{\alpha}$ for $t > 1$. If $0 < t \le 1$, then $f_t(x)$ is not continuous at $x = 0$. If $n < t \le n+1$ with a positive integer n, then $f_t(x)$ is of class C^{n-1} on $\mathbb{R}$ but not of class C^n on $\mathbb{R}$. This example shows that the properties (2) and (8) in (a) are time dependent.

EXAMPLE 23.4 (Wolfe [**577, 581**]). Let $\{Y_t\}$ be the Lévy process (subordinator) such that the characteristic function of P_{Y_t} is the right-hand side of (23.1). Let μ^t be the Γ-distribution (23.2) with $\alpha = 1$ with density $f_t(x)$. Since $\{Y_t\}$ is compound Poisson, we have

$$P_{Y_t} = \mathrm{e}^{-t}\delta_0 + \sum_{n=1}^{\infty} \mathrm{e}^{-t} \frac{t^n}{n!} \mu^n.$$

Hence P_{Y_t} has a point mass at 0 and is absolutely continuous on $(0, \infty)$ with density

$$g_t(x) = \sum_{n=1}^{\infty} \mathrm{e}^{-t} \frac{t^n}{n!} f_n(x).$$

Since $f_n'(x) = -f_n(x) + f_{n-1}(x)$ for $n = 1, 2, \ldots$ with $f_0(x) = 0$, we have

$$g_t'(x) = \sum_{n=1}^{\infty} \mathrm{e}^{-t} \frac{t^n}{n!} \Big(\frac{t}{n+1} - 1 \Big) f_n(x).$$

Hence, if $0 < t \le 2$, then $g_t'(x) < 0$ and P_{Y_t} is unimodal with mode 0. If $t > 2$, then P_{Y_t} is not unimodal, because $g_t'(0+) = \mathrm{e}^{-t} t (\frac{t}{2} - 1) > 0$. Further, it will be shown in Proposition 54.12 that $[P_{Y_t}]_{\{x>0\}}$ is unimodal; in this sense P_{Y_t} is bimodal for $t > 2$.

24. Supports

For any measure ρ on $\mathbb{R}^d$, its *support* $S_\rho = S(\rho)$ is defined to be the set of $x \in \mathbb{R}^d$ such that $\rho(G) > 0$ for any open set G containing x. The support S_ρ is a closed set. We say that ρ is supported on a set B, meaning that $S_\rho \subset B$. For any random variable X on $\mathbb{R}^d$, the support of P_X is called the support of X and denoted by $S_X = S(X)$. It is the smallest closed set F satisfying $P[X \in F] = 1$. The following simple lemma is basic in studying supports of random variables.

LEMMA 24.1. *If X and Y are independent random variables on $\mathbb{R}^d$, then S_{X+Y} is the closure of $\{x+y: x \in S_X,\ y \in S_Y\}$, that is, $S_{X+Y} = \overline{S_X + S_Y}$.*

Proof. If $x \in S_X$ and $y \in S_Y$, then $x + y \in S_{X+Y}$, since, for any $\varepsilon > 0$,

$$P[\,|X+Y-x-y| < \varepsilon\,] \geq P[\,|X-x| < \varepsilon/2]P[\,|Y-y| < \varepsilon/2\,] > 0.$$

Hence $S_{X+Y} \supset \overline{S_X + S_Y}$. If K_1 and K_2 are both compact, then $K_1 + K_2$ is compact. Consequently, $S_X + S_Y$ is the union of a countable number of compact sets, hence it is a Borel set. We have

$$P[\,X+Y \in S_X + S_Y\,] \geq P[\,X \in S_X\,]P[\,Y \in S_Y\,] = 1.$$

Hence $\overline{S_X + S_Y}$ is a closed set with P_{X+Y}-measure 1. Therefore it contains S_{X+Y}. □

COROLLARY 24.2. *Let ρ be a non-zero finite measure on $\mathbb{R}^d$. Suppose that ρ has a finite measure ρ_1 as a convolution factor, that is, $\rho = \rho_1 * \rho_2$ with some finite measure ρ_2. If S_{ρ_1} is unbounded, then S_ρ is unbounded.*

THEOREM 24.3 (Unboundedness). *Let $\{X_t: t \geq 0\}$ be a non-trivial Lévy process on $\mathbb{R}^d$. Then, for any $t > 0$, $S(X_t)$ is unbounded.*

Proof. Let (A, ν, γ) be the generating triplet of $\{X_t\}$. Since $\{X_t\}$ is non-trivial (Definition 13.6), we have $A \neq 0$ or $\nu \neq 0$.

Case 1. Suppose $A \neq 0$. Then A has rank $l \geq 1$. Let ρ_t be the Gaussian distribution with variance matrix tA and mean 0. Then $S(\rho_t)$ is an l-dimensional linear subspace, which is unbounded. The distribution P_{X_t} has ρ_t as a convolution factor. Hence $S(X_t)$ is unbounded by Corollary 24.2.

Case 2. Suppose $\nu \neq 0$. Choose $\varepsilon > 0$ such that $\int_{|x|>\varepsilon} \nu(\mathrm{d}x) = c > 0$ and let $\{Y_t\}$ be the compound Poisson process with Lévy measure $\nu_1 = [\nu]_{\{|x|>\varepsilon\}}$. Then P_{Y_t} is a convolution factor of P_{X_t}. Since

$$P_{Y_t} = \mathrm{e}^{-ct} \sum_{n=0}^{\infty} (n!)^{-1} t^n \nu_1^n, \tag{24.1}$$

$S(\nu_1^n) \subset S(Y_t)$ for every n. If $x \in S(\nu_1)$, then $nx \in S(\nu_1^n)$ by Lemma 24.1. Hence $S(Y_t)$ is unbounded. Then $S(X_t)$ is unbounded by Corollary 24.2. □

COROLLARY 24.4. *If μ is infinitely divisible on $\mathbb{R}^d$ and if μ is not a δ-distribution, then S_μ is unbounded.*

THEOREM 24.5 (Compound Poisson process). *Let $\{X_t\}$ be a compound Poisson process on $\mathbb{R}^d$ with Lévy measure ν. Let $F_0 = \{0\}$, $F_1 = S_\nu$, and $F_{n+1} = F_n + F_1$ for $n \geq 1$. Then, for every $t > 0$, $S(X_t)$ is the closure of $\bigcup_{n=0}^{\infty} F_n$.*

Proof. We have $S(\nu^n) = \overline{F_n}$. In fact, this is obvious for $n = 0$ and 1, and, if it is true for a given n, then, by Lemma 24.1, $S(\nu^{n+1}) = \overline{\overline{F_n} + F_1} = \overline{F_{n+1}}$. As P_{X_t} is expressed by the right-hand side of (24.1) with $c = \nu(\mathbb{R}^d)$ and with ν in place of ν_1, $S(X_t)$ equals the closure of $\bigcup_{n=0}^{\infty} \overline{F_n}$, which is no other than the closure of $\bigcup_{n=0}^{\infty} F_n$. □

COROLLARY 24.6 (Support in $\mathbb{Z}$). *Let μ be an infinitely divisible distribution on $\mathbb{R}$ with generating triplet (A, ν, γ). Then, $S_\mu \subset \mathbb{Z}$ if and only if $A = 0$, $S_\nu \subset \mathbb{Z}$, and the drift γ_0 is in $\mathbb{Z}$.*

Proof. Let us show the 'only if' part. Assume $S_\mu \subset \mathbb{Z}$. If $A \neq 0$, then μ has a nondegenerate Gaussian distribution as a convolution factor, which implies $S_\mu = \mathbb{R}$, contrary to the assumption. Hence $A = 0$. Next we claim that $\nu(\mathbb{R}) < \infty$. Suppose that, on the contrary, $\nu(\mathbb{R}) = \infty$. Then $\nu(\{x \colon 0 < |x| < 1\}) > 0$, and hence there is $x_0 \in S_\nu$ with $0 < |x_0| < 1$. Choose ε with $0 < \varepsilon < |x_0|$ and let U be the ε-neighborhood of x_0, ν_0 be the restriction of ν to U, and μ_0 be the compound Poisson distribution with Lévy measure ν_0. Then, 0 and x_0 belong to $S(\mu_0)$ by Theorem 24.5. Since μ_0 is a convolution factor of μ, it follows from Lemma 24.1 that there is $y \in \mathbb{R}$ such that y and $y + x_0$ are in S_μ, which contradicts that $S_\mu \subset \mathbb{Z}$. Hence it is shown that $\nu(\mathbb{R}) < \infty$. Let μ_1 be the compound Poisson distribution with Lévy measure ν. Then we have $\mu = \mu_1 * \delta_{\gamma_0}$. Since $\mu_1(\{0\}) > 0$, we have $\mu(\{\gamma_0\}) > 0$. Hence $\gamma_0 \in \mathbb{Z}$ and $S_{\mu_1} \subset \mathbb{Z}$. Therefore, Theorem 24.5 implies that $S_\nu \subset \mathbb{Z}$. This finishes the proof of the 'only if' part. The 'if' part is a direct consequence of Theorem 24.5. Note that $S_\nu \subset \mathbb{Z}$ implies that the total mass of ν is finite. □

Let us study the support of μ^t for a Lévy process on $\mathbb{R}$. We use the types A, B, C introduced in Definition 11.9. The following theorem gives a condition for μ^t to be bounded below.

THEOREM 24.7 (Support bounded below). *Let $\{X_t\}$ be a Lévy process on $\mathbb{R}$ with Lévy measure ν. Then, $S(X_t)$ is bounded below for every t if and only if $S_\nu \subset [0, \infty)$ and $\{X_t\}$ is of type A or B. Boundedness from below of the support is not a time dependent distributional property.*

Notice that the condition that $A = 0$ and $S_\nu \subset [0, \infty)$ does not imply boundedness below of $S(X_t)$ if $\{X_t\}$ is of type C.

Proof of theorem. Assume that $S_\nu \subset [0, \infty)$ and $\{X_t\}$ is of type A or B. Then

$$E[\mathrm{e}^{\mathrm{i}zX(t)}] = \exp\left[t\left(\int_{(0,\infty)} (\mathrm{e}^{\mathrm{i}zx} - 1)\nu(\mathrm{d}x) + \mathrm{i}\gamma_0 z\right)\right]. \tag{24.2}$$

Let $\{X^n(t)\}$ be a Lévy process satisfying

$$E[\mathrm{e}^{\mathrm{i}zX^n(t)}] = \exp\left[t\left(\int_{(1/n,\infty)} (\mathrm{e}^{\mathrm{i}zx} - 1)\nu(\mathrm{d}x) + \mathrm{i}\gamma_0 z\right)\right].$$

The support of $X^n(t) - \gamma_0 t$ is in $[0, \infty)$, since it is a compound Poisson process with positive jumps. Hence $S(X^n(t)) \subset [\gamma_0 t, \infty)$. As $n \to \infty$, the distribution of $X^n(t)$ tends to the distribution of $X(t)$. Therefore, $S(X(t)) \subset [\gamma_0 t, \infty)$, which shows the 'if' part.

Assume that, for some $t_0 > 0$, the support of $X(t_0)$ is bounded below. If we show that $S_\nu \subset [0, \infty)$ and that $\{X_t\}$ is of type A or B, then all the assertion of the theorem is true. Let $c(x)$ be a nonnegative bounded continuous function on $\mathbb{R}$ satisfying (8.3) and (8.4). Let $(t_0 A, t_0\nu, \beta')_c$ be the generating triplet of the infinitely divisible distribution $P_{X(t_0)}$ (see Remark 8.4). If $A \neq 0$, then $S(X(t_0)) = \mathbb{R}$ by Lemma 24.1, contrary to the assumption. Hence $A = 0$. For some $\beta'' \in \mathbb{R}$, $X(t_0) + \beta''$ has support in $[0, \infty)$. The generating triplet of $\rho = P_{X(t_0)+\beta''}$ is $(0, t_0\nu, \beta)$ with $\beta = \beta' + \beta''$. By Lemma 24.1, $\rho^{1/n}$ has support in $[0, \infty)$ for any $n \in \mathbb{N}$. Let ρ_n be a compound Poisson distribution defined by

$$\widehat{\rho}_n(z) = \exp[n(\widehat{\rho}(z)^{1/n} - 1)].$$

As is shown in the proof of Theorem 8.1(i) before Corollary 8.8, ρ_n tends to ρ as $n \to \infty$. Since

$$\widehat{\rho}_n(z) = \exp\left[n\int(\mathrm{e}^{\mathrm{i}zx} - 1 - \mathrm{i}zxc(x))\rho^{1/n}(\mathrm{d}x) + \mathrm{i}nz\int xc(x)\rho^{1/n}(\mathrm{d}x)\right],$$

Theorem 8.7 says that

$$n\int f(x)\rho^{1/n}(\mathrm{d}x) \to t_0\int f(x)\nu(\mathrm{d}x) \tag{24.3}$$

for every bounded continuous function f vanishing on a neighborhood of 0, and

$$n\int xc(x)\rho^{1/n}(\mathrm{d}x) \to \beta. \tag{24.4}$$

It follows from (24.3) and $S(\rho^{1/n}) \subset [0,\infty)$ that $S_\nu \subset [0,\infty)$. For any $b>0$ satisfying $\nu(\{b\})=0$, we have

$$n\int_{(b,\infty)} xc(x)\rho^{1/n}(\mathrm{d}x) \to t_0 \int_{(b,\infty)} xc(x)\nu(\mathrm{d}x)$$

by (24.3). Hence

$$\int_{(b,\infty)} xc(x)\nu(\mathrm{d}x) \le t_0^{-1}\beta$$

by (24.4). As b can be chosen arbitrarily small,

$$\int_{(0,\infty)} xc(x)\nu(\mathrm{d}x) \le t_0^{-1}\beta.$$

Therefore, $\int_{(0,1]} x\nu(\mathrm{d}x) < \infty$. It follows that $\{X_t\}$ is of type A or B. □

COROLLARY 24.8. *Let $\{X_t\}$ be a Lévy process on $\mathbb{R}$ of type A or B with Lévy measure ν and drift γ_0. If $S_\nu \subset [0,\infty)$, then $S(X_t)$ has infimum $\gamma_0 t$.*

Proof. The characteristic function of $X(t)$ is expressed by (24.2). Let a_t be the infimum of $S(X_t)$. The proof of the 'if' part of the theorem above shows that $a_t \ge \gamma_0 t$. Let ρ_t be the distribution of $X_t - a_t$. Then $S(\rho_t) \subset [0,\infty)$. Let $c(x)$ be as in the proof above and let $(0, t\nu, \beta_t)_c$ be the generating triplet of ρ_t. The argument in the proof above shows that

$$\int_{(0,\infty)} xc(x)\nu(\mathrm{d}x) \le t^{-1}\beta_t.$$

We get

$$\beta_t = \gamma_0 t - a_t + t\int xc(x)\nu(\mathrm{d}x)$$

from the representation (8.5) with 0, $t\nu$, and β_t in place of A, ν, and γ_c. Therefore $\gamma_0 t - a_t \ge 0$. □

REMARK 24.9. Our proof of Theorem 24.7 avoided the use of the Lévy–Itô decomposition of sample functions. But, if we do use it, the proof is simplified as follows.

Assume that $S_\nu \subset [0,\infty)$ and $\{X_t\}$ is of type A or B. Let γ_0 be the drift of $\{X_t\}$. Then $\{X_t - \gamma_0 t\}$ has drift 0. Hence, by Theorem 21.5, $\{X_t - \gamma_0 t\}$ is a subordinator (that is, an increasing Lévy process) and $S(X_t - \gamma_0 t) \subset [0,\infty)$, that is, $S(X_t) \subset [\gamma_0 t, \infty)$.

Conversely, assume that, for some $t_0 > 0$, $S(X_{t_0})$ is bounded below. Let (A,ν,γ) be the generating triplet of $\{X_t\}$. For some $\gamma' \ge 0$, $X_{t_0} + \gamma' t_0$ is supported on $[0,\infty)$. Let $Y_t = X_t + \gamma' t$. Then $\{Y_t\}$ is a Lévy process with $S(Y_{t_0}) \subset [0,\infty)$. Its generating triplet is $(A,\nu,\gamma+\gamma')$. We have $S(Y_{t_0/n}) \subset [0,\infty)$ for every $n \in \mathbb{N}$ by Lemma 24.1, since $P_{Y(t_0)}$ is the n-fold convolution of $P_{Y(t_0/n)}$. Hence $S(Y_{kt_0/n}) \subset [0,\infty)$ for every n and $k \in \mathbb{N}$ again by Lemma 24.1. By the stochastic continuity, $S(Y_t) \subset [0,\infty)$ for every $t \ge 0$. Hence, if $t \le t'$, then $P[\,Y_t \le Y_{t'}\,] = 1$. It follows that $\{Y_t\}$ is a subordinator. Now, by Theorem 21.5, $A = 0$, $S_\nu \subset [0, \infty)$, and $\int_{(0,1]} x\nu(\mathrm{d}x) < \infty$.

In the reverse direction, Theorem 21.5 readily follows from Theorem 24.7 and Corollary 24.8 without the use of the Lévy–Itô decomposition.

THEOREM 24.10 (Lévy process on $\mathbb{R}$). *Let $\{X_t\}$ be a Lévy process on $\mathbb{R}$ with Gaussian variance A and Lévy measure ν. The support of X_t for $t > 0$ is described as follows, except the case of type A with $0 \notin S_\nu$.*

(i) *Assume type C. Then $S(X_t) = \mathbb{R}$.*

(ii) *Suppose that $0 \in S_\nu$, $S_\nu \cap (0, \infty) \neq \emptyset$, and $S_\nu \cap (-\infty, 0) \neq \emptyset$. Then $S(X_t) = \mathbb{R}$.*

(iii) *Suppose that $0 \in S_\nu$. Assume type A or B and let γ_0 be the drift. If $S_\nu \subset [0, \infty)$, then $S(X_t) = [t\gamma_0, \infty)$. If $S_\nu \subset (-\infty, 0]$, then $S(X_t) = (-\infty, t\gamma_0]$.*

Proof. If $\{X_t\}$ is of type B, then $0 \in S_\nu$. Hence, the case not covered by (i)–(iii) is type A with $0 \notin S_\nu$. Let us give the proof in the order (iii), (i), and (ii).

(iii) Assume that $S_\nu \subset [0, \infty)$. Assume, further, $\gamma_0 = 0$. Let us prove that $S(X_t) = [0, \infty)$ for $t > 0$. The case $\gamma_0 \neq 0$ is reduced to this case by translation. We have $S(X_t) \subset [0, \infty)$ by Corollary 24.8. We claim that $P[\,X_t \in (a, b)\,] > 0$ for arbitrary $0 \le a < b < \infty$. This will give $S(X_t) = [0, \infty)$. Since $0 \in S_\nu$ and $\nu\{0\} = 0$, the point 0 is a cluster point of S_ν. Choose $\varepsilon \in S_\nu$ such that $0 < \varepsilon < b - a$. Let $\{Y_t\}$ and $\{Z_t\}$ be independent Lévy processes such that

$$E[\mathrm{e}^{\mathrm{i}zY(t)}] = \exp\left[t \int_{(\varepsilon/2, \infty)} (\mathrm{e}^{\mathrm{i}zx} - 1)\nu(\mathrm{d}x)\right],$$

$$E[\mathrm{e}^{\mathrm{i}zZ(t)}] = \exp\left[t \int_{(0, \varepsilon/2]} (\mathrm{e}^{\mathrm{i}zx} - 1)\nu(\mathrm{d}x)\right].$$

Then $X_t \stackrel{\mathrm{d}}{=} Y_t + Z_t$. Since $\{Y_t\}$ is compound Poisson, Theorem 24.5 says that $n\varepsilon \in S(Y_t)$ for every $n \in \mathbb{N}$. On the other hand, by Corollary 24.8, $0 \in S(Z_t)$. Hence $n\varepsilon \in S(X_t)$ for $n \in \mathbb{N}$ by Lemma 24.1. Since $a < n\varepsilon < b$ for some n, we have $P[\,X_t \in (a, b)\,] > 0$. The case $S_\nu \subset (-\infty, 0]$ is similar.

(i) Since $\{X_t\}$ is of type C, we have $A \neq 0$, or $\int_{(0,1)} x\nu(\mathrm{d}x) = \infty$, or $\int_{(-1,0)} |x|\nu(\mathrm{d}x) = \infty$. If $A \neq 0$, then P_{X_t} has a nondegenerate Gaussian factor, and $S(X_t) = \mathbb{R}$ by Lemma 24.1. Assume that $\int_{(0,1)} x\nu(\mathrm{d}x) = \infty$. Let $\nu_1(\mathrm{d}x) = x1_{(0,1)}(x)\nu(\mathrm{d}x)$ and $\nu_2 = \nu - \nu_1$. Then, $\int_{(0,1)} x\nu_1(\mathrm{d}x) < \infty$ and $\int_{(0,1)} x\nu_2(\mathrm{d}x) = \infty$. There are independent Lévy processes $\{Y_t\}$ and $\{Z_t\}$ such that $X_t \stackrel{\mathrm{d}}{=} Y_t + Z_t$ and

$$E[\mathrm{e}^{\mathrm{i}zY(t)}] = \exp\left[t \int (\mathrm{e}^{\mathrm{i}zx} - 1)\nu_1(\mathrm{d}x)\right].$$

Since we have $0 \in S_{\nu_1}$, it follows from the assertion (iii) that $S(Y_t) = [0,\infty)$ for every $t > 0$. On the other hand, $S(Z_t)$ is unbounded below by virtue of Theorem 24.7. Thus, by Lemma 24.1, we see that $S(X_t) = \mathbb{R}$. The remaining case, that $\int_{(-1,0)} |x|\nu(\mathrm{d}x) = \infty$, is reduced to the above by reflection.

(ii) The type C case is treated in (i). So we assume that $\{X_t\}$ is of type A or B. Since $0 \in S_\nu$ and $\nu(\{0\}) = 0$, the origin belongs to $S([\nu]_{(0,\infty)})$ or $S([\nu]_{(-\infty,0)})$. Assume that $0 \in S([\nu]_{(0,\infty)})$. Choose independent Lévy processes $\{Y_t\}$ and $\{Z_t\}$ such that $X_t \overset{\mathrm{d}}{=} Y_t + Z_t$ and

$$E[\mathrm{e}^{\mathrm{i}zY(t)}] = \exp\left[t\int_{(0,\infty)} (\mathrm{e}^{\mathrm{i}zx} - 1)\nu(\mathrm{d}x)\right].$$

Then $S(Y_t) = [0,\infty)$ for any $t > 0$ by virtue of the assertion (iii). Since $[\nu]_{(-\infty,0)} \neq 0$, $S(Z_t)$ is unbounded below for every $t > 0$ by Theorem 24.7. Hence $S(X_t) = \mathbb{R}$ by Lemma 24.1. The case that $0 \in S([\nu]_{(-\infty,0)})$ is reduced to this case by reflection. □

THEOREM 24.11 (Subordinator). *Let $\{X_t\}$ be a Lévy process on $\mathbb{R}$. Then the following four conditions are equivalent to each other: $\{X_t\}$ is a subordinator; $S(X_t) \subset [0,\infty)$ for every $t > 0$; $S(X_t) \subset [0,\infty)$ for some $t > 0$; $A = 0$, $S_\nu \subset [0,\infty)$, $\int_{(0,1]} x\nu(\mathrm{d}x) < \infty$, and $\gamma_0 \geq 0$. If $\{X_t\}$ is a subordinator, then*

$$E[\mathrm{e}^{-uX_t}] = \exp\left[t\left(\int_{(0,\infty)} (\mathrm{e}^{-ux} - 1)\nu(\mathrm{d}x) - \gamma_0 u\right)\right], \qquad u \geq 0. \tag{24.5}$$

Proof. The equivalence of the four conditions is a consequence of Theorem 24.7, Corollary 24.8, and the discussion in Remark 24.9. The Laplace transform (24.5) of P_{X_t} is given in Remark 21.6. But we give here a proof independent of Chapter 4. We claim that

$$E[\mathrm{e}^{wX_t}] = \exp\left[t\left(\int_{(0,\infty)} (\mathrm{e}^{wx} - 1)\nu(\mathrm{d}x) + \gamma_0 w\right)\right] \tag{24.6}$$

for any $w \in \mathbb{C}$ with $\mathrm{Re}\, w \leq 0$. Since $S(X_t) \subset [0,\infty)$ and $\gamma_0 \geq 0$, both sides of (24.6) are finite and continuous in w with $\mathrm{Re}\, w \leq 0$. Let $\Phi_1(w)$ and $\Phi_2(w)$ be the left-hand and the right-hand side of (24.6), respectively. They are analytic in $\{w \in \mathbb{C}\colon \mathrm{Re}\, w < 0\}$, which is proved as in the proof of Proposition 2.6. We have $\Phi_1(w) - \Phi_2(w) = 0$ when $\mathrm{Re}\, w = 0$. Hence, we can extend the range of analyticity of $\Phi_1 - \Phi_2$ by using H. A. Schwarz's principle of reflection. Now the uniqueness theorem shows that $\Phi_1(w) - \Phi_2(w) = 0$ for all w we are considering. This gives (24.6) and, in particular, (24.5). □

EXAMPLE 24.12. A stable subordinator, that is, a stable process on $\mathbb{R}$ with increasing sample functions, is given by a stable distribution μ with support in $[0,\infty)$. Its characteristic function in the nondegenerate case is expressed as

$$\widehat{\mu}(z) = \exp\left[c_1 \int_0^\infty (e^{izx} - 1)x^{-1-\alpha}dx + i\gamma_0 z\right] \tag{24.7}$$

with $0 < \alpha < 1$, $\gamma_0 \geq 0$, and $c_1 > 0$ or, equivalently,

$$\widehat{\mu}(z) = \exp[-c|z|^\alpha(1 - i\tan\tfrac{\pi\alpha}{2}\operatorname{sgn} z) + i\gamma_0 z] \tag{24.8}$$

with

$$c = \tfrac{1}{\alpha}\Gamma(1-\alpha)\bigl(\cos\tfrac{\pi\alpha}{2}\bigr)c_1. \tag{24.9}$$

That is, $\widehat{\mu}(z)$ is given by the formula (14.24) with $0 < \alpha < 1$, $\beta = 1$, and $\tau = \gamma_0 \geq 0$. This is a consequence of Theorem 24.11 combined with Theorems 14.3 and 14.15. As to (24.9), see the solution of Exercise 18.8. The representation (24.8) was already discussed in Definition 14.16 and Example 21.7. The Laplace transform of μ is written as

$$L_\mu(u) = \exp[-c'u^\alpha - \gamma_0 u] \quad \text{with} \quad c' = \tfrac{1}{\alpha}\Gamma(1-\alpha)c_1. \tag{24.10}$$

This is obtained from (24.5), since

$$\begin{aligned}\int_0^\infty (e^{-ux} - 1)x^{-1-\alpha}dx &= -\int_0^\infty x^{-1-\alpha}dx \int_0^x ue^{-uy}dy \\ &= -u\int_0^\infty e^{-uy}dy \int_y^\infty x^{-1-\alpha}dx = -\tfrac{1}{\alpha}\Gamma(1-\alpha)u^\alpha.\end{aligned}$$

The distribution μ has a connection with the Mittag–Leffler function $E_\alpha(x)$ defined by

$$E_\alpha(x) = \sum_{n=0}^\infty \frac{x^n}{\Gamma(n\alpha+1)}. \tag{24.11}$$

If a random variable X has the distribution μ with $c' = 1$ and $\gamma_0 = 0$, then the distribution of $X^{-\alpha}$ is called the *Mittag–Leffler distribution* with parameter α, as it has Laplace transform $E_\alpha(-u)$. See Exercise 29.18.

We introduce the concept of the support of a Lévy process.

DEFINITION 24.13. Let $\{X_t\}$ be a Lévy process on $\mathbb{R}^d$ defined on a probability space $(\Omega, \mathcal{F}, P)$. Let Ω_0 be the subset of Ω in Definition 1.6 of a Lévy process. The *support* Σ of $\{X_t\}$ is a closed set such that

$$P[\,\{X_t \in \Sigma \text{ for all } t \geq 0\} \cap \Omega_0\,] = 1 \tag{24.12}$$

and such that, if (24.12) holds with another closed set F in place of Σ, then $F \supset \Sigma$. Note that

$$\{X_t \in \Sigma \text{ for all } t \geq 0\} \cap \Omega_0 = \{X_t \in \Sigma \text{ for all } t \in \mathbb{Q}_+\} \cap \Omega_0 \in \mathcal{F}.$$

Existence of the support Σ is proved below. Its uniqueness comes from the definition.

PROPOSITION 24.14. *Any Lévy process $\{X_t\}$ on $\mathbb{R}^d$ has its support Σ. It satisfies the following.*

(i) *Σ is closed under addition and contains* 0.

(ii) *Σ is the closure of $\bigcup_{t\geq 0} S(X_t)$.*

(iii) *Σ is the set of points x such that, for every $\varepsilon > 0$, there is $t \geq 0$ satisfying $P[\,|X_t - x| < \varepsilon\,] > 0$.*

(iv) *Σ is the set of points x such that, for every $\varepsilon > 0$,*

$$P[\,\{|X_t - x| < \varepsilon \text{ for some } t \geq 0\} \cap \Omega_0\,] > 0.$$

Another characterization of Σ will be given in Exercise 44.1.

Proof of proposition. Denote by Σ_1, Σ_2, and Σ_3 the sets that are asserted to be equal to Σ in (ii), (iii), and (iv), respectively. Note that, in (iv),

$$\{|X_t - x| < \varepsilon \text{ for some } t \geq 0\} \cap \Omega_0 = \{|X_t - x| < \varepsilon \text{ for some } t \in \mathbb{Q}_+\} \cap \Omega_0 \in \mathcal{F}.$$

Let us see that Σ_2 and Σ_3 are closed. Let $\{x_n\}$ be a sequence in Σ_2 or Σ_3 such that $x_n \to x$. For any $\varepsilon > 0$, choose n_0 such that $|x_{n_0} - x| < \varepsilon/2$. If $\{x_n\}$ is in Σ_2, then, choosing $t_0 \geq 0$ satisfying $P[\,|X_{t_0} - x_{n_0}| < \varepsilon/2\,] > 0$, we have $P[\,|X_{t_0} - x| < \varepsilon\,] > 0$, and hence $x \in \Sigma_2$. If $\{x_n\}$ is in Σ_3, then

$$\{|X_t - x_{n_0}| < \varepsilon \text{ for some } t\} \cap \Omega_0 \supset \{|X_t - x_{n_0}| < \varepsilon/2 \text{ for some } t\} \cap \Omega_0,$$

which has positive probability. Hence $x \in \Sigma_3$. Thus Σ_2 and Σ_3 are closed. Let F be a closed set such that (24.12) holds with F in place of Σ. We claim that

$$\Sigma_1 \subset \Sigma_2 \subset \Sigma_3 \subset F. \tag{24.13}$$

It follows from the definition of Σ_2 that $S(X_t) \subset \Sigma_2$ for every $t \geq 0$. Hence $\Sigma_1 \subset \Sigma_2$. The relation $\Sigma_2 \subset \Sigma_3$ is obvious. We have

$$P[\,\{X_t \in F^c \text{ for some } t \geq 0\} \cap \Omega_0\,] = 0,$$

where $F^c = \mathbb{R}^d \setminus F$. If $x \notin F$, then $x \notin \Sigma_3$, since there is $\varepsilon > 0$ such that $\{y\colon |y - x| < \varepsilon\} \subset F^c$. Hence $\Sigma_3 \subset F$, and (24.13) is proved. We have $P[X_t \in \Sigma_1] = 1$ for each t, and hence $P[X_t \in \Sigma_1 \text{ for all } t \in \mathbb{Q}_+] = 1$. Hence we have (24.12) with Σ_1 in place of Σ. Now, by (24.13), the support Σ of $\{X_t\}$ exists and Σ, Σ_1, Σ_2, and Σ_3 are identical. The origin obviously belongs to Σ. As to the other assertion in (i), let x and y be in Σ_2. For any $\varepsilon > 0$, there are t and s such that $P[\,|X_t - x| < \varepsilon/2\,] > 0$ and $P[\,|X_s - y| < \varepsilon/2\,] > 0$. Then,

$$\begin{aligned} P[\,|X_{t+s} - x - y| < \varepsilon\,] &\geq P[\,|X_t - x| < \varepsilon/2,\ |X_{t+s} - X_t - y| < \varepsilon/2\,] \\ &= P[\,|X_t - x| < \varepsilon/2\,]\, P[\,|X_s - y| < \varepsilon/2\,], \end{aligned}$$

which is positive. This shows that $x + y \in \Sigma_2$. □

EXAMPLE 24.15. Let $\{X_t\}$ be a Lévy process on $\mathbb{R}$ such that $1 \in S_\nu \subset \mathbb{N}$ and $A = 0$. Then, $S(X_t) = \gamma_0 t + \mathbb{Z}_+$. By (ii) of the proposition above, $\Sigma = \mathbb{R}_+$, $\mathbb{Z}_+$, or $\mathbb{R}$ according as $\gamma_0 > 0$, $= 0$, or < 0.

DEFINITION 24.16. A measure ρ on $\mathbb{R}^d$ is *degenerate* if there are $a \in \mathbb{R}^d$ and a proper linear subspace V of $\mathbb{R}^d$ (that is, a linear subspace with $\dim V \leq d-1$) such that $S_\rho \subset a + V$. Otherwise, ρ is called *nondegenerate.*

PROPOSITION 24.17. *Let $\{X_t\}$ be a Lévy process on $\mathbb{R}^d$ generated by (A, ν, γ).*

(i) *For a linear subspace V of $\mathbb{R}^d$ the following three statements are equivalent.*

(1) *$S(X_t) \subset V$ for each $t > 0$.*
(2) *$S(X_{t_0}) \subset V$ for some $t_0 > 0$.*
(3) *$A(\mathbb{R}^d) \subset V$, $S_\nu \subset V$, and $\gamma \in V$, where $A(\mathbb{R}^d) = \{Ax \colon x \in \mathbb{R}^d\}$, called the range of the matrix A.*

(ii) *The following statements are equivalent.*

(1) *The distribution of X_{t_0} is degenerate for some $t_0 > 0$.*
(2) *There are $a \in \mathbb{R}^d$ and a proper linear subspace V such that $S(X_t) \subset ta + V$ for every $t > 0$.*
(3) *There is a proper linear subspace V such that $A(\mathbb{R}^d) \subset V$ and $S_\nu \subset V$.*

Proof. (i) Let T be the orthogonal projector (hence symmetric) to V. Let $\{Y_t\}$ be the Lévy process defined by $Y_t = TX_t$. Assume that (2) holds. Then $P[\,X_{t_0} \in V\,] = 1$ and $Y_{t_0} = X_{t_0}$ a. s., hence $\{Y_t\} \stackrel{\mathrm{d}}{=} \{X_t\}$ and $S(X_t) = S(Y_t) \subset V$. This shows (1). The condition (1) trivially implies (2).

Next assume (1) and let us show (3). Since $\{Y_t\} \stackrel{\mathrm{d}}{=} \{X_t\}$ we have $TAT = A$ and $[\nu T^{-1}]_{\mathbb{R}^d \setminus \{0\}} = \nu$ by Proposition 11.10. Thus $A(\mathbb{R}^d) \subset V$ and $S_\nu \subset V$. Further, the γ_T in (11.10) coincides with $T\gamma$, and hence $\gamma = T\gamma \in V$.

To show that (3) implies (1), first notice that $TA = A$ by $A(\mathbb{R}^d) \subset V$. Since A is symmetric, the null space $\{x \colon Ax = 0\}$ is the orthogonal complement of $A(\mathbb{R}^d)$. Hence $A(I - T) = 0$, and thus $TAT = A$. Since $S_\nu \subset V$, we have $\nu(T^{-1}(B)) = \nu(V \cap T^{-1}(B)) = \nu(B)$ for any $B \in \mathcal{B}(V)$. Since $\gamma \in V$, we have $\gamma_T = T\gamma = \gamma$. Therefore $\{Y_t\} \stackrel{\mathrm{d}}{=} \{X_t\}$ by Proposition 11.10.

(ii) The condition (2) trivially implies (1). Suppose that (1) holds. Then we can find $a \in \mathbb{R}^d$ and V with $\dim V \leq d-1$ such that $S(X_{t_0}) \subset t_0 a + V$. Let $Y_t = X_t - ta$. Then $\{Y_t\}$ is a Lévy process with generating triplet

$(A, \nu, \gamma - a)$. Hence, applying the assertion (i), we get (3). To see that (3) implies (2), again apply (i) to the Lévy process $Z_t = X_t - t\gamma$. □

DEFINITION 24.18. Let $\{X_t\}$ be a Lévy process on $\mathbb{R}^d$. We say that $\{X_t\}$ is *degenerate* if P_{X_t} is degenerate for every t. Otherwise $\{X_t\}$ is *nondegenerate*. Each of the conditions (1), (2), and (3) in Proposition 24.17(ii) is equivalent to degeneracy of $\{X_t\}$. We say that $\{X_t\}$ is *genuinely d-dimensional* if no proper linear subspace of $\mathbb{R}^d$ contains the support Σ of $\{X_t\}$. If $\{X_t\}$ is not genuinely d-dimensional, then there is a proper linear subspace V such that $S(X_t) \subset V$ for every t (see Proposition 24.14), and $\{X_t\}$ is degenerate. If $\{X_t\}$ is degenerate, it is possible that $\{X_t\}$ is genuinely d-dimensional.

By Proposition 24.17(i), $\{X_t\}$ is genuinely d-dimensional if and only if there is no proper linear subspace V satisfying (3) of (i).

The following facts will be useful in the future.

PROPOSITION 24.19. *Let μ be a probability measure on $\mathbb{R}^d$. Then there are $\varepsilon > 0$ and $c > 0$ such that*

$$|\widehat{\mu}(z)| \leq 1 - c|z|^2 \quad \text{for } |z| < \varepsilon, \tag{24.14}$$

if and only if μ is nondegenerate.

Proof. The 'if' part is proved in two steps.

Step 1. Let $X = (X_j)_{1 \leq j \leq d}$ be a random variable with nondegenerate distribution μ. Assume that there is a compact set K with $P[\, X \in K\,] = 1$ and that $EX = 0$. Then $\widehat{\mu}(z)$ is of class C^∞ and

$$\widehat{\mu}(z) = 1 - \frac{1}{2} \sum_{j,k=1}^{d} a_{jk} z_j z_k + o(|z|^2),$$

where $a_{jk} = -\dfrac{\partial^2 \widehat{\mu}}{\partial z_j \partial z_k}(0) = E[X_j X_k]$. Let $F(z) = \sum_{j,k=1}^{d} a_{jk} z_j z_k$. Then $F(z) = E\big[\big(\sum_{j=1}^{d} X_j z_j\big)^2\big] \geq 0$. If $F(z^0) = 0$ for some $z^0 \neq 0$, then X is orthogonal to z^0 with probability 1, which is impossible by the nondegeneracy of μ. Hence $F(z)$ is a positive-definite quadratic form. Therefore $F(z) \geq 4c|z|^2$ with some $c > 0$. Thus $|\widehat{\mu}(z)| \leq 1 - c|z|^2$ in a neighborhood of 0.

Step 2. Let μ be a general nondegenerate distribution. Then we can choose $b > 0$ such that $[\mu]_{\{|x| \leq b\}}$ is nondegenerate (Exercise 29.3). Let $k = \mu(\{|x| \leq b\})$ and let $\mu' = k^{-1}[\mu]_{\{|x| \leq b\}}$. Denote by γ the mean of μ' and let $\mu'' = \mu' * \delta_{-\gamma}$. Then μ'' is a nondegenerate probability measure and, by Step 1, there are $c > 0$ and $\varepsilon > 0$ such that

$$|\widehat{\mu''}(z)| \leq 1 - c|z|^2 \quad \text{for } |z| < \varepsilon.$$

Since $|\widehat{\mu''}(z)| = k^{-1}\left|\int_{|z|\le b} e^{i\langle z,x\rangle}\mu(dx)\right|$, we have

$$|\widehat{\mu}(z)| \le \left|\int_{|z|\le b} e^{i\langle z,x\rangle}\mu(dx)\right| + (1-k) \le k(1-c|z|^2) + (1-k)$$
$$= 1 - kc|z|^2$$

for $|z| < \varepsilon$.

To see the 'only if' part, let μ be degenerate. Then $S_\mu \subset a+V$ for some a and V with $\dim V \le d-1$. Since $\widehat{\mu}(z)e^{-i\langle a,z\rangle} = 1$ for any z orthogonal to V, we cannot find ε and c satisfying (24.14). □

PROPOSITION 24.20. *Let μ be a nondegenerate α-semi-stable distribution on $\mathbb{R}^d$, $0 < \alpha \le 2$. Then there is a constant $K > 0$ such that*

$$|\widehat{\mu}(z)| \le e^{-K|z|^\alpha} \quad \text{on } \mathbb{R}^d. \tag{24.15}$$

Proof. We have $-|z|^{-\alpha}\log|\widehat{\mu}(z)| = \eta_1(z)$ for $z \ne 0$ in Proposition 14.9. Since $\eta_1(bz) = \eta_1(z)$ with a span $b > 1$,

$$\inf_{z\in\mathbb{R}^d\setminus\{0\}} \eta_1(z) = \inf_{b^n<|z|\le b^{n+1}} \eta_1(z) \quad \text{for } n \in \mathbb{Z}.$$

By Proposition 24.19, there is $\varepsilon > 0$ such that $\eta_1(z) > 0$ for $0 < |z| \le \varepsilon$. Hence, using the continuity of $\eta_1(z)$ on $\mathbb{R}^d \setminus \{0\}$, we see that there is $K > 0$ such that $\eta_1(z) \ge K$. □

DEFINITION 24.21. Let $\{X_t\}$ be a Lévy process on $\mathbb{R}^d$ with support Σ. The smallest closed additive subgroup $\mathfrak{G}$ of $\mathbb{R}^d$ that contains Σ is called *the group of* $\{X_t\}$.

PROPOSITION 24.22. *Let Σ and $\mathfrak{G}$ be as in Definition 24.21. Then $\mathfrak{G}$ is the closure of $\Sigma - \Sigma$.*

Proof. Let $B = \overline{\Sigma - \Sigma}$. That $\mathfrak{G} \supset B$ is immediate from the definition. Since $0 \in \Sigma$, B contains Σ. For any x and y in B, there are x_n, x'_n, y_n, and y'_n in Σ such that $x_n - x'_n \to x$ and $y_n - y'_n \to y$. We have $(x_n - x'_n) - (y_n - y'_n) = (x_n + y'_n) - (x'_n + y_n)$, which is in $\Sigma - \Sigma$ by Proposition 24.14(i). Hence $x - y \in B$. Thus B is an additive group. Hence $B \supset \mathfrak{G}$. □

DEFINITION 24.23. A measure ρ on $\mathbb{R}^d$ is called an *invariant measure* of a Lévy process $\{X_t\}$ if

$$\rho(B) = \int_{\mathbb{R}^d} P[\,x + X_t \in B\,]\,\rho(dx) \quad \text{for every } B \in \mathcal{B}(\mathbb{R}^d). \tag{24.16}$$

THEOREM 24.24 (Lebesgue measure). *If $\{X_t\}$ is a Lévy process on $\mathbb{R}^d$, then the Lebesgue measure is an invariant measure of $\{X_t\}$.*

See Exercises 29.4–29.6 for further information.

Proof of theorem. Use the translation invariance of the Lebesgue measure. Then

$$\int P[\,x+X_t\in B\,]\mathrm{d}x=\int \mathrm{d}x\int 1_B(x+y)\mu^t(\mathrm{d}y)=\int \mu^t(\mathrm{d}y)\,\mathrm{Leb}(B-y)$$
$$=\int \mu^t(\mathrm{d}y)\,\mathrm{Leb}(B)=\mathrm{Leb}(B).$$

This is (24.16) for the Lebesgue measure. □

25. Moments

We define the g-moment of a random variable and discuss finiteness of the g-moment of X_t for a Lévy process $\{X_t\}$.

DEFINITION 25.1. Let $g(x)$ be a nonnegative measurable function on $\mathbb{R}^d$. We call $\int g(x)\mu(\mathrm{d}x)$ the *g-moment* of a measure μ on $\mathbb{R}^d$. We call $E[g(X)]$ the *g-moment* of a random variable X on $\mathbb{R}^d$.

DEFINITION 25.2. A function $g(x)$ on $\mathbb{R}^d$ is called *submultiplicative* if it is nonnegative and there is a constant $a>0$ such that

$$g(x+y)\le ag(x)g(y)\quad \text{for } x,y\in\mathbb{R}^d. \tag{25.1}$$

A function bounded on every compact set is called *locally bounded.*

THEOREM 25.3 (g-Moment). *Let g be a submultiplicative, locally bounded, measurable function on $\mathbb{R}^d$. Then, finiteness of the g-moment is not a time dependent distributional property in the class of Lévy processes. Let $\{X_t\}$ be a Lévy process on $\mathbb{R}^d$ with Lévy measure ν. Then, X_t has finite g-moment for every $t>0$ if and only if $[\nu]_{\{|x|>1\}}$ has finite g-moment.*

The following facts indicate the wide applicability of the theorem.

PROPOSITION 25.4. (i) *The product of two submultiplicative functions is submultiplicative.*

(ii) *If $g(x)$ is submultiplicative on $\mathbb{R}^d$, then so is $g(cx+\gamma)^\alpha$ with $c\in\mathbb{R}$, $\gamma\in\mathbb{R}^d$, and $\alpha>0$.*

(iii) *Let $0<\beta\le 1$. Then the following functions are submultiplicative:*

$$|x|\vee 1,\ |x_j|\vee 1,\ x_j\vee 1,\ \exp(|x|^\beta),\ \exp(|x_j|^\beta),$$
$$\exp((x_j\vee 0)^\beta),\ \log(|x|\vee \mathrm{e}),\ \log(|x_j|\vee \mathrm{e}),\ \log(x_j\vee \mathrm{e}),$$
$$\log\log(|x|\vee \mathrm{e}^{\mathrm{e}}),\ \log\log(|x_j|\vee \mathrm{e}^{\mathrm{e}}),\ \log\log(x_j\vee \mathrm{e}^{\mathrm{e}}).$$

Here x_j is the jth component of x.

Proof. (i) Immediate from the definition.

(ii) Let $g_1(x) = g(cx)$, $g_2(x) = g(x+\gamma)$, and $g_3(x) = g(x)^\alpha$. Then it follows from (25.1) that $g_1(x+y) \le a g_1(x) g_1(y)$, $g_2(x+y) \le a^2 g(-\gamma) g_2(x) g_2(y)$, and $g_3(x+y) \le a^\alpha g_3(x) g_3(y)$.

(iii) Let $h(u)$ be a positive increasing function on $\mathbb{R}$ such that, for some $b \ge 0$, $h(u)$ is flat on $(-\infty, b]$ and $\log h(u)$ is concave on $[b, \infty)$. Then $h(u)$ is submultiplicative on $\mathbb{R}$. In fact, for $u, v \ge b$, the function $f(u) = \log h(u)$ satisfies

$$f(u+b) - f(u) \le f(2b) - f(b),$$
$$f(u+v) - f(v) \le f(u+b) - f(b),$$

and hence

$$f(u+v) \le f(u+b) - f(b) + f(v) \le f(2b) - 2f(b) + f(u) + f(v),$$

which shows

$$h(u+v) \le \text{const}\, h(u)h(v). \tag{25.2}$$

It follows that (25.2) holds for all $u, v \in \mathbb{R}$. The functions $u \vee 1$, $\exp((u \vee 0)^\beta)$, $\log(u \vee \mathrm{e})$, and $\log\log(u \vee \mathrm{e}^{\mathrm{e}})$ fulfill the conditions on $h(u)$. By (25.2) and by the increasingness of h, the functions $h(|x|)$, $h(|x_j|)$, and $h(x_j)$ are submultiplicative on $\mathbb{R}^d$. □

We prove Theorem 25.3 after three lemmas.

LEMMA 25.5. *If $g(x)$ is submultiplicative and locally bounded, then*

$$g(x) \le b\mathrm{e}^{c|x|} \tag{25.3}$$

with some $b > 0$ and $c > 0$.

Proof. Choose b in such a way that $\sup_{|x| \le 1} g(x) \le b$ and $ab > 1$. If $n - 1 < |x| \le n$, then

$$g(x) \le a^{n-1} g(\tfrac{1}{n}x)^n \le a^{n-1} b^n \le b(ab)^{|x|},$$

which shows (25.3). □

LEMMA 25.6. *Let μ be an infinitely divisible distribution on $\mathbb{R}$ with Lévy measure ν supported on a bounded set. Then $\widehat{\mu}(z)$ can be extended to an entire function on $\mathbb{C}$.*

Proof. There is a finite $a > 0$ such that $S_\nu \subset [-a, a]$. The Lévy–Khintchine representation of $\widehat{\mu}(z)$ is written as

$$\widehat{\mu}(z) = \exp\left[-\frac{1}{2}Az^2 + \int_{[-a,a]} (\mathrm{e}^{\mathrm{i}zx} - 1 - \mathrm{i}zx)\nu(\mathrm{d}x) + \mathrm{i}\gamma' z\right]$$

with some $\gamma' \in \mathbb{R}$. The right-hand side is meaningful even if z is complex. Denote this function by $\Phi(z)$. Then $\Phi(z)$ is an entire function, since we can exchange the order of integration and differentiation. □

LEMMA 25.7. *If μ is a probability measure on $\mathbb{R}$ and $\widehat{\mu}(z)$ is extendible to an entire function on $\mathbb{C}$, then μ has finite exponential moments, that is, it has finite $\mathrm{e}^{c|x|}$-moment for every $c > 0$.*

Proof. It follows from Proposition 2.5(x) that $\alpha_n = \int x^n \mu(\mathrm{d}x)$ and $\beta_n = \int |x|^n \mu(\mathrm{d}x)$ are finite for any $n \geq 1$. Since $\frac{\mathrm{d}^n \widehat{\mu}}{\mathrm{d}z^n}(0) = \mathrm{i}^n \alpha_n$, we have

$$\widehat{\mu}(z) = \sum_{n=0}^{\infty} \frac{1}{n!} \mathrm{i}^n \alpha_n z^n,$$

the radius of convergence of the right-hand side being infinite. Notice that $\beta_{2k} = \alpha_{2k}$ and $\beta_{2k+1} \leq \frac{1}{2}(\alpha_{2k+2} + \alpha_{2k})$, since $|x|^{2k+1} \leq \frac{1}{2}(x^{2k+2} + x^{2k})$. It follows that

$$\int \mathrm{e}^{c|x|} \mu(\mathrm{d}x) = \sum_{n=0}^{\infty} \frac{1}{n!} \beta_n c^n < \infty,$$

completing the proof. □

Proof of Theorem 25.3. Let $\nu_0 = [\nu]_{\{|x| \leq 1\}}$ and $\nu_1 = [\nu]_{\{|x|>1\}}$. Construct independent Lévy processes $\{X_t^0\}$ and $\{X_t^1\}$ on $\mathbb{R}^d$ such that $\{X_t\} \stackrel{\mathrm{d}}{=} \{X_t^0 + X_t^1\}$ and $\{X_t^1\}$ is compound Poisson with Lévy measure ν_1. Let μ_0 and μ_1 be the distributions of X_1^0 and X_1^1, respectively.

Suppose that X_t has finite g-moment for some $t > 0$. It follows from

$$E[g(X_t)] = \iint g(x+y) {\mu_0}^t(\mathrm{d}x) {\mu_1}^t(\mathrm{d}y)$$

that $\int g(x+y) {\mu_1}^t(\mathrm{d}y) < \infty$ for some x. This means

$$\sum_{n=0}^{\infty} \frac{t^n}{n!} \int g(x+y) {\nu_1}^n(\mathrm{d}y) < \infty.$$

Since $g(y) \leq a g(-x) g(x+y)$, we get

$$\sum_{n=0}^{\infty} \frac{t^n}{n!} \int g(y) {\nu_1}^n(\mathrm{d}y) < \infty. \tag{25.4}$$

Hence $\int g(y) \nu_1(\mathrm{d}y) < \infty$.

Conversely, suppose that $\int g(y) \nu_1(\mathrm{d}y) < \infty$. Let us prove that $E[g(X_t)] < \infty$ for every t. By the submultiplicativity,

$$\begin{aligned} \int g(y) {\nu_1}^n(\mathrm{d}y) &= \int \ldots \int g(y_1 + \cdots + y_n) \nu_1(\mathrm{d}y_1) \ldots \nu_1(dy_n) \\ &\leq a^{n-1} \left(\int g(y) \nu_1(\mathrm{d}y) \right)^n. \end{aligned}$$

Hence we have (25.4) for every t. That is, X_t^1 has finite g-moment. Since

$$E[g(X_t)] \le abE[\mathrm{e}^{c|X_t^0|}]\,E[g(X_t^1)]$$

by (25.1) and (25.3), it remains only to show that $E[\mathrm{e}^{c|X_t^0|}] < \infty$. Let $X_j^0(t)$, $1 \le j \le d$, be the components of X_t^0. Then

$$E[\mathrm{e}^{c|X_t^0|}] \le E\left[\exp\left(c\sum_{j=1}^d |X_j^0(t)|\right)\right] \le E\left[\prod_{j=1}^d (\mathrm{e}^{cX_j^0(t)} + \mathrm{e}^{-cX_j^0(t)})\right],$$

which is written as a sum of a finite number of terms of the form $E[\exp X_t^\sharp]$ with $X_t^\sharp$ being a linear combination of $X_j^0(t)$, $1 \le j \le d$. Since $\{X_t^\sharp\}$ is a Lévy process on $\mathbb{R}$ with Lévy measure supported on a bounded set (use Proposition 11.10), $E[\exp X_t^\sharp]$ is finite by virtue of Lemmas 25.6 and 25.7. This proves all statements in the theorem. □

COROLLARY 25.8. *Let $\alpha > 0$, $0 < \beta \le 1$, and $\gamma \ge 0$. None of the properties $\int |x|^\alpha \mu(\mathrm{d}x) < \infty$, $\int (0 \vee \log|x|)^\alpha \mu(\mathrm{d}x) < \infty$, and $\int |x|^\gamma \mathrm{e}^{\alpha|x|^\beta}\mu(\mathrm{d}x) < \infty$ is time dependent in the class of Lévy processes. For a Lévy process on $\mathbb{R}^d$ with Lévy measure ν, each of the properties is expressed by the corresponding property of $[\nu]_{\{|x|>1\}}$.*

This follows from Theorem 25.3 and Proposition 25.4.

REMARK 25.9. There is a nonnegative measurable function $g(x)$ satisfying (25.3) such that finiteness of the g-moment is a time dependent distributional property in the class of Lévy processes. For example, let $g(x) = (1 \wedge |x|^{-\alpha})\mathrm{e}^{|x|}$ with $\alpha > 0$. Consider a Γ-process $\{X_t\}$ with $EX_1 = 1$. Then it is easy to see that $E[g(X_t)] < \infty$ if and only if $t < \alpha$. This process has $\nu = x^{-1}\mathrm{e}^{-x}1_{(0,\infty)}(x)\mathrm{d}x$, so that ν_1 has finite g-moment (Example 8.10).

EXAMPLE 25.10. Let $\{X_t\}$ be a non-trivial semi-stable process on $\mathbb{R}^d$ with index $\alpha \in (0,2)$. Then, for every $t > 0$, $E[|X_t|^\eta]$ is finite or infinite according as $0 < \eta < \alpha$ or $\eta \ge \alpha$, respectively. To see this, notice that the argument in the proofs of Theorem 13.15 and Proposition 14.5 gives

$$\int_{S_n(b)} |x|^\eta \nu(\mathrm{d}x) = b^{n(\eta-\alpha)} \int_{S_0(b)} |x|^\eta \nu(\mathrm{d}x),$$

and hence $\int_{|x|>1} |x|^\eta \nu(\mathrm{d}x) < \infty$ if and only if $\eta < \alpha$; apply Corollary 25.8. In particular, for a stable process on $\mathbb{R}$ with parameters (α, β, τ, c) (Definition 14.16), $E[X_t] = \tau t$ if $1 < \alpha < 2$ (use Proposition 2.5(ix)). The following explicit results are known. If $0 < \alpha < 1$ and $\{X_t\}$ is a stable subordinator with $E[\mathrm{e}^{-uX_t}] = \mathrm{e}^{-tc'u^\alpha}$ (Example 24.12), then, for $-\infty < \eta < \alpha$,

$$E[X_t{}^\eta] = (tc')^{\eta/\alpha}\frac{\Gamma(1-\frac{\eta}{\alpha})}{\Gamma(1-\eta)}, \tag{25.5}$$

which is shown by Wolfe [**579**] and Shanbhag and Sreehari [**475**] (Exercise 29.17). If $0 < \alpha < 2$ and $\{X_t\}$ is symmetric and α-stable on $\mathbb{R}$ with $E[\mathrm{e}^{\mathrm{i}zX_t}] = \mathrm{e}^{-tc|z|^\alpha}$ (Theorem 14.14), then, for $-1 < \eta < \alpha$,

$$E[|X_t|^\eta] = (tc)^{\eta/\alpha} \frac{2^\eta \Gamma(\frac{1+\eta}{2})\Gamma(1-\frac{\eta}{\alpha})}{\sqrt{\pi}\Gamma(1-\frac{\eta}{2})}, \tag{25.6}$$

as is shown in [**475**].

EXAMPLE 25.11. If $\{X_t\}$ is a Lévy process on $\mathbb{R}$ with Lévy measure supported on $(-\infty, 0]$, then $E[\mathrm{e}^{cX_t}] < \infty$ for every $c > 0$ and $t > 0$. Use Theorem 25.3 for $g(x) = \mathrm{e}^{cx}$. For instance, a stable process on $\mathbb{R}$ with $1 \leq \alpha < 2$ and $\beta = -1$ satisfies this assumption (it has support $\mathbb{R}$ for every $t > 0$ as is shown in Theorem 24.10(i)).

EXAMPLE 25.12. Let $\{X_t\}$ be a Lévy process on $\mathbb{R}^d$ generated by (A, ν, γ). In components, $X_t = (X_j(t))$, $\gamma = (\gamma_j)$, and $A = (A_{jk})$. Then X_t has finite mean for $t > 0$ if and only if $\int_{|x|>1} |x|\nu(\mathrm{d}x) < \infty$. When this condition is met, we can find $m_j(t) = E[X_j(t)]$ expressed as

$$m_j(t) = t\left(\int_{|x|>1} x_j \nu(\mathrm{d}x) + \gamma_j\right) = t\gamma_{1,j}, \qquad j = 1, \ldots, d, \tag{25.7}$$

differentiating $\widehat{\mu}(z)$ (Proposition 2.5(ix)). Here $\gamma_{1,j}$ is the jth component of the center γ_1 in (8.8). Similarly, $E[|X_t|^2] < \infty$ for all $t > 0$ if and only if $\int_{|x|>1} |x|^2\nu(\mathrm{d}x) < \infty$. In this case,

$$v_{jk}(t) = E[(X_j(t) - m_j(t))(X_k(t) - m_k(t))], \qquad j, k = 1, \ldots, d,$$

the (j, k) elements of the covariance matrix of $X(t)$, are expressed as

$$v_{jk}(t) = t\left(A_{jk} + \int_{\mathbb{R}^d} x_j x_k \nu(\mathrm{d}x)\right). \tag{25.8}$$

Theorem 25.3 shows that, for a Lévy process $\{X_t\}$ with Lévy measure ν, the tails of P_{X_t} and ν have a kind of similarity. Are they actually equivalent in some class of Lévy processes? This question was answered by Embrechts, Goldie, and Veraverbeke [**123**] for subordinators. We state their result without proof in two remarks below.

DEFINITION 25.13. A probability measure μ on $[0, \infty)$ is called *subexponential* if $\mu(x, \infty) > 0$ for every x and

$$\lim_{x\to\infty} \frac{\mu^n(x, \infty)}{\mu(x, \infty)} = n \quad \text{for } n = 2, 3, \ldots. \tag{25.9}$$

The class of probability measures satisfying (25.9) above was introduced by Chistyakov [**75**]. The condition can be weakened. Specifically, if

$$\limsup_{x\to\infty} \frac{\mu^2(x, \infty)}{\mu(x, \infty)} \leq 2,$$

then μ is subexponential. The meaning of (25.9) is as follows. Let $\{Z_j\}$ be independent nonnegative random variables each with distribution μ and let $S_n = \sum_{j=1}^n Z_j$ and $M_n = \max_{1\le j\le n} Z_j$. Then μ satisfies (25.9) if and only if

$$P[\, S_n > x\,] \sim P[\, M_n > x\,], \quad x \to \infty, \quad \text{for } n = 2, 3, \ldots.$$

In fact, $P[\, S_n > x\,] = \mu^n(x, \infty)$ and

$$\begin{aligned} P[\, M_n > x\,] &= \textstyle\sum_{j=1}^n P[\, Z_1 \le x, \ldots, Z_{j-1} \le x, Z_j > x\,] \\ &= \textstyle\sum_{j=1}^n \mu(0, x]^{j-1} \mu(x, \infty) \sim n\mu(x, \infty) \quad \text{as } x \to \infty. \end{aligned}$$

REMARK 25.14. A basic result on subexponentiality is as follows. If $\{X_t\}$ is a subordinator with Lévy measure ν, then the following conditions are equivalent [**123**]:

(1) $\nu(1, \infty) > 0$ and $\frac{1}{\nu(1,\infty)}[\nu]_{(1,\infty)}$ is subexponential;
(2) P_{X_t} is subexponential for every $t > 0$;
(3) P_{X_t} is subexponential for some $t > 0$;
(4) $P[\, X_t > x\,] \sim t\nu(x, \infty)$, $x \to \infty$, for every $t > 0$;
(5) $P[\, X_t > x\,] \sim t\nu(x, \infty)$, $x \to \infty$, for some $t > 0$.

Some of the consequences of subexponentiality are as follows. Let μ be a subexponential probability measure on $[0, \infty)$. Then,

(1) for any $y \in \mathbb{R}$, $\mu(x - y, \infty)/\mu(x, \infty) \to 1$ as $x \to \infty$;
(2) for every $\varepsilon > 0$, $\int_{[0,\infty)} e^{\varepsilon x} \mu(dx) = \infty$;
(3) if μ' is a probability measure on $[0, \infty)$ satisfying $\lim_{x\to\infty} \frac{\mu'(x,\infty)}{\mu(x,\infty)} = c$ for some $c \in (0, \infty)$, then μ' is subexponential.

A function $L(x)$ is called *slowly varying* at ∞ if $L(x)$ is positive, measurable, and $L(cx) \sim L(x)$, $x \to \infty$, for any $c > 0$. A function $f(x)$ is called *regularly varying of index* η at ∞ if $f(x) = x^\eta L(x)$ with $L(x)$ slowly varying at ∞.

REMARK 25.15. A sufficient condition for subexponentiality is as follows. If μ is a probability measure on $[0, \infty)$ such that $\mu(x, \infty)$ is regularly varying of index $-\alpha$ at ∞ with some $\alpha \ge 0$, then μ is subexponential. In the case of an infinitely divisible distribution with Lévy measure ν, we can also apply this to $\frac{1}{\nu(1,\infty)}[\nu]_{(1,\infty)}$. For example, the Pareto distribution (Remark 8.12) and one-sided stable distributions (by the form of the Lévy measures in Remark 14.4) are subexponential. As examples not covered by this sufficient condition, the Weibull distribution with parameter $0 < \alpha < 1$ and the log-normal distribution in Remark 8.12 are subexponential.

For related results and references on subexponentiality, see the recent book [**124**] of Embrechts, Klüppelberg, and Mikosch.

REMARK 25.16. Grübel [**175**] extends a part of the assertions in Remark 25.14 as follows. Let $h(x)$ be a nonnegative continuous function on $[0, \infty)$ decreasing to 0 as $x \to \infty$ such that

$$-\textstyle\int_0^x h(x - y) dh(y) = O(h(x)), \quad x \to \infty. \tag{25.10}$$

Let μ be an infinitely divisible distribution on $\mathbb{R}$ and let ν be its Lévy measure. Then the following hold as $x \to \infty$: $\mu(x, \infty) = O(h(x))$ if and only if $\nu(x, \infty) = O(h(x))$; $\mu(x, \infty) = o(h(x))$ if and only if $\nu(x, \infty) = o(h(x))$. Examples of functions $h(x)$ satisfying the conditions above are $h(x) = (1+x)^{-\alpha}(1 + \log(1 + x))^{-\beta}$ with $\alpha > 0$, $\beta \geq 0$ or with $\alpha = 0$, $\beta > 0$, and $h(x) = \mathrm{e}^{-cx^{\alpha}}$ with $c > 0$, $0 < \alpha < 1$. A sufficient condition for (25.10) is that $\sup_x \frac{h(x)}{h(2x)} < \infty$.

When $g(x) = \mathrm{e}^{\langle c,x \rangle}$, the g-moment of a Lévy process is explicitly expressible. We define, for $w = (w_j)_{1 \leq j \leq d}$ and $v = (v_j)_{1 \leq j \leq d}$ in $\mathbb{C}^d$, the inner product $\langle w, v \rangle = \sum_{j=1}^{d} w_j v_j$ (not the Hermitian inner product $\sum_{j=1}^{d} w_j \overline{v_j}$). We write $\mathrm{Re}\, w = (\mathrm{Re}\, w_j)_{1 \leq j \leq d} \in \mathbb{R}^d$. Let $D = \{x \in \mathbb{R}^d \colon |x| \leq 1\}$.

THEOREM 25.17 (Exponential moment). *Let $\{X_t\}$ be a Lévy process on $\mathbb{R}^d$ generated by (A, ν, γ). Let*

$$C = \left\{ c \in \mathbb{R}^d \colon \int_{|x|>1} \mathrm{e}^{\langle c,x \rangle} \nu(\mathrm{d}x) < \infty \right\}.$$

(i) *The set C is convex and contains the origin.*

(ii) *$c \in C$ if and only if $E\mathrm{e}^{\langle c,X_t \rangle} < \infty$ for some $t > 0$ or, equivalently, for every $t > 0$.*

(iii) *If $w \in \mathbb{C}^d$ is such that $\mathrm{Re}\, w \in C$, then*

$$\Psi(w) = \frac{1}{2}\langle w, Aw \rangle + \int_{\mathbb{R}^d} (\mathrm{e}^{\langle w,x \rangle} - 1 - \langle w, x \rangle 1_D(x)) \nu(\mathrm{d}x) + \langle \gamma, w \rangle \tag{25.11}$$

is definable, $E|\mathrm{e}^{\langle w,X_t \rangle}| < \infty$, and

$$E[\mathrm{e}^{\langle w,X_t \rangle}] = \mathrm{e}^{t\Psi(w)}. \tag{25.12}$$

Proof. (i) Obviously C contains the origin. If c_1 and c_2 are in C, then, for any $0 < r < 1$ and $s = 1 - r$,

$$\int_{|x|>1} \mathrm{e}^{\langle rc_1 + sc_2, x \rangle} \nu(\mathrm{d}x) \leq \left(\int_{|x|>1} \mathrm{e}^{\langle c_1,x \rangle} \nu(\mathrm{d}x) \right)^r \left(\int_{|x|>1} \mathrm{e}^{\langle c_2,x \rangle} \nu(\mathrm{d}x) \right)^s < \infty$$

by Hölder's inequality. Hence C is convex.

(ii) The function $g(x) = \mathrm{e}^{\langle c,x \rangle}$ is clearly submultiplicative. Hence Theorem 25.3 gives the assertion.

(iii) Any linear transformation U of $\mathbb{R}^d$ to $\mathbb{R}^d$ can be uniquely extended to a linear transformation of $\mathbb{C}^d$ to $\mathbb{C}^d$. Regarding U as a $d \times d$ matrix, it is easy to see that $\langle w, Uv \rangle = \langle U'w, v \rangle$ for $w, v \in \mathbb{C}^d$, where U' is the transpose of U. Now let $\mathrm{Re}\, w \in C$. Then $\int_{|x|>1} |\mathrm{e}^{\langle w,x \rangle}| \nu(\mathrm{d}x) = \int_{|x|>1} \mathrm{e}^{\langle \mathrm{Re}\, w,x \rangle} \nu(\mathrm{d}x) < \infty$, which shows that $\Psi(w)$ of (25.11) is definable and finite. Also, $E|\mathrm{e}^{\langle w,X_t \rangle}| = E\mathrm{e}^{\langle \mathrm{Re}\, w,X_t \rangle} < \infty$ by (ii). Let us show (25.12) in three steps.

Step 1. Let e_1 be the unit vector with first component 1. Assume that $e_1 \in C$. Let us prove (25.12) for all $w = (w_j)_{1 \leq j \leq d}$ with $\mathrm{Re}\, w_1 \in [0, 1]$ and

$\operatorname{Re} w_j = 0$, $2 \leq j \leq d$. Fix $t > 0$ and $w_2, \ldots, w_d \in \mathbb{C}$ with $\operatorname{Re} w_j = 0$, $2 \leq j \leq d$, and regard w_1 as variable in $F = \{w_1 \in \mathbb{C}\colon \operatorname{Re} w_1 \in [0,1]\}$. Consider $f(w_1) = E\mathrm{e}^{\langle w, X_t \rangle}$. Then $f(w_1)$ is continuous on F, since

$$|\mathrm{e}^{\langle w, X(t)\rangle}| = \mathrm{e}^{(\operatorname{Re} w_1) X_1(t)} \leq (\operatorname{Re} w_1)\mathrm{e}^{X_1(t)} + (1 - \operatorname{Re} w_1) \leq \mathrm{e}^{X_1(t)} + 1$$

by the convexity of $\mathrm{e}^{uX_1(t)}$ in u, where $X_1(t)$ is the first component of $X(t)$. Moreover, $f(w_1)$ is analytic in the interior of F, since it is the limit of the analytic functions $E[\mathrm{e}^{\langle w, X_t\rangle};\ |X_t| \leq n]$ as $n \to \infty$. Similarly, $h(w_1) = \mathrm{e}^{t\Psi(w)}$ is continuous on F and analytic in the interior of F. If $\operatorname{Re} w_1 = 0$, then $f(w_1) = h(w_1)$, which is the Lévy–Khintchine representation of $P_{X_1(t)}$. Therefore, as in the proof of Theorem 24.11, the principle of reflection and the uniqueness theorem yield (25.12) when $\operatorname{Re} w_1 \in [0,1]$.

Step 2. Let U be a linear transformation from $\mathbb{R}^d$ onto $\mathbb{R}^d$. Let $Y_t = UX_t$. Then $\{Y_t\}$ is a Lévy process with generating triplet (A_U, ν_U, γ_U) by Proposition 11.10. Write

$$C_U = \left\{c \in \mathbb{R}^d\colon \int_{|x|>1} \mathrm{e}^{\langle c, x\rangle} \nu_U(\mathrm{d}x) < \infty\right\}.$$

Since $\nu_U = \nu U^{-1}$, we have $C_U = (U')^{-1} C$. Given $w \in \mathbb{C}^d$ satisfying $\operatorname{Re} w \in C$, let $v = (U')^{-1} w$. Then $\operatorname{Re} v \in C_U$. Define

$$\Psi_U(v) = \frac{1}{2}\langle v, A_U v\rangle + \int_{\mathbb{R}^d} (\mathrm{e}^{\langle v, x\rangle} - 1 - \langle v, x\rangle 1_D(x))\nu_U(\mathrm{d}x) + \langle \gamma_U, v\rangle.$$

We claim that if

$$E[\mathrm{e}^{\langle v, Y_t\rangle}] = \mathrm{e}^{t\Psi_U(v)}, \tag{25.13}$$

then w satisfies (25.12). In fact, $\langle v, Y_t\rangle = \langle (U^{-1})' w, Y_t\rangle = \langle w, X_t\rangle$ and

$$\begin{aligned}\Psi_U(v) &= \frac{1}{2}\langle U'v, AU'v\rangle + \int (\mathrm{e}^{\langle v, Ux\rangle} - 1 - \langle v, Ux\rangle 1_D(x))\nu(\mathrm{d}x) + \langle U\gamma, v\rangle \\ &= \Psi(w)\end{aligned}$$

by (11.8)–(11.10). That is, (25.13) is identical with (25.12).

Step 3. Given $w \in \mathbb{C}^d$ satisfying $\operatorname{Re} w \in C$, we shall show (25.12). If $\operatorname{Re} w = 0$, there is nothing to prove. Assume $\operatorname{Re} w \neq 0$. Choose a linear transformation U from $\mathbb{R}^d$ onto $\mathbb{R}^d$ such that $\operatorname{Re} w = U' e_1$. Consider the Lévy process $Y_t = UX_t$. Since $C_U = (U')^{-1} C$, we have $e_1 \in C_U$. We know, by Step 1, that, if $v \in \mathbb{C}^d$ satisfies $\operatorname{Re} v = e_1$, then (25.13) holds. Hence, by the result of Step 2, w satisfies (25.12). $\square$

We close this section with a discussion of the g-moments of $\sup_{s\in[0,t]} |X_s|$.

THEOREM 25.18. *Let $\{X_t\}$ be a Lévy process on $\mathbb{R}^d$. Define*

$$X_t^* = \sup_{s\in[0,t]} |X_s|. \tag{25.14}$$

Let $g(r)$ be a nonnegative continuous submultiplicative function on $[0,\infty)$, increasing to ∞ as $r \to \infty$. Then the following four statements are equivalent.

(1) $E[g(X_t^*)] < \infty$ *for some* $t > 0$.
(2) $E[g(X_t^*)] < \infty$ *for every* $t > 0$.
(3) $E[g(|X_t|)] < \infty$ *for some* $t > 0$.
(4) $E[g(|X_t|)] < \infty$ *for every* $t > 0$.

Proof. Since $g(|x|)$ is submultiplicative on $\mathbb{R}^d$, (3) and (4) are equivalent by Theorem 25.3. As $|X_t| \le X_t^*$, all we have to show is that, for any fixed $t > 0$, $E[g(|X_t|)] < \infty$ implies $E[g(X_t^*)] < \infty$. We claim that, for any $a > 0$ and $b > 0$,

$$P[X_t^* > a + b] \le P[|X_t| > a]/P[X_t^* \le b/2]. \tag{25.15}$$

Fix t and let $t_{n,j} = jt/2^n$ for $j = 1, \ldots, 2^n$ and $X_{(n)}^* = \max_{1\le j\le 2^n} |X_{t_{n,j}}|$. Choosing $Z_j(s) = Z_j = X_{t_{n,j}} - X_{t_{n,j-1}}$ in Lemma 20.2 and using Remark 20.3, we have

$$P[X_{(n)}^* > a + b] \le P[|X_t| > a]/P[X_{(n)}^* \le b/2]$$

in (20.2). Hence, letting $n \to \infty$, we get (25.15). Choose $b > 0$ such that $P[X_t^* \le b/2] > 0$. Let $\widetilde{g}(r)$ be a continuous increasing function on $[0,\infty)$ such that $\widetilde{g}(0) = 0$ and $\widetilde{g}(r) = g(r)$ for $r \ge 1$. Apply Lemma 17.6 to $k(r) = 1 - P[|X_t| \le r]$ and $l(r) = \widetilde{g}(r)$. Then

$$\int_{0+}^{\infty} P[|X_t| > r] \mathrm{d}\widetilde{g}(r) = \int_{(0,\infty)} \widetilde{g}(r)\, P[|X_t| \in \mathrm{d}r] = E[\widetilde{g}(|X_t|)].$$

It follows from (25.15) that

$$\int_{0+}^{\infty} P[X_t^* > r + b] \mathrm{d}\widetilde{g}(r) \le E[\widetilde{g}(|X_t|)]/P[X_t^* \le b/2].$$

The integral in the left-hand side equals

$$\int_{(0,\infty)} \widetilde{g}(r) P[X_t^* - b \in \mathrm{d}r] = E[\widetilde{g}(X_t^* - b);\, X_t^* > b],$$

similarly. Hence, if $E[g(|X_t|)] < \infty$, then $E[g(X_t^* - b);\, X_t^* > b] < \infty$ and, by the submultiplicativity of g, $E[g(X_t^*)] < \infty$. □

REMARK 25.19. Let $d = 1$. Doob [**106**], p. 337, shows an explicit bound:

$$E[(X_t^*)^\alpha] \le 8E[|X_t|^\alpha] \quad \text{for } \alpha \ge 1, \tag{25.16}$$

provided that $E|X_t| < \infty$ and $EX_t = 0$. This is true not only for Lévy processes but also for additive processes on $\mathbb{R}$. Define the supremum process $M_t = \sup_{0\le s\le t} X_s$. If $\frac{1}{\nu(1,\infty)}[\nu]_{(1,\infty)}$ is subexponential, then $P[M_t > x]/\nu(x,\infty) \to t$ as $x \to \infty$. This is due to Berman [**22**] and Rosinski and Samorodnitsky [**423**]. Extension of this result in the case where the right tail of ν is lighter is studied in Braverman and Samorodnitsky [**64**] and Braverman [**63**].

26. Lévy measures with bounded supports

The size of the support S_ν of the Lévy measure ν of a Lévy process $\{X_t\}$ is connected with the $\mathrm{e}^{\alpha|x|\log|x|}$-moment of X_t.

THEOREM 26.1. *Let* $\{X_t\}$ *be a Lévy process on* $\mathbb{R}^d$ *with Lévy measure* ν. *Let*

$$c = \inf\{a > 0\colon S_\nu \subset \{x\colon |x| \le a\}\}. \tag{26.1}$$

If $\nu = 0$, *then let* $c = 0$. *If* S_ν *is unbounded, then let* $c = \infty$. *We understand* $1/\infty = 0$ *and* $1/0 = \infty$ *in the following.*

(i) *For any* α *with* $0 < \alpha < 1/c$ *and for any* $t > 0$,

$$E[\mathrm{e}^{\alpha|X_t|\log|X_t|}] < \infty \tag{26.2}$$

and

$$P[\,|X_t| > r\,] = o(\mathrm{e}^{-\alpha r\log r}), \qquad r \to \infty. \tag{26.3}$$

(ii) *For any* α *with* $\alpha > 1/c$ *and for any* $t > 0$,

$$E[\mathrm{e}^{\alpha|X_t|\log|X_t|}] = \infty \tag{26.4}$$

and

$$P[\,|X_t| > r\,]/\mathrm{e}^{-\alpha r\log r} \to \infty, \qquad r \to \infty. \tag{26.5}$$

REMARK 26.2. Let $\alpha_0 \in (0,\infty)$. Theorem 26.1 implies that the property that $\int \mathrm{e}^{\alpha|x|\log|x|}\mu(\mathrm{d}x) < \infty$ for all $\alpha \in (0,\alpha_0)$ is not time dependent in the class of Lévy processes. But it does not imply that the property that $\int \mathrm{e}^{\alpha_0|x|\log|x|}\mu(\mathrm{d}x) < \infty$ is not time dependent.

REMARK 26.3. If $\nu = 0$, then P_{X_t} is Gaussian with covariance matrix tA and, using the maximum eigenvalue a of A, we have

$$P[\,|X_t| > r\,] = o(\mathrm{e}^{-\alpha r^2}), \qquad r \to \infty,$$

for $0 < \alpha < 1/(2at)$, and

$$P[\,|X_t| > r\,]/\mathrm{e}^{-\alpha r^2} \to \infty, \qquad r \to \infty,$$

for $\alpha > 1/(2at)$. This follows from the form of the density on the support (affine subspace) of P_{X_t}. Theorem 26.1 tells us that, if $\nu \ne 0$, the rate of decay of the tail of P_{X_t} is much slower than in the Gaussian case. Thus there is a gap in the rate of decay of the tail between Gaussian and non-Gaussian infinitely divisible distributions. But keep it in mind that here we are looking at the tail defined by $P[\,|X_t| > r\,]$. If we consider the one-sided tail in one-dimensional case, the situation is different. For example, the tail $P[\,X_t > r\,]$ in the positive direction for a stable process of parameters (α, β, τ, c) with $1 \le \alpha < 2$ and $\beta = -1$, which is described in Remark 14.18(vi), is thinner than the tail of the Gaussian distribution.

We shall prove Theorem 26.1 after the preparation of several lemmas. As the theorem can be considered as a result on an infinitely divisible distribution, let μ be an infinitely divisible distribution on $\mathbb{R}^d$ with generating triplet (A, ν, γ) and define c by (26.1).

LEMMA 26.4. *Assume $d = 1$ and μ is non-trivial. Let C be the interval defined by*

(26.6)
$$C = \left\{ u \in \mathbb{R} \colon \int_{\mathbb{R}} e^{ux} \mu(dx) < \infty \right\} = \left\{ u \in \mathbb{R} \colon \int_{|x|>1} e^{ux} \nu(dx) < \infty \right\}$$

and define

$$\Psi(u) = \frac{1}{2} A u^2 + \int_{\mathbb{R}} (e^{ux} - 1 - ux 1_{[-1,1]}(x)) \nu(dx) + \gamma u \tag{26.7}$$

for $u \in C$. Then $\Psi(u)$ is of class C^∞ and $\Psi''(u) > 0$ in the interior of C. Let b be the supremum of C. If $b > 0$, then, letting $\lim_{u \downarrow 0} \Psi'(u) = \xi_0 \geq -\infty$ and letting $u = \theta(\xi)$ be the inverse function of $\xi = \Psi'(u)$ defined on the interval $(\xi_0, \Psi'(b-))$, we have

$$\mu[x, \infty) \leq \exp\left[- \int_{\xi_0}^{x} \theta(\xi) d\xi \right] \tag{26.8}$$

for all $x \in (\xi_0, \Psi'(b-))$.

Proof. We use the method of Cramér [**87**] in the estimating of large deviations. The set C is the same as that of Theorem 25.17. There we have seen that C is convex (an interval for $d = 1$), the second equality in (26.6) holds, and hence $\Psi(u)$ is definable on C. It follows from (26.6) and (26.7) that Ψ is of class C^∞ and

$$\Psi'(u) = Au + \int (x e^{ux} - x 1_{[-1,1]}(x)) \nu(dx) + \gamma, \tag{26.9}$$

$$\Psi''(u) = A + \int x^2 e^{ux} \nu(dx) > 0. \tag{26.10}$$

Now, assume $b > 0$. Since $\int e^{ux} \mu(dx) = e^{\Psi(u)}$ by Theorem 25.17, we get

$$\mu[x, \infty) \leq \int e^{u(y-x)} \mu(dy) = e^{\Psi(u) - ux}.$$

We want to make this bound as good as we can. Let $x \in (\xi_0, \Psi'(b-))$. As $\Psi'(u) - x$ changes from negative to positive at $u = \theta(x)$, we have

$$\min_{0<u<b} (\Psi(u) - ux) = \Psi(\theta(x)) - x\theta(x).$$

Since $\Psi(0) = 0$,

$$\Psi(\theta(x)) = \int_0^{\theta(x)} \Psi'(u) du = \int_{\xi_0}^{x} \Psi'(\theta(\xi)) d\theta(\xi) = \int_{\xi_0}^{x} \xi d\theta(\xi),$$

and hence

$$\Psi(\theta(x)) - x\theta(x) = -\lim_{\xi\downarrow\xi_0} \xi\theta(\xi) - \int_{\xi_0}^{x} \theta(\xi)\mathrm{d}\xi.$$

Noting that

$$\lim_{\xi\downarrow\xi_0} \xi\theta(\xi) = \lim_{u\downarrow 0} \Psi'(u)u = \lim_{u\downarrow 0} u \int (y\mathrm{e}^{uy} - y1_{[-1,1]}(y))\nu(\mathrm{d}y) = 0,$$

we obtain (26.8). □

LEMMA 26.5. *Assume $c < \infty$. Then, for $0 < \alpha < 1/(c\sqrt{d})$,*

$$\int_{|x|>r} \mu(\mathrm{d}x) = o(\mathrm{e}^{-\alpha r \log r}), \qquad r \to \infty. \tag{26.11}$$

Proof. Step 1. Let $d = 1$. If $\nu = 0$, then μ is Gaussian and (26.11) is evident. Let $\nu \neq 0$. Let us show

$$\mu[x,\infty) = o(\mathrm{e}^{-\alpha x \log x}), \qquad x \to \infty. \tag{26.12}$$

Since $\int_{|y|>c} \nu(\mathrm{d}y) = 0$, the set C in Lemma 26.4 is the whole line $\mathbb{R}$ and $b = \infty$. It follows from (26.9) that $\lim_{u\to\infty} \Psi'(u) < \infty$ if and only if $A = 0$, $\nu(0,c] = 0$, and $\int_{[-c,0)} |y|\nu(\mathrm{d}y) < \infty$. In this case, the support of μ is bounded above by virtue of Theorem 24.7 and (26.12) is evident. So we assume that $\lim_{u\to\infty} \Psi'(u) = \infty$. Then we have (26.8) for $\xi_0 < x < \infty$. Let us estimate $\theta(\xi)$. Since

$$\xi = A\theta(\xi) + \int_{[-c,c]} (\mathrm{e}^{\theta(\xi)y} - 1)y\nu(\mathrm{d}y) + \gamma_1$$

with some constant γ_1 by (26.9), we have

$$\xi \leq A\theta(\xi) + \mathrm{e}^{\theta(\xi)c}\theta(\xi) \int_{[-c,c]} y^2\nu(\mathrm{d}y) + \gamma_1.$$

Hence, for $0 < \alpha < \alpha' < 1/c$, we have $\xi\mathrm{e}^{-\theta(\xi)/\alpha'} \to 0$ as $\xi \to \infty$. Thus, there is $\xi_1 > 0$ such that $-\theta(\xi)/\alpha' < -\log\xi$ for $\xi > \xi_1$. Therefore,

$$\mu[x,\infty) \leq \text{const} \exp\left[-\alpha' \int_{\xi_1}^{x} \log\xi \mathrm{d}\xi\right] \leq \text{const}\, \mathrm{e}^{-\alpha' x(\log x - 1)},$$

which shows (26.12). As $\mu(-\infty,-x]$ is similarly estimated, we get (26.11) in the case $d = 1$.

Step 2. Let $d \geq 2$. Let $X = (X_j)_{1\leq j\leq d}$ be a random variable with distribution μ. Then each X_j is infinitely divisible with Lévy measure supported on $[-c,c]$ (Proposition 11.10). Using the result of Step 1, we see that

$$P[\,|X| > r\,] \leq \sum_{j=1}^{d} P[\,|X_j| > r/\sqrt{d}\,] = o(\mathrm{e}^{-\alpha'\sqrt{d}(r/\sqrt{d})\log(r/\sqrt{d})}) = o(\mathrm{e}^{-\alpha r \log r})$$

for $0 < \alpha < \alpha' < 1/(c\sqrt{d})$, completing the proof. □

LEMMA 26.6. *Assume that μ is compound Poisson on $\mathbb{R}^d$. Then (26.11) holds for $0 < \alpha < 1/c$ and*

$$\int_{|x|>r} \mu(\mathrm{d}x) \Big/ \mathrm{e}^{-\alpha r \log r} \to \infty, \qquad r \to \infty, \tag{26.13}$$

for $\alpha > 1/c$.

Proof. Let $a = \nu(\mathbb{R}^d)$ and $a_1 = a \vee 1$. Suppose that $c < \infty$ and $0 < \alpha < 1/c$. Since $S(\nu^n) \subset \{x \colon |x| \le nc\}$, we have

$$\int_{|x|>r} \mu(\mathrm{d}x) = \mathrm{e}^{-a} \sum_{n=1}^{\infty} \frac{1}{n!} \int_{|x|>r} \nu^n(\mathrm{d}x) \le \mathrm{e}^{-a} \sum_{n>r/c} \frac{a^n}{n!} \le \frac{a^{n_r}}{n_r!},$$

where n_r is the integer such that $n_r > r/c \ge n_r - 1$. By Stirling's formula,

$$\begin{aligned} \frac{a^{n_r}}{n_r!} \Big/ \mathrm{e}^{-\alpha r \log r} &\le \frac{a_1^{r/c+1}}{\Gamma(r/c+1)} \mathrm{e}^{\alpha r \log r} \\ &\sim \text{const} \frac{\mathrm{e}^{(r/c)\log a_1 + \alpha r \log r}}{(r/c)^{1/2} \mathrm{e}^{(r/c)\log(r/c) - r/c}} = o(1), \qquad r \to \infty. \end{aligned}$$

Hence we get (26.11).

To show the latter half, suppose that $0 < c \le \infty$ and $\alpha > 1/c$. Choose c' with $1/\alpha < c' < c$. By making an orthogonal transformation of the state space if necessary, we may and do assume that $\nu\{x \colon x_1 > c'\} > 0$, where x_1 is the first component of x. Using the integer n_r with $n_r > r/c' \ge n_r - 1$, we get

$$\begin{aligned} \int_{|x|>r} \mu(\mathrm{d}x) &\ge \frac{\mathrm{e}^{-a}}{n_r!} \int_{|x|>r} \nu^{n_r}(\mathrm{d}x) \ge \frac{\mathrm{e}^{-a}}{n_r!} \left(\int_{\{x_1 > r/n_r\}} \nu(\mathrm{d}x) \right)^{n_r} \\ &\ge \frac{\mathrm{e}^{-a}}{n_r!} a_2^{n_r} \ge \frac{\mathrm{e}^{-a}}{\Gamma(r/c'+2)} a_2^{r/c'+1}, \end{aligned}$$

where $a_2 = 1 \wedge \nu\{x \colon x_1 > c'\}$. Again by Stirling's formula,

$$\frac{a_2^{r/c'+1}}{\Gamma(r/c'+2)} \Big/ \mathrm{e}^{-\alpha r \log r} \to \infty, \qquad r \to \infty,$$

which proves (26.13). □

LEMMA 26.7. *Let ρ be a finite measure on $[0, \infty)$.*

(i) *If $g(x)$ is a positive increasing function and if ρ has finite g-moment, then $\rho(x, \infty) = o(1/g(x))$ as $x \to \infty$.*

(ii) *Let $g_\alpha(x) = \mathrm{e}^{\alpha x \log x}$ for $\alpha > 0$. If $\rho(x, \infty) = O(1/g_\alpha(x))$ as $x \to \infty$, then, for any $\beta \in (0, \alpha)$, ρ has finite g_β-moment.*

Proof. (i) Use $g(x)\rho(x,\infty) \le \int_{(x,\infty)} g(y)\rho(\mathrm{d}y) \to 0$.

(ii) Let $G(x) = \rho(x,\infty)$. Then

$$\begin{aligned}\int_{(0,\infty)} \mathrm{e}^{\beta x \log x}\rho(\mathrm{d}x) &= -\int_0^\infty \mathrm{e}^{\beta x\log x}\mathrm{d}G(x)\\ &= G(0) + \beta\int_0^\infty \mathrm{e}^{\beta x\log x}(\log x + 1)G(x)\mathrm{d}x < \infty,\end{aligned}$$

as asserted. □

Proof of Theorem 26.1. It is enough to prove the assertions for an infinitely divisible distribution μ on $\mathbb{R}^d$ with generating triplet (A,ν,γ). That is, we prove them for $t=1$. If $\nu = 0$, then μ is Gaussian, $c = 0$, and the result is known. Assume $\nu \ne 0$, that is, $c > 0$. First, suppose $c < \infty$ and $0 < \alpha < 1/c$, and let us prove (26.3). Choose $0 < \delta < 1$ with $\alpha < (1-\delta)/c$ and, then, $0 < c' < c$ with $c'\alpha\sqrt{d} < \delta$. Let Y and Z be independent random variables satisfying $Y + Z \overset{\mathrm{d}}{=} X_1$ and

$$E\mathrm{e}^{\mathrm{i}\langle z,Z\rangle} = \exp\left[\int_{c'<|x|\le c} (\mathrm{e}^{\mathrm{i}\langle z,x\rangle} - 1)\nu(\mathrm{d}x)\right].$$

Then

$$P[\,|X_1| > r\,] \le P[\,|Y| + |Z| > r\,] \le P[\,|Y| > r\delta\,] + P[\,|Z| > r(1-\delta)\,].$$

Choose α' such that $\alpha < \alpha' < (1-\delta)/c$ and $c'\alpha'\sqrt{d} < \delta$. Since Y is infinitely divisible and its Lévy measure is supported on $\{|x| \le c'\}$, Lemma 26.5 says that

$$P[\,|Y| > r\delta\,] = o(\mathrm{e}^{-(\alpha'/\delta)r\delta\log r\delta}) = o(\mathrm{e}^{-\alpha r\log r}), \qquad r\to\infty.$$

On the other hand, since Z has a compound Poisson distribution, we have

$$P[\,|Z| > r(1-\delta)\,] = o(\mathrm{e}^{-(\alpha'/(1-\delta))r(1-\delta)\log(r(1-\delta))}) = o(\mathrm{e}^{-\alpha r\log r})$$

by Lemma 26.6. Hence (26.3) follows.

Next, let us suppose $c \le \infty$ and $\alpha > 1/c$ and prove (26.5). Choosing $0 < c' < c$, define Y and Z as above. Choose α' and $a > 1$ such that $\alpha > \alpha' > 1/c$ and $\alpha' a < \alpha$. We have

$$P[\,|X_1| > r\,] \ge P[\,|Y| \le r(a-1)\,]\,P[\,|Z| > ra\,]$$

and $P[\,|Y| \le r(a-1)\,] \to 1$ as $r\to\infty$. Since, by Lemma 26.6,

$$\frac{P[\,|Z| > ra\,]}{\mathrm{e}^{-\alpha r\log r}} = \frac{P[\,|Z| > ra\,]}{\mathrm{e}^{-\alpha' ra\log ra}}\mathrm{e}^{(\alpha r-\alpha' ra)\log r - \alpha' ra\log a} \to \infty,$$

we obtain (26.5).

The assertions (26.2) and (26.4) concerning $E[\mathrm{e}^{\alpha|X_1|\log|X_1|}]$ follow from (26.3) and (26.5) by Lemma 26.7. □

An assertion similar to Theorem 26.1 remains true when we consider a fixed direction. We give it in one dimension.

THEOREM 26.8. *Let* $\{X_t\}$ *be a Lévy process on* $\mathbb{R}$ *with Lévy measure* ν *and let* c *be the infimum of* $a \geq 0$ *such that* $S_\nu \subset (-\infty, a]$. *If* S_ν *is unbounded above, let* $c = \infty$.

(i) *For any* α *with* $0 < \alpha < 1/c$ *and for any* $t > 0$,

$$E[\mathrm{e}^{\alpha X_t \log X_t};\ X_t > 0] < \infty, \tag{26.14}$$

$$P[\,X_t > r\,] = o(\mathrm{e}^{-\alpha r \log r}), \qquad r \to \infty. \tag{26.15}$$

(ii) *For any* $\alpha > 1/c$ *and* $t > 0$,

$$E[\mathrm{e}^{\alpha X_t \log X_t};\ X_t > 0] = \infty, \tag{26.16}$$

$$P[\,X_t > r\,]/\mathrm{e}^{-\alpha r \log r} \to \infty, \qquad r \to \infty. \tag{26.17}$$

Proof. It suffices to prove these for $t = 1$. The assertions (26.14) and (26.16) follow from (26.15) and (26.17) by Lemma 26.7. To prove (26.15), let $c < \infty$ and $0 < \alpha < 1/c$. Choose $c' > 0$ with $\alpha < 1/c' \leq 1/c$, let Y and Z be independent and satisfy $Y + Z \stackrel{\mathrm{d}}{=} X_1$ and

$$E\mathrm{e}^{\mathrm{i}zZ} = \exp\left[\int_{(-\infty,-c']} (\mathrm{e}^{\mathrm{i}zx} - 1)\nu(\mathrm{d}x)\right].$$

Then Z has a compound Poisson distribution and $Z \leq 0$ a. s. Apply Theorem 26.1 to Y and get

$$P[\,X_1 > r\,] \leq P[\,Y > r\,] \leq P[\,|Y| > r\,] = o(\mathrm{e}^{-\alpha r \log r}), \qquad r \to \infty.$$

This shows (26.15).

To show (26.17), let $c > 0$ and $\alpha > 1/c$. This time choose $0 < c' < c$ and independent Y and Z such that $Y + Z \stackrel{\mathrm{d}}{=} X_1$ and

$$E\mathrm{e}^{\mathrm{i}zZ} = \exp\left[\int_{(c',c]} (\mathrm{e}^{\mathrm{i}zx} - 1)\nu(\mathrm{d}x)\right].$$

Then Z is compound Poisson and $Z \geq 0$ a. s. Using α' and a that satisfy $\alpha > \alpha' > 1/c$ and $\alpha/\alpha' > a > 1$, we see that

$$P[\,X_1 > r\,] \geq P[\,Z > ra\,]\,P[\,|Y| \leq r(a-1)\,],$$
$$P[\,Z > ra\,]/\mathrm{e}^{-\alpha' ra \log ra} \to \infty, \qquad r \to \infty,$$

by Lemma 26.6. It follows that

$$P[\,Z > ra\,]/\mathrm{e}^{-\alpha r \log r} \to \infty$$

and hence (26.17). □

27. Continuity properties

A measure ρ on $\mathcal{B}(\mathbb{R}^d)$ is said to be *discrete* if there is a countable set C such that $\rho(\mathbb{R}^d \setminus C) = 0$; ρ is said to be *continuous* if $\rho\{x\} = 0$ for every $x \in \mathbb{R}^d$; ρ is said to be *singular* if there is a set $B \in \mathcal{B}(\mathbb{R}^d)$ such that $\rho(\mathbb{R}^d \setminus B) = 0$ and $\mathrm{Leb}(B) = 0$ (Leb denotes the Lebesgue measure on $\mathbb{R}^d$); ρ is said to be *absolutely continuous* if $\rho(B) = 0$ for every $B \in \mathcal{B}(\mathbb{R}^d)$ satisfying $\mathrm{Leb}(B) = 0$. If ρ is discrete, absolutely continuous, or continuous singular, then ρ is said to be *pure*.

Let ρ be σ-finite. Then there are measures ρ_d, ρ_{ac}, and ρ_{cs} such that $\rho = \rho_d + \rho_{ac} + \rho_{cs}$, ρ_d is discrete, ρ_{ac} is absolutely continuous, and ρ_{cs} is continuous singular. The set of ρ_d, ρ_{ac}, and ρ_{cs} is uniquely determined by ρ (Lebesgue decomposition). The measures ρ_d, ρ_{ac}, and ρ_{cs} are respectively called the discrete, absolutely continuous, and continuous singular parts of ρ. If $\rho_d \neq 0$ (or $\rho_{ac} \neq 0$ or $\rho_{cs} \neq 0$), then we say that ρ has a discrete (or absolutely continuous or continuous singular, respectively) part. The measure $\rho_{ac} + \rho_{cs}$ is called the continuous part of ρ. The discrete part is sometimes called the point masses of ρ.

Let us study discreteness, continuity, absolute continuity, and continuous singularity of P_{X_t} for a Lévy process $\{X_t\}$ on $\mathbb{R}^d$ in relation to its generating triplet (A, ν, γ). In general probability measures μ, these properties are reflected by asymptotic behavior at ∞ of the characteristic functions $\widehat{\mu}(z)$. For example, the Riemann–Lebesgue theorem tells us that $\widehat{\mu}(z) \to 0$ as $|z| \to \infty$ if μ is absolutely continuous on $\mathbb{R}^d$ (see [**34, 66, 139, 151, 268**] for $d = 1$ and [**51**], p. 57, [**497**], p. 2 for general d). On the other hand, we know that μ is absolutely continuous with bounded continuous density if $\int |\widehat{\mu}(z)| \mathrm{d}z < \infty$ (Proposition 2.5(xii)). (As a digression we add the fact that μ is absolutely continuous with square integrable density if and only if $\widehat{\mu}(z)$ is square integrable, that is, $\int |\widehat{\mu}(z)|^2 \mathrm{d}z < \infty$; see [**51, 268, 497**].) In distributions of Lévy processes, we usually need more analysis pertaining to infinite divisibility than to apply general theorems.

We begin with a simple lemma.

LEMMA 27.1. *Let ρ_1 and ρ_2 be non-zero finite measures on $\mathbb{R}^d$. Let $\rho = \rho_1 * \rho_2$;*

(i) *ρ is continuous if and only if ρ_1 or ρ_2 is continuous;*

(ii) *ρ is discrete if and only if ρ_1 and ρ_2 are discrete;*

(iii) *ρ is absolutely continuous if ρ_1 or ρ_2 is absolutely continuous;*

(vi) *ρ_1 or ρ_2 is continuous singular if ρ is continuous singular.*

Proof. (i), (ii). If ρ_1 is continuous, then ρ is continuous, because $\rho\{x\} = \int \rho_1\{x - y\}\rho_2(\mathrm{d}y) = 0$. If ρ_1 and ρ_2 are discrete, then ρ is discrete, since $\rho_1(\mathbb{R}^d \setminus C_1) = \rho_2(\mathbb{R}^d \setminus C_2) = 0$ with some countable sets C_1 and C_2 and $\rho(\mathbb{R}^d \setminus C) = 0$ for $C = C_1 + C_2$, which is countable. Thus we get the 'if' parts of (i) and (ii). The 'only if' part of (i) follows from the 'if' part of (ii), because, if both ρ_1 and ρ_2 have discrete parts, then ρ has a discrete part. Similarly, the 'only if' part of (ii) follows from the 'if' part of (i).

(iii) Suppose that ρ_1 is absolutely continuous. If $B \in \mathcal{B}(\mathbb{R}^d)$ satisfies $\mathrm{Leb}(B) = 0$, then $\mathrm{Leb}(B - y) = 0$ for every y and $\rho(B) = \int \rho_1(B - y)\rho_2(\mathrm{d}y) = 0$.

(iv) Suppose that neither ρ_1 nor ρ_2 is continuous singular. Then $(\rho_1)_d + (\rho_1)_{ac} \neq 0$ and $(\rho_2)_d + (\rho_2)_{ac} \neq 0$. It follows from (ii) and (iii) that $((\rho_1)_d + (\rho_1)_{ac}) * ((\rho_2)_d + (\rho_2)_{ac})$ has a discrete or absolutely continuous part. □

REMARK 27.2. There is a continuous singular ρ_1 such that ${\rho_1}^2$ is absolutely continuous. Hence, neither the assertion (iii) nor (iv) can be strengthened to 'if and only if'. On the other hand, there is a continuous singular ρ_2 such that ${\rho_2}^n$ is continuous singular for every $n \in \mathbb{N}$. Later Remark 27.22, Theorem 27.23 and Remark 27.24 will furnish such ρ_1 and ρ_2 in the class of infinitely divisible probability measures on $\mathbb{R}$. Other examples are found in Kawata [**268**], p. 558 and Lukacs [**329**], p. 20.

REMARK 27.3. If $\{X_t\}$ is a compound Poisson process on $\mathbb{R}^d$ with Lévy measure ν, then

$$P_{X_t} = \mathrm{e}^{-t\nu(\mathbb{R}^d)} \textstyle\sum_{k=0}^{\infty} (k!)^{-1} t^k \nu^k, \tag{27.1}$$

which is not continuous, as $P[\,X_t = 0\,] > 0$. Many properties of P_{X_t} are derived from those of ν by Lemma 27.1 and (27.1). For example, for $t > 0$, $[P_{X_t}]_{\mathbb{R}^d \setminus \{0\}}$ is continuous if and only if ν is continuous.

THEOREM 27.4 (Continuity). *For a Lévy process $\{X_t\}$ on $\mathbb{R}^d$ with generating triplet (A, ν, γ), the following three statements are equivalent.*

(1) *P_{X_t} is continuous for every $t > 0$.*
(2) *P_{X_t} is continuous for some $t > 0$.*
(3) *$\{X_t\}$ is of type B or C (that is, $A \neq 0$ or $\nu(\mathbb{R}^d) = \infty$).*

Proof. The statement (2) implies (3) by Remark 27.3. That is, if $A = 0$ and $\nu(\mathbb{R}^d) < \infty$, then $\{X_t - \gamma_0 t\}$ is a compound Poisson process and P_{X_t} has a point mass at $\gamma_0 t$. Let us prove that (3) implies (1). If $A \neq 0$, then P_{X_t}, $t > 0$, has a non-trivial Gaussian distribution as a convolution factor and P_{X_t} is continuous by Lemma 27.1(i), since a non-trivial Gaussian has a density on an affine subspace. In the following, suppose that $\nu(\mathbb{R}^d) = \infty$.

Case 1. The case that ν is discrete. Let $x_1, x_2, \ldots$ be the points with positive ν-measure and let $m_j = \nu\{x_j\}$ and $m'_j = m_j \wedge 1$. We have $\sum_{j=1}^{\infty} m_j = \infty$, and hence $\sum_{j=1}^{\infty} m'_j = \infty$. Let $\{Y_n(t)\}$ be the compound Poisson process with Lévy measure $\nu_n = \sum_{j=1}^{n} m'_j \delta_{x_j}$. In general define, for a probability measure μ on $\mathbb{R}^d$,

$$D(\mu) = \sup_{x \in \mathbb{R}^d} \mu\{x\}.$$

When μ is the distribution of X, we write $D(X) = D(\mu)$. If $\mu = \mu_1 * \mu_2$, then $D(\mu) \leq D(\mu_1)$, since

$$\mu\{x\} = \int \mu_1\{x-y\}\mu_2(\mathrm{d}y) \leq D(\mu_1).$$

It follows that

$$D(X(t)) \leq D(Y_n(t)). \tag{27.2}$$

Let $c_n = \nu_n(\mathbb{R}^d)$ and $\sigma_n = c_n^{-1}\nu_n$. Since $D({\sigma_n}^k) \leq D(\sigma_n) \leq c_n^{-1}$, we have, by (27.1),

$$P[\,Y_n(t) = x\,] = \mathrm{e}^{-tc_n} \sum_{k=0}^{\infty} \frac{t^k}{k!} c_n^k {\sigma_n}^k\{x\} \leq \mathrm{e}^{-tc_n} + {c_n}^{-1}.$$

Consequently, $D(Y_n(t)) \leq \mathrm{e}^{-tc_n} + {c_n}^{-1}$. Now it follows from $c_n \to \infty$ that $D(X(t)) = 0$, that is, $X(t)$ has a continuous distribution.

Case 2. The case that ν is continuous. Let $\{Y_n(t)\}$ be the compound Poisson with Lévy measure $\nu_n = [\nu]_{\{|x|>1/n\}}$. Then we have (27.2). For any $k \geq 1$, ${\nu_n}^k$ is continuous by Lemma 27.1(i). Hence, by (27.1), the distribution of $Y_n(t)$ has a point mass only at 0. It follows that $D(Y_n(t)) = \mathrm{e}^{-tc_n}$, where $c_n = \nu_n(\mathbb{R}^d)$. Since $c_n \to \infty$, we have $D(X(t)) = 0$.

Remaining case. Let ν_d and ν_c be the discrete and the continuous part of ν, respectively. Then ν_d or ν_c has infinite total measure. If $\nu_d(\mathbb{R}^d) = \infty$, then, letting $\{Y_t\}$ be the Lévy process with generating triplet $(0, \nu_d, 0)$, we see that Y_t has a continuous distribution for any $t > 0$ by Step 1, and hence so does X_t again by Lemma 27.1(i). If $\nu_c(\mathbb{R}^d) = \infty$, then, similarly, use Step 2. □

COROLLARY 27.5. *The following statements are equivalent for a Lévy process $\{X_t\}$ on $\mathbb{R}^d$ generated by (A, ν, γ).*

(1) *P_{X_t} is discrete for every $t > 0$.*
(2) *P_{X_t} is discrete for some $t > 0$.*
(3) *$\{X_t\}$ is of type A and ν is discrete.*

Proof. The statement (3) implies (1) because of Lemma 27.1(ii) and (27.1). To see that (2) implies (3), notice that $A = 0$ and $\nu(\mathbb{R}^d) < \infty$ by Theorem 27.4 and that ν is discrete by (27.1). □

When ρ is a discrete measure on $\mathbb{R}^d$, let us define $C_\rho = \{x \in \mathbb{R}^d \colon \rho\{x\} > 0\}$ and call it the *carrier* of ρ. It is a countable set, but it can be a dense set. If a random variable X on $\mathbb{R}^d$ has a discrete distribution, then the carrier of P_X is written as C_X.

PROPOSITION 27.6. *Let $\{X_t\}$ be a Lévy process of type A on $\mathbb{R}^d$ with P_{X_t} discrete for every t. Let ν and γ_0 be its Lévy measure and drift. Then,* *for every* $t > 0$,

$$C_{X_t} = \left(\{0\} \cup \bigcup_{n=1}^{\infty} \{x_1 + \cdots + x_n \colon x_1, \ldots, x_n \in C_\nu\} \right) + \gamma_0 t. \tag{27.3}$$

Proof. By Corollary 27.5, ν is discrete and $\{X_t - \gamma_0 t\}$ is compound Poisson. Hence (27.1) gives (27.3). □

If the Gaussian covariance matrix A for a Lévy process $\{X_t\}$ on $\mathbb{R}^d$ has rank d, then P_{X_t}, $t > 0$, has a nondegenerate Gaussian as a convolution factor and hence, by Lemma 27.1(iii), it is absolutely continuous. In the case $A = 0$, to find conditions for absolute continuity of X_t, $t > 0$, is a hard problem.

THEOREM 27.7 (Sufficiency for absolute continuity). *Let* $\{X_t\}$ *be a Lévy process on* $\mathbb{R}^d$ *generated by* (A, ν, γ) *with* $\nu(\mathbb{R}^d) = \infty$. *Define* $\widetilde{\nu}$ *by*

$$\widetilde{\nu}(B) = \int_B (|x|^2 \wedge 1) \nu(\mathrm{d}x). \tag{27.4}$$

If $(\widetilde{\nu})^l$ *is absolutely continuous for some* $l \in \mathbb{N}$, *then, for every* $t > 0$, P_{X_t} *is absolutely continuous.*

Proof. Let $\nu_n = [\nu]_{\{|x|>1/n\}}$ and $c_n = \nu_n(\mathbb{R}^d)$. Let $\{Y_n(t)\}$ be compound Poisson with Lévy measure ν_n. Fix $t > 0$ and let μ and μ_n be the distributions of $X(t)$ and $Y_n(t)$, respectively. We have

$$\mu_n = \sum_{k=0}^{l-1} \mathrm{e}^{-tc_n} \frac{t^k}{k!} {\nu_n}^k + \sum_{k=l}^{\infty} \mathrm{e}^{-tc_n} \frac{t^k}{k!} {\nu_n}^k$$

by (27.1). Since $(\nu_n)^l$ is absolutely continuous, the second term in the right-hand side is absolutely continuous by Lemma 27.1(iii). Let $\mu = \mu_d + \mu_{ac} + \mu_{cs}$ and $\mu_n = \mu_{n,d} + \mu_{n,ac} + \mu_{n,cs}$ be the Lebesgue decompositions. Then we have

$$\mu_{n,d}(\mathbb{R}^d) + \mu_{n,cs}(\mathbb{R}^d) \leq \sum_{k=0}^{l-1} \mathrm{e}^{-tc_n} \frac{t^k}{k!} {c_n}^k.$$

The right-hand side tends to 0 as $n \to \infty$, since $c_n \to \infty$. We have

$$\mu_d(\mathbb{R}^d) + \mu_{cs}(\mathbb{R}^d) \leq \mu_{n,d}(\mathbb{R}^d) + \mu_{n,cs}(\mathbb{R}^d),$$

since μ has μ_n as a convolution factor. Therefore $\mu_d + \mu_{cs} = 0$, completing the proof. □

EXAMPLE 27.8. Absolute continuity of non-trivial selfdecomposable distributions on $\mathbb{R}$ is now proved. In fact, if μ is non-trivial and selfdecomposable on $\mathbb{R}$, then, by Corollary 15.11, either $A > 0$ or ν is absolutely continuous with total mass infinite. So, if $A = 0$, Theorem 27.7 with $l = 1$ applies.

Using Theorem 27.7, we give another sufficient condition for absolute continuity.

DEFINITION 27.9. We say that a measure ν on $\mathbb{R}^d$ is *radially absolutely continuous*, if there are a finite measure λ on the unit sphere S and a nonnegative measurable function $g(\xi, r)$ on $S \times (0, \infty)$ such that

$$\nu(B) = \int_S \lambda(\mathrm{d}\xi) \int_0^\infty g(\xi, r) 1_B(r\xi)\mathrm{d}r \quad \text{for } B \in \mathcal{B}(\mathbb{R}^d \setminus \{0\}). \tag{27.5}$$

A radially absolutely continuous measure ν is said to satisfy *the divergence condition,* if the λ and the $g(\xi, r)$ can be chosen to satisfy an additional condition

$$\int_0^\infty g(\xi, r)\mathrm{d}r = \infty \quad \text{for } \lambda\text{-a.e. } \xi. \tag{27.6}$$

THEOREM 27.10 (Sufficiency for absolute continuity). *If $\{X_t\}$ is a nondegenerate Lévy process on $\mathbb{R}^d$ with Lévy measure ν being radially absolutely continuous with the divergence condition satisfied, then P_{X_t} is absolutely continuous for each $t > 0$.*

We need two lemmas.

LEMMA 27.11. *Let ν be a measure on $\mathbb{R}^d$ such that $\nu\{0\} = 0$ and the measure $\widetilde{\nu}$ defined by (27.4) is finite. If ν is radially absolutely continuous and if $\nu(V) = 0$ for every proper linear subspace V of $\mathbb{R}^d$, then $(\widetilde{\nu})^d$ is absolutely continuous.*

Proof. Let C be the set of $\xi \in S$ such that $\int_0^\infty g(\xi, r)\mathrm{d}r = 0$. We may and do assume that $\lambda(C) = 0$. For any proper linear subspace V of $\mathbb{R}^d$, we have $\lambda(S \cap V) = 0$ from $\nu(V) = 0$. Let $B \in \mathcal{B}(\mathbb{R}^d)$ be such that $\mathrm{Leb}(B) = 0$. Since (27.5) holds for all $B \in \mathcal{B}(\mathbb{R}^d)$, we have

$$\begin{aligned}(\widetilde{\nu})^d(B) &= \int \cdots \int_{(\mathbb{R}^d)^d} 1_B(x_1 + \cdots + x_d)\widetilde{\nu}(\mathrm{d}x_1) \ldots \widetilde{\nu}(\mathrm{d}x_d) \\ &= \int \cdots \int_{S^d} I(\xi_1, \ldots, \xi_d)\lambda(\mathrm{d}\xi_1) \ldots \lambda(\mathrm{d}\xi_d),\end{aligned}$$

$$I(\xi_1, \ldots, \xi_d) = \int_0^\infty \cdots \int_0^\infty 1_B(r_1\xi_1 + \cdots + r_d\xi_d) \prod_{j=1}^d g(\xi_j, r_j)(r_j^2 \wedge 1)\mathrm{d}r_j.$$

If $\xi_1, \ldots, \xi_d$ are linearly independent, then change of variables in the multiple integral gives $I(\xi_1, \ldots, \xi_d) = 0$ from $\mathrm{Leb}(B) = 0$. Let us denote by $V(\xi_1, \ldots, \xi_s)$ the linear subspace spanned by $\xi_1, \ldots, \xi_s$. Let

$$K_r = \{(\xi_1, \ldots, \xi_d) \in S^d \colon \dim V(\xi_1, \ldots, \xi_d) = r\} = \bigcup_{(i_1, \ldots, i_r)} K_r(i_1, \ldots, i_r),$$

$$K_r(i_1, \ldots, i_r) = \{(\xi_1, \ldots, \xi_d) \in K_r \colon \xi_{i_1}, \ldots, \xi_{i_r} \text{ are linearly independent}\}.$$

Given $i_1, \ldots, i_r$ with $1 \le r \le d-1$, we choose i_0 different from $i_1, \ldots, i_r$ and obtain

$$\int \cdots \int_{K_r(i_1,\ldots,i_r)} \lambda(\mathrm{d}\xi_1) \ldots \lambda(\mathrm{d}\xi_d)$$
$$\le \int \cdots \int_{S^{d-1}} \prod_{j \ne i_0} \lambda(\mathrm{d}\xi_j) \int_S 1_{V(\xi(i_1),\ldots,\xi(i_r))}(\xi_{i_0}) \lambda(\mathrm{d}\xi_{i_0}) = 0.$$

It follows that $(\widetilde{\nu})^d(B) = 0$. □

LEMMA 27.12. *Let V_1 be a linear subspace of $\mathbb{R}^d$ with $1 \le \dim V_1 \le d-1$ and let T_1 be the orthogonal projector to V_1. If ν is a radially absolutely continuous measure on $\mathbb{R}^d$, then the measure νT_1^{-1} is radially absolutely continuous on V_1. If, moreover, ν satisfies the divergence condition, then νT_1^{-1} also satisfies the divergence condition.*

Proof. Denote by V_2 the orthogonal complement of V_1 in $\mathbb{R}^d$ and by T_2 the orthogonal projector to V_2. The measure ν has the representation (27.5). Let $c = \lambda(S \setminus V_2)$. If $c = 0$, then ν and νT_1^{-1} are concentrated on V_2 and $\{0\}$, respectively, and our assertion is trivial. Assume $c > 0$. We consider $(S \setminus V_2, c^{-1}[\lambda]_{S \setminus V_2})$ as a probability space and $Y(\xi) = |T_1\xi|^{-1}T_1\xi$ and $Z(\xi) = T_2\xi$ as random variables defined there. Let $P_Y(\mathrm{d}\eta)$ be the distribution of Y on $S \cap V_1$, and let $P_Z^\eta(\mathrm{d}\zeta)$ be the conditional distribution of Z given $Y = \eta$. This $P_Z^\eta(\mathrm{d}\zeta)$ is a probability measure on $\{\zeta \in V_2 \colon |\zeta| < 1\}$, uniquely determined up to η of P_Y-measure 0 [**34, 151**]. Using $\xi = T_1\xi + T_2\xi = (1-|Z|^2)^{1/2}Y + Z$, we have, for any $B \in \mathcal{B}(V_1)$ with $0 \notin B$,

$$\begin{aligned}
\nu T_1^{-1}(B) &= \int_{S \setminus V_2} \lambda(\mathrm{d}\xi) \int_0^\infty g(\xi, r) 1_B(rT_1\xi)\mathrm{d}r \\
&= c \int P_Y(\mathrm{d}\eta) \int P_Z^\eta(\mathrm{d}\zeta) \int g((1-|\zeta|^2)^{1/2}\eta + \zeta, r) 1_B(r(1-|\zeta|^2)^{1/2}\eta)\mathrm{d}r \\
&= \int_{S \cap V_1} \Lambda(\mathrm{d}\eta) \int_0^\infty G(\eta, r) 1_B(r\eta)\mathrm{d}r
\end{aligned}$$

with $\Lambda(\mathrm{d}\eta) = cP_Y(\mathrm{d}\eta)$ and

$$G(\eta, r) = \int (1-|\zeta|^2)^{-1/2} g((1-|\zeta|^2)^{1/2}\eta + \zeta, (1-|\zeta|^2)^{-1/2}r) P_Z^\eta(\mathrm{d}\zeta).$$

Thus νT_1^{-1} is radially absolutely continuous on V_1.

Suppose that, in addition, ν satisfies the divergence condition. A statement equivalent to (27.6) is that, for any $C \in \mathcal{B}(S)$, $\nu(\widetilde{C}) = 0$ or ∞, where $\widetilde{C} = \{r\xi \colon \xi \in C, 0 < r < \infty\}$. If $C \in \mathcal{B}(S \cap V_1)$, then $\nu T_1^{-1}(\widetilde{C}) = \nu\{x \colon T_1x \ne 0, |T_1x|^{-1}T_1x \in C\}$, which is 0 or ∞. Hence νT_1^{-1} satisfies the divergence condition. □

Proof of Theorem 27.10. Let (A, ν, γ) be the generating triplet of $\{X_t\}$. If $d = 1$, then $A \neq 0$ or ν is absolutely continuous with total mass infinite, and hence P_{X_t} is absolutely continuous by Theorem 27.7. Suppose that $d \geq 2$ and that our theorem is true for dimensions less than d. Let $\mu = P_{X_1}$. It suffices to prove that μ is absolutely continuous. This is because, for any $c > 0$, the Lévy process generated by $(cA, c\nu, c\gamma)$ satisfies the assumption of the theorem. If A has rank d, then μ is absolutely continuous, having a nondegenerate Gaussian as a convolution factor. So we assume that A has rank $\leq d - 1$. Then $\nu \neq 0$. If ν vanishes on every $(d-1)$-dimensional linear subspace, then μ is absolutely continuous by virtue of Theorem 27.7 and Lemma 27.11. So we assume that $\nu(V) > 0$ for some linear subspace V with $\dim V = d-1$. Define a linear subspace V_2 and a distribution μ_2 on V_2 as follows. If $A \neq 0$, then let $\widehat{\mu}_2(z) = \mathrm{e}^{-\langle z, Az\rangle/2}$ and let V_2 be the support of μ_2. If $A = 0$, then let V_2 be the smallest linear subspace that contains the support of $[\nu]_V$ and let

$$\widehat{\mu}_2(z) = \exp\left[\int_{V_2} (\mathrm{e}^{\mathrm{i}\langle z,x\rangle} - 1 - \mathrm{i}\langle z, x\rangle 1_{\{|x|\leq 1\}}(x))\nu(\mathrm{d}x)\right].$$

Let V_1 be the orthogonal complement of V_2 and let T_1 and T_2 be the orthogonal projectors to V_1 and V_2, respectively. The dimensions of V_1 and V_2 are l_1 and l_2 with $l_1 + l_2 = d$. We write $x_1 = T_1 x$ and $x_2 = T_2 x$ for $x \in \mathbb{R}^d$. Let μ_1 be the distribution such that $\mu = \mu_1 * \mu_2$. Since μ_2 is absolutely continuous with respect to the l_2-dimensional Lebesgue measure $\mathrm{d}x_2$, we have $\mu_2(\mathrm{d}x_2) = f(x_2)\mathrm{d}x_2$ with a nonnegative measurable function f. Here we have used, in the case $A = 0$, the induction hypothesis. Given $B \in \mathcal{B}(\mathbb{R}^d)$ with $\mathrm{Leb}(B) = 0$, let us prove that $\mu(B) = 0$. We have

$$\mu(B) = \int_{\mathbb{R}^d} \mu_1(\mathrm{d}y) \int_{V_2} 1_B(y_1, x_2 + y_2) f(x_2)\mathrm{d}x_2.$$

Let

$$h(y_1, y_2) = \int_{V_2} 1_B(y_1, x_2 + y_2) f(x_2)\mathrm{d}x_2,$$

which is measurable in (y_1, y_2). Since $\int_{V_1} \mathrm{d}y_1 \int_{V_2} 1_B(y_1, x_2)\mathrm{d}x_2 = 0$, there is $B_1 \in \mathcal{B}(V_1)$ such that the l_1-dimensional Lebesgue measure of B_1 is 0 and, for every $y_1 \notin B_1$, $\int_{V_2} 1_B(y_1, x_2)\mathrm{d}x_2 = 0$. Hence, for every $y_1 \notin B_1$ and $y_2 \in V_2$, we have $\int_{V_2} 1_B(y_1, x_2 + y_2)\mathrm{d}x_2 = 0$. Thus $h(y_1, y_2) = 1_{B_1}(y_1)h(y_1, y_2)$. Let Y be a random variable on $\mathbb{R}^d$ with distribution μ_1 and let $T_1 Y = Y_1$ and $T_2 Y = Y_2$. Let ρ_1 be the distribution of Y_1 and $\rho_2(\mathrm{d}y_2|y_1)$ be the conditional distribution of Y_2 given $Y_1 = y_1$. Then

$$\mu(B) = \int 1_{B_1}(y_1)h(y_1, y_2)\mu_1(\mathrm{d}y) = \int_{B_1} \rho_1(dy_1) \int_{V_2} h(y_1, y_2)\rho_2(\mathrm{d}y_2|y_1).$$

The ρ_1 is an infinitely divisible measure on V_1. The μ_1 is infinitely divisible without Gaussian part. Denote the Lévy measure of μ_1 by ν_1. By Proposition 11.10 the Lévy measure of ρ_1 is $[\nu_1 T_1^{-1}]_{V_1 \setminus \{0\}}$. If this is supported on a proper linear subspace V_1^0 of V_1, then ν_1 is supported on $V_1^0 + V_2$ and, hence, ν is also supported on $V_1^0 + V_2$, which is impossible by the nondegeneracy of μ by Proposition 24.17. It follows that ρ_1 is nondegenerate in V_1, again by Proposition 24.17. Now, by Lemma 27.12, we can apply the induction hypothesis to ρ_1. Thus ρ_1 is absolutely continuous with respect to the l_1-dimensional Lebesgue measure on V_1. Hence $\rho_1(B_1) = 0$. Therefore $\mu(B) = 0$, completing the proof. □

The absolute continuity of selfdecomposable distributions is an important application of the theorem just proved.

THEOREM 27.13. *Any nondegenerate selfdecomposable distribution on* $\mathbb{R}^d$ *is absolutely continuous.*

Proof. The Lévy measure ν of a selfdecomposable distribution has the representation (15.8) in Theorem 15.10. There we can choose the measure λ to vanish on the set $\{\xi\colon k_\xi(0+) = 0\}$ (Remark 15.12(ii)). Thus we see that ν is radially absolutely continuous, satisfying the divergence condition. Now the assumption of Theorem 27.10 is fulfilled. □

REMARK 27.14. Using Theorem 27.10, it is easy to give, on $\mathbb{R}^d$, $d \geq 2$, an absolutely continuous, purely non-Gaussian infinitely divisible distribution with singular Lévy measure. It is enough to choose λ with support $\{\xi_1, \ldots, \xi_d\}$, using linearly independent $\xi_1, \ldots, \xi_d$.

We can prove a weaker result for semi-selfdecomposable distributions. See Exercise 29.13 for an extension to non-infinitely-divisible case.

THEOREM 27.15 (Wolfe). *If* μ *is a non-trivial semi-selfdecomposable distribution on* $\mathbb{R}^d$, *then it is either absolutely continuous or continuous singular.*

Proof. Let μ be non-trivial, semi-selfdecomposable with span $b > 1$, generated by (A, ν, γ). Define $S_n(b)$ and $T_{b^{-1}}$ as in (13.10) and (13.8), respectively. We have $A \neq 0$ or $\nu \neq 0$. If $A = 0$, then $\nu(\mathbb{R}^d) = \infty$, because $\mathbb{R}^d \setminus \{0\}$ is the disjoint union of $S_n(b)$, $n \in \mathbb{Z}$, and $\nu(S_n(b)) \geq \nu(S_{n+1}(b))$ by Theorem 15.8. Hence μ is continuous by Theorem 27.4. We have

$$\mu = T_{b^{-1}}\mu * \rho \tag{27.7}$$

with some distribution ρ. By the equation (27.7), ρ determines μ uniquely, because the induction procedure gives

$$\mu = T_{b^{-n}}\mu * \rho_n \quad \text{with } \rho_n = T_{b^{-(n-1)}}\rho * \cdots * T_{b^{-1}}\rho * \rho \tag{27.8}$$

and $\rho_n \to \mu$ as $n \to \infty$. Let $\mu = \mu_{ac} + \mu_{cs}$, where μ_{ac} is absolutely continuous and μ_{cs} is continuous singular. Let c_1 and c_2 be the total measures of μ_{ac}

and μ_{cs}, respectively. We have $\mu_{ac}+\mu_{cs}=(T_{b^{-1}}\mu_{ac})*\rho+(T_{b^{-1}}\mu_{cs})*\rho$. Since $(T_{b^{-1}}\mu_{ac})*\rho$ is absolutely continuous (Lemma 27.1(iii)) with total mass c_1, $(T_{b^{-1}}\mu_{cs})*\rho$ does not have an absolutely continuous part. Hence $\mu_{ac}=(T_{b^{-1}}\mu_{ac})*\rho$. If $c_1>0$, then (27.7) holds with $c_1^{-1}\mu_{ac}$ in place of μ, and thus $c_1^{-1}\mu_{ac}=\mu$, implying $c_1=1$. Therefore, either $c_1=0$ or $c_2=0$. □

Let us consider the case of discrete Lévy measures.

THEOREM 27.16 (Hartman–Wintner). *Suppose that μ is infinitely divisible on $\mathbb{R}^d$ with generating triplet (A,ν,γ) such that $A=0$ and ν is discrete with total measure infinite. Then, μ is either absolutely continuous or continuous singular.*

We need the following two facts.

PROPOSITION 27.17. *Let X_n, $n=1,2,\ldots$, be independent random variables on $\mathbb{R}^d$. If the distribution of $\sum_{j=1}^n X_j$ is convergent as $n\to\infty$, then $\sum_{j=1}^n X_j$ converges a. s. as $n\to\infty$.*

This is one of the basic facts on sums of independent random variables and proofs in the univariate case are found in [**327**], Vol. 1, p. 263 and [**151**], p. 272. The d-variate case is proved by applying the result of the univariate case to each component.

PROPOSITION 27.18 (Jessen–Wintner law of pure types). *Let $X_1,X_2,\ldots$ be independent random variables on $\mathbb{R}^d$ such that $\sum_{j=1}^n X_j$ is convergent a. s. as $n\to\infty$. Let X be the limit random variable. If each X_j has a discrete distribution, then the distribution of X is pure, that is, discrete, absolutely continuous, or continuous singular.*

Proof. Let N be the countable set of possible values of X_n, $n=1,2,\ldots$. Choose Ω_0 with $P[\Omega_0]=1$ such that, for all $\omega\in\Omega_0$, $X_n(\omega)\in N$, $n=1,2,\ldots$, and $\sum_{n=1}^\infty X_n(\omega)=X(\omega)$. Let M be the set of points x of the form $x=\sum_{j=1}^n m_jx_j$ with $n\in\mathbb{N}$, $x_j\in N$, and $m_j\in\mathbb{Z}$. Then M is a countable set.

Case 1. Assume that P_X is not singular. Let us show that P_X is then absolutely continuous. Let B be a Borel set with $\mathrm{Leb}(B)=0$. Then $\mathrm{Leb}(B+M)=0$. Hence $P_X(B+M)<1$. Let $C=\{\omega\in\Omega_0\colon X(\omega)\in B+M\}$. Then, by the definition of M, $C=\{\omega\in\Omega_0\colon \sum_{j=n}^\infty X_j(\omega)\in B+M\}$ for each n. Hence, by Kolmogorov's 0–1 law (Theorem 1.14), $P[C]=0$ or 1. Since $P[C]=P_X(B+M)$, it follows that $P_X(B+M)=0$. Hence $P_X(B)=0$, that is, P_X is absolutely continuous.

Case 2. Assume that P_X is singular. We claim that P_X is discrete or continuous singular. Suppose that P_X is not discrete. Let B be a one-point set. Then $B+M$ is a countable set. Hence $P_X(B+M)<1$. By the same reasoning as in Case 1, we have $P_X(B+M)=0$. Hence $P_X(B)=0$. Thus P_X is continuous. □

Proof of Theorem 27.16. We may and do assume $\gamma = 0$. Let $\nu_1 = [\nu]_{\{|x|>1\}}$ and $\nu_n = [\nu]_{\{n^{-1}<|x|\le(n-1)^{-1}\}}$ for $n \ge 2$. Let μ_n be the infinitely divisible distribution with generating triplet $(0, \nu_n, 0)$. Let X_n, $n = 1, 2, \ldots$, be independent random variables with $P_{X_n} = \mu_n$. Each μ_n is discrete by Corollary 27.5. The distribution of $\sum_{j=1}^{n} X_j$ is convergent to μ as $n \to \infty$. Hence, by Proposition 27.17, $\sum_{j=1}^{\infty} X_n$ is convergent a. s. Now, using Proposition 27.18, we see that μ is pure. Since $\nu(\mathbb{R}^d) = \infty$, μ cannot be discrete, by Theorem 27.4. □

THEOREM 27.19 (Watanabe). *Let b be an integer with $b \ge 2$. Let $\{X_t\}$ be a Lévy process on $\mathbb{R}$ of type B with Lévy measure*

$$\nu = \sum_{n=-\infty}^{\infty} k_{1,n}\delta_{b^n} + \sum_{n=-\infty}^{\infty} k_{2,n}\delta_{-b^n},$$

where $k_{j,n} \ge 0$ and $\sup_{j,n} k_{j,n} < \infty$. Then, P_{X_t} is continuous singular for any $t > 0$.

Proof. We have, for $\mu = P_{X_1}$,

$$\log|\widehat{\mu}(z)^t| = t\sum_{j=1}^{2}\sum_{n=-\infty}^{\infty}(\cos b^n z - 1)k_{j,n}.$$

Let $z_m = 2\pi b^m$ for $m = 1, 2, \ldots$. Since b is an integer with $b \ge 2$, we get, using $1 - \cos u \le \frac{1}{2}u^2$ for $u \in \mathbb{R}$,

$$\log|\widehat{\mu}(z_m)^t| = -t\sum_{j=1}^{2}\sum_{n=-\infty}^{-m-1}(1 - \cos 2\pi b^{m+n})k_{j,n} \ge -2\pi^2 t(b^2-1)^{-1}\sup_{j,n} k_{j,n}.$$

If P_{X_t} is absolutely continuous, then the Riemann–Lebesgue theorem implies that $\log|\widehat{\mu}(z_m)^t|$ tends to $-\infty$, contrary to the above. Therefore P_{X_t} is not absolutely continuous. Hence, by Theorem 27.16, P_{X_t} is continous singular for any $t > 0$. □

REMARK 27.20. Choosing $k_{j,n}$ decreasing in n in Theorem 27.19, we get a semi-selfdecomposable process $\{X_t\}$ on $\mathbb{R}$. If $\{X_t^{(j)}\}$, $j = 1, \ldots, d$, are semi-selfdecomposable processes of this type and if they are independent, then the process $\{X_t\} = \{(X_t^{(j)})_{j=1,\ldots,d}\}$ on $\mathbb{R}^d$ is semi-selfdecomposable and nondegenerate. The distribution of X_t on $\mathbb{R}^d$ is continuous singular for $t > 0$, since it is the direct product of uni-variate continuous singular distributions.

REMARK 27.21. Prior to Theorem 27.19 we have not shown any example of continuous singular infinitely divisible distributions. Other examples are given in Theorem 27.23 and Exercise 29.12. Orey [**364**] shows that no condition of the form $\int_{|x|\le 1}|x|^{\alpha}\nu(dx) = \infty$ with $0 < \alpha < 2$ guarantees absolute continuity of μ on $\mathbb{R}$ (Exercise 29.12).

REMARK 27.22. Theorem 27.19 is a special case of the following fact (i) proved by Watanabe [**560**] using the idea of Erdős [**127**].

(i) Theorem 27.19 is true not only for integers b with $b \geq 2$ but also for any b in the set B_1 defined as follows: $b \in B_1$ if and only if b is in $(1, \infty)$ and there is a polynomial $F(u)$ with integer coefficients and the leading coefficient 1 such that $F(b) = 0$ and all other roots of $F(u) = 0$ satisfy $|u| < 1$. Numbers in B_1 are called Pisot–Vijayaraghavan numbers. For example, $b = n \in \mathbb{N} \setminus \{1\}$ is in B_1 with $F(u) = u - n$ and $b = (\sqrt{5}+1)/2 \in B_1$ with $F(u) = u^2 - u - 1$.

Further he shows the following. Let $\{X_t\}$ be a semi-selfdecomposable process of type B on $\mathbb{R}$ with span b and Lévy measure $\nu = \sum_{n\leq 0} k_n \delta_{b^n}$ (hence $k_n \geq k_{n+1}$ for $n \leq 0$). Let $c = \lim_{n\to-\infty} k_n \leq \infty$.

(ii) Let b be arbitrary in $(1, \infty)$. If $c < \infty$, then P_{X_t} is continuous singular for any t satisfying $0 < t < h^{-1}(\log b)/c$, where h^{-1} is the inverse function of the entropy function (Exercise 29.24) of the Poisson process with parameter 1.

(iii) Let $b \in B_1$. Then P_{X_t} is absolutely continuous with a bounded continuous density for every $t > 0$ if and only if $\sum_{n\leq 0} \mathrm{e}^{-tk_n} < \infty$ for every $t > 0$.

(iv) Define a subset B_2 of $(1, \infty)$ as follows. Let $Y_0, Y_1, \ldots$ be independent random variables such that $P[\,Y_n = 0\,] = p$ and $P[\,Y_n = b^{-n}\,] = 1-p$ with $0 < p < 1$. Let $\mu_{b,p}$ be the distribution of $Y = \sum_{n=0}^{\infty} Y_n$. The set B_2 is the totality of $b > 1$ such that there exist $p \in (\frac{1}{2}, 1)$ and $n \in \mathbb{N}$ satisfying $\int_{-\infty}^{\infty} |\widehat{\mu}_{b,p}(z)|^n \mathrm{d}z < \infty$. If $b \in B_2$ and $c < \infty$, then P_{X_t} is absolutely continuous with a bounded continuous density for all sufficiently large t. If $b \in B_2$ and $c = \infty$, then P_{X_t} is absolutely continuous with a bounded continuous density for every $t > 0$. Watanabe calls b^{-1} a Peres–Solomyak number if $b \in B_2$. We have $B_1 \cap B_2 = \emptyset$. By using [**369**] it is shown that $(1, \infty) \setminus B_2$ has Lebesgue measure zero.

The facts in Remark 27.22 show that absolute continuity and continuous singularity are time dependent distributional properties in the class of Lévy processes. Let us construct another example and prove this assertion. It is a subordinator having a discrete Lévy measure of a special type with infinite total mass. According to Tucker [**540**], this is H. Rubin's construction. We follow the paper [**540**] but, since Tucker is not dealing with Lévy processes, a slight modification related to the condition (27.10) is necessary. Let $a_1, a_2, \ldots$ be a sequence of positive reals satisfying

$$\sum_{n=1}^{\infty} a_n = \infty. \tag{27.9}$$

Choose positive integers $b_1, b_2, \ldots$ satisfying

$$\sum_{n=1}^{\infty} a_n b_n^{-1} < \infty, \tag{27.10}$$

and then choose integers $c_0 = 0 < c_1 < c_2 < \cdots$ such that

$$\sum_{n=0}^{\infty} 2^{c_n} \sum_{j=n+1}^{\infty} b_{j+1} 2^{-c_j} < \infty. \tag{27.11}$$

For $n = 1, 2, \ldots$, let ρ_n be the discrete uniform distribution on the set

$$E_n = \{k 2^{-c_n} \colon k = 0, 1, \ldots, 2^{c_n - c_{n-1}} - 1\}.$$

Let ν be the restriction of $\sum_{n=1}^{\infty} a_n \rho_n$ to $(0,\infty)$. Then ν is a discrete measure on $(0,1)$ such that $\int x\nu(dx) < \infty$ and $\nu(0,1) = \infty$. In fact,

$$\begin{aligned}\textstyle\int x\nu(dx) &= \textstyle\sum_{n=1}^{\infty} a_n \sum_{k=1}^{2^{c_n-c_{n-1}}-1} k 2^{-c_n} 2^{-c_n+c_{n-1}} \\ &\le \textstyle\frac{1}{2}\sum_{n=1}^{\infty} a_n 2^{c_{n-1}-2c_n}(2^{c_n-c_{n-1}})^2 \le \mathrm{const}\sum_{n=1}^{\infty} b_n 2^{-c_{n-1}} < \infty\end{aligned}$$

by (27.10) and (27.11), and

$$\nu(0,1) = \textstyle\sum_{n=1}^{\infty} a_n(1-2^{-c_n+c_{n-1}}) \ge \frac{1}{2}\sum_{n=1}^{\infty} a_n = \infty.$$

THEOREM 27.23. *Let $\{X_t\}$ be the subordinator having the measure ν given above as its Lévy measure. Then, P_{X_t} is continuous singular for any $t > 0$ satisfying $\sum_{n=1}^{\infty} e^{-ta_n} = \infty$; P_{X_t} is absolutely continuous for any $t > 0$ satisfying $\sum_{n=1}^{\infty} e^{-ta_n} < \infty$.*

REMARK 27.24. For any sequence $\{a_n\}$ of positive reals satisfying (27.9), we can find $\{b_n\}$ and $\{c_n\}$ satisfying (27.10) and (27.11). In fact, it is obvious that $\{b_n\}$ can be chosen to satisfy (27.10). Then, fix a sequence $\varepsilon_1, \varepsilon_2, \ldots$ of positive reals with $\sum_{n=1}^{\infty} \varepsilon_n < \infty$. Choose integers $0 < c_1^{(1)} < c_2^{(1)} < \cdots$ such that $\sum_{j=1}^{\infty} b_{j+1} 2^{-c_j^{(1)}} \le \varepsilon_1$ and let $c_1 = c_1^{(1)}$. Next choose integers $0 < c_2^{(2)} < c_3^{(2)} < \cdots$ such that $c_n^{(2)} \ge c_n^{(1)}$ $(n = 2,3,\ldots)$ and $2^{c_1}\sum_{j=2}^{\infty} b_{j+1} 2^{-c_j^{(2)}} \le \varepsilon_2$ and let $c_2 = c_2^{(2)}$. Next choose integers $0 < c_3^{(3)} < c_4^{(3)} < \cdots$ such that $c_n^{(3)} \ge c_n^{(2)}$ $(n = 3,4,\ldots)$ and $2^{c_2}\sum_{j=3}^{\infty} b_{j+1} 2^{-c_j^{(3)}} \le \varepsilon_3$ and let $c_3 = c_3^{(3)}$. Continue this procedure.

Special choices of $\{a_n\}$ give the following processes.

1. Take $a_n = \log\log(n \vee 3)$. Then P_{X_t} is continuous singular for any $t > 0$. In fact, $\sum_{n=3}^{\infty} e^{-ta_n} = \sum_{n=3}^{\infty} (\log n)^{-t} = \infty$ for $t > 0$.

2. Take $a_n = \log(n \vee 2)$. Then, P_{X_t} is continuous singular for $0 < t \le 1$, and is absolutely continuous for $t > 1$. We have $\sum_{n=2}^{\infty} e^{-ta_n} = \sum_{n=2}^{\infty} n^{-t} = \infty$ or $< \infty$ according as $0 < t \le 1$ or $t > 1$.

3. Take $a_n = \log(n \vee 2) + 2\log\log(n \vee 3)$. Then, P_{X_t} is continuous singular for $0 < t < 1$, and is absolutely continuous for $t \ge 1$. In fact, $\sum_{n=3}^{\infty} e^{-ta_n} = \sum_{n=3}^{\infty} (n(\log n)^2)^{-t} = \infty$ or $< \infty$ according as $0 < t < 1$ or $t \ge 1$.

4. Take $a_n = n$. Then, P_{X_t} is absolutely continuous for any $t > 0$, since $\sum_{n=1}^{\infty} e^{-ta_n} = \sum_{n=1}^{\infty} e^{-tn} < \infty$ for $t > 0$.

REMARK 27.25. The subordinator $\{X_t\}$ in Theorem 27.23 is given as follows. Let V_{jk} $(j,k = 1,2,\ldots)$ be independent identically distributed random variables with $P[V_{jk} = 0] = P[V_{jk} = 1] = 1/2$. Let $\{Y_t^{(n)} \colon t \ge 0\}$ $(n = 1,2,\ldots)$ be independent Poisson processes with $E[Y_t^{(n)}] = a_n t$. Suppose, further, that $\{Y_t^{(n)} \colon t \ge 0\}$ $(n = 1,2,\ldots)$ and V_{jk} $(j,k = 1,2,\ldots)$ are independent. Let

$$Z_{jn} = \textstyle\sum_{k=c_{n-1}+1}^{c_n} 2^{-k} V_{jk}, \quad X_t^{(n)} = \sum_{j=1}^{Y_t^{(n)}} Z_{jn},$$

where $\sum_{j=1}^{0}$ is understood to be zero. Considering the distribution function, it is not hard to see that each Z_{jn} has distribution ρ_n. The process $\{X_t^{(n)} \colon t \ge 0\}$ is an increasing compound Poisson process with Lévy measure $[a_n\rho_n]_{(0,1)}$, that is,

$$E[\exp(-uX_t^{(n)})] = \exp\left[ta_n \textstyle\int_{\mathbb{R}}(e^{-ux}-1)\rho_n(dx)\right], \quad u \geq 0.$$

Now let

$$X_t = \textstyle\sum_{n=1}^{\infty} X_t^{(n)}.$$

The infinite series in the right-hand side converges a. s. and we get the process $\{X_t\}$ with drift 0 in Theorem 27.23. Notice that $\sum_{n=1}^{\infty} Z_{jn}$ converges a. s. and has uniform distribution on $[0,1]$, because, if $\Omega_0 = [0,1)$, P_0 =Lebesgue measure, and $U_n(\omega) = 1_{B_n}(\omega)$, $\omega \in \Omega_0$, with B_n being the union of $[\frac{2k-1}{2^n}, \frac{2k}{2^n})$ for $k = 1, \ldots, 2^{n-1}$, then $\omega = \sum_{n=1}^{\infty} 2^{-n}U_n(\omega)$ is the binary expansion of ω and this has the same distribution as $\sum_{n=1}^{\infty} Z_{jn}$.

We prepare two lemmas.

LEMMA 27.26. *Let X and Y be random variables on $\mathbb{R}$ with distributions μ and ρ, respectively. Let $\mu = \mu_d + \mu_{ac} + \mu_{cs}$ and $\rho = \rho_d + \rho_{ac} + \rho_{cs}$ be their Lebesgue decompositions. If $Y-X$ has a discrete distribution, then $\mu_i(\mathbb{R}) = \rho_i(\mathbb{R})$ for $i = d$, ac, cs.*

Proof. When $F(x)$ is a distribution function on $\mathbb{R}$, let us denote by $M_i(F)$ the total measure of the i-part ($i = d$, ac, cs). Let $Z = Y - X$ and let $\{z_n\}$ be the possible values of Z. Then,

$$P[X \leq x] = \textstyle\sum_n P[X \leq x \mid Z = z_n]\, P[Z = z_n],$$
$$P[Y \leq y] = \textstyle\sum_n P[X \leq y - z_n \mid Z = z_n]\, P[Z = z_n].$$

Hence,

$$\begin{aligned}\rho_i(\mathbb{R}) &= \textstyle\sum_n M_i(P[X \leq \cdot - z_n \mid Z = z_n])\, P[Z = z_n] \\ &= \textstyle\sum_n M_i(P[X \leq \cdot \mid Z = z_n])\, P[Z = z_n] = \mu_i(\mathbb{R})\end{aligned}$$

for $i = d$, ac, cs. □

LEMMA 27.27. *Suppose that X_n and Y_n, $n = 1, 2, \ldots$, are random variables on $\mathbb{R}$ such that $\sum_{n=1}^{\infty} P[X_n \neq Y_n] < 1$ and $\sum_{n=1}^{\infty} X_n$ is convergent a. s. If $\sum_{n=1}^{\infty} X_n$ has a singular distribution, then the distribution of $\sum_{n=1}^{\infty} Y_n$ is not absolutely continuous.*

Proof. By the Borel–Cantelli lemma $X_n = Y_n$ for all large n with probability 1. Hence $\sum_{n=1}^{\infty} Y_n$ converges a. s. Let $B = \{X_n = Y_n \text{ for all } n\}$. Then

$$P[B] = 1 - P[X_n \neq Y_n \text{ for some } n] \geq 1 - \textstyle\sum_n P[X_n \neq Y_n] > 0.$$

If the distribution of $\sum_{n=1}^{\infty} X_n$ is singular, then we can find $C \in \mathcal{B}(\mathbb{R})$ with $\mathrm{Leb}(C) = 0$ such that $P[\sum_n X_n \in C] = 1$, and hence

$$P[\textstyle\sum_n Y_n \in C] \geq P[B \cap \{\sum_n Y_n \in C\}] = P[B \cap \{\sum_n X_n \in C\}] = P[B] > 0,$$

completing the proof. □

Proof of Theorem 27.23. Use the notation in Remark 27.25. Since the Lévy measure ν is discrete with infinite total mass, the distribution of X_t for any fixed

$t > 0$ is absolutely continuous or continuous singular by the Hartman–Wintner theorem.

Step 1. Fix $t > 0$ such that $\sum_n \mathrm{e}^{-ta_n} < \infty$. We claim that X_t has an absolutely continuous distribution. Since $\sum_n P[\,Y_t^{(n)} = 0\,] = \sum_n \mathrm{e}^{-ta_n} < \infty$, the Borel–Cantelli lemma says that $Y_t^{(n)} > 0$ for all large n with probability 1. Define

$$\Omega_0 = \bigcap_{k=1}^{\infty}\{Y_t^{(k)} > 0\},$$
$$\Omega_n = \{Y_t^{(n)} = 0\} \cap \bigcap_{k=n+1}^{\infty}\{Y_t^{(k)} > 0\} \quad \text{for } n = 1, 2, \ldots.$$

These are disjoint events and the sum of their probabilities is 1. Let $S_n = \sum_{k=n+1}^{\infty} Z_{1k}$ and $T_n = X_t - S_n$. As $\sum_{k=1}^{\infty} Z_{1k}$ has an absolutely continuous distribution, so does S_n by Lemma 27.26. Let $n \geq 0$ and $B, C \in \mathcal{B}(\mathbb{R})$ be arbitrary. Obviously, $\{S_n \in B\}$ and Ω_n are independent events. We have

$$\Omega_n \cap \{T_n \in C\} = \Omega_n \cap \Big\{\sum_{k=1}^{n-1}\sum_{j=1}^{Y_t^{(k)}} Z_{jk} + \sum_{k=n+1}^{\infty}\sum_{j=2}^{Y_t^{(k)}} Z_{jk} \in C\Big\},$$

where $\sum_{j=2}^{1}$ is understood to be 0. Hence, $\{S_n \in B\}$ and $\Omega_n \cap \{T_n \in C\}$ are independent. Thus

$$\begin{aligned}
P[\,S_n \in B,\, T_n \in C \,|\, \Omega_n] &= P[\,S_n \in B,\, T_n \in C,\, \Omega_n]/P[\Omega_n] \\
&= P[\,S_n \in B\,]\, P[\,T_n \in C,\, \Omega_n]/P[\Omega_n] \\
&= P[\,S_n \in B \,|\, \Omega_n]\, P[\,T_n \in C \,|\, \Omega_n],
\end{aligned}$$

that is, S_n and T_n are independent conditionally on Ω_n. Since $P[\,S_n \in \cdot \,|\, \Omega_n] = P[\,S_n \in \cdot\,]$ is absolutely continuous, $P[\,S_n + T_n \in \cdot \,|\, \Omega_n]$ is absolutely continuous by Lemma 27.1(iii). As

$$P[\,X_t \in \cdot\,] = \sum_{n=0}^{\infty} P[\,X_t \in \cdot \,|\, \Omega_n]\, P[\Omega_n],$$

the distribution of X_t is absolutely continuous.

Step 2. Fix $t > 0$ with $\sum_n \mathrm{e}^{-ta_n} = \infty$. We prove that P_{X_t} is continuous singular. Since $\nu(0,1) = \infty$, the continuity follows from Theorem 27.4. Define

$$\overline{Y}_t^{(n)} = b_n \wedge Y_t^{(n)}, \quad \overline{X}_t^{(n)} = \sum_{j=1}^{\overline{Y}_t^{(n)}} Z_{jn}, \quad \overline{X}_t = \sum_{n=1}^{\infty} \overline{X}_t^{(n)}.$$

Then

$$P[\,\overline{X}_t^{(n)} \neq X_t^{(n)}\,] \leq P[\,Y_t^{(n)} > b_n\,] \leq b_n^{-1} E[Y_t^{(n)}] = ta_n b_n^{-1}.$$

By the conditions (27.10) and (27.11), there is $m \geq 1$ such that $\sum_{n=m}^{\infty} ta_n b_n^{-1} < 1$ and $\sum_{j=m-1}^{\infty} b_{j+1} 2^{-c_j} < 1$. Write $W = \sum_{n=m}^{\infty} X_t^{(n)}$ and $\overline{W} = \sum_{n=m}^{\infty} \overline{X}_t^{(n)}$, suppressing m and t. It is enough to prove that $\overline{W}$ has a singular distribution. For, if $P_{\overline{W}}$ is singular, then P_W is not absolutely continuous, by Lemma 27.27, and, since each $X_t^{(n)}$ has a discrete distribution, the distribution of $X_t = W + \sum_{n=1}^{m-1} X_t^{(n)}$ is not absolutely continuous, by Lemma 27.26. Let us construct, for any $\eta > 0$, a Borel set B_η such that $\mathrm{Leb}(B_\eta) \leq \eta$ and $P[\,\overline{W} \in B_\eta\,] = 1$. This will finish the proof, as $B = \bigcap_{n=1}^{\infty} B_{1/n}$ has Lebesgue measure 0 and $P[\,\overline{W} \in B\,] = 1$.

Let $r_n = \sum_{j=n}^{\infty} b_{j+1} 2^{-c_j}$ and $\varepsilon_n = 2^{c_{n-1}} r_n$. Then $\sum_{n=1}^{\infty} \varepsilon_n < \infty$ by the condition (27.11). Fix $N \geq m$ such that $\sum_{n=N}^{\infty} \varepsilon_n < \eta$, and let

$$\Omega' = \{\overline{Y}_t^{(n)} = 0 \text{ for some } n \geq N\},$$

$$\Omega_n = \{\overline{Y}_t^{(l)} > 0 \text{ for } N \leq l \leq n-1,\ \overline{Y}_t^{(n)} = 0\} \quad \text{for } n = N, N+1, \ldots.$$

Then Ω_n, $n \geq N$, are disjoint and their union is Ω'. Since $\sum_{n=1}^{\infty} P[\,Y_t^{(n)} = 0\,] = \sum_{n=1}^{\infty} \mathrm{e}^{-ta_n} = \infty$, $Y_t^{(n)} = 0$ infinitely often with probability 1 by the Borel–Cantelli lemma. Thus $P[\Omega'] = 1$. We have

$$\overline{W} 1_{\Omega_n} = (\overline{X}_t^{(m)} + \cdots + \overline{X}_t^{(n-1)}) 1_{\Omega_n} + (\overline{X}_t^{(n+1)} + \overline{X}_t^{(n+2)} + \cdots) 1_{\Omega_n}.$$

We can check that

$$(27.12) \qquad (\overline{X}_t^{(m)} + \cdots + \overline{X}_t^{(n-1)}) 1_{\Omega_n} \in \{k 2^{-c_{n-1}} \colon k = 0, \ldots, 2^{c_{n-1}} - 1\} \quad \text{a. s.},$$

$$(27.13) \quad (\overline{X}_t^{(n+1)} + \overline{X}_t^{(n+2)} + \cdots) 1_{\Omega_n} \in [0, r_n) \quad \text{a. s.}$$

In fact, since $Z_{jl} \in E_l$ a. s. for every j and l, we have

$$\overline{X}_t^{(l)} \in \{k 2^{-c_l} \colon k = 0, \ldots, (2^{c_l - c_{l-1}} - 1) b_l\} \quad \text{a. s.}$$

Thus $(\overline{X}_t^{(m)} + \cdots + \overline{X}_t^{(n-1)}) 1_{\Omega_n}$ takes values in $2^{-c_{n-1}} \mathbb{Z}_+ \cap [0, 1)$, since $2^{-c_{m-1}} b_m + \cdots + 2^{-c_{n-2}} b_{n-1} < \sum_{j=m-1}^{\infty} b_{j+1} 2^{-c_j} < 1$. Likewise, $\overline{X}_t^{(n+1)} + \overline{X}_t^{(n+2)} + \cdots < 2^{-c_n} b_{n+1} + 2^{-c_{n+1}} b_{n+2} + \cdots = r_n$. Hence (27.12) and (27.13) are true. Therefore

$$\overline{W} 1_{\Omega_n} \in F_n \text{ a. s. with } F_n = \bigcup_{k=0}^{2^{c_{n-1}}-1} [k 2^{-c_{n-1}}, k 2^{-c_{n-1}} + r_n),$$

and $\mathrm{Leb}(F_n) \leq r_n 2^{c_{n-1}} = \varepsilon_n$. Let $B_\eta = \bigcup_{n=N}^{\infty} F_n$. Then $\mathrm{Leb}(B_\eta) \leq \sum_{n=N}^{\infty} \varepsilon_n < \eta$ and $\overline{W} = \sum_{n=N}^{\infty} \overline{W} 1_{\Omega_n} \in B_\eta$ a. s. The proof is complete. □

We add a property of discrete distributions.

PROPOSITION 27.28. *If μ is a discrete probability measure on $\mathbb{R}^d$, then* $\limsup_{r \to \infty} |\widehat{\mu}(r z_0)| = 1$ *for any* $z_0 \in \mathbb{R}^d$.

Proof. If X is a random variable with a discrete distribution μ, then $\langle z_0, X \rangle$ is discrete and $\widehat{\mu}(r z_0) = E[\mathrm{e}^{\mathrm{i} r \langle z_0, X \rangle}]$. Hence, it is enough to prove the proposition for $d = 1$. If the assertion is proved for probability measures on $\mathbb{R}$ with finite supports, then it is true for all discrete probability measures on $\mathbb{R}$. In fact, letting x_j, $j = 1, 2, \ldots$, be all points with positive mass $\mu\{x_j\} = c_j$, we have

$$\left|\widehat{\mu}(z) - \sum_{j=1}^{n} c_j \mathrm{e}^{\mathrm{i} z x_j}\right| \leq \sum_{j=n+1}^{\infty} c_j,$$

which is arbitrarily small, and $\limsup_{z \to \infty} \left|\sum_{j=1}^{n} c_j \mathrm{e}^{\mathrm{i} z x_j}\right| = \sum_{j=1}^{n} c_j$. The proof of the assertion in the case of finite support is divided into three steps.

Step 1. Following Bohr [**52**], we say that a complex-valued continuous function $\varphi(z)$ on $\mathbb{R}$ is *almost periodic* if, for any $\varepsilon > 0$, there is an $L(\varepsilon) > 0$ such that every interval of length $L(\varepsilon)$ contains a point τ satisfying

$$(27.14) \qquad \sup_{z \in \mathbb{R}} |\varphi(z + \tau) - \varphi(z)| \leq \varepsilon.$$

Such a τ is called an *ε-translation number* for φ. If φ is periodic with period τ, then it is almost periodic, because $\sup_{z \in \mathbb{R}} |\varphi(z + n\tau) - \varphi(z)| = 0$ for any $n \in \mathbb{Z}$.

Any almost periodic function is uniformly continuous and bounded. But we do not need this fact.

Step 2. If $\varphi(z)$ is uniformly continuous and almost periodic, then, for any $\varepsilon > 0$, there are $\widetilde{L}(\varepsilon) > 0$ and $\delta(\varepsilon) > 0$ such that every interval of length $\widetilde{L}(\varepsilon)$ contains an interval of length $\delta(\varepsilon)$ all points of which are ε-translation numbers for φ. To see this, choose $\delta(\varepsilon)$ such that $|\varphi(z_1) - \varphi(z_2)| < \varepsilon/2$ whenever $|z_1 - z_2| < \delta(\varepsilon)/2$. Let $\widetilde{L}(\varepsilon) = L(\frac{\varepsilon}{2}) + \delta(\varepsilon)$. Given an interval $(a, a + \widetilde{L}(\varepsilon))$, let τ be an $\frac{\varepsilon}{2}$-translation number in $(a + \frac{\delta(\varepsilon)}{2}, a + \frac{\delta(\varepsilon)}{2} + L(\frac{\varepsilon}{2}))$. If $|b| < \frac{\delta(\varepsilon)}{2}$, then $\tau + b$ is an ε-translation number, since

$$|\varphi(z + \tau + b) - \varphi(z)| \le |\varphi(z + \tau + b) - \varphi(z + \tau)| + |\varphi(z + \tau) - \varphi(z)| \le \varepsilon.$$

Step 3. Let us prove that, if φ_1 and φ_2 are uniformly continuous, almost periodic functions, then $\varphi_1 + \varphi_2$ is again uniformly continuous and almost periodic. The uniform continuity of $\varphi_1 + \varphi_2$ is obvious. To see the almost-periodicity, it is enough to show that, for each $\varepsilon > 0$, there is $L > 0$ such that every interval of length L contains a common 2ε-translation number τ for φ_1 and φ_2. In fact,

$$\begin{aligned}&|(\varphi_1(z + \tau) + \varphi_2(z + \tau)) - (\varphi_1(z) + \varphi_2(z))|\\ &\qquad \le |\varphi_1(z + \tau) - \varphi_1(z)| + |\varphi_2(z + \tau) + \varphi_2(z))| \le 4\varepsilon,\end{aligned}$$

that is, τ is a 4ε-translation number for $\varphi_1 + \varphi_2$. Denote $\widetilde{L}(\varepsilon)$ and $\delta(\varepsilon)$ in Step 2 for φ_k by L_k and δ_k, where $k = 1, 2$. Let $L_0 = L_1 \vee L_2$. Fix $0 < \eta < \delta_1 \wedge \delta_2$. Any interval of length L_0 contains, by Step 2, an ε-translation number τ_k for φ_k expressible as an integer multiple of η. Writing $\tau_1 = n'\eta$ and $\tau_2 = n''\eta$, we have $\tau_1 - \tau_2 = (n' - n'')\eta = n\eta$ with $|n|\eta < L_0$. There can be only finitely many distinct values of $n\eta$. Denote them by $n_1\eta, \ldots, n_N\eta$. For $j = 1, \ldots, N$ choose and fix $\tau_1^{(j)}$ and $\tau_2^{(j)}$ such that $\tau_1^{(j)} - \tau_2^{(j)} = n_j\eta$. Let $\max_{j=1,\ldots,N} |\tau_1^{(j)}| = T$ and let $L = L_0 + 2T$. We claim that every interval $(a, a + L)$ contains a common 2ε-translation number τ for φ_1 and φ_2. The interval $(a + T, a + L_0 + T)$ contains ε-translation numbers $\tau_1 = n'\eta$ and $\tau_2 = n''\eta$ for φ_1 and φ_2, respectively. Then, for some j, $\tau_1 - \tau_2 = n_j\eta = \tau_1^{(j)} - \tau_2^{(j)}$. Look at the number $\tau = \tau_1 - \tau_1^{(j)} = \tau_2 - \tau_2^{(j)}$. First, τ is a 2ε-translation number for φ_k, $k = 1, 2$, because

$$|\varphi_k(z+\tau) - \varphi_k(z)| \le |\varphi_k(z+\tau_k - \tau_k^{(j)}) - \varphi_k(z - \tau_k^{(j)})| + |\varphi_k(z - \tau_k^{(j)}) - \varphi_k(z)| \le 2\varepsilon.$$

Second, $a < \tau < a + L = a + L_0 + 2T$, because $a + T < \tau_1 < a + L_0 + T$ and $|\tau_1^{(j)}| \le T$. This finishes Step 3.

Now, consider $\widehat{\mu}(z) = \sum_{j=1}^{n} c_j e^{izx_j}$. Each term $c_j e^{izx_j}$ is uniformly continuous and periodic. Therefore, applying Step 3, we see that $\widehat{\mu}(z)$ is almost periodic. For any $\varepsilon > 0$ and $l \in \mathbb{N}$ we can find $\tau \in (l, l + L(\varepsilon))$ such that $1 - |\widehat{\mu}(\tau)| \le |\widehat{\mu}(0) - \widehat{\mu}(\tau)| \le \varepsilon$. It follows that $\limsup_{z\to\infty} |\widehat{\mu}(z)| = 1$. □

28. Smoothness

If the characteristic function $\widehat{\mu}(z)$ of a probability measure μ on $\mathbb{R}^d$ is integrable, then μ has a continuous density $g(x)$ that tends to 0 as $|x| \to \infty$

(Proposition 2.5(xii) and the Riemann–Lebesgue theorem). Rapid decrease of $\widehat{\mu}(z)$ as $|z| \to \infty$ implies smoothness of $g(x)$.

PROPOSITION 28.1. *If a probability measure μ on $\mathbb{R}^d$ satisfies*

$$\int_{\mathbb{R}^d} |\widehat{\mu}(z)||z|^n \mathrm{d}z < \infty \tag{28.1}$$

for some $n \in \mathbb{Z}_+$, then μ has a density $g(x)$ of class C^n and the partial derivatives of $g(x)$ of orders $0, \ldots, n$ tend to 0 as $|x| \to \infty$.

Proof. By Proposition 2.5(xii), the function $g(x)$ defined by

$$g(x) = (2\pi)^{-d} \int_{\mathbb{R}^d} \mathrm{e}^{-\mathrm{i}\langle x,z\rangle} \widehat{\mu}(z)\mathrm{d}z$$

is the density of μ. The right-hand side is n times differentiable with respect to x and differentiation under the integral sign is possible by virtue of (28.1). Thus g is of class C^n and the Riemann–Lebesgue theorem applies. □

EXAMPLE 28.2. If μ is a nondegenerate semi-stable distribution on $\mathbb{R}^d$, then it has a density $g(x)$ of class C^∞ and the partial derivative of $g(x)$ of any order tends to 0 as $|x| \to \infty$. A nondegenerate stable distribution is a special case. To see this, let α be the index. By Proposition 24.20 there is $K > 0$ such that $|\widehat{\mu}(z)| \le \mathrm{e}^{-K|z|^\alpha}$ on $\mathbb{R}^d$. Hence (28.1) follows for every $n \in \mathbb{N}$. As a special case, if μ is semi-stable on $\mathbb{R}$ with support $[0, \infty)$, then it has a C^∞ density, all derivatives of which vanish at $x = 0$.

The following result is known for $d = 1$.

PROPOSITION 28.3 (Orey [**364**]). *Let μ be an infinitely divisible distribution on $\mathbb{R}$ with Lévy measure ν satisfying*

$$\liminf_{r\downarrow 0} \frac{\int_{[-r,r]} x^2 \nu(\mathrm{d}x)}{r^{2-\alpha}} > 0 \tag{28.2}$$

for some $0 < \alpha < 2$. Then μ has a density of class C^∞ and all derivatives of the density tend to 0 as $|x| \to \infty$.

Proof. There is $c_1 > 0$ such that $\int_{[-r,r]} x^2 \nu(\mathrm{d}x) \ge c_1 r^{2-\alpha}$ for all small $r > 0$. Hence, using $1 - \cos u = 2(\sin \frac{u}{2})^2 \ge 2(\frac{u}{\pi})^2$ for $|u| \le \pi$, we see that there are constants c_2, c_3 such that

$$\begin{aligned} |\widehat{\mu}(z)| &\le \exp \int_{\mathbb{R}} (\cos zx - 1)\nu(\mathrm{d}x) \\ &\le \exp\Big[-c_2 \int_{|x|\le\pi/|z|} z^2 x^2 \nu(\mathrm{d}x)\Big] \le \exp(-c_3 |z|^\alpha) \end{aligned}$$

for all large $|z|$. Thus (28.1) holds for every $n \in \mathbb{Z}_+$. □

Smoothness of selfdecomposable distributions on $\mathbb{R}$ can be fully analyzed.

THEOREM 28.4. *Let μ be a selfdecomposable distribution on $\mathbb{R}$ with*

$$(28.3)\qquad \widehat{\mu}(z) = \exp\left[\int_{\mathbb{R}} (\mathrm{e}^{\mathrm{i}zx} - 1 - \mathrm{i}zx 1_{[-1,1]}(x)) \frac{k(x)}{|x|} \mathrm{d}x + \mathrm{i}\gamma z\right],$$

where $k(x)$ is a nonnegative function right-continuous and increasing on $(-\infty, 0)$, and left-continuous and decreasing on $(0, \infty)$. Let $c = k(0+) + k(0-) > 0$.

(i) *Suppose $c < \infty$. Define $N \in \mathbb{Z}_+$ by $N < c \le N + 1$. Let γ_0 be the drift of μ. Then μ has a density $f(x)$ continuous on $\{x \neq \gamma_0\}$ and the function $g(x)$ defined by*

$$(28.4)\qquad g(x) = \begin{cases} (x - \gamma_0) f(x), & x \neq \gamma_0, \\ 0, & x = \gamma_0, \end{cases}$$

is continuous on $\mathbb{R}$. If $c \le 1$, then $f(x)$ cannot be extended to a continuous function on $\mathbb{R}$. If $c > 1$, then $f(x)$ is extended to a C^{N-1} function on $\mathbb{R}$ and $g(x)$ is a C^N function on $\mathbb{R}$.

(ii) *If $c = \infty$, then μ has a C^∞ density on $\mathbb{R}$.*

Note that (28.3) is a general form of a purely non-Gaussian selfdecomposable distribution on $\mathbb{R}$ (Corollary 15.11).

LEMMA 28.5. *Let μ be as in Theorem 28.4. Then $|\widehat{\mu}(z)| = o(|z|^{-\alpha})$ as $|z| \to \infty$ for any α with $0 < \alpha < c$.*

Proof. Step 1. Assume that $\widehat{\mu}(z) = \exp\left[\int_0^b (\mathrm{e}^{\mathrm{i}zx} - 1) c x^{-1} \mathrm{d}x\right]$ with some $0 < b < \infty$. If $z > 1/b$, then

$$|\widehat{\mu}(z)| = \exp\left[\int_0^{bz} (\cos y - 1) c y^{-1} \mathrm{d}y\right] \le \exp\left[\int_1^{bz} c y^{-1} \cos y \, \mathrm{d}y - \int_1^{bz} c y^{-1} \mathrm{d}y\right].$$

Since $\int_1^u y^{-1} \cos y \, \mathrm{d}y$ is convergent as $u \to \infty$, we get $|\widehat{\mu}(z)| \le \text{const}\, z^{-c}$ for $z > 1/b$. If $z < -1/b$, then $|\widehat{\mu}(z)| = |\widehat{\mu}(-z)| \le \text{const}\, |z|^{-c}$.

Step 2. Assume that $k(0-) = 0$. Then $k(0+) = c > \alpha > 0$. Let $\alpha < \beta < c$. Choose ε such that $k(x) > \beta$ for $0 < x < \varepsilon$. Define μ_1 by $\widehat{\mu}_1(z) = \exp\left[\int_0^\varepsilon (\mathrm{e}^{\mathrm{i}zx} - 1) \beta x^{-1} \mathrm{d}x\right]$. Then μ_1 is a convolution factor of μ. Hence, using Step 1 for μ_1,

$$|\widehat{\mu}(z)| \le |\widehat{\mu}_1(z)| \le \text{const}\, |z|^{-\beta} = o(|z|^{-\alpha}).$$

The case that $k(0+) = 0$ is handled in the same way.

Step 3. Suppose $k(0+) > 0$ and $k(0-) > 0$. Letting μ_1 and μ_2 be defined by restricting the Lévy measure to $(0, \infty)$ and $(-\infty, 0)$, respectively, we have $\mu = \mu_1 * \mu_2$. Apply the result of Step 2 to μ_1 and μ_2. □

Proof of Theorem 28.4. We know, by Example 27.8, that μ is absolutely continuous. If $1 < c < \infty$, then, by Proposition 28.1 and Lemma 28.5, μ has a C^{N-1} density. Moreover, if $c = \infty$, then μ has a C^∞ density for the same reason. Define a measure ρ on $\mathbb{R}$ by $\rho(-\infty, x] = k(x)$ for $x < 0$, $\rho[x, \infty) = k(x)$ for $x > 0$, and $\rho\{0\} = 0$.

Step 1. Let us show that

(28.5) $$\widehat{\mu}(z) = \exp\bigl[\textstyle\int_{\mathbb{R}}\bigl(\int_0^{zx}(\mathrm{e}^{\mathrm{i}u}-1)u^{-1}\mathrm{d}u - \mathrm{i}z\tau(x)\bigr)\rho(\mathrm{d}x) + \mathrm{i}\gamma z\bigr],$$

where

$$\tau(x) = x1_{[-1,1]}(x) - 1_{(-\infty,-1)}(x) + 1_{(1,\infty)}(x).$$

Since $\int_{|x|\le 1}|x|k(x)\mathrm{d}x < \infty$ and $\int_{|x|>1}|x|^{-1}k(x)\mathrm{d}x < \infty$, it follows from Lemma 17.6 that $\lim_{x\to 0} x^2k(x) = 0$ and $\lim_{|x|\to\infty}(\log|x|)k(x) = 0$. Fix $z \ne 0$ and let

$$h(x) = \textstyle\int_0^x(\mathrm{e}^{\mathrm{i}zu} - 1 - \mathrm{i}zu1_{[-1,1]}(u))u^{-1}\mathrm{d}u = \int_0^{zx}(\mathrm{e}^{\mathrm{i}u}-1)u^{-1}\mathrm{d}u - \mathrm{i}z\tau(x).$$

As $x \to 0$, $h(x) \sim -\frac{1}{4}z^2x^2$ and hence $|h(x)|k(x) \to 0$. As $|x| \to \infty$, $h(x) = O(\log|x|)$ and hence $|h(x)|k(x) \to 0$. Thus, integration by parts gives

$$\begin{aligned}\textstyle\int_0^\infty(\mathrm{e}^{\mathrm{i}zx} - 1 - \mathrm{i}zx1_{[-1,1]}(x))k(x)x^{-1}\mathrm{d}x &= \textstyle\lim_{a\downarrow 0,\, b\uparrow\infty}\int_a^b k(x)\mathrm{d}h(x)\\ &= \textstyle -\lim\int_a^b h(x)\mathrm{d}k(x+) = \int_{(0,\infty)}h(x)\rho(\mathrm{d}x).\end{aligned}$$

Similarly

$$\textstyle\int_{-\infty}^0(\mathrm{e}^{\mathrm{i}zx} - 1 - \mathrm{i}zx1_{[-1,1]}(x))k(x)|x|^{-1}\mathrm{d}x = \int_{(-\infty,0)}h(x)\rho(\mathrm{d}x).$$

Hence we have (28.5).

Step 2. Differentiating (28.5), we see that $\widehat{\mu}(z)$ is of class C^1 on $\{z \ne 0\}$ and satisfies

(28.6) $$\tfrac{\mathrm{d}}{\mathrm{d}z}\widehat{\mu}(z) = \widehat{\mu}(z)\bigl[z^{-1}\textstyle\int_{\mathbb{R}}(\mathrm{e}^{\mathrm{i}zx} - 1 - \mathrm{i}z\tau(x))\rho(\mathrm{d}x) + \mathrm{i}\gamma\bigr].$$

Differentiation under the integral sign is justified, since the integrals $\int_{|x|\le 2}x^2\rho(\mathrm{d}x)$, $\int_{|x|>2}\log|x|\rho(\mathrm{d}x)$ are finite by Lemma 17.6.

Step 3. Let $F(x) = \mu(-\infty, x]$, the distribution function of μ. It is continuous in x. It follows from Lemma 28.5 and the inversion formula (Proposition 2.5(xi)) that

(28.7) $$\begin{aligned}F(x+\gamma_0) - F(\gamma_0) &= (2\pi)^{-1}\textstyle\int_{\mathbb{R}}(-\mathrm{i}z)^{-1}(\mathrm{e}^{-\mathrm{i}zx}-1)\widehat{\mu}(z)\mathrm{e}^{-\mathrm{i}\gamma_0 z}\mathrm{d}z\\ &= (2\pi)^{-1}(\operatorname{sgn}x)\textstyle\int_{\mathbb{R}}(-\mathrm{i}u)^{-1}(\mathrm{e}^{-\mathrm{i}u}-1)\widehat{\mu}(\frac{u}{x})\mathrm{e}^{-\mathrm{i}\gamma_0 u/x}\mathrm{d}u\end{aligned}$$

for $x \ne 0$. Suppose that $c < \infty$. Then, ρ is a finite measure and

$$\gamma_0 = \gamma - \textstyle\int_{-1}^1(\operatorname{sgn}x)k(x)\mathrm{d}x = \gamma - \int_{\mathbb{R}}\tau(x)\rho(\mathrm{d}x).$$

Hence we have, for $z \ne 0$,

$$\tfrac{\mathrm{d}}{\mathrm{d}z}(\widehat{\mu}(z)\mathrm{e}^{-\mathrm{i}\gamma_0 z}) = \widehat{\mu}(z)\mathrm{e}^{-\mathrm{i}\gamma_0 z}z^{-1}\textstyle\int_{\mathbb{R}}(\mathrm{e}^{\mathrm{i}zx}-1)\rho(\mathrm{d}x)$$

by (28.6). It follows that, for $u \ne 0$ and $x \ne 0$,

(28.8) $$\tfrac{\mathrm{d}}{\mathrm{d}x}(\widehat{\mu}(\tfrac{u}{x})\mathrm{e}^{-\mathrm{i}\gamma_0 u/x}) = -\tfrac{1}{x}\widehat{\mu}(\tfrac{u}{x})\mathrm{e}^{-\mathrm{i}\gamma_0 u/x}\textstyle\int_{\mathbb{R}}(\mathrm{e}^{\mathrm{i}yu/x}-1)\rho(\mathrm{d}y).$$

If $\alpha < c$ and if K is a compact set not containing 0, then the right-hand side of (28.8) is of order $o(|u|^{-\alpha})$ as $|u| \to \infty$, uniformly in $x \in K$ by Lemma 28.5. Therefore the last integral in (28.7) is differentiable with respect to $x \ne 0$, and $F'(x+\gamma_0)$ equals

$$-(2\pi)^{-1}|x|^{-1}\textstyle\int_{\mathbb{R}}(-\mathrm{i}u)^{-1}(\mathrm{e}^{-\mathrm{i}u}-1)\widehat{\mu}(\frac{u}{x})\mathrm{e}^{-\mathrm{i}\gamma_0 u/x}\mathrm{d}u\int_{\mathbb{R}}(\mathrm{e}^{\mathrm{i}yu/x}-1)\rho(\mathrm{d}y)$$

for $x \neq 0$. Thus $F(x)$ has a continuous density $f(x)$ on $\{x \neq \gamma_0\}$ and

$$g(x) = -(2\pi)^{-1} \int_{\mathbb{R}} (-\mathrm{i}z)^{-1} (\mathrm{e}^{-\mathrm{i}zx} - \mathrm{e}^{-\mathrm{i}\gamma_0 z}) \widehat{\mu}(z) \mathrm{d}z \int_{\mathbb{R}} (\mathrm{e}^{\mathrm{i}zy} - 1) \rho(\mathrm{d}y) \tag{28.9}$$

on $\mathbb{R}$. It follows that $g(x)$ is continuous on $\mathbb{R}$. Using Lemma 28.5 again, we see from (28.9) that $g(x)$ is of class C^N and

$$g^{(n)}(x) = -(2\pi)^{-1} \int_{\mathbb{R}} (-\mathrm{i}z)^{n-1} \mathrm{e}^{-\mathrm{i}zx} \widehat{\mu}(z) \mathrm{d}z \int_{\mathbb{R}} (\mathrm{e}^{\mathrm{i}zy} - 1) \rho(\mathrm{d}y) \tag{28.10}$$

on $\mathbb{R}$ for $n = 1, \ldots, N$. A consequence is that f is of class C^N on $\{x \neq \gamma_0\}$.

We postpone to Chapter 10 the proof that $f(x)$ cannot be extended to a continuous function on $\mathbb{R}$ if $c \leq 1$ (Theorem 53.8 and Remark 53.10). □

REMARK 28.6. It is known that, in Theorem 28.4, the density $f(x)$ cannot be extended to a C^N function on $\mathbb{R}$ if $c < \infty$. Theorem 28.4 and this result are proved by Zolotarev [**604**] and Wolfe [**578**]. Further, the order of singularity of $f^{(N)}(x)$ as $x \to \gamma_0$ is studied by Sato and Yamazato [**466, 467**]. See Section 53.

REMARK 28.7. Let $\{X_t\}$ be a Lévy process on $\mathbb{R}$ with $P_{X_1} = \mu$, where μ is a selfdecomposable distribution as in Theorem 28.4. Assume that $c < \infty$. From Theorem 28.4 and a result mentioned in Remark 28.6, we obtain time evolution of the smoothness of P_{X_t}. When $0 < t \leq \frac{1}{c}$, P_{X_t} has a density $f_t(x)$ continuous on $\{x \neq t\gamma_0\}$ but discontinuous at $x = t\gamma_0$. When $\frac{n}{c} < t \leq \frac{n+1}{c}$ with $n \geq 1$, P_{X_t} has a C^{n-1} density $f_t(x)$ on $\mathbb{R}$ and $f_t(x)$ is of class C^n on $\{x \neq t\gamma_0\}$ but not of class C^n on $\mathbb{R}$.

REMARK 28.8. Another problem on the density $p(t, x)$ of P_{X_t} of a Lévy process $\{X_t\}$ is whether it is positive or 0. For $d = 1$ Sharpe [**478**] proves that, if $p(t, x)$ is continuous with respect to $(t, x) \in (0, \infty) \times \mathbb{R}$, then the set $\{x\colon p(t, x) > 0\}$ is the whole line or a half line for each $t > 0$. All non-trivial semi-stable processes on $\mathbb{R}$ are the case. Also, the unimodality in Chapter 10 will give positivity of a density, except possibly at the endpoint of its support. For multi-dimensional stable processes this positivity problem was studied in Taylor [**526**] and Port and Vitale [**388**].

29. Exercises 5

E 29.1. Let X and Y be independent random variables on $\mathbb{R}^d$. Show that S_{X+Y} is not necessarily equal to $S_X + S_Y$. Show that, if S_X is compact, then $S_{X+Y} = S_X + S_Y$.

E 29.2. Let us say that a measure μ on $\mathbb{R}^d$ is *genuinely d-dimensional* if no proper linear subspace of $\mathbb{R}^d$ contains S_μ. Show that μ is genuinely d-dimensional if and only if there are linearly independent vectors $x_1, \ldots, x_d$ in S_μ.

E 29.3. Show that μ is nondegenerate if and only if there are vectors $x_0, x_1, \ldots, x_d$ in S_μ such that $x_1 - x_0, \ldots, x_d - x_0$ are linearly independent.

E 29.4. Let $\mathfrak{G}$ be the group of a Lévy process $\{X_t\}$ on $\mathbb{R}$. Prove the following.
(i) $\mathfrak{G}$ equals either $\mathbb{R}$ or $a\mathbb{Z}$ with a positive real a.
(ii) $\mathfrak{G} \subset a\mathbb{Z}$ with $a > 0$ if and only if $A = 0$, $S_\nu \subset a\mathbb{Z}$, and $\gamma_0 = 0$.

(iii) $\mathfrak{G} = a\mathbb{Z}$ with $a > 0$ if and only if $A = 0$, $\gamma_0 = 0$, and a is the largest positive real such that $a\mathbb{Z} \supset S_\nu$.

(iv) If $\mathfrak{G} = a\mathbb{Z}$, then the counting measure on $a\mathbb{Z}$ is an invariant measure.

E 29.5. Give a Lévy process $\{X_t\}$ for which $\mathfrak{G} = \mathbb{Z}$ and which has an invariant measure supported on $\mathfrak{G}$, not equal to a multiple of the counting measure.

E 29.6. Show that any non-zero Lévy process on $\mathbb{R}^d$ has no invariant distribution.

E 29.7 (Zolotarev [**602, 608**]). Let μ be α-stable on $\mathbb{R}$ with parameters $1 < \alpha < 2$, $\beta = -1$, $\tau = 0$, and $c = |\cos\frac{\pi\alpha}{2}|$ in Theorem 14.15. Show that $\mu[0,\infty) = 1/\alpha$ and the distribution μ_+ on $[0,\infty)$ defined by $\mu_+(B) = \alpha\mu(B)$, $B \in \mathcal{B}([0,\infty))$, is a Mittag–Leffler distribution with Laplace transform $E_{1/\alpha}(-u)$ (Example 24.12).

E 29.8. This is an example of the g-moment of a Lévy process when $g(x)$ does not satisfy (25.3). Let $g(x) = \mathrm{e}^{|x|h(|x|)}$, where $h(u)$ is a nonnegative function on $[0,\infty)$ increasing to ∞ with u. Show that there exists a Lévy process $\{X_t\}$ on $\mathbb{R}$ with Lévy measure ν such that $[\nu]_{\{|x|>1\}}$ has finite g-moment but $E[g(X_t)] = \infty$ for every $t > 0$.

E 29.9. Show that $\mu(\mathrm{d}x) = 1_{[0,\infty)}(x)(2/\pi)^{1/2}\mathrm{e}^{-x^2/2}\mathrm{d}x$, the half-Gaussian distribution, is not infinitely divisible.

E 29.10. Let $\alpha > 1$. Show that, if a probability measure μ on $\mathbb{R}^d$ satisfies $\int_{|x|>r}\mu(\mathrm{d}x) \sim c\mathrm{e}^{-r^\alpha}$ $(r \to \infty)$ with $c > 0$, then μ is not infinitely divisible. Such is the Weibull distribution (extreme value distribution of type 3 in [**241**]) with parameter $\alpha > 1$ (that is, $d = 1$ and $\mu(-\infty, x] = 1_{(0,\infty)}(x)(1 - \mathrm{e}^{-x^\alpha})$).

E 29.11 (Millar [**347**]). Let $\{X_t\}$ be a Lévy process on $\mathbb{R}^d$ and let $0 < \eta \le 2$. Show that $E[|X_t|^\eta] = O(t)$ as $t \downarrow 0$, assuming that $\int_{\mathbb{R}^d}|x|^\eta\nu(\mathrm{d}x) < \infty$.

E 29.12 (Orey [**364**]). Let $0 < \alpha < 2$ and let c be an integer $> 2/(2-\alpha)$. Let $\nu = \sum_{n=1}^\infty {a_n}^{-\alpha}\delta_{a_n}$ with $a_n = 2^{-c^n}$. Show that $\int_{(0,1)} x^\alpha\nu(\mathrm{d}x) = \infty$ and $\int_{(0,1)} x^\beta\nu(\mathrm{d}x) < \infty$ for $\beta > \alpha$ and that the Lévy process $\{X_t\}$ on $\mathbb{R}$ generated by $(0,\nu,0)$ has a continuous singular distribution for any $t > 0$.

E 29.13 (Wolfe [**583**]). (Continued from E 18.14) Let μ be a non-trivial probability measure on $\mathbb{R}^d$ satisfying $\widehat{\mu}(z) = \widehat{\mu}(b^{-1}z)\widehat{\rho}(z)$ with some $b > 1$ and some (not necessarily infinitely divisible) ρ. Show that μ is either absolutely continuous or continuous singular. When $d = 1$ and $\rho = p\delta_0 + (1-p)\delta_1$ with some $p \in (0,1)$, the corresponding distribution μ is called the *Bernoulli convolution* and many papers (see Peres and Solomyak [**369**] and the references therein) are devoted to the study of its continuity properties.

E 29.14. Let $\{X_t\}$ be a Lévy process on $\mathbb{R}^d$ generated by (A,ν,γ). Show that if A has rank d, then, for each $t > 0$, P_{X_t} has a C^∞ density, all derivatives of which tend to 0 as $|x| \to \infty$.

E 29.15. Let μ be α-stable on $\mathbb{R}$ with $0 < \alpha < 2$ and $\beta = 1$. Show that, for any $\eta < 0$,

$$\int_{\mathbb{R}} e^{\eta x}\mu(dx) = \begin{cases} \exp(-c|\eta|^{\alpha}/\cos\frac{\pi\alpha}{2} + \tau\eta) & \text{for } \alpha \neq 1, \\ \exp(2\pi^{-1}c|\eta|\log|\eta| + \tau\eta) & \text{for } \alpha = 1, \end{cases}$$

where c and τ are the parameters in Theorem 14.15.

E 29.16 (Shanbhag and Sreehari [**475**]). Let X and Z be independent positive random variables. Suppose that X has strictly α-stable distribution, $0 < \alpha < 1$, with $E[e^{-uX}] = e^{-u^{\alpha}}$, $u \geq 0$, and that Z has exponential distribution with parameter 1. Show that $(Z/X)^{\alpha} \stackrel{d}{=} Z$.

E 29.17. Prove (25.5).

E 29.18. Show that, if X is a positive random variable satisfying $E[e^{-uX}] = e^{-u^{\alpha}}$, $u \geq 0$, with $0 < \alpha < 1$, then $Y = X^{-\alpha}$ satisfies $E[e^{-uY}] = E_{\alpha}(-u)$, $u \geq 0$ (Mittag–Leffler distribution with parameter α).

E 29.19. Show that Mittag–Leffler distributions are not infinitely divisible.

E 29.20. Let $\{\mu_n\}$ be a sequence of probability measures on $\mathbb{R}$ such that μ_n is unimodal with mode a_n. Suppose that μ_n tends to a probability measure μ as $n \to \infty$. Let $b_0 = \liminf_{n\to\infty} a_n$ and $b_1 = \limsup_{n\to\infty} a_n$. Show that b_0 and b_1 are finite and that, for any $a \in [b_0, b_1]$, μ is unimodal with mode a.

E 29.21 (Khintchine [**281**]). Prove the following. Let X and U be independent random variables on $\mathbb{R}$. Suppose U is uniformly distributed on $[0,1]$. Then P_{UX} is unimodal with mode 0. Conversely, any distribution unimodal with mode 0 can be expressed in this way.

E 29.22 (Wintner [**575**]). Show that, if μ_1 and μ_2 are symmetric unimodal probability measures on $\mathbb{R}$, then $\mu_1 * \mu_2$ is symmetric and unimodal.

E 29.23. For any discrete probability measure ρ on $\mathbb{R}^d$ with carrier C_{ρ}, define $H(\rho) = -\sum_{a\in C_{\rho}} \rho\{a\}\log\rho\{a\}$ and call it the *entropy* of ρ. If a random variable on $\mathbb{R}^d$ has a discrete distribution, then $H(P_X)$ is written as $H(X)$ and called the entropy of X. Give an example that $H(X) = \infty$. Show the following.

(i) Let ρ_n, $n = 0, 1, \ldots$, be discrete probability measures on $\mathbb{R}^d$. Let σ be a probability measure on $\mathbb{Z}_+$ with $\sigma\{n\} = p_n$. Then

$$\textstyle\sum_{n=0}^{\infty} p_n H(\rho_n) \leq H(\sum_{n=0}^{\infty} p_n\rho_n) \leq \sum_{n=0}^{\infty} p_n H(\rho_n) + H(\sigma).$$

(ii) Let X_1 and X_2 be independent discrete random variables on $\mathbb{R}^d$. Then

$$H(X_1) \vee H(X_2) \leq H(X_1 + X_2) \leq H(X_1) + H(X_2).$$

We have $H(X_1) = H(X_1 + X_2)$ if and only if X_2 is trivial.

E 29.24. Let $\{X_t\}$ be a non-trivial Lévy process of type A on $\mathbb{R}^d$ with Lévy measure ν. Assume that ν is discrete and let $c = \nu(\mathbb{R}^d)$ and $\sigma = c^{-1}\nu$. Let $h(t) = H(X_t)$ and call it the *entropy function* of $\{X_t\}$. Obviously $h(0) = 0$. Prove the following.

(i) If $H(\sigma) = \infty$, then $h(t) = \infty$ for every $t > 0$.

(ii) If $H(\sigma) < \infty$, then $h(t)$ is a finite, continuous, strictly increasing function of $t \in [0, \infty)$ and there are positive constants b_1 and b_2 such that $b_1 \log t \leq h(t) \leq b_2 t$ for all large t.

Notes

The naming of "time dependent distributional property (in the class of Lévy processes)" was suggested by Frank Knight. The notion was introduced in Sato [**445**] under the name of "properties of time-evolution type". The proof of Theorem 24.7 and Corollary 24.8 independent of the Lévy–Itô decomposition is by Baxter and Shapiro [**19**]. Theorem 24.10 is pointed out by Tucker [**541**]. Proposition 24.19 is given by Hoeffding [**205**]; the present proof is taken from Petrov's book [**371**].

Elementary treatment of infinitely divisible distributions on $\mathbb{Z}_+$ (Corollary 24.6) by generating functions is possible, as in Feller's book [**138**].

Theorem 25.3 on g-moments is by Kruglov [**296**]. Related papers are Kruglov [**298**] and Sato [**430**]. When $d = 1$ and $g(x) = (|x| \vee 1)^\alpha$, it was proved by Ramachandran [**408**], but Lévy [**317**], p.176, pointed it out in the case of compound Poisson processes. Remark 25.9 is from [**296**]. An essential part of Theorem 25.18 was suggested by Kwapień and Woyczyński [**302**].

Theorems 26.1 and 26.8 are by Sato [**430**]. Another proof using an expression like E 22.2–22.4 is given in Csörgő and Mason [**90**]. A weak form of part (i) of Theorem 26.1 is given by Kruglov [**296**]. The important Lemma 26.4 is proved by Zolotarev [**607**].

Theorem 27.4 on continuity was announced by Dœblin [**100**], p. 28–29. Various proofs by Hartman and Wintner [**184**], Blum and Rosenblatt [**40**], Itô [**225**], and others are known. The present proof is by Itô [**225**]. Theorem 27.7 is the extension by Sato [**433**] of a result of Tucker [**538**] and Fisz and Varadarajan [**142**] given in the case that $d = 1$ and $l = 1$. Theorem 27.13 is proved in [**433**]. A generalization of the method of its proof is given by Theorem 27.10. Extension of the absolute continuity results is made by Yamazato [**590, 593**]. Theorems 27.15 and 27.19 are by Wolfe [**583**] and Watanabe [**560**], respectively. Theorem 27.16 and Proposition 27.18 are by Hartman and Wintner [**184**] and Jessen and Wintner [**240**], respectively. Further sufficient conditions for absolute continuity and continuous singularity are given by Hartman and Wintner [**184**], Tucker [**539, 540**], and Orey [**364**]. Time evolution from continuous singular to absolutely continuous in Theorem 27.23 can be shown also for some symmetric Lévy processes and some non-symmetric Lévy processes with jumps in both directions. See Sato [**444**]. Given an arbitrary increasing function $f(t)$ from $[0, \infty)$ to $[0, 1] \cup \{\infty\}$, Rubin [**425**] describes the construction of a Lévy process $\{X_t\}$ on $\mathbb{R}$ such that $\dim_{\mathrm{R}} P_{X_t} = f(t)$. Here, for any singular distribution μ, $\dim_{\mathrm{R}} \mu$ is equal to the Hausdorff dimension $\dim_{\mathrm{H}} \mu$ defined as the infimum of the Hausdorff dimensions of all Borel sets B with $\mu(B) = 1$. If μ is not singular, then $\dim_{\mathrm{R}} \mu$ is defined to be ∞. The proof of Proposition 27.28 is taken from Bohr [**52**].

Exercises 29.23 and 29.24 are from Watanabe [**560**].

CHAPTER 6

Subordination and density transformation

30. Subordination of Lévy processes

Subordination is a transformation of a stochastic process to a new stochastic process through random time change by an increasing Lévy process (subordinator) independent of the original process. The new process is called subordinate to the original one. Subordination is a transformation of a temporally homogeneous Markov process to another temporally homogeneous Markov process, and of a Lévy process to another Lévy process. The idea of subordination was introduced by Bochner [**48**] in 1949 and expounded in his book [**49**]. Subordination can be carried out also on a semigroup of linear operators on a Banach space, producing a new semigroup. We enter the theory of semigroups induced by temporally homogeneous Markov processes and, especially, by Lévy processes. Another procedure to obtain a second Lévy process from a Lévy process is to create a density on the probability space of the original process on every finite time interval. We call this density transformation and give its characterization in the last substantive section of this chapter.

In this chapter, when we talk about a Lévy process $\{X_t\}$, we always assume that, for *every* ω, $X_t(\omega)$ is right-continuous with left limits in t and $X_0(\omega) = 0$. When we talk about a subordinator $\{Z_t\}$, we assume that, for *every* ω, $Z_t(\omega)$ is increasing, right-continuous in t and $Z_0(\omega) = 0$. This is slightly stronger than Definitions 1.6 and 21.4, but it does not restrict generality, as we can use a probability space $(\Omega_0, [\mathcal{F}]_{\Omega_0}, [P]_{\Omega_0})$ instead of $(\Omega, \mathcal{F}, P)$ whenever $\Omega_0 \in \mathcal{F}$ has probability 1. This caution is needed, for instance, when we prove that $Y_t(\omega)$ given by (30.4) is a random variable.

We begin with subordination of Lévy processes.

THEOREM 30.1. *Let $\{Z_t\colon t \geq 0\}$ be a subordinator (an increasing Lévy process on $\mathbb{R}$) with Lévy measure ρ, drift β_0, and $P_{Z_1} = \lambda$. That is,*

$$E[\mathrm{e}^{-uZ_t}] = \int_{[0,\infty)} \mathrm{e}^{-us}\lambda^t(\mathrm{d}s) = \mathrm{e}^{t\Psi(-u)}, \qquad u \geq 0, \tag{30.1}$$

where, for any complex w with $\operatorname{Re} w \leq 0$,

$$\Psi(w) = \beta_0 w + \int_{(0,\infty)} (\mathrm{e}^{ws} - 1)\rho(\mathrm{d}s) \tag{30.2}$$

with

$$\beta_0 \geq 0 \quad \textit{and} \quad \int_{(0,\infty)} (1 \wedge s)\rho(\mathrm{d}s) < \infty. \tag{30.3}$$

Let $\{X_t\}$ *be a Lévy process on* $\mathbb{R}^d$ *with generating triplet* (A, ν, γ) *and let* $\mu = P_{X_1}$. *Suppose that* $\{X_t\}$ *and* $\{Z_t\}$ *are independent. Define*

$$Y_t(\omega) = X_{Z_t(\omega)}(\omega), \qquad t \geq 0. \tag{30.4}$$

Then $\{Y_t\}$ *is a Lévy process on* $\mathbb{R}^d$ *and*

$$P[Y_t \in B] = \int_{[0,\infty)} \mu^s(B)\lambda^t(\mathrm{d}s), \qquad B \in \mathcal{B}(\mathbb{R}^d), \tag{30.5}$$

$$E[\mathrm{e}^{\mathrm{i}\langle z, Y_t\rangle}] = \mathrm{e}^{t\Psi(\log \widehat{\mu}(z))}, \qquad z \in \mathbb{R}^d. \tag{30.6}$$

The generating triplet $(A^\sharp, \nu^\sharp, \gamma^\sharp)$ *of* $\{Y_t\}$ *is as follows:*

$$A^\sharp = \beta_0 A, \tag{30.7}$$

$$\nu^\sharp(B) = \beta_0 \nu(B) + \int_{(0,\infty)} \mu^s(B)\rho(\mathrm{d}s), \qquad B \in \mathcal{B}(\mathbb{R}^d \setminus \{0\}), \tag{30.8}$$

$$\gamma^\sharp = \beta_0 \gamma + \int_{(0,\infty)} \rho(\mathrm{d}s) \int_{|x| \leq 1} x\mu^s(\mathrm{d}x). \tag{30.9}$$

If $\beta_0 = 0$ *and* $\int_{(0,1]} s^{1/2}\rho(\mathrm{d}s) < \infty$, *then* $\{Y_t\}$ *is of type A or B and has drift* 0.

DEFINITION 30.2. The transformation above of $\{X_t\}$ to $\{Y_t\}$ is called *subordination* by the subordinator $\{Z_t\}$. Any Lévy process identical in law with $\{Y_t\}$ is said to be *subordinate* to $\{X_t\}$. Sometimes $\{Z_t\}$ is called the *directing process*.

LEMMA 30.3. *Let* $\{X_t\}$ *be a Lévy process on* $\mathbb{R}^d$. *For any* $\varepsilon > 0$ *there is* $C = C(\varepsilon)$ *such that, for any* t,

$$P[|X_t| > \varepsilon] \leq Ct. \tag{30.10}$$

There are C_1, C_2, *and* C_3 *such that, for any* t,

$$E[|X_t|^2; |X_t| \leq 1] \leq C_1 t, \tag{30.11}$$

$$|E[X_t; |X_t| \leq 1]| \leq C_2 t, \tag{30.12}$$

$$E[|X_t|; |X_t| \leq 1] \leq C_3 t^{1/2}. \tag{30.13}$$

Proof. Let $D = \{x\colon |x| \leq 1\}$. Let $\{X_1(t)\}$ and $\{X_2(t)\}$ be independent Lévy processes generated by $(A, [\nu]_D, \gamma)$ and $(0, [\nu]_{D^c}, 0)$, respectively. Then $\{X_1(t) + X_2(t)\} \stackrel{\mathrm{d}}{=} \{X_t\}$. We have $E[|X_1(t)|^2] < \infty$ by Theorem 25.3. Let

$X_j(t)$ and $X_{1,j}(t)$ be the jth components of X_t and $X_1(t)$, respectively. Noting that $\{X_2(t)\}$ is a compound Poisson process, we have

$$\begin{aligned} P[|X_t| > \varepsilon] &\le P[X_2(t) \neq 0] + P[X_2(t) = 0, |X_1(t)| > \varepsilon] \\ &\le 1 - \mathrm{e}^{-t\nu(D^c)} + \varepsilon^{-2} E[|X_1(t)|^2] \\ &\le t\nu(D^c) + \varepsilon^{-2} \sum_{j=1}^{d} (t^2 |E[X_{1,j}(1)]|^2 + t\mathrm{Var}(X_{1,j}(1))) \end{aligned}$$

by Example 25.12. Thus we get (30.10), as it is enough to consider $t \le 1$. Similarly (30.11) follows from

$$E[|X_t|^2; |X_t| \le 1] \le P[X_2(t) \neq 0] + E[|X_1(t)|^2].$$

Since

$$E[|X_t|; |X_t| \le 1] \le (E[|X_t|^2; |X_t| \le 1])^{1/2}$$

by the Schwarz inequality, (30.13) also follows. Since

$$\begin{aligned} E[\mathrm{i}X_j(t); |X_t| \le 1] = E[\mathrm{e}^{\mathrm{i}X_j(t)} - 1] &- E[\mathrm{e}^{\mathrm{i}X_j(t)} - 1; |X_t| > 1] \\ &- E[\mathrm{e}^{\mathrm{i}X_j(t)} - 1 - \mathrm{i}X_j(t); |X_t| \le 1], \end{aligned}$$

we have

$$|E[X_j(t); |X_t| \le 1]| \le |\widehat{\mu}(e_j)^t - 1| + 2P[|X_t| > 1] + \tfrac{1}{2} E[X_j(t)^2; |X_t| \le 1],$$

where e_j is the jth unit vector. Each term of the right-hand side is bounded by constant multiples of t. As $E[X_t; |X_t| \le 1]$ is the vector with components $E[X_j(t); |X_t| \le 1]$, (30.12) is now proved. □

Proof of Theorem 30.1. Since $X_t(\omega)$ is measurable in ω and right-continuous in t, it is measurable as a function of (t, ω). Hence, for each t, $Y_t(\omega) = X_{Z_t(\omega)}(\omega)$ is a random variable on $\mathbb{R}^d$. Let $f(x)$ be bounded and continuous on $\mathbb{R}^d$. Then, the independence of $\{X_t\}$ and $\{Z_t\}$ yields

$$E[f(Y_t)] = E[g(Z_t)] \quad \text{with } g(s) = E[f(X_s)]. \tag{30.14}$$

In fact, for $k_n(s) = \sum_{j=1}^{n^2} (j/n) 1_{[(j-1)/n, j/n)}(s)$, $f(X_{k_n(Z_t)})$ is a measurable function of $X_{1/n}(\omega), X_{2/n}(\omega), \ldots, X_n(\omega), Z(t, \omega)$, and we have

$$E[f(X_{k_n(Z_t)})] = E[g_{(n)}(Z_t)] \quad \text{with } g_{(n)}(s) = E[f(X_{k_n(s)})]$$

by Proposition 1.16. Letting $n \to \infty$, we get (30.14). Thus we get (30.5). Similarly, for $0 \le t_1 < t_2$,

$$E[f(Y_{t_2} - Y_{t_1})] = E[h(Z_{t_1}, Z_{t_2})] \quad \text{with } h(s_1, s_2) = E[f(X_{s_2} - X_{s_1})].$$

Since Z_t is increasing in t and since $h(s_1, s_2) = g(s_2 - s_1)$ for $s_1 \le s_2$,

$$E[f(Y_{t_2} - Y_{t_1})] = E[g(Z_{t_2} - Z_{t_1})] = E[g(Z_{t_2 - t_1})]. \tag{30.15}$$

Let $f_1(x), \ldots, f_n(x)$ be bounded and continuous and let $0 \le t_1 < \cdots < t_n$. Define $g_j(s)$ and $h_j(s_1, s_2)$ similarly from $f_j(x)$. We get, in the same way,

$$E\left[\prod_{j=1}^{n} f_j(Y_{t_{j+1}} - Y_{t_j})\right] = E[G(Z_{t_1}, \ldots, Z_{t_n})],$$

where

$$G(s_1, \ldots, s_n) = E\left[\prod_{j=1}^{n} f_j(X_{s_{j+1}} - X_{s_j})\right] \quad \text{for } 0 \le s_1 \le \cdots \le s_n.$$

Hence

$$E\left[\prod_{j=1}^{n} f_j(Y_{t_{j+1}} - Y_{t_j})\right] = E\left[\prod_{j=1}^{n} h_j(Z_{t_j}, Z_{t_{j+1}})\right] = E\left[\prod_{j=1}^{n} g_j(Z_{t_{j+1}} - Z_{t_j})\right]$$
$$= \prod_{j=1}^{n} E[g_j(Z_{t_{j+1}} - Z_{t_j})] = \prod_{j=1}^{n} E[f_j(Y_{t_{j+1}} - Y_{t_j})],$$

that is, $\{Y_t\}$ has independent increments. We have

$$E[f(Y_{t_2} - Y_{t_1})] = E[f(Y_{t_2 - t_1})]$$

by (30.14) and (30.15), that is, $\{Y_t\}$ has stationary increments. Evidently, $Y_0 = 0$ and Y_t is right-continuous with left limits in t. Therefore $\{Y_t\}$ is a Lévy process on $\mathbb{R}^d$.

We have, for any w with $\operatorname{Re} w \le 0$,

$$E[\mathrm{e}^{wZ_t}] = \int_{[0,\infty)} \mathrm{e}^{sw} \lambda^t(\mathrm{d}s) = \mathrm{e}^{t\Psi(w)}. \tag{30.16}$$

This is shown in the proofs of Theorems 24.11 and 25.17. Since

$$E[\mathrm{e}^{\mathrm{i}\langle z, Y_t\rangle}] = \int_{[0,\infty)} E[\mathrm{e}^{\mathrm{i}\langle z, X_s\rangle}]\lambda^t(\mathrm{d}s) = \int_{[0,\infty)} \mathrm{e}^{s \log \widehat{\mu}(z)} \lambda^t(\mathrm{d}s) = \mathrm{e}^{t\Psi(\log \widehat{\mu}(z))}$$

by (30.16), we obtain (30.6). To calculate $A^\sharp$, $\nu^\sharp$, and $\gamma^\sharp$, we use Lemma 30.3. By (30.2) we have

$$\Psi(\log \widehat{\mu}(z)) = \beta_0 \log \widehat{\mu}(z) + \int_{(0,\infty)} (\widehat{\mu}(z)^s - 1)\rho(\mathrm{d}s). \tag{30.17}$$

Define a measure ν_1 by $\nu_1(\{0\}) = 0$ and

$$\nu_1(B) = \int_{(0,\infty)} \mu^s(B)\rho(\mathrm{d}s) \quad \text{for } B \in \mathcal{B}(\mathbb{R}^d \setminus \{0\}).$$

Let $D = \{x\colon |x| \le 1\}$. Then Lemma 30.3 yields

$$\int_D |x|^2 \nu_1(\mathrm{d}x) = \int_{(0,\infty)} \rho(\mathrm{d}s) \int_D |x|^2 \mu^s(\mathrm{d}x) < \infty,$$

$$\int_{|x|>1} \nu_1(\mathrm{d}x) = \int_{(0,\infty)} P[|X_s| > 1]\rho(\mathrm{d}s) < \infty,$$

$$\int_{(0,\infty)} \rho(\mathrm{d}s) \left| \int_D x\mu^s(\mathrm{d}x) \right| < \infty.$$

Hence, with $g(z, x) = \mathrm{e}^{\mathrm{i}\langle z,x\rangle} - 1 - \mathrm{i}\langle z, x\rangle 1_D(x)$,

$$\begin{aligned}\int_{(0,\infty)} (\widehat{\mu}(z)^s - 1)\rho(\mathrm{d}s) &= \int \rho(\mathrm{d}s) \int (\mathrm{e}^{\mathrm{i}\langle z,x\rangle} - 1)\mu^s(\mathrm{d}x) \\ &= \int \rho(\mathrm{d}s) \int g(z, x)\mu^s(\mathrm{d}x) + \mathrm{i} \int \rho(\mathrm{d}s) \int \langle z, x\rangle 1_D(x)\mu^s(\mathrm{d}x) \\ &= \int g(z, x)\nu_1(\mathrm{d}x) + \mathrm{i}\Big\langle z, \int \rho(\mathrm{d}s) \int_D x\mu^s(\mathrm{d}x)\Big\rangle.\end{aligned}$$

Therefore, using (30.6) and (30.17), we get (30.7), (30.8), and (30.9).

Suppose that $\beta_0 = 0$ and $\int_{(0,1]} s^{1/2}\rho(\mathrm{d}s) < \infty$. Then it follows from (30.7), (30.8), and (30.13) that $A^\sharp = 0$ and

$$\int_{|x|\le 1} |x|\nu^\sharp(\mathrm{d}x) = \int \rho(\mathrm{d}s) \int_{|x|\le 1} |x|\mu^s(\mathrm{d}x) < \infty.$$

Hence $\{Y_t\}$ is of type A or B. Its drift $\gamma_0^\sharp = \gamma^\sharp - \int_{|x|\le 1} x\nu^\sharp(\mathrm{d}x)$ vanishes by (30.9). This finishes the proof. □

Iteration of subordination is again subordination.

THEOREM 30.4. *Let $\{Z_1(t)\}$ and $\{Z_2(t)\}$ be independent subordinators and $Z_3(t) = Z_1(Z_2(t))$. Then $\{Z_3(t)\}$ is a subordinator. Define $\Psi_j(-u)$ by*

$$E[\mathrm{e}^{-uZ_j(t)}] = \mathrm{e}^{t\Psi_j(-u)}, \qquad u \ge 0. \tag{30.18}$$

Then

$$\Psi_3(-u) = \Psi_2(\Psi_1(-u)) \qquad u \ge 0. \tag{30.19}$$

Let $\{X_t\}$, $\{Y_t\}$, and $\{W_t\}$ be Lévy processes on $\mathbb{R}^d$. If $\{Y_t\}$ is subordinate to $\{X_t\}$ by the subordinator $\{Z_1(t)\}$ and $\{W_t\}$ is subordinate to $\{Y_t\}$ by $\{Z_2(t)\}$, then $\{W_t\}$ is subordinate to $\{X_t\}$ by $\{Z_3(t)\}$.

Proof. We have shown in the previous theorem that $\{Z_3(t)\}$ is a Lévy process. It is obviously increasing, so it is a subordinator. If $u \ge 0$, then $\Psi_j(-u) \le 0$ and, by (30.14),

$$\begin{aligned}E[\mathrm{e}^{-uZ_3(t)}] &= \int E[\mathrm{e}^{-uZ_1(s)}]P[Z_2(t) \in \mathrm{d}s] \\ &= \int \mathrm{e}^{s\Psi_1(-u)}P[Z_2(t) \in \mathrm{d}s] = \mathrm{e}^{t\Psi_2(\Psi_1(-u))}.\end{aligned}$$

Hence (30.19) follows. To see the second half, choose a probability space so that $\{X_t\}$, $\{Z_1(t)\}$, and $\{Z_2(t)\}$ are defined on it and independent. Let $Y'_t = X_{Z_1(t)}$ and $W'_t = Y'_{Z_2(t)}$. Then $\{Y'_t\} \stackrel{\mathrm{d}}{=} \{Y_t\}$ and $\{W'_t\} \stackrel{\mathrm{d}}{=} \{W_t\}$. That is, $\{W_t\} \stackrel{\mathrm{d}}{=} \{X_{Z_1(Z_2(t))}\}$. □

EXAMPLE 30.5. If $\{Z_1(t)\}$ and $\{Z_2(t)\}$ in the theorem above are strictly stable subordinators with indices α_1 and α_2, respectively, then $\{Z_3(t)\}$ is a strictly stable subordinator with index $\alpha_1\alpha_2$. In fact, $\Psi_1(-u) = -c_1 u^{\alpha_1}$ and $\Psi_2(-u) = -c_2 u^{\alpha_2}$ with some $c_1 > 0$ and $c_2 > 0$, which imply $\Psi_3(-u) = -c_2 {c_1}^{\alpha_2} u^{\alpha_1\alpha_2}$. Here α_1 and α_2 are in $(0,1)$. See Example 24.12.

EXAMPLE 30.6. Subordination of the Brownian motion $\{X_t\}$ on $\mathbb{R}^d$ by a strictly α-stable subordinator (Example 24.12) yields a rotation invariant 2α-stable process on $\mathbb{R}^d$. In fact, setting $\Psi(-u) = -c'u^\alpha$, $c' > 0$, and $\log\widehat{\mu}(z) = -\frac{1}{2}|z|^2$ in (30.6), we get

$$E[\mathrm{e}^{\mathrm{i}\langle z, Y_t\rangle}] = \exp(-t2^{-\alpha}c'|z|^{2\alpha}). \tag{30.20}$$

Using Theorem 14.14, we see that all rotation invariant 2α-stable processes are obtained in this way. This example continues in Example 32.7.

EXAMPLE 30.7. Let $\{X_t\}$ be the Brownian motion on $\mathbb{R}^2$. Its components $\{X_1(t)\}$ and $\{X_2(t)\}$ are independent one-dimensional Brownian motions (Proposition 5.2). Let, for $s \geq 0$, $T_{D(s)}$ be the hitting time of $D(s) = (s, \infty) \times \mathbb{R}$ by $\{X_t\}$, that is,

$$T_{D(s)}(\omega) = \inf\{t > 0 \colon X_t(\omega) \in D(s)\},$$

with the convention that the infimum of the empty set is ∞. This is the hitting time of (s, ∞) by $\{X_1(t)\}$. By Theorem 5.5, $T_{D(s)}$ is almost surely finite. We will show in Example 40.14 and again in Example 46.5 that $\{T_{D(s)} \colon s \geq 0\}$ is a strictly $\frac{1}{2}$-stable subordinator and

$$E[\mathrm{e}^{-uT_{D(s)}}] = \mathrm{e}^{-s\sqrt{2u}}.$$

We have $X_1(T_{D(s)}) = s$ by the continuity of X_t. Since $\{X_2(t)\}$ and $\{T_{D(s)}\}$ are independent, $\{X_2(T_{D(s)})\}$ is subordinate to $\{X_2(t)\}$. By the previous example, $\{X_2(T_{D(s)})\}$ is a Cauchy process on $\mathbb{R}$ and

$$E[\mathrm{e}^{\mathrm{i}zX_2(T_{D(s)})}] = \mathrm{e}^{-s|z|}.$$

This explains that, in the theory of harmonic functions, the half plane has harmonic measure equal to the Cauchy distribution.

If we consider the Brownian motion on $\mathbb{R}^d$, $X_t = (X_j(t))_{1\leq j\leq d}$, then, using $(X_j(t))_{2\leq j\leq d}$ in place of the $X_2(t)$ above, we have the same conclusion and get a Cauchy process on $\mathbb{R}^{d-1}$.

EXAMPLE 30.8. Let $\{Z_t\}$ be the Γ-process with $E[Z_1] = 1/q$, $q > 0$. Then

$$\Psi(-u) = \int_0^\infty (\mathrm{e}^{-ux} - 1)x^{-1}\mathrm{e}^{-qx}\mathrm{d}x = -\log(1 + q^{-1}u), \tag{30.21}$$

as in Example 8.10. Then the Lévy process $\{Y_t\}$ on $\mathbb{R}^d$ subordinate to $\{X_t\}$ by $\{Z_t\}$ in Theorem 30.1 satisfies

$$P[Y_t \in B] = \frac{q^t}{\Gamma(t)} \int_0^\infty P[X_s \in B] s^{t-1}\mathrm{e}^{-qs}\mathrm{d}s, \tag{30.22}$$

$$E[\mathrm{e}^{\mathrm{i}\langle z, Y_t\rangle}] = (1 - q^{-1}\log\widehat{\mu}(z))^{-t}. \tag{30.23}$$

In particular, if $d = 1$ and $\{X_t\}$ is a Poisson process with parameter $c > 0$, then, for each $t > 0$, Y_t has negative binomial distribution with parameters t and $p = q/(c + q)$ (Example 2.16), since $E[\mathrm{e}^{-uX_t}] = \mathrm{e}^{tc(\mathrm{e}^{-u}-1)}$ and

$$E[\mathrm{e}^{-uY_t}] = \mathrm{e}^{-t\log(1-q^{-1}c(\mathrm{e}^{-u}-1))} = p^t(1 - (1-p)\mathrm{e}^{-u})^{-t}.$$

If $d = 1$ and μ is α-stable with $\widehat{\mu}(z) = \mathrm{e}^{-|z|^\alpha}$, $0 < \alpha \leq 2$, then the distribution of Y_1 has characteristic function

$$(1 + q^{-1}|z|^\alpha)^{-1}.$$

This is called Linnik distribution or geometric stable distribution; see Erdoğan and Ostrovskii [**126**].

In the theory of temporally homogeneous Markov processes, the Laplace transform with respect to t of the transition function is important. It is given by the following in the case of Lévy processes.

DEFINITION 30.9. Let $\{X_t\}$ be a Lévy process on $\mathbb{R}^d$ with $\mu = P_{X_1}$. The *q-potential measure* $V^q(B)$, $B \in \mathcal{B}(\mathbb{R}^d)$, for $q \geq 0$ is

$$V^q(B) = \int_0^\infty \mathrm{e}^{-qt}\mu^t(B)\mathrm{d}t = E\left[\int_0^\infty \mathrm{e}^{-qt}1_B(X_t)\mathrm{d}t\right]. \tag{30.24}$$

The existence of the integrals comes from the measurability of $X_t(\omega)$ in (t, ω) and Fubini's theorem. The 0-potential measure is simply called the *potential measure* and written as $V(B)$. If $q > 0$, then the q-potential measure has total mass $1/q$. If $q = 0$, then $V^q(B) = V(B)$ may be infinite.

THEOREM 30.10. *Let $\{X_t\}$ be a Lévy process on $\mathbb{R}^d$ with $\mu = P_{X_1}$. Let $q > 0$.*

(i) *The probability measure qV^q is infinitely divisible and purely non-Gaussian with Lévy measure $\nu_q^\sharp$ equal to*

$$\nu_q^\sharp(B) = \int_0^\infty \mathrm{e}^{-qt}\mu^t(B)\frac{\mathrm{d}t}{t}, \qquad B \in \mathcal{B}(\mathbb{R}^d \setminus \{0\}), \tag{30.25}$$

and satisfying $\int(1 \wedge |x|)\nu_q^\sharp(\mathrm{d}x) < \infty$. The drift of qV^q is 0.

(ii) *The total mass of* $\nu_q^\sharp$ *is finite if and only if* $\{X_t\}$ *is a compound Poisson process or the zero process.*

Proof. (i) Looking at (30.22) and (30.24), we see that qV^q is the distribution of Y_1 in Example 30.8. Therefore qV^q is infinitely divisible. Theorem 30.1 and (30.21) give the generating triplet $(A_q^\sharp, \nu_q^\sharp, \gamma_q^\sharp)$ of qV^q, using $\rho(\mathrm{d}s) = \mathrm{e}^{-qs}\frac{\mathrm{d}s}{s}$ and $\beta_0 = 0$. Thus $A_q^\sharp = 0$ and $\nu_q^\sharp$ is given by (30.25). Since $\int_{(0,1]} s^{1/2}\rho(\mathrm{d}s) < \infty$ and $\beta_0 = 0$, Theorem 30.1 tells us that $\{Y_t\}$ is of type A or B and has drift 0.

(ii) Let a be the total mass of the Lévy measure ν of $\{X_t\}$. If $\{X_t\}$ is the zero process, then $\nu_q^\sharp = 0$. Suppose that $\{X_t\}$ is not the zero process. If $\{X_t\}$ is a compound Poisson process, then $a < \infty$ and $P[\,X_t \neq 0\,] \leq 1 - \mathrm{e}^{-at}$, which implies $\nu_q^\sharp(\mathbb{R}^d \setminus \{0\}) < \infty$ by (30.25). If $\{X_t\}$ is of type B or C, then $P[\,X_t \neq 0\,] = 1$ for each $t > 0$ by Theorem 27.4 and hence $\nu_q^\sharp(\mathbb{R}^d \setminus \{0\}) = \infty$ by (30.25). If $\{X_t\}$ is of type A and not a compound Poisson process, then it has drift $\gamma_0 \neq 0$ and we can prove $\nu_q^\sharp(\mathbb{R}^d \setminus \{0\}) = \infty$. In fact,

$$P[\,X_t = 0\,] = \sum_{n=0}^{\infty} \mathrm{e}^{-at}\frac{(at)^n}{n!} P[\,S_n = -\gamma_0 t\,],$$

using the random walk $\{S_n\}$ with $P_{S_1} = a^{-1}\nu$. Since $P[\,S_n = x\,] = 0$ except for a countable number of points x, we have $P[\,X_t = 0\,] = 0$ except for a countable number of t, and (30.25) shows that $\nu_q^\sharp(\mathbb{R}^d \setminus \{0\}) = \infty$. □

EXAMPLE 30.11. Let $\{X_t\}$ be the Brownian motion on $\mathbb{R}^d$. Then the distribution of X_t, $t > 0$, has density

$$p(t,x) = (2\pi t)^{-d/2}\mathrm{e}^{-|x|^2/(2t)}. \tag{30.26}$$

Hence, for $q > 0$, the q-potential measure has density

$$v^q(x) = \int_0^\infty \mathrm{e}^{-qt}p(t,x)\mathrm{d}t = \int_0^\infty \mathrm{e}^{-qt}(2\pi t)^{-d/2}\mathrm{e}^{-|x|^2/(2t)}\mathrm{d}t. \tag{30.27}$$

We use the following formula in 8.432.6 of [**169**] for the modified Bessel function $K_p(x)$ of (4.9) and (4.10):

$$K_p(x) = \frac{1}{2}\left(\frac{x}{2}\right)^p \int_0^\infty \mathrm{e}^{-t-x^2/(4t)}t^{-p-1}\mathrm{d}t, \quad x > 0,\ p \in \mathbb{R}. \tag{30.28}$$

The derivation of this formula with historical remark is given in Watson [**569**], pp. 181–183. Using this, we get

$$v^q(x) = 2(2\pi)^{-d/2}(2q)^{(d-2)/4}|x|^{-(d-2)/2}K_{(d-2)/2}((2q)^{1/2}|x|) \tag{30.29}$$

for $q > 0$, $x \neq 0$. Thus $qv^q(x)\mathrm{d}x$ is an example of an infinitely divisible distribution expressible by Bessel functions. The Lévy measure $\nu_q^\sharp = w^q(x)\mathrm{d}x$ is similarly calculated from (30.25):

$$w^q(x) = 2(2\pi)^{-d/2}(2q)^{d/4}|x|^{-d/2}K_{d/2}((2q)^{1/2}|x|) \tag{30.30}$$

for $q > 0$, $x \neq 0$. In the case $d = 1$ or 3, calculation of (30.27) as in Example 2.13 gives

$$v^q(x) = \begin{cases} (2q)^{-1/2} \exp(-(2q)^{1/2}|x|) & \text{for } d = 1, \\ (2\pi|x|)^{-1} \exp(-(2q)^{1/2}|x|) & \text{for } d = 3. \end{cases} \tag{30.31}$$

In general, if d is odd, then $v^q(x)$ and $w^q(x)$ are expressed by elementary functions. The one-dimensional case will be further discussed in Example 45.4.

31. Infinitesimal generators of Lévy processes

The generating triplet of a Lévy process gives a description of the infinitesimal generator of its transition semigroup. In order to formulate this, we use the theory of semigroups of linear operators. We introduce basic definitions. Let $\mathbf{B}$ be a real (or complex) Banach space. That is, $\mathbf{B}$ is a vector space over the real (or complex) scalar field equipped with a mapping $\|f\|$ from $\mathbf{B}$ into $\mathbb{R}$, called the norm, satisfying

$$\|af\| = |a| \, \|f\| \quad \text{for } f \in \mathbf{B},\ a \in \mathbb{R} \text{ (or } a \in \mathbb{C}),$$
$$\|f + g\| \leq \|f\| + \|g\| \quad \text{for } f, g \in \mathbf{B},$$
$$\|f\| = 0 \text{ if and only if } f = 0,$$

such that, if a sequence $\{f_n\}$ in $\mathbf{B}$ satisfies $\lim_{n,m\to\infty} \|f_n - f_m\| = 0$, then there is $f \in \mathbf{B}$ with $\lim_{n\to\infty} \|f_n - f\| = 0$. Convergence of f_n, $n = 1, 2, \ldots$, to g in norm, that is, $\|f_n - g\| \to 0$ as $n \to \infty$, is called *strong convergence* and denoted by

$$\lim_{n\to\infty} f_n = g \text{ (strong)} \quad \text{or} \quad f_n \to g \text{ (strong)}.$$

A *linear operator* L in $\mathbf{B}$ is a mapping from a linear subspace $\mathfrak{D}(L)$ of $\mathbf{B}$ into $\mathbf{B}$ such that

$$L(af + bg) = aLf + bLg \quad \text{for } f, g \in \mathfrak{D}(L),\ a, b \in \mathbb{R} \text{ (or } a, b \in \mathbb{C}).$$

The set $\mathfrak{D}(L)$ is called the *domain* of L. A linear operator L is called *bounded* if $\mathfrak{D}(L) = \mathbf{B}$ and $\sup_{\|f\|\leq 1} \|Lf\|$, called the *norm* of L and denoted by $\|L\|$, is finite. A linear operator L with $\mathfrak{D}(L) = \mathbf{B}$ is bounded if and only if L is continuous in the sense that $f_n \to f$ (strong) implies $Lf_n \to Lf$ (strong). A linear operator L is said to be *closed* if $f_n \in \mathfrak{D}(L)$, $f_n \to f$ (strong), and $Lf_n \to g$ (strong) imply $f \in \mathfrak{D}(L)$ and $Lf = g$, in other words, if the graph of L, $\{(f, Lf) \colon f \in \mathfrak{D}(L)\}$, is a closed set in $\mathbf{B} \times \mathbf{B}$. A linear operator L_2 is said to be an *extension* of a linear operator L_1 if $\mathfrak{D}(L_1) \subset \mathfrak{D}(L_2)$ and $L_1 f = L_2 f$ for $f \in \mathfrak{D}(L_1)$. A linear operator L is called *closable* if it has a closed extension. A linear operator L is closable if and only if $f_n \in \mathfrak{D}(L)$, $f_n \to 0$ (strong), and $Lf_n \to g$ (strong) imply $g = 0$. If a linear operator

L is closable, then the *smallest closed extension* (or *closure*) $\overline{L}$ exists; $\overline{L}$ is a closed extension of L, and every closed extension of L is an extension of $\overline{L}$. A description of $\overline{L}$ is as follows: $f \in \mathfrak{D}(\overline{L})$ and $\overline{L}f = g$ if and only if there is a sequence $f_n \in \mathfrak{D}(L)$ such that $f_n \to f$ (strong) and $Lf_n \to g$ (strong). A linear subspace $\mathfrak{D}_0$ of $\mathbf{B}$ is said to be a *core* of a closed operator L if $\mathfrak{D}_0 \subset \mathfrak{D}(L)$ and if the smallest closed extension of $[L]_{\mathfrak{D}_0}$, the restriction of L to $\mathfrak{D}_0$, equals L. If a description of the domain $\mathfrak{D}(L)$ is not known, to find an easily describable core of L is important. The set $\{Lf : f \in \mathfrak{D}(L)\}$, called the *range* of L, is denoted by $\mathfrak{R}(L)$. The identity operator on $\mathbf{B}$ is denoted by I. A subset $\mathfrak{D}_1$ of $\mathbf{B}$ is said to be *dense* in $\mathbf{B}$ if, for any $f \in \mathbf{B}$, there is a sequence $\{f_n\}$ in $\mathfrak{D}_1$ such that $f_n \to f$ (strong).

DEFINITION 31.1. A family $\{P_t : t \geq 0\}$ of bounded linear operators on $\mathbf{B}$ is called a *strongly continuous semigroup* if

$$P_t P_s = P_{t+s} \quad \text{for } t, s \in [0, \infty), \tag{31.1}$$

$$P_0 = I \tag{31.2}$$

$$\lim_{t \downarrow 0} P_t f = f \text{ (strong)} \quad \text{for any } f \in \mathbf{B}. \tag{31.3}$$

It is called a strongly continuous *contraction* semigroup if, moreover,

$$\|P_t\| \leq 1. \tag{31.4}$$

DEFINITION 31.2. The *infinitesimal generator* L of a strongly continuous contraction semigroup $\{P_t\}$ is defined by

$$Lf = \lim_{t \downarrow 0} t^{-1}(P_t f - f) \text{ (strong)} \tag{31.5}$$

with $\mathfrak{D}(L)$ being the set of f such that the right-hand side of (31.5) exists.

A major theorem of the theory of semigroups of operators is as follows. It was independently proved by Hille [**201**] and Yosida [**597**]. Proofs are found also in Dunford and Schwartz [**111**] and Ethier and Kurtz [**133**].

THEOREM 31.3 (Hille–Yosida). (i) *If L is the infinitesimal generator of a strongly continuous contraction semigroup $\{P_t\}$, then L is closed, $\mathfrak{D}(L)$ is dense, and, for any $q > 0$, $\mathfrak{R}(qI - L) = \mathbf{B}$, $qI - L$ is one-to-one, $\|(qI - L)^{-1}\| \leq 1/q$, and*

$$(qI - L)^{-1} f = \int_0^\infty \mathrm{e}^{-qt} P_t f \, \mathrm{d}t \quad \textit{for } f \in \mathbf{B}. \tag{31.6}$$

(ii) *The infinitesimal generator determines the semigroup. That is, two strongly continuous contraction semigroups coincide if their infinitesimal generators coincide.*

(iii) *If a linear operator L in $\mathbf{B}$ has a dense domain $\mathfrak{D}(L)$ and, for any $q > 0$, $\mathfrak{R}(qI - L) = \mathbf{B}$, $qI - L$ is one-to-one, $\|(qI - L)^{-1}\| \leq 1/q$, then L is the infinitesimal generator of a strongly continuous contraction semigroup on $\mathbf{B}$.*

The integral on $[0, \infty)$ in the right-hand side of (31.6) is defined to be the strong limit of the integral on $[0, s]$ as $s \to \infty$ and the integral on $[0, s]$ is the Riemann type integral in strong convergence, that is, for $u(t) = \mathrm{e}^{-qt} P_t f$,

$$\int_0^s u(t)\mathrm{d}t = \lim_{\mathrm{mesh}(\Delta)\to 0} \sum_{j=1}^{n} u(t_{j-1})(t_j - t_{j-1}) \text{ (strong)},$$

where Δ is a partition $0 = t_0 < t_1 < \cdots < t_n = s$. The operator $U^q = (qI - L)^{-1}$ for $q > 0$ is called the *resolvent operator* of L. It satisfies the *resolvent equation*,

$$U^q - U^r + (q - r)U^q U^r = 0, \tag{31.7}$$

and

$$\lim_{q\to\infty} qU^q f = f \text{ (strong)} \quad \text{for } f \in \mathbf{B}. \tag{31.8}$$

REMARK 31.4. For any bounded linear operator L on $\mathbf{B}$, the exponential e^L is defined to be a bounded linear operator such that

$$\mathrm{e}^L f = \textstyle\sum_{n=0}^{\infty} (n!)^{-1} L^n f.$$

If a strongly continuous contraction semigroup $\{P_t\}$ has a bounded operator L as its infinitesimal generator, then we can prove that $P_t = \mathrm{e}^{tL}$. Extending this, any strongly continuous contraction semigroup $\{P_t\}$ with infinitesimal generator L (not necessarily bounded) is sometimes denoted by $P_t = \mathrm{e}^{tL}$.

Now let $C_0 = C_0(\mathbb{R}^d)$ be the real Banach space of continuous functions f from $\mathbb{R}^d$ into $\mathbb{R}$ satisfying $\lim_{|x|\to\infty} f(x) = 0$ with norm $\|f\| = \sup_x |f(x)|$. Let C_0^n be the set of $f \in C_0$ such that f is n times differentiable and the partial derivatives of f with order $\leq n$ belong to C_0. Let $C_0^\infty = \bigcap_{n=1}^\infty C_0^n$. The *support* of a function f is the closure of the set $\{x \colon f(x) \neq 0\}$. Let C_c^∞ be the set of $f \in C_0^\infty$ with compact support. A bounded linear operator L on C_0 is said to have *positivity* if $Lf \geq 0$ for every $f \in C_0$ satisfying $f \geq 0$.

Suppose that $\{X_t\}$ is a Lévy process on $\mathbb{R}^d$ corresponding to an infinitely divisible distribution $\mu = P_{X_1}$. The transition function $P_t(x, B)$ is defined by

$$P_t(x, B) = \mu^t(B - x) \quad \text{for } t \geq 0, x \in \mathbb{R}^d, B \in \mathcal{B}(\mathbb{R}^d), \tag{31.9}$$

as in (10.8). Define, for $f \in C_0$,

$$(P_t f)(x) = \int_{\mathbb{R}^d} P_t(x, \mathrm{d}y) f(y) \tag{31.10}$$

$$= \int_{\mathbb{R}^d} \mu^t(\mathrm{d}y)\, f(x+y) = E[f(x+X_t)].$$

Then $P_t f \in C_0$ by the Lebesgue convergence theorem. The following is a major result of this section.

THEOREM 31.5. *The family of operators $\{P_t \colon t \geq 0\}$ defined above from a Lévy process $\{X_t\}$ on $\mathbb{R}^d$ is a strongly continuous semigroup on $C_0(\mathbb{R}^d)$ with norm $\|P_t\| = 1$. Let L be its infinitesimal generator. Then C_c^∞ is a core of L, $C_0^2 \subset \mathfrak{D}(L)$, and*

$$\text{(31.11)} \qquad Lf(x) = \frac{1}{2}\sum_{j,k=1}^{d} A_{jk}\frac{\partial^2 f}{\partial x_j \partial x_k}(x) + \sum_{j=1}^{d} \gamma_j \frac{\partial f}{\partial x_j}(x)$$
$$+ \int_{\mathbb{R}^d}\Big(f(x+y) - f(x) - \sum_{j=1}^{d} y_j \frac{\partial f}{\partial x_j}(x) 1_D(y)\Big)\nu(\mathrm{d}y)$$

for $f \in C_0^2$, where (A, ν, γ) is the generating triplet of $\{X_t\}$, $A = (A_{jk})$, $\gamma = (\gamma_j)$, and $D = \{x \colon |x| \leq 1\}$.

The semigroup $\{P_t\}$ on $C_0(\mathbb{R}^d)$ is called the *transition semigroup* of $\{X_t\}$. In order to prove the theorem, we need two lemmas.

LEMMA 31.6. *Let $\{P_t\}$ be a strongly continuous contraction semigroup on $\mathbf{B}$ with infinitesimal generator L. If $\mathfrak{D}_0$ and $\mathfrak{D}_1$ are linear subspaces of $\mathbf{B}$ such that*

$$\text{(31.12)} \qquad \mathfrak{D}_0 \subset \mathfrak{D}_1 \subset \mathfrak{D}(L) \text{ and } \mathfrak{D}_0 \text{ is dense in } \mathbf{B}$$

and

$$\text{(31.13)} \qquad f \in \mathfrak{D}_0 \text{ implies } P_t f \in \mathfrak{D}_1 \text{ for any } t > 0,$$

then $\mathfrak{D}_1$ is a core of L.

Proof. Fix $q > 0$. Let $\mathfrak{R} = \{(qI - L)u \colon u \in \mathfrak{D}_1\}$. It suffices to show that $\mathfrak{R}$ is dense in $\mathbf{B}$. In fact, if $\mathfrak{R}$ is dense, then, for any $u \in \mathfrak{D}(L)$, we let $f = (qI - L)u$, choose $f_n \in \mathfrak{R}$ satisfying $f_n \to f$, and find that $U^q f_n \to U^q f = u$ and $LU^q f_n = qU^q f_n - f_n \to qU^q f - f = Lu$, which, combined with $U^q f_n \in \mathfrak{D}_1$, shows that L is the closure of $[L]_{\mathfrak{D}_1}$. All convergences here are strong. To show the denseness of $\mathfrak{R}$, we see that any $g \in \mathfrak{D}_0$ is approximated by elements of $\mathfrak{R}$. Since

$$g = U^q(qI - L)g = \lim_{s\to\infty}\int_0^s \mathrm{e}^{-qt} P_t (qI - L) g\, \mathrm{d}t$$
$$= \lim_{s\to\infty}\lim_{n\to\infty} \frac{s}{n}\sum_{j=0}^{n-1} \mathrm{e}^{-qjs/n} P_{js/n}(qI - L) g$$

in strong convergence and since $P_{js/n}(qI - L)g = (qI - L)P_{js/n}g \in \mathfrak{R}$ by (31.13), an element of $\mathfrak{R}$ can be chosen as close to g as we want. $\square$

LEMMA 31.7. *Suppose that $\{P_t\}$ is a strongly continuous contraction semigroup on C_0 with infinitesimal generator L and that P_t has positivity. If $f \in C_0$, $g \in C_0$, and*

$$\lim_{t \downarrow 0} \frac{1}{t}(P_t f(x) - f(x)) = g(x) \text{ pointwise on } \mathbb{R}^d, \tag{31.14}$$

then $f \in \mathfrak{D}(L)$ and $Lf = g$.

Proof. Define $L^\sharp f = g$ whenever $f \in C_0$, $g \in C_0$, and (31.14) holds. Then $L^\sharp$ is an extension of L. We claim that, if $f \in \mathfrak{D}(L^\sharp)$ and $f(x_0) = \max_x f(x) > 0$, then $(L^\sharp f)(x_0) \leq 0$; we call this property *dispersiveness* of $L^\sharp$. Let $f^+(x) = f(x) \vee 0$ and $f^-(x) = -(f(x) \wedge 0)$. Then $f \leq f^+$ and $P_t f \leq P_t(f^+)$. Hence, if f takes a positive maximum at x_0, then

$$(P_t f)(x_0) - f(x_0) \leq P_t(f^+)(x_0) - \|f^+\| \leq \|P_t(f^+)\| - \|f^+\| \leq 0,$$

which implies $(L^\sharp f)(x_0) \leq 0$. That is, $L^\sharp$ is dispersive. Now we see that, if $(qI - L^\sharp)f = 0$ with some $q > 0$, then $f = 0$. Indeed, if $f^+ \neq 0$, then f takes a positive maximum at some point x_0 and

$$(qI - L^\sharp)f(x_0) \geq f(x_0) > 0,$$

which contradicts $(qI - L^\sharp)f = 0$; if $f^- \neq 0$, then we have the same absurdity because $f^- = (-f)^+$. The operator $qI - L^\sharp$ is an extension of $qI - L$, but $\mathfrak{R}(qI - L)$ is already the whole space C_0 by Theorem 31.3. So the one-to-one property of $qI - L^\sharp$ shows that $qI - L^\sharp = qI - L$. Hence $L^\sharp = L$. $\square$

Proof of Theorem 31.5. Let us see that $\{P_t\}$ is a strongly continuous semigroup with $\|P_t\| = 1$. The property (31.1) comes from the Chapman–Kolmogorov identity (10.2), (31.2) is evident, and (31.3) is proved as follows. Let $f \in C_0$. It is easy to see that f is uniformly continuous on $\mathbb{R}^d$. Given $\varepsilon > 0$, choose $\delta > 0$ so that $|f(x+y) - f(x)| < \varepsilon$ whenever $|y| \leq \delta$. Then,

$$\begin{aligned}
&|P_t f(x) - f(x)| \\
&\quad \leq \left| \int_{|y| \leq \delta} \mu^t(\mathrm{d}y)(f(x+y) - f(x)) \right| + \left| \int_{|y| > \delta} \mu^t(\mathrm{d}y)(f(x+y) - f(x)) \right| \\
&\quad \leq \varepsilon + 2\|f\| \mu^t\{y \colon |y| > \delta\} \\
&\quad \leq \varepsilon + 2\|f\|\varepsilon
\end{aligned}$$

for small t, because of the stochastic continuity. Hence $\|P_t f - f\| \to 0$ as $t \downarrow 0$. This is (31.3). Also $\|P_t\| \leq 1$ is evident from (31.10). Choose $f_n \in C_0$ such that $0 \leq f_n \leq 1$ and $f_n(x) = 1$ for $|x| \leq n$. Then $\lim_{n \to \infty} P_t f_n(x) = 1$. Hence $\|P_t\| = 1$.

Now let $f \in C_0^2$. We use Lemma 31.7 to prove that $f \in \mathfrak{D}(L)$ and that Lf is of the form (31.11). Define $L_0 f$ by the right-hand side of (31.11). Write $D_j = \frac{\partial}{\partial x_j}$ and $D_{jk} = \frac{\partial^2}{\partial x_j \partial x_k}$. By Taylor's theorem

$$\left| f(x+y) - f(x) - \sum_{j=1}^{d} y_j D_j f(x) \right| \leq \frac{1}{2}|y|^2 \sum_{j,k=1}^{d} \|D_{jk} f\|. \tag{31.15}$$

Hence $L_0 f$ belongs to C_0. We claim that (31.14) holds with $g = L_0 f$. Let $t_n \downarrow 0$. We have

$$\exp(t_n^{-1}(\widehat{\mu}(z)^{t_n} - 1)) = \exp(t_n^{-1}(\mathrm{e}^{t_n \log \widehat{\mu}(z)} - 1)) \to \widehat{\mu}(z)$$

as $n \to \infty$. In order to apply Theorem 8.7, we use a bounded continuous function $c(x)$ from $\mathbb{R}^d$ to $\mathbb{R}$ satisfying (8.3) and (8.4). Then

$$\begin{aligned}\exp(t_n^{-1}(\widehat{\mu}(z)^{t_n} - 1)) &= \exp\left[t_n^{-1} \int_{\mathbb{R}^d \setminus \{0\}} (\mathrm{e}^{\mathrm{i}\langle z,y\rangle} - 1)\, \mu^{t_n}(\mathrm{d}y)\right] \\ &= \exp\left[i\langle \beta_n, z\rangle + \int (\mathrm{e}^{\mathrm{i}\langle z,y\rangle} - 1 - \mathrm{i}\langle z, y\rangle c(y))\nu_n(\mathrm{d}y)\right],\end{aligned}$$

where $\nu_n = [t_n^{-1}\mu^{t_n}]_{\mathbb{R}^d\setminus\{0\}}$ and $\beta_n = \int y c(y) \nu_n(\mathrm{d}y)$. On the other hand $\widehat{\mu}(z)$ has the Lévy–Khintchine representation $(A, \nu, \beta)_c$ with $\beta = \gamma + \int y(c(y) - 1_D(y))\, \nu(\mathrm{d}y)$. Thus, we have convergence of ν_n, $A_{n,\varepsilon}$, and β_n to ν, A, and β as described in (1), (2), and (3) of Theorem 8.7. Now choose $\varepsilon_m \downarrow 0$ such that $\int_{|y|=\varepsilon_m} \nu(\mathrm{d}y) = 0$. Then, decomposing

$$\begin{aligned} t_n^{-1}(P_{t_n} f(0) - f(0)) &= \int (f(y) - f(0))\, \nu_n(\mathrm{d}y) = I_1 + I_2 + I_3, \\ I_1 &= \int_{|y|\leq\varepsilon_m} \Big(f(y) - f(0) - \sum_j y_j D_j f(0) c(y)\Big)\, \nu_n(\mathrm{d}y), \\ I_2 &= \int_{|y|>\varepsilon_m} \Big(f(y) - f(0) - \sum_j y_j D_j f(0) c(y)\Big)\, \nu_n(\mathrm{d}y), \\ I_3 &= \sum_j \int y_j c(y)\, \nu_n(\mathrm{d}y)\, D_j f(0), \end{aligned}$$

we get

$$\lim_{n\to\infty} I_2 = \int_{|y|>\varepsilon_m} \Big(f(y) - f(0) - \sum_j y_j D_j f(0) c(y)\Big)\, \nu(\mathrm{d}y)$$

from (1), and

$$\lim_{n\to\infty} I_3 = \sum_j (\beta)_j D_j f(0)$$

from (3). Here $(\beta)_j$ is the jth component of β. The statement (2) is, in our case, equivalent to

$$\lim_{\varepsilon\downarrow 0}\limsup_{n\to\infty}\int_{|y|\le\varepsilon} y_j y_k\,\nu_n(\mathrm{d}y)=\lim_{\varepsilon\downarrow 0}\liminf_{n\to\infty}\int_{|y|\le\varepsilon} y_j y_k\,\nu_n(\mathrm{d}y)=A_{jk}.$$

Hence, noting that

$$f(y)-f(0)-\sum_j y_j D_j f(0)=\frac{1}{2}\sum_{j,k} y_j y_k D_{jk} f(0)+o(|y|^2)$$

and $c(y)=1+o(|y|)$, we get

$$\lim_{m\to\infty}\limsup_{n\to\infty} I_1=\lim_{m\to\infty}\liminf_{n\to\infty} I_1=\frac{1}{2}\sum_{j,k} A_{jk} D_{jk} f(0).$$

Therefore

$$\lim_{n\to\infty} {t_n}^{-1}(P_{t_n} f(0)-f(0))=L_0 f(0).$$

Since $t_n\downarrow 0$ is arbitrary, we get

$$\lim_{t\downarrow 0} t^{-1}(P_t f(0)-f(0))=L_0 f(0).$$

To deal with an arbitrary x, let M_x be the translation operator

$$(M_x f)(y)=f(x+y).$$

Then $f(x)=(M_x f)(0)$ and $(P_t f)(x)=(M_x P_t f)(0)=(P_t M_x f)(0)$. Hence

$$\lim_{t\downarrow 0} t^{-1}(P_t f(x)-f(x))=(L_0 M_x f)(0)=(L_0 f)(x).$$

It remains to consider cores. If $f\in C_0^1$, then, from (31.10),

$$(D_j P_t f)(x)=\int \mu^t(\mathrm{d}y) D_j f(x+y)$$

and $P_t f\in C_0^1$. Repeating this argument, we see that P_t maps C_0^n into itself. Hence P_t maps C_0^∞ into itself. It is easy to see that C_0^∞ is dense in C_0. Hence, by Lemma 31.6, C_0^∞ is a core of L. Let $\psi(r)$ be a C^∞ function on $[0,\infty)$ such that $\psi(r)=1$ on $[0,1]$ and $\psi(r)=0$ on $[2,\infty)$. Given $f\in C_0^\infty$, let $f_n(x)=f(x)\psi(|x|^2/n^2)$. Then $f_n\in C_c^\infty$ and f_n, $D_j f_n$, and $D_{jk} f_n$ strongly converge to f, $D_j f$, and $D_{jk} f$, respectively, as $n\to\infty$. It follows that

$$L_0 f_n\to L_0 f\quad\text{(strong)},$$

since

$$\|L_0 g\|\le \mathrm{const}\Big(\|g\|+\sum_j\|D_j g\|+\sum_{j,k}\|D_{jk} g\|\Big)$$

for any $g\in C_0^\infty$ by (31.15). This implies that not only C_0^∞ but also C_c^∞ is a core of L. □

EXAMPLE 31.8. If $\{X_t\}$ is the Brownian motion on $\mathbb{R}^d$, then, for $f \in C_0^2$, $Lf = \frac{1}{2}\Delta f$, where Δ is the *Laplacian*, $\Delta = \sum_{j=1}^d (\frac{\partial}{\partial x_j})^2$. If $\{X_t\}$ is a compound Poisson process on $\mathbb{R}^d$ with Lévy measure ν, then L is a bounded operator and

$$Lf(x) = \int_{\mathbb{R}^d} (f(x+y) - f(x))\,\nu(\mathrm{d}y). \tag{31.16}$$

In fact, (31.16) is true for $f \in C_0^2$ by Theorem 31.5 and this shows $\|Lf\| \leq 2\|f\|\nu(\mathbb{R}^d)$. Hence, by the closedness of L, $\mathfrak{D}(L)$ is the whole space and (31.16) holds for $f \in C_0$.

REMARK 31.9. Let $P_t(x, B)$ be a temporally homogeneous transition function on $\mathbb{R}^d$ and define $(P_t f)(x) = \int P_t(x, \mathrm{d}y) f(y)$. Suppose that, if $f \in C_0$, then $P_t f \in C_0$ and $P_t f(x) \to f(x)$ for each x as $t \downarrow 0$. Then we can prove that $\{P_t\}$ is a strongly continuous semigroup on C_0 with $\|P_t\| = 1$, although the proof above no longer works. See Loève [**327**].

REMARK 31.10. Let L be the infinitesimal generator of the semigroup $\{P_t\}$ on $C_0(\mathbb{R}^d)$ induced by a Lévy process $\{X_t\}$ on $\mathbb{R}^d$. Assume that $\{X_t\}$ is not the zero process. Then, it can be proved that L is one-to-one and that the range of L is dense in $C_0(\mathbb{R}^d)$. The operator $V = -L^{-1}$ is the *potential operator* of $\{P_t\}$ in the sense of Yosida [**598**]. By general theory, f is in $\mathfrak{D}(V)$ and $Vf = g$ if and only if $\lim_{q \downarrow 0} U^q f$ (strong) exists and equals g.

32. Subordination of semigroups of operators

Let $\mathbf{B}$ be a real (or complex) Banach space. Combining Bochner's idea and the Hille–Yosida theorem, Phillips [**373**] proved the following theorem. The assertion on cores was made explicit in [**439**].

THEOREM 32.1. *Let $\{Z_t : t \geq 0\}$ be a subordinator with Lévy measure ρ, drift β_0, and $P_{Z_1} = \lambda$. Let $\{P_t : t \geq 0\}$ be a strongly continuous contraction semigroup of linear operators on $\mathbf{B}$ with infinitesimal generator L. Define*

$$Q_t f = \int_{[0,\infty)} P_s f \lambda^t(\mathrm{d}s), \qquad f \in \mathbf{B}. \tag{32.1}$$

Then $\{Q_t : t \geq 0\}$ is a strongly continuous contraction semigroup of linear operators on $\mathbf{B}$. Denote its infinitesimal generator by M. Then $\mathfrak{D}(L)$ is a core of M and

$$Mf = \beta_0 Lf + \int_{(0,\infty)} (P_s f - f)\rho(\mathrm{d}s), \qquad f \in \mathfrak{D}(L). \tag{32.2}$$

The definition of the integral in the right-hand side of (32.1) is as follows. First define

$$\int_{[0,r]} P_s f \lambda^t(\mathrm{d}s) = \lim_{\mathrm{mesh}(\Delta)\to 0} \left(P_0 f \lambda^t\{0\} + \sum_{j=1}^{n} P_{s_{j-1}} f \, \lambda^t(s_{j-1}, s_j] \right) \text{ (strong)},$$

where Δ is a partition $0 = s_0 < s_1 < \cdots < s_n = r$. It is definable because $P_s f$ is strongly continuous in s. Then

$$\int_{[0,\infty)} P_s f \, \lambda^t(\mathrm{d}s) = \lim_{r\to\infty} \int_{[0,r]} P_s f \, \lambda^t(\mathrm{d}s) \text{ (strong)}.$$

The right-hand side exists, since $\|P_s f\| \le \|f\|$ and $\lambda^t(r,\infty) \to 0$ as $r \to \infty$. Similarly the integral in the right-hand side of (32.2) is defined to be the strong limit of $\int_{r_1}^{r_2} (P_s f - f)\rho(\mathrm{d}s)$ as $r_1 \downarrow 0$ and $r_2 \to \infty$. It is definable since $f \in \mathfrak{D}(L)$ implies $\|P_s f - f\| = O(s)$ as $s \downarrow 0$ and since $\int_{(0,\infty)} (1 \wedge s)\rho(\mathrm{d}s) < \infty$.

DEFINITION 32.2. The transformation of $\{P_t\}$ to $\{Q_t\}$ in Theorem 32.1 is called *subordination* by the subordinator $\{Z_t\}$. The semigroup $\{Q_t\}$ and its infinitesimal generator M are said to be *subordinate* to $\{P_t\}$ and L, respectively. Using the function Ψ in (30.2), we denote M by $\Psi(L)$.

In the proof of the theorem above, we use the notion of weak convergence. A mapping l from $\mathbf{B}$ to $\mathbb{R}$ (or $\mathbb{C}$) is called a *linear functional* if $l(af + bg) = al(f) + bl(g)$. A linear functional l is said to be *continuous* if $f_n \to f$ (strong) implies $l(f_n) \to l(f)$ or, equivalently, if $\sup_{\|f\|\le 1} |l(f)|$ is finite. Denote by $\mathbf{B}'$ the space of continuous linear functionals of $\mathbf{B}$. It is a real (or complex) Banach space with the norm $\|l\| = \sup_{\|f\|\le 1} |l(f)|$, called the *dual space* of $\mathbf{B}$. A sequence $\{f_n \colon n = 1, 2, \dots\}$ in $\mathbf{B}$ is called *weakly convergent* to $f \in \mathbf{B}$, written $f_n \to f$ (weak), if $l(f_n) \to l(f)$ for any $l \in \mathbf{B}'$. If $\{f_n\}$ is weakly convergent to f and to g, then $f = g$, since $l(f) = l(g)$ for all $l \in \mathbf{B}'$ implies $f = g$ by the Hahn–Banach theorem.

LEMMA 32.3. *Let $\{P_t\}$ be a strongly continuous contraction semigroup of linear operators on $\mathbf{B}$. Let L be its infinitesimal generator. If f and g in $\mathbf{B}$ satisfy*

$$t^{-1}(P_t f - f) \to g \text{ (weak) as } t \downarrow 0, \tag{32.3}$$

then $f \in \mathfrak{D}(L)$ and $Lf = g$.

Proof. Let U^q be the resolvent of L. Let $g = L^\sharp f$ if (32.3) holds. The linear operator $L^\sharp$ thus defined is an extension of L. Let $f \in \mathfrak{D}(L^\sharp)$ and $q > 0$. We claim that

$$f = 0 \quad \text{if } qf - L^\sharp f = 0. \tag{32.4}$$

Suppose that this is proved. Then, for any $f \in \mathfrak{D}(L^\sharp)$, the element $h = qf - L^\sharp f$ satisfies $(qI - L)U^q h = h$ and hence $(qI - L^\sharp)U^q h = h$, which implies

$f = U^q h \in \mathfrak{D}(L)$ by (32.4). Therefore $L^\sharp = L$. To show (32.4), assume $qf - L^\sharp f = 0$. Write $t^{-1}(P_t f - f) - L^\sharp f = k_t$. Then $P_t f = tk_t + tqf + f$ and thus $|l(P_t f)| \geq (tq+1)|l(f)| - t|l(k_t)|$ for $l \in \mathbf{B}'$. Choose l satisfying $|l(f)| = \|f\|$ and $\|l\| = 1$, which is possible by the Hahn–Banach theorem. Then we have $0 \geq tq\|f\| - t|l(k_t)|$. Since $l(k_t) \to 0$ as $t \downarrow 0$, it follows that $\|f\| = 0$. □

Proof of Theorem 32.1. It is evident from (32.1) that Q_t is a linear operator with $\|Q_t\| \leq 1$ and $Q_0 = I$. As $t \downarrow 0$,

$$\|Q_t f - f\| \leq \int_{[0,\infty)} \|P_s f - f\| \lambda^t(\mathrm{d}s) \to \|P_0 f - f\| = 0,$$

since $\lambda^t \to \delta_0$. We have

$$P_r Q_t f = \int_{[0,\infty)} P_{r+s} f \lambda^t(\mathrm{d}s) \quad \text{for } r \geq 0 \text{ and } t \geq 0, \tag{32.5}$$

and, for any $l \in \mathbf{B}'$,

$$\begin{aligned} l(Q_u Q_t f) &= \int l(P_r Q_t f) \lambda^u(\mathrm{d}r) = \iint l(P_{r+s} f) \lambda^t(\mathrm{d}s) \lambda^u(\mathrm{d}r) \\ &= \int l(P_s f) \lambda^{t+u}(\mathrm{d}s) = l(Q_{u+t} f), \end{aligned}$$

using (32.5). Hence $Q_u Q_t = Q_{u+t}$. Thus $\{Q_t\}$ is a strongly continuous contraction semigroup. Recall that we have (30.1), (30.2), and (30.3) for the subordinator $\{Z_t\}$. Let $c(x)$ be a nonnegative continuous function on $\mathbb{R}$ satisfying (8.3) and (8.4). As in the proof of Theorem 24.7, it follows from $\exp[t^{-1}(\widehat{\lambda}(z)^t - 1)] \to \widehat{\lambda}(z)$ as $t \downarrow 0$ that

$$t^{-1} \int_{[0,\infty)} g(s) \lambda^t(\mathrm{d}s) \to \int_{(0,\infty)} g(s) \rho(\mathrm{d}s) \quad \text{as } t \downarrow 0 \tag{32.6}$$

for any bounded continuous function $g(s)$ on $(0,\infty)$ which vanishes on a neighborhood of 0 and that

$$t^{-1} \int_{[0,\infty)} s\, c(s) \lambda^t(\mathrm{d}s) \to \beta_0 + \int_{(0,\infty)} s\, c(s) \rho(\mathrm{d}s) \quad \text{as } t \downarrow 0. \tag{32.7}$$

Hence (32.6) holds for any bounded continuous function $g(s)$ satisfying $g(s) = o(s)$ as $s \downarrow 0$. Let $f \in \mathfrak{D}(L)$. We get, for any $l \in \mathbf{B}'$,

$$\begin{aligned} l(t^{-1}(Q_t f - f)) &= t^{-1} \int l(P_s f - f) \lambda^t(\mathrm{d}s) \\ &= t^{-1} \int l(P_s f - f)(1 - c(s)) \lambda^t(\mathrm{d}s) \\ &\quad + t^{-1} \int l(P_s f - f - sLf) c(s) \lambda^t(\mathrm{d}s) + l(Lf) t^{-1} \int s\, c(s) \lambda^t(\mathrm{d}s) \end{aligned}$$

$$\begin{aligned}&\to \int l(P_sf-f)(1-c(s))\rho(\mathrm{d}s)\\&\quad+\int l(P_sf-f-sLf)c(s)\rho(\mathrm{d}s)+l(Lf)\left(\beta_0+\int s\,c(s)\rho(\mathrm{d}s)\right)\\&=l\left(\beta_0Lf+\int l(P_sf-f)\rho(\mathrm{d}s)\right)\end{aligned}$$

as $t\downarrow 0$. Hence, by Lemma 32.3, $f\in\mathfrak{D}(M)$ and (32.2) holds.

Let $f\in\mathfrak{D}(L)$. In order to prove that $\mathfrak{D}(L)$ is a core of M, it suffices to show that $Q_tf\in\mathfrak{D}(L)$ for any $t>0$ (Lemma 31.6). Let $g=\int P_sLf\lambda^t(\mathrm{d}s)$. Then, by (32.5),

$$\begin{aligned}\|r^{-1}(P_rQ_tf-Q_tf)-g\|&\le\int\|r^{-1}(P_{r+s}f-P_sf)-P_sLf\|\lambda^t(\mathrm{d}s)\\&\le\|r^{-1}(P_rf-f)-Lf\|\to 0,\qquad r\downarrow 0.\end{aligned}$$

Hence $Q_tf\in\mathfrak{D}(L)$ and $LQ_tf=g$. □

REMARK 32.4. Consider the case $\mathbf{B}=C_0(\mathbb{R}^d)$. For $f\in C_0(\mathbb{R}^d)$, P_sf is a real-valued continuous function, so that $\int_{[0,\infty)}P_sf(x)\lambda^t(\mathrm{d}s)$ is defined as an ordinary integral. It coincides with Q_tf defined in (32.1). Hence, if $\{P_t\}$ has positivity, so does $\{Q_t\}$. If $\{P_t\}$ is the semigroup induced by a temporally homogeneous transition function as in Remark 31.9, so is the semigroup $\{Q_t\}$.

Let us prove some properties of subordinate infinitesimal generators.

PROPOSITION 32.5. *Let L and M be as in Theorem 32.1.*
(i) *If $f\in\mathfrak{D}(L)$, then*

$$\|Mf\|\le\left(\beta_0+\int_{(0,1]}s\rho(\mathrm{d}s)\right)\|Lf\|+2\int_{(1,\infty)}\rho(\mathrm{d}s)\|f\|. \tag{32.8}$$

(ii) *If a linear subspace $\mathfrak{D}_0$ is a core of L, then $\mathfrak{D}_0$ is a core of M.*
(iii) *Assume that $\beta_0=0$. Then, the collection of all $f\in\mathbf{B}$ satisfying $\|P_tf-f\|=O(t)$, $t\downarrow 0$, is a core of M and, for such f,*

$$Mf=\int_{(0,\infty)}(P_sf-f)\rho(\mathrm{d}s). \tag{32.9}$$

Proof. (i) If $f\in\mathfrak{D}_0$, then

$$P_tf-f=\int_0^tP_sLf\,\mathrm{d}s, \tag{32.10}$$

because we have

$$h^{-1}(P_h-I)\int_0^tP_sf\,\mathrm{d}s=h^{-1}\left(\int_t^{t+h}P_sf\,\mathrm{d}s-\int_0^hP_sf\,\mathrm{d}s\right)\to P_tf-f,$$

$$h^{-1}(P_h - I)\int_0^t P_s f\, \mathrm{d}s = \int_0^t P_s(h^{-1}(P_h - I)f)\mathrm{d}s \to \int_0^t P_s Lf\, \mathrm{d}s,$$

both strongly. It follows from (32.10) that $\|P_t f - f\| \le (t\|Lf\|) \wedge (2\|f\|)$. Recalling (32.2), we get (32.8).

(ii) Let $\mathfrak{D}_0$ be a core of L. Write $M_0 = [M]_{\mathfrak{D}_0}$. For any $f \in \mathfrak{D}(L)$, there are $f_n \in \mathfrak{D}_0$ such that $f_n \to f$ and $Lf_n \to Lf$, both strongly. Therefore, by (i), $M_0 f_n \to Mf$ (strong). It follows that the closure $\overline{M_0}$ of M_0 is an extension of $[M]_{\mathfrak{D}(L)}$. Since $\mathfrak{D}(L)$ is a core of M by Theorem 32.1, we see that $\overline{M_0} = M$.

(iii) Let $\beta_0 = 0$ and $\|P_t f - f\| = O(t)$, $t \downarrow 0$. Let U^q be the resolvent operator of L. Then $qU^q f \in \mathfrak{D}(L) \subset \mathfrak{D}(M)$ and $qU^q f \to f$, $q \to \infty$. Further

$$M(qU^q f) \to \int_{(0,\infty)} (P_s f - f)\rho(\mathrm{d}s) \text{ (strong)},$$

since

$$M(qU^q f) = \int_{(0,\infty)} (P_s - I)(qU^q f)\rho(\mathrm{d}s)$$

and $\|(P_s - I)(qU^q f)\| = \|q\int_0^\infty \mathrm{e}^{-qt}(P_{s+t}f - P_t f)\mathrm{d}t\| \le \|P_s f - f\|$. This shows that $f \in \mathfrak{D}(M)$ and (32.9), since M is a closed operator. The collection of such f is a core of M because it includes $\mathfrak{D}(L)$. □

An important example of subordination is to make fractional powers of the negatives of infinitesimal generators.

EXAMPLE 32.6. Let $\{Z_t\}$ be a strictly α-stable subordinator. Then $0 < \alpha < 1$ and $\Psi(-u) = -c'u^\alpha$ for $u \ge 0$ with $c' > 0$. Assume that $c' = 1$. Then it is natural to write the operator $\Psi(L)$ as $\Psi(L) = -(-L)^\alpha$ for the infinitesimal generator L of $\{P_t\}$. We have

$$\Psi(-u) = -u^\alpha = \frac{\alpha}{\Gamma(1-\alpha)}\int_0^\infty (\mathrm{e}^{-us} - 1)s^{-1-\alpha}\mathrm{d}s, \qquad u \ge 0$$

(see Example 24.12). Thus, if $f \in \mathfrak{D}(L)$, or more generally, if $\|P_t f - f\| = O(t)$, $t \downarrow 0$, then $f \in \mathfrak{D}((-L)^\alpha)$ and

$$-(-L)^\alpha f = \frac{\alpha}{\Gamma(1-\alpha)}\int_0^\infty (P_s f - f)s^{-1-\alpha}\mathrm{d}s \tag{32.11}$$

by Theorem 32.1 or Proposition 32.5.

We can prove that

$$((-L)^\alpha)^\beta = (-L)^{\alpha\beta} \quad \text{for } 0 < \alpha < 1,\ 0 < \beta < 1, \tag{32.12}$$

$$(-L)^\beta(-L)^\alpha = (-L)^{\alpha+\beta} \quad \text{for } 0 < \alpha < 1,\ 0 < \beta \le 1 - \alpha, \tag{32.13}$$

$$\mathfrak{D}((-L)^\alpha) \subset \mathfrak{D}((-L)^\beta) \quad \text{for } 0 < \beta \le \alpha < 1. \tag{32.14}$$

The equality of two linear operators means that they are identical mappings with identical domains, and the products of linear operators are defined as in Exercise 34.6. The equality (32.12) follows from (30.19); (32.14) follows from (32.12).

EXAMPLE 32.7. Let $0 < \alpha < 1$. The rotation invariant 2α-stable process $\{Y_t\}$ on $\mathbb{R}^d$ is subordinate to the Brownian motion on $\mathbb{R}^d$ by a strictly α-stable subordinator, and we have (30.20) in Example 30.6. Hence, the infinitesimal generator M of $\{Y_t\}$ in $C_0(\mathbb{R}^d)$ is expressible as

$$M = -2^{-\alpha}c'(-\overline{\Delta})^\alpha,$$

where the Laplacian Δ is defined on C_c^∞ and $\overline{\Delta}$ is the closure of Δ (Theorem 31.5 and Example 31.8). Let $f \in C_0^2$. Expressing the infinitesimal generator by Theorem 31.5, Remark 14.6, and Theorem 14.7 and using the constant c_0 in Exercise 18.9, we get the following formulas. Let λ_0 be the uniform probability measure on the unit sphere S. If $0 < \alpha < 1/2$, then

$$-(-\overline{\Delta})^\alpha f(x) = \frac{1}{c_0}\int_S \lambda_0(\mathrm{d}\xi)\int_0^\infty (f(x+r\xi) - f(x))r^{-1-2\alpha}\mathrm{d}r.$$

If $1/2 \le \alpha < 1$, then

$$-(-\overline{\Delta})^\alpha f(x) = \frac{1}{2c_0}\int_S \lambda_0(\mathrm{d}\xi)\int_0^\infty (f(x+r\xi) + f(x-r\xi) - 2f(x))r^{-1-2\alpha}\mathrm{d}r.$$

EXAMPLE 32.8. Let $\{Z_t\}$ be a Γ-process. Then $\Psi(-u) = -\log(1+\frac{u}{r})$ as in Example 30.8. Thus it is natural to write the operator $\Psi(L)$ as $-\log(I - \frac{1}{r}L)$. This is a way to define the logarithm of an operator.

33. Density transformation of Lévy processes

Fix $d \ge 1$ and let $\mathbf{D} = D([0,\infty),\mathbb{R}^d)$ be the space of mappings ξ from $[0,\infty)$ into $\mathbb{R}^d$ right-continuous with left limits. Write $x_t(\xi) = x(t,\xi) = \xi(t)$. Let $\mathcal{F}_\mathbf{D}$ be the smallest σ-algebra that makes x_t, $t \in [0,\infty)$, measurable. These are already familiar objects from Chapter 4. Further let $\mathcal{F}_t$ be the smallest σ-algebra that makes x_s, $s \in [0,t]$, measurable. In this section we denote a Lévy process $\{X_t : t \ge 0\}$ defined on a probability space with a probability measure P^0 by $(\{X_t\}, P^0)$, stressing the probability measure. Any Lévy process $(\{X_t\}, P^0)$ induces a probability measure $P^\mathbf{D}$ on $(\mathbf{D}, \mathcal{F}_\mathbf{D})$ such that $(\{x_t\}, P^\mathbf{D})$ is a Lévy process identical in law with $(\{X_t\}, P^0)$. The Lévy processes that we consider in this section are of the form $(\{x_t\}, P)$, where P is a probability measure on $(\mathbf{D}, \mathcal{F}_\mathbf{D})$. For two Lévy processes $(\{x_t\}, P)$ and $(\{x_t\}, P^\sharp)$, what is the condition for the mutual absolute continuity of $[P]_{\mathcal{F}_t}$ and $[P^\sharp]_{\mathcal{F}_t}$ for every t? This problem was solved by Skorohod [**490, 492**], Kunita and Watanabe [**300**], and Newman [**360, 361**]. It is connected with the work of Kakutani [**252**]. We reformulate and prove their results.

Two measures ρ_1, ρ_2 on a common measurable space $(M, \mathcal{F}_M)$ are called *mutually absolutely continuous,* written $\rho_1 \approx \rho_2$, if the collection $\{B \in \mathcal{F}_M \colon \rho_1(B) = 0\}$ is identical with $\{B \in \mathcal{F}_M \colon \rho_2(B) = 0\}$. The Radon–Nikodým derivative of ρ_2 with respect to ρ_1 is denoted by $\frac{\mathrm{d}\rho_2}{\mathrm{d}\rho_1}$. If $\rho_1 \approx \rho_2$, then $\frac{\mathrm{d}\rho_2}{\mathrm{d}\rho_1}$ is positive and finite ρ_1-almost everywhere or, equivalently, ρ_2-almost everywhere. For a $d \times d$ matrix A we write $\mathfrak{R}(A) = \{Ax \colon x \in \mathbb{R}^d\}$. The expectation under the measure P is denoted by E^P.

Our main results are the following two theorems. The first one describes a necessary and sufficient condition in terms of generating triplets. The second one gives an expression for the densities on the path space.

THEOREM 33.1. *Let $(\{x_t\}, P)$ and $(\{x_t\}, P^\sharp)$ be Lévy processes on $\mathbb{R}^d$ with generating triplets (A, ν, γ) and $(A^\sharp, \nu^\sharp, \gamma^\sharp)$, respectively. Then the following two statements (1) and (2) are equivalent.*

(1) *$[P]_{\mathcal{F}_t} \approx [P^\sharp]_{\mathcal{F}_t}$ for every $t \in (0, \infty)$.*
(2) *The generating triplets satisfy*

$$A = A^\sharp, \tag{33.1}$$

$$\nu \approx \nu^\sharp \tag{33.2}$$

with the function $\varphi(x)$ defined by $\frac{\mathrm{d}\nu^\sharp}{\mathrm{d}\nu} = \mathrm{e}^{\varphi(x)}$ satisfying

$$\int_{\mathbb{R}^d} (\mathrm{e}^{\varphi(x)/2} - 1)^2 \nu(\mathrm{d}x) < \infty, \tag{33.3}$$

and

$$\gamma^\sharp - \gamma - \int_{|x| \le 1} x(\nu^\sharp - \nu)(\mathrm{d}x) \in \mathfrak{R}(A). \tag{33.4}$$

The condition (33.3) is equivalent to finiteness of the Hellinger–Kakutani distance between ν and $\nu^\sharp$, but we do not introduce that. The integral in (33.3) is symbolically written as $\int_{\mathbb{R}^d} (\sqrt{\mathrm{d}\nu^\sharp} - \sqrt{\mathrm{d}\nu})^2$. Finiteness of the integral appearing in (33.4) follows from (33.3) (Remark 33.3).

Given a Lévy process $(\{x_t\}, P)$ with generating triplet (A, ν, γ), we can define the jump part

$$x_t^\nu(\xi) = \lim_{\varepsilon \downarrow 0} \Bigg(\sum_{(s, x_s - x_{s-}) \in (0,t] \times \{|x| > \varepsilon\}} (x_s - x_{s-}) - t \int_{\varepsilon < |x| \le 1} x \nu(\mathrm{d}x) \Bigg), \tag{33.5}$$

where the convergence in the right-hand side is uniform in t on any bounded time interval, P-a. s. The Lévy–Itô decomposition in Theorem 19.2 guarantees this fact and, moreover, $(\{x_t^\nu\}, P)$ is a Lévy process with generating triplet $(0, \nu, 0)$. The process $(\{x_t - x_t^\nu\}, P)$ is the continuous part of $(\{x_t\}, P)$. It is a Lévy process with generating triplet $(A, 0, \gamma)$. The two processes $(\{x_t^\nu\}, P)$ and $(\{x_t - x_t^\nu\}, P)$ are independent.

THEOREM 33.2. *Let $(\{x_t\}, P)$ and $(\{x_t\}, P^\sharp)$ be Lévy processes on $\mathbb{R}^d$ with generating triplets (A, ν, γ) and $(A^\sharp, \nu^\sharp, \gamma^\sharp)$, respectively. Suppose that the equivalent conditions (1) and (2) in the previous theorem are satisfied. Choose $\eta \in \mathbb{R}^d$ such that*

$$\gamma^\sharp - \gamma - \int_{|x|\le 1} x(\nu^\sharp - \nu)(\mathrm{d}x) = A\eta. \tag{33.6}$$

Then we can define, P-a. s.,

$$\begin{aligned} U_t = \langle \eta, x_t - x_t^\nu \rangle - \frac{t}{2}\langle \eta, A\eta \rangle - t\langle \gamma, \eta \rangle \\ + \lim_{\varepsilon \downarrow 0} \Bigg(\sum_{(s, x_s - x_{s-}) \in (0,t] \times \{|x| > \varepsilon\}} \varphi(x_s - x_{s-}) - t\int_{|x|>\varepsilon} (\mathrm{e}^{\varphi(x)} - 1)\nu(\mathrm{d}x) \Bigg), \end{aligned} \tag{33.7}$$

where φ is the function in (2) and $(\{x_t - x_t^\nu\}, P)$ is the continuous part of $(\{x_t\}, P)$. The convergence in the right-hand side of (33.7) is uniform in t on any bounded interval, P-a. s. We have, for every $t \in [0, \infty)$,

$$E^P[\mathrm{e}^{U_t}] = E^{P^\sharp}[\mathrm{e}^{-U_t}] = 1 \tag{33.8}$$

and

$$\frac{\mathrm{d}[P^\sharp]_{\mathcal{F}_t}}{\mathrm{d}[P]_{\mathcal{F}_t}} = \mathrm{e}^{U_t}, \quad P\text{-a. s.} \tag{33.9}$$

The process $(\{U_t\}, P)$ is a Lévy process on $\mathbb{R}$ with generating triplet (A_U, ν_U, γ_U) expressed by

$$A_U = \langle \eta, A\eta \rangle, \tag{33.10}$$

$$\nu_U = [\nu\varphi^{-1}]_{\mathbb{R}\setminus\{0\}}, \tag{33.11}$$

$$\gamma_U = -\frac{1}{2}\langle \eta, A\eta \rangle - \int_{\mathbb{R}} (\mathrm{e}^y - 1 - y\,1_{\{0<|y|\le 1\}}(y))(\nu\varphi^{-1})(\mathrm{d}y). \tag{33.12}$$

Notice that, if two Lévy processes satisfy the condition (1) of mutual absolute continuity in Theorem 33.1, then they have any almost sure local behavior of sample functions in common.

REMARK 33.3. The property (33.3) is equivalent to the following three properties combined:

$$\int_{\{x:\,|\varphi(x)|\le 1\}} \varphi(x)^2\nu(\mathrm{d}x) < \infty, \tag{33.13}$$

$$\int_{\{x:\,\varphi(x)>1\}} \mathrm{e}^{\varphi(x)}\nu(\mathrm{d}x) < \infty, \tag{33.14}$$

$$\int_{\{x:\,\varphi(x)<-1\}} \nu(\mathrm{d}x) < \infty. \tag{33.15}$$

To see this, denote by $f(u) \asymp g(u)$ the existence of two positive constants c_1 and c_2 such that $c_1 g(u) \le f(u) \le c_2 g(u)$. Then $|\mathrm{e}^{u/2} - 1| \asymp |u|$ for $|u| \le 1$, $(\mathrm{e}^{u/2} - 1)^2 \asymp \mathrm{e}^u$ for $u > 1$, and $(\mathrm{e}^{u/2} - 1)^2 \asymp 1$ for $u < -1$, hence the equivalence that we have asserted. Let us further note the finiteness of some integrals. Assume (33.3) and hence (33.13)–(33.15). Denote some sets in $\mathbb{R}^d$ as follows: $B_0 = \{|x| \le 1\}$, $B_1 = \{|x| > 1\}$, $B_\varepsilon = \{|x| > \varepsilon\}$, $C_0 = \{|\varphi(x)| \le 1\}$, $C_1 = \{|\varphi(x)| > 1\}$, and

$$B_{pq} = B_p \cap C_q \quad \text{for } p = 0, 1, \varepsilon \text{ and } q = 0, 1.$$

Then $\int_{C_1} \nu(\mathrm{d}x)$, $\int_{B_1} \mathrm{e}^{\varphi(x)}\nu(\mathrm{d}x)$, and $\int_{C_1} \mathrm{e}^{\varphi(x)}\nu(\mathrm{d}x)$ are finite,

$$\int_{B_0} |x|^2 \mathrm{e}^{\varphi(x)}\nu(\mathrm{d}x) \le \mathrm{e}\int_{B_{00}} |x|^2\nu(\mathrm{d}x) + \int_{C_1} \mathrm{e}^{\varphi(x)}\nu(\mathrm{d}x) < \infty,$$

and

$$\begin{aligned}\int_{B_0} |x|\,|\mathrm{e}^{\varphi(x)} - 1|\nu(\mathrm{d}x) &\le \int_{B_{00}} |x|\,|\mathrm{e}^{\varphi(x)} - 1|\nu(\mathrm{d}x) + \int_{B_{01}} |\mathrm{e}^{\varphi(x)} - 1|\nu(\mathrm{d}x)\\ &\le \left(\int_{B_0} |x|^2\nu(\mathrm{d}x)\right)^{1/2}\left(\int_{C_0} (\mathrm{e}^{\varphi(x)} - 1)^2\nu(\mathrm{d}x)\right)^{1/2} + \int_{C_1} (\mathrm{e}^{\varphi(x)} + 1)\nu(\mathrm{d}x)\\ &< \infty.\end{aligned}$$

Moreover,

$$\int_{C_0} |\mathrm{e}^{\varphi(x)} - 1 - \varphi(x)|\nu(\mathrm{d}x) < \infty.$$

In particular, finiteness of the integrals appearing in (33.4), (33.6), (33.7), and (33.12) follows from (33.3).

DEFINITION 33.4. Let $(\{x_t\}, P)$ be a Lévy process on $\mathbb{R}^d$ with generating triplet (A, ν, γ). Given $\varphi(x)$ satisfying (33.3) and $-\infty < \varphi(x) < \infty$ and given $\eta \in \mathbb{R}^d$, define $A^\sharp = A$, $\nu^\sharp(\mathrm{d}x) = \mathrm{e}^{\varphi(x)}\nu(\mathrm{d}x)$, and $\gamma^\sharp$ by (33.6). Then we can define U_t as in Theorem 33.2 and a probability measure $P_t^\sharp$ on $\mathcal{F}_t$ by

$$P_t^\sharp[B] = E^P[\mathrm{e}^{U_t} 1_B] \quad \text{for } B \in \mathcal{F}_t. \tag{33.16}$$

The family $\{P_t^\sharp\colon t \geq 0\}$ is consistent and has a unique extension $P^\sharp$ on $\mathcal{F}$. Then $(\{x_t\}, P^\sharp)$ is a Lévy process, which has generating triplet $(A^\sharp, \nu^\sharp, \gamma^\sharp)$. These are consequences of Theorems 33.1 and 33.2. Let us call this construction of $(\{x_t\}, P^\sharp)$ from $(\{x_t\}, P)$ *density transformation* by $\varphi(x)$ and η.

REMARK 33.5. The condition (1) in Theorem 33.1 is equivalent to $[P]_{\mathcal{F}_t} \approx [P^\sharp]_{\mathcal{F}_t}$ for some $t \in (0, \infty)$. Actually, we shall prove that this property implies the condition (2). This fact is by the temporal homogeneity of Lévy processes.

We prove the two theorems after a series of lemmas.

LEMMA 33.6. *Let $(\{x_t\}, P)$ be a Lévy process on $\mathbb{R}^d$ with generating triplet (A, ν, γ). Let $\varphi(x)$ be a finite measurable function on $\mathbb{R}^d$ satisfying (33.3). Let $\eta \in \mathbb{R}^d$. Then the limit in (33.7) exists uniformly in t on any bounded interval, P-a. s. The process $(\{U_t\}, P)$ thus defined is a Lévy process on $\mathbb{R}$ with generating triplet (A_U, ν_U, γ_U) satisfying (33.10), (33.11), and (33.12), and*

$$E^P[\mathrm{e}^{U_t}] = 1. \tag{33.17}$$

Proof. Define

$$J(B, \xi) = \#\{s\colon (s, x_s(\xi) - x_{s-}(\xi)) \in B\} \tag{33.18}$$

for $B \in \mathcal{B}((0, \infty) \times (\mathbb{R}^d \setminus \{0\}))$. Define

$$\begin{aligned}
V_\varepsilon^1(t) &= \sum_{(s, x_s - x_{s-}) \in (0,t] \times B_{\varepsilon 0}} \varphi(x_s - x_{s-}) - t \int_{B_{\varepsilon 0}} \varphi(x) \nu(\mathrm{d}x) \\
&= \int_{(0,t] \times B_{\varepsilon 0}} \varphi(x) J(\mathrm{d}(s, x)) - t \int_{B_{\varepsilon 0}} \varphi(x) \nu(\mathrm{d}x), \\
V_\varepsilon^2(t) &= \sum_{(s, x_s - x_{s-}) \in (0,t] \times B_{\varepsilon 1}} \varphi(x_s - x_{s-}) = \int_{(0,t] \times B_{\varepsilon 1}} \varphi(x) J(\mathrm{d}(s, x)), \\
c_\varepsilon &= -\frac{1}{2}\langle \eta, A\eta \rangle - \langle \gamma, \eta \rangle - \int_{B_\varepsilon} (\mathrm{e}^{\varphi(x)} - 1 - \varphi(x) 1_{[-1,1]}(\varphi(x))) \nu(\mathrm{d}x), \\
U_\varepsilon(t) &= V_\varepsilon^1(t) + V_\varepsilon^2(t) + \langle \eta, x_t - x_t^\nu \rangle + t c_\varepsilon.
\end{aligned}$$

Then the definition of U_t by (33.7) is the same as

$$U_t = \lim_{\varepsilon \downarrow 0} U_\varepsilon(t). \tag{33.19}$$

By Theorem 19.2 and Proposition 19.5 we have, for $u \in \mathbb{R}$,

$$\begin{aligned}
E^P[\mathrm{e}^{\mathrm{i}uU_\varepsilon(t)}] &= E^P[\mathrm{e}^{\mathrm{i}u(V_\varepsilon^1(t) + V_\varepsilon^2(t))}] E^P[\mathrm{e}^{\mathrm{i}u\langle \eta, x_t - x_t^\nu \rangle}] \mathrm{e}^{\mathrm{i}utc_\varepsilon} \\
&= \exp\Big[t\Big(\int_{B_{\varepsilon 0}} (\mathrm{e}^{\mathrm{i}u\varphi(x)} - 1 - \mathrm{i}u\varphi(x)) \nu(\mathrm{d}x) + \int_{B_{\varepsilon 1}} (\mathrm{e}^{\mathrm{i}u\varphi(x)} - 1) \nu(\mathrm{d}x)
\end{aligned}$$

$$- \frac{1}{2}\langle u\eta, A(u\eta)\rangle + \mathrm{i}\langle \gamma, u\eta\rangle + \mathrm{i}uc_\varepsilon\Big)\Big],$$

which tends to

$$\exp\Big[t\Big(\int_{\mathbb{R}^d}(\mathrm{e}^{\mathrm{i}u\varphi(x)} - 1 - \mathrm{i}u\varphi(x)1_{[-1,1]}(\varphi(x)))\nu(\mathrm{d}x) - \mathrm{i}u\int_{\mathbb{R}^d}(\mathrm{e}^{\varphi(x)} - 1 - \varphi(x)1_{[-1,1]}(\varphi(x)))\nu(\mathrm{d}x) - \frac{1}{2}\langle\eta, A\eta\rangle u^2 - \frac{\mathrm{i}}{2}\langle\eta, A\eta\rangle u\Big)\Big] \tag{33.20}$$

as $\varepsilon \downarrow 0$. Here we can apply the argument in Lemmas 20.6 and 20.7 to $V_\varepsilon^1(t)$. Proposition 19.5 is directly utilized on $V_\varepsilon^2(t)$. The nonrandom term tc_ε is convergent. Thus we see that $U_\varepsilon(t)$ converges uniformly in t on any bounded interval, P-a. s. Hence U_t is defined by (33.19) P-a. s. and $E[\mathrm{e}^{\mathrm{i}uU_t}]$ equals the expression (33.20) for $u \in \mathbb{R}$. We see that $(\{U_t\}, P)$ is a Lévy process on $\mathbb{R}$, because $(\{U_\varepsilon(t)\}, P)$ has stationary independent increments. As is seen from (33.20), the generating triplet (A_U, ν_U, γ_U) of the process $(\{U_t\}, P)$ is given by (33.10), (33.11), and (33.12). Since $\int_{|y|>1}\mathrm{e}^y(\nu\varphi^{-1})(\mathrm{d}y)$ is finite, $E^P[\mathrm{e}^{U_t}]$ is finite by Theorem 25.3. Now Theorem 25.17 on exponential moments shows that $E^P[\mathrm{e}^{U_t}]$ is evaluated as (33.20) with $\mathrm{i}u$ replaced by 1. Thus we obtain (33.17). □

LEMMA 33.7. *In the same setting as in Lemma 33.6, define* $A^\sharp$, $\nu^\sharp$, *and* $\gamma^\sharp$ *by* $A^\sharp = A$, $\nu^\sharp(\mathrm{d}x) = \mathrm{e}^{\varphi(x)}\nu(\mathrm{d}x)$, *and*

$$\gamma^\sharp = \gamma + \int_{|x|\le 1} x(\mathrm{e}^{\varphi(x)} - 1)\nu(\mathrm{d}x) + A\eta.$$

Then, the Lévy process $(\{x_t\}, P^\sharp)$ *generated by* $(A^\sharp, \nu^\sharp, \gamma^\sharp)$ *satisfies*

$$P^\sharp[B] = E^P[\mathrm{e}^{U_t}1_B] \quad \textit{for } B \in \mathcal{F}_t,\ t \in [0,\infty). \tag{33.21}$$

Proof. Define the $(d+1)$-dimensional process $x_t^*(\xi) = (x_j^*(t,\xi))_{1\le j\le d+1}$, $\xi \in \mathbf{D}$, by $(x_j^*(t,\xi))_{1\le j\le d} = x_t(\xi)$ and $x_{d+1}^*(t,\xi) = U_t(\xi)$. Then $(\{x_t^*\}, P)$ is a Lévy process on $\mathbb{R}^{d+1}$, for it has stationary independent increments and its sample functions are right-continuous with left limits P-a. s. and start at 0. Let us calculate its characteristic function. We claim that, for $z \in \mathbb{R}^d$ and $u \in \mathbb{R}$,

$$E^P[\mathrm{e}^{\mathrm{i}\langle z, x_t\rangle + \mathrm{i}uU_t}] = \exp\Big[t\Big(I - \frac{1}{2}\langle z + u\eta, A(z+u\eta)\rangle + \mathrm{i}\langle\gamma, z\rangle - \frac{\mathrm{i}}{2}\langle\eta, A\eta\rangle u - \mathrm{i}u\int_{\mathbb{R}^d}(\mathrm{e}^{\varphi(x)} - 1 - \varphi(x)1_{[-1,1]}(\varphi(x)))\nu(\mathrm{d}x)\Big)\Big], \tag{33.22}$$

where

$$I = \int_{\mathbb{R}^d} (\mathrm{e}^{\mathrm{i}\langle z,x\rangle + \mathrm{i}u\varphi(x)} - 1 - \mathrm{i}\langle z,x\rangle 1_{B_0}(x) - \mathrm{i}u\varphi(x)1_{[-1,1]}(\varphi(x)))\nu(\mathrm{d}x).$$

First note that

$$\int_{\mathbb{R}^d} |\mathrm{e}^{\mathrm{i}\langle z,x\rangle + \mathrm{i}u\varphi(x)} - 1 - \mathrm{i}\langle z,x\rangle 1_{B_0}(x) - \mathrm{i}u\varphi(x)1_{[-1,1]}(\varphi(x))|\nu(\mathrm{d}x) < \infty.$$

In fact, the integrals over B_{00}, B_{01}, B_{10}, and B_{11} are finite as follows:

$$\begin{aligned}
\int_{B_{00}} &\le \frac{1}{2}\int_{B_{00}} (\langle z,x\rangle + u\varphi(x))^2\nu(\mathrm{d}x) \\
&\le |z|^2 \int_{B_0} |x|^2\nu(\mathrm{d}x) + u^2 \int_{C_0} \varphi(x)^2\nu(\mathrm{d}x), \\
\int_{B_{01}} &= \int_{B_{01}} |(\mathrm{e}^{\mathrm{i}\langle z,x\rangle} - 1 - \mathrm{i}\langle z,x\rangle)\mathrm{e}^{\mathrm{i}u\varphi(x)} + (\mathrm{e}^{\mathrm{i}u\varphi(x)} - 1) \\
&\qquad + \mathrm{i}\langle z,x\rangle(\mathrm{e}^{\mathrm{i}u\varphi(x)} - 1)|\nu(\mathrm{d}x) \\
&\le \frac{|z|^2}{2}\int_{B_0} |x|^2\nu(\mathrm{d}x) + 2\int_{C_1} \nu(\mathrm{d}x) \\
&\qquad + 2|z|\Big(\int_{B_0} |x|^2\nu(\mathrm{d}x)\Big)^{1/2}\Big(\int_{C_1} \nu(\mathrm{d}x)\Big)^{1/2}, \\
\int_{B_{10}} &\le \frac{u^2}{2}\int_{C_0} \varphi(x)^2\nu(\mathrm{d}x) + 2\int_{B_1} \nu(\mathrm{d}x) \\
&\qquad + 2|u|\Big(\int_{C_0} \varphi(x)^2\nu(\mathrm{d}x)\Big)^{1/2}\Big(\int_{B_1} \nu(\mathrm{d}x)\Big)^{1/2}, \\
\int_{B_{11}} &\le 2\int_{B_1} \nu(\mathrm{d}x).
\end{aligned}$$

To prove (33.22), we go back to the ε-approximation of $E^P[\mathrm{e}^{\mathrm{i}uU_t + \mathrm{i}\langle z,x_t\rangle}]$, using (33.5) and (33.19). Let

$$x_\varepsilon^\nu(t,\xi) = \sum_{(s,x_s - x_{s-})\in(0,t]\times B_\varepsilon} (x_s - x_{s-}) - t\int_{B_\varepsilon\cap B_0} x\nu(\mathrm{d}x).$$

We have

$$\begin{aligned}
&E^P[\mathrm{e}^{\mathrm{i}\langle z,x_\varepsilon^\nu(t)\rangle + \mathrm{i}\langle z,x_t - x_t^\nu\rangle + \mathrm{i}uU_\varepsilon(t)}] \\
&\quad = E^P\Big[\exp\Big(\mathrm{i}\int_{(0,t]\times B_\varepsilon} \langle z,x\rangle J(\mathrm{d}(s,x)) - \mathrm{i}t\int_{B_\varepsilon\cap B_0} \langle z,x\rangle\nu(\mathrm{d}x) \\
&\qquad\qquad + \mathrm{i}\langle z,x_t - x_t^\nu\rangle + \mathrm{i}uV_\varepsilon^1(t) + \mathrm{i}uV_\varepsilon^2(t)
\end{aligned}$$

$$+ \mathrm{i}u\langle \eta, x_t - x_t^\nu\rangle + \mathrm{i}utc_\varepsilon\Big)\Big]$$
$$= k_\varepsilon l_\varepsilon m,$$

where

$$k_\varepsilon = E^P\Big[\exp\Big(\mathrm{i}\int_{(0,t]\times B_\varepsilon}(\langle z,x\rangle + u\varphi(x))J(\mathrm{d}(s,x))\Big)\Big],$$
$$l_\varepsilon = \exp\Big(-\mathrm{i}t\int_{B_\varepsilon\cap B_0}\langle z,x\rangle\nu(\mathrm{d}x) - \mathrm{i}ut\int_{B_{\varepsilon 0}}\varphi(x)\nu(\mathrm{d}x) + \mathrm{i}utc_\varepsilon\Big),$$
$$m = E^P[\mathrm{e}^{\mathrm{i}\langle z+u\eta, x_t - x_t^\nu\rangle}].$$

It follows from Proposition 19.5 that

$$k_\varepsilon = \exp\Big(t\int_{B_\varepsilon}(\mathrm{e}^{\mathrm{i}(\langle z,x\rangle + u\varphi(x))} - 1)\nu(\mathrm{d}x)\Big),$$

while $m = \mathrm{e}^{-(t/2)\langle z+u\eta, A(z+u\eta)\rangle + \mathrm{i}\langle\gamma, z+u\eta\rangle}$. Hence, letting $\varepsilon\downarrow 0$, we get (33.22).

The identity (33.22) shows that the generating triplet (A^*,ν^*,γ^*) of the Lévy process $(\{x_t^*\},P)$ is such that $\langle z^*, A^*z^*\rangle = \langle z+u\eta, A(z+u\eta)\rangle$ for $z^* = (z_j^*)_{1\le j\le d+1}\in\mathbb{R}^{d+1}$ with $(z^*)_{1\le j\le d} = z$ and $z_{d+1}^* = u$ and $\nu^* = \nu f^{-1}$ with $f(x) = (x_j^*)_{1\le j\le d+1}$, $(x_j^*)_{1\le j\le d} = x$, and $x_{d+1}^* = \varphi(x)$. We apply Theorem 25.17 to the $(d+1)$-dimensional process. Recall that $E^P[\mathrm{e}^{U_t}]$ is finite. This means that $E^P[\exp\langle c^*, x_t^*\rangle] < \infty$ for $c^* = (c_j^*)_{1\le j\le d+1}$ with $(c_j^*)_{1\le j\le d} = 0$ and $c_{d+1}^* = 1$. Now we can get the expression for $E^P[\mathrm{e}^{\mathrm{i}\langle z,x_t\rangle + U_t}]$ by analytic continuation. The result is exactly the right-hand side of (33.22) with $\mathrm{i}u$ replaced by 1. That is,

$$E^P[\mathrm{e}^{\mathrm{i}\langle z,x_t\rangle + U_t}] = \exp\Big[t\Big(\frac{1}{2}\langle \mathrm{i}z+\eta, A(\mathrm{i}z+\eta)\rangle + \mathrm{i}\langle\gamma,z\rangle$$
$$+ \int_{\mathbb{R}^d}(\mathrm{e}^{\mathrm{i}\langle z,x\rangle + \varphi(x)} - 1 - \mathrm{i}\langle z,x\rangle 1_{B_0}(x) - \varphi(x)1_{[-1,1]}(\varphi(x)))\nu(\mathrm{d}x)$$
$$- \frac{1}{2}\langle\eta, A\eta\rangle - \int_{\mathbb{R}^d}(\mathrm{e}^{\varphi(x)} - 1 - \varphi(x)1_{[-1,1]}(\varphi(x)))\nu(\mathrm{d}x)\Big)\Big].$$

Therefore,

$$E^P[\mathrm{e}^{\mathrm{i}\langle z,x_t\rangle + U_t}] = \exp\Big[t\Big(-\frac{1}{2}\langle z, Az\rangle + \mathrm{i}\langle\gamma,z\rangle + \mathrm{i}\langle A\eta, z\rangle$$
$$+ \int_{\mathbb{R}^d}(\mathrm{e}^{\mathrm{i}\langle z,x\rangle} - 1 - \mathrm{i}\langle z,x\rangle 1_{B_0}(x))\mathrm{e}^{\varphi(x)}\nu(\mathrm{d}x)$$
$$+ \mathrm{i}\int_{B_0}\langle z,x\rangle(\mathrm{e}^{\varphi(x)} - 1)\nu(\mathrm{d}x)\Big)\Big]. \tag{33.23}$$

Finiteness of the integrals that appear is now easily checked.

Now let $(\{x_t\}, P^\sharp)$ be the Lévy process generated by $(A^\sharp, \nu^\sharp, \gamma^\sharp)$ in the statement of the lemma. Fix t. Define a probability measure $P_t^\sharp$ on $\mathcal{F}_t$ by (33.16). Then the two processes given by $\{x_s\colon 0 \leq s \leq t\}$ under the measures $P^\sharp$ and $P_t^\sharp$ are identical in law, since both have stationary independent increments and, by (33.23), they have a common characteristic function. Consequently, (33.21) is true. □

PROPOSITION 33.8. *Let P and $P^\sharp$ be two probability measures on a measurable space $(\Theta, \mathcal{F}_\Theta)$. Let $(\Theta', \mathcal{F}_{\Theta'})$ be another measurable space and let X be a mapping from Θ to Θ' measurable with respect to $\mathcal{F}_\Theta$ and $\mathcal{F}_{\Theta'}$. Denote by P_X and $P_X^\sharp$ the probability measures on Θ' induced by X under P and $P^\sharp$, respectively. If $P \approx P^\sharp$, then $P_X \approx P_X^\sharp$.*

Proof. Note that if $P \approx P^\sharp$, then, for $B \in \mathcal{F}_{\Theta'}$, $P[\,X \in B\,] = 0$ and $P^\sharp[\,X \in B\,] = 0$ are equivalent. □

LEMMA 33.9. *Let $(\{x_t\}, P)$ and $(\{x_t\}, P^\sharp)$ be Lévy processes with generating triplets (A, ν, γ) and $(A^\sharp, \nu^\sharp, \gamma^\sharp)$, respectively. Fix $t \in (0, \infty)$ and assume that $[P]_{\mathcal{F}_t} \approx [P^\sharp]_{\mathcal{F}_t}$. Then $\nu \approx \nu^\sharp$ and the function $\varphi(x)$ defined by $\mathrm{e}^{\varphi(x)} = \frac{\mathrm{d}\nu^\sharp}{\mathrm{d}\nu}$ satisfies, for any $\varepsilon > 0$,*

$$\nu\{x\colon |\varphi(x)| > \varepsilon\} < \infty, \tag{33.24}$$

$$\nu^\sharp\{x\colon |\varphi(x)| > \varepsilon\} < \infty. \tag{33.25}$$

Proof. We have, by Theorem 19.2(i),

$$P[\,J((0,t] \times B) = 0\,] = \mathrm{e}^{-t\nu(B)}, \tag{33.26}$$

$$P^\sharp[\,J((0,t] \times B) = 0\,] = \mathrm{e}^{-t\nu^\sharp(B)} \tag{33.27}$$

for $B \in \mathcal{B}(\mathbb{R}^d)$. Here $J(\cdot)$ is defined by (33.18). If $\nu(B) = 0$, then $\nu^\sharp(B) = 0$, because the left-hand side of (33.26) is 1 and, by the mutual absolute continuity, the left-hand side of (33.27) is 1. Similarly, if $\nu^\sharp(B) = 0$, then $\nu(B) = 0$. Hence $\nu \approx \nu^\sharp$. Hence $\varphi(x)$ is defined and $0 < \varphi(x) < \infty$, ν-a. e. Similarly $\nu(B) < \infty$ and $\nu^\sharp(B) < \infty$ are equivalent. Therefore (33.24) and (33.25) are equivalent.

Let us show (33.24). The following argument is based on Kakutani [**252**]. Suppose that, for some $\varepsilon > 0$,

$$\nu\{x\colon \varphi(x) > \varepsilon\} = \infty.$$

Since $\mathrm{d}s\nu(\mathrm{d}x)$ is a continuous measure, we can choose disjoint sets $B_n \in \mathcal{B}((0,\infty) \times (\mathbb{R}^d \setminus \{0\}))$, $n = 1, 2, \ldots$, such that

$$\bigcup_{n=1}^{\infty} B_n = (0,t] \times \{x\colon \varphi(x) > \varepsilon\} \quad \text{and} \quad \iint_{B_n} \mathrm{d}s\, \nu(\mathrm{d}x) = 1.$$

Define $\{0,1\}$-valued measurable mappings $Y_n(\xi) = \operatorname{sgn} J(B_n, \xi)$, $\xi \in \mathbf{D}$. As $J(\cdot)$ is a Poisson random measure under P and $P^\sharp$, $\{Y_n\}$ are independent both under P and under $P^\sharp$. We have

$$P[Y_n = 0] = P[J(B_n) = 0] = \exp\left(- \iint_{B_n} \mathrm{d}s\, \nu(\mathrm{d}x) \right) = \mathrm{e}^{-1},$$
$$P^\sharp[Y_n = 0] = P^\sharp[J(B_n) = 0] = \mathrm{e}^{-c_n},$$

where $0 < c_n = \iint_{B_n} \mathrm{d}s\, \nu^\sharp(\mathrm{d}x) < \infty$. Let $\Theta = \{0,1\}^{\mathbb{N}}$ and let $\mathcal{F}_\Theta$ be the σ-algebra generated by the cylinder sets. Define $y_n(\theta) = \theta(n)$ for $\theta = (\theta(n))_{n \in \mathbb{N}} \in \Theta$. Let Q and $Q^\sharp$ be the probability measures on $\mathcal{F}_\Theta$ induced from P and $P^\sharp$, respectively, by the mapping ψ defined by $\psi(\xi) = (Y_n(\xi))_{n\in\mathbb{N}}$. It follows from $[P]_{\mathcal{F}_t} \approx [P^\sharp]_{\mathcal{F}_t}$, by Proposition 33.8, that $Q \approx Q^\sharp$. The $\{y_n\}$ are independent under Q and $Q^\sharp$. Thus Q and $Q^\sharp$ are the product measures of $\{P_{Y_n}\}$ and $\{P^\sharp_{Y_n}\}$, respectively, where P_{Y_n} and $P^\sharp_{Y_n}$ are the distributions on $\{0,1\}$ of Y_n under P and $P^\sharp$. Both P_{Y_n} and $P^\sharp_{Y_n}$ have positive measures at 0 and 1, and

$$\frac{\mathrm{d}P^\sharp_{Y_n}}{\mathrm{d}P_{Y_n}}(k) = 1_{\{0\}}(k)\frac{\mathrm{e}^{-c_n}}{\mathrm{e}^{-1}} + 1_{\{1\}}(k)\frac{1-\mathrm{e}^{-c_n}}{1-\mathrm{e}^{-1}} \quad \text{for } k = 0, 1.$$

Define

$$z_n(\theta) = (1 - c_n) 1_{\{y_n = 0\}}(\theta) + \left(\log \frac{1-\mathrm{e}^{-c_n}}{1-\mathrm{e}^{-1}}\right) 1_{\{y_n=1\}}(\theta), \quad \theta \in \Theta.$$

Then

$$\frac{\mathrm{d}P^\sharp_{Y_n}}{\mathrm{d}P_{Y_n}}(y_n(\theta)) = \mathrm{e}^{z_n(\theta)}.$$

Write

$$r_n = \int_{\{0,1\}} \left(\frac{\mathrm{d}P^\sharp_{Y_n}}{\mathrm{d}P_{Y_n}}\right)^{1/2} \mathrm{d}P_{Y_n} = \int_\Theta \mathrm{e}^{z_n/2} \mathrm{d}Q.$$

Then $0 < r_n \le 1$ by the concavity of $\sqrt{x}$. Let us show that

$$\prod_{n=1}^{\infty} r_n > 0. \tag{33.28}$$

Suppose that $\prod_{n=1}^\infty r_n = 0$. Then, for each n, there is k_n such that $\prod_{k=1}^{k_n} r_k < 2^{-n}$. Let $F_n = \{\theta\colon \prod_{k=1}^{k_n} \mathrm{e}^{z_k} > 1\}$. We have

$$Q[F_n] \le \int_{F_n} \prod_{k=1}^{k_n} \mathrm{e}^{z_k/2} \mathrm{d}Q \le \prod_{k=1}^{k_n} r_k < 2^{-n},$$
$$Q^\sharp[F_n^{\mathrm{c}}] = \int_{F_n^{\mathrm{c}}} \prod_{k=1}^{k_n} \mathrm{e}^{z_k} \mathrm{d}Q \le \int_{F_n^{\mathrm{c}}} \prod_{k=1}^{k_n} \mathrm{e}^{z_k/2} \mathrm{d}Q \le \prod_{k=1}^{k_n} r_k < 2^{-n}.$$

Therefore, by the Borel–Cantelli lemma,

$$Q\left[\limsup_{n\to\infty} F_n\right] = 0 \quad \text{and} \quad Q^\sharp\left[\limsup_{n\to\infty} F_n^c\right] = 0.$$

The latter implies $Q^\sharp[\limsup_{n\to\infty} F_n] = 1$. This contradicts $Q \approx Q^\sharp$. Hence (33.28) is true.

Define $w_n = \prod_{k=1}^n \mathrm{e}^{z_k(\theta)/2}$. Consider $L^2 = L^2(\Theta, Q)$. Let $n < m$. Then

$$\begin{aligned}
\|w_n - w_m\|_{L^2}^2 &= \int_\Theta \left(\prod_{k=1}^n \mathrm{e}^{z_k/2}\right)^2 \left(1 - \prod_{k=n+1}^m \mathrm{e}^{z_k/2}\right)^2 \mathrm{d}Q \\
&= \int_\Theta \left(1 - \prod_{k=n+1}^m \mathrm{e}^{z_k/2}\right)^2 \mathrm{d}Q \\
&= \int_\Theta \left(1 + \prod_{k=n+1}^m \mathrm{e}^{z_k} - 2\prod_{k=n+1}^m \mathrm{e}^{z_k/2}\right) \mathrm{d}Q = 2 - 2\prod_{k=n+1}^m r_k,
\end{aligned}$$

which tends to 0 as $n, m \to \infty$ by (33.28). Hence there is $w \in L^2$ such that $\|w - w_n\|_{L^2} \to 0$. We have $Q^\sharp[B] = \int_B w^2 \mathrm{d}Q$ for $B \in \mathcal{F}_\Theta$. For, if B is in the σ-algebra generated by $y_1, \ldots, y_n$, then this equality is true as $Q^\sharp[B] = \int_B w_n^2 \mathrm{d}Q \to \int_B w^2 \mathrm{d}Q$, $n \to \infty$, and the equality is extended to all $B \in \mathcal{F}_\Theta$. Hence $w^2 = \frac{\mathrm{d}Q^\sharp}{\mathrm{d}Q}$. Hence, by $Q \approx Q^\sharp$,

$$w > 0 \quad Q\text{-a.s.} \tag{33.29}$$

It follows from the L^2-convergence that $w_n \to w$ in probability (Q). Thus, using (33.29) and Exercise 6.6, we see that $\log w_n \to \log w$ in probability (Q). Since $\log w_n = \frac{1}{2}\sum_{k=1}^n z_k$, it follows that $z_n \to 0$ in probability (Q). Since $B_n \subset (0, t] \times \{\varphi(x) > \varepsilon\}$,

$$c_n = \iint_{B_n} \mathrm{e}^{\varphi(x)} \mathrm{d}s\, \nu(\mathrm{d}x) > \mathrm{e}^\varepsilon \iint_{B_n} \mathrm{d}s\, \nu(\mathrm{d}x) = \mathrm{e}^\varepsilon.$$

Therefore, $Q[z_n = 1 - c_n] \le Q[z_n < 1 - \mathrm{e}^\varepsilon] \to 0$ as $n \to \infty$. But $Q[z_n = 1 - c_n] = Q[y_n = 0] = \mathrm{e}^{-1}$. This is absurd. This finishes the proof that $\nu\{\varphi(x) > \varepsilon\} < \infty$. It also follows that $\nu^\sharp\{\varphi(x) > \varepsilon\} < \infty$.

Since $\frac{\mathrm{d}\nu}{\mathrm{d}\nu^\sharp} = \mathrm{e}^{-\varphi}$, we get finiteness of $\nu\{-\varphi(x) > \varepsilon\}$ and $\nu^\sharp\{-\varphi(x) > \varepsilon\}$, interchanging the roles of ν and $\nu^\sharp$. The proof of (33.24) and (33.25) is now complete. □

We shall use a part of the following general result of Kolmogorov.

PROPOSITION 33.10 (Three-series theorem). *Let $\{X_n, n = 1, 2, \ldots\}$ be independent random variables on $\mathbb{R}$. Fix a positive real number c. Then*

$\sum_{n=1}^{\infty} X_n$ *converges a. s. if and only if the following three series converge:*

$$\sum_n P[|X_n| > c], \quad \sum_n E[X_n 1_{\{|X_n| \le c\}}], \quad \sum_n \mathrm{Var}(X_n 1_{\{|X_n| \le c\}}).$$

Proofs are found in [**34**], p. 299, [**80**], p. 118, and [**151**], p. 203.

LEMMA 33.11. *In the same setting as in Lemma 33.9, we have (33.3).*

Proof. Let $\varepsilon > 0$. We have

$$\int_{|\varphi(x)|>\varepsilon} (\mathrm{e}^{\varphi(x)/2} - 1)^2 \nu(\mathrm{d}x) \le \int_{|\varphi(x)|>\varepsilon} (\mathrm{e}^{\varphi(x)} + 1)\nu(\mathrm{d}x)$$
$$= \nu^{\sharp}\{|\varphi(x)| > \varepsilon\} + \nu\{|\varphi(x)| > \varepsilon\},$$

which is finite by Lemma 33.9. Let us show that

$$\int_{0<\varphi(x)\le\varepsilon} (\mathrm{e}^{\varphi(x)/2} - 1)^2 \nu(\mathrm{d}x) < \infty. \tag{33.30}$$

In the case $\nu\{0 < \varphi(x) \le \varepsilon\} < \infty$, (33.30) is evident. Consider the case that

$$\nu\{0 < \varphi(x) \le \varepsilon\} = \infty. \tag{33.31}$$

Using Lemma 33.9 and the continuity of $\mathrm{d}s\,\nu(\mathrm{d}x)$, we can choose disjoint sets B_n, $n = 1, 2, \ldots$, in $\mathcal{B}((0,t] \times \{|x| > 0\})$ and a sequence $\{b_n\}$ decreasing to 0 such that $\iint_{B_n} \mathrm{d}s\,\nu(\mathrm{d}x) = 1$,

$$B_n \subset (0,t] \times \{x\colon b_n \ge \mathrm{e}^{\varphi(x)/2} - 1 \ge b_{n+1}\} \tag{33.32}$$

and

$$\bigcup_{n=1}^{\infty} B_n = (0,t] \times \{x\colon 0 < \varphi(x) \le \varepsilon\}. \tag{33.33}$$

Let

$$c_n = \iint_{B_n} \mathrm{d}s\,\nu^{\sharp}(\mathrm{d}x) = \iint_{B_n} \mathrm{e}^{\varphi(x)} \mathrm{d}s\,\nu(\mathrm{d}x).$$

We have $(1 + b_n)^2 \ge \mathrm{e}^{\varphi(x)} \ge (1 + b_{n+1})^2$ on B_n by (33.32). Hence

$$(1 + b_n)^2 \ge c_n \ge (1 + b_{n+1})^2.$$

Define $Y_n(\xi)$, Θ, $\mathcal{F}_\Theta$, Q, $y_n(\theta)$, and $z_n(\theta)$ in the same way as in the proof of Lemma 33.9. Then $\{z_n\}$ are independent and we see that $\sum_{n=1}^{\infty} z_n$ is convergent in $L^2(Q)$. Since there is c such that $1 < c_n \le c$, we have $1 - \mathrm{e}^{-1} < 1 - \mathrm{e}^{-c_n} \le 1 - \mathrm{e}^{-c}$. Thus $z_n(\theta)$ are uniformly bounded. Therefore, by Propositions 27.17 and 33.10, $\sum_{n=1}^{\infty} z_n$ is convergent Q-a. s. and

$$\sum_{n=1}^{\infty} E^Q[{z_n}^2] < \infty. \tag{33.34}$$

It follows that $\sum_{n=1}^{\infty} b_n{}^2 < \infty$, since

$$E^Q[z_n{}^2] \geq E^Q[z_n{}^2 1_{\{y_n=0\}}] = (1-c_n)^2 \mathrm{e}^{-1}$$
$$\geq ((1+b_{n+1})^2-1)\mathrm{e}^{-1} = (b_{n+1}{}^2 + 2b_{n+1})^2\mathrm{e}^{-1} \geq 4\mathrm{e}^{-1} b_{n+1}{}^2.$$

On the other hand,

$$\int_{0<\varphi(x)\leq\varepsilon} (\mathrm{e}^{\varphi(x)/2}-1)^2 \nu(\mathrm{d}x) = \frac{1}{t}\sum_{n=1}^{\infty} \iint_{B_n} (\mathrm{e}^{\varphi(x)/2}-1)^2 \mathrm{d}s\, \nu(\mathrm{d}x)$$
$$\leq \frac{1}{t}\sum_{n=1}^{\infty} b_n{}^2$$

by (33.32) and (33.33). So we have (33.30).

Interchanging the roles of P and $P^\sharp$ and noting that $\frac{\mathrm{d}\nu}{\mathrm{d}\nu^\sharp} = \mathrm{e}^{-\varphi}$, we get

$$\int_{0<-\varphi(x)\leq\varepsilon} (\mathrm{e}^{-\varphi(x)/2}-1)^2 \nu^\sharp(\mathrm{d}x) < \infty,$$

that is,

$$\int_{0>\varphi(x)\geq-\varepsilon} (\mathrm{e}^{\varphi(x)/2}-1)^2 \nu(\mathrm{d}x) < \infty. \tag{33.35}$$

We obtain (33.3) by (33.30) and (33.35). □

LEMMA 33.12. *Let $(\{x_t\}, P)$ and $(\{x_t\}, P^\sharp)$ be Lévy processes with generating triplets $(A, 0, \gamma)$ and $(A^\sharp, 0, \gamma^\sharp)$, respectively. Assume that $[P]_{\mathcal{F}_t} \approx [P^\sharp]_{\mathcal{F}_t}$ for some $t \in (0,\infty)$. Then $A = A^\sharp$ and $\gamma^\sharp - \gamma \in \mathfrak{R}(A)$.*

Proof. Step 1. Consider the case $d = 1$. Then $A, \gamma, A^\sharp$, and $\gamma^\sharp$ are real numbers. Fix t for which $[P]_{\mathcal{F}_t} \approx [P^\sharp]_{\mathcal{F}_t}$. Suppose that $A > 0$. It follows that $A^\sharp > 0$. Let $y_s = x_{s/A} - s\gamma/A$. Then $(\{y_s\}, P)$ is a Brownian motion. Hence, by Theorem 5.8, $\sum_{k=1}^n (y_{ks/n} - y_{(k-1)s/n})^2 \to s$ in $L^2(P)$ for every fixed s. Choose $s = tA$. Then, since $\sum_{k=1}^n (y_{ks/n} - y_{(k-1)s/n})^2 = \sum_{k=1}^n (x_{kt/n} - x_{(k-1)t/n})^2 - 2t\gamma x_t/n + t^2\gamma^2/n$, we have $\sum_{k=1}^n (x_{kt/n} - x_{(k-1)t/n})^2 \to tA$ in $L^2(P)$ and hence in probability ($[P]_{\mathcal{F}_t}$). Using the assumption and Exercise 6.6, we see that $\sum_{k=1}^n (x_{kt/n} - x_{(k-1)t/n})^2 \to tA$ in probability ($[P^\sharp]_{\mathcal{F}_t}$). On the other hand, we have similarly $\sum_{k=1}^n (x_{kt/n} - x_{(k-1)t/n})^2 \to tA^\sharp$ in probability ($[P^\sharp]_{\mathcal{F}_t}$). Therefore $A = A^\sharp > 0$. In the case $A = 0$, we have $A^\sharp = 0$ and the processes are γs and $\gamma^\sharp s$, from which follows $\gamma = \gamma^\sharp$, by the mutual absolute continuity. The lemma is proved for $d = 1$.

Step 2. Let $z \in \mathbb{R}^d$. Then $(\{\langle z, x_t\rangle\}, P)$ and $(\{\langle z, x_t\rangle\}, P^\sharp)$ are Lévy processes with generating triplets $(\langle z, Az\rangle, 0, \langle\gamma, z\rangle)$ and $(\langle z, A^\sharp z\rangle, 0, \langle\gamma^\sharp, z\rangle)$, respectively. Write $\mathbf{D}_1 = D([0,\infty), \mathbb{R})$ and $x_t(\xi) = \xi(t)$ for $\xi \in \mathbf{D}_1$. Let $\mathcal{F}_{\mathbf{D}_1}$ be the σ-algebra generated by x_t, $t \in [0,\infty)$, and $\mathcal{F}_t(\mathbf{D}_1)$ be that generated by x_s, $s \in [0,t]$. The processes $(\{\langle z, x_t\rangle\}, P)$ and $(\{\langle z, x_t\rangle\}, P^\sharp)$ induce probability measures Q and $Q^\sharp$ on $(\mathbf{D}_1, \mathcal{F}_{\mathbf{D}_1})$, respectively. By Proposition

33.8, $[Q]_{\mathcal{F}_t(\mathbf{D}_1)} \approx [Q^\sharp]_{\mathcal{F}_t(\mathbf{D}_1)}$ for some t. Therefore, by Step 1, either we have $\langle z, Az\rangle = \langle z, A^\sharp z\rangle > 0$, or we have $\langle z, Az\rangle = \langle z, A^\sharp z\rangle = 0$ and $\langle \gamma, z\rangle = \langle \gamma^\sharp, z\rangle$. Thus $\langle z, Az\rangle = \langle z, Az\rangle$ for every z. Hence $A = A^\sharp$. If $\mathfrak{R}(A) = \mathbb{R}^d$, then $\gamma^\sharp - \gamma \in \mathfrak{R}(A)$ is trivial. Suppose that $\mathfrak{R}(A) \neq \mathbb{R}^d$. Let T be the orthogonal projector to $\mathfrak{R}(A)^\perp$. Note that $TAT' = 0$. Proposition 11.10 yields that $(\{Tx_t\}, P)$ and $(\{Tx_t\}, P^\sharp)$ are Lévy processes generated by $(0, 0, T\gamma)$ and $(0, 0, T\gamma^\sharp)$, respectively. By the mutual absolute continuity, $T\gamma = T\gamma^\sharp$, that is, $\gamma^\sharp - \gamma \in \mathfrak{R}(A)$. □

Proof of Theorem 33.1 and Theorem 33.2. Let $(\{x_t\}, P)$ and $(\{x_t\}, P^\sharp)$ be Lévy processes and let (A, ν, γ) and $(A^\sharp, \nu^\sharp, \gamma^\sharp)$ be their generating triplets.

Assume that the condition (2) in Theorem 33.1 is satisfied. Using the function $\varphi(x)$ in the condition (2), apply Lemmas 33.6 and 33.7. Then we have (33.21). This means that the condition (1) is satisfied.

Next, assuming the condition (2) (hence also the condition (1)) in Theorem 33.1, let us see the conclusion of Theorem 33.2. Use Lemmas 33.6 and 33.7. Then all assertions except $E^{P^\sharp}[\mathrm{e}^{-U_t}] = 1$ follow. But this one is also evident, as (33.21) implies $P[B] = E^{P^\sharp}[\mathrm{e}^{-U_t} 1_B]$ for $B \in \mathcal{F}_t$.

Assume that the condition (1) is satisfied or, more generally, that $[P]_{\mathcal{F}_t} \approx [P^\sharp]_{\mathcal{F}_t}$ for some $t \in (0, \infty)$. Let us show that the condition (2) is then satisfied. First, use Lemmas 33.9 and 33.11. Then, $\nu \approx \nu^\sharp$, and (33.3) holds for the function $\varphi(x)$ defined by $\frac{\mathrm{d}\nu^\sharp}{\mathrm{d}\nu} = \mathrm{e}^{\varphi(x)}$. Consider the continuous parts $(\{x_s - x_s^\nu\}, P)$ and $(\{x_s - x_s^{\nu^\sharp}\}, P^\sharp)$. They have the generating triplets $(A, 0, \gamma)$ and $(A^\sharp, 0, \gamma^\sharp)$, respectively. Denote by P_c and $P_c^\sharp$ the measures induced on $(\mathbf{D}, \mathcal{F}_{\mathbf{D}})$ by $(\{x_s - x_s^\nu\}, P)$ and $(\{x_s - x_s^\nu\}, P^\sharp)$, respectively. By Proposition 33.8, $[P_c]_{\mathcal{F}_t} \approx [P_c^\sharp]_{\mathcal{F}_t}$. The generating triplet of $(\{x_s - x_s^\nu\}, P^\sharp)$ is $(A^\sharp, 0, \gamma^\sharp - \int_{|x|\le 1} x(\nu^\sharp - \nu)(\mathrm{d}x))$. Therefore, by Lemma 33.12, $A = A^\sharp$ and $\gamma^\sharp - \gamma - \int_{|x|\le 1} x(\nu^\sharp - \nu)(\mathrm{d}x) \in \mathfrak{R}(A)$. This finishes the proof. □

REMARK 33.13. In Theorem 33.2 assume, in addition, that the function $\varphi(x)$ satisfies

$$\int_{|\varphi(x)|\le 1} |\varphi(x)|\nu(\mathrm{d}x) < \infty. \tag{33.36}$$

Then we have $\int_{\mathbb{R}^d} |\mathrm{e}^{\varphi(x)} - 1|\nu(\mathrm{d}x) < \infty$, and

$$\begin{aligned} U_t = &\sum_{(s, x_s - x_{s-}) \in (0,t] \times \{|x|>0\}} \varphi(x_s - x_{s-}) - t \int_{\mathbb{R}^d} (\mathrm{e}^{\varphi(x)} - 1)\nu(\mathrm{d}x) \\ &+ \langle \eta, x_t - x_t^\nu\rangle - \frac{1}{2}\langle \eta, A\eta\rangle - t\langle \gamma, \eta\rangle, \quad P\text{-a.s.} \end{aligned} \tag{33.37}$$

Note that, in this case,

$$\begin{aligned} E^P\Big[\exp\Big(-u \sum_{(s,x_s-x_{s-})\in(0,t]\times\{|x|>0\}} |\varphi(x_s - x_{s-})|\Big)\Big] \\ = E^P\Big[\exp\Big(-u \int_{(0,t]\times\{|x|>0\}} |\varphi(x)| J(\mathrm{d}(s,x))\Big)\Big] \\ = \exp\Big(t \int_{\mathbb{R}^d} (\mathrm{e}^{-u|\varphi(x)|} - 1)\nu(\mathrm{d}x)\Big) \end{aligned}$$

for $u > 0$ by Proposition 19.5 as in the proof of Theorem 19.3 (end of Section 20). The last member tends to 1 as $u \downarrow 0$. Hence the sum in (33.37) is absolutely convergent P-a. s.

EXAMPLE 33.14. Given a Lévy process $(\{x_t\}, P)$ with (A, ν, γ), let $\eta \neq 0$ and $\varphi(x) = \langle \eta, x\rangle$. Then (33.13) and (33.15) are satisfied, since

$$\int_{|\varphi(x)|\leq 1} \varphi(x)^2 \nu(\mathrm{d}x) \leq |\eta|^2 \int_{|x|\leq 1} |x|^2 \nu(\mathrm{d}x) + \int_{|x|>1} \nu(\mathrm{d}x) < \infty$$

and

$$\int_{\varphi(x)<-1} \nu(\mathrm{d}x) \leq \int_{|x|\geq 1/|\eta|} \nu(\mathrm{d}x) < \infty.$$

Thus (33.3) is equivalent to

$$\int_{\langle \eta,x\rangle>1} \mathrm{e}^{\langle \eta,x\rangle} \nu(\mathrm{d}x) < \infty, \tag{33.38}$$

by Remark 33.3. This condition (33.38) is equivalent also to

$$\int_{|x|>1} \mathrm{e}^{\langle \eta,x\rangle} \nu(\mathrm{d}x) < \infty.$$

Hence, as we have shown in Theorem 25.17, (33.38) holds if and only if $E^P[\mathrm{e}^{\langle \eta,x_t\rangle}] < \infty$. Now, assume that (33.38) is satisfied. We can make a new Lévy process $(\{x_t\}, P^\sharp)$ by the density transformation by our $\varphi(x)$ and η. This transformation is studied in [**439**]. It follows from the definition of U_t by (33.7) and from the Lévy–Itô decomposition that

$$(33.39)\qquad \begin{aligned} U_t =&\langle \eta, x_t\rangle - t\Big(\int_{\mathbb{R}^d}(\mathrm{e}^{\langle \eta, x\rangle} - 1 - \langle \eta, x\rangle 1_{\{|x|\le 1\}}(x))\nu(\mathrm{d}x) \\ &\qquad + \frac{1}{2}\langle \eta, A\eta\rangle + \langle \gamma, \eta\rangle\Big) \\ =&\langle \eta, x_t\rangle - t\Psi(\eta), \end{aligned}$$

where Ψ is defined by (25.11). Thus the equality $E[\mathrm{e}^{U_t}] = 1$ is the same as (25.12). We can write (33.21) as

$$(33.40)\qquad P^\sharp[B] = \mathrm{e}^{-t\Psi(\eta)}E^P[\mathrm{e}^{\langle \eta, x_t\rangle}1_B] \quad \text{for } B \in \mathcal{F}_t.$$

In particular,

$$(33.41)\qquad P^\sharp[\, x_t \in B\,] = \mathrm{e}^{-t\Psi(\eta)}\int_B \mathrm{e}^{\langle \eta, x\rangle}P[\, x_t \in \mathrm{d}x\,] \quad \text{for } B \in \mathcal{B}(\mathbb{R}^d).$$

This is the transformation utilized by Cramér [**87**] in 1938.

If $(\{x_t\}, P)$ is the Brownian motion on $\mathbb{R}^d$, then (33.41) is reduced to

$$(33.42)\qquad P^\sharp[\, x_t \in B\,] = \mathrm{e}^{-t|\eta|^2/2}\int_B \mathrm{e}^{\langle \eta, x\rangle}P[\, x_t \in \mathrm{d}x\,],$$

which is the simplest case of the drift transformation in the theory of diffusion processes discovered by Cameron and Martin [**72**] and Maruyama [**339**] (see Ikeda and Watanabe [**215**] and Fukushima, Oshima, and Takeda [**156**]).

EXAMPLE 33.15. As a special case of Example 33.14, let $d = 1$ and suppose that $(\{x_t\}, P)$ is a subordinator with

$$\Psi(w) = \int_{(0,\infty)}(\mathrm{e}^{ws} - 1)\nu(\mathrm{d}s) + \gamma_0 w \quad \text{for } w \le 0.$$

Here $\gamma_0 \ge 0$. Let $\eta < 0$. Then η satisfies (33.38), since ν is concentrated on the positive axis. Let $\varphi(x) = \eta x$. The density transformation by $\varphi(x)$ and η gives a subordinator $(\{x_t\}, P^\sharp)$ and

$$(33.43)\qquad E^{P^\sharp}[\mathrm{e}^{-ux_t}] = \mathrm{e}^{t\widetilde{\Psi}(-u)} \quad \text{with } \widetilde{\Psi}(-u) = \Psi(-u+\eta) - \Psi(\eta)$$

for $u \ge 0$. This follows from (33.40). Hence

$$(33.44)\qquad \widetilde{\Psi}(-u) = \int_{(0,\infty)}(\mathrm{e}^{-us} - 1)\mathrm{e}^{\eta s}\nu(\mathrm{d}s) - \gamma_0 u.$$

Given a strongly continuous contraction semigroup $\{P_t\colon t \ge 0\}$ of linear operators on a Banach space $\mathbf{B}$, let $\{Q_t\colon t \ge 0\}$ be the semigroup subordinate to $\{P_t\}$ by the subordinator $(\{x_t\}, P^\sharp)$ above. Then we have

$$Q_t f = \mathrm{e}^{-t\Psi(\eta)}\int_{(0,\infty)} \mathrm{e}^{\eta s}P_s f\, P[\, x_t \in \mathrm{d}s\,] \quad \text{for } f \in \mathbf{B}$$

by (32.1) and (33.41). Let L and M be the infinitesimal generators of $\{P_t\}$ and $\{Q_t\}$, respectively. We have, by Theorem 32.1 and Definition 32.2,

$$M = \widetilde{\Psi}(L) = \Psi(L + \eta I) - \Psi(\eta) I. \tag{33.45}$$

Suppose, in particular, that $(\{x_t\}, P)$ is a strictly α-stable subordinator with $\Psi(-u) = -c'u^\alpha$ for $u \geq 0$. Here $0 < \alpha < 1$ and $c' > 0$ (Example 32.6). Then

$$E^{P^\sharp}[\mathrm{e}^{-ux_t}] = \mathrm{e}^{-tc'((u-\eta)^\alpha - (-\eta)^\alpha)} \quad \text{for } u \geq 0,$$

and (33.45) becomes

$$M = -c'(-L - \eta I)^\alpha + c'(-\eta)^\alpha I.$$

The case $\alpha = 1/2$ appears in Bochner [**48**]. Some limit theorems as $\eta \to -\infty$ are discussed in Sato [**439**]. If $\alpha = 1/2$, then Example 2.13 gives an explicit density:

$$P^\sharp[\, x_t \in B\,] = \frac{tc'}{2\sqrt{\pi}} \mathrm{e}^{tc'(-\eta)^{1/2}} \int_{B\cap(0,\infty)} \mathrm{e}^{\eta x - (tc')^2/(4x)} x^{-3/2} \mathrm{d}x \quad \text{for } B \in \mathcal{B}(\mathbb{R}^d).$$

This is called *inverse Gaussian distribution*. If μ is an inverse Gaussian, then its cumulant generating function, $\log \int \mathrm{e}^{vx}\mu(\mathrm{d}x)$, is the inverse function of that of a Gaussian distribution. The naming comes from this. We shall see the following in Example 46.6. Let T_x be the hitting time of a point $x \geq 0$ for a Brownian motion with drift $\gamma > 0$ (that is, a Lévy process on $\mathbb{R}$ generated by $(1, 0, \gamma)$). If we regard x as time parameter, $\{T_t \colon t \geq 0\}$ is a Lévy process identical in law with $(\{x_t\}, P^\sharp)$ with $c' = \sqrt{2}$ and $\eta = -\gamma^2/2$. Thus, for each $x > 0$, T_x has an inverse Gaussian distribution. For other properties see Seshadri [**474**].

34. Exercises 6

E 34.1 (Feller [**139**]). Let $\{X_t\}$ be the Γ-process with $E[X_1] = 1$ and $\{Z_t^0\}$ be the Poisson process with $E[Z_1^0] = 1$. Let $\{Y_t\}$ be subordinate to $\{X_t\}$ by the directing process $\{Z_t^0 + \beta t\}$ with $\beta \geq 0$. Show the following for the distribution of Y_t, $t > 0$. If $\beta > 0$, then

$$P[\, Y_t \in B\,] = \textstyle\int_{B\cap(0,\infty)} \mathrm{e}^{-t-x} (\sqrt{x/t})^{\beta t - 1} I_{\beta t - 1}(2\sqrt{tx}) \mathrm{d}x.$$

If $\beta = 0$, then

$$P[\, Y_t \in B\,] = \mathrm{e}^{-t}\delta_0(B) + \textstyle\int_{B\cap(0,\infty)} \mathrm{e}^{-t-x} \sqrt{t/x}\, I_1(2\sqrt{tx}) \mathrm{d}x.$$

Here I_ν is the modified Bessel function (4.11). When $\beta = n/a$ with $n \in \mathbb{N}$ and $a \in (0, \infty)$, the distribution of $2Y_{a/2}$ is identical with the noncentral χ^2 distribution with n degrees of freedom and noncentrality parameter a in statistics.

E 34.2 (Feller [**139**]). The distribution λ_t^0 defined by

$$\lambda_t^0(\mathrm{d}x) = \sum_{k=0}^{\infty} \frac{t}{2k+t}\binom{2k+t}{k} 2^{-2k-t}\delta_{2k}$$

has Laplace transform

$$\int_{[0,\infty)} \mathrm{e}^{-ux}\lambda_t^0(\mathrm{d}x) = (1+\sqrt{1-\mathrm{e}^{-2u}})^{-t}.$$

A proof is given in Feller [**137**]. Let $\{Z_t^0\}$ be the subordinator with distribution λ_t^0. Let $\{X_t\}$ be the Γ-process with $EX_1 = 1$. Let $\mu_t^\sharp$ be the distribution of the Lévy process $\{Y_t\}$ subordinate to $\{X_t\}$ by the directing process $\{Z_t^0 + t\}$. Show that

$$\mu_t^\sharp(\mathrm{d}x) = \mathrm{e}^{-x} t x^{-1} I_t(x) 1_{(0,\infty)}(x)\mathrm{d}x$$

and

$$\int_0^\infty \mathrm{e}^{-ux}\mu_t^\sharp(\mathrm{d}x) = (u+1-\sqrt{(u+1)^2-1})^t, \qquad u>0,$$

and that the Lévy measures ρ and $\nu^\sharp$ of, respectively, $\{Z_t^0\}$ and $\{Y_t\}$ are

$$\rho = \sum_{k=1}^{\infty} \frac{(2k-1)!}{(k!)^2} 2^{-2k}\delta_{2k} \quad \text{and} \quad \nu^\sharp(\mathrm{d}x) = \mathrm{e}^{-x}x^{-1}I_0(x)1_{(0,\infty)}(x)\mathrm{d}x.$$

E 34.3 (Ismail and Kelker [**218**] and Halgreen [**178**]). Let $\{Y_t\}$ be a Lévy process on $\mathbb{R}^d$ subordinate to the Brownian motion on $\mathbb{R}^d$ by a selfdecomposable subordinator $\{Z_t\}$. Show that $\{Y_t\}$ is a selfdecomposable process.

E 34.4 (Pillai [**375**]). Let $\{Z_t\}$ be the Γ-process with $E[Z_1] = 1$ and let $\{X_t\}$ be the α-stable subordinator with $E[\mathrm{e}^{-uX_t}] = \mathrm{e}^{-tu^\alpha}$, $0<\alpha<1$. Show that, if $\{Y_t\}$ is the process subordinate to $\{X_t\}$ by $\{Z_t\}$, then $P[Y_1 \le x] = 1 - E_\alpha(-x^\alpha)$, $x \ge 0$, where E_α is the Mittag–Leffler function of (24.11). It follows that $1 - E_\alpha(-x^\alpha)$ is a selfdecomposable distribution function. Further, show that

$$P[\,Y_t \le x\,] = \sum_{n=0}^{\infty} (-1)^n \frac{\Gamma(t+n)}{\Gamma(t)n!\Gamma(1+\alpha(t+n))} x^{\alpha(t+n)}.$$

E 34.5. Let $\{S_n\}$ be a random walk on $\mathbb{R}^d$ and $\{Z_t\}$ be an integer-valued subordinator. Assume that they are independent. Define $Y_t = S_{Z_t}$. Show that $\{Y_t\}$ is a compound Poisson process with Lévy measure $\nu^\sharp(B) = \sum_{n=1}^{\infty} \mu^n(B)\rho\{n\}$, $B \in \mathcal{B}(\mathbb{R}^d \setminus \{0\})$, where $\mu = P_{S_1}$ and ρ is the Lévy measure of $\{Z_t\}$. The transformation is called *compounding* of $\{S_n\}$ by $\{Z_t\}$. Discuss the case where $\{Z_t\}$ is a Poisson process and the case where Z_1 has a geometric distribution with parameter $p \in (0,1)$. Show that, in the latter case,

$$E[\mathrm{e}^{\mathrm{i}\langle z, Y_t\rangle}] = p^t(1-q\widehat{\mu}(z))^{-t}, \quad q = 1-p.$$

If μ is infinitely divisible, then compounding is a special case of subordination.

E 34.6. If L and M are linear operators in a Banach space $\mathbf{B}$, then ML is defined by $(ML)f = M(Lf)$ with domain $\mathfrak{D}(ML) = \{f\colon f \in \mathfrak{D}(L)$ and $Lf \in \mathfrak{D}(M)\}$. Thus we define L^n for $n \in \mathbb{N}$ by $L^1 = L$ and $L^{n+1} = LL^n$. Show that, if L is the infinitesimal generator of a strongly continuous contraction semigroup on $\mathbf{B}$, then, for any $n \in \mathbb{N}$, $\mathfrak{D}(L^n)$ is a core of L.

E 34.7. Let $\{P_t\colon t \ge 0\}$ be a strongly continuous contraction semigroup on a Banach space $\mathbf{B}$ and let L be its infinitesimal generator. Show that L is a bounded operator if and only if $\|P_t - I\| \to 0$ as $t \downarrow 0$.

E 34.8. Let L be the infinitesimal generator as in Theorem 31.5 of a Lévy process $\{X_t\}$ on $\mathbb{R}^d$. Show that, if L is a bounded operator, then $\{X_t\}$ is a compound Poisson process or the zero process.

E 34.9. Let $\{P_t\}$ and $\{Q_t\}$ be the semigroups on $C_0(\mathbb{R}^d)$ determined by Lévy processes $\{X_t\}$ and $\{Y_t\}$, respectively. Let a and b be positive reals and let $R_t = Q_{bt}P_{at}$. Show that $\{R_t\}$ is the semigroup determined by another Lévy process $\{Z_t\}$ and that $L_3 f = aL_1 f + bL_2 f$ for $f \in C_0^2$, where L_1, L_2, and L_3 are the infinitesimal generators of $\{P_t\}$, $\{Q_t\}$, and $\{R_t\}$, respectively. Show that, if $\{X_t\}$ and $\{Y_t\}$ are independent, then $\{Z_t\} \stackrel{\mathrm{d}}{=} \{X_{at} + Y_{bt}\}$.

E 34.10. Let $1 \leq p < \infty$. Let $L^p(\mathbb{R}^d)$ be the real Banach space of measurable functions f from $\mathbb{R}^d$ to $\mathbb{R} \cup \{+\infty, -\infty\}$ satisfying $\int_{\mathbb{R}^d} |f(x)|^p \mathrm{d}x < \infty$ with norm $\|f\| = \left(\int |f(x)|^p \mathrm{d}x\right)^{1/p}$, where two functions equal almost everywhere are identified. Let $P_t(x, B) = \mu^t(B - x)$ be the temporally homogeneous transition function associated with a Lévy process $\{X_t\}$. Show that, for $f \in L^p(\mathbb{R}^d)$, $(P_t f)(x) = \int \mu^t(\mathrm{d}y) f(x + y)$ is defined as an element of $L^p(\mathbb{R}^d)$ and that $\{P_t\}$ is a strongly continuous contraction semigroup.

E 34.11. Let $\{X_t\}$ and $\{\widetilde{X}_t\}$ be Lévy processes on $\mathbb{R}^d$ such that $\{\widetilde{X}_t\} \stackrel{\mathrm{d}}{=} \{-X_t\}$. Let $\{P_t\}$ and $\{\widetilde{P}_t\}$ be the semigroups in $L^2(\mathbb{R}^d)$ induced by $\{X_t\}$ and $\{\widetilde{X}_t\}$, respectively, as in E 34.10. Show that, for any f and g in $L^2(\mathbb{R}^d)$,

$$\int_{\mathbb{R}^d} (P_t f)(x) g(x) \mathrm{d}x = \int_{\mathbb{R}^d} f(x) (\widetilde{P}_t g)(x) \mathrm{d}x.$$

E 34.12. Let $(\{x_t\}, P)$ be a selfdecomposable subordinator. Let $\varphi(x)$ be decreasing for $x > 0$ and satisfying $\varphi(x) = O(x)$, $x \downarrow 0$, and let η be arbitrary. Show that density transformation of $(\{x_t\}, P)$ by $\varphi(x)$ and η gives another selfdecomposable subordinator.

E 34.13 (Barndorff-Nielsen and Halgreen [**14**], Halgreen [**178**]). The following probability measure is called *generalized inverse Gaussian* with parameters (λ, χ, ψ):

$$\mu(\mathrm{d}x) = cx^{\lambda-1} \exp\left(-\tfrac{1}{2}(\chi x^{-1} + \psi x)\right) 1_{(0,\infty)}(x) \mathrm{d}x,$$

where c is a normalizing positive constant. The domain of the parameters is given by $\{\lambda < 0, \chi > 0, \psi \geq 0\}$, $\{\lambda = 0, \chi > 0, \psi > 0\}$, and $\{\lambda > 0, \chi \geq 0, \psi > 0\}$. Show that $c = (\psi/\chi)^{\lambda/2}/(2K_\lambda(\sqrt{\chi\psi}))$ if $\chi > 0$ and $\psi > 0$, where K_λ is the modified Bessel function (4.9), (4.10). If $\chi = 0$ or $\psi = 0$, then c is given by the limit value in the same formula. Show, for the Laplace transform $L_\mu(u)$, $u \geq 0$, of μ, that

$$L_\mu(u) = \begin{cases} \left(\frac{\psi}{\psi + 2u}\right)^{\lambda/2} \frac{K_\lambda(\sqrt{\chi(\psi+2u)})}{K_\lambda(\sqrt{\chi\psi})} & \text{if } \chi > 0 \text{ and } \psi > 0, \\ \frac{2^{1+\lambda/2} K_\lambda(\sqrt{2\chi u})}{\Gamma(-\lambda)(\chi u)^{\lambda/2}} & \text{if } \lambda < 0,\ \chi > 0, \text{ and } \psi = 0. \end{cases}$$

Show that a Γ-distribution is μ with $\chi = 0$ and that the distribution of $1/X$ where X is Γ-distributed is μ with $\psi = 0$. Show that all generalized inverse Gaussians are infinitely divisible and, furthermore, selfdecomposable.

E 34.14 (Barndorff-Nielsen [**12**] and Halgreen [**178**]). A distribution on $\mathbb{R}$ having density function $g(x) = c\exp(-a\sqrt{1+x^2}+bx)$ with $a > 0$, $|b| < a$, and a normalizing constant $c > 0$ is called a *hyperbolic distribution,* since the graph of $\log g(x)$ is a branch of a hyperbola. Show that it is infinitely divisible and, moreover, selfdecomposable. The result is generalized to distributions on $\mathbb{R}$ with density $c(\sqrt{1+x^2})^{\lambda-1/2}K_{\lambda-1/2}(a\sqrt{1+x^2})e^{bx}$, $\lambda \in \mathbb{R}$, $a > 0$, $|b| < a$ ($|b| = a$ is also permitted if $\lambda < 0$).

E 34.15. Let $\mu = qv^q(x)dx$, $q > 0$, with $v^q(x)$ of (30.29). Show that, for any $t > 0$, μ^t has density

$$2(2\pi)^{-d/2}(\Gamma(t))^{-1}q^t(2q)^{d/4-t/2}|x|^{t-d/2}K_{d/2-t}((2q)^{1/2}|x|).$$

Notes

Subordination was introduced by Bochner, as we have mentioned at the beginning of the chapter. It is treated in Feller's book [**139**]. Theorem 30.1 is obtained by Zolotarev [**603**], Bochner [**50**], Ikeda and Watanabe [**213**], and Rogozin [**418**]. Theorem 30.10(i) is pointed out by Rogozin [**419**]. Example 30.7 is by Spitzer [**495**]. The same result is obtained by using the inverse of the local time at 0 for $X_1(t)$. The latter is extended by Molchanov and Ostrovskii [**354**] to get a representation of rotation invariant stable processes.

Theorem 31.5 is essentially given by Itô [**221**] and Hunt [**208**]. The assertion on cores there is given by Sato [**428**]. Lemma 31.6 is by Watanabe [**548**]. The proof in this book follows [**133**]. Lemma 31.7 is by Itô [**223**]. Existence of potential operators in the sense of Yosida for Lévy processes is proved by [**428**]. Their cores are studied in [**429**]. Berg and Forst [**21**] has some exposition. Compounding in E 34.5 was found by Zolotarev [**603**].

The works connected with Sections 32 and 33 are mentioned in the text. Concerning the density transformation, all of Skorohod [**490, 492**], Kunita and Watanabe [**300**], and Newman [**361**] treat not only Lévy processes but also additive processes. Moreover, Kunita and Watanabe [**300**] prove similar results for semimartingales. For two Lévy processes $(\{x_t\}, P)$ and $(\{x_t\}, P^\sharp)$ Newman [**360, 361**] obtains the condition that $[P^\sharp]_{\mathcal{F}_t}$ is singular with respect to $[P]_{\mathcal{F}_t}$. In general, there are cases where $[P^\sharp]_{\mathcal{F}_t}$ is neither absolutely continuous nor singular with respect to $[P]_{\mathcal{F}_t}$. But, as Newman [**361**] and Brockett and Tucker [**69**] point out, such cases do not exist if $\nu \approx \nu^\sharp$. Some related later works are Memin and Shiryaev [**344**], Jacod and Shiryaev [**230**], Takahashi [**512**], and Inoue [**216**].

Halgreen's result [**178**], of which E 34.14 is a special case, has a multivariate generalization discussed by Takano [**515**].

CHAPTER 7

Recurrence and transience

35. Dichotomy of recurrence and transience

The Lévy processes on $\mathbb{R}^d$ are divided into two classes, called recurrent and transient, according to large time behavior of sample functions. We give the dichotomy theorem, prove some criteria in terms of characteristic functions, and discuss important recurrent and transient cases in this chapter. The analogue of the law of large numbers is also given.

DEFINITION 35.1. A Lévy process $\{X_t : t \geq 0\}$ on $\mathbb{R}^d$ defined on a probability space $(\Omega, \mathcal{F}, P)$ is called *recurrent* if

$$\liminf_{t\to\infty} |X_t| = 0 \quad \text{a. s.} \tag{35.1}$$

It is called *transient* if

$$\lim_{t\to\infty} |X_t| = \infty \quad \text{a. s.} \tag{35.2}$$

Note that the events $\Omega_0 \cap \{\liminf_{t\to\infty} |X_t| = 0\}$ and $\Omega_0 \cap \{\lim_{t\to\infty} |X_t| = \infty\}$ are in $\mathcal{F}$, since they equal

$$\Omega_0 \cap \bigcap_{k=1}^{\infty} \bigcap_{n=1}^{\infty} \bigcup_{t\in\mathbb{Q}\cap(n,\infty)} \{|X_t| < 1/k\} \text{ and } \Omega_0 \cap \bigcap_{k=1}^{\infty} \bigcup_{n=1}^{\infty} \bigcap_{t\in\mathbb{Q}\cap(n,\infty)} \{|X_t| > k\},$$

respectively. Here Ω_0 is the event in Definition 1.6.

First we consider random walks. They are simpler than Lévy processes.

DEFINITION 35.2. A random walk $\{S_n : n \in \mathbb{Z}_+\}$ on $\mathbb{R}^d$ is called *recurrent* if

$$\liminf_{n\to\infty} |S_n| = 0 \quad \text{a. s.} \tag{35.3}$$

It is called *transient* if

$$\lim_{n\to\infty} |S_n| = \infty \quad \text{a. s.} \tag{35.4}$$

An important quantity in recurrence and transience of a Lévy process is its *potential measure* $V(B)$, $B \in \mathcal{B}(\mathbb{R}^d)$, as in Definition 30.9, that is,

$$V(B) = \int_0^\infty P[X_t \in B] \mathrm{d}t = E\Big[\int_0^\infty 1_B(X_t) \mathrm{d}t\Big]. \tag{35.5}$$

Sometimes we call $\int_0^\infty 1_B(X_t)dt$ the *sojourn time* on B and $V(B)$ the *mean sojourn time* on B. The analogue of $V(B)$ for a random walk is

$$W(B) = \sum_{n=1}^{\infty} P[S_n \in B] = E\left[\sum_{n=1}^{\infty} 1_B(S_n)\right] \tag{35.6}$$

for $B \in \mathcal{B}(\mathbb{R}^d)$. Note that we allow infinite values for $V(B)$ and $W(B)$. Let

$$B_a = \{x \in \mathbb{R}^d \colon |x| < a\}, \tag{35.7}$$

the open ball with radius $a > 0$ and center at the origin, that is, the a-neighborhood of the origin.

THEOREM 35.3 (Dichotomy for random walks). *Let $\{S_n\}$ be a random walk on $\mathbb{R}^d$. Then:*

(i) *It is either recurrent or transient.*

(ii) *It is recurrent if and only if*

$$W(B_a) = \infty \quad \text{for every } a > 0. \tag{35.8}$$

(iii) *It is transient if and only if*

$$W(B_a) < \infty \quad \text{for every } a > 0. \tag{35.9}$$

Proof. Step 1. Assume (35.9). Then $\{S_n\}$ is transient. In fact, we have $P[\limsup_{n\to\infty}\{|S_n| < a\}] = 0$ from (35.9) by the Borel–Cantelli lemma (Proposition 1.11). Thus, almost surely, there exists m such that $|S_n| \geq a$ for all $n \geq m$. Since a is arbitrary, this shows transience of $\{S_n\}$.

Step 2. Suppose that $W(B_a) = \infty$ for some a. Let us prove that $\{S_n\}$ is recurrent. Let $K = \{|x| \leq a\}$. Then $W(K) = \infty$. Let $\eta > 0$. Since K is covered by a finite number of open balls with radii $\eta/2$, there is an open ball B with radius $\eta/2$ such that $W(B) = \infty$. Since

$$\begin{aligned} 1 &\geq \sum_{k=1}^{\infty} P[\,S_k \in B \text{ and } S_{k+n} \notin B,\ n \geq 1\,] \\ &\geq \sum_{k=1}^{\infty} P[\,S_k \in B \text{ and } |S_{k+n} - S_k| \geq \eta,\ n \geq 1\,] \\ &= P[\,|S_n| \geq \eta,\ n \geq 1\,] \sum_{k=1}^{\infty} P[\,S_k \in B\,] \end{aligned}$$

by the stationary independent increments property, we get

$$P[\,|S_n| \geq \eta,\ n \geq 1\,] = 0 \tag{35.10}$$

from $W(B) = \infty$. Now, for any $\varepsilon > 0$,

$$P[\,\exists\, m \geq 1 \text{ such that } |S_n| \geq \varepsilon \text{ for all } n \geq m\,] = \sum_{k=1}^{\infty} p_k, \tag{35.11}$$

where $p_k = P[\,|S_k| < \varepsilon \text{ and } |S_{k+n}| \geq \varepsilon,\, n \geq 1\,]$. This follows from (35.10) with η replaced by ε. We have, for $0 < \eta < \varepsilon$,

$$\begin{aligned} &P[\,|S_k| < \varepsilon - \eta \text{ and } |S_{k+n}| \geq \varepsilon,\, n \geq 1\,] \\ &\quad \leq P[\,|S_k| < \varepsilon - \eta \text{ and } |S_{k+n} - S_k| \geq \eta,\, n \geq 1\,] \\ &\quad = P[\,|S_k| < \varepsilon - \eta\,]\, P[\,|S_n| \geq \eta,\, n \geq 1\,] \\ &\quad = 0, \end{aligned}$$

using the stationary independent increments property and (35.10). Letting $\eta \downarrow 0$, we see that $p_k = 0$ for all $k \geq 1$. Hence the probability on the left-hand side of (35.11) is 0. That is,

$$P[\,|S_n| < \varepsilon \text{ for infinitely many } n\,] = 1$$

for every $\varepsilon > 0$. Hence $\{S_n\}$ is recurrent.

Steps 1 and 2 combined give the proof of (i). Also, (iii) has been proved by Steps 1 and 2. The 'if' part of (ii) follows from Step 2, too. If $W(B_a) < \infty$ for some $a > 0$, then the argument in Step 1 shows that $P[\,\exists\, m \text{ such that } |S_n| \geq a \text{ for all } n \geq m\,] = 1$ for this a, which implies that $\{S_n\}$ is not recurrent. Thus (ii) is shown. □

THEOREM 35.4 (Dichotomy for Lévy processes). *Let $\{X_t\}$ be a Lévy process on $\mathbb{R}^d$. Then:*

(i) *It is either recurrent or transient.*

(ii) *It is recurrent if and only if*

$$V(B_a) = \infty \quad \textit{for every } a > 0. \tag{35.12}$$

(iii) *It is recurrent if and only if*

$$\int_0^\infty 1_{B_a}(X_t)\mathrm{d}t = \infty \quad \textit{a. s.} \quad \textit{for every } a > 0. \tag{35.13}$$

(iv) *It is transient if and only if*

$$V(B_a) < \infty \quad \textit{for every } a > 0. \tag{35.14}$$

(v) *It is transient if and only if*

$$\int_0^\infty 1_{B_a}(X_t)\mathrm{d}t < \infty \quad \textit{a. s.} \quad \textit{for every } a > 0. \tag{35.15}$$

(vi) *Fix $h > 0$ arbitrarily. The process $\{X_t\}$ is recurrent if and only if the random walk $\{X_{nh}\colon n = 0, 1, \ldots\}$ is recurrent.*

This theorem gives not only the dichotomy but also a criterion by potential measures of recurrence/transience. We prepare a lemma for the proof.

LEMMA 35.5 (Kingman [**284**]). *Fix $a > 0$. For any Lévy process $\{X_t\}$ there is a function $\gamma(\varepsilon)$ satisfying $\gamma(\varepsilon) \to 1$, $\varepsilon \downarrow 0$, such that, for every $t > 0$ and $\varepsilon > 0$,*

$$P\Big[\int_t^\infty 1_{B_{2a}}(X_s)ds > \varepsilon\Big] \geq \gamma(\varepsilon) P[\,|X_{t+s}| < a \text{ for some } s > 0\,]. \tag{35.16}$$

Proof. Denote by $\mathcal{F}_t^0$ the σ-algebra generated by $\{X_s \colon s \in [0,t]\}$. Let $\Lambda \in \mathcal{F}_t^0$ with $\Lambda \subset \{|X_t| < a\}$ and $P[\Lambda] > 0$. Let

$$Y = \frac{1}{2\varepsilon}\int_t^{t+2\varepsilon} 1_{B_{2a}}(X_s)ds.$$

Then $0 \leq Y \leq 1$ and, using the conditional probability (3.3) given Λ, we have

$$2E[Y \mid \Lambda] \leq 2P[Y > \tfrac{1}{2} \mid \Lambda] + P[Y \leq \tfrac{1}{2} \mid \Lambda] = P[Y > \tfrac{1}{2} \mid \Lambda] + 1.$$

Hence,

$$\begin{aligned} P\Big[\int_t^{t+2\varepsilon} 1_{B_{2a}}(X_s)ds > \varepsilon \;\Big|\; \Lambda\Big] &\geq \frac{1}{\varepsilon} E\Big[\int_t^{t+2\varepsilon} 1_{B_{2a}}(X_s)ds \;\Big|\; \Lambda\Big] - 1 \\ &= \frac{1}{\varepsilon}\int_0^{2\varepsilon} P[\,|X_{t+s}| < 2a \mid \Lambda\,]ds - 1 \\ &\geq \frac{1}{\varepsilon}\int_0^{2\varepsilon} P[\,|X_{t+s} - X_t| < a \mid \Lambda\,]ds - 1 = \gamma(\varepsilon), \end{aligned}$$

where $\gamma(\varepsilon) = \frac{1}{\varepsilon}\int_0^{2\varepsilon} P[\,|X_s| < a\,]ds - 1$. Since $P[\,|X_s| < a\,] \to 1$ as $s \downarrow 0$, $\gamma(\varepsilon)$ tends to 1 as $\varepsilon \downarrow 0$. We have

$$P\Big[\int_t^\infty 1_{B_{2a}}(X_s)ds > \varepsilon \;\Big|\; \Lambda\Big] \geq \gamma(\varepsilon). \tag{35.17}$$

For each $\eta > 0$, we have

$$\begin{aligned} &P\Big[\int_t^\infty 1_{B_{2a}}(X_s)ds > \varepsilon\Big] \\ &\quad\geq \sum_{n=0}^\infty P\Big[|X_{t+j\eta}| \geq a \text{ for } 0 \leq j < n,\ |X_{t+n\eta}| < a,\ \int_{t+n\eta}^\infty 1_{B_{2a}}(X_s)ds > \varepsilon\Big]. \end{aligned}$$

By use of (35.17) with $t + n\eta$ in place of t, this is

$$\begin{aligned} &\geq \gamma(\varepsilon)\sum_{n=0}^\infty P[\,|X_{t+j\eta}| \geq a \text{ for } 0 \leq j < n,\ |X_{t+n\eta}| < a\,] \\ &= \gamma(\varepsilon) P[\,|X_{t+n\eta}| < a \text{ for some } n \in \mathbb{Z}_+\,]. \end{aligned}$$

Choosing $\eta = 2^{-k}$ and letting $k \to \infty$, we get (35.16). $\square$

Proof of Theorem 35.4. Fix $a > 0$. Let us prove the equivalence of the following statements.

(1) Almost surely there are $t_n = t_n(\omega) \to \infty$ such that $X_{t_n} \in B_a$.
(2) Almost surely $\int_0^\infty 1_{B_{2a}}(X_t)dt = \infty$.
(3) $\int_0^\infty P[X_t \in B_{2a}]dt = \infty$.
(4) There is $h_0 > 0$ such that, for every $h \in (0, h_0]$, $\sum_{n=1}^\infty P[X_{nh} \in B_{3a}] = \infty$.
(5) There is $h_0 > 0$ such that, for every $h \in (0, h_0]$, $\{X_{nh}: n = 0, 1, \dots\}$ is recurrent.

(1) $\Rightarrow$ (2). Let $0 < \varepsilon < 1$. Use Lemma 35.5. For every t, the probability on the right-hand side of (35.16) is 1, by virtue of (1). Hence, letting $t \to \infty$, we get

$$P\Big[\int_0^\infty 1_{B_{2a}}(X_s)ds = \infty\Big] \geq \gamma(\varepsilon).$$

Then, letting $\varepsilon \to 0$, we obtain (2).

(2) $\Rightarrow$ (3). This is clear, as $\int_0^\infty P[X_t \in B_{2a}]dt = E\big[\int_0^\infty 1_{B_{2a}}(X_t)dt\big]$.

(3) $\Rightarrow$ (4). Choose $h_0 > 0$ such that, for every $s \leq h_0$, $P[X_s \in B_a] > \frac{1}{2}$. Let $0 < h \leq h_0$. We have, for every $x \in B_{2a}$ and $s \leq h$, $P[x + X_s \in B_{3a}] > \frac{1}{2}$. If $(n-1)h \leq t \leq nh$, then

$$\begin{aligned} P[X_{nh} \in B_{3a}] &\geq P[X_t \in B_{2a}, X_t + (X_{nh} - X_t) \in B_{3a}] \\ &= E[f(X_t)\, 1_{\{X_t \in B_{2a}\}}] \quad \text{with } f(x) = P[x + X_{nh-t} \in B_{3a}], \end{aligned}$$

and hence

$$P[X_{nh} \in B_{3a}] \geq \tfrac{1}{2} P[X_t \in B_{2a}].$$

Therefore we get

$$P[X_{nh} \in B_{3a}] \geq \frac{1}{2h} \int_{(n-1)h}^{nh} P[X_t \in B_{2a}]dt. \tag{35.18}$$

Using (3), we get (4).

(4) $\Rightarrow$ (5). This follows from Theorem 35.3.

(5) $\Rightarrow$ (1). Almost surely there are integers $k_n \uparrow \infty$ such that $|X_{k_n h}| \to 0$. Hence we have (1).

This completes the proof of the equivalence of (1)–(5). The condition (5) does not involve a. Hence, each of the conditions (1), (2), (3), and (4) is independent of a. Thus, (1) holds for some a if and only if (1) holds for every a, which, in turn, is equivalent to recurrence of $\{X_t\}$. The assertions (ii) and (iii) now follow.

If (35.12) does not hold, then (35.14) holds. In fact, if $V(B_a) = \infty$ for some $a > 0$, then, by the independence of (3) from a, $V(B_a) = \infty$ for every $a > 0$.

Now let us prove the assertion (iv). If $\{X_t\}$ is transient, then it is not recurrent, and hence (35.12) fails to hold, which implies (35.14). Conversely, assume (35.14). Choose $\varepsilon > 0$ in such a way that $\gamma(\varepsilon) > 1/2$ for the function γ in Lemma 35.5. Then

$$E\Big[\int_t^\infty 1_{B_{2a}}(X_s)\mathrm{d}s\Big] \geq \varepsilon P\Big[\int_t^\infty 1_{B_{2a}}(X_s)\mathrm{d}s > \varepsilon\Big]$$
$$\geq (\varepsilon/2)P[\,|X_{t+s}| < a \text{ for some } s > 0\,].$$

Let $t \uparrow \infty$ and use (35.14). Then the left-hand side goes to 0 and

$$\lim_{t\to\infty} P[\,|X_{t+s}| < a \text{ for some } s > 0\,] = 0.$$

Since a and t are arbitrary, we obtain transience, noting that

$$\{\lim_{t\to\infty} |X_t| = \infty\} = \bigcap_{k=1}^{\infty}\bigcup_{n=1}^{\infty}\{|X_{n+s}| \geq k \text{ for all } s > 0\}.$$

Now we prove (i). Assume that $\{X_t\}$ is not recurrent. Then, by (ii), (35.12) does not hold and hence (35.14) does hold. Hence the process is transient by (iv).

To show (v), if (35.15) holds, then the process is transient by (iii) and (i). If the process is transient, then (35.14) holds by (iv) and (35.14) implies (35.15).

The proof of (vi) is as follows. If the random walk $\{X_{nh}\}$ is recurrent, then we see that the process $\{X_t\}$ is recurrent, like the proof that (5) implies (1). Conversely, suppose that $\{X_t\}$ is recurrent. Then we have (35.12). For the given h we can find $a > 0$ such that $P[\,\sup_{s\in[0,h]}|X_s(\omega)| \leq a\,] > \frac{1}{2}$, because $\sup_{s\in[0,h]}|X_s(\omega)|$ is finite a. s. by the right-continuity with left limits of the sample functions. Now we have $\inf_{s\leq h} P[\,X_s \in B_a\,] > \frac{1}{2}$ for this a. Recall the proof that (3) implies (4). We see that (35.18) holds for this a. Hence $\sum_{n=1}^{\infty} P[\,X_{nh} \in B_{3a}\,] = \infty$ and $\{X_{nh}\}$ is recurrent by Theorem 35.3.

The proof of Theorem 35.4 is complete. □

We shall frequently use the recurrence/transience criterion by potential measures given by (ii) and (iv). The q-potential measure for $q \geq 0$ of a Lévy process is defined by Definition 30.9. If V^q is absolutely continuous, then we call the density $v^q(x)$ the *q-potential density.* As V^0 is denoted by V and called the potential measure, we write $v^0(x) = v(x)$ and call it the potential density. If the distribution μ^t of X_t, $t > 0$, has density $p(t,x)$ measurable in (t,x), then

$$v^q(x) = \int_0^\infty \mathrm{e}^{-qt} p(t,x)\mathrm{d}t.$$

Therefore, in this case, we can determine recurrence/transience if we can find $v(x)$. Let us consider Brownian motions and Cauchy processes.

EXAMPLE 35.6 (Brownian motion). Let $\{X_t\}$ be the Brownian motion on $\mathbb{R}^d$. Then the distribution of X_t, $t > 0$, has density $p(t,x)$ of (30.26).

If $x \neq 0$, then $p(t,x) \sim (2\pi t)^{-d/2}$ as $t \to \infty$. Hence $v(x) = \infty$ for $x \neq 0$ if $d = 1$ or 2. It follows that $V(B) = \infty$ for any nonempty open set B if $d = 1$ or 2. Hence, by Theorem 35.4, *the Brownian motion for $d = 1$ or 2 is recurrent.* If $d \geq 3$, then, by the change of variable $|x|^2/(2t) = s$,

$$v(x) = \frac{1}{2}\pi^{-d/2}|x|^{2-d}\int_0^\infty \mathrm{e}^{-s}s^{d/2-2}\mathrm{d}s = \frac{1}{2}\pi^{-d/2}\Gamma\Big(\frac{d}{2}-1\Big)|x|^{2-d} \tag{35.19}$$

for $x \neq 0$. Since this $v(x)$ is integrable on any bounded Borel set, *the Brownian motion for $d \geq 3$ is transient*, by Theorem 35.4. The function $|x|^{2-d}$ for $d \geq 3$ is the basic density in the Newtonian potential in potential theory. Theory of Newtonian potential is thus connected with transient Brownian motions.

The random walk analogue of the Brownian motion on $\mathbb{R}^d$ is the simple random walk $\{S_n\}$ with $P[\,S_1 = x\,] = 1/(2d)$ when x is one of the $2d$ adjacent lattice points to the origin. It is recurrent for $d = 1$, 2 and transient for $d \geq 3$. Proofs are found in [**34**], [**138**], [**151**]. Such dependence on the dimension of the large time behavior was discovered by Pólya [**378**] and led to the concepts of recurrence and transience.

EXAMPLE 35.7 (Cauchy process). Let $\{X_t\}$ be the Cauchy process on $\mathbb{R}^d$ having the Cauchy distribution with $c = 1$ and $\gamma = 0$ (Example 2.12) at time 1. Then

$$\begin{aligned} v(x) &= \pi^{-(d+1)/2}\Gamma(\tfrac{d+1}{2})\int_0^\infty t(|x|^2+t^2)^{-(d+1)/2}\mathrm{d}t \\ &= \begin{cases} \infty & \text{for } d = 1, \\ 2^{-1}\pi^{-(d+1)/2}\Gamma(\tfrac{d-1}{2})|x|^{1-d} & \text{for } d \geq 2,\ x \neq 0. \end{cases} \end{aligned}$$

Hence $\{X_t\}$ is recurrent for $d = 1$ and transient for $d \geq 2$. Notice that the two-dimensional Cauchy process is transient, although it is subordinate to the recurrent Brownian motion.

We add a property of recurrent Lévy processes. The supports of Lévy processes are defined in Definition 24.13.

THEOREM 35.8. *Let $\{X_t\}$ be a recurrent Lévy process on $\mathbb{R}^d$. Let Σ be the support of $\{X_t\}$. Then:*

(i) *Σ is a closed additive subgroup of $\mathbb{R}^d$. If $d = 1$, then Σ equals either $\mathbb{R}$ or $a\mathbb{Z}$ with some $a > 0$.*

(ii) *Almost surely the set of $x \in \mathbb{R}^d$ such that $\liminf_{t\to\infty}|X_t(\omega) - x| = 0$ coincides with Σ.*

(iii) *$V(B) = \infty$ for any open set B with $B \cap \Sigma \neq \emptyset$.*

The assertion (i) says that, in the recurrent case, the group $\mathfrak{G}$ of the process as in Definition 24.21 is identical with the support Σ itself.

Proof of theorem. Let R be the set of x such that, almost surely, $\liminf_{t\to\infty}|X_t - x| = 0$. We have $R \subset \Sigma$ by Proposition 24.14(iii) and $0 \in R$ by the recurrence. Let us prove that

$$\text{if } x \in \Sigma \text{ and } y \in R, \text{ then } y - x \in R. \tag{35.20}$$

In fact, suppose that $x \in \Sigma$ and $y - x \notin R$. Then, for some ε and t_0,

$$P[\,|X_s - y + x| \geq \varepsilon \text{ for all } s \geq t_0\,] > 0.$$

We can pick t_1 such that $P[\,|X_{t_1} - x| < \varepsilon/2\,] > 0$ again by Proposition 24.14(iii). Then,

$$\begin{aligned}
&P[\,|X_s - y| \geq \varepsilon/2 \text{ for all } s \geq t_0 + t_1\,]\\
&\qquad \geq P[\,|X_{t_1} - x| < \varepsilon/2,\ |X_s - X_{t_1} - y + x| \geq \varepsilon \text{ for all } s \geq t_0 + t_1\,]\\
&\qquad \geq P[\,|X_{t_1} - x| < \varepsilon/2]\, P[\,|X_s - y + x| \geq \varepsilon \text{ for all } s \geq t_0\,]\\
&\qquad > 0,
\end{aligned}$$

which implies that $y \notin R$. This proves (35.20). If $x \in \Sigma$, then $-x \in \Sigma$ by (35.20) and by $0 \in R$. The support Σ is closed under addition by Proposition 24.14(i). Hence Σ is a closed additive subgroup of $\mathbb{R}^d$. If $d = 1$ and $a = \inf\{x > 0\colon x \in \Sigma\}$, then $\Sigma = \mathbb{R}$ in the case $a = 0$ and $\Sigma = a\mathbb{Z}$ in the case $a > 0$. This proves (i).

We have $\Sigma \subset R$, because, if $x \in \Sigma$, then $-(-x) \in R$ by $0 \in R$ and $-x \in \Sigma$. Hence $\Sigma = R$. Let $\mathcal{O}$ be the collection of open balls B with rational radii and with centers at rational points such that $B \cap \Sigma \neq \emptyset$. Since $\mathcal{O}$ is countable, it follows from $\Sigma = R$ that

$$P\left[\bigcap_{B\in\mathcal{O},t\geq 0} \{X_s \in B,\ \exists s > t\}\right] = 1.$$

Now, recalling Definition 24.13 of the support Σ, we obtain (ii).

The proof of (iii) is as follows. Let $x \in \Sigma$, $\varepsilon > 0$, and B be the ε-neighborhood of x. It is enough to show $V(B) = \infty$ in this case. Let C be the $(\varepsilon/2)$-neighborhood of x. There is $t \geq 0$ such that $P[\,X_t \in C\,] > 0$. We have

$$\begin{aligned}
V(B) &\geq E\left[1_{\{X_t\in C\}} \int_t^\infty 1_B(X_t + (X_s - X_t))\mathrm{d}s\right]\\
&= E[1_{\{X_t\in C\}} f(X_t)] \quad \text{with } f(y) = E\left[\int_0^\infty 1_B(y + X_s)\mathrm{d}s\right]
\end{aligned}$$

by Proposition 10.7. Since

$$f(y) \geq E\left[\int_0^\infty 1_{B_{\varepsilon/2}}(X_s)\mathrm{d}s\right] = \infty \quad \text{for } y \in C$$

by the recurrence, $V(B)$ is infinite. $\square$

36. Laws of large numbers

We use the concept of uniform integrability.

DEFINITION 36.1. A family $\{X_\lambda\colon \lambda \in \Lambda\}$ of real random variables on a probability space $(\Omega, \mathcal{F}, P)$, where Λ is a parameter set, is said to be *uniformly integrable* if $\sup_{\lambda\in\Lambda} E[|X_\lambda|;\ |X_\lambda| > a] \to 0$ as $a \to \infty$.

PROPOSITION 36.2. *If $\{X_n\colon n = 1, 2, \ldots\}$ is a sequence of real random variables such that $X = \lim_{n\to\infty} X_n$ exists almost surely, then the following three statements are equivalent.*

(1) *$\{X_n\}$ is uniformly integrable.*
(2) *$E|X| < \infty$ and $\lim_{n\to\infty} E|X_n - X| = 0$.*
(3) *$E|X| < \infty$ and $\lim_{n\to\infty} E|X_n| = E|X|$.*

Each of (1)–(3) *implies that* $\lim_{n\to\infty} EX_n = EX$.

This proposition is given in Doob [**106**], p. 629, Meyer [**345**], p. 38, Chung [**80**], p. 97, and Fristedt and Gray [**151**], p. 108.

THEOREM 36.3 (Strong law of large numbers). *Let $\{S_n\}$ be a random walk on $\mathbb{R}^d$. If $E|S_1| < \infty$ and $ES_1 = \gamma$, then*

$$\lim_{n\to\infty} n^{-1} S_n = \gamma \quad a.\,s. \tag{36.1}$$

and

$$\lim_{n\to\infty} E|n^{-1} S_n - \gamma| = 0. \tag{36.2}$$

If $E|S_1| = \infty$, then

$$\limsup_{n\to\infty} n^{-1} |S_n| = \infty \quad a.\,s. \tag{36.3}$$

Proof. Suppose $d = 1$ and $ES_1 = \gamma$, finite. The a. s. convergence (36.1) is Kolmogorov's theorem and the proof is found in [**34**], [**66**], [**80**], [**139**], [**151**], [**327**]. Let $\{Z_n\}$ be independent identically distributed random variables such that $S_n = Z_1 + \cdots + Z_n$. Write $Z_n^+ = Z_n \vee 0$, $Z_n^- = (-Z_n) \vee 0$, $S_n^+ = Z_1^+ + \cdots + Z_n^+$, $S_n^- = Z_1^- + \cdots + Z_n^-$, $EZ_1^+ = \gamma^+$, and $EZ_1^- = \gamma^-$. Then $Z_n = Z_n^+ - Z_n^-$ and $S_n = S_n^+ - S_n^-$. Since $\{Z_n^+\}$ is independent and identically distributed, $n^{-1}S_n^+ \to \gamma^+$ a. s. Since $E[n^{-1}S_n^+] = EZ_1^+ = \gamma^+$, Proposition 36.2 tells us that $\{n^{-1}S_n^+\}$ is uniformly integrable. Hence $E|n^{-1}S_n^+ - \gamma^+| \to 0$. Similarly we have $E|n^{-1}S_n^- - \gamma^-| \to 0$. Together these imply (36.2).

Suppose $d = 1$ and $E|S_1| = \infty$. We have, for each $a > 0$,

$$\sum_{n=1}^{\infty} P[|S_1| > na] \geq \int_1^{\infty} P[a^{-1}|S_1| > u]du = \infty,$$

since $E[a^{-1}|S_1|] = \infty$. Noticing that $P[|Z_n| > na] = P[|S_1| > na]$, we have $\sum_{n=1}^{\infty} P[|Z_n| > na] = \infty$. By the Borel–Cantelli lemma, it follows that

$$P[\, n^{-1}|Z_n| > a \text{ for infinitely many } n\,] = 1.$$

Hence $\limsup_{n\to\infty} n^{-1}|Z_n| = \infty$ a. s. Since $Z_n = S_n - S_{n-1}$, (36.3) follows.

In the d-variate case denote the jth component by the subscript j. For each $j = 1, \ldots, d$, $\{(S_n)_j\}$ is a random walk on $\mathbb{R}$. So, if $E|S_1| < \infty$ and $ES_1 = \gamma$, then we obtain (36.1) and (36.2), applying the one-dimensional result to each j. If $E|S_1| = \infty$, then $E|(S_1)_j| = \infty$ for some j and we can apply the one-dimensional result to this j to get (36.3). □

THEOREM 36.4 (Weak law of large numbers). *Let $\{S_n\}$ be a random walk on $\mathbb{R}^d$ and let $\gamma \in \mathbb{R}^d$. Then*

$$n^{-1}S_n \to \gamma \quad \text{in prob. as } n \to \infty \tag{36.4}$$

if and only if

$$\lim_{r\to\infty} rP[|S_1| > r] = 0 \quad \text{and} \quad \lim_{r\to\infty} E[S_1;\, |S_1| \le r] = \gamma. \tag{36.5}$$

Proof. In the case $d = 1$, the proof is given in Feller [**139**], pp. 235 and 565. Let $d \ge 2$. Suppose that (36.4) holds. Then $n^{-1}(S_n)_j \to \gamma_j$ in prob. for each j, and we get (36.5), noticing that

$$rP[|S_1| > r] \le r\sum_{j=1}^{d} P\left[|(S_1)_j| > \frac{r}{d}\right] \to 0$$

and

$$\begin{aligned} &|E[(S_1)_j; |S_1| \le r] - E[(S_1)_j; |(S_1)_j| \le r]| \\ &\qquad \le E[|(S_1)_j|;\, |S_1| > r,\, |(S_1)_j| \le r] \le rP[|S_1| > r] \to 0. \end{aligned}$$

The converse is similar. □

The analogue of the strong law of large numbers for Lévy processes is given by the following two theorems.

THEOREM 36.5. *Let $\{X_t\}$ be a Lévy process on $\mathbb{R}^d$. If $E|X_1| < \infty$ and $EX_1 = \gamma$, then*

$$\lim_{t\to\infty} t^{-1}X_t = \gamma \quad \text{a. s.} \tag{36.6}$$

and

$$\lim_{t\to\infty} E|t^{-1}X_t - \gamma| = 0. \tag{36.7}$$

If $E|X_1| = \infty$, then

$$\limsup_{t\to\infty} t^{-1}|X_t| = \infty \quad \text{a. s.} \tag{36.8}$$

Proof. We apply Theorem 36.3 to the random walk $\{X_n\}$. Suppose that $E|X_1| < \infty$ and $EX_1 = \gamma$. Then $n^{-1}X_n \to \gamma$ a. s. Since

$$t^{-1}X_t = (t^{-1}n)(n^{-1}X_n + n^{-1}(X_t - X_n)),$$

it is enough to show

$$n^{-1} \sup_{t\in[n,n+1]} |X_t - X_n| \to 0 \quad \text{a. s.,} \tag{36.9}$$

in order to prove (36.6). Let $Y_n = \sup_{t\in[n,n+1]} |X_t - X_n|$. Then $\{Y_n\}$ is independent and identically distributed. By Theorem 25.18 EY_1 is finite. Thus, by Theorem 36.3 again,

$$\lim_{n\to\infty} \frac{1}{n} \sum_{k=1}^{n} Y_k \to EY_1 \quad \text{a. s.}$$

Hence

$$\lim_{n\to\infty} \frac{1}{n} \sum_{k=1}^{n-1} Y_k \to EY_1 \quad \text{a. s.}$$

by $\frac{n-1}{n} \to 1$. It follows that $n^{-1}Y_n \to 0$ a. s., that is, (36.9). To see (36.7), let $n = n(t)$ be the smallest integer such that $n \geq t$. Then, by Exercise 6.14,

$$E|t^{-1}X_t - \gamma| = t^{-1}E|X_t - t\gamma| \leq t^{-1}E|X_n - n\gamma|,$$

which tends to 0 by (36.2).

If $E|X_1| = \infty$, then $\limsup_{n\to\infty} n^{-1}|X_n| = \infty$ a. s. by (36.3), and hence (36.8). □

THEOREM 36.6. ($d = 1$) *Let $\{X_t\}$ be a Lévy process on $\mathbb{R}$. If $EX_1 = \infty$, then*

$$\lim_{t\to\infty} t^{-1}X_t = \infty \quad \textit{a. s.} \tag{36.10}$$

If $EX_1 = -\infty$, then

$$\lim_{t\to\infty} t^{-1}X_t = -\infty \quad \textit{a. s.} \tag{36.11}$$

Proof. Assume that $EX_1 = \infty$. Let (A, ν, γ) be the generating triplet. Then $\int_{(1,\infty)} x\nu(\mathrm{d}x) = \infty$ and $\int_{(-\infty,-1)} |x|\nu(\mathrm{d}x) < \infty$ by Corollary 25.8. Let $a > 1$ and let

$$Y_t = X_t - \sum_{s\leq t} (X_s - X_{s-}) 1_{(a,\infty)}(X_s - X_{s-}).$$

Then, by the Lévy–Itô decomposition, $\{Y_t\}$ is a Lévy process with Lévy measure $1_{(-\infty,a]}(x)\nu(\mathrm{d}x)$. Since

$$EY_1 = \gamma + \int_{(-\infty,-1)} x\nu(\mathrm{d}x) + \int_{(1,a]} x\nu(\mathrm{d}x)$$

by Example 25.12, $EY_1 \to \infty$ as $a \to \infty$. Since $X_t \geq Y_t$, we have $\liminf_{t\to\infty} t^{-1}X_t \geq EY_1$ a. s. by Theorem 36.5 and we obtain (36.10). The latter half of the theorem is reduced to the first half. $\square$

THEOREM 36.7. ($d = 1$) *Let $\{X_t\}$ be a Lévy process on $\mathbb{R}$. If*

$$\lim_{r\to\infty} rP[|X_1| > r] = 0 \quad \text{and} \quad \lim_{r\to\infty} E[X_1;\ |X_1| \leq r] = 0, \tag{36.12}$$

then $\{X_t\}$ is recurrent. In particular, if $E|X_1| < \infty$ and $EX_1 = 0$, then $\{X_t\}$ is recurrent. If $0 < EX_1 \leq \infty$ or if $0 > EX_1 \geq -\infty$, then $\{X_t\}$ is transient.

Proof. Assume (36.12). This is equivalent to $n^{-1}X_n \to 0$ in prob. by Theorem 36.4. Define $W(B) = E\sum_{n=1}^{\infty} 1_B(X_n)$. If B is an interval of length 1, then

$$\begin{aligned}
W(B) &= \sum_{m=1}^{\infty} E\left[\sum_{k=0}^{\infty} 1_B(X_{m+k});\ X_n \notin B \text{ for } 0 < n < m,\ X_m \in B\right] \\
&\leq \sum_{m=1}^{\infty} E\left[\sum_{k=0}^{\infty} 1_{[-1,1]}(X_{m+k} - X_m);\ X_n \notin B \text{ for } 0 < n < m,\ X_m \in B\right] \\
&= \sum_{m=1}^{\infty} P[X_n \notin B \text{ for } 0 < n < m,\ X_m \in B]\, E\left[1 + \sum_{k=1}^{\infty} 1_{[-1,1]}(X_k)\right] \\
&\leq 1 + W[-1,1].
\end{aligned}$$

Let $\varepsilon > 0$. There is n_0 such that $P[|X_n| < n\varepsilon] > \frac{1}{2}$ for every $n \geq n_0$. Let a be an integer such that $a > \varepsilon n_0$. Then

$$W[-a,a] \geq \sum_{n_0 \leq n \leq a/\varepsilon} P[|X_n| < n\varepsilon] \geq \frac{1}{2}\Big(\frac{a}{\varepsilon} - n_0\Big).$$

On the other hand,

$$W[-a,a] \leq \sum_{j=-a+1}^{a} W[j-1,j] \leq 2a(1 + W[-1,1]).$$

Hence

$$1 + W[-1,1] \geq \frac{1}{4}\Big(\frac{1}{\varepsilon} - \frac{n_0}{a}\Big).$$

Letting $a \to \infty$, we get $1 + W[-1,1] \geq 1/(4\varepsilon)$. Then, letting $\varepsilon \downarrow 0$, we see that $W[-1,1] = \infty$. Hence the random walk $\{X_n\}$ is recurrent by Theorem 35.3. Hence the Lévy process $\{X_t\}$ is recurrent.

If $0 < EX_1 \leq \infty$, then $t^{-1}X_t \to EX_1$ a. s. by Theorems 36.5 and 36.6, which implies $X_t \to \infty$ a. s. Similarly, if $0 > EX_1 \geq -\infty$, then $X_t \to -\infty$ a. s. $\square$

THEOREM 36.8. *Let $\{X_t\}$ be a Lévy process on $\mathbb{R}^d$ with $E|X_1| < \infty$ and $EX_1 = \gamma$. Fix $t_0 \in (0, \infty)$. Then, for every $\eta > 0$,*

$$\lim_{\varepsilon \downarrow 0} P\left[\sup_{t\in[0,t_0]} |\varepsilon X(t/\varepsilon) - t\gamma| > \eta\right] = 0. \tag{36.13}$$

Proof. Let $Y_\varepsilon(t) = \varepsilon X(t/\varepsilon) - t\gamma$. Then $EY_\varepsilon(t) = 0$. The process $\{Y_\varepsilon(t) \colon t \in [0, t_0]\}$ has independent increments, since, for $t_1 < t_2$,

$$Y_\varepsilon(t_2) - Y_\varepsilon(t_1) = \varepsilon(X(t_2/\varepsilon) - X(t_1/\varepsilon)) - (t_2 - t_1)\gamma,$$

which is independent of $\{X(t/\varepsilon) \colon t \le t_1\}$. Let $t_{n,k} = kt_0/2^n$ and $Z_k(s) = Z_k = Y_\varepsilon(t_{n,k}) - Y_\varepsilon(t_{n,k-1})$ in Lemma 20.2. Then

$$P\left[\max_{1\le k\le 2^n} |Y_\varepsilon(t_{n,k})| > 3\eta\right] \le 3 \max_{1\le k\le 2^n} P[|Y_\varepsilon(t_{n,k})| > \eta]$$

for $\eta > 0$. Letting $n \to \infty$, we get

$$P\left[\sup_{t\in[0,t_0]} |Y_\varepsilon(t)| > 3\eta\right] \le 3 \sup_{t\in[0,t_0]} P[|Y_\varepsilon(t)| > \eta].$$

We have

$$P[|Y_\varepsilon(t)| > \eta] \le \frac{1}{\eta} E|Y_\varepsilon(t)| = \frac{\varepsilon}{\eta} E\left|X\left(\frac{t}{\varepsilon}\right) - \frac{t}{\varepsilon}\gamma\right|.$$

Notice that $E|X(\frac{t}{\varepsilon}) - \frac{t}{\varepsilon}\gamma|$ is increasing in t by Exercise 6.14 and use (36.7) in Theorem 36.5. Then

$$\sup_{t\in[0,t_0]} P[|Y_\varepsilon(t)| > \eta] \le \frac{\varepsilon}{\eta} E\left|X\left(\frac{t_0}{\varepsilon}\right) - \frac{t_0}{\varepsilon}\gamma\right| = \frac{t_0}{\eta} E\left|\frac{\varepsilon}{t_0} X\left(\frac{t_0}{\varepsilon}\right) - \gamma\right| \to 0$$

as $\varepsilon \downarrow 0$, which was to be shown. □

REMARK 36.9. Theorem 36.8 can be considered as the convergence of the probability measure induced by $\{\varepsilon X(t/\varepsilon) \colon t \in [0, t_0]\}$ to the probability measure concentrated at a single point represented by the path $\{t\gamma \colon t \in [0, t_0]\}$, as $\varepsilon \downarrow 0$, on the function space $D([0, t_0], \mathbb{R}^d)$ equipped with the metric of uniform convergence. In the case of the Brownian motion, large deviations in this convergence are studied by Schilder [**471**], which is related to Cameron and Martin [**72**]. The study is extended to the case of Lévy processes to some extent by Borovkov [**58**], Lynch and Sethuraman [**330**], Mogulskii [**353**], de Acosta [**3**], Dobrushin and Pechersky [**99**], and Jain [**231**].

REMARK 36.10. Kesten [**273**] makes a deep study of the limit points of $n^{-1}S_n$ for random walks $\{S_n\}$ on $\mathbb{R}$. He proves the following. There exists a nonrandom closed set F in $[-\infty, \infty]$ such that

$$P[\text{the set of limit points of } n^{-1}S_n \text{ as } n \to \infty \text{ is } F\,] = 1. \tag{36.14}$$

If

$$E[S_1 \vee 0] = \infty \quad \text{and} \quad E[S_1 \wedge 0] = -\infty, \tag{36.15}$$

then one of the following three cases necessarily occurs:

Case 1: $\lim_{n\to\infty} n^{-1}S_n = \infty$ a. s.;

Case 2: $\lim_{n\to\infty} n^{-1}S_n = -\infty$ a. s.;

Case 3: $\limsup_{n\to\infty} n^{-1}S_n = \infty$ and $\liminf_{n\to\infty} n^{-1}S_n = -\infty$ a. s.;

each case is non-void. Further, for any closed set F in $[-\infty,\infty]$ containing $-\infty$ and ∞, there is a random walk $\{S_n\}$ satisfying (36.14) and (36.15). If, moreover, $0 \in F$, then one can even take $\{S_n\}$ recurrent.

Erickson [**130**] proves the following. Let $\{S_n\}$ be a random walk on $\mathbb{R}$ satisfying (36.15). Let $\rho = P_{S_1}$ and define

$$J^+ = \textstyle\int_{(0,\infty)} x \big(\int_{-x}^0 \rho(-\infty,y)\mathrm{d}y\big)^{-1}\rho(\mathrm{d}x),$$
$$J^- = \textstyle\int_{(-\infty,0)} |x| \big(\int_0^{|x|} \rho(y,\infty)\mathrm{d}y\big)^{-1}\rho(\mathrm{d}x).$$

Then,

$$J^+ = \infty \quad \text{and} \quad J^- < \infty \quad \Longleftrightarrow \quad \text{Case 1},$$
$$J^+ < \infty \quad \text{and} \quad J^- = \infty \quad \Longleftrightarrow \quad \text{Case 2},$$
$$J^+ = \infty \quad \text{and} \quad J^- = \infty \quad \Longleftrightarrow \quad \text{Case 3}.$$

A related result is obtained by Mori [**356**]. This remark continues to Remark 37.13.

37. Criteria and examples

We give criteria for recurrence and transience of a Lévy process on $\mathbb{R}^d$ in terms of the characteristic function of the corresponding infinitely divisible distribution, and then discuss the dependence on the dimension. Stable and semi-stable processes and other examples are analyzed. The oscillating property in one dimension is also discussed.

A measurable function $f(x)$ on $\mathbb{R}^d$ is called *integrable* if $\int_{\mathbb{R}^d} |f(x)|\mathrm{d}x < \infty$. We use the Fourier transforms of integrable functions.

DEFINITION 37.1. Let $f(x)$ be a complex-valued integrable function on $\mathbb{R}^d$. The *Fourier transform* Ff of f is a complex-valued function on $\mathbb{R}^d$ defined by

$$(Ff)(z) = \int_{\mathbb{R}^d} \mathrm{e}^{\mathrm{i}\langle z,x\rangle} f(x)\mathrm{d}x, \qquad z \in \mathbb{R}^d. \tag{37.1}$$

PROPOSITION 37.2 (Fourier inversion formula). *Let $f(x)$ be a complex-valued integrable function on $\mathbb{R}^d$. Then, $(Ff)(z)$ is continuous and bounded. If Ff is integrable, then*

$$f(x) = (2\pi)^{-d} \int_{\mathbb{R}^d} \mathrm{e}^{-\mathrm{i}\langle x,z\rangle} (Ff)(z)\mathrm{d}z \tag{37.2}$$

for almost every $x \in \mathbb{R}^d$ and the function on the right-hand side is continuous and bounded on $\mathbb{R}^d$.

The continuity of Ff is proved by Lebesgue's dominated convergence theorem. The boundedness of Ff is clear. Considering the real and imaginary parts, we reduce the proof of (37.2) to the case that $f(x)$ is real-valued. Then, considering the positive and negative parts, we reduce it to the case $f(x) \geq 0$. Then, this is the result (xii) of Proposition 2.5.

PROPOSITION 37.3. *Let $a > 0$. There exists a bounded, continuous, nonnegative, integrable function $f(x)$ on $\mathbb{R}^d$, not identically 0, such that $(Ff)(z)$ is nonnegative, and vanishes on $\{z \colon |z| \geq a\}$.*

Proof. Let $b > 0$. Let

$$g(x) = \frac{1}{2b}\Big(1 - \frac{|x|}{2b}\Big) 1_{[-2b,2b]}(x),$$

a triangular density function on $\mathbb{R}$. Then the distribution $g(x)\mathrm{d}x$ is the convolution of the uniform distribution on $[-b, b]$ with itself. Hence, by Example 2.18,

$$(Fg)(z) = \frac{\sin^2 bz}{(bz)^2}, \qquad z \in \mathbb{R}.$$

Let $f(x) = (Fg)(x_1) \ldots (Fg)(x_d)$ for $x = (x_j)_{1 \leq j \leq d} \in \mathbb{R}^d$. We have

$$\begin{aligned}(Ff)(z) &= \prod_{j=1}^{d} \int_{\mathbb{R}} \mathrm{e}^{\mathrm{i}z_j x_j} (Fg)(x_j)\mathrm{d}x_j = \prod_{j=1}^{d} \int_{\mathbb{R}} \mathrm{e}^{-\mathrm{i}z_j x_j} (Fg)(x_j)\mathrm{d}x_j \\ &= (2\pi)^d g(z_1) \ldots g(z_d),\end{aligned}$$

using Proposition 37.2 in one dimension. Thus we get the desired function if we choose b small enough. □

Now suppose that we are given a Lévy process $\{X_t \colon t \geq 0\}$ on $\mathbb{R}^d$ with distribution μ at time 1. Let $\psi(z) = \log \widehat{\mu}(z)$, the distinguished logarithm of the characteristic function $\widehat{\mu}(z)$. For $q \geq 0$ let $V^q(\mathrm{d}y)$ be the q-potential measure of $\{X_t\}$ defined in Definition 30.9. If $f(x)$ is bounded and measurable on $\mathbb{R}^d$, then, for $q > 0$, we define $(U^q f)(x)$ by

$$\begin{aligned}(U^q f)(x) &= \int_{\mathbb{R}^d} f(x + y) V^q(\mathrm{d}y) = E\Big[\int_0^\infty \mathrm{e}^{-qt} f(x + X_t)\mathrm{d}t\Big] \\ &= \int_0^\infty \mathrm{e}^{-qt}\mathrm{d}t \int_{\mathbb{R}^d} f(x + y)\mu^t(\mathrm{d}y).\end{aligned} \tag{37.3}$$

If $f(x)$ is nonnegative and measurable, then we define $(Uf)(x) = (U^0 f)(x)$ by (37.3) with $q = 0$, allowing the value ∞.

PROPOSITION 37.4. *Let $q > 0$. The probability measure qV^q has characteristic function*

$$\widehat{(qV^q)}(z) = \frac{q}{q - \psi(z)}, \qquad z \in \mathbb{R}^d. \tag{37.4}$$

If $f(x)$ is continuous and integrable on $\mathbb{R}^d$ and if $(Ff)(z)$ is integrable on $\mathbb{R}^d$, then

$$(U^q f)(x) = (2\pi)^{-d} \int_{\mathbb{R}^d} (Ff)(-z) \frac{e^{i\langle x,z\rangle}}{q-\psi(z)} dz. \tag{37.5}$$

Proof. We have

$$\int_{\mathbb{R}^d} e^{i\langle z,x\rangle} V^q(dx) = \int_0^\infty e^{-qt} dt \int_{\mathbb{R}^d} e^{i\langle z,x\rangle} \mu^t(dx) = \int_0^\infty e^{-qt+t\psi(z)} dt$$
$$= \left[\frac{-e^{-t(q-\psi(z))}}{q-\psi(z)} \right]_{t=0}^\infty = \frac{1}{q-\psi(z)},$$

that is, (37.4). Here note that, since $|e^{\psi(z)}| = |\widehat{\mu}(z)| \le 1$, we have $\operatorname{Re}\psi(z) \le 0$ and $|q-\psi(z)| \ge \operatorname{Re}(q-\psi(z)) \ge q$. Since $f(x)$ is continuous, Proposition 37.2 tells us that (37.2) holds everywhere and that $f(x)$ is bounded. Hence, by (37.3), Fubini's theorem, and (37.4),

$$(U^q f)(x) = (2\pi)^{-d} \int_{\mathbb{R}^d} V^q(dy) \int_{\mathbb{R}^d} e^{-i\langle x+y,z\rangle} (Ff)(z) dz$$
$$= (2\pi)^{-d} \int_{\mathbb{R}^d} e^{-i\langle x,z\rangle} (Ff)(z) \frac{1}{q-\psi(-z)} dz,$$

which equals the right-hand side of (37.5). □

These propositions give the following criterion, which is proved for random walks by Chung and Fuchs [**84**].

THEOREM 37.5 (Criterion of Chung–Fuchs type). *Fix an a-neighborhood B_a of the origin. Then the following three statements are equivalent.*

(1) *$\{X_t\}$ is recurrent.*

(2) $\lim_{q\downarrow 0} \int_{B_a} \operatorname{Re}\left(\frac{1}{q-\psi(z)}\right) dz = \infty.$

(3) $\limsup_{q\downarrow 0} \int_{B_a} \operatorname{Re}\left(\frac{1}{q-\psi(z)}\right) dz = \infty.$

Proof. First let us prove that (1) implies (2). Assume that $\{X_t\}$ is recurrent. For the given a choose the function $f(x)$ described in Proposition 37.3. Let $q > 0$. Since $U^q f$ is real-valued, we have

$$(U^q f)(0) = (2\pi)^{-d} \int_{B_a} (Ff)(-z) \operatorname{Re}\left(\frac{1}{q-\psi(z)}\right) dz.$$

by Proposition 37.4. Since

$$\operatorname{Re}\left(\frac{1}{q-\psi(z)}\right) = \frac{q-\operatorname{Re}\psi(z)}{|q-\psi(z)|^2} > 0, \tag{37.6}$$

there is a constant $b > 0$ such that

$$(U^q f)(0) \le b \int_{B_a} \mathrm{Re}\left(\frac{1}{q-\psi(z)}\right) \mathrm{d}z.$$

Since $f \ge 0$ and $f(0) > 0$ by (37.2), we have $(U^q f)(0) \to \infty$ as $q \to 0$, using Theorem 35.4(ii). This proves (2). Trivially (2) implies (3).

Next, let us show that (3) implies (1). Let $f(x)$ be the function in Proposition 37.3 with 1 in place of a. Let $g(x) = (Ff)(-x)$. Then $(Fg)(z) = (2\pi)^d f(z)$ by (37.2), and $(Fg)(0) > 0$. Let $g_c(x) = g(cx)$ for $c > 0$. Choose c large enough. Then we have $\inf_{z \in B_a} (Fg_c)(z) > 0$, since $(Fg_c)(z) = c^{-d}(Fg)(c^{-1}z)$. Hence there is a constant $b > 0$ such that

$$(U^q g_c)(0) \ge b \int_{B_a} \mathrm{Re}\left(\frac{1}{q-\psi(z)}\right) \mathrm{d}z.$$

It follows from (3) that $\limsup_{q \downarrow 0} (U^q g_c)(0) = \infty$. But $(U^q g_c)(0)$ increases as q decreases, since $g_c \ge 0$. Hence $(Ug_c)(0) = \infty$. This implies the recurrence of $\{X_t\}$ by Theorem 35.4. □

COROLLARY 37.6. *Fix B_a. Let us understand that $\mathrm{Re}\left(\frac{1}{-\psi(z)}\right) = \infty$ and $\frac{1}{|\psi(z)|} = \infty$ for any z such that $\psi(z) = 0$. If*

$$\int_{B_a} \mathrm{Re}\left(\frac{1}{-\psi(z)}\right) \mathrm{d}z = \infty, \tag{37.7}$$

then $\{X_t\}$ is recurrent. If

$$\int_{B_a} \frac{\mathrm{d}z}{|\psi(z)|} < \infty, \tag{37.8}$$

then $\{X_t\}$ is transient. If $\{X_t\}$ is symmetric, then it is necessary and sufficient for recurrence that

$$\int_{B_a} \frac{\mathrm{d}z}{-\psi(z)} = \infty. \tag{37.9}$$

Proof. By (37.6) and Fatou's lemma

$$\int_{B_a} \mathrm{Re}\left(\frac{1}{-\psi(z)}\right) \mathrm{d}z \le \liminf_{q \downarrow 0} \int_{B_a} \mathrm{Re}\left(\frac{1}{q-\psi(z)}\right) \mathrm{d}z.$$

Hence (37.7) implies (2) of Theorem 37.5 and the recurrence. Since $|q-\psi|^2 = (q - \mathrm{Re}\,\psi)^2 + (\mathrm{Im}\,\psi)^2 \ge (\mathrm{Re}\,\psi)^2 + (\mathrm{Im}\,\psi)^2 = |\psi|^2$, we have

$$\mathrm{Re}\left(\frac{1}{q-\psi(z)}\right) \le \left|\frac{1}{q-\psi(z)}\right| \le \frac{1}{|\psi(z)|}.$$

Hence, if (37.8) holds, then (3) of Theorem 37.5 does not hold and $\{X_t\}$ is transient. If $\{X_t\}$ is symmetric, then $\widehat{\mu}(z)$ is real and hence $-\psi(z) \ge 0$. In this case, (37.7), (37.9), and the negation of (37.8) are identical. □

REMARK 37.7. It is known that the condition (37.7) is not only sufficient but also necessary for recurrence. That is, $\{X_t\}$ *is recurrent if and only if (37.7) holds.* We call this *the criterion of Spitzer type.* The 'only if' part is connected with the existence of the so-called recurrent potential operators and its proof has the following history. Originally Spitzer [**496**] found a similar fact for $\mathbb{Z}^d$-valued random walks. Extension to $\mathbb{R}^d$-valued random walks needed involved argument and was done by Ornstein [**366**] and Stone [**505**] by different methods. The case of Lévy processes was reduced to random walks by Port and Stone [**386**]. Another proof was given by M. Itô [**229**].

The criteria just proved give the following important result. The definition of the genuine d-dimensionality of a Lévy process is given in Definition 24.18.

THEOREM 37.8. *Let $d \geq 3$. Any genuinely d-dimensional Lévy process on $\mathbb{R}^d$ is transient.*

Proof. Let $\{X_t\}$ be genuinely d-dimensional. There are two cases. 1: $\{X_t\}$ is nondegenerate. 2: $\{X_t\}$ is degenerate.

Case 1. We can use Proposition 24.19. Thus, there are $c > 0$ and $a > 0$ such that $|\widehat{\mu}(z)| \leq 1 - c|z|^2$ on B_a. Since $|\widehat{\mu}(z)| = \mathrm{e}^{\mathrm{Re}\,\psi(z)}$, we can choose $c' > 0$ and $a' > 0$ such that

$$-\mathrm{Re}\,\psi(z) \geq -\log(1 - c|z|^2) \geq c'|z|^2 \quad \text{on } B_{a'}.$$

Hence, using the surface measure c_d of the unit sphere, we get

$$\int_{B_{a'}} \frac{\mathrm{d}z}{|\psi(z)|} \leq \int_{B_{a'}} \frac{\mathrm{d}z}{-\mathrm{Re}\,\psi(z)} \leq \frac{1}{c'} \int_{B_{a'}} \frac{\mathrm{d}z}{|z|^2} = \frac{c_d}{c'} \int_0^{a'} r^{d-3}\mathrm{d}r < \infty.$$

Hence $\{X_t\}$ is transient by Corollary 37.6.

Case 2. The process $\{X_t\}$ is degenerate but genuinely d-dimensional. Let (A, ν, γ) be the generating triplet of $\{X_t\}$ and let S_ν be the support of ν. In this case, by Definition 24.18 and Proposition 24.17, there is a proper linear subspace M of $\mathbb{R}^d$ such that $A(\mathbb{R}^d) \subset M$, $S_\nu \subset M$, and $\gamma \notin M$. Hence $\{X_t - t\gamma\}$ is a Lévy process on M. Decompose γ as $\gamma = \gamma_1 + \gamma_2$ with $\gamma_1 \in M$ and γ_2 is in the orthogonal complement of M. Then

$$|X_t|^2 = |X_t - t\gamma + t\gamma_1|^2 + |t\gamma_2|^2 \geq t^2|\gamma_2|^2 \to \infty$$

as $t \to \infty$. Therefore $\{X_t\}$ is transient. □

A sufficient condition for recurrence for $d = 1$ is given in Theorem 36.7. We give a related remark.

REMARK 37.9. ($d = 1$) Let $\{X_t\}$ be a Lévy process on $\mathbb{R}$. Let $X_t^+ = X_t \vee 0$ and $X_t^- = (-X_t) \vee 0$. If $EX_1^+ < \infty$ or $EX_1^- < \infty$, then a necessary and sufficient condition for recurrence is that $EX_1 = 0$. This is contained in

Theorem 36.7. The theorem also says that (36.12) is a sufficient condition for recurrence. Suppose that

$$EX_1^+ = \infty \quad \text{and} \quad EX_1^- = \infty. \tag{37.10}$$

Then there are three cases:

Case 1: (36.12) holds (hence $\{X_t\}$ is recurrent).

Case 2: (36.12) does not hold and $\{X_t\}$ is recurrent.

Case 3: (36.12) does not hold and $\{X_t\}$ is transient.

Let us show that each of the three cases can actually occur.

To get a process in Case 1, choose a symmetric finite measure ν on $\mathbb{R}$ such that $\nu\{0\} = 0$, $\lim_{x\to\infty} x \int_{|y|>x} \nu(\mathrm{d}y) = 0$, and $\int_{\mathbb{R}} |x|\nu(\mathrm{d}x) = \infty$ (for example, let $\nu(\mathrm{d}x) = 1_{\{|x|>2\}}(x)|x|^{-2}(\log|x|)^{-1}\mathrm{d}x$). Let $\{X_t\}$ be a compound Poisson process with Lévy measure ν. Then it satisfies (37.10) by Theorem 25.3 and Proposition 25.4. The second property in (36.12) is evident, since $\{X_t\}$ is symmetric. To see the first property in (36.12), let $c = \nu(\mathbb{R})$ and let $\{S_n\}$ be a random walk such that $P_{S_1} = c^{-1}\nu$. Since

$$P[\,|S_n| > x\,] \le nP[\,|S_1| > x/n\,],$$

we have

$$\begin{aligned} xP[\,|X_1| > x\,] &= x\mathrm{e}^{-c}\sum_{n=0}^{\infty}(n!)^{-1}c^n P[\,|S_n| > x\,] \\ &\le \mathrm{e}^{-c}\sum_{n=0}^{\infty} n^2(n!)^{-1}c^n(x/n)P[\,|S_1| > x/n\,], \end{aligned}$$

which tends to 0 as $x \to \infty$ by Lebesgue's dominated convergence theorem. Hence $\{X_t\}$ is in Case 1.

A Cauchy process is in Case 2 and any stable process with $0 < \alpha < 1$ and $|\beta| < 1$ is an example of Case 3. Their recurrence/transience will be shown in Corollary 37.17. They satisfy (37.10) by Theorem 25.3 and Proposition 25.4 and by the explicit form of their Lévy measures in Remark 14.6 combined with Theorem 14.15. The condition (36.12) is equivalent to $n^{-1}X_n \to 0$ in prob. (Theorem 36.4) and hence to $\widehat{\mu}(n^{-1}z)^n \to 1$. The characteristic functions $\widehat{\mu}(z)$ for α-stable processes with $\alpha \le 1$ on $\mathbb{R}$ do not have this property, as is seen from Theorem 14.15.

PROPOSITION 37.10. *Let $\{X_t\}$ be a non-zero Lévy process on $\mathbb{R}$. Then it satisfies one of the following three conditions:*

(1) $\lim_{t\to\infty} X_t = \infty$ *a. s.;*

(2) $\lim_{t\to\infty} X_t = -\infty$ *a. s.;*

(3) $\limsup_{t\to\infty} X_t = \infty$ *and* $\liminf_{t\to\infty} X_t = -\infty$ *a. s.*

Proof. If $\{X_t\}$ is recurrent, then (3) holds. In fact, Theorem 35.8 makes a much stronger assertion. Suppose that $\{X_t\}$ is transient. Let $M = \limsup_{t\to\infty} X_t$ and $N = \liminf_{t\to\infty} X_t$. Since $|X_t(\omega)| \to \infty$, no finite point is a limit point of $X_t(\omega)$ as $t \to \infty$. Hence $P[\, M = \infty \text{ or } -\infty\,] = 1$. Kolmogorov's 0–1 law (Theorem 1.14) tells us that $P[\, M = \infty\,] = 1$ or 0. It follows that either $P[\, M = \infty\,] = 1$ or $P[\, M = -\infty\,] = 1$. Similarly, either $P[\, N = \infty\,] = 1$ or $P[\, N = -\infty\,] = 1$. Hence one of (1), (2), and (3) holds. □

DEFINITION 37.11. We use the following terminology for the properties in the preceding proposition. A non-zero Lévy process $\{X_t\}$ on $\mathbb{R}$ is *drifting to* ∞ if (1) holds; it is *drifting to* $-\infty$ if (2) holds; it is *oscillating* if (3) holds.

REMARK 37.12. Necessary and sufficient conditions for the three properties above in terms of $P[\, X_t > 0\,]$ and $P[\, X_t < 0\,]$ will be given in Theorem 48.1. In the class of Lévy processes on $\mathbb{R}$ that satisfy (37.10) and belong to Case 3 of Remark 37.9, there are processes drifting to ∞, drifting to $-\infty$, and oscillating. A stable process with $0 < \alpha < 1$ and $|\beta| < 1$, which is an example satisfying (37.10) and belonging to Case 3 given in Remark 37.9, is oscillating. This will be shown in Theorem 48.6. The process $X_t - X'_t$ in Example 48.5 will be shown to be drifting to ∞ in addition to satisfying (37.10) and being in Case 3. The negative of this process is drifting to $-\infty$.

If $\{X_t\}$ is an oscillating transient Lévy process, then

$$P[\text{the set of limit points of } X_t \text{ as } t \to \infty \text{ is } \{\infty, -\infty\}] = 1. \tag{37.11}$$

This is a remarkable almost sure behavior of sample functions; as t grows large, $X_t(\omega)$ does not have any finite limit point, but by jumps of large size it goes up and down infinitely often between any neighborhoods of ∞ and $-\infty$. A recurrent Lévy process on $\mathbb{R}$ is oscillating, but it has the contrary property descried in Theorem 35.8.

Any symmetric non-zero Lévy process on $\mathbb{R}$ is oscillating, because drifting to ∞ or to $-\infty$ contradicts the symmetry.

REMARK 37.13. The results of Kesten [**273**] and Erickson [**130**] on Lévy processes are as follows. They correspond to those on random walks in Remark 36.10. Let $\{X_t\}$ be a Lévy process on $\mathbb{R}$ satisfying (37.10). Then it satisfies one of the following three:

(1) $\lim_{t\to\infty} t^{-1}X_t = \infty$ a. s.;

(2) $\lim_{t\to\infty} t^{-1}X_t = -\infty$ a. s.;

(3) $\limsup_{t\to\infty} t^{-1}X_t = \infty$ and $\liminf_{t\to\infty} t^{-1}X_t = -\infty$ a. s.

Further, let ν be its Lévy measure and define

$$K^+ = \textstyle\int_{(2,\infty)} x \big(\int_{-x}^{-1} \nu(-\infty, y)\mathrm{d}y\big)^{-1} \nu(\mathrm{d}x),$$

$$K^- = \textstyle\int_{(-\infty,-2)} |x| \big(\int_{1}^{|x|} \nu(y, \infty)\mathrm{d}y\big)^{-1} \nu(\mathrm{d}x).$$

Then $K^+ + K^- = \infty$ and the following equivalences are true:

$$(1) \iff K^+ = \infty \quad \text{and} \quad K^- < \infty;$$
$$(2) \iff K^+ < \infty \quad \text{and} \quad K^- = \infty;$$
$$(3) \iff K^+ = \infty \quad \text{and} \quad K^- = \infty.$$

See [**130**], p. 373, where these are proved by the reduction to the results on random walks.

It follows that, under the condition (37.10), the properties (1), (2), and (3) are respectively equivalent to drifting to ∞, drifting to $-\infty$, and oscillating. Indeed this is obvious if 'are respectively equivalent to' is replaced by 'respectively imply'; then note that (1), (2), and (3) are exhaustive. We now have a criterion of drifting to ∞, drifting to $-\infty$, and oscillating for Lévy processes on $\mathbb{R}$ in terms of Lévy measure and parameter γ.

Let us consider Lévy processes on $\mathbb{R}^2$.

THEOREM 37.14. *Let $d = 2$. If $E[|X_1|^2] < \infty$ and $EX_1 = 0 \in \mathbb{R}^2$, then $\{X_t\}$ is recurrent.*

Proof. Let (A, ν, γ) and Σ be the generating triplet and the support of $\{X_t\}$. If $\{X_t\}$ is not genuinely two-dimensional, then Σ is in a straight line through the origin and $\{X_t\}$ is recurrent by Theorem 36.7. Suppose that $\{X_t\}$ is genuinely two-dimensional. By Proposition 24.17, no straight line through the origin contains $A(\mathbb{R}^2)$, S_ν, and γ. If $A(\mathbb{R}^2)$ and S_ν lie in a straight line M through the origin, then $\gamma \notin M$ and $X_1 - \gamma \in M$, hence also $EX_1 - \gamma \in M$, contradicting the assumption $EX_1 = 0$. Hence $\{X_t\}$ must be nondegenerate by Proposition 24.17. Now, using Proposition 24.19, we can find $c' > 0$ and $a' > 0$ such that $-\mathrm{Re}\,\psi(z) \geq c'|z|^2$ on $B_{a'}$. On the other hand, the assumptions $E[|X_1|^2] < \infty$ and $EX_1 = 0$ imply that $\psi(z)$ is of class C^2 and that $\partial\psi/\partial z_1$ and $\partial\psi/\partial z_2$ vanish at the origin. Hence $|\psi(z)| \leq c''|z|^2$ with some $c'' > 0$. Therefore,

$$\mathrm{Re}\left(\frac{1}{-\psi(z)}\right) = \frac{-\mathrm{Re}\,\psi(z)}{|\psi(z)|^2} \geq \frac{c'}{(c'')^2|z|^2}$$

in a neighborhood of the origin. Hence we have (37.7) for some a and the process is recurrent. □

REMARK 37.15. Let $d = 2$. If $E|X_1| < \infty$ and $EX_1 \neq 0$, then $\{X_t\}$ is transient, which follows from the Lévy process analogue of the strong law of large numbers (Theorem 36.5). If $E[|X_1|^2] = \infty$, $E|X_1| < \infty$, and $EX_1 = 0$, then there are both recurrent and transient cases. Even under an additional condition that $E[|X_1|^\eta] < \infty$ for every $\eta \in (0, 2)$, we can construct a recurrent one and a transient one in the following way.

Let $g(r)$ be a nonnegative, measurable function on $(1, \infty)$ such that $\int_1^\infty g(r)dr < \infty$. Let ν be the finite measure supported on $\{|x| \geq 1\}$

defined by

$$\nu(B) = \int_S \lambda(\mathrm{d}\xi) \int_1^\infty 1_B(r\xi) g(r) \mathrm{d}r, \quad B \in \mathcal{B}(\mathbb{R}^2),$$

where λ is the uniform measure on the unit circle S with total measure 2π. Let $\{X_t\}$ be the compound Poisson process with Lévy measure ν. Then $\{X_t\}$ is rotation invariant and $-\psi(z)$ is real-valued and nonnegative, and depends only on $|z|$. We have $-\psi(z) = -\psi(z')$, where z' has first component $|z|$ and second component 0. Hence

$$\begin{aligned}
-\psi(z) &= \int (1 - \mathrm{e}^{\mathrm{i}\langle z,x\rangle})\nu(\mathrm{d}x) = \int_0^{2\pi} \mathrm{d}\theta \int_1^\infty (1 - \mathrm{e}^{\mathrm{i}|z|r\cos\theta}) g(r) \mathrm{d}r \\
&= \int_0^{2\pi} \mathrm{d}\theta \int_1^\infty (1 - \cos(|z| r \cos\theta)) g(r) \mathrm{d}r \\
&= \frac{1}{|z|} \int_0^{2\pi} \mathrm{d}\theta \int_{|z|}^\infty (1 - \cos(r\cos\theta)) g\Big(\frac{r}{|z|}\Big) \mathrm{d}r.
\end{aligned}$$

Let us show the following.

(i) If $g(r) = r^{-3}$ on $(1, \infty)$, then $\{X_t\}$ is recurrent, $E[|X_1|^\eta] < \infty$ for $0 < \eta < 2$, $E[|X_1|^2] = \infty$, and $E[X_1] = 0$.

(ii) If $g(r) = r^{-3} \log r$ on $(1, \infty)$, then $\{X_t\}$ is transient and possesses the other properties above.

Denote positive constants by $c_1, c_2, \ldots$. Let $g(r) = r^{-3}$. Using $1 - \cos a \le a^2/2$, we have

$$-\psi(z) \le |z|^2 \Big(c_1 + c_2 \int_{|z|}^1 \frac{\mathrm{d}r}{r} \Big) \le c_3 |z|^2 \log \frac{1}{|z|}$$

for $|z| \le 1/2$. Hence

$$\int_{|z| \le 1/2} \frac{\mathrm{d}z}{-\psi(z)} \ge c_4 \int_0^{1/2} \frac{\mathrm{d}r}{r \log(1/r)} = \infty,$$

which implies recurrence by (37.7). Since $\int |x|^\eta \nu(\mathrm{d}x) < \infty$ for $0 < \eta < 2$ and $\int |x|^2 \nu(\mathrm{d}x) = \infty$, we have $E[|X_1|^\eta] < \infty$ for $0 < \eta < 2$ and $E[|X_1|^2] = \infty$ by Theorem 25.3. Since $\{X_t\}$ is symmetric, $E[X_1] = 0$.

Consider the case that $g(r) = r^{-3} \log r$ on $(1, \infty)$. Using $1 - \cos a \ge c_5 a^2$ for $0 \le a \le 1$, we have, for $|z| \le 1$,

$$\begin{aligned}
-\psi(z) &\ge \int_0^{2\pi} \mathrm{d}\theta \int_1^{1/|z|} (1 - \cos(|z| r \cos\theta)) r^{-3} \log r \, \mathrm{d}r \\
&\ge c_6 \int_0^{2\pi} \mathrm{d}\theta \int_1^{1/|z|} |z|^2 (\cos\theta)^2 r^{-1} \log r \, \mathrm{d}r = c_7 |z|^2 (\log|z|)^2.
\end{aligned}$$

Therefore

$$\int_{|z|\le 1/2} \frac{\mathrm{d}z}{-\psi(z)} \le \frac{1}{c_7}\int_{|z|\le 1/2} \frac{\mathrm{d}z}{|z|^2(\log|z|)^2} = c_8\int_0^{1/2} \frac{\mathrm{d}r}{r(\log r)^2} < \infty,$$

which proves transience by (37.8). The other properties are shown similarly to (i).

Let us determine recurrence and transience of semi-stable processes.

THEOREM 37.16. ($d = 1$) *Let $\{X_t\}$ be a non-trivial α-semi-stable process on $\mathbb{R}$.*

(i) *Suppose that $1 \le \alpha \le 2$. Then $\{X_t\}$ is recurrent if and only if it is strictly α-semi-stable.*

(ii) *Suppose that $0 < \alpha < 1$. Then $\{X_t\}$ is transient.*

A necessary and sufficient condition for an α-semi-stable distribution having b as a span to be strictly α-semi-stable having b as a span is given in Theorem 14.7. The theorem above says that, in the case $1 \le \alpha \le 2$, the recurrence condition is reduced to this condition, since we know the result in Exercise 18.11.

Proof of theorem. (i) If $1 < \alpha \le 2$, then the process has a finite mean and the assertion follows from Theorems 14.7 and 36.7.

Consider the case $\alpha = 1$. Suppose that $\{X_t\}$ is not strictly 1-semi-stable. Then $\widehat{\mu}(z)^b = \widehat{\mu}(bz)\mathrm{e}^{icz}$ with some $b > 1$ and $c \ne 0$. It follows that

$$b^n\psi(z) = \psi(b^n z) + inb^{n-1}cz \quad \text{for } n \in \mathbb{Z}.$$

Let us show that $\{X_t\}$ is transient, using Theorem 37.5. Let $\psi_1 = \operatorname{Re}\psi$ and $\psi_2 = \operatorname{Im}\psi$. Since $\operatorname{Re}\left(\frac{1}{q-\psi}\right)$ is even,

$$\int_{-1}^{1} \operatorname{Re}\left(\frac{1}{q-\psi}\right)\mathrm{d}z = 2(I_1 + I_2)$$

for $q > 0$, where

$$I_1 = \int_0^1 \frac{q\mathrm{d}z}{(q-\psi_1)^2 + {\psi_2}^2} \quad \text{and} \quad I_2 = \int_0^1 \frac{-\psi_1\mathrm{d}z}{(q-\psi_1)^2 + {\psi_2}^2}.$$

Since non-triviality and nondegeneracy are equivalent for distributions on $\mathbb{R}$, Proposition 24.20 shows that there is $K > 0$ such that $-\psi_1(z) \ge K|z|$ for $z \in \mathbb{R}$. Thus

$$I_1 \le \int_0^1 \frac{q\mathrm{d}z}{q^2 + {\psi_1}^2} \le \int_0^1 \frac{q\mathrm{d}z}{q^2 + K^2z^2} = \int_0^{1/q} \frac{\mathrm{d}z}{1 + K^2z^2},$$

which is bounded in q. We have

$$I_2 \le \int_0^1 \frac{-\psi_1\mathrm{d}z}{{\psi_1}^2 + {\psi_2}^2} = \sum_{n=0}^{\infty} J_n \quad \text{where } J_n = \int_{b^{-n-1}}^{b^{-n}} \frac{-\psi_1\mathrm{d}z}{{\psi_1}^2 + {\psi_2}^2}.$$

Noticing that

$$b^n\psi_1(z) = \psi_1(b^n z) \quad \text{and} \quad b^n\psi_2(z) = \psi_2(b^n z) + nb^{n-1}cz \quad \text{for } n \in \mathbb{Z},$$

we get

$$J_n = \int_{1/b}^{1} \frac{-\psi_1 \mathrm{d}z}{{\psi_1}^2 + (\psi_2 + nb^{-1}cz)^2} = O\Big(\frac{1}{n^2}\Big) \quad \text{as } n \to \infty.$$

Hence

$$\limsup_{q\downarrow 0} \int_{-1}^{1} \mathrm{Re}\Big(\frac{1}{q-\psi}\Big)\mathrm{d}z < \infty.$$

That is, $\{X_t\}$ is transient.

Conversely suppose that $\{X_t\}$ is strictly 1-semi-stable. Then $\widehat{\mu}(z)^b = \widehat{\mu}(bz)$ for some $b > 1$, and hence $b^n\psi(z) = \psi(b^n z)$ for $n \in \mathbb{Z}$. Therefore

$$\int_{b^n}^{b^{n+1}} \mathrm{Re}\Big(\frac{1}{-\psi(z)}\Big)\mathrm{d}z = \int_{1}^{b} \mathrm{Re}\Big(\frac{1}{-\psi(z)}\Big)\mathrm{d}z.$$

The integrand is nonnegative. Hence $\int_{-1}^{1} \mathrm{Re}\big(\frac{1}{-\psi}\big)\mathrm{d}z = 0$ or ∞. Since $\{X_t\}$ is non-trivial, $0 < \mathrm{Re}\big(\frac{1}{-\psi(z)}\big) < \infty$ for some z with $0 < |z| < 1$. It follows that the integral is infinite, which shows recurrence by Corollary 37.6.

(ii) By Proposition 24.20, we can find a constant $K > 0$ such that $-\mathrm{Re}\,\psi(z) \geq K|z|^\alpha$ on $\mathbb{R}$. Hence (37.8) holds and the process is transient. □

COROLLARY 37.17. ($d = 1$) *Let $\{X_t\}$ be a non-trivial α-stable process on $\mathbb{R}$. If $1 \leq \alpha \leq 2$, then its recurrence is equivalent to strict α-stability. If $0 < \alpha < 1$, then it is transient. If $0 < \alpha < 2$, then another expression is as follows: a non-trivial stable process on $\mathbb{R}$ with parameters (α, β, τ, c) is recurrent if either $1 < \alpha < 2$ and $\tau = 0$, or $\alpha = 1$ and $\beta = 0$; otherwise it is transient.*

Proof. This can be shown in a way similar to (and simpler than) Theorem 37.16. But we derive it from the theorem. For this we have only to note that $\{X_t\}$ is α-stable or strictly α-stable, respectively, if and only if it is α-semi-stable or strictly α-semi-stable and every $b \in (1, \infty)$ is a span. To verify the last part of the corollary, use Theorem 14.15. □

Noticing Theorem 37.16 and the result on moments in Example 25.10, one might think the possibility to give a sufficient condition for transience in the form $E[|X_1|^\eta] = \infty$ for small $\eta > 0$. But this cannot be done. That is, however small $\eta > 0$ we choose, we can find a recurrent Lévy process $\{X_t\}$ on $\mathbb{R}$ with $E[|X_1|^\eta] = \infty$. This will be shown by Theorem 38.4 combined with Corollary 25.8.

THEOREM 37.18. ($d = 2$) *Let $0 < \alpha \leq 2$. A genuinely two-dimensional α-semi-stable process on $\mathbb{R}^2$ is recurrent if and only if it is strictly 2-stable.*

Proof. If $\{X_t\}$ is 2-semi-stable, then it is Gaussian distributed and Theorem 37.14 and Remark 37.15 apply. If $\{X_t\}$ is α-semi-stable with $0 < \alpha < 2$ and nondegenerate, then $-\mathrm{Re}\,\psi(z) \geq K|z|^{\alpha}$ with some $K > 0$ by Proposition 24.20 and, since $d = 2$, this implies (37.8) and transience. If $\{X_t\}$ is degenerate but genuinely two-dimensional, then its transience is shown as in Case 2 of the proof of Theorem 37.8. □

EXAMPLE 37.19. Any nondegenerate α-semi-stable process on $\mathbb{R}^d$ has distribution density $p(t,x)$ for $t > 0$ continuous in (t,x), since we have $p(t,x) = (2\pi)^{-d}\int \mathrm{e}^{-\mathrm{i}\langle x,z\rangle}\widehat{\mu}(z)^t \mathrm{d}z$ by Proposition 24.20. Thus it has q-potential density $v^q(x)$. Explicit expressions for $v(x) = v^0(x)$ are known in some cases.

(i) Let $\{X_t\}$ be a stable subordinator of index $0 < \alpha < 1$ with $E[\mathrm{e}^{-uX_t}] = \mathrm{e}^{-tc'u^{\alpha}}$, $c' > 0$, for $u \geq 0$. See Example 24.12. Then

$$v(x) = \frac{1}{c'\Gamma(\alpha)}|x|^{\alpha-1}1_{(0,\infty)}(x). \tag{37.12}$$

In fact, letting $f_t(x)$ and $v^q(x)$ be the distribution density at time $t > 0$ and the q-potential density for $q > 0$, respectively, we have

$$\int_0^{\infty} \mathrm{e}^{-ux}v^q(x)\mathrm{d}x = \int_0^{\infty} \mathrm{e}^{-ux}\mathrm{d}x \int_0^{\infty} \mathrm{e}^{-qt}f_t(x)\mathrm{d}t$$
$$= \int_0^{\infty} \mathrm{e}^{-qt-tc'u^{\alpha}}\mathrm{d}t = (q + c'u^{\alpha})^{-1}$$

for $u > 0$. Let $q \downarrow 0$. Then

$$\int_0^{\infty} \mathrm{e}^{-ux}v(x)\mathrm{d}x = \frac{1}{c'}u^{-\alpha} = \frac{1}{c'\Gamma(\alpha)}\int_0^{\infty} \mathrm{e}^{-ux}x^{\alpha-1}\mathrm{d}x, \quad u > 0.$$

Applying Proposition 2.6 to constant multiples of $\mathrm{e}^{-x}v(x)$ and $\mathrm{e}^{-x}x^{\alpha-1}$, we get (37.12).

(ii) Let $\{X_t\}$ be a rotation invariant transient α-stable process on $\mathbb{R}^d$. There are three cases: $d = 1$ and $0 < \alpha < 1$; $d = 2$ and $0 < \alpha < 2$; $d \geq 3$ and $0 < \alpha \leq 2$. In each case,

$$v(x) = \text{const}\,|x|^{\alpha-d}. \tag{37.13}$$

In potential theory the potential with this kernel is called the Riesz potential after M. Riesz. When $\alpha = 1$ or 2, we have already shown (37.13) in Examples 35.6 and 35.7. A proof of it for general $0 < \alpha < 2$ is as follows. The process has the characteristic function

$$E[\mathrm{e}^{\mathrm{i}\langle z,X_t\rangle}] = \exp(-tc|z|^{\alpha}), \quad z \in \mathbb{R}^d$$

with some $c > 0$ (Theorem 14.14). It is subordinate to the Brownian motion $\{X_t^0\}$ on $\mathbb{R}^d$ by the $\frac{\alpha}{2}$-stable subordinator $\{Z_t\}$ with $E[\mathrm{e}^{-uZ_t}] = \mathrm{e}^{-tc2^{\alpha/2}u^{\alpha/2}}$,

$u \geq 0$ (Example 30.6). Write $P_{X_1^0} = \mu_0$, $P_{Z_1} = \mu_Z$, and denote the q-potential measure of $\{Z_t\}$ by V_Z^q. Then, for $q > 0$,

$$V^q(B) = \int_0^\infty \mathrm{e}^{-qt}\mathrm{d}t \int_0^\infty \mu_0^s(B)\mu_Z^t(\mathrm{d}s) = \int_0^\infty \mu_0^s(B)V_Z^q(\mathrm{d}s)$$

by (30.5). Letting $q \downarrow 0$, we have

$$V(B) = \int_0^\infty \mu_0^s(B)V_Z^0(\mathrm{d}s) = \frac{(2\pi)^{-d/2}2^{-\alpha/2}}{c\Gamma(\alpha/2)} \int_B \mathrm{d}x \int_0^\infty \mathrm{e}^{-|x|^2/(2s)} s^{(\alpha-d)/2-1}\mathrm{d}s$$

by (i). Evaluating the integral in s by change of variable, we get (37.13) with

$$\text{const} = c^{-1}c_{d,\alpha}, \quad c_{d,\alpha} = \pi^{-d/2}2^{-\alpha}\Gamma\left(\frac{d-\alpha}{2}\right) \Big/ \Gamma\left(\frac{\alpha}{2}\right). \tag{37.14}$$

In the one-dimensional case the potential densities for the stable process with parameters $(\alpha, \beta, 0, c)$ with $0 < \alpha < 1$ and the Brownian motion with drift γ added are calculated in Exercises 39.1 and 39.2.

REMARK 37.20. For a general Lévy process $\{X_t\}$ on $\mathbb{R}^d$ define the *last exit time* from a set B by

$$L_B(\omega) = \sup\{t > 0 \colon X_t(\omega) \in B\},$$

where the supremum of the empty set is 0. If B is open, then L_B is a random variable, since $\Omega_0 \cap \{L_B > t\} = \Omega_0 \cap \bigcup_{s \in \mathbb{Q} \cap (t,\infty)} \{X_s \in B\}$, where Ω_0 is the event in Definition 1.6. Let $B_a = \{x \colon |x| < a\}$ as before. Transience of $\{X_t\}$ is equivalent to the finiteness a. s. of L_{B_a} for every $a > 0$. This is a mere rephrasing of (35.2). A transient Lévy process is called *weakly transient* if $E[L_{B_a}] = \infty$ for every $a > 0$; *strongly transient* if $E[L_{B_a}] < \infty$ for every $a > 0$. Then we can prove the following. A transient Lévy process is either weakly transient or strongly transient. It is weakly transient if and only if

$$\textstyle\int_0^\infty \mathrm{d}s \int_s^\infty P[X_t \in B_a]\mathrm{d}t = \infty \quad \text{for every } a > 0.$$

It is strongly transient if and only if

$$\textstyle\int_0^\infty \mathrm{d}s \int_s^\infty P[X_t \in B_a]\mathrm{d}t < \infty \quad \text{for every } a > 0.$$

The d-dimensional Brownian motion is strongly transient if and only if $d \geq 5$. We can derive this fact from (16.1). Every genuinely d-dimensional Lévy process on $\mathbb{R}^d$ is strongly transient if $d \geq 5$. An α-stable process on $\mathbb{R}$ with $0 < \alpha < 2$ and parameters (α, β, τ, c) or with $\alpha = 2$ and center γ is strongly transient in the following four cases:

(1) $\alpha = 2$ and $\gamma \neq 0$,
(2) $1 < \alpha < 2$, $|\beta| = 1$, $\tau \neq 0$, and $\beta\tau > 0$,
(3) $1/2 \leq \alpha \leq 1$ and $|\beta| = 1$,
(4) $0 < \alpha < 1/2$;

otherwise it is either weakly transient or recurrent. An analogue of the Chung–Fuchs type criterion is known under a slight restriction called strongly non-lattice. Analogues of Theorems 38.2–38.4 below for weak and strong transience of symmetric transient Lévy processes on $\mathbb{R}$ are also known [**448**]. Papers related to weak and strong transience are Getoor [**158**], Port [**380, 383, 384**], Port and Stone [**386**], Sato [**447, 448**], and Yamamuro [**586**].

REMARK 37.21. A notion stronger than recurrence is point recurrence. A Lévy process $\{X_t\}$ on $\mathbb{R}^d$ is called *point recurrent* if $\limsup_{t\to\infty} 1_{\{0\}}(X_t) = 1$ a. s. The Brownian motion on $\mathbb{R}$ is point recurrent, which follows from the oscillating property (Theorem 35.8) and the continuity of sample functions. The two-dimensional Brownian motion is recurrent but not point recurrent, as will be proved in Example 43.7. See Remark 43.12 on the condition for point recurrence.

38. The symmetric one-dimensional case

The preceding section gives recurrence criteria for a Lévy process $\{X_t\}$ in terms of $\psi(z)$, the distinguished logarithm of the characteristic function $\widehat{\mu}(z)$. But we want to decide recurrence/transience of $\{X_t\}$ directly from its generating triplet (A, ν, γ), not through the function $\psi(z)$. Let $d = 1$. When X_1 has mean m, $\{X_t\}$ is recurrent if and only if $m = 0$ (Theorem 36.7). However, in the case where both EX_1^+ and EX_1^- are infinite, no general criterion in terms of the generating triplet is known. Only in the symmetric case do we have some results. They are Lévy process analogues of Shepp's theory [**480, 481**] on symmetric random walks. It might seem plausible that, if a transient Lévy process has Lévy measure with fat tails in some sense, then any second Lévy process with Lévy measure having fatter tails is transient. But this is not true even among symmetric ones, as will be shown below.

DEFINITION 38.1. Let ρ and ρ' be symmetric measures on $\mathbb{R}$ finite outside of any neighborhood of the origin. We say that ρ has a *bigger tail* than ρ' or an *identical tail* with ρ' if there is $x_0 > 0$ such that $\rho(x,\infty) \geq \rho'(x,\infty)$ for $x > x_0$ or $\rho(x,\infty) = \rho'(x,\infty)$ for $x > x_0$, respectively. We say that ρ is *quasi-unimodal* if there is $x_0 > 0$ such that $\rho(x,\infty)$ is convex for $x > x_0$.

We give three theorems. Recall that a Lévy process generated by (A, ν, γ) is symmetric if and only if ν is symmetric and $\gamma = 0$ (Exercise 18.1). Thus, in the symmetric case,

$$\psi(z) = -2^{-1}Az^2 - 2\int_{(0,\infty)} (1 - \cos zx)\nu(\mathrm{d}x). \tag{38.1}$$

When we consider two symmetric Lévy processes $\{X_t\}$ and $\{Y_t\}$, their Lévy measures are denoted by ν_X and ν_Y, respectively. The measure of total variation of a signed measure σ is denoted by $|\sigma|$.

THEOREM 38.2. *Let* $\{X_t\}$ *and* $\{Y_t\}$ *be symmetric Lévy processes on* $\mathbb{R}$.
(i) *If*

$$\int_{(0,\infty)} x^2 |\nu_X - \nu_Y|(\mathrm{d}x) < \infty, \tag{38.2}$$

then recurrence of $\{X_t\}$ *is equivalent to that of* $\{Y_t\}$.

(ii) *If* ν_Y *has a bigger tail than* ν_X *and if* ν_Y *is quasi-unimodal, then transience of* $\{X_t\}$ *implies that of* $\{Y_t\}$.

Without quasi-unimodality of ν_Y, the process $\{Y_t\}$ may possibly be recurrent even if ν_Y has a bigger tail than ν_X of a transient $\{X_t\}$. An explicit example for this fact is given by Exercise 39.16. A much stronger fact than this will be shown in Theorem 38.4.

THEOREM 38.3. *Let* $\{X_t\}$ *be a symmetric Lévy process on* $\mathbb{R}$ *with Lévy measure* ν. *Define*

$$R(r,x) = \nu\left(\bigcup_{n=0}^{\infty}(2nr + x, 2(n+1)r - x] \cap (1,\infty)\right) \tag{38.3}$$

for $r \geq x \geq 0$, *and*

$$N(x) = \nu(x \vee 1, \infty) \tag{38.4}$$

for $x \geq 0$. *Let* $c > 0$ *be fixed. Then recurrence of* $\{X_t\}$ *is equivalent to*

$$\int_c^\infty \left(\int_0^r xR(r,x)\mathrm{d}x\right)^{-1} \mathrm{d}r = \infty. \tag{38.5}$$

If ν *is quasi-unimodal, then recurrence of* $\{X_t\}$ *is equivalent to*

$$\int_c^\infty \left(\int_0^r xN(x)\mathrm{d}x\right)^{-1} \mathrm{d}r = \infty. \tag{38.6}$$

Even without quasi-unimodality of ν, the condition (38.6) implies recurrence, since $N(x) \geq R(r,x)$. But recurrence does not imply (38.6), because, if it does, the remark after Theorem 38.2 would not be true.

THEOREM 38.4. *For an arbitrarily given symmetric finite measure* ρ *on* $\mathbb{R}$, *there exists a recurrent symmetric Lévy process* $\{X_t\}$ *on* $\mathbb{R}$ *such that its Lévy measure* ν *has a bigger tail than* ρ.

Proof of Theorem 38.2(i). Assume (38.2). Then, by Theorem 25.3, $E[X_1^2] < \infty$ if and only if $E[Y_1^2] < \infty$. Thus, if $E[X_1^2] < \infty$, then both $\{X_t\}$ and $\{Y_t\}$ are recurrent by Theorem 36.7, since $E[X_1] = E[Y_1] = 0$ by symmetry. Suppose that $E[X_1^2] = \infty$. By Theorem 25.3, $\int_{(0,\infty)} x^2 \nu_X(\mathrm{d}x) = \infty$. We have

$$z^{-2} \int_{(0,\infty)} (1 - \cos zx)\nu_X(\mathrm{d}x) \to \infty \quad \text{as } z \downarrow 0, \tag{38.7}$$

using Fatou's lemma in

$$z^{-2}\int_{(0,\infty)}(1-\cos zx)\nu_X(\mathrm{d}x) = 2^{-1}\int_{(0,\infty)}\left(\frac{\sin(zx/2)}{zx/2}\right)^2 x^2\,\nu_X(\mathrm{d}x).$$

It follows from (38.7) that

$$-z^{-2}\psi_X(z) \to \infty \quad \text{as } z \downarrow 0. \tag{38.8}$$

Since

$$\psi_X(z) - \psi_Y(z) = -2^{-1}(A_X - A_Y)z^2 - 2\int_{(0,\infty)}(1-\cos zx)(\nu_X - \nu_Y)(\mathrm{d}x),$$

we have

$$|\psi_X(z) - \psi_Y(z)| \le c_1 z^2 + 4\int_{(0,\infty)}\left(\sin\tfrac{zx}{2}\right)^2 |\nu_X - \nu_Y|(\mathrm{d}x) \le c_2 z^2$$

with some constants c_1, c_2 by (38.2). Hence

$$(-\psi_Y(z))/(-\psi_X(z)) = 1 + (\psi_X(z) - \psi_Y(z))/(-\psi_X(z)) \to 1 \quad \text{as } z \downarrow 0$$

by virtue of (38.8). Thus the condition (37.9) for ψ_X is equivalent to that for ψ_Y. □

We introduce the notion of unimodal correspondent after Shepp [**480**]. We also use the notion of "more peaked" of Birnbaum [**39**].

DEFINITION 38.5. Let X and U be independent random variables on $\mathbb{R}$ and suppose that U is uniformly distributed on $[0,1]$. The unimodal distribution ρ of UX (see Exercise 29.21) is called the *unimodal correspondent* of the distribution μ of X. We denote it by $\rho = \Lambda\mu$.

LEMMA 38.6. *For any $a \le 0 \le b$, $(\Lambda\mu)[a,b] \ge \mu[a,b]$.*

Proof. Since $0 \le U \le 1$, the event $\{a \le X \le b\}$ is included in the event $\{a \le UX \le b\}$. □

DEFINITION 38.7. Let μ_1 and μ_2 be symmetric probability measures on $\mathbb{R}$. We say that μ_1 is *more peaked than* μ_2 if $\mu_1[-x,x] \ge \mu_2[-x,x]$ for every $x \ge 0$.

Proof of Theorem 38.2(ii). Assume that ν_Y is quasi-unimodal and has a bigger tail than ν_X and that $\{X_t\}$ is transient. We will show that $\{Y_t\}$ is also transient.

Step 1. Suppose that $A_X = A_Y = 0$, $\nu_X(\mathbb{R}) < \infty$, $\nu_Y(\mathbb{R}) < \infty$, that both ν_X and ν_Y are unimodal with mode 0, and that

$$\nu_Y(x,\infty) \ge \nu_X(x,\infty) \quad \text{for every } x > 0.$$

Choose a symmetric measure $\nu^\sharp$ with $\nu^\sharp\{0\} = 0$, unimodal with mode 0, such that $\nu^\sharp$ has an identical tail with ν_X,

$$\nu_Y(x,\infty) \ge \nu^\sharp(x,\infty) \quad \text{for every } x > 0,$$

and $\nu^\sharp(0,\infty) = c/2$, where $c = \nu_Y(\mathbb{R})$. With normalizing to probability measures, this means that $\nu^\sharp$ is more peaked than ν_Y. Let $\{X_t^\sharp\}$ be the symmetric Lévy process generated by $(0,\nu^\sharp,0)$. It is transient by (i), since $\{X_t\}$ is transient. By

Exercise 39.15, we obtain $\nu_Y{}^n(x,\infty) \geq (\nu^\sharp)^n(x,\infty)$ for $x > 0$, using unimodality and symmetry of ν_Y and $\nu^\sharp$. Thus

$$P[\, Y_t > x\,] = \textstyle\sum_{n=0}^\infty e^{-ct}\frac{t^n}{n!}\nu_Y{}^n(x,\infty) \geq \sum_{n=0}^\infty e^{-ct}\frac{t^n}{n!}(\nu^\sharp)^n(x,\infty) = P[\, X_t^\sharp > x\,].$$

Hence

$$P[\, Y_t \in [-x,x]\,] \leq P[\, X_t^\sharp \in [-x,x]\,] \quad \text{for } t > 0 \text{ and } x > 0.$$

Hence $\{Y_t\}$ is transient by virtue of Theorem 35.4.

Step 2. Suppose that $A_X = A_Y = 0$, $\nu_X(\mathbb{R}) = \nu_Y(\mathbb{R}) = 1$, and that $\nu_Y = \Lambda\nu_X$, the unimodal correspondent of ν_X. By symmetry

$$\textstyle\int_{(0,\infty)} \cos zx\, \nu_Y(dx) = \int_0^1 du \int_{(0,\infty)} \cos uzx\, \nu_X(dx) = \int_{(0,\infty)} \frac{\sin zx}{zx}\nu_X(dx).$$

Hence

$$-\psi_X(z) = 2\textstyle\int_{(0,\infty)}(1-\cos zx)\nu_X(dx),$$
$$-\psi_Y(z) = 2\textstyle\int_{(0,\infty)}(1-\frac{\sin zx}{zx})\nu_X(dx).$$

There is a positive constant C such that $1-\cos u \leq C(1-\frac{\sin u}{u})$ for $u \in \mathbb{R}$. Then, $-\psi_Y(z) \geq -C^{-1}\psi_X(z)$. Hence $\{Y_t\}$ is transient by Corollary 37.6.

Step 3. Suppose that $A_X = A_Y = 0$, $\nu_X(\mathbb{R}) = 1$, $\nu_Y(\mathbb{R}) < \infty$, ν_Y unimodal with mode 0, and that

$$\nu_Y(x,\infty) \geq \nu_X(x,\infty) \quad \text{for every } x > 0.$$

Let $\{Z_t\}$ be the Lévy process generated by $(0,\Lambda\nu_X,0)$. Then $\{Z_t\}$ is transient, by the result of Step 2. We have

$$\nu_Y(x,\infty) \geq (\Lambda\nu_X)(x,\infty) \quad \text{for every } x > 0,$$

since $\nu_X(x,\infty) \geq (\Lambda\nu_X)(x,\infty)$ by Lemma 38.6. By Step 1 $\{Y_t\}$ is transient.

Step 4. General case. By the assumption there is $x_0 > 0$ such that

$$\nu_Y(x,\infty) \geq \nu_X(x,\infty) \quad \text{for } x \geq x_0.$$

Since ν_Y is quasi-unimodal, we can choose a symmetric measure $\nu_Y^\sharp$ unimodal with mode 0 with $\nu_Y^\sharp\{0\} = 0$ such that there is $x_1 \geq x_0$ satisfying

$$\nu_Y^\sharp(x,\infty) = \nu_Y(x,\infty) \quad \text{for } x \geq x_1,$$

and $\nu_Y^\sharp(x_1,\infty) < 1/2 \leq \nu_Y^\sharp(0,\infty) < \infty$. Then choose a symmetric measure $\nu_X^\sharp$ such that

$$\nu_X^\sharp(x,\infty) = \nu_X(x,\infty) \quad \text{for } x \geq x_1,$$
$$\nu_X^\sharp(x,\infty) \leq \nu_Y^\sharp(x,\infty) \quad \text{for } x > 0,$$

and $\nu_X^\sharp(0,\infty) = 1/2$, $\nu_X^\sharp\{0\} = 0$. Let $\{X_t^\sharp\}$ and $\{Y_t^\sharp\}$ be the Lévy processes generated by $(0,\nu_X^\sharp,0)$ and $(0,\nu_Y^\sharp,0)$, respectively. By (i), $\{X_t^\sharp\}$ is transient. Hence, by Step 3, $\{Y_t^\sharp\}$ is transient. Then, by (i), $\{Y_t\}$ is transient. □

Proof of Theorem 38.3. We may assume that $A = 0$ and $\nu[-1, 1] = 0$. This does not change $R(r, x)$ and $N(x)$ and it does not affect recurrence and transience, as Theorem 38.2(i) says. Then

$$\begin{aligned}-\psi(z) &= 2\int_1^\infty (1-\cos zx)\mathrm{d}(-N(x)) \\ &= 2(1-\cos z)N(1) + \lim_{y\to\infty} 2\int_1^y N(x)z\sin zx\,\mathrm{d}x \\ &= 2z\lim_{y\to\infty}\int_0^y N(x)\sin zx\,\mathrm{d}x = 2z\sum_{n=0}^\infty \int_0^{2\pi/z} N(\tfrac{2\pi n}{z}+x)\sin zx\,\mathrm{d}x \\ &= 2z\sum_{n=0}^\infty (I_{n,1}+I_{n,2}+I_{n,3}+I_{n,4}),\end{aligned}$$

where

$$\begin{aligned}I_{n,1} &= \int_0^{\pi/2z} N(\tfrac{2\pi n}{z}+x)\sin zx\,\mathrm{d}x, \\ I_{n,2} &= \int_{\pi/2z}^{\pi/z} N(\tfrac{2\pi n}{z}+x)\sin zx\,\mathrm{d}x = \int_0^{\pi/2z} N(\tfrac{2\pi n}{z}+\tfrac{\pi}{z}-x)\sin zx\,\mathrm{d}x, \\ I_{n,3} &= \int_{\pi/z}^{3\pi/2z} N(\tfrac{2\pi n}{z}+x)\sin zx\,\mathrm{d}x = -\int_0^{\pi/2z} N(\tfrac{2\pi n}{z}+\tfrac{\pi}{z}+x)\sin zx\,\mathrm{d}x, \\ I_{n,4} &= \int_{3\pi/2z}^{2\pi/z} N(\tfrac{2\pi n}{z}+x)\sin zx\,\mathrm{d}x = -\int_0^{\pi/2z} N(\tfrac{2\pi n}{z}+\tfrac{2\pi}{z}-x)\sin zx\,\mathrm{d}x.\end{aligned}$$

We have

$$\begin{aligned}I_{n,1}+I_{n,4} &= \int_0^{\pi/2z} \nu(\tfrac{2\pi n}{z}+x, \tfrac{2\pi(n+1)}{z}-x]\sin zx\,\mathrm{d}x, \\ I_{n,2}+I_{n,3} &= \int_0^{\pi/2z} \nu(\tfrac{2\pi n}{z}+\tfrac{\pi}{z}-x, \tfrac{2\pi n}{z}+\tfrac{\pi}{z}+x]\sin zx\,\mathrm{d}x.\end{aligned}$$

Now we can change the order of summation and integration, as the integrands are nonnegative. Thus, defining

$$R^0(r,x) = \nu\big(\textstyle\bigcup_{n=0}^\infty ((2n+1)r-x, (2n+1)r+x]\big),$$

we have

$$-\psi(z) = 2z\big(\int_0^{\pi/2z} R(\tfrac{\pi}{z},x)\sin zx\,\mathrm{d}x + \int_0^{\pi/2z} R^0(\tfrac{\pi}{z},x)\sin zx\,\mathrm{d}x\big).$$

Note that

$$R(\tfrac{\pi}{z},x) \geq R^0(\tfrac{\pi}{z},x) \geq 0 \quad \text{for } 0 < x < \tfrac{\pi}{2z}$$

and use $2u/\pi \leq \sin u \leq u$ for $0 \leq u \leq \pi/2$. Then, for $z > 0$,

$$-\psi(z) \geq 2z\int_0^{\pi/2z} R(\tfrac{\pi}{z},x)\sin zx\,\mathrm{d}x \geq \tfrac{4}{\pi}z^2\int_0^{\pi/2z} xR(\tfrac{\pi}{z},x)\mathrm{d}x \tag{38.9}$$

and

$$-\psi(z) \leq 4z\int_0^{\pi/2z} R(\tfrac{\pi}{z},x)\sin zx\,\mathrm{d}x \leq 4z^2\int_0^{\pi/2z} xR(\tfrac{\pi}{z},x)\mathrm{d}x. \tag{38.10}$$

Suppose that (38.5) is satisfied. Then

$$\int_0^{\pi/c}\big(z^2\int_0^{\pi/z} xR(\tfrac{\pi}{z},x)\mathrm{d}x\big)^{-1}\mathrm{d}z = \infty. \tag{38.11}$$

Hence, by (38.10),

$$\int_0^{\pi/c}\frac{1}{-\psi(z)}\mathrm{d}z = \infty, \tag{38.12}$$

which means recurrence by Corollary 37.6. Conversely, suppose that $\{X_t\}$ is recurrent. Then (38.12) holds, again by Corollary 37.6. Hence

$$\int_0^{\pi/c}\big(z^2\int_0^{\pi/2z} xR(\tfrac{\pi}{z},x)\mathrm{d}x\big)^{-1}\mathrm{d}z = \infty$$

by (38.9). We have

$$\int_{r/2}^{r} xR(r,x)\mathrm{d}x \leq 4\int_{r/4}^{r/2} xR(r,x)\mathrm{d}x,$$

since

$$\begin{aligned}\int_{r/2}^{r} x\nu(2nr+x, 2(n+1)r-x]\,\mathrm{d}x &= 4\int_{r/4}^{r/2} x\nu(2nr+2x, 2(n+1)r-2x]\,\mathrm{d}x\\ &\leq 4\int_{r/4}^{r/2} x\nu(2nr+x, 2(n+1)r-x]\,\mathrm{d}x.\end{aligned}$$

Hence

$$\int_0^r xR(r,x)\mathrm{d}x \leq 5\int_0^{r/2} xR(r,x)\mathrm{d}x.$$

Hence we obtain (38.11), which is the same as (38.5).

It remains to prove the assertions related to $N(x)$. If (38.6) holds, then $\{X_t\}$ is recurrent, since $N(x) \geq R(x,r)$ for $0 \leq x \leq r$ and (38.5) holds. Conversely, suppose that $\{X_t\}$ is recurrent and that ν is quasi-unimodal. We can find a symmetric probability measure $\nu^\sharp$ unimodal with mode 0 with $\nu^\sharp\{0\} = 0$ such that, for some $x_0 > 0$,

$$\nu^\sharp(x,\infty) = \nu(x,\infty) \quad \text{for } x \geq x_0.$$

In order to construct such a $\nu^\sharp$, first, from the quasi-unimodality, find a symmetric measure ν_1 unimodal with mode 0, $\nu_1\{0\} = 0$, such that, for some $x_1 > 0$, $\nu_1(x,\infty) = \nu(x,\infty)$ for $x \geq x_1$, and then get the graph of $\nu^\sharp(x,\infty)$ by drawing a straight line from the point $(0,1/2)$ tangent to the curve $y = \nu_1(x,\infty)$ if $\nu_1(0,\infty) > 1/2$, or through a point $(x_2, \nu_1(x_2,\infty))$ with $x_2 > 0$ if $\nu_1(0,\infty) < 1/2$. Let $\{X_t^\sharp\}$ be the Lévy process generated by $(0,\nu^\sharp,0)$. Then $\{X_t^\sharp\}$ is recurrent by Theorem 38.2(i). Let $\{Y_t\}$ be the Lévy process generated by $(0,\Lambda\nu^\sharp,0)$. Since $\nu^\sharp(x,\infty) \geq \Lambda\nu^\sharp(x,\infty)$ for $x > 0$ by Lemma 38.6, it follows from Theorem 38.2(ii) that $\{Y_t\}$ is recurrent and hence $\int_0^\varepsilon \frac{\mathrm{d}z}{-\psi_Y(z)} = \infty$ for any $\varepsilon > 0$. As in Step 2 of the proof of Theorem 38.2(ii),

$$-\psi_Y(z) = 2\int_{(0,\infty)} \left(1 - \tfrac{\sin zx}{zx}\right) \nu^\sharp(\mathrm{d}x).$$

Hence

$$-\psi_Y(z) \geq 2C\int_{(0,\infty)} ((zx)^2 \wedge 1)\nu^\sharp(\mathrm{d}x),$$

where $C > 0$ is a constant such that $1 - \frac{\sin u}{u} \geq C(u^2 \wedge 1)$ for $u \in \mathbb{R}$. Thus

$$-\psi_Y(z) \geq 2C\int_0^\infty N^\sharp(x)\mathrm{d}((zx)^2 \wedge 1) = 4Cz^2\int_0^{1/z} xN^\sharp(x)\mathrm{d}x,$$

for $N^\sharp(x) = \nu^\sharp(x \vee 1,\infty)$. Therefore

$$\int_0^{1/c} \left(z^2\int_0^{1/z} xN^\sharp(x)\mathrm{d}x\right)^{-1}\mathrm{d}z = \infty,$$

that is,

$$\int_c^\infty \left(\int_0^r xN^\sharp(x)\mathrm{d}x\right)^{-1}\mathrm{d}r = \infty.$$

This implies (38.6) because, in the case $\int_0^\infty xN(x)\mathrm{d}x < \infty$, (38.6) is evident and, in the case $\int_0^\infty xN(x)\mathrm{d}x = \infty$,

$$\frac{\int_0^r xN^\sharp(x)\mathrm{d}x}{\int_0^r xN(x)\mathrm{d}x} = \frac{\int_{x_0}^r xN(x)\mathrm{d}x}{\int_0^r xN(x)\mathrm{d}x} + o(1) = 1 + o(1) \quad \text{as } r \to \infty.$$

This completes the proof. □

Theorem 38.4 is essentially contained in the following lemma.

LEMMA 38.8 (Shepp [**481**]). *Suppose that $y_n > 0$, increasing to ∞, and $p_n > 0$ with $\sum_{n=1}^{\infty} p_n < \infty$ are given. Then, for some $x_n \geq y_n$, $n = 1, 2, \ldots$,*

$$\int_0^1 \left(\sum_{n=1}^{\infty} p_n(1 - \cos x_n z)\right)^{-1} dz = \infty. \tag{38.13}$$

Proof. Assume that $n_0 = 0 < n_1 < \cdots < n_k$ and x_n for $n = 1, 2, \ldots, n_k$ have been chosen to satisfy $x_n \geq y_n$ for $n = 1, \ldots, n_k$ and

$$\int_0^1 \left(\sum_{n \leq n_k} p_n(1 - \cos x_n z) + 2\sum_{n > n_k} p_n\right)^{-1} dz \geq k. \tag{38.14}$$

Let us show that then we can choose $n_{k+1} > n_k$ and x_n for $n = n_k + 1, \ldots, n_{k+1}$ so that $x_n \geq y_n$ and so that (38.14) is satisfied with $k+1$ in place of k. As (38.14) is evident for $k = 0$, these inductively defined x_n satisfy (38.13), since

$$\left(\sum_n p_n(1 - \cos x_n z)\right)^{-1} \geq \left(\sum_{n \leq n_k} p_n(1 - \cos x_n z) + 2\sum_{n > n_k} p_n\right)^{-1}.$$

We will show that, choosing $m > n_k$ suitably large and then choosing $x \geq y_m$ suitably large, we can fulfill the requirement by $n_{k+1} = m$ and $x_n = x$ for $n = n_k + 1, \ldots, n_{k+1}$. For that purpose it is enough to show that

$$\limsup_{m \to \infty} \limsup_{N \to \infty} \int_0^1 \left\{\sum_{n \leq n_k} p_n(1 - \cos x_n z) + \left(\sum_{n=n_k+1}^{m} p_n\right)(1 - \cos 2\pi N z) + 2\sum_{n > m} p_n\right\}^{-1} dz = \infty. \tag{38.15}$$

Let $\varepsilon_m = \left(\sum_{n>m} p_n\right)^{1/2}$. Then $\varepsilon_m \downarrow 0$. Since the integrand in (38.15) is greater than

$$c(z^2 + (1 - \cos 2\pi N z) + {\varepsilon_m}^2)^{-1}$$

with some constant $c > 0$, it is enough to show that

$$\limsup_{m \to \infty} \limsup_{N \to \infty} \int_0^1 (z^2 + 1 - \cos 2\pi N z + {\varepsilon_m}^2)^{-1} dz = \infty. \tag{38.16}$$

Denote the integral in (38.16) by $I(m, N)$. Then

$$I(m, N) = \frac{1}{N} \sum_{n=1}^{N} \int_0^1 \left(\left(\frac{n-1}{N} + \frac{r}{N}\right)^2 + 1 - \cos 2\pi r + {\varepsilon_m}^2\right)^{-1} dr.$$

Using $1 - \cos 2\pi r \leq \frac{1}{2}(2\pi r)^2$, we can find another constant $c > 0$ such that

$$I(m, N) \geq \frac{c}{N} \sum_{n=1}^{N} \int_0^1 \left(\left(\frac{n}{N}\right)^2 + r^2 + {\varepsilon_m}^2\right)^{-1} dr.$$

Hence

$$\begin{aligned} I(m, N) &\geq \frac{c}{N} \sum_{n=1}^{N} \int_0^1 \left(\frac{n}{N} + r + \varepsilon_m\right)^{-2} dr \\ &= \frac{c}{N} \sum_{n=1}^{N} \left(\frac{n}{N} + \varepsilon_m\right)^{-1} \left(\frac{n}{N} + 1 + \varepsilon_m\right)^{-1}. \end{aligned}$$

Let m be so large that $\varepsilon_m < 1$. Then

$$\begin{aligned} I(m, N) &\geq \frac{c}{3N} \sum_{n=1}^{N} \left(\frac{n}{N} + \varepsilon_m\right)^{-1} \geq \frac{c}{3} \int_1^{N+1} \frac{dr}{r + N\varepsilon_m} \\ &= \frac{c}{3} \log \frac{N+1+N\varepsilon_m}{1+N\varepsilon_m} = \frac{c}{3} \log \frac{1+\varepsilon_m}{\varepsilon_m} + o(1) \end{aligned}$$

as $N \to \infty$. Hence (38.16) is true. The lemma is now proved. □

Proof of Theorem 38.4. We are given a symmetric finite measure ρ. We will construct a symmetric finite measure ν, $\nu\{0\} = 0$, with a bigger tail than ρ so that $\int_0^1 \frac{1}{-\psi(z)} dz = \infty$, where $-\psi(z) = 2\int_{(0,\infty)}(1 - \cos zx)\nu(dx)$. Then the symmetric Lévy process with Lévy measure ν is recurrent. We may assume that $\rho = \sum_{n=1}^{\infty} p_n(\delta_{y_n} + \delta_{-y_n})$ with $p_n > 0$, $\sum_n p_n < \infty$, $y_n > 0$, and $y_n \uparrow \infty$, because we can choose a measure of this form with a bigger tail than the original ρ. Now apply Lemma 38.8 and let $\nu = \sum_{n=1}^{\infty} p_n(\delta_{x_n} + \delta_{-x_n})$. Then, by (38.13), ν has the property required. □

39. Exercises 7

E 39.1. Let $\{X_t\}$ be the Brownian motion on $\mathbb{R}$ with drift $\gamma > 0$ added. Show that its potential density is

$$v(x) = \gamma^{-1}[1_{[0,\infty)}(x) + e^{2\gamma x} 1_{(-\infty,0)}(x)].$$

It is remarkable that, on $[0, \infty)$, the potential measure is a constant multiple of the Lebesgue measure, while the density tends to 0 as $x \to -\infty$. Compare this to a general result in E 39.14.

E 39.2. Let $\{X_t\}$ be the stable process on $\mathbb{R}$ having parameters $(\alpha, \beta, 0, c)$ with $0 < \alpha < 1$, $-1 \le \beta \le 1$, and $c > 0$. Show that

$$v(x) = C(1 + \beta \operatorname{sgn} x)|x|^{\alpha-1}, \quad C = \left[2c\Gamma(\alpha)\left(1 + \beta^2 \left(\tan \tfrac{\pi\alpha}{2}\right)^2\right) \cos \tfrac{\pi\alpha}{2}\right]^{-1}.$$

E 39.3. Show the following. A non-trivial 1-stable process on $\mathbb{R}$ is recurrent if and only if its Lévy measure is symmetric or, in other words, if and only if it is a Cauchy process. A non-trivial 1-semi-stable process on $\mathbb{R}$ is recurrent if its Lévy measure is symmetric. But there are recurrent 1-semi-stable processes on $\mathbb{R}$ with non-symmetric Lévy measures.

E 39.4. Let $\{X_t\}$ be a Lévy process on $\mathbb{R}^d$. Show that $\{X_t\}$ is recurrent if, for every nonempty open set G, $P[X_t \in G \text{ for some } t > 0] = 1$. Show that the converse is also true, provided that its group $\mathfrak{G}$ is identical with $\mathbb{R}^d$.

E 39.5. Show that, if $\{X_t\}$ is a recurrent Lévy process on $\mathbb{R}^d$ with $d \ge 3$, then the support Σ of $\{X_t\}$ is contained in a two-dimensional linear subspace of $\mathbb{R}^d$.

E 39.6. Let $\{X_t\}$ and $\{Y_t\}$ be Lévy processes on $\mathbb{R}^d$ with Lévy measures ν_X and ν_Y, respectively. Assume that they have a common Gaussian part. Show that, if $\{X_t\}$ is transient and symmetric and if $\nu_Y(B) \ge \nu_X(B)$ for every $B \in \mathcal{B}(\mathbb{R}^d)$, then $\{Y_t\}$ is transient.

E 39.7. Let $\{X_t\}$ and $\{Y_t\}$ be independent Lévy processes on $\mathbb{R}^d$ identical in law. Show that, if $\{X_t\}$ is recurrent, then its symmetrization $\{X_t - Y_t\}$ is recurrent, too.

E 39.8. Let $\{X_t\}$ and $\{Y_t\}$ be independent Lévy processes on $\mathbb{R}$. Suppose that $\{X_t\}$ is symmetric and generated by $(0, \nu_X, 0)$. Consider the condition

$$\text{(C)} \qquad \liminf_{z \downarrow 0} z^{-1} \int_0^\infty (1 - \cos zx) \nu_X(\mathrm{d}x) > 0.$$

(i) Suppose that $E|Y_t| < \infty$ for $t > 0$. Show that, if $\{X_t\}$ is recurrent and satisfies the condition (C), then $\{X_t + Y_t\}$ is recurrent.

(ii) Show that the condition (C) is determined only by the tail of ν_X.

(iii) Show that the condition (C) implies $E|X_t| = \infty$ for $t > 0$.

(iv) Show that a Cauchy process with $\gamma = 0$ satisfies the condition (C).

E 39.9. Let $\{X_t\}$ be a symmetric Lévy process on $\mathbb{R}$ with Lévy measure ν. Suppose that there are $\alpha \in \mathbb{R}$ and $a > 1$ such that $\nu(\mathrm{d}x) = |x|^{-2}(\log|x|)^\alpha \mathrm{d}x$ on $\{x \colon |x| > a\}$. Show that $\{X_t\}$ is recurrent or transient according as $\alpha \le 1$ or $\alpha > 1$, respectively. Consider also the case where $\nu(\mathrm{d}x) = |x|^{-2}(\log|x|)(\log\log|x|)^\alpha \mathrm{d}x$ on $\{x \colon |x| > a\}$.

E 39.10. Show that (37.8) is not a necessary condition for transience of a Lévy process.

E 39.11. Let $\{X_t\}$ be a non-zero Lévy process on $\mathbb{R}$. Show that the cases (1), (2), and (3) in Proposition 37.10 are respectively described as follows:

(1) $\sup_t X_t = \infty$ and $\inf_t X_t > -\infty$ a. s.
(2) $\sup_t X_t < \infty$ and $\inf_t X_t = -\infty$ a. s.
(3) $\sup_t X_t = \infty$ and $\inf_t X_t = -\infty$ a. s.

E 39.12 ([**519**]). Let $\{X_t\}$ be a transient Lévy process on $\mathbb{R}^d$ such that μ^t, $t > 0$, has density $p_t(x)$ measurable in (t, x). For $0 < \alpha < 1$ let $\{Z_t^\alpha\}$ be the α-stable subordinator with $E[\mathrm{e}^{-uZ_t^\alpha}] = \mathrm{e}^{-tu^\alpha}$, $u \ge 0$. Let $\{X_t^\alpha\}$ be the Lévy process on $\mathbb{R}^d$ subordinate to $\{X_t\}$ by $\{Z_t^\alpha\}$. Show that $\{X_t^\alpha\}$ is transient with 0-potential density $v_\alpha(x) = \frac{1}{\Gamma(\alpha)} \int_0^\infty p_t(x) t^{\alpha-1} \mathrm{d}t$. Further show that

$$\int_{\mathbb{R}^d} v_\alpha(y - x) v_\beta(z - y) \mathrm{d}y = v_{\alpha+\beta}(z - x) \qquad \text{a. e. } x \text{ and } z$$

for $\alpha > 0$ and $\beta > 0$ with $\alpha + \beta < 1$. This is an expression of the identity (32.13). A special case is the identity

$$\int_{\mathbb{R}^d} c_{d,\alpha} |y - x|^{\alpha-d} c_{d,\beta} |z - y|^{\beta-d} \mathrm{d}y = c_{d,\alpha+\beta} |z - x|^{\alpha+\beta-d}$$

for $\alpha > 0$ and $\beta > 0$ with $\alpha + \beta < d \wedge 2$. Here the constants are those in (37.14).

E 39.13. Let $\{X_t\}$ be a non-zero subordinator. Show that, for any $u > 0$,

$$\int_0^\infty \mathrm{e}^{-ux} V(\mathrm{d}x) = \left\{ \gamma_0 u + \int_{(0,\infty)} (1 - \mathrm{e}^{-ux})\, \nu(\mathrm{d}x) \right\}^{-1}.$$

E 39.14 (Renewal theorem for Lévy processes). Let $\{X_t\}$ be a Lévy process on $\mathbb{R}$ with the group $\mathfrak{G}$ being $\mathbb{R}$. Prove the following. If $E|X_t| < \infty$ and $EX_1 = \gamma_1 > 0$, then $\int f(y - x)\, V(\mathrm{d}y)$ tends to $\gamma_1{}^{-1} \int f(y) \mathrm{d}y$ as $x \to \infty$ and to 0 as $x \to -\infty$ for any $f \in C_c$, a continuous function with compact support. If $E|X_t| < \infty$ and $EX_1 < 0$, then we have the dual situation. In the remaining transient case, $\int f(y - x)\, V(\mathrm{d}y) \to 0$ as $|x| \to \infty$ for any $f \in C_c$. (In the case $d \ge 2$, $\int_{\mathbb{R}^d} f(y - x)\, V(\mathrm{d}y) \to 0$ as $|x| \to \infty$ for any $f \in C_c$ for any transient Lévy process with $\mathfrak{G} = \mathbb{R}^d$.)

E 39.15 (Birnbaum [**39**]). Let μ_1, μ_2, ρ_1, ρ_2 be continuous symmetric unimodal probability measures on $\mathbb{R}$. Show that, if μ_1 and μ_2 are more peaked than ρ_1 and ρ_2, respectively, then $\mu_1 * \mu_2$ is more peaked than $\rho_1 * \rho_2$.

E 39.16 (Shepp [**480**]). Let $\nu = \sum_{n=1}^{\infty} p_n(\delta_{a_n} + \delta_{-a_n})$ be a probability measure with $a_n = 2^{n^2}$ and $p_n = c2^{n-n^2}n^2$, where c is a normalizing constant. Show that the Lévy process on $\mathbb{R}$ generated by $(0, \nu, 0)$ is recurrent while that generated by $(0, \Lambda\nu, 0)$ is transient. Note that ν has a bigger tail than $\Lambda\nu$. Here $\Lambda\nu$ is the unimodal correspondent of ν.

Notes

Recurrence and transience are defined and studied in temporally homogeneous Markov processes. Many books, for example Chung [**80, 81**] and Resnick [**413**], treat them.

The proof of Theorem 35.4 follows Kingman [**284**], except the proof that (3) implies (4) and the proof of (iv). The convergence (36.6) is from Doob [**106**], p. 364. Theorem 36.8 is by de Acosta [**3**]. Theorems 36.7, 37.8, and 37.14 are analogues of results on random walks in Chung and Fuchs [**84**] and Chung [**80**]. The results on semi-stable processes in Theorems 37.16 and 37.18 are by Choi [**76**]. Choi and Sato [**77**] extend them to operator-semi-stable processes. The extension of Shepp's theory [**480, 481**] in Section 38 is taken from [**448**]. E 39.6–E 39.8 are also from [**448**]. The quasi-unimodality in Definition 38.1 is called convex at infinity by Shepp [**480**].

The two-dimensional Brownian motion is recurrent, but we can subtract an appropriate divergent part from $V^q f$ as $q \downarrow 0$, to get a finite function for f of a suitable class. This corresponds to defining the logarithmic potential operator in potential theory. Port and Stone [**386**] define potential operators for recurrent Lévy processes on $\mathbb{R}$ and $\mathbb{R}^2$. The potential operators in the sense of Yosida are also definable, as is mentioned in Remark 31.10 and Notes in Chapter 6.

A recurrence criterion for the processes of Ornstein–Uhlenbeck type on $\mathbb{R}^d$ defined in Section 17 is given by Shiga [**483**] and Sato, Watanabe, and Yamazato [**462**]. Its extension to the Markov process on $\mathbb{R}^d$ defined by

$$X_t = x + Z_t - \int_0^t QX_s \mathrm{d}s$$

in place of (17.1) is made by Sato, Watanabe, Yamamuro, and Yamazato [**461**] and Watanabe [**559**]. Here Q is a $d \times d$ matrix such that all of its eigenvalues have positive real parts. The process has infinitely divisible distribution at each t whenever it starts at a single point. If a process of Ornstein–Uhlenbeck type on $\mathbb{R}^d$ satisfies the condition (17.11), then it is recurrent. But the converse does not hold.

Study of recurrence and transience of selfsimilar additive processes is initiated by Sato and Yamamuro [**465**] and Yamamuro [**587**].

CHAPTER 8

Potential theory for Lévy processes

40. The strong Markov property

A remarkable development of potential theory for temporally homogeneous Markov processes was made in the middle of the twentieth century. In this chapter we treat the elementary part of the potential theory in relation to Lévy processes. It is the contribution of Hunt, Blumenthal, Getoor, Kesten, Kanda, Port, Stone, and many others. The strong Markov property is the key to open this field. Now we need to consider filtrations of σ-algebras. Using them, we introduce stopping times and the strong Markov property in this section.

In this chapter let $\Omega = D([0,\infty),\mathbb{R}^d)$, the collection of functions $\omega(t)$ from $[0,\infty)$ into $\mathbb{R}^d$, right-continuous with left limits. For $\omega \in \Omega$, let $X_t(\omega) = \omega(t)$ and let $\mathcal{F}_t^0 = \sigma(X_s \colon s \in [0,t])$ and $\mathcal{F}^0 = \sigma(X_s \colon s \in [0,\infty))$. (In Section 20 we have written $\Omega = \mathbf{D}$ and $\mathcal{F}^0 = \mathcal{F}_{\mathbf{D}}$.) We consider a probability measure P on $\mathcal{F}^0$ such that $\{X_t \colon t \geq 0\}$ is a Lévy process under P. The process $\{X_t\}$ under P is denoted by $(\{X_t\}, P)$. Any Lévy process on $\mathbb{R}^d$ can be realized in this way. Now fix such a Lévy process. Define, for $x \in \mathbb{R}^d$,

$$P^x[X_{t_1} \in B_1, \ldots, X_{t_n} \in B_n] = P[x + X_{t_1} \in B_1, \ldots, x + X_{t_n} \in B_n] \tag{40.1}$$

for $0 \leq t_1 < \cdots < t_n$ and $B_1, \ldots, B_n \in \mathcal{B}(\mathbb{R}^d)$. This P^x can be uniquely extended to a probability measure on $\mathcal{F}^0$; the extension is denoted by the same symbol. Thus $P^0 = P$. For any $H \in \mathcal{F}^0$, $P^x[H]$ is measurable (that is, Borel-measurable) in x. To show this, note that (40.1) is measurable in x since $E[f(x + X_{t_1}, \ldots, x + X_{t_n})]$ is continuous in x for any bounded continuous function $f(x_1, \ldots, x_n)$, and then use Proposition 1.15. For any probability measure ρ on $\mathcal{B}(\mathbb{R}^d)$ define a probability measure P^ρ on $\mathcal{F}^0$ by

$$P^\rho[H] = \int_{\mathbb{R}^d} P^x[H]\rho(\mathrm{d}x) \tag{40.2}$$

for $H \in \mathcal{F}^0$. We have $P^x[X_0 = x] = 1$ and $P^\rho[X_0 \in B] = \rho(B)$ for $B \in \mathcal{B}(\mathbb{R}^d)$. Thus $\{X_t\}$ is, under P^x, a process starting at x and, under P^ρ, a process with initial distribution ρ. The expectations with respect to P^x and P^ρ are denoted by E^x and E^ρ, respectively.

PROPOSITION 40.1. *The process $(\{X_t\}, P^\rho)$ has stationary independent increments and is stochastically continuous. Its increments have the same distributions as those of the original process.*

Proof. Let $0 \le t_0 < t_1 < \cdots < t_n$ and $B_0, \ldots, B_n \in \mathcal{B}(\mathbb{R}^d)$. Then

$$\begin{aligned} P^\rho[X_{t_0} \in B_0,\ X_{t_1} - X_{t_0} \in B_1, \ldots, X_{t_n} - X_{t_{n-1}} \in B_n] \\ &= \int \rho(\mathrm{d}x) P^x[X_{t_0} \in B_0,\ X_{t_1} - X_{t_0} \in B_1, \ldots, X_{t_n} - X_{t_{n-1}} \in B_n] \\ &= \int \rho(\mathrm{d}x) P^0[x + X_{t_0} \in B_0,\ X_{t_1} - X_{t_0} \in B_1, \ldots, X_{t_n} - X_{t_{n-1}} \in B_n] \\ &= P^\rho[X_{t_0} \in B_0] P^0[X_{t_1} - X_{t_0} \in B_1] \ldots P^0[X_{t_n} - X_{t_{n-1}} \in B_n]. \end{aligned}$$

We also have, for any s, t, and B,

$$P^\rho[X_{s+t} - X_s \in B] = \int \rho(\mathrm{d}x) P^x[X_{s+t} - X_s \in B] = P^0[X_{s+t} - X_s \in B].$$

Thus our assertions follow. □

Let $(\mathcal{F}^0)^{P^\rho}$ be the completion of $\mathcal{F}^0$ by P^ρ. That is, $H \in (\mathcal{F}^0)^{P^\rho}$ if and only if there are H_1 and H_2 in $\mathcal{F}^0$ such that $H_1 \subset H \subset H_2$ and $P^\rho[H_2 \setminus H_1] = 0$. The measure P^ρ can be uniquely extended to $(\mathcal{F}^0)^{P^\rho}$. Define

$$\mathcal{F} = \bigcap_\rho (\mathcal{F}^0)^{P^\rho}.$$

Next, define $\mathcal{F}_t$ as the collection of sets H in Ω such that, for any probability measure ρ on $\mathcal{B}(\mathbb{R}^d)$, there are $H_\rho \in \mathcal{F}_t^0$, $G_\rho \in \mathcal{F}$, and $G'_\rho \in \mathcal{F}$ satisfying $H_\rho \setminus G_\rho \subset H \subset H_\rho \cup G'_\rho$ and $P^\rho[G_\rho] = P^\rho[G'_\rho] = 0$. In other words, $\mathcal{F}_t$ is the collection of $H \in \mathcal{F}$ such that, for any ρ, there is $H_\rho \in \mathcal{F}_t^0$ satisfying $P^\rho[H \setminus H_\rho] = P^\rho[H_\rho \setminus H] = 0$. $\mathcal{F}$ and $\mathcal{F}_t$ are σ-algebras. Further, define the σ-algebra $\mathcal{F}_{t+}$ by

$$\mathcal{F}_{t+} = \bigcap_{\varepsilon > 0} \mathcal{F}_{t+\varepsilon}.$$

Let $\mathcal{B}^*(\mathbb{R}^d)$ be the σ-algebra defined by $\mathcal{B}^*(\mathbb{R}^d) = \bigcap_\rho \mathcal{B}(\mathbb{R}^d)^\rho$, where the superscript of a probability measure ρ denotes the completion by ρ. Sets in $\mathcal{B}^*(\mathbb{R}^d)$ and $\mathcal{B}^*(\mathbb{R}^d)$-measurable functions are called *universally measurable.*

PROPOSITION 40.2. *For any $H \in \mathcal{F}$, $P^x[H]$ is universally measurable in x and satisfies (40.2) for any probability measure ρ.*

Proof. By the definition of $\mathcal{F}$, there are $H_{\rho,1}$ and $H_{\rho,2}$ in $\mathcal{F}^0$ such that $H_{\rho,1} \subset H \subset H_{\rho,2}$ and $P^\rho[H_{\rho,2} \setminus H_{\rho,1}] = 0$. Hence $P^x[H_{\rho,1}] \le P^x[H] \le P^x[H_{\rho,2}]$ and $\int (P^x[H_{\rho,2}] - P^x[H_{\rho,1}])\rho(\mathrm{d}x) = 0$. Hence $P^x[H]$ is $\mathcal{B}(\mathbb{R}^d)^\rho$-measurable and (40.2) holds. □

We have two propositions on $\{\mathcal{F}_{t+}\}$.

PROPOSITION 40.3. *$\mathcal{F}_t = \mathcal{F}_{t+}$ for every $t \in [0, \infty)$.*

Proof. For every $t \geq 0$ and $s \geq 0$, $X_{t+s} - X_t$ is independent of $\mathcal{F}_t^0$ under P^ρ by Proposition 40.1. It follows that it is independent of $\mathcal{F}_t$ under P^ρ. Hence, for every $\varepsilon > 0$, $X_{t+s+\varepsilon} - X_{t+\varepsilon}$ is independent of $\mathcal{F}_{t+}$ under P^ρ. Letting $\varepsilon \downarrow 0$, we see that $X_{t+s} - X_t$ is independent of F_{t+} under P^ρ. Similarly, for every choice of n and $s_j \geq 0$ $(j = 1, \ldots, n)$, $\{X_{t+s_j} - X_t \colon j = 1, \ldots, n\}$ is independent of $\mathcal{F}_{t+}$ under P^ρ. Therefore, if $f(x_1, \ldots, x_m)$ is bounded and measurable and, if $0 \leq t_1 < \cdots < t_r = t < t_{r+1} < \cdots < t_m$, then, for every $H \in \mathcal{F}_{t+}$,

$$
\begin{aligned}
&E^\rho[f(X_{t_1}, \ldots, X_{t_m})1_H] \\
&\quad = E^\rho[f(X_{t_1}, \ldots, X_{t_r}, X_{t_r} + (X_{t_{r+1}} - X_{t_r}), \ldots, X_{t_r} + (X_{t_m} - X_{t_r}))1_H] \\
&\quad = E^\rho[g(X_{t_1}, \ldots, X_{t_r})1_H]
\end{aligned}
$$

by an extension of Proposition 1.16, where $g(x_1, \ldots, x_r) = E[f(x_1, \ldots, x_r, x_r + X_{t_{r+1}-t_r}, \ldots, x_r + X_{t_m - t_r})]$ is measurable. Now, using Proposition 1.15, we see that, for every $G \in \mathcal{F}^0$, there exists a bounded $\mathcal{F}_t^0$-measurable function $h(\omega)$ such that

$$E^\rho[1_G 1_H] = E^\rho[h 1_H] \quad \text{for every } H \in \mathcal{F}_{t+}.$$

If there are two such functions h_1 and h_2 for a given $G \in \mathcal{F}^0$, then $h_1(\omega) = h_2(\omega)$ P^ρ-a. s. We have $0 \leq h \leq 1$ P^ρ-a. s. By the definition of $\mathcal{F}$, the same is true for $G \in \mathcal{F}$.

Now, suppose that $G \in \mathcal{F}_{t+}$. Using the function h for G, let $G' = \{\omega \colon h(\omega) = 1\}$. Then we have $P^\rho[G \setminus G'] = 0$ and $P^\rho[G' \setminus G] = 0$, using $E^\rho[(1_G - h)1_H] = 0$ for $H = G \setminus G'$ and $H = G' \setminus G$, respectively. Hence we get $G \in \mathcal{F}_t$ from $G' \in \mathcal{F}_t^0$. Thus $\mathcal{F}_{t+} \subset \mathcal{F}_t$. The reverse inclusion is trivial. □

The following fact is called Blumenthal's 0–1 law.

PROPOSITION 40.4. *If $H \in \mathcal{F}_{0+}$, then, for each x, $P^x[H] = 0$ or 1.*

Proof. Let $H \in \mathcal{F}_{0+}$. Then $H \in \mathcal{F}_0$ by the proposition above. Hence, for each x, there is $H_x \in \mathcal{F}_0^0$ such that $P^x[H \setminus H_x] = P^x[H_x \setminus H] = 0$. Since $X_0 = x$ P^x-a. s., $P^x[H_x]$ is 0 or 1. □

DEFINITION 40.5. A mapping T from Ω into $[0, \infty]$ is a *stopping time* if $\{T \leq t\} \in \mathcal{F}_t$ for every $t \in [0, \infty)$. If T is a stopping time, denote by $\mathcal{F}_T$ the class of $H \in \mathcal{F}$ such that $H \cap \{T \leq t\} \in \mathcal{F}_t$ for every $t \in [0, \infty)$, and write $\mathcal{F}_T' = \{H \in \mathcal{F}_T \colon H \subset \{T < \infty\}\}$. (Intuitively, T is a stopping time if one can determine whether $T \leq t$ or not by the observation of the motion up to time t; $H \in \mathcal{F}_T$ if H is an event expressible by the observation of the motion up to time T.)

DEFINITION 40.6. Let $B \subset \mathbb{R}^d$. The *hitting time* T_B of B is defined by

$$T_B(\omega) = \inf\{t > 0\colon X_t(\omega) \in B\}, \tag{40.3}$$

where the infimum of the empty set is defined to be $+\infty$.

EXAMPLE 40.7. If B is open, then T_B is a stopping time, because

$$\{T_B < t\} = \bigcup_{s \in \mathbb{Q} \cap (0,t)} \{X_s \in B\} \in \mathcal{F}_t^0 \subset \mathcal{F}_t$$

by the right-continuity of ω, and $\{T_B \le t\} = \bigcap_n \{T_B < t + 1/n\} \in \mathcal{F}_{t+} = \mathcal{F}_t$ by Proposition 40.3. Hitting times of more general sets are treated later.

Let $\varepsilon > 0$. The time T of the first jump of size $> \varepsilon$ is defined by

$$T(\omega) = \inf\{t > 0\colon |X_t(\omega) - X_{t-}(\omega)| > \varepsilon\}.$$

This is a stopping time, as is proved similarly to Lemma 20.9.

PROPOSITION 40.8. (i) *If T is a stopping time, then $\mathcal{F}_T$ is a σ-algebra.*

(ii) *If $T = s =$ const, then T is a stopping time and $\mathcal{F}_T = \mathcal{F}_s$.*

(iii) *If T_1 and T_2 are stopping times, then $T_1 \wedge T_2$ and $T_1 \vee T_2$ are stopping times.*

(iv) *If T_n, $n = 1, 2, \ldots$, are stopping times, then $\sup_n T_n$ is a stopping time.*

(v) *If T is a stopping time and $s =$ const ≥ 0, then $T + s$ is a stopping time.*

(vi) *If T_1 and T_2 are stopping times and $T_1 \le T_2$, then $\mathcal{F}_{T_1} \subset \mathcal{F}_{T_2}$.*

(vii) *If T_n, $n = 1, 2, \ldots$, are stopping times and $T = \inf_n T_n$, then T is a stopping time and $\mathcal{F}_T = \bigcap_n \mathcal{F}_{T_n}$.*

(viii) *If T_1 and T_2 are stopping times, then $\{T_1 < T_2\}$ and $\{T_1 \le T_2\}$ belong to $\mathcal{F}_{T_1} \cap \mathcal{F}_{T_2}$.*

Proof. (i) If $H \in \mathcal{F}_T$, then $H^c \in \mathcal{F}_T$, since $H^c \cap \{T \le t\} = \{T \le t\} \setminus (H \cap \{T \le t\}) \in \mathcal{F}_t$. If $H_n \in \mathcal{F}_T$, $n = 1, 2, \ldots$, then $\bigcup_n H_n \in \mathcal{F}_T$, since $(\bigcup_n H_n) \cup \{T \le t\} = \bigcup_n (H_n \cap \{T \le t\}) \in \mathcal{F}_t$. It is evident that $\Omega \in \mathcal{F}_T$. Hence F_T is a σ-algebra.

(ii) $\{T \le t\} = \Omega$ or $\emptyset$ and $H \cap \{T \le t\} = H$ or $\emptyset$, according as $t \ge s$ or $t < s$, respectively.

(iii) $\{T_1 \wedge T_2 \le t\} = \{T_1 \le t\} \cup \{T_2 \le t\} \in \mathcal{F}_t$ and $\{T_1 \vee T_2 \le t\} = \{T_1 \le t\} \cap \{T_2 \le t\} \in \mathcal{F}_t$.

(iv) $\{\sup_n T_n \le t\} = \bigcap_n \{T_n \le t\} \in \mathcal{F}_t$.

(v) $\{T + s \le t\} = \{T \le t - s\} \in \mathcal{F}_{(t-s) \vee 0} \subset \mathcal{F}_t$.

(vi) If $H \in \mathcal{F}_{T_1}$, then $H \cap \{T_2 \le t\} = (H \cap \{T_1 \le t\}) \cap \{T_2 \le t\} \in \mathcal{F}_t$.

(vii) In general, if S is a stopping time, then $\{S < t\} = \bigcup_k \{S \le t - 1/k\} \in \mathcal{F}_t$. Thus $\{T < t\} = \bigcup_n \{T_n < t\} \in \mathcal{F}_t$. Hence $\{T \le t\} = \bigcap_k \{T < t + 1/k\} \in \mathcal{F}_{t+} = \mathcal{F}_t$ by Proposition 40.3, and T is a stopping time. It follows from (vi) that $\mathcal{F}_T \subset \bigcap_n \mathcal{F}_{T_n}$. Conversely, if $H \in \bigcap_n \mathcal{F}_{T_n}$, then

$H \cap \{T < t\} = \bigcup_n (H \cap \{T_n < t\}) \in \mathcal{F}_t$ and hence $H \cap \{T \le t\} \in \mathcal{F}_{t+} = \mathcal{F}_t$, that is, $H \in \mathcal{F}_T$.

(viii) $\{T_1 < T_2\} \cap \{T_2 \le t\} = \bigcup_{r \in \mathbb{Q} \cap (0,t)} \{T_1 < r < T_2 \le t\}$, which is in $\mathcal{F}_t$. $\{T_1 < T_2\} \cap \{T_1 \le t\}$ belongs to $\mathcal{F}_t$, since

$$\begin{aligned}\{T_1 < T_2\} \cap \{T_1 \le t\} &= \{T_1 < T_2,\, T_1 = t\} \cup \{T_1 < T_2,\, T_1 < t\} \\ &= \{t < T_2,\, T_1 = t\} \cup \bigcup_{r \in \mathbb{Q} \cap (0,t)} \{T_1 < r < T_2\}.\end{aligned}$$

Hence $\{T_1 < T_2\}$ belongs to $\mathcal{F}_{T_1} \cap \mathcal{F}_{T_2}$. So does its complement $\{T_1 \ge T_2\}$. Exchanging the roles of T_1 and T_2, we see $\{T_1 \le T_2\} \in \mathcal{F}_{T_1} \cap \mathcal{F}_{T_2}$. □

PROPOSITION 40.9. *Let T be a stopping time. Then T is $\mathcal{F}_T$-measurable and $X_T = X(T(\omega), \omega)$ defined on $\{T < \infty\}$ is $\mathcal{F}'_T$-measurable. Moreover, for any $B \in \mathcal{B}^*(\mathbb{R}^d)$, $\{\omega \colon X_T \in B\}$ is in $\mathcal{F}'_T$.*

Proof. Since $\{T \le s\} \cap \{T \le t\} \in \mathcal{F}_{s \wedge t}$, we have $\{T \le s\} \in \mathcal{F}_T$, that is, T is $\mathcal{F}_T$-measurable. Let us prove the $\mathcal{F}'_T$-measurability of $X(T(\omega), \omega)$. Consider first the case where T takes values only in a countable set $D = \{t_1, t_2, \ldots, \infty\}$. Then, for any $B \in \mathcal{B}(\mathbb{R}^d)$,

$$\{T < \infty,\, X_T \in B\} \cap \{T \le t\} = \bigcup_{t_n \le t} \{T = t_n,\, X_{t_n} \in B\} \in \mathcal{F}_t,$$

which implies that X_T defined on $\{T < \infty\}$ is $\mathcal{F}'_T$-measurable.

For a general T, define T_n by $T_n = (k+1)/2^n$ for $k/2^n \le T < (k+1)/2^n$ and $T_n = \infty$ for $T = \infty$. Then T_n, $n = 1, 2, \ldots$, are stopping times and $T_n \downarrow T$ as $n \to \infty$. As shown above, X_{T_n} defined on $\{T < \infty\} = \{T_n < \infty\}$ is $\mathcal{F}'_{T_n}$-measurable. By (vi) of the preceding proposition, X_{T_n} is $\mathcal{F}'_{T_m}$-measurable if $m < n$. Since $X_{T_n} \to X_T$ on $\{T < \infty\}$ by the right-continuity, X_T is $\mathcal{F}'_{T_m}$-measurable for each m. Hence, by (vii), X_T is $\mathcal{F}'_T$-measurable.

In general, $H \in \mathcal{F}_t$ if, for any ρ, there are H_1 and H_2 in $\mathcal{F}_t$ satisfying $H_1 \subset H \subset H_2$ and $P^\rho[H_2 \setminus H_1] = 0$. To prove this, first we see that $H \in \mathcal{F}$. In fact, since $H_1, H_2 \in \mathcal{F}$, there are $G_1, G_2 \in \mathcal{F}^0$ such that $G_1 \subset H_1$, $G_2 \supset H_2$, $P^\rho[G_1] = P^\rho[H_1]$, and $P^\rho[G_2] = P^\rho[H_2]$. Hence $H \in \mathcal{F}$. Since $H_1 \in \mathcal{F}_t$, we can find $H'_1 \in \mathcal{F}^0_t$ such that $P^\rho[H_1 \setminus H'_1] = P^\rho[H'_1 \setminus H_1] = 0$. Then $P^\rho[H \setminus H'_1] = P^\rho[H'_1 \setminus H] = 0$, because $H \setminus H'_1 \subset (H \setminus H_1) \cup (H_1 \setminus H'_1) \subset (H_2 \setminus H_1) \cup (H_1 \setminus H'_1)$ and $H'_1 \setminus H \subset H'_1 \setminus H_1$. This shows that $H \in \mathcal{F}_t$.

Now let us prove the last sentence in the proposition. Let $B \in \mathcal{B}^*(\mathbb{R}^d)$ and let $H = \{X_T \in B,\, T \le t\}$. We claim that $H \in \mathcal{F}_t$. Given a probability measure ρ on $\mathbb{R}^d$, define a finite measure η by $\eta(C) = P^\rho[X_T \in C,\, T \le t]$ for $C \in \mathcal{B}_{\mathbb{R}^d}$. Choose $B_1, B_2 \in \mathcal{B}_{\mathbb{R}^d}$ such that $B_1 \subset B \subset B_2$ and $\eta(B_2 \setminus B_1) = 0$. Let $H_j = \{X_T \in B_j,\, T \le t\}$ for $j = 1, 2$. Then $H_1, H_2 \in \mathcal{F}_t$, $H_1 \subset H \subset H_2$, and $P^\rho[H_2 \setminus H_1] = \eta(B_2 \setminus B_1) = 0$. Hence $H \in \mathcal{F}_t$. □

The following theorem is an extension of the stationary independent increments property to the increments after a stopping time.

THEOREM 40.10. *Let T be a stopping time and ρ be a probability measure on $\mathbb{R}^d$. Suppose that $P^\rho[T<\infty]>0$. Define $\Omega'=\{T<\infty\}$, $\mathcal{F}'=\{H\in\mathcal{F}\colon H\subset\Omega'\}$, and $P'^\rho[H]=P^\rho[H]/P^\rho[\Omega']$ for $H\in\mathcal{F}'$. Then, $\{X_{T+t}-X_T, t\geq 0\}$ and $\mathcal{F}'_T$ are independent under P'^ρ, and the process $\{X_{T+t}-X_T, t\geq 0\}$ under P'^ρ is a Lévy process identical in law with $(\{X_t\},P^0)$.*

Proof. Write $X'_t=X_{T+t}-X_T$. The assertion of the theorem is equivalent to saying that, for any bounded continuous function $f(x_1,\ldots,x_k)$ and $H\in\mathcal{F}'_T$,

$$E'^\rho[f(X'_{s_1},\ldots,X'_{s_k})1_H]=E^0[f(X_{s_1},\ldots,X_{s_k})]\,P'^\rho[H], \tag{40.4}$$

where E'^ρ is the expectation with respect to P'^ρ. In the first step let us prove (40.4) under the assumption that the value of T is restricted to a countable set $D=\{t_1,t_2,\ldots,t_\infty\}$, where $t_\infty=\infty$ and $t_n<\infty$ for $n<\infty$. We have

$$\begin{aligned}
&E'^\rho[f(X'_{s_1},\ldots,X'_{s_k})1_H]\\
&\quad=\sum_{n<\infty}E^\rho[f(X_{t_n+s_1}-X_{t_n},\ldots,X_{t_n+s_k}-X_{t_n})1_{H\cap\{T=t_n\}}]/P^\rho[\Omega']\\
&\quad=\sum_{n<\infty}E^\rho[f(X_{t_n+s_1}-X_{t_n},\ldots,X_{t_n+s_k}-X_{t_n})]\,P^\rho[H\cap\{T=t_n\}]/P^\rho[\Omega']\\
&\quad=\sum_{n<\infty}E^0[f(X_{s_1},\ldots,X_{s_k})]\,P'^\rho[H\cap\{T=t_n\}]\\
&\quad=E^0[f(X_{s_1},\ldots,X_{s_k})]\,P'^\rho[H].
\end{aligned}$$

In the second equality above we have used Proposition 40.1. Thus we get (40.4) in this case.

In the second step, consider a general T. Define T_n as in the proof of Proposition 40.9. Let $H\in\mathcal{F}'_T$. Noting that $\mathcal{F}_T\subset\mathcal{F}_{T_n}$ and using the first step, we get

$$E'^\rho[f(X_{T_n+s_1}-X_{T_n},\ldots,X_{T_n+s_k}-X_{T_n})1_H]=E^0[f(X_{s_1},\ldots,X_{s_k})]\,P'^\rho[H].$$

Let $n\to\infty$ and use the right-continuity of $X_t(\omega)$. We obtain (40.4). $\square$

The following property is called the *strong Markov property*. When T is a constant time, it is the Markov property.

COROLLARY 40.11. *Let T be a stopping time and ρ be a probability measure. For any $H\in\mathcal{F}'_T$, $0\leq s_1<\cdots<s_n$, and bounded measurable $f(x_1,\ldots,x_n)$,*

$$E^\rho[1_H f(X_{T+s_1},\ldots,X_{T+s_n})]=E^\rho[1_H\,E^{X_T}[f(X_{s_1},\ldots,X_{s_n})]]. \tag{40.5}$$

Proof. Using an extension of Proposition 1.16, we have from the theorem

$$\begin{aligned}\text{left-hand side} &= E'^{\rho}[1_H f(X_T + X'_{s_1}, \dots, X_T + X'_{s_n})]\, P^{\rho}[T < \infty] \\ &= E'^{\rho}\big[1_H \big(E'^{\rho}[f(x + X'_{s_1}, \dots, x + X'_{s_n})]\big)_{x=X_T}\big]\, P^{\rho}[T < \infty] \\ &= E'^{\rho}\big[1_H \big(E^{0}[f(x + X_{s_1}, \dots, x + X_{s_n})]\big)_{x=X_T}\big]\, P^{\rho}[T < \infty] \\ &= E'^{\rho}[1_H E^{X_T}[f(X_{s_1}, \dots, X_{s_n})]]\, P^{\rho}[T < \infty].\end{aligned}$$

That is, (40.5) is true. □

There is another important property, called *quasi-left-continuity.*

THEOREM 40.12. *Let $\{T_n\}$ be a sequence of stopping times increasing to T as $n \to \infty$. Then, for any probability measure ρ, $X_{T_n} \to X_T$ P^{ρ}-almost surely on $\{T < \infty\}$.*

Proof. Assume that $\{T < \infty\} = \Omega$. Let $Y = \lim_{n\to\infty} X_{T_n}$ and $Y_t = \lim_{n\to\infty} X_{T_n + t}$ for $t > 0$. These are definable since X_t has left limits in t. We must prove that $Y = X_T$ P^{ρ}-a. s. As $Y_t(\omega)$ equals either $X_{(T+t)-}(\omega)$ or $X_{T+t}(\omega)$, the right-continuity of X_t gives $\lim_{t\downarrow 0} Y_t = X_T$. Let $f, g \in C_0(\mathbb{R}^d)$. By the strong Markov property

$$E^{\rho}[f(X_{T_n}) g(X_{T_n+t})] = E^{\rho}[f(X_{T_n}) g_t(X_{T_n})],$$

where $g_t(x) = E^x[g(X_t)]$. A part of Theorem 31.5 tells us that $g_t \in C_0(\mathbb{R}^d)$ and $\|g_t - g\| \to 0$ as $t \to 0$. Therefore, letting $n \to \infty$, we get $E^{\rho}[f(Y)g(Y_t)] = E^{\rho}[f(Y)g_t(Y)]$, and then, by letting $t \downarrow 0$, $E^{\rho}[f(Y)g(X_T)] = E^{\rho}[f(Y)g(Y)]$. It follows that, for any bounded measurable function $h(x, y)$,

$$E^{\rho}[h(Y, X_T)] = E^{\rho}[h(Y, Y)].$$

Choose the indicator function of $\{(x, y) \colon x = y\}$ as the function $h(x, y)$. Then we get $P^{\rho}[Y = X_T] = 1$.

For a general T without the assumption $\{T < \infty\} = \Omega$, consider $T_n \wedge t$ and $T \wedge t$. These are stopping times by Proposition 40.8 and $T_n \wedge t \uparrow T \wedge t$. Hence, $X(T_n \wedge t) \to X(T \wedge t)$ P^{ρ}-a. s. This means that $X_{T_n} \to X_T$ P^{ρ}-a. s. on $\{T \le t\}$. As t is arbitrary, this holds P^{ρ}-a. s. on $\{T < \infty\}$. □

Using the quasi-left-continuity, let us prove that hitting times of a larger class are stopping times. A set B in $\mathbb{R}^d$ is called an F_σ set if it is the union of a countable number of closed sets. An open set G is an F_σ set, since it is the union of $F_n = \{x \colon \operatorname{dis}(x, \mathbb{R}^d \setminus G) \ge 1/n\}$.

THEOREM 40.13. *Let B be an F_σ set. Then the hitting time T_B of B defined by (40.3) is a stopping time.*

Proof. In general, if S is a stopping time and if T is a mapping from Ω into $[0, \infty]$ such that $\{T \ne S\} \in \mathcal{F}$ and $P^{\rho}[T \ne S] = 0$ for any ρ, then T is a stopping time. In fact, for any t, $H = \{S \le t\} \setminus \{T \le t\}$ and

$G = \{T \le t\} \setminus \{S \le t\}$ are subsets of $\{T \ne S\}$, and hence H and G are in $(\mathcal{F}^0)^{P^\rho}$ and $P^\rho[H] = P^\rho[G] = 0$. Hence $\{S \le t\} \in \mathcal{F}_t$ implies $\{T \le t\} \in \mathcal{F}_t$.

Assume that B is a closed set. Let D_n be the set of points x having distance from B smaller than $1/n$. Then D_n is open and includes B. Let, for $\varepsilon > 0$, $T_B^\varepsilon = \inf\{t \ge \varepsilon \colon X_t \in B\}$ and $T_{D_n}^\varepsilon = \inf\{t \ge \varepsilon \colon X_t \in D_n\}$. Then $\{T_{D_n}^\varepsilon < t\}$ is empty for $t \le \varepsilon$ and equals $\{X_s \in D_n \text{ for some } s \in \mathbb{Q} \cap (\varepsilon, t)\}$ for $t > \varepsilon$. Thus $T_{D_n}^\varepsilon$ is a stopping time, since $\mathcal{F}_{t+} = \mathcal{F}_t$. They are increasing in n and bounded by T_B^ε. Let $T = \lim_{n\to\infty} T_{D_n}^\varepsilon$. Then T is a stopping time satisfying $\varepsilon \le T \le T_B^\varepsilon$. Let ρ be an arbitrary probability measure. By the quasi-left-continuity $X(T_{D_n}^\varepsilon) \to X_T$ P^ρ-a.s. on $\{T < \infty\}$. Hence $X_T \in B$ P^ρ-a.s. on $\{T < \infty\}$, because $X(T_{D_n}^\varepsilon) \in \overline{D_n}$. We have $\{T \ne T_B^\varepsilon\} = \{\varepsilon \le T < T_B^\varepsilon\} \subset \{T < \infty \text{ and } X_T \notin B\}$, which is in $\mathcal{F}$ and has P^ρ-measure 0. Therefore T_B^ε is a stopping time by the remark at the beginning of the proof. Let $\varepsilon_n \downarrow 0$. Then $T_B^{\varepsilon_n}$ decreases to T_B by the definition (40.3). Hence T_B is a stopping time by Proposition 40.8(vii), in the case where B is closed.

Now let B be a general F_σ set. Then $T_B = \inf T_{B_n}$, where B_n are closed sets with $B = \bigcup_{n=1}^\infty B_n$. Hence T_B is a stopping time, again by Proposition 40.8(vii). □

EXAMPLE 40.14. Suppose that $d = 1$ and $(\{X_t\}, P^0)$ is the Brownian motion. Let us write $T_a = T_{\{a\}}$ for $a \in \mathbb{R}$. For any $a > 0$, $b > 0$, and $t > 0$ we have

$$P^0[T_a < t \text{ and } X_t < a - b] = P^0[T_a < t \text{ and } X_t > a + b] = P^0[X_t > a + b]. \tag{40.6}$$

This is called the *reflection principle* for the Brownian motion. The hitting time T_a of $a > 0$ has, under P^0, the one-sided strictly $\frac{1}{2}$-stable distribution with parameter c in Example 2.13 written as a. Further, $(\{T_{s+}, s \ge 0\}, P^0)$ is a strictly $\frac{1}{2}$-stable subordinator. Note that $T_{s+}(\omega) = T_{(s,\infty)}(\omega)$ if $X_t(\omega)$ is continuous in t and $X_0(\omega) = 0$.

The proof is as follows. Let $\Omega' = \{X_t \text{ is continuous in } t\}$. Notice that $\Omega' \in \mathcal{F}$ and $P^\rho[\Omega'] = 1$ for any ρ. Since $\{X_t > a + b,\ X_0 = 0\} \cap \Omega' \subset \{T_a < t\}$, the second equality in (40.6) follows. Since T_a is a stopping time by the foregoing theorem, letting $X'_s = X_{T_a+s} - X_{T_a} = X_{T_a+s} - a$, we obtain

$$\begin{aligned} P^0[T_a < t,\ X_t < a - b] &= P^0[T_a < t,\ (X'_s)_{s=t-T_a} < -b] \\ &= E^0[(P^0[X_s < -b])_{s=t-T_a};\ T_a < t] \end{aligned}$$

by Theorem 40.10 and by an extension of Proposition 1.16. Similarly,

$$P^0[T_a < t,\ X_t > a + b] = E^0[(P^0[X_s > b])_{s=t-T_a};\ T_a < t].$$

Since $P^0[X_s < -b] = P^0[X_s > b]$ by symmetry, the first equality in (40.6) holds.

Letting $b \downarrow 0$ in (40.6), we get

$$P^0[T_a < t,\, X_t < a] = P^0[T_a < t,\, X_t > a] = P^0[X_t > a].$$

Since $P^0[X_t = a] = 0$, it follows that

$$P^0[T_a < t] = 2P^0[X_t > a].$$

Hence

$$P^0[T_a < t] = \sqrt{\frac{2}{\pi t}} \int_a^\infty \mathrm{e}^{-x^2/(2t)} \mathrm{d}x = \frac{a}{\sqrt{2\pi}} \int_0^t \mathrm{e}^{-a^2/(2s)} s^{-3/2} \mathrm{d}s$$

by the substitution $x = a\sqrt{t/s}$. Hence T_a has the one-sided strictly $\frac{1}{2}$-stable distribution as asserted. In particular, $P^0[T_a < \infty] = 1$. Define $Y_s = T_{s+}$ for $s \geq 0$. Then Y_s is right-continuous in s. It is increasing in s on $\Omega' \cap \{X_0 = 0\}$. We have $Y_0 = 0$ P^0-a. s., because $T_{0+}(\omega) > 0$ implies $X_t(\omega) \leq 0$ for all sufficiently small t in contradiction to Theorem 5.6. Since T_{s+} is a stopping time, we have, for $0 \leq s_0 < s_1 < \cdots < s_n$ and $B_1, \ldots, B_n \in \mathcal{B}(\mathbb{R})$,

$$\begin{aligned} P^0[Y_{s_k} - Y_{s_{k-1}} \in B_k,\, k = 1, \ldots, n] &= P^0[T_{s_k+} - T_{s_{k-1}+} \in B_k,\, k = 1, \ldots, n] \\ &= E^0[P^0[T_{(s_n - s_{n-1})+} \in B_n];\, T_{s_k+} - T_{s_{k-1}+} \in B_k,\, k = 1, \ldots, n-1] \end{aligned}$$

by Theorem 40.10. Repeating this, we get

$$P^0[Y_{s_k} - Y_{s_{k-1}} \in B_k,\, k = 1, \ldots, n] = \prod_{k=1}^n P^0[Y_{s_k - s_{k-1}} \in B_k].$$

Hence $(\{Y_s\}, P^0)$ is a strictly $\frac{1}{2}$-stable subordinator. We have $Y_s = T_{s+} = T_{(s,\infty)}$ on $\Omega' \cap \{X_0 = 0\}$.

41. Potential operators

We continue to fix a Lévy process $(\{X_t\}, P)$. We introduce the notion of excessive functions and consider absolute continuity conditions for potential measures. We cannot avoid subtleties in measurability.

PROPOSITION 41.1. *Let f be a bounded universally measurable function and $q > 0$. Let λ and ρ be probability measures on $[0, \infty)$ and $\mathbb{R}^d$, respectively. Then:*

(i) *$f(X_t(\omega))$ is $(\mathcal{B}_{[0,\infty)} \times \mathcal{F}^0)^{\lambda \times P^\rho}$-measurable in (t, ω), $\mathcal{F}_t$-measurable in ω, and $\mathcal{B}^\lambda_{[0,\infty)}$-measurable in t.*

(ii) *$E^x[f(X_t)]$ is $(\mathcal{B}_{[0,\infty)} \times \mathcal{B}_{\mathbb{R}^d})^{\lambda \times \rho}$-measurable in (t, x), universally measurable in x, and $\mathcal{B}^\lambda_{[0,\infty)}$-measurable in t.*

(iii) *$\int_0^\infty \mathrm{e}^{-qt} E^x[f(X_t)] \mathrm{d}t$ is universally measurable in x.*

(iv) $\int_0^\infty e^{-qt} f(X_t(\omega))dt$ *is* $\mathcal{F}$*-measurable in* ω *and*

$$\int_0^\infty e^{-qt} E^x[f(X_t)]dt = E^x\left[\int_0^\infty e^{-qt} f(X_t)dt\right].$$

Proof. Let us denote a mapping of x to y by $x \mapsto y$. If a mapping from a measurable space $(\mathcal{X}, \mathcal{F}_{\mathcal{X}})$ to a measurable space $(\mathcal{Y}, \mathcal{F}_{\mathcal{Y}})$ is measurable with respect to those σ-algebras, we say that it is $(\mathcal{F}_{\mathcal{X}}/\mathcal{F}_{\mathcal{Y}})$-measurable.

(i) The mapping $\varphi(t,\omega) = X_t(\omega)$ is $(\mathcal{B}_{[0,\infty)} \times \mathcal{F}^0/\mathcal{B}_{\mathbb{R}^d})$-measurable, since $X_t(\omega)$ is right-continuous in t. Let $B \in \mathcal{B}_{\mathbb{R}^d}{}^*$. Define η by $\eta(C) = (\lambda \times P^\rho)(\varphi^{-1}(C))$ for $C \in \mathcal{B}_{\mathbb{R}^d}$. We can find $B_1, B_2 \in \mathcal{B}_{\mathbb{R}^d}$ such that $B_1 \subset B \subset B_2$ and $\eta(B_2 \setminus B_1) = 0$. Hence $\varphi^{-1}(B_1) \subset \varphi^{-1}(B) \subset \varphi^{-1}(B_2)$ and $(\lambda \times P^\rho)(\varphi^{-1}(B_2) \setminus \varphi^{-1}(B_1)) = 0$. Hence $\varphi^{-1}(B) \in (\mathcal{B}_{[0,\infty)} \times \mathcal{F}^0)^{\lambda \times P^\rho}$. Hence φ is $((\mathcal{B}_{[0,\infty)} \times \mathcal{F}^0)^{\lambda\times P^\rho}/\mathcal{B}_{\mathbb{R}^d}{}^*)$-measurable. Since f is $(\mathcal{B}_{\mathbb{R}^d}{}^*/\mathcal{B}_{\mathbb{R}})$-measurable, it follows that $(t,\omega) \mapsto f(X_t(\omega))$ is $((\mathcal{B}_{[0,\infty)} \times \mathcal{F}^0)^{\lambda\times P^\rho}/\mathcal{B}_{\mathbb{R}})$-measurable. Similarly, $(\mathcal{B}_{[0,\infty)}/\mathcal{B}_{\mathbb{R}^d})$-measurability of the mapping $t \mapsto X_t(\omega)$ for fixed ω yields its $(\mathcal{B}^\lambda_{[0,\infty)}/\mathcal{B}_{\mathbb{R}^d}{}^*)$-measurability; $(\mathcal{F}^0_t/\mathcal{B}_{\mathbb{R}^d})$-measurability of the mapping $\omega \mapsto X_t(\omega)$ for fixed t yields its $(\mathcal{F}_t/\mathcal{B}_{\mathbb{R}^d}{}^*)$-measurability.

(ii) Using the right-continuity of $X_t(\omega)$ in t, we can prove that $(t,x) \mapsto E^x[1_B(X_t)]$ is $(\mathcal{B}_{[0,\infty)} \times \mathcal{B}_{\mathbb{R}^d}/\mathcal{B}_{\mathbb{R}})$-measurable if $B \in \mathcal{B}_{\mathbb{R}^d}$. Define a probability measure η by $\eta(B) = \iint E^x[1_B(X_t)]\lambda(dt)\rho(dx)$. Since f is $\mathcal{B}_{\mathbb{R}^d}{}^*$-measurable, there are $\mathcal{B}_{\mathbb{R}^d}$-measurable functions f_1, f_2 such that $f_1 \le f \le f_2$ and $\int (f_2(x) - f_1(x))\eta(dx) = 0$. Thus $E^x[f_1(X_t)] \le E^x[f(X_t)] \le E^x[f_2(X_t)]$ and $\iint (E^x[f_2(X_t)] - E^x[f_1(X_t)])\lambda(dt)\rho(dx) = \int (f_2(x) - f_1(x))\eta(dx) = 0$. Hence $(t,x) \mapsto E^x[f(X_t)]$ is $((\mathcal{B}_{[0,\infty)} \times \mathcal{B}_{\mathbb{R}^d})^{\lambda\times\rho}/\mathcal{B}_{\mathbb{R}})$-measurable. The $(\mathcal{B}^\lambda_{[0,\infty)}/\mathcal{B}_{\mathbb{R}})$-measurability of $t \mapsto E^x[f(X_t)]$ is shown similarly. The $(\mathcal{B}_{\mathbb{R}^d}{}^*/\mathcal{B}_{\mathbb{R}})$-measurability of $x \mapsto E^x[f(X_t)]$ is a consequence of (i) and Proposition 40.2.

(iii) For $B \in \mathcal{B}_{\mathbb{R}^d}$, $x \mapsto \int_0^\infty e^{-qt} E^x[1_B(X_t)]dt$ is $(\mathcal{B}_{\mathbb{R}^d}/\mathcal{B}_{\mathbb{R}})$-measurable. Given a probability measure σ on $\mathbb{R}^d$, define a finite measure η by $\eta(B) = \int \sigma(dx) \times \int_0^\infty e^{-qt} E^x[1_B(X_t)]dt$. Then, by the same argument as above, we see that $x \mapsto \int_0^\infty e^{-qt} E^x[f(X_t)]dt$ is $(\mathcal{B}^\sigma_{\mathbb{R}^d}/\mathcal{B}_{\mathbb{R}})$-measurable. Hence it is $(\mathcal{B}_{\mathbb{R}^d}{}^*/\mathcal{B}_{\mathbb{R}})$-measurable. Note that the existence of $\int_0^\infty e^{-qt} E^x[f(X_t)]dt$ is a consequence of (ii) with λ taken to be $qe^{-qt}dt$.

(iv) For $B \in \mathcal{B}_{\mathbb{R}^d}$, $\omega \mapsto \int_0^\infty e^{-qt} 1_B(X_t(\omega))dt$ is $(\mathcal{F}^0/\mathcal{B}_{\mathbb{R}})$-measurable. Given σ on $\mathbb{R}^d$, define η by $\eta(B) = E^\sigma\left[\int_0^\infty e^{-qt} 1_B(X_t(\omega))dt\right]$. The same argument shows that $\omega \mapsto \int_0^\infty e^{-qt} f(X_t)dt$ is $((\mathcal{F}^0)^{P^\sigma}/\mathcal{B}_{\mathbb{R}})$-measurable. Hence it is $(\mathcal{F}/\mathcal{B}_{\mathbb{R}})$-measurable. The equality in the last assertion is obtained from Fubini's theorem. □

DEFINITION 41.2. For $t \ge 0$ and $q \ge 0$, the *transition kernel* $P_t(x,B)$ and the q*-potential kernel* $U^q(x,B)$ are defined by

$$P_t(x,B) = P^x[X_t \in B], \qquad U^q(x,B) = \int_0^\infty e^{-qt} P_t(x,B)dt,$$

where $x \in \mathbb{R}^d$, $B \in \mathcal{B}_{\mathbb{R}^d}{}^*$. They are measures with respect to B and universally measurable functions with respect to x. The *transition operator* P_t, the *q-potential operator* U^q, and the *q-balayage operator* P_B^q are given by

$$P_t f(x) = \int_{\mathbb{R}^d} P_t(x, \mathrm{d}y) f(y), \qquad U^q f(x) = \int_{\mathbb{R}^d} U^q(x, \mathrm{d}y) f(y),$$
$$P_B^q f(x) = E^x[\mathrm{e}^{-qT_B} f(X_{T_B})]$$

for f $\mathcal{B}_{\mathbb{R}^d}{}^*$-measurable, whenever the integrals are defined. Here B is an F_σ set and T_B is the hitting time of B. We understand $\mathrm{e}^{-qT_B} = 0$ for $q \geq 0$ whenever $T_B = \infty$. Write $U = U^0$ and $P_B = P_B^0$. Sometimes we write $P_t^q = \mathrm{e}^{-qt} P_t$.

In the notation of the previous chapters,

$$P_t(x, B) = P_t(0, B - x) = \mu^t(B - x),$$
$$U^q(x, B) = U^q(0, B - x) = V^q(B - x),$$

where μ is the distribution of X_1 under P^0 and V^q is the q-potential measure of Definition 30.9. For f nonnegative and universally measurable, we have

$$P_t f(x) = E^x[f(X_t)] = \int_{\mathbb{R}^d} P_t(0, \mathrm{d}y) f(x + y) = \int_{\mathbb{R}^d} \mu^t(\mathrm{d}y) f(x + y),$$
$$U^q f(x) = \int_0^\infty \mathrm{e}^{-qt} E^x[f(X_t)]\mathrm{d}t = E^x\left[\int_0^\infty \mathrm{e}^{-qt} f(X_t)\mathrm{d}t\right]$$
$$= \int_{\mathbb{R}^d} U^q(0, \mathrm{d}y) f(x + y) = \int_{\mathbb{R}^d} V^q(\mathrm{d}y) f(x + y).$$

We have used Proposition 41.1.

PROPOSITION 41.3. *Let f be a nonnegative universally measurable function and T be a stopping time. Then*

$$(41.1) \quad P_t P_s f = P_{t+s} f, \qquad t \geq 0,\ s \geq 0$$

$$(41.2) \quad U^q f = U^r f + (r - q) U^r U^q f = U^r f + (r - q) U^q U^r f, \quad 0 \leq q < r,$$

$$(41.3) \quad U^q f(x) = E^x\left[\int_0^T \mathrm{e}^{-qt} f(X_t)\mathrm{d}t\right] + E^x[\mathrm{e}^{-qT} U^q f(X_T)], \qquad q \geq 0.$$

Here (41.1) is the semigroup property; (41.2) is called the *resolvent equation.* (41.3) is from the strong Markov property.

Proof of proposition. The first relation (41.1) is familiar to us. We have $U^r(U^q f) = (U^r U^q) f$, where $(U^r U^q)(x, B) = \int U^r(x, \mathrm{d}y) U^q(y, B)$. We write $(U^r U^q)(x, B)$ as $U^r U^q(x, B)$. Using (41.1), we get

$$U^r U^q(x, B) = \int_0^\infty \mathrm{e}^{-rt}\mathrm{d}t \int_{\mathbb{R}^d} P_t(x, \mathrm{d}y) \int_0^\infty \mathrm{e}^{-qs}\mathrm{d}s P_s(y, B)$$

$$= \int_0^\infty e^{-rt}dt \int_0^\infty e^{-qs}ds P_{t+s}(x,B) = \int_0^\infty e^{-(r-q)t}dt \int_t^\infty e^{-qs} P_s(x,B)ds$$
$$= \int_0^\infty e^{-qs} P_s(x,B)ds \int_0^s e^{-(r-q)t}dt = \int_0^\infty e^{-qs} \frac{1-e^{-(r-q)s}}{r-q} P_s(x,B)ds$$
$$= \frac{1}{r-q}(U^q(x,B) - U^r(x,B))$$

and

$$U^r U^q(x,B) = \int_0^\infty e^{-rt}dt \int_0^\infty e^{-qs}ds P_{t+s}(x,B) = U^q U^r(x,B).$$

Thus (41.2) follows.

Similarly to Proposition 41.1(iv), we see that, for each t, $\int_0^t e^{-qs} f(X_s)ds$ is $\mathcal{F}$-measurable in ω. It is continuous in t. Hence it is $(\mathcal{B}_{[0,\infty)} \times \mathcal{F})$-measurable in (t,ω). Hence $\int_0^{T(\omega)} e^{-qs} f(X_s(\omega))ds$ is $\mathcal{F}$-measurable in ω. It follows that

$$U^q f(x) = E^x\left[\int_0^T e^{-qt} f(X_t)dt\right] + E^x\left[\int_T^\infty e^{-qt} f(X_t)dt\right].$$

The second term in the right-hand side equals

$$E^x\left[e^{-qT}\int_0^\infty e^{-qt} f(X_{T+t})dt\right] = E^x\left[e^{-qT} E^{X_T}\left[\int_0^\infty e^{-qt} f(X_t)dt\right]\right]$$
$$= E[e^{-qT} U^q f(X_T)]$$

by Corollary 40.11. □

DEFINITION 41.4. Let $0 \le q < \infty$. A function $f(x)$ on $\mathbb{R}^d$ taking values in $[0,\infty]$ is *q-excessive* if it is universally measurable, $e^{-qt}P_t f \le f$ for all $t > 0$ and $e^{-qt}P_t f(x) \to f(x)$ for all x as $t \downarrow 0$.

Note that, if f is q-excessive, then $e^{-qt}P_t f(x)$ increases as t decreases, since $P_t^q f = P_s^q(P_{t-s}^q f) \le P_s^q f$ for $0 < s < t$. The following are basic properties of q-excessive functions. Further properties are proved in Blumenthal and Getoor [**45**] and Chung [**81**].

PROPOSITION 41.5. *Let $0 \le q < \infty$.*

(i) *If $\{f_n,\ n = 1,2,\dots\}$ is an increasing sequence of q-excessive functions, then the limit function $f(x)$ is q-excessive.*

(ii) *If f is universally measurable, taking values in $[0,\infty]$, then $U^q f$ is q-excessive.*

(iii) *A function f is q-excessive if and only if f is r-excessive for every $r > q$.*

(iv) *Let f be universally measurable, taking values in $[0,\infty]$. Then, f is q-excessive if and only if $rU^{q+r} f \le f$ for $r > 0$ and $rU^{q+r} f(x) \to f(x)$ for every x as $r \to \infty$.*

(v) *If $q > 0$ and f is q-excessive, then there is a sequence of bounded nonnegative universally measurable functions $\{g_n\colon n = 1, 2, \dots\}$ such that $U^q g_n(x)$ increases to $f(x)$ for every x as $n \uparrow \infty$.*

(vi) *If f is q-excessive, then $E^x[\mathrm{e}^{-qT} f(X_T)] \le f(x)$ for every stopping time T.*

(vii) *If f is q-excessive, then $P^q_B f$ is q-excessive for every F_σ set B.*

Proof. (i) We have $P^q_t f \le f$, passing from the same relation for f_n. Hence, $f \ge \lim_{t\downarrow 0} P^q_t f \ge \lim_{t\downarrow 0} P^q_t f_n = f_n$. Letting $n \to \infty$, we get $f = \lim_{t\downarrow 0} P^q_t f$.

(ii) Notice that

$$P^q_t U^q f = P^q_t \int_0^\infty P^q_s f \mathrm{d}s = \int_0^\infty P^q_{t+s} f \mathrm{d}s = \int_t^\infty P^q_s f \mathrm{d}s,$$

which increases to $U^q f$ as $t \downarrow 0$.

(iii) If f is q-excessive, then, for $r > q$, $\mathrm{e}^{-rt} P_t f = \mathrm{e}^{-(r-q)t} \mathrm{e}^{-qt} P_t f$ increases to f as $t \downarrow 0$. If f is r-excessive for $r > q$, then $\mathrm{e}^{-qt} P_t f = \lim_{r\downarrow q} \mathrm{e}^{-rt} P_t f \le f$ and $\mathrm{e}^{-qt} P_t f = \mathrm{e}^{-(q-r)t} \mathrm{e}^{-rt} P_t f \to f$ as $t \downarrow 0$.

(iv) If f is q-excessive, then

$$rU^{q+r} f = r \int_0^\infty \mathrm{e}^{-(q+r)t} P_t f \mathrm{d}t = \int_0^\infty \mathrm{e}^{-t-qt/r} P_{t/r} f \mathrm{d}t \ \uparrow \ \int_0^\infty \mathrm{e}^{-t} f \mathrm{d}t = f.$$

Conversely, suppose that $rU^{q+r} f \le f$ and $\lim_{r\to\infty} rU^{q+r} f = f$. Assume $q > 0$. Let $f_n = f \wedge n$. Then $rU^{q+r} f_n \le (rU^{q+r} f) \wedge (nrU^{q+r} 1) \le f_n$. Hence, using (41.2) and $q > 0$, we see

$$U^{q+r} f_n = U^q (f_n - rU^{q+r} f_n),$$

which is q-excessive by (ii). If $r < r'$, then, by (41.2), $rU^{q+r} f_n = rU^{q+r'} f_n + (r' - r) U^{q+r'} (rU^{q+r} f_n) \le rU^{q+r'} f_n + (r' - r) U^{q+r'} f_n = r' U^{q+r'} f_n$. Thus $rU^{q+r} f_n$ increases to some h_n as $r \to \infty$. As $n \to \infty$, h_n increases to some h. By virtue of (i), h is q-excessive. We have $h_n \le f_n$ and $h \le f$ on one hand, and $rU^{q+r} f_n \le h_n$ and $rU^{q+r} f \le h$ on the other. Now, using $\lim_{r\to\infty} rU^{q+r} f = f$, we get $f = h$, and q-excessiveness of f is shown. In the case $q = 0$, notice that $rU^{q'+r} f \le f$ and $rU^{q'+r} f = \frac{r}{q'+r} (q' + r) U^{q'+r} f \to f$ as $r \to \infty$ for $q' > 0$, and see that f is 0-excessive by (iii).

(v) By (iv), $rU^{q+r} f \uparrow f$ as $r \uparrow \infty$. Hence we can use f_n and h_n in the proof of (iv). Let $g_n = n(f_n - nU^{q+n} f_n)$. Then g_n is nonnegative, bounded, and $U^q g_n = nU^{q+n} f_n \le (n+1) U^{q+n+1} f_n \le (n+1) U^{q+n+1} f_{n+1} = U^q g_{n+1}$. Furthermore $h_k = \lim_{n\to\infty} nU^{q+n} f_k \le \lim_{n\to\infty} nU^{q+n} f_n \le \lim_{n\to\infty} f_n = f$. Since $h_k \uparrow f$ as $k \uparrow \infty$, $U^q g_n = nU^{q+n} f_n \uparrow f$ as $n \uparrow \infty$.

(vi) Let $q > 0$. Use g_n in (v). By (41.3), $E^x[\mathrm{e}^{-qT} U^q g_n(X_T)] \le U^q g_n(x)$. Letting $n \uparrow \infty$, we get $E^x[\mathrm{e}^{-qT} f(X_T)] \le f(x)$. If $q = 0$, use (iii).

(vii) Assume $q > 0$. Using g_n in (v), we have $P_B^q U^q g_n \uparrow P_B^q f$ as $n \uparrow \infty$. Looking at the proof of (41.3), we get

$$P_t^q P_B^q U^q g_n(x)$$
$$= E^x\left[\mathrm{e}^{-qt} E^{X_T}\left[\int_{T_B}^{\infty} \mathrm{e}^{-qs} g_n(X_s)\mathrm{d}s\right]\right] = E^x\left[\int_{T_{B,t}}^{\infty} \mathrm{e}^{-qs} g_n(X_s)\mathrm{d}s\right],$$

where $T_{B,t} = \inf\{s > t \colon X_s \in B\}$. As $t \downarrow 0$, the last member increases to $E^x\left[\int_{T_B}^{\infty} \mathrm{e}^{-qs} g_n(X_s)\mathrm{d}s\right]$, which equals $P_B^q U^q g_n(x)$. This shows that $P_B^q U^q g_n$ is q-excessive. Hence $P_B^q f$ is q-excessive by (i).

Next consider the case $q = 0$. Since f is q'-excessive for $q' > 0$ by (iii), $P_B^{q'} f$ is q'-excessive. Thus $P_t^{q'} P_B^{q'} f \le P_B^{q'} f$, and hence $P_t P_B f \le P_B f$ follows. Further we get $\lim_{t\downarrow 0} P_t P_B f = P_B f$, by letting $q' \downarrow 0$ in $P_B^{q'} f = \lim_{t\downarrow 0} P_t^{q'} P_B^{q'} f \le \lim_{t\downarrow 0} P_t P_B f \le P_B f$. □

Let us introduce the dual process. We have started with a fixed Lévy process $(\{X_t\}, P)$ and defined the probability measures P^x, P^ρ, the σ-algebras $\mathcal{F}$, $\mathcal{F}_t$, and the operators P_t, U^q with respect to this process. Here $X_t(\omega) = \omega(t)$ for $\omega \in \Omega = D([0,\infty), \mathbb{R}^d)$ and P is a probability measure on $\mathcal{F}^0$. The probability measure $\widetilde{P}$ on $\mathcal{F}^0$ satisfying $\widetilde{P}[X_{t_k} \in B_k,\ k = 1,\dots,n] = P[-X_{t_k} \in B_k,\ k = 1,\dots,n]$ for any n, t_k, and B_k defines another Lévy process $(\{X_t\}, \widetilde{P})$. Using $\widetilde{P}$, we define $\widetilde{P}^x$, $\widetilde{P}^\rho$, $\widetilde{\mathcal{F}}$, $\widetilde{\mathcal{F}}_t$, $\widetilde{P}_t$, and $\widetilde{U}^q$ in parallel. Also $\widetilde{\mu}$ and $\widetilde{V}^q$ are defined in parallel to μ and V^q.

DEFINITION 41.6. The Lévy process $(\{X_t\}, \widetilde{P})$ is called the *dual process* of $(\{X_t\}, P)$. A function q-excessive with respect to the dual process is called *q-co-excessive.*

We have

$$\widetilde{\mu}^t(B) = \mu^t(-B), \qquad \widetilde{V}^q(B) = V^q(-B), \tag{41.4}$$

$$\widetilde{P}_t f(x) = \int \widetilde{P}_t(x,\mathrm{d}y) f(y) = \int \widetilde{\mu}^t(\mathrm{d}y) f(x+y) = \int \mu^t(\mathrm{d}y) f(x-y), \tag{41.5}$$

$$\widetilde{U}^q f(x) = \int \widetilde{U}^q(x,\mathrm{d}y) f(y) = \int \widetilde{V}^q(\mathrm{d}y) f(x+y) = \int V^q(\mathrm{d}y) f(x-y) \tag{41.6}$$

for $B \in \mathcal{B}_{\mathbb{R}^d}{}^*$ and f nonnegative and universally measurable.

PROPOSITION 41.7. *Let f and g be nonnegative and universally measurable. Then*

$$\int_{\mathbb{R}^d} P_t f(x) g(x)\mathrm{d}x = \int_{\mathbb{R}^d} f(x) \widetilde{P}_t g(x)\mathrm{d}x, \qquad t \ge 0, \tag{41.7}$$

$$\int_{\mathbb{R}^d} U^q f(x)g(x)\mathrm{d}x = \int_{\mathbb{R}^d} f(x)\widetilde{U}^q g(x)\mathrm{d}x, \qquad q \geq 0. \tag{41.8}$$

Proof. Use the expressions for P_t and $\widetilde{P}_t$ in terms of μ^t and $\widetilde{\mu}^t$. Then

$$\begin{aligned}\int P_t f(x)g(x)\mathrm{d}x &= \iint \mu^t(\mathrm{d}y)f(x+y)g(x)\mathrm{d}x \\ &= \int \mu^t(\mathrm{d}y)\int f(x)g(x-y)\mathrm{d}x = \int f(x)\widetilde{P}_t g(x)\mathrm{d}x,\end{aligned}$$

that is, (41.7). The second identity is proved in the same way. □

The pathwise meaning of the dual process is given by time reversal.

PROPOSITION 41.8. *Fix $t > 0$ and a probability measure ρ. Let*

$$Y_s(\omega) = X_{(t-s)-}(\omega) - X_{t-}(\omega) \quad \textit{for } 0 \leq s < t.$$

Then Y_s is right-continuous in $s \in [0,t)$ with left limits in $s \in (0,t]$ and

$$(\{Y_s,\, 0 \leq s < t\},\, P^\rho) \stackrel{\mathrm{d}}{=} (\{X_s,\, 0 \leq s < t\},\, \widetilde{P}^0).$$

Proof. If $s' > s$ and $s' \to s$, then $t - s' < t - s$, $t - s' \to t - s$, and $Y_{s'} \to Y_s$. If $s' < s$ and $s' \to s$, then $t - s' > t - s$, $t - s' \to t - s$, and $Y_{s'} \to X_{t-s} - X_{t-}$. Let $Z_s = X_{t-s} - X_t$ for $0 \leq s < t$. We have $P^\rho[Y_s = Z_s] = 1$ for each s by (1.10). Hence

$$(\{Y_s,\, 0 \leq s < t\},\, P^\rho) \stackrel{\mathrm{d}}{=} (\{Z_s,\, 0 \leq s < t\},\, P^\rho).$$

For $0 \leq s_1 < \cdots < s_n < t$,

$$(Z_{s_2} - Z_{s_1}, \ldots, Z_{s_n} - Z_{s_{n-1}}) = (-(X_{t-s_1} - X_{t-s_2}), \ldots, -(X_{t-s_{n-1}} - X_{t-s_n}))$$

P^ρ-a. s. For $0 \leq s_1 < s_2 < t$,

$$(Z_{s_2} - Z_{s_1},\, P^\rho) = (X_{t-s_2} - X_{t-s_1},\, P^\rho) \stackrel{\mathrm{d}}{=} (X_0 - X_{s_2-s_1},\, P^\rho) \stackrel{\mathrm{d}}{=} (X_{s_2-s_1},\, \widetilde{P}^0).$$

Hence $(\{Z_s,\, 0 \leq s < t\},\, P^\rho)$ has stationary independent increments, starts at 0, and is identical in law with $(\{X_s,\, 0 \leq s < t\},\, \widetilde{P}^0)$. □

PROPOSITION 41.9. *Let $q \geq 0$. A Borel set B has Lebesgue measure 0 if and only if $U^q(x, B) = 0$ for almost every x.*

Proof. If $\operatorname{Leb} B = 0$, then $U^q(x, B) = 0$ for a. e. x, since

$$\int U^q 1_B(x)g(x)\mathrm{d}x = \int_B \widetilde{U}^q g(x)\mathrm{d}x = 0$$

for any nonnegative Borel-measurable g by (41.8). To see the converse, choose $g = 1$. □

PROPOSITION 41.10. *Let $g(x) = f(-x)$. Then $f(x)$ is q-co-excessive if and only if $g(x)$ is q-excessive.*

Proof. By (41.5)

$$\widetilde{P}_t f(x) = \int \mu^t(\mathrm{d}y) f(x-y) = \int \mu^t(\mathrm{d}y) g(y-x) = P_t g(-x).$$

Hence $\mathrm{e}^{-qt}\widetilde{P}_t f \leq f$ if and only if $\mathrm{e}^{-qt}P_t g \leq g$. As $t \downarrow 0$, $\mathrm{e}^{-qt}\widetilde{P}_t f \to f$ if and only if $\mathrm{e}^{-qt}P_t g \to g$. □

Now we discuss absolute continuity conditions.

DEFINITION 41.11. *Condition* (ACP), or absolute continuity of potential measures, holds if, for every $q \geq 0$, V^q is absolutely continuous. *Condition* (ACT), or absolute continuity of transition measures, holds if, for every $t > 0$, μ^t is absolutely continuous.

REMARK 41.12. Condition (ACP) is equivalent to saying that V^q is absolutely continuous for some $q \geq 0$. In fact, let V^q be absolutely continuous and let B be a Borel set with $\mathrm{Leb}\, B = 0$. If $r > q$, then $V^r(B) \leq V^q(B) = 0$. If $0 \leq r < q$, then $U^q 1_B(x) = V^q(B - x) = 0$ and $U^r 1_B = U^q 1_B + (q-r)U^r U^q 1_B = 0$ by (41.2). The situation for Condition (ACT) is different; absolute continuity of μ^t for some $t > 0$ does not imply (ACT). See Theorem 27.23 and Remark 27.24.

(ACP) for the original process is equivalent to (ACP) for the dual process. Likewise, (ACT) for the original process is equivalent to (ACT) for the dual process. These are obvious from (41.4), since $\mathrm{Leb}\, B = 0$ is equivalent to $\mathrm{Leb}(-B) = 0$.

REMARK 41.13. (ACT) implies (ACP). This follows from the definition of V^q. However, (ACP) does not imply (ACT). A simple example is the trivial process $P^x[X_t = x + \gamma t] = 1$ for $d = 1$ with $\gamma \neq 0$. For this process $\mu^t = \delta_{\gamma t}$ and $V^0(a, b] = \int_0^\infty \mu^t(a, b]\mathrm{d}t = (b-a)/\gamma$ for $0 \leq a < b$ if $\gamma > 0$. See Exercise 44.2 for another example. Fukushima [**154, 155**] proves that (ACP) does imply (ACT), provided that $(\{X_t\}, P^0)$ is symmetric. His result is for general temporally homogeneous Markov processes with symmetry.

DEFINITION 41.14. An F_σ set B is *polar* if $P^x[T_B = \infty] = 1$ for every $x \in \mathbb{R}^d$, that is, if it cannot be hit from any starting point. It is *essentially polar* if $P^x[T_B = \infty] = 1$ for almost every $x \in \mathbb{R}^d$. Polar and essentially polar sets relative to the dual process are called *co-polar* and *essentially co-polar*, respectively.

A function $f(x)$ on $\mathbb{R}^d$ taking values in $[-\infty, \infty]$ is said to be *lower semi-continuous*, if $\liminf_{y \to x} f(y) \geq f(x)$ for every x. If $\{f_n\}$ is an increasing sequence of lower semi-continuous functions, then its limit f is lower semi-continuous. To see this, notice that $\liminf_{y\to x} f(y) \geq \liminf_{y \to x} f_n(y) \geq f_n(x)$ and let $n \to \infty$.

The support S_f of a function f on $\mathbb{R}^d$ is the closure of the set $\{x \in \mathbb{R}^d \colon f(x) \neq 0\}$.

THEOREM 41.15. *The following statements are equivalent.*

(1) *Condition* (ACP) *holds.*
(2) *If f is a bounded Borel-measurable function with compact support, then, for $q > 0$, $U^q f$ is continuous.*
(3) *If f is bounded and universally measurable, then, for $q > 0$, $U^q f$ is continuous.*
(4) *For every $q \geq 0$ any q-excessive function is lower semi-continuous.*
(5) *If f and g are q-excessive for some $q \geq 0$ and if $f \geq g$ almost everywhere, then $f \geq g$ everywhere.*
(6) *Any essentially polar F_σ set is polar.*

Proof. Let us show that (1) $\Rightarrow$ (3) $\Rightarrow$ (2) $\Rightarrow$ (1), (3) $\Rightarrow$ (4) $\Rightarrow$ (6) $\Rightarrow$ (1), and (1) $\Rightarrow$ (5) $\Rightarrow$ (1).

(1) $\Rightarrow$ (3). Let $q > 0$. We have $V^q(\mathrm{d}y) = v^q(y)\mathrm{d}y$ with some nonnegative Borel-measurable function v^q. Let g be a Borel-measurable function with $|g| \leq M$. For any $\varepsilon > 0$, choose a continuous function w^q with compact support such that $\int |v^q(y) - w^q(y)|\mathrm{d}y < \varepsilon$. Then

$$\begin{aligned}|U^q g(x) - U^q g(x')| &= \left|\int g(y)(v^q(y-x) - v^q(y-x'))\mathrm{d}y\right| \\ &\leq M \int (|v^q(y-x) - w^q(y-x)| + |w^q(y-x) - w^q(y-x')| \\ &\qquad + |w^q(y-x') - v^q(y-x')|)\mathrm{d}y \\ &\leq 3M\varepsilon\end{aligned}$$

if $|x - x'|$ is small enough. Hence $U^q g$ is continuous. Let $\{x_1, x_2, \ldots\}$ be dense in $\mathbb{R}^d$. Define, for $q > 0$, a probability measure ρ by $\rho(B) = \int qV^q(\mathrm{d}y) \sum_{n=1}^{\infty} 2^{-n} 1_B(x_n + y)$. Given a bounded universally measurable function f, we can find bounded Borel-measurable g_1 and g_2 such that $g_1 \leq f \leq g_2$ and $\int (g_2 - g_1)\rho(\mathrm{d}x) = 0$. Then $U^q g_1 \leq U^q f \leq U^q g_2$ and $U^q g_k$ is continuous for $k = 1, 2$ by (3). Since

$$\int g_k(x)\rho(\mathrm{d}x) = \int qV^q(\mathrm{d}y) \sum_n 2^{-n} g_k(x_n + y) = \sum_n 2^{-n} qU^q g_k(x_n)$$

for $k = 1, 2$, we see $U^q g_1(x_n) = U^q g_2(x_n)$ for all n. It follows from the denseness of $\{x_n\}$ and from the continuity that $U^q g_1 = U^q g_2$. Hence $U^q f = U^q g_1$ and $U^q f$ is continuous.

Proof that (3) $\Rightarrow$ (2) is trivial.

(2) $\Rightarrow$ (1). Let B be a Borel set with $\mathrm{Leb}\, B = 0$. Let $q > 0$. Let $B_n = B \cap \{x \colon |x| \leq n\}$. Then $U^q 1_B = U^q 1_{B_n} = 0$ a. e. by Proposition 41.9. Since $U^q 1_{B_n}$ is continuous by (2), $U^q 1_{B_n} = 0$ everywhere. Letting $n \to \infty$, we get $U^q 1_B = 0$.

(3) $\Rightarrow$ (4). Let f be q-excessive with $q > 0$. By Proposition 41.5(v) there are bounded nonnegative universally measurable functions g_n such

that $U^q g_n \uparrow f$ as $n \uparrow \infty$. By (3) each $U^q g_n$ is continuous. Hence f is lower semi-continuous. In the case where f is 0-excessive, it is q-excessive for $q > 0$ by Proposition 41.5(iii) and hence lower semi-continuous.

(4) $\Rightarrow$ (6). Let B be an essentially polar F_σ set. Let $f(x) = P^x[T_B < \infty] = P_B 1(x)$. Then f is 0-excessive by Proposition 41.5(vii). Since $f = 0$ a. e., the lower semi-continuity implies $f = 0$ everywhere. Hence, B is polar.

(6) $\Rightarrow$ (1). Let B be a Borel set with $\mathrm{Leb}\, B = 0$. Then $U(x, B) = 0$ for a. e. x by Proposition 41.9. Let $C = \{x \colon U(x, B) > 0\}$. Then C is a Borel set. We will prove that $C = \emptyset$, which shows (1). Suppose that $C \neq \emptyset$. Let $x_0 \in C$. There is $t > 0$ such that $E^{x_0}\left[\int_t^\infty 1_B(X_s)\mathrm{d}s\right] > 0$. Thus

$$0 < E^{x_0}\left[E^{X_t}\left[\int_0^\infty 1_B(X_s)\mathrm{d}s\right]\right] = E^{x_0}[U(X_t, B)].$$

Hence $P^{x_0}[X_t \in C] > 0$. Therefore we can find a compact set $K \subset C$ such that $P^{x_0}[X_t \in K] > 0$. Since $U(x, B)$ is 0-excessive by Proposition 41.5(ii), $U(x, B) \geq E^x[U(X_{T_K}, B);\, T_K < \infty]$ by Proposition 41.5(vi). Since $U(X_{T_K}, B) > 0$, we have $P^x[T_K < \infty] = 0$ for a. e. x. Thus K is essentially polar. Hence, by (6), K is polar, which contradicts the positivity of $P^{x_0}[X_t \in K]$. Hence $C = \emptyset$.

(1) $\Rightarrow$ (5). Suppose that f and g are q-excessive and $f \geq g$ a. e. It follows from (1) that $rU^{q+r} f \geq rU^{q+r} g$ on $\mathbb{R}^d$. Letting $r \to \infty$, we get $f \geq g$ by Proposition 41.5(iv).

(5) $\Rightarrow$ (1). Let B be a Borel set with $\mathrm{Leb}\, B = 0$. Then, as in the proof that (6) $\Rightarrow$ (1), $U(x, B)$ is 0-excessive and vanishes a. e. It follows from (5) that $0 \geq U(x, B)$ for every x. Hence $U(x, B) = 0$. □

THEOREM 41.16. *Suppose that* (ACP) *holds. Then, for any* $q > 0$, *there is a unique* q*-co-excessive function* u^q *such that*

$$U^q f(x) = \int_{\mathbb{R}^d} u^q(y - x) f(y)\mathrm{d}y \tag{41.9}$$

for any nonnegative universally measurable function f. *If the process is transient, then the same assertion is true also for* $q = 0$.

In the following, u^q always denotes the function in this theorem under Condition (ACP).

Proof of theorem. Let $v^q(x)$ be a density of V^q for $q > 0$. Then $v^q(-x)$ is a density of $\widetilde{V}^q$. We have $U^q f(x) = \int v^q(y - x) f(y)\mathrm{d}y$. By the resolvent equation, $U^q f(0) = U^{q+r} f(0) + rU^{q+r} U^q f(0)$ for $r > 0$. Hence

$$\begin{aligned} v^q(y) &= v^{q+r}(y) + r\int v^{q+r}(z) v^q(y - z)\mathrm{d}z \\ &= v^{q+r}(y) + r\widetilde{U}^{q+r} v^q(y) \qquad \text{for a. e. } y. \end{aligned} \tag{41.10}$$

It follows that

$$v^q(y) \geq r\widetilde{U}^{q+r}v^q(y) \qquad \text{for a. e. } y. \tag{41.11}$$

If $r < r'$, then $r\widetilde{U}^{q+r}v^q = r\widetilde{U}^{q+r'}v^q + (r'-r)\widetilde{U}^{q+r'}(r\widetilde{U}^{q+r}v^q) \leq r\widetilde{U}^{q+r'}v^q + (r'-r)\widetilde{U}^{q+r'}v^q = r'\widetilde{U}^{q+r'}v^q$ everywhere by (41.11). Thus we can define

$$u^q(x) = \lim_{r\to\infty} r\widetilde{U}^{q+r}v^q(x) \qquad \text{for every } x. \tag{41.12}$$

The sequence v^{q+n}, $n = 1, 2, \ldots$, is decreasing a. e. Using Fatou's lemma in $\int v^{q+n}(y)\mathrm{d}y = 1/(q+n)$, we see that $\lim_{n\to\infty} v^{q+n}(y) = 0$ a. e. Hence (41.10) shows that $v^q = u^q$ a. e. Hence we have (41.9). Let us see that u^q is q-co-excessive. It follows from (41.11) that $r'U^{q+r'}v^q \geq rr'\widetilde{U}^{q+r'}\widetilde{U}^{q+r}v^q = r\widetilde{U}^{q+r}(r'\widetilde{U}^{q+r'}v^q)$ everywhere. Letting $r' \to \infty$, we get $u^q \geq r\widetilde{U}^{q+r}u^q$. We have (41.12) with v^q in the right-hand side replaced by u^q, since $v^q = u^q$ a. e. Thus u^q is q-co-excessive by virtue of Proposition 41.5(iv) applied to the dual process.

To see the uniqueness, suppose that u_1^q and u_2^q both satisfy the requirements for u^q. Then (41.9) with $x = 0$ shows that $u_1^q = u_2^q$ a. e. It follows from (5) of Theorem 41.15 for the dual process that $u_1^q = u_2^q$ everywhere.

In the transient case, all the argument is valid even when $q = 0$. □

PROPOSITION 41.17. *Suppose that* (ACP) *holds. Let $\widetilde{u}^q$ be the unique q-excessive function such that $\widetilde{U}^q f(x) = \int_{\mathbb{R}^d} \widetilde{u}^q(y-x)f(y)\mathrm{d}y$ for every nonnegative universally measurable f. Then we have $\widetilde{u}^q(x) = u^q(-x)$.*

Proof. Consequence of (41.6) and Proposition 41.10. □

Let us consider the recurrent case.

PROPOSITION 41.18. *Recurrence of the process is equivalent to recurrence of its dual process.*

Proof. Evident from Definition 35.1, of recurrence, since $(\{X_t\}, \widetilde{P})$ is identical in law with $(\{-X_t\}, P)$. □

THEOREM 41.19. *Suppose that $(\{X_t\}, P)$ is recurrent and let Σ be the support of the process.*

(i) *If $\Sigma = \mathbb{R}^d$, then, for every Borel set B with* Leb $B > 0$, *$U(x, B) = \infty$ for a. e. x.*

(ii) *If* (ACP) *holds, then $\Sigma = \mathbb{R}^d$ and, for every Borel set B with* Leb $B > 0$ *and for every x, $U(x, B) = \infty$.*

Proof. (i) Let $\Sigma = \mathbb{R}^d$. If G is nonempty and open, then $V(G) = \infty$ by Theorem 35.8. Let B and C be bounded Borel sets with positive Lebesgue measures. Let $f(y) = \int_C 1_B(x+y)\mathrm{d}x$. We have

$$\int_C U(x, B)\mathrm{d}x = \int_C \mathrm{d}x \int_{\mathbb{R}^d} V(\mathrm{d}y)1_B(x+y) = \int_{\mathbb{R}^d} f(y)V(\mathrm{d}y).$$

The function f is not identically zero, since

$$\int f(y)\mathrm{d}y = \iint 1_B(x+y)1_C(x)\mathrm{d}x\mathrm{d}y = \mathrm{Leb}(B)\,\mathrm{Leb}(C) > 0.$$

Moreover f is continuous, as is proved by approximation of 1_B by continuous functions as in the proof that (1) $\Rightarrow$ (3) in Theorem 41.15. It follows that $\int_C U(x,B)\mathrm{d}x = \infty$. Hence $U(x,B) = \infty$ for a. e. x.

(ii) Let (ACP) hold. Since Σ is the support of V^q for $q > 0$ by Exercise 44.1 and since Σ is a closed additive subgroup of $\mathbb{R}^d$ by Theorem 35.8, we have $\Sigma = \mathbb{R}^d$. Otherwise $\mathrm{Leb}\,\Sigma = 0$, contradictory with (ACP). The function $U(x,B)$ is 0-excessive by Proposition 41.5(ii). Since $U(x,B) = \infty$ a. e., it is identically infinite by (5) of Theorem 41.15. □

REMARK 41.20. Assume that, for each $t > 0$, there is a nonnegative bounded continuous function $p_t(x)$ such that $P^0[X_t \in B] = \int_B p_t(x)\mathrm{d}x$ for any $B \in \mathcal{B}(\mathbb{R}^d)$. Then,

$$P_t f(x) = \textstyle\int_{\mathbb{R}^d} p_t(y-x) f(y)\mathrm{d}y, \quad t > 0,$$

for any nonnegative universally measurable function f;

$$\textstyle\int_{\mathbb{R}^d} p_t(y-x) p_s(z-y)\mathrm{d}y = p_{t+s}(z-x)$$

for any $t > 0$, $s > 0$, x and $z \in \mathbb{R}^d$; and

$$u^q(y) = \textstyle\int_0^\infty \mathrm{e}^{-qt} p_t(y)\mathrm{d}t, \quad y \in \mathbb{R}^d,$$

for $q > 0$ (and for $q \geq 0$ in the transient case). To see the last equality, we can check that the right-hand side is a q-co-excessive function satisfying (41.9). The Brownian motion and nondegenerate stable or semi-stable processes satisfy this assumption (Example 28.2). Hence, for the Brownian motion and some stable processes, the functions $u^q(y)$ ($q > 0$ or $q \geq 0$) are equal, not only almost everywhere but everywhere, to those calculated in Examples 30.11, 35.6, 35.7, and 37.19 and Exercises 39.1 and 39.2.

EXAMPLE 41.21. Any compound Poisson process does not satisfy (ACP), since the one-point set $\{0\}$ has positive V^q-measure. On the other hand, nondegenerate selfdecomposable processes satisfy (ACP), since they satisfy (ACT). See Theorem 27.13.

Consider a compound Poisson process $(\{X_t\}, P)$ on $\mathbb{R}$ with Lévy measure ν concentrated on $\mathbb{Q} \setminus \{0\}$. Let x_n, $n = 1, 2, \ldots$, be an enumeration of $\mathbb{Q} \setminus \{0\}$. Let ν_0 be a measure such that $0 < \nu_0\{x_n\} \leq (|x_n|^{-1} \wedge 1) n^{-2}$. Then $\sum_n \nu_0\{x_n\}$ and $\sum_n |x_n| \nu_0\{x_n\}$ are finite. Suppose that $\nu(B) = \nu_0(B \cap (0,\infty)) + a\nu_0(B \cap (-\infty, 0))$ for $B \in \mathcal{B}_{\mathbb{R}}$, where $a > 0$ is chosen to satisfy $\int x\nu(\mathrm{d}x) = 0$. Then $E^0 X_t = t\int x\nu(\mathrm{d}x) = 0$ by Example 25.12. Hence $(\{X_t\}, P)$ is recurrent by Theorem 36.7. We have $\Sigma = \mathfrak{G} = \mathbb{R}$, since X_1 under P^0 has support $\mathbb{R}$. Since $X_t \in \mathbb{Q}$ for all t P^0-almost surely, we have $U(0, \mathbb{R} \setminus \mathbb{Q}) = V(\mathbb{R} \setminus \mathbb{Q}) = 0$. But, for any Borel subset B of $\mathbb{R} \setminus \mathbb{Q}$ with $\mathrm{Leb}\,B > 0$, we have $U(x,B) = \infty$ for a. e. x by Theorem 41.19.

EXAMPLE 41.22. Semi-selfdecomposable processes do not necessarily satisfy (ACP). We will show more. Let $d = 1$. Semi-selfdecomposable processes of type

B not satisfying (ACP) exist in each of the four classes given by combination of recurrent or transient and symmetric or non-symmetric. In the following, given ν_k, $k = 1, 2, \ldots$, we define μ_k by $\widehat{\mu}_k(z) = \mathrm{e}^{\psi_k(z)}$ with $\psi_k(z) = \int (\mathrm{e}^{izx} - 1)\nu_k(\mathrm{d}x)$ and denote the Lévy process corresponding to μ_k by $(\{X_t\}, P_k)$. The expectation with respect to P_k is denoted by E_k.

1. Let $\nu_1 = \sum_{n=1}^{\infty} \delta_{2^{-n}} + \sum_{n=0}^{\infty} c_n \delta_{2^n}$ with $1 \geq c_0 \geq c_1 \geq \cdots > 0$. Then $(\{X_t\}, P_1)$ is semi-selfdecomposable of type B. Let $z_k = 2^k \pi$, $k = 1, 2, \ldots$. Then,

$$0 \leq -\mathrm{Re}\,\psi_1(z_k) = \sum_{n=k}^{\infty} (1 - \cos 2^{k-n}\pi) \leq \tfrac{\pi^2}{2} \sum_{n=k}^{\infty} 2^{2(k-n)} = \tfrac{2\pi^2}{3},$$
$$|\mathrm{Im}\,\psi_1(z_k)| = \left|\sum_{n=k+1}^{\infty} \sin 2^{k-n}\pi\right| \leq \pi \sum_{n=k+1}^{\infty} 2^{k-n} = \pi.$$

Hence, along a subsequence, $\psi_1(z_k)$ tends to a complex number a with $-\mathrm{Re}\, a \geq 0$. Thus, for the q-potential measure V_1^q of the process,

$$\int \mathrm{e}^{izx} V_1^q(\mathrm{d}x) = (q - \psi_1(z))^{-1} \to (q - a)^{-1}$$

for $q > 0$ as z goes to ∞ along the subsequence of $\{z_k\}$. Therefore, by the Riemann–Lebesgue theorem, V_1^q is not absolutely continuous. The process is a subordinator and hence transient and non-symmetric.

2. Let $\nu_2 = \sum_{n=1}^{\infty} \delta_{2^{-n}} + \sum_{n=0}^{\infty} c'_n \delta_{2^n}$ with $1 \geq c'_0 \geq c'_1 \geq \cdots > 0$. Let $\widetilde{\nu}_2$ be the dual measure of ν_2, that is, $\widetilde{\nu}_2(B) = \nu_2(-B)$, and let $\nu_3 = \nu_1 + a\widetilde{\nu}_2$ with $a > 0$. Then the process $(\{X_t\}, P_3)$ is semi-selfdecomposable and of type B. It does not satisfy (ACP) for the same reason. Assume that $\sum_{n=0}^{\infty} 2^n c_n < \infty$ and $\sum_{n=0}^{\infty} 2^n c'_n < \infty$. Thus we have $E_3[|X_t|] < \infty$. If $\{c_k\} = \{c'_k\}$ and $a = 1$, then the process is symmetric, $E_3[X_t] = 0$, and it is recurrent by Theorem 36.7. If $\{c_k\}$ and $\{c'_k\}$ are not identical and if $a > 0$ is chosen so that $\int x\nu(\mathrm{d}x) = 0$, then $E_3[X_t] = 0$ and the process is recurrent and non-symmetric.

3. Let $\nu_4 = \sum_{n=-\infty}^{\infty} 2^{-n\alpha}(\delta_{2^n} + \delta_{-2^n})$ with $0 < \alpha < 1$. Then $(\{X_t\}, P_4)$ is a symmetric α-semi-stable process. It is transient by Theorem 37.16. Since $\psi_4(z)$ is real, we have $\psi_4(z) \leq -K|z|^{\alpha}$ with some $K > 0$ by Proposition 24.20. We have

$$\left|\int_{|x|<1} (\mathrm{e}^{izx} - 1)\nu_4(\mathrm{d}x)\right| = \int_{|x|<1} (1 - \cos zx)\nu_4(\mathrm{d}x) \leq \tfrac{1}{2} \int_{|x|<1} |zx|^2 \nu_4(\mathrm{d}x) = K'|z|^2$$

with $K' > 0$. Let $\nu_5 = \sum_{n=0}^{\infty} 2^{-n\alpha}(\delta_{2^n} + \delta_{-2^n})$. Then

$$|\psi_5(z)| \geq |\psi_4(z)| - \left|\int_{|x|<1} (\mathrm{e}^{izx} - 1)\nu_4(\mathrm{d}x)\right| \geq K|z|^{\alpha} - K'|z|^2 \geq K''|z|^{\alpha}$$

for small $|z|$ with some $K'' > 0$. Hence $\int_{|z|<\varepsilon} |\psi_5(z)|^{-1} \mathrm{d}z < \infty$ for some $\varepsilon > 0$. It follows from Corollary 37.6 that $(\{X_t\}, P_5)$ is transient. Let $\nu_6 = \sum_{n=1}^{\infty} (\delta_{2^{-n}} + \delta_{-2^{-n}}) + \nu_5$. We see, from Exercise 39.6, that $(\{X_t\}, P_6)$ is symmetric and transient. It is semi-selfdecomposable, of type B, not satisfying (ACP).

EXAMPLE 41.23. Using the infinitely divisible distributions of Orey [**364**], let us construct other processes not satisfying (ACP). Let $d = 1$ and $0 < \alpha < 2$. Let c be an integer with $c > 2/(2 - \alpha)$. Let $a_n = 2^{-c^n}$ and $\nu_1 = \sum_{n=1}^{\infty} a_n{}^{-\alpha}(\delta_{a_n} + \delta_{-a_n})$. Then ν_1 has support in $(-1, 1)$ and $\int_{(-1,1)} |x|^{\alpha} \nu_1(\mathrm{d}x) = \infty$, while $\int_{(-1,1)} |x|^{\beta} \nu_1(\mathrm{d}x) < \infty$ for $\beta > \alpha$, as we have seen in Exercise 29.12. Consider the Lévy process $(\{X_t\}, P_1)$ generated by $(0, \nu_1, 0)$. Then $\widehat{\mu}_1(z) = \mathrm{e}^{\psi_1(z)}$

with $\psi_1(z) = 2\sum_{n=1}^{\infty}(\cos za_n - 1){a_n}^{-\alpha}$. For $z_k = 2\pi {a_k}^{-1}$ we have

$$\begin{aligned}|\psi_1(z_k)| &= 4\textstyle\sum_{n=1}^{\infty}\sin^2(z_k a_n/2)\,{a_n}^{-\alpha} = 4\sum_{n=1}^{\infty}\sin^2(\pi 2^{c^k-c^n})\,{a_n}^{-\alpha}\\ &\le 4\pi^2\textstyle\sum_{n=k+1}^{\infty}2^{2c^k-2c^n+\alpha c^n} = 4\pi^2\sum_{n=1}^{\infty}2^{-c^k((2-\alpha)c^n-2)} \to 0\end{aligned}$$

as $k \to \infty$. Consequently, the q-potential measure V_1^q, $q > 0$, satisfies

$$\textstyle\int e^{izx}V_1^q(dx) = (q - \psi_1(z))^{-1} \to q^{-1} = V_1^q(\mathbb{R}),$$

when z goes to ∞ along the sequence $\{z_k\}$. By virtue of the Riemann–Lebesgue theorem it follows that, for any $q > 0$, V_1^q does not have absolutely continuous part. As a matter of fact, it is continuous singular, since it is continuous by Theorem 30.10.

If we choose $1 \le \alpha < 2$, then this process is of type C, symmetric and recurrent. The recurrence follows from $E_1[X_t] = 0$. If $1 \le \alpha < 2$ and if we make $\nu_2 = \nu_1 + a_2\delta_{a_1} + a_1\delta_{-a_2}$ and

$$\psi_2(z) = \psi_1(z) + a_2(e^{ia_1 z} - 1) + a_1(e^{-ia_2 z} - 1),$$

then the associated process $(\{X_t\}, P_2)$ is type C, recurrent and non-symmetric, with continuous singular V_2^q for $q > 0$. Note that $E_2[X_t] = t(a_2a_1 - a_1a_2) = 0$ and that $\psi_2(z_k) = \psi_1(z_k)$ for $k \ge 2$.

To construct a transient process of type C with continuous singular q-potential measure for $q > 0$, use $\nu_3 = \nu_1 + \delta_{a_1}$ with $1 \le \alpha < 2$ and $\psi_3(z) = \psi_1(z) + e^{ia_1 z} - 1$. We have $\psi_3(z_k) = \psi_1(z_k)$ for $k \ge 1$ and the associated process $(\{X_t\}, P_3)$ satisfies $E_3[X_t] = ta_1$. This is non-symmetric.

A transient symmetric process of type C with continuous singular q-potential measure for $q > 0$ is obtained from $\nu_4 = \nu_1 + \sum_{n=0}^{\infty}2^{-n\kappa}(\delta_{2^n} + \delta_{-2^n})$ with $1 \le \alpha < 2$ and $0 < \kappa < 1$ and $\psi_4(z) = \int(\cos zx - 1)\nu_4(dx)$. The transience of the associated process $(\{X_t\}, P_4)$ is proved as in the process $(\{X_t\}, P_6)$ of Example 41.22. The continuous singularity comes from $\psi_4(z_k) = \psi_1(z_k)$ for $k \ge 1$.

It is also possible to construct a subordinator with continuous singular potential measure by this method. Assume $0 < \alpha < 1$ and let c be an integer with $c > 1/(1-\alpha)$. Let $(\{X_t\}, P_5)$ be the subordinator with $\psi_5(z) = \sum_{n=1}^{\infty}(e^{iza_n} - 1){a_n}^{-\alpha}$. For the same sequence $\{z_k\}$ as above, we have

$$\begin{aligned}|\mathrm{Re}\,\psi_5(z_k)| &= 2\textstyle\sum_{n=1}^{\infty}{a_n}^{-\alpha}\sin^2(z_k a_n/2) \to 0,\\ |\mathrm{Im}\,\psi_5(z_k)| &= \big|\textstyle\sum_{n=1}^{\infty}{a_n}^{-\alpha}\sin z_k a_n\big| = \big|\sum_{n=1}^{\infty}{a_n}^{-\alpha}\sin(2\pi 2^{c^k-c^n})\big|\\ &\le 2\pi\textstyle\sum_{n=k+1}^{\infty}2^{c^k-c^n+\alpha c^n} = 2\pi\sum_{n=1}^{\infty}2^{-c^k((1-\alpha)c^n-1)} \to 0\end{aligned}$$

as $k \to \infty$. Hence $\psi_5(z_k) \to 0$ and V_5^q is continuous singular for any $q > 0$.

See Berg [**20**] for some other examples.

42. Capacity

We continue to fix a Lévy process $(\{X_t\}, P)$ as in Sections 40 and 41. After proving Hunt's switching formula, we introduce capacity and energy, study their relations to essential polarity, and make comparison of the classes of essentially polar sets for two Lévy processes.

DEFINITION 42.1. Let B be an F_σ set. The transition operator P_t^B and potential operator U_B^q for the part of the process up to hitting the set B are defined by

$$P_t^B f(x) = E^x[f(X_t);\, t < T_B], \qquad t \geq 0,$$
$$U_B^q f(x) = E^x\left[\int_0^{T_B} \mathrm{e}^{-qt} f(X_t)\mathrm{d}t\right] = \int_0^\infty \mathrm{e}^{-qt} P_t^B f(x)\mathrm{d}t, \qquad q \geq 0,$$

for nonnegative universally measurable f. By the dual process $(\{X_t\}, \widetilde{P})$ in place of $(\{X_t\}, P)$ the operators $\widetilde{P}_t^B$ and $\widetilde{U}_B^q$ are similarly defined. Write $P_t^B(x, C) = P_t^B 1_C(x)$, $U_B^q(x, C) = U_B^q 1_C(x)$, and similarly $\widetilde{P}_t^B(x, C)$, $\widetilde{U}_B^q(x, C)$ for $C \in \mathcal{B}_{\mathbb{R}^d}{}^*$. For a measure ρ on $\mathbb{R}^d$ we define

$$\rho U^q(C) = \int_{\mathbb{R}^d} \rho(\mathrm{d}x) U^q(x, C), \qquad \rho P_B^q(C) = \int_{\mathbb{R}^d} \rho(\mathrm{d}x) P_B^q(x, C),$$
$$\rho \widetilde{U}^q(C) = \int_{\mathbb{R}^d} \rho(\mathrm{d}x) \widetilde{U}^q(x, C), \qquad \rho \widetilde{P}_B^q(C) = \int_{\mathbb{R}^d} \rho(\mathrm{d}x) \widetilde{P}_B^q(x, C)$$

for $C \in \mathcal{B}_{\mathbb{R}^d}{}^*$.

In the definition above we have used some of the following facts.

PROPOSITION 42.2. *Let T be a stopping time. Then all assertions in Proposition 41.1 remain true if we replace $f(X_t(\omega))$ by $f(X_t(\omega))1_{\{t<T\}}(\omega)$.*

Proof. To see the $(\mathcal{B}_{[0,\infty)} \times \mathcal{F}^0)^{\lambda \times P^\rho}$-measurability of $f(X_t(\omega))1_{\{t<T\}}(\omega)$ in (t, ω), it is enough to use (i) of Proposition 41.1 and to note that $\{(t, \omega) \colon t < T(\omega)\} = \bigcup_{r \in \mathbb{Q}} \{(t, \omega) \colon t < r < T(\omega)\} \in \mathcal{B}_{[0,\infty)} \times \mathcal{F} \subset (\mathcal{B}_{[0,\infty)} \times \mathcal{F}^0)^{\lambda \times P^\rho}$. We can prove the other assertions, modifying the proof of Proposition 41.1. As a typical modification, let us show the analogue of (iii). First note that if f is Borel-measurable, then $(t, x) \mapsto E^x[f(X_t)1_{\{t<T\}}]$ is $(\mathcal{B}_{[0,\infty)} \times \mathcal{B}_{\mathbb{R}^d}{}^*/\mathcal{B}_{\mathbb{R}})$-measurable, and hence $x \mapsto g(x) = \int_0^\infty \mathrm{e}^{-qt} E^x[f(X_t)1_{\{t<T\}}]\mathrm{d}t$ is $(\mathcal{B}_{\mathbb{R}^d}{}^*/\mathcal{B}_{\mathbb{R}})$-measurable. Given a probability measure σ on $\mathbb{R}^d$, let $\eta(B) = q \int \sigma(\mathrm{d}x) \int_0^\infty \mathrm{e}^{-qt} E^x[1_B(X_t)1_{\{t<T\}}]\mathrm{d}t$. Let f be universally measurable. We can find Borel-measurable functions f_1, f_2 such that $f_1 \leq f \leq f_2$ and $\int (f_2 - f_1)\eta(\mathrm{d}x) = 0$. For $j = 1, 2$ let $g_j(x) = \int_0^\infty \mathrm{e}^{-qt} E^x[f_j(X_t)1_{\{t<T\}}]\mathrm{d}t$. Then $g_1 \leq g \leq g_2$ and $\int (g_2 - g_1)\sigma(\mathrm{d}x) = 0$. Since g_1 and g_2 are universally measurable, we can find Borel-measurable h_1 and h_2 such that $h_1 \leq g_1$, $g_2 \leq h_2$, and $\int (g_1 - h_1)\sigma(\mathrm{d}x) = \int (h_2 - g_2)\sigma(\mathrm{d}x) = 0$. Thus g is $(\mathcal{B}_{\mathbb{R}^d})^\sigma$-measurable, and hence g is universally measurable. $\square$

PROPOSITION 42.3 (Hunt's switching formula). *Let B be an F_σ set and f and g be nonnegative universally measurable. Then,*

$$\int_{\mathbb{R}^d} P_t^B f(x)g(x)\mathrm{d}x = \int_{\mathbb{R}^d} f(x)\widetilde{P}_t^B g(x)\mathrm{d}x, \qquad t \geq 0, \tag{42.1}$$

$$\int_{\mathbb{R}^d} U_B^q f(x)g(x)\mathrm{d}x = \int_{\mathbb{R}^d} f(x)\widetilde{U}_B^q g(x)\mathrm{d}x, \qquad q \geq 0. \tag{42.2}$$

Proof. The second formula is obtained from the first by going to the Laplace transforms. To see the first, let f and g be nonnegative and continuous with compact supports. Let B be open set and let $F = \mathbb{R}^d \setminus B$. Let $t > 0$ and $t_{nk} = k2^{-n}t$, $k = 0, 1, \ldots, 2^n$. Then, using (41.7), we have

$$\begin{aligned}
\int P_t^B f(x)g(x)\mathrm{d}x &= \int g(x)\mathrm{d}xE^x[f(X_t);\, t < T_B] \\
&\leq \int g(x)\mathrm{d}xE^x\left[\prod_{k=1}^{2^n} 1_F(X_{t_{nk}})f(X_t)\right] \\
&= \int g(x)\mathrm{d}xE^x\left[1_F(X_{t_{n1}})E^{X(t_{n1})}\left[\prod_{k=2}^{2^n} 1_F(X_{t_{nk}-t_{n1}})f(X_{t-t_{n1}})\right]\right] \\
&= \int \widetilde{E}^x[g(X_{t_{n1}})]\mathrm{d}x 1_F(x)E^x\left[\prod_{k=2}^{2^n} 1_F(X_{t_{nk}-t_{n1}})f(X_{t-t_{n1}})\right] \\
&= \int \widetilde{E}^x[1_F(X_{t_{n2}-t_{n1}})g(X_{t_{n2}})]\mathrm{d}x 1_F(x)E^x\left[\prod_{k=3}^{2^n} 1_F(X_{t_{nk}-t_{n2}})f(X_{t-t_{n2}})\right] \\
&= \cdots \\
&= \int \widetilde{E}^x\left[\prod_{k=1}^{2^n-1} 1_F(X_{t-t_{nk}})g(X_t)\right] 1_F(x)f(x)\mathrm{d}x \\
&\to \int \widetilde{E}^x[g(X_t); X_s \in F \text{ for } 0 < s < t]1_F(x)f(x)\mathrm{d}x
\end{aligned}$$

as $n \to \infty$. Thus

$$\int g(x)\mathrm{d}xE^x[f(X_{t'});\, t' < T_B] \leq \int \widetilde{E}^x[g(X_{t'});\, t' \leq T_B]1_F(x)f(x)\mathrm{d}x$$

for any t'. Let $t' > t$ and $t' \to t$. Then, since f and g are continuous with compact supports and B is open,

$$\begin{aligned}
\int g(x)\mathrm{d}xE^x[f(X_t);\, t < T_B] &\leq \int \widetilde{E}^x[g(X_t);\, t < T_B]1_F(x)f(x)\mathrm{d}x \\
&= \int \widetilde{E}^x[g(X_t);\, t < T_B]f(x)\mathrm{d}x.
\end{aligned}$$

We get the reverse inequality by interchanging the roles of the original and dual processes. Thus (42.1) is shown in this case. It is extended to f and g Borel-measurable with compact supports, and then to f and g universally measurable with compact supports. Now we can relax the assumption that B is open. If B is closed, then (42.1) is shown by approximation of B by $\frac{1}{n}$-neighborhoods and by use of quasi-left-continuity. Now let $B = \bigcup_{n=1}^{\infty} F_n$ with $\{F_n\}$ being an increasing sequence of closed sets. Then $T_{F_n} \downarrow T_B$, but T_{F_n} is not necessarily strictly decreasing. Let f and g be continuous with compact supports. It follows from (42.1) for F_n that

$$\int E^x[f(X_{t-}); t \le T_{F_n}]g(x)\mathrm{d}x = \int f(x)\widetilde{E}^x[g(X_{t-}); t \le T_{F_n}]\mathrm{d}x.$$

By (1.10) we can replace X_{t-} by X_t. Let $n \to \infty$. Then

$$\int E^x[f(X_t); t \le T_B]g(x)\mathrm{d}x = \int f(x)\widetilde{E}^x[g(X_t); t \le T_B]\mathrm{d}x.$$

Now, shifting t from above, we get the identity with $t < T_B$ in place of $t \le T_B$. □

COROLLARY 42.4. *Assume* (ACP). *Let* $q > 0$. *Then, for any* F_σ *set* B,

$$\int_{\mathbb{R}^d} P_B^q(x, \mathrm{d}y)u^q(z - y) = \int_{\mathbb{R}^d} u^q(y - x)\widetilde{P}_B^q(z, \mathrm{d}y), \qquad x, z \in \mathbb{R}^d. \tag{42.3}$$

If the process is transient, the same is true also for $q = 0$.

Proof. Let f and g be universally measurable, nonnegative and bounded, with compact supports. By (41.8), (42.2), (41.3) and its dual version, we get

$$\int P_B^q U^q f(x)g(x)\mathrm{d}x = \int f(x)\widetilde{P}_B^q\widetilde{U}^q g(x)\mathrm{d}x. \tag{42.4}$$

Hence, for any such g,

$$\iint g(x)\mathrm{d}x P_B^q(x, \mathrm{d}y)u^q(z - y) = \iint g(x)\mathrm{d}x \widetilde{P}_B^q(z, \mathrm{d}y)u^q(y - x) \tag{42.5}$$

for a. e. z. Here we have used Proposition 41.17. Since $u^q(z - y)$ is q-co-excessive in z for any fixed y, the left-hand side is q-co-excessive in z. The right-hand side is also q-co-excessive by Proposition 41.5(vii). Therefore (42.5) is true for every z, by the dual version of (5) of Theorem 41.15. Hence, for every x, (42.3) is true for a. e. z. Similar discussion shows that there is no exceptional point. □

THEOREM 42.5. *Let* B *be an* F_σ *set and let* $q > 0$. *There exists a unique measure* ρ *on* $\mathbb{R}^d$ *such that*

$$\rho\widetilde{U}^q(C) = \int_C E^x[\mathrm{e}^{-qT_B}]\mathrm{d}x, \qquad C \in \mathcal{B}(\mathbb{R}^d). \tag{42.6}$$

This measure ρ is expressed as

$$\rho(C) = q\int_{\mathbb{R}^d} \widetilde{E}^x[\mathrm{e}^{-qT_B};\ X_{T_B}\in C]\mathrm{d}x, \qquad C\in\mathcal{B}(\mathbb{R}^d), \tag{42.7}$$

and is supported on $\overline{B}$ and $\rho(C)<\infty$ for any compact set C.

DEFINITION 42.6. The measure ρ in the preceding theorem is the *q-capacitary measure* of B, denoted by m_B^q. Its total mass is the *q-capacity* of B and denoted by $C^q(B)$. That is, $C^q(B) = m_B^q(\overline{B})$. The q-capacitary measure of B for the dual process is the *q-co-capacitary measure* of B and denoted by $\widetilde{m}_B^q$.

In the proof of the theorem we use the following fact.

PROPOSITION 42.7. *Let $q>0$ and let ρ be a measure on $\mathbb{R}^d$ such that $\rho\widetilde{P}_B^q(B)<\infty$ for any compact B. Then $\rho\widetilde{U}^q$ determines ρ.*

Proof. Let ρ and ρ' satisfy $\rho\widetilde{U}^q = \rho'\widetilde{U}^q$.

Step 1. Suppose that $\rho(\mathbb{R}^d)<\infty$. We have $\rho\widetilde{U}^q(\mathbb{R}^d) = q^{-1}\rho(\mathbb{R}^d)$ and similarly for ρ'. Hence $\rho'(\mathbb{R}^d)<\infty$. By the resolvent equation (41.2) we have $\rho\widetilde{U}^r = \rho'\widetilde{U}^r$ for all $r>0$. For any bounded continuous function f we have $r\int\rho(\mathrm{d}x)\widetilde{U}^r f(x) = r\int\rho'(\mathrm{d}x)\widetilde{U}^r f(x)$ and, letting $r\to\infty$, $\int\rho(\mathrm{d}x)f(x) = \int\rho'(\mathrm{d}x)f(x)$. Thus $\rho=\rho'$.

Step 2. Suppose that $\rho(\mathbb{R}^d)=\infty$. Then $\rho'(\mathbb{R}^d)=\infty$. Let $D_n = \{x\colon |x|<n\}$. Since $\rho(D_n) = \int_{D_n}\rho(\mathrm{d}x)\widetilde{P}_B^q(x,B)\le\rho\widetilde{P}_B^q(B)$ for $B=\overline{D}_n$, $\rho(D_n)$ is finite. Denote by ρ^n the restriction of ρ to $D_n\setminus D_{n-1}$. Write $\rho_k = \rho\widetilde{P}_{D_k}^q$ and $\rho_k^n = \rho^n\widetilde{P}_{D_k}^q$. Finiteness of ρ_k follows from $\rho_k(\mathbb{R}^d)\le\rho P_B^q(\mathbb{R}^d)<\infty$ for $B=\overline{D}_k$. We have $\rho=\sum_{n=1}^\infty\rho^n$ and $\rho_k=\sum_{n=1}^\infty\rho_k^n$. Further $\rho_k^n=\rho^n$ for $k>n$, since $\widetilde{P}_{D_k}^q(x,\cdot)=\delta_x(\cdot)$ for $x\in D_n$. For any $x\in\mathbb{R}^d$ and $C\in\mathcal{B}_{\mathbb{R}^d}$,

$$\begin{aligned}&\int\widetilde{P}_{D_{k+1}}^q(x,\mathrm{d}y)\widetilde{P}_{D_k}^q(y,C)\\ &\quad=\widetilde{E}^x[\mathrm{e}^{-qT(D_{k+1})}\widetilde{E}^{X(T(D_{k+1}))}[\mathrm{e}^{-qT(D_k)};X_{T(D_k)}\in C]] = \widetilde{E}^x[\mathrm{e}^{-qS};X_S\in C],\end{aligned}$$

where $S=\inf\{t>T_{D_{k+1}}\colon X_t\in D_k\}$. But $S=T_{D_k}$ since D_k is open. Hence

$$\rho_k^n(C) = \rho^n\widetilde{P}_{D_k}^q(C) = \rho^n\widetilde{P}_{D_{k+1}}^q\widetilde{P}_{D_k}^q(C)\ge\rho^n\widetilde{P}_{D_{k+1}}^q(C) = \rho_{k+1}^q(C)$$

if $C\subset D_k$. Thus, if $C\subset D_l$ for some l, then we can apply Lebesgue's dominated convergence theorem and get

$$\rho_k(C) = \sum_{n=1}^\infty\rho_k^n(C)\to\sum_{n=1}^\infty\rho^n(C) = \rho(C), \qquad k\to\infty.$$

Likewise $\rho'(D_n)$ is finite and $\rho_k'(C) = \rho'\widetilde{P}_{D_k}^q(C)$ tends to $\rho'(C)$ for bounded C. Note that the assumption $\rho\widetilde{U}^q = \rho'\widetilde{U}^q$ implies $\rho\widetilde{P}_B^q\widetilde{U}^q = \rho'\widetilde{P}_B^q\widetilde{U}^q$ for any

F_σ set B, which is proved by use of Proposition 41.5(v). Thus, by Step 1, $\rho_k = \rho'_k$. Hence $\rho = \rho'$. □

Proof of Theorem 42.5. Let us prove the existence of ρ. Define $\rho_0(C)$ by the right-hand side of (42.7). Suppose that C is a bounded Borel set and let $f(x) = 1_C(x)$. Integrating (41.3) for $T = T_B$ over $\mathbb{R}^d$, we get

$$\int \widetilde{U}^q f(x)\mathrm{d}x = \int \widetilde{U}^q_B f(x)\mathrm{d}x + q^{-1} \int \rho_0(\mathrm{d}x)\widetilde{U}^q f(x).$$

The left-hand side equals $q^{-1}\int f(x)\mathrm{d}x$ by (41.8), while the first term of the right-hand side equals $\int f(x)U^q_B 1(x)\mathrm{d}x = q^{-1}\int f(x)(1 - E^x[\mathrm{e}^{-qT_B}])\mathrm{d}x$ by Hunt's switching formula. It follows that

$$\int f(x)E^x[\mathrm{e}^{-qT_B}]\mathrm{d}x = \int \rho_0(\mathrm{d}x)\widetilde{U}^q f(x),$$

that is, ρ_0 satisfies (42.6) for bounded Borel C, and hence for all Borel C. It is evident that ρ_0 is supported on $\overline{B}$. Writing $B_a = \{x\colon |x| \le a\}$ for $a > 0$, we have

$$\widetilde{U}^q(x, B_a) = \widetilde{V}^q(B_a - x) \ge \widetilde{V}^q(B_{a/2}) \quad \text{for } x \in B_{a/2}. \tag{42.8}$$

Note that $\widetilde{V}^q(B_{a/2}) = V^q(B_{a/2})$. Hence, for any $x \in \mathbb{R}^d$,

$$\widetilde{U}^q(x, B_a) \ge \widetilde{E}^x[\mathrm{e}^{-qT(B_{a/2})}\widetilde{U}^q 1_{B_a}(X_{T(B_{a/2})})] \ge V^q(B_{a/2})\widetilde{E}^x[\mathrm{e}^{-qT(B_{a/2})}].$$

Using this, we have

$$\rho_0(B_{a/2}) = q\int \widetilde{E}^x[\mathrm{e}^{-qT_B}; X_{T_B} \in B_{a/2}]\mathrm{d}x \le q\,\mathrm{Leb}(B_{a/2}) + I,$$

where

$$\begin{aligned} I &= q\int_{|x|>a/2} \widetilde{E}^x[\mathrm{e}^{-qT_B}; T_B \ge T_{B_{a/2}}]\mathrm{d}x \le q\int_{\mathbb{R}^d} \widetilde{E}^x[\mathrm{e}^{-qT(B_{a/2})}]\mathrm{d}x \\ &\le \frac{q}{V^q(B_{a/2})}\int_{\mathbb{R}^d} \widetilde{U}^q(x, B_a)\mathrm{d}x = \frac{\mathrm{Leb}(B_a)}{V^q(B_{a/2})}. \end{aligned}$$

Hence $\rho_0(C)$ is finite for any compact C.

The uniqueness of ρ is reduced to Proposition 42.7. Notice that finiteness of $\rho\widetilde{U}^q(B)$ for all compact B implies finiteness of $\rho\widetilde{P}^q_B(B)$ for all compact B. Indeed, $\rho\widetilde{P}^q_B\widetilde{U}^q(B_a) \le \rho\widetilde{U}^q(B_a) < \infty$ and

$$\rho\widetilde{P}^q_B\widetilde{U}^q(B_a) \ge \int_B \rho\widetilde{P}^q_B(\mathrm{d}x)V^q(B_{a/2}) = \rho\widetilde{P}^q_B(B)V^q(B_{a/2})$$

by (42.8), if $B \subset B_{a/2}$. □

THEOREM 42.8. *Assume that the process is transient. For any bounded F_σ set B there is a unique measure ρ such that*

$$\rho\widetilde{U}(C) = \int_C P^x[T_B < \infty]\mathrm{d}x, \qquad C \in \mathcal{B}(\mathbb{R}^d). \tag{42.9}$$

This measure ρ is finite and supported on $\overline{B}$.

DEFINITION 42.9. The measure ρ above is the 0-*capacitary measure*, or the *equilibrium measure*, of B and denoted by m_B. The total mass of m_B is the 0-*capacity*, or the *capacity*, of B, denoted by $C^0(B)$ or $C(B)$. The 0-capacitary measure of B for the dual process is the 0-*co-capacitary measure*, or *co-equilibrium measure*, of B, denoted by $\widetilde{m}_B$.

Proof of Theorem 42.8. Let B be an F_σ set with $\overline{B}$ compact. Let $K = \{x\colon \operatorname{dis}(x, \overline{B}) \le 1\}$ and $D = \{x\colon |x| > 1\}$. If $x \in \overline{B}$, then

$$\begin{aligned}\widetilde{U}^q(x, K) &= \widetilde{E}^0\left[\int_0^\infty \mathrm{e}^{-qt} 1_{K-x}(X_t)\mathrm{d}t\right] \ge \widetilde{E}^0\left[\int_0^{T_D} \mathrm{e}^{-qt}\mathrm{d}t\right] \\ &= q^{-1}\widetilde{E}^0[1 - \mathrm{e}^{-qT_D}] \ge k \quad \text{for } 0 < q \le 1,\end{aligned}$$

where $k = \widetilde{E}^0[1 - \mathrm{e}^{-T_D}] > 0$. Hence

$$k m_B^q(\overline{B}) \le \int_{\overline{B}} m_B^q(\mathrm{d}x)\widetilde{U}^q(x, K) = \int_K E^x[\mathrm{e}^{-qT_B}]\mathrm{d}x \le \operatorname{Leb} K.$$

Thus $\{m_B^q\colon 0 < q \le 1\}$ is uniformly bounded. Hence we can choose a sequence $m_B^{q_n}$, $q_n \to 0$, convergent to a finite measure ρ on $\overline{B}$. Let f be a nonnegative continuous function with compact support. Then $\widetilde{U}^q f$ and $\widetilde{U} f$ are continuous. As $q \downarrow 0$, $\widetilde{U}^q f$ tends to $\widetilde{U} f$ uniformly on any compact set, since the increasing convergence on a compact set of a sequence of continuous functions to a continuous function is uniform (Dini's theorem). Hence

$$\int_{\overline{B}} (\widetilde{U} f(x) - \widetilde{U}^q f(x)) m_B^q(\mathrm{d}x) \to 0, \qquad q \to 0.$$

On the other hand,

$$\int_{\overline{B}} \widetilde{U}^q f(x) m_B^q(\mathrm{d}x) = \int_{\mathbb{R}^d} f(x) E^x[\mathrm{e}^{-qT_B}]\mathrm{d}x \to \int_{\mathbb{R}^d} f(x) P^x[T_B < \infty]\mathrm{d}x.$$

Thus $\int_{\overline{B}} \widetilde{U} f(x) m_B^q(\mathrm{d}x)$ has the same limit when $q \to 0$. It follows that

$$\int \rho(\mathrm{d}x)\widetilde{U} f(x) = \int_{\mathbb{R}^d} f(x) P^x[T_B < \infty]\mathrm{d}x.$$

This shows (42.9).

To show the uniqueness, let ρ and ρ' satisfy (42.9). Then $\rho\widetilde{U}$ and $\rho'\widetilde{U}$ are identical and they are finite for compact sets. It follows from the resolvent equation (41.2) that $\rho\widetilde{U}^q = \rho'\widetilde{U}^q$ for $q > 0$ and they are finite for compact sets. As in the last part of the proof of Theorem 42.5, Proposition 42.7 applies and $\rho = \rho'$. □

PROPOSITION 42.10. *Let B be an F_σ set. Then, for any $q > 0$,*

$$C^q(B) = \widetilde{C}^q(B), \tag{42.10}$$

where $\widetilde{C}^q(B) = \widetilde{m}^q_B(\overline{B})$. If the process is transient and if B is a bounded F_σ set, then

$$C^0(B) = \lim_{q \downarrow 0} C^q(B) \tag{42.11}$$

and (42.10) is true also for $q = 0$.

Proof. We have, for $q > 0$,

$$C^q(B) = q \int_{\mathbb{R}^d} \widetilde{E}^x[\mathrm{e}^{-qT_B}]\mathrm{d}x \tag{42.12}$$

by (42.7). Hence

$$C^q(B) = q^2 \int \widetilde{P}^q_B \widetilde{U}^q 1(x)\mathrm{d}x = q^2 \int P^q_B U^q 1(x)\mathrm{d}x = \widetilde{C}^q(B)$$

by (42.4). Assume the transience of the process and compactness of $\overline{B}$. It is shown in the proof of Theorem 42.8 that

$$\lim_{q \downarrow 0} \int_{\overline{B}} f(x) m^q_B(\mathrm{d}x) = \int_{\overline{B}} f(x) m_B(\mathrm{d}x)$$

for any continuous function f on $\overline{B}$. Letting $f = 1$ on $\overline{B}$, we get (42.11). (42.10) for $q = 0$ is a consequence. □

REMARK 42.11. Let the process be transient and B be a bounded F_σ set. If the process has continuous sample functions, then m_B is concentrated on the boundary ∂B of B. Indeed, for the interior B^0 of B,

$$\begin{aligned} m^q_B(B^0) &= q \textstyle\int_{\mathbb{R}^d} \widetilde{E}^x[\mathrm{e}^{-qT_B}; X_{T_B} \in B^0]\mathrm{d}x = q \int_{\overline{B}} \widetilde{E}^x[\mathrm{e}^{-qT_B}; X_{T_B} \in B^0]\mathrm{d}x \\ &\le q \,\mathrm{Leb}(\overline{B}) \to 0 \quad \text{as } q \downarrow 0, \end{aligned}$$

which shows that $m_B(B^0) = 0$. It follows that m_B is concentrated on ∂B.

PROPOSITION 42.12. *Let $q > 0$. Let B, B', and B_n be F_σ sets.*

(i) *If B is bounded, then $C^q(B) < \infty$.*

(ii) *If $B \subset B'$, then $C^q(B) \le C^q(B')$.*

(iii) $C^q(B \cup B') + C^q(B \cap B') \le C^q(B) + C^q(B')$.

(iv) *If B_n, $n = 1, 2, \ldots$, are increasing and $\bigcup_{n=1}^\infty B_n = B$, then $C^q(B_n) \uparrow C^q(B)$.*

(v) $C^q(B) = \inf\{C^q(D) \colon D \text{ open and } D \supset B\}$.

(vi) $C^q(B) = C^q(-B) = C^q(B + x)$ *for $x \in \mathbb{R}^d$.*

Proof. The assertion (i) is a part of Theorem 42.5. The assertion (ii) follows from (42.12) and from $T_B \geq T_{B'}$ for $B \subset B'$. To show (iii), note that

$$\begin{aligned}\widetilde{P}^x[T_{B\cap B'} \leq t] &\leq \widetilde{P}^x[T_B \leq t \text{ and } T_{B'} \leq t]\\ &= \widetilde{P}^x[T_B \leq t] + \widetilde{P}^x[T_{B'} \leq t] - \widetilde{P}^x[T_B \leq t \text{ or } T_{B'} \leq t]\\ &= \widetilde{P}^x[T_B \leq t] + \widetilde{P}^x[T_{B'} \leq t] - \widetilde{P}^x[T_{B\cup B'} \leq t]\end{aligned}$$

and that, for $h(t) = \widetilde{P}^x[0 < T_B \leq t]$,

$$\begin{aligned}\widetilde{E}^x[\mathrm{e}^{-qT_B}] &= \widetilde{P}^x[T_B = 0] + \int_0^\infty \mathrm{e}^{-qt}\mathrm{d}h(t)\\ &= \widetilde{P}^x[T_B = 0] + q\int_0^\infty \mathrm{e}^{-qt}h(t)\mathrm{d}t = q\int_0^\infty \widetilde{P}^x[T_B \leq t]\mathrm{e}^{-qt}\mathrm{d}t\end{aligned}$$

and use (42.12). The proof of (iv) is from $T_{B_n} \downarrow T_B$ and (42.12).

Let us prove (v). First consider the case that B is compact. Let $D_n = \{x\colon \mathrm{dis}(x,B) < n^{-1}\}$, an open set. Then $\widetilde{P}^x[T_{D_n} \uparrow T_B] = 1$ for any x by quasi-left-continuity. Hence $\widetilde{E}^x[\mathrm{e}^{-qT(D_n)}] \downarrow \widetilde{E}^x[\mathrm{e}^{-qT(B)}]$. Since $q\int \widetilde{E}^x[\mathrm{e}^{-qT(D_1)}]\mathrm{d}x = C^q(D_1) < \infty$, we can use the dominated convergence theorem to get $C^q(D_n) \downarrow C^q(B)$. Next consider the case of a general F_σ set B. Let $\varepsilon > 0$. It is enough to find an open set $D \supset B$ such that $C^q(D) \leq C^q(B) + \varepsilon$. Choose compact sets B_n increasing to B. We can find an open set $D_n \supset B_n$ such that $C^q(D_n) < C^q(B_n) + \varepsilon_n$, where $\varepsilon_n = 2^{-n}\varepsilon$. Then

$$C^q\left(\bigcup_{k=1}^n D_k\right) < C^q(B_n) + \sum_{k=1}^n \varepsilon_k. \tag{42.13}$$

Indeed, this is true for $n = 1$. Assuming (42.13) for n, we have

$$\begin{aligned}C^q\left(\bigcup_{k=1}^{n+1} D_k\right) &\leq C^q\left(\bigcup_{k=1}^n D_k\right) + C^q(D_{n+1}) - C^q\left(\bigcup_{k=1}^n D_k \cap D_{n+1}\right)\\ &\leq C^q\left(\bigcup_{k=1}^n D_k\right) + C^q(D_{n+1}) - C^q(B_n)\\ &\leq C^q(D_{n+1}) + \sum_{k=1}^n \varepsilon_k < C^q(B_{n+1}) + \sum_{k=1}^{n+1} \varepsilon_k\end{aligned}$$

by (ii) and (iii). Hence (42.13) is true for all n. Let $D = \bigcup_{n=1}^\infty D_n$. Then $D \supset B$ and (42.13) and (iv) show that $C^q(D) \leq C^q(B) + \varepsilon$.

Since $\widetilde{E}^x[\mathrm{e}^{-qT_B}] = \widetilde{E}^0[\mathrm{e}^{-qT_{B-x}}] = E^0[\mathrm{e}^{-qT_{-B+x}}] = E^{-x}[\mathrm{e}^{-qT_{-B}}]$, we have

$$C^q(B) = q\int E^{-x}[\mathrm{e}^{-qT_{-B}}]\mathrm{d}x = q\int E^x[\mathrm{e}^{-qT_{-B}}]\mathrm{d}x = \widetilde{C}^q(-B) = C^q(-B),$$

$$C^q(B+x) = q\int \widetilde{E}^y[\mathrm{e}^{-qT_{B+x}}]\mathrm{d}y = q\int \widetilde{E}^{y-x}[\mathrm{e}^{-qT_B}]\mathrm{d}y$$
$$= q\int \widetilde{E}^y[\mathrm{e}^{-qT_B}]\mathrm{d}y = C^q(B).$$

This is (vi). □

PROPOSITION 42.13. *Suppose that* (ACP) *holds. For an F_σ set B and $q>0$, m_B^q is characterized as follows: it is a unique measure such that*

$$\int u^q(y-x)m_B^q(\mathrm{d}y) = E^x[\mathrm{e}^{-qT_B}], \qquad x\in\mathbb{R}^d. \tag{42.14}$$

Suppose, further, that the process is transient and that B is bounded. Then m_B is a unique measure such that

$$\int u^0(y-x)m_B(\mathrm{d}y) = P^x[T_B<\infty], \qquad x\in\mathbb{R}^d. \tag{42.15}$$

Proof. Recall that $\widetilde{u}^q(x-y) = u^q(y-x)$. It follows from Theorems 42.5 and 42.8 that (42.14) and (42.15) are true for a. e. x. Since both sides are q-excessive, there are no exceptional points, by Theorem 41.15. The uniqueness follows from that in the two theorems. □

REMARK 42.14. The equilibrium measure m_B is given a meaning related to the last exit time L_B from B in Remark 37.20. Assume (ACP) and transience and let B be a bounded F_σ set. Chung [**81**], p. 212, proves that, if, in addition, $u^0(x)$ is continuous, $u^0(0)=\infty$, and $0<u^0(x)<\infty$ for $x\neq 0$, then

$$\textstyle\int_C u^0(y-x)m_B(\mathrm{d}y) = P^x[L_B>0 \text{ and } X_{L_B-}\in C] \tag{42.16}$$

for $C\in\mathcal{B}_{\mathbb{R}^d}$. See also Getoor and Sharpe [**162**].

PROPOSITION 42.15. *Assume* (ACP). *Let B and D be F_σ sets such that $\overline{B}\subset D^0$, where D^0 is the interior of D. If $q>0$, then*

$$m_B^q(C) = \int m_D^q(\mathrm{d}x)\widetilde{P}_B^q(x,C) \tag{42.17}$$

for any Borel set C and

$$C^q(B) = \int m_D^q(\mathrm{d}x)\widetilde{E}^x[\mathrm{e}^{-qT_B}]. \tag{42.18}$$

If the process is transient and B and D are bounded, then

$$m_B(C) = \int m_D(\mathrm{d}x)\widetilde{P}^x[X_{T_B}\in C] \tag{42.19}$$

for any Borel set C and

$$C(B) = \int m_D(\mathrm{d}x)\widetilde{P}^x[T_B<\infty]. \tag{42.20}$$

Proof. Let $m(C) = \int m_D^q(\mathrm{d}x)\widetilde{P}_B^q(x,C)$. Then, by Corollary 42.4 and Proposition 42.13,

$$\begin{aligned}\int m(\mathrm{d}y)u^q(y-x) &= \iint m_D^q(\mathrm{d}z)\widetilde{P}_B^q(z,\mathrm{d}y)u^q(y-x)\\ &= \int_{\overline{D}} m_D^q(\mathrm{d}z)\int_{\overline{B}} u^q(z-y)P_B^q(x,\mathrm{d}y) = \int_{\overline{B}} E^y[\mathrm{e}^{-qT_D}]P_B^q(x,\mathrm{d}y)\\ &= P_B^q(x,\overline{B}) = E^x[\mathrm{e}^{-qT_B}].\end{aligned}$$

Hence, again by Proposition 42.13, $m = m_B^q$. The remaining assertion in the transient case is proved in the same way. □

REMARK 42.16. Let $(\{X_t\},P)$ be the Brownian motion on $\mathbb{R}^d$, $d \geq 3$. For any compact set B, we have $C(B) = C(\partial B)$. This is seen from (42.20), since m_D is concentrated on ∂D by Remark 42.11 and $P^x[T_B < \infty] = P^x[T_{\partial B} < \infty]$ for $x \in \partial D$.

EXAMPLE 42.17. Suppose that the process $(\{X_t\},P^0)$ is strictly α-stable with $0 < \alpha \leq 2$. Then, for any F_σ set B,

$$C^q(aB) = a^{d-\alpha}C^{a^\alpha q}(B) \quad \text{for } q > 0,\ a > 0. \tag{42.21}$$

If $d > \alpha$ and the process is genuinely d-dimensional, then the process is transient (Theorems 37.8, 37.16, and 37.18), and

$$C(aB) = a^{d-\alpha}C(B) \quad \text{for } a > 0. \tag{42.22}$$

In fact, we have

$$C^q(aB) = q\textstyle\int \widetilde{E}^x[\mathrm{e}^{-qT_{aB}}]\mathrm{d}x = q\int \widetilde{E}^0[\mathrm{e}^{-qT_{aB-x}}]\mathrm{d}x.$$

Since $T_{aB-x} = \inf\{t > 0\colon X_t \in aB - x\} = \inf\{t > 0\colon a^{-1}X_t \in B - a^{-1}x\}$, the strict α-stability implies

$$T_{aB-x} \overset{\mathrm{d}}{=} \inf\{t > 0\colon X_{a^{-\alpha}t} \in B - a^{-1}x\} \quad \text{under } P^0.$$

Thus $T_{aB-x} \overset{\mathrm{d}}{=} a^\alpha T_{B-a^{-1}x}$ under P^0. Hence

$$\begin{aligned}C^q(aB) &= q\textstyle\int \widetilde{E}^0[\mathrm{e}^{-qa^\alpha T(B-a^{-1}x)}]\mathrm{d}x = q\int \widetilde{E}^{a^{-1}x}[\mathrm{e}^{-qa^\alpha T_B}]\mathrm{d}x\\ &= qa^d\textstyle\int \widetilde{E}^x[\mathrm{e}^{-qa^\alpha T_B}]\mathrm{d}x = a^{d-\alpha}C^{a^\alpha q}(B).\end{aligned}$$

In the transient case, (42.21) tends to (42.22) by (42.11).

In particular, for the Brownian motion on $\mathbb{R}^d$, $d \geq 3$, $B_a = \{x\colon |x| \leq a\}$ has capacity

$$C(B_a) = a^{d-2}C(B_1). \tag{42.23}$$

See Exercise 44.12 for the evaluation of $C(B_1)$.

REMARK 42.18. Let $(\{X_t\},P)$ be a rotation invariant α-stable process on $\mathbb{R}^d$ with $\widehat{\mu}(z) = \mathrm{e}^{-|z|^\alpha}$, $0 < \alpha < 2$. The explicit form of $u^0(x)$ is given in Example 37.19(ii). Blumenthal, Getoor, and Ray [**47**] calculate the following quantities.

Let $B_0 = \{x\colon |x| \le 1\}$ and $B_1 = \{x\colon |x| \ge 1\}$. Let $b_{d,\alpha} = \pi^{-d/2-1}\Gamma(\frac{d}{2})\sin(\frac{\pi\alpha}{2}) = \pi^{-d/2}\Gamma(\frac{d}{2})/(\Gamma(\frac{\alpha}{2})\Gamma(\frac{2-\alpha}{2}))$. For $|x| < 1$ and C Borel in B_1,

$$P^x[X_{T(B_1)} \in C] = b_{d,\alpha}(1-|x|^2)^{\alpha/2}\int_C (|y|^2-1)^{-\alpha/2}|x-y|^{-d}\mathrm{d}y.$$

For $d=1$ and $|x|<1$,

$$P^x[X_{T(B_1)} \ge 1] = 2^{1-\alpha}\tfrac{\Gamma(\alpha)}{(\Gamma(\alpha/2))^2}\int_{-1}^{x}(1-r^2)^{\alpha/2-1}\mathrm{d}r.$$

In particular, for $d=\alpha=1$ and $|x|<1$,

$$P^x[X_{T(B_1)} \ge 1] = \tfrac{1}{2} + \tfrac{1}{\pi}\arcsin x,$$

which is found by Spitzer [**495**].

For $|x|>1$ and C Borel in B_0,

$$P^x[X_{T(B_0)} \in C] = b_{d,\alpha}(|x|^2-1)^{\alpha/2}\int_C (1-|y|^2)^{-\alpha/2}|x-y|^{-d}\mathrm{d}y$$

whenever $\alpha < d\wedge 2$ or $\alpha = d = 1$, and

$$P^x[X_{T(B_0)} \in C] = b_{1,\alpha}(x^2-1)^{\alpha/2}\int_C (1-y^2)^{-\alpha/2}|x-y|^{-1}\mathrm{d}y$$
$$- b_{1,\alpha}(\alpha-1)\int_1^{|x|}(r^2-1)^{2/\alpha-1}\mathrm{d}r\int_C (1-y^2)^{-\alpha/2}\mathrm{d}y$$

whenever $d=1<\alpha<2$.

In the transient case (that is, $\alpha<d$), the 0-potential density is calculated in Example 37.19, and

$$m_{B_0}(\mathrm{d}y) = \tfrac{2^\alpha\Gamma(d/2)}{\Gamma((d-\alpha)/2)\Gamma((2-\alpha)/2)}(1-|y|^2)^{-\alpha/2}\mathrm{d}y,$$

$$P^x[T_{B_0}=\infty] = \tfrac{\Gamma(d/2)}{\Gamma((d-\alpha)/2)\Gamma(\alpha/2)}\int_0^{|x|^2-1}(r+1)^{-d/2}r^{\alpha/2-1}\mathrm{d}r \qquad \text{for } |x|>1.$$

In one dimension ($d=1$) for general $0<\alpha<2$, Ray [**410**] finds that, for $b>0$ and for C Borel in $[b,\infty)$,

$$P^0[X(T_{[b,\infty)}) \in C] = \tfrac{\sin(\pi\alpha/2)}{\pi}\int_C y^{-1}(b/(y-b))^{\alpha/2}\mathrm{d}y.$$

Nullity and positivity of the capacity are expressed by sample function behavior.

THEOREM 42.19. *Let B be an F_σ set. The following are equivalent.*

(1) $C^q(B)=0$ *for some* $q>0$.

(2) $C^q(B)=0$ *for all* $q>0$.

(3) *B is essentially polar.*

If the process is transient and B is bounded, then the following condition is also equivalent.

(4) $C^0(B)=0$.

Proof. We see from (42.10) and from the dual of (42.12) that $C^q(B)=0$ if and only if $E^x[\mathrm{e}^{-qT_B}]=0$ for a. e. x. Hence, $C^q(B)=0$ if and only if $P^x[T_B=\infty]=1$ for a. e. x. Thus (1), (2), and (3) are equivalent. Assume transience and boundedness of B. If (2) holds, then (4) holds by (42.11). Conversely, if (4) holds, then $m_B=0$ and hence $P^x[T_B<\infty]=0$ for a. e. x by (42.9), that is, B is essentially polar. □

COROLLARY 42.20. *An F_σ set B is essentially polar if and only if it is essentially co-polar.*

Proof. Consequence of the theorem above and (42.10).

PROPOSITION 42.21. *Suppose that $\Sigma = \mathbb{R}^d$, where Σ is the support of the process. If B is a Borel set with* $\mathrm{Leb}\, B > 0$, *then, for any $q > 0$,*

$$U^q(x, B) > 0 \quad \textit{for a. e. } x. \tag{42.24}$$

The conclusion holds for all x under Condition (ACP).

Proof. Assume that $\mathrm{Leb}\, B > 0$. Suppose that, for some $q > 0$, there is a Borel set C with $\mathrm{Leb}\, C > 0$ such that $U^q(x, B) = 0$ for $x \in C$. As in the proof of Theorem 41.19(i), we have $\int_C U^q(x, B)\mathrm{d}x = \int_{\mathbb{R}^d} f(y)V^q(\mathrm{d}y)$ for a nonnegative continuous function f not identically 0. Since V^q has support $\mathbb{R}^d$ by Exercise 44.1, $\int f(y)V^q(\mathrm{d}y) > 0$, which is a contradiction. Hence (42.24) is true. Assume (ACP). Given $q > 0$, choose $q' > q$ and note that $U^q(x, B) \geq (q' - q)U^q U^{q'} 1_B(x)$ by (41.2). Since $U^{q'} 1_B$ is positive a. e. and $U^q(x, \cdot)$ is absolutely continuous, we have $U^q(x, B) > 0$. □

THEOREM 42.22. *Suppose that $\Sigma = \mathbb{R}^d$. Let B be an F_σ set. The following are equivalent.*

(1) *$C^q(B) > 0$ for some $q > 0$.*
(2) *$C^q(B) > 0$ for all $q > 0$.*
(3) *$P^x[T_B < \infty] > 0$ for a. e. x.*
(4) *$\widetilde{P}^x[T_B < \infty] > 0$ for a. e. x.*

If (ACP) *is satisfied, then the conditions (3) and (4) with "a. e. x" replaced by "all x" are also equivalent.*

Proof. By Theorem 42.19 and (42.10), we already know that (1) $\Leftrightarrow$ (2), (3) $\Rightarrow$ (1), and (4) $\Rightarrow$ (1). Let us prove that (1) $\Rightarrow$ (3). Theorem 42.19 says that (1) implies that there is a Borel set D with $\mathrm{Leb}\, D > 0$ such that $P^x[T_B < \infty] > 0$ for $x \in D$. Hence there are a Borel subset C of D and $k > 0$ such that $\mathrm{Leb}\, C > 0$ and $P^x[T_B < \infty] \geq k$ for $x \in C$. For every x and s,

$$\begin{aligned} P^x[T_B < \infty] &\geq P^x[X_t \in B \text{ for some } t > s] \\ &= E^x[P^{X_s}[T_B < \infty]] \geq kP^x[X_s \in C]. \end{aligned}$$

Hence

$$P^x[T_B < \infty] \geq kq \int_0^\infty \mathrm{e}^{-qs} P^x[X_s \in C]\mathrm{d}s = kqU^q(x, C).$$

Therefore we get (3) by Proposition 42.21. By (42.10) (1) implies (4) for the same reason. Under (ACP) "a. e. x" can be replaced by "all x", as is proved in Proposition 42.21. □

As before, the distribution of X_t at $t = 1$ under P^0 is denoted by μ. The distinguished logarithm of the characteristic function $\widehat{\mu}(z)$ is denoted by $\psi(z) = \log \widehat{\mu}(z)$.

LEMMA 42.23. *Let ρ be a probability measure on $\mathbb{R}^d$ with integrable characteristic function $\widehat{\rho}(z)$. Let f be the bounded continuous density of ρ (see Proposition 2.5(xii)). Then*

$$\int_{\mathbb{R}^d} f(x)U^q f(x)\mathrm{d}x = (2\pi)^{-d} \int_{\mathbb{R}^d} |\widehat{\rho}(z)|^2 \mathrm{Re}\left(\frac{1}{q-\psi(z)}\right)\mathrm{d}z \tag{42.25}$$

for $q > 0$.

Note that $0 < \mathrm{Re}\left(\frac{1}{q-\psi(z)}\right) \leq q^{-1}$, since $\mathrm{Re}\left(\frac{1}{q-\psi(z)}\right) = \frac{\mathrm{Re}\,(q-\psi(z))}{|q-\psi(z)|^2} \leq \frac{1}{\mathrm{Re}\,(q-\psi(z))}$ and since $\mathrm{Re}\,\psi(z) \leq 0$.

Proof of lemma. Let Ff be the Fourier transform of f given in Definition 37.1. Then $\widehat{\rho}(z) = Ff(z)$. Using Proposition 37.4 and Fubini's theorem, we get

$$\begin{aligned}\int f(x)U^q f(x)\mathrm{d}x &= (2\pi)^{-d} \int f(x)\mathrm{d}x \int \widehat{\rho}(-z)\frac{\mathrm{e}^{\mathrm{i}\langle x,z\rangle}}{q-\psi(z)}\mathrm{d}z \\ &= (2\pi)^{-d} \int \widehat{\rho}(z)\widehat{\rho}(-z)\frac{1}{q-\psi(z)}\mathrm{d}z,\end{aligned}$$

which is the right-hand side of (42.25). ☐

Based on this lemma, we define the q-energy integral of a general probability measure ρ and q-energy of a set.

DEFINITION 42.24. Let $q > 0$. The *q-energy integral* $I^q(\rho)$ of a probability measure ρ on $\mathbb{R}^d$ is

$$I^q(\rho) = (2\pi)^{-d} \int_{\mathbb{R}^d} |\widehat{\rho}(z)|^2 \mathrm{Re}\left(\frac{1}{q-\psi(z)}\right)\mathrm{d}z. \tag{42.26}$$

The *q-energy* $e^q(B)$ of a nonempty Borel set B is

$$e^q(B) = \inf\{I^q(\rho)\colon \rho \text{ is a probability measure with } \rho(B) = 1\}. \tag{42.27}$$

$I^q(\rho)$ and $e^q(B)$ are nonnegative and possibly infinite. The following are some simple properties of them:

$$I^q(\rho * \delta_x) = I^q(\rho) \quad \text{for } x \in \mathbb{R}^d; \tag{42.28}$$

$$I^q(\rho) \leq I^q(\delta_0) = (2\pi)^{-d} \int_{\mathbb{R}^d} \mathrm{Re}\left(\frac{1}{q-\psi(z)}\right)\mathrm{d}z \leq \infty; \tag{42.29}$$

$$\text{if } A \subset B, \text{ then } e^q(A) \geq e^q(B); \tag{42.30}$$

$$e^q(B) = e^q(-B) = e^q(B+x) \quad \text{for } x \in \mathbb{R}^d; \tag{42.31}$$

$$e^q(\{x\}) = e^q(\{0\}) = (2\pi)^{-d} \int_{\mathbb{R}^d} \mathrm{Re}\left(\frac{1}{q-\psi(z)}\right)\mathrm{d}z \leq \infty. \tag{42.32}$$

PROPOSITION 42.25. *If B is a bounded Borel set, then $e^q(B) > 0$.*

Proof. Suppose that $e^q(B) = 0$. Then there are probability measures ρ_n with $\rho_n(B) = 1$ and $I^q(B_n) \to 0$. A subsequence $\{\rho_{n_k}\}$ tends to a probability measure ρ on $\overline{B}$. Thus

$$\begin{aligned} I^q(\rho) &= (2\pi)^{-d} \int |\widehat{\rho}(z)|^2 \mathrm{Re}\left(\frac{1}{q-\psi(z)}\right)\mathrm{d}z \\ &\leq \liminf_{k\to\infty} (2\pi)^{-d} \int |\widehat{\rho}_{n_k}(z)|^2 \mathrm{Re}\left(\frac{1}{q-\psi(z)}\right)\mathrm{d}z = \liminf_{k\to\infty} I^q(\rho_{n_k}) = 0. \end{aligned}$$

Hence $\widehat{\rho}(z) = 0$ a. e. This is absurd. □

PROPOSITION 42.26. *Let B be open and bounded. Then, for $q > 0$,*

$$e^q(B) = \inf \int_{\mathbb{R}^d} f(x) U^q f(x) \mathrm{d}x, \tag{42.33}$$

where the infimum is taken over all nonnegative continuous functions f on $\mathbb{R}^d$ such that $f = 0$ on $\mathbb{R}^d \setminus B$ and $\int f \mathrm{d}x = 1$.

Proof. Denote the right-hand side of (42.33) by I. Let σ^n be a probability measure supported by $\{x\colon |x| \leq 1/n\}$ with $\widehat{\sigma^n}$ being nonnegative and integrable. Such $\sigma^n = a^n(x)\mathrm{d}x$ is given by choosing a^n as a constant multiple of the Fourier transform of the function in Proposition 37.3. Let $B_k = \{x \in B\colon \mathrm{dis}(x, \mathbb{R}^d \setminus B) > 1/k\}$. Let b_k be a continuous function, $0 \leq b_k \leq 1$, $b_k = 1$ on B_{k-1}, and $b_k = 0$ on $\mathbb{R}^d \setminus B_k$.

Step 1. Let f be nonnegative, continuous on $\mathbb{R}^d$ with $\int f \mathrm{d}x = 1$ and $f = 0$ on $\mathbb{R}^d \setminus B$. Write $\rho = f\mathrm{d}x$. For large k, let $\rho_k = f_k \mathrm{d}x$, where $f_k = \left(\int f(y) b_k(y) \mathrm{d}y\right)^{-1} f(x) b_k(x)$. Let $\rho_k^n = \sigma^n * \rho_k$ with $n > k$. Since $\widehat{\rho_k^n}(z) = \widehat{\sigma^n}(z)\widehat{\rho_k}(z)$ is integrable, the density f_k^n of ρ_k^n satisfies $\int f_k^n U^q f_k^n \mathrm{d}x = I^q(\rho_k^n)$ by Lemma 42.23. Since $\rho_k^n(B) = 1$, we have $e^q(B) \leq \int f_k^n U^q f_k^n \mathrm{d}x$. For each k, $f_k^n(x) = \int \sigma^n(\mathrm{d}y) f_k(x-y) \to f_k(x)$ boundedly as $n \to \infty$. Hence $\int f_k^n U^q f_k^n \mathrm{d}x \to \int f_k U^q f_k \mathrm{d}x$ as $n \to \infty$. Since $f_k \to f$ boundedly as $k \to \infty$, we have $\int f_k U^q f_k \mathrm{d}x \to \int f U^q f \mathrm{d}x$. Hence $e^q(B) \leq I$.

Step 2. Given $0 < \varepsilon < 1$, choose a probability measure ρ such that $\rho(B) = 1$ and $I^q(\rho) \leq e^q(B) + \varepsilon$. Then choose k such that $\rho(B_{k-1}) > 1 - \varepsilon$. Define $\rho_k = c_k{}^{-1} b_k(x) \rho(\mathrm{d}x)$ with $c_k = \int b_k(y) \rho(\mathrm{d}y)$ and $\rho^n = f^n \mathrm{d}x = \sigma^n * \rho$. Let $\rho_k^n = f_k^n \mathrm{d}x = \sigma^n * \rho_k$ with $n > k$. Since $|\widehat{\rho^n}| \leq |\widehat{\rho}|$, we have $I^q(\rho^n) \leq I^q(\rho)$. Lemma 42.23 tells us that $\int f^n U^q f^n \mathrm{d}x = I^q(\rho^n)$. Since $f^n(x) = \int a^n(x-y)\rho(\mathrm{d}y)$ and $f_k^n(x) = \int a^n(x-y)\rho_k(\mathrm{d}y) = c_k{}^{-1} \int a^n(x-y) b_k(y) \rho(\mathrm{d}y)$, we have $f^n \geq c_k f_k^n \geq \rho(B_{k-1}) f_k^n \geq (1-\varepsilon) f_k^n$. It follows that $I^q(\rho^n) \geq (1-\varepsilon)^2 \int f_k^n U^q f_k^n \mathrm{d}x$. Noting $f_k^n = 0$ on $\mathbb{R}^d \setminus B$, we see that $I \leq (1-\varepsilon)^{-2}(e^q(B) + \varepsilon)$. Hence $I \leq e^q(B)$. □

THEOREM 42.27. *Let $q > 0$. If B is a bounded open set or a compact set, then*

$$\frac{1}{4e^q(B)} \le C^q(B) \le \frac{1}{e^q(B)}. \tag{42.34}$$

Proof. Step 1. Upper bound. Let B be bounded and open. Use B_k and σ^n in the proof of the preceding proposition. Let $\rho = C^q(B)^{-1}m_B^q$, $\rho_k = C^q(B_k)^{-1}m_{B_k}^q$, and $\rho_k^n = f_k^n \mathrm{d}x = \sigma^n * \rho_k$ for $n > k$. Then, by Theorem 42.5,

$$\int m_B^q(\mathrm{d}x)\widetilde{U}^q f_k^n = \int f_k^n \mathrm{d}x E^x[\mathrm{e}^{-qT_B}] = \int_B f_k^n \mathrm{d}x = 1,$$

while

$$\begin{aligned}\int f_k^n \mathrm{d}x E^x[\mathrm{e}^{-qT_B}] &\ge \int f_k^n \mathrm{d}x E^x[\mathrm{e}^{-qT(B_k)}] = \int m_{B_k}^q(\mathrm{d}x)\widetilde{U}^q f_k^n \\ &= (2\pi)^{-d}\int m_{B_k}^q(\mathrm{d}x)\int \widetilde{U}^q(x,\mathrm{d}y)\int \mathrm{e}^{\mathrm{i}\langle y,z\rangle}Ff_k^n(-z)\mathrm{d}z \\ &= (2\pi)^{-d}C^q(B_k)\int |\widehat{\rho_k}(z)|^2\widehat{\sigma^n}(-z)\frac{\mathrm{d}z}{\overline{q-\psi(z)}} \\ &= (2\pi)^{-d}C^q(B_k)\int |\widehat{\rho_k}(z)|^2\widehat{\sigma^n}(-z)\mathrm{Re}\left(\frac{1}{q-\psi(z)}\right)\mathrm{d}z,\end{aligned}$$

since $Ff_k^n = \widehat{\rho_k^n} = \widehat{\sigma^n}\widehat{\rho_k}$ is integrable. As $n \to \infty$, $\widehat{\sigma^n}(-z) \to 1$ and, by Fatou's lemma,

$$\begin{aligned}1 &\ge (2\pi)^{-d}C^q(B_k)\int |\widehat{\rho_k}(z)|^2\mathrm{Re}\left(\frac{1}{q-\psi(z)}\right)\mathrm{d}z \\ &= C^q(B_k)I^q(\rho_k) \ge C^q(B_k)e^q(B).\end{aligned}$$

Letting $k \to \infty$, we get $1 \ge C^q(B)e^q(B)$.

Let B be compact and $D_n = \{x\colon \mathrm{dis}(x,B) < 1/n\}$, an open set. Given $\varepsilon > 0$, choose a probability measure ρ_n on D_n such that $I^q(\rho_n) \le e^q(D_n) + \varepsilon \le C^q(D_n)^{-1} + \varepsilon \le C^q(B)^{-1} + \varepsilon$. Choose a subsequence $\{\rho_{n_k}\}$ convergent to some ρ. Then $S_\rho \subset B$, since B is compact. By Fatou's lemma $I^q(\rho) \le \liminf_{k\to\infty} I^q(\rho_{n_k})$. Hence $e^q(B) \le C^q(B)^{-1} + \varepsilon$.

Step 2. Lower bound. Let B be bounded and open. For any $\varepsilon > 0$ we can choose f nonnegative and continuous, with $\int f\mathrm{d}x = 1$ such that $f = 0$ on $\mathbb{R}^d \setminus B$ and

$$\int fU^q f\mathrm{d}x \le e^q(B)(1+\varepsilon),$$

using Propositions 42.25 and 42.26. Let $D = \{x \in B\colon \widetilde{U}^q f(x) < 2e^q(B)\}$. Since $\widetilde{U}^q f$ is continuous, D is open. Since $\int f\widetilde{U}^q f\mathrm{d}x \ge 2e^q(B)\int_{B\setminus D} f\mathrm{d}x$, we have $\int_{B\setminus D} f\mathrm{d}x \leq (1+\varepsilon)/2$. Hence $\int_D f\mathrm{d}x \geq (1-\varepsilon)/2$. Now

$$\int m_D^q(\mathrm{d}x)\widetilde{U}^q f = \int f\mathrm{d}x E^x[\mathrm{e}^{-qT_D}] \geq \int_D f\mathrm{d}x \geq (1-\varepsilon)/2,$$

while

$$\int m_D^q(\mathrm{d}x)\widetilde{U}^q f = \int_{\overline{D}} m_D^q(\mathrm{d}x)\widetilde{U}^q f \leq 2e^q(B)m_D^q(\overline{D}) \leq 2e^q(B)C^q(B).$$

Hence $C^q(B) \geq (1-\varepsilon)/(4e^q(B))$. Hence $C^q(B) \geq 1/(4e^q(B))$.

If B is compact, choose bounded open sets $D_n \supset B$ with $C^q(D_n) \to C^q(B)$, using Proposition 42.12(v). Notice that $C^q(D_n) \geq 1/(4e^q(D_n)) \geq 1/(4e^q(B))$. We have $C^q(B) \geq 1/(4e^q(B))$. □

REMARK 42.28. Suppose that the process is symmetric and satisfies (ACT). Then the following fact is known (see Chung [**81**], p. 226). Let $q > 0$ and let B be a compact set with $C^q(B) > 0$. Then, for any probability measure ρ with $\rho(B) = 1$,

$$1/C^q(B) \leq \iint \rho(\mathrm{d}x)u^q(y-x)\rho(\mathrm{d}y). \tag{42.35}$$

The equality holds if and only if $\rho = C^q(B)^{-1}m_B^q$. This is an extension of the classical result of Gauss and Frostman. It follows from (42.35) that (42.34) is strengthened to

$$C^q(B) = 1/e^q(B) \tag{42.36}$$

in this case. Brownian motion, nondegenerate symmetric stable processes and, more generally, nondegenerate symmetric semi-stable processes are examples.

On the other hand, it is known that (42.36) does not hold in general. Actually, Hawkes [**185, 190**] shows that, for any $\varepsilon > 0$, there is a strictly α-stable process with $0 < \alpha < 1$ on $\mathbb{R}$ such that $C^q(B) < (\frac{1}{2}+\varepsilon)/e^q(B)$ for some $q > 0$ and a bounded interval B.

THEOREM 42.29. *For $j = 1$ and 2 let $(\{X_t\}, P_j)$ be Lévy processes on $\mathbb{R}^d$. The quantities related to $(\{X_t\}, P_j)$ are denoted by the subscript j. Fix $q > 0$. Assume that there is a constant $k > 0$ such that*

$$\operatorname{Re}\left(\frac{1}{q-\psi_2(z)}\right) \geq k\operatorname{Re}\left(\frac{1}{q-\psi_1(z)}\right), \qquad z \in \mathbb{R}^d. \tag{42.37}$$

Then,

$$C_1^q(B) \geq (k/4)C_2^q(B) \tag{42.38}$$

for all F_σ sets B. If B is an essentially polar F_σ set for $(\{X_t\}, P_1)$, then it is essentially polar for $(\{X_t\}, P_2)$.

Recall that the functions $\operatorname{Re}\left(\frac{1}{q-\psi_j(z)}\right)$ are positive, bounded, and continuous. The existence of k satisfying (42.37) depends only on their behavior outside any compact set.

Proof of theorem. The assumption (42.37) implies $I_2^q(\rho) \geq kI_1^q(\rho)$ for any probability measure ρ. Hence $e_2^q(B) \geq ke_1^q(B)$ for any nonempty Borel set B. Hence, by Theorem 42.27,

$$C_1^q(B) \geq 1/(4e_1^q(B)) \geq k/(4e_2^q(B)) \geq (k/4)C_2^q(B)$$

for B either bounded and open, or compact. The resulting inequality (42.38) is extended to F_σ sets by virtue of Proposition 42.12(v). The last sentence of the theorem is a consequence of Theorem 42.19. □

We apply the preceding theorem to semi-stable processes.

THEOREM 42.30. *Let $(\{X_t\}, P_1)$ and $(\{X_t\}, P_2)$ be nondegenerate Lévy processes on $\mathbb{R}^d$ that satisfy one of the following three assumptions.*

(1) $1 < \alpha \leq 2$. *Both processes are α-semi-stable.*
(2) $0 < \alpha \leq 1$. *Both processes are strictly α-semi-stable.*
(3) $d = 1$, $0 < \alpha \leq 1$. *Both processes are α-semi-stable and neither of them is strictly α-semi-stable.*

For each j let C_j^q be the q-capacity associated with $(\{X_t\}, P_j)$. Then, for each fixed $q > 0$, there are positive constants k and k' such that

$$kC_1^q(B) \leq C_2^q(B) \leq k'C_1^q(B) \quad \textit{for } F_\sigma \textit{ sets } B. \tag{42.39}$$

Consequently, an F_σ set B is polar for $(\{X_t\}, P_1)$ if and only if it is polar for $(\{X_t\}, P_2)$.

Proof. Let $(\{X_t\}, P)$ be either $(\{X_t\}, P_1)$ or $(\{X_t\}, P_2)$. We use the functions $\eta_1(z)$ and $\eta_2(z)$ for $(\{X_t\}, P)$ in Proposition 14.9.

Case (1). Let $1 < \alpha < 2$. We have $\operatorname{Re}\psi(z) = -|z|^\alpha\eta_1(z)$ and $\operatorname{Im}\psi(z) = -|z|^\alpha\eta_2(z) + \langle\gamma_1, z\rangle$. Hence

$$|z|^\alpha\operatorname{Re}\left(\frac{1}{q-\psi(z)}\right) = \frac{|z|^{-\alpha}q + \eta_1(z)}{(|z|^{-\alpha}q + \eta_1(z))^2 + (\eta_2(z) - |z|^{-\alpha}\langle\gamma_1, z\rangle)^2}. \tag{42.40}$$

There are positive constants k_1, k_2, and k_3 such that, for $z \neq 0$, $k_1 \leq \eta_1(z) \leq k_2$ and $|\eta_2(z)| \leq k_3$. The existence of k_2 and k_3 is because of the continuity combined with $\eta_1(bz) = \eta_1(z)$ and $\eta_2(bz) = \eta_2(z)$, where b is a span. The existence of k_1 is by nondegeneracy as in Proposition 24.20. Thus

$$\limsup_{|z|\to\infty} |z|^\alpha\operatorname{Re}\left(\frac{1}{q-\psi(z)}\right) \leq k_1^{-2}k_2$$

and the lim inf is bounded from below by $k_1(k_2^2+k_3^2)^{-1}$. Note that $|z|^{-\alpha}\langle\gamma_1, z\rangle \to 0$ by the assumption $\alpha > 1$. Therefore the two processes $(\{X_t\}, P_1)$ and $(\{X_t\}, P_2)$ satisfy (42.37) as well as the inequality with ψ_1 and ψ_2 interchanged. The case $\alpha = 2$ is similar, as $\operatorname{Re}\psi(z) = -2^{-1}\langle z, Az\rangle$ with A nondegenerate and $\operatorname{Im}\psi(z) = \langle\gamma, z\rangle$.

Case (2). We have (42.40) with $-|z|^{-\alpha}\langle\gamma_1, z\rangle$ deleted. Hence we get the same conclusion. We need to be careful in the case $\alpha = 1$, as we do not have

$\eta_2(bz) = \eta_2(z)$ for general 1-semi-stable processes. But strict 1-semi-stable processes satisfy this relation. See Theorem 14.7 and (14.14).

Case (3). Let $0 < \alpha < 1$. We have (42.40) with γ_1 replaced by $\gamma_0 \neq 0$. Hence

$$|z|^{2-\alpha}\mathrm{Re}\left(\frac{1}{q-\psi(z)}\right) = \frac{|z|^{-\alpha}q + \eta_1(z)}{(|z|^{-1}q + |z|^{\alpha-1}\eta_1(z))^2 + (|z|^{\alpha-1}\eta_2(z) - |z|^{-1}z\gamma_0)^2}.$$

Using the constants k_1 and k_2 as in case (1), we have

$$\limsup_{|z|\to\infty} |z|^{2-\alpha}\mathrm{Re}\left(\frac{1}{q-\psi(z)}\right) \leq |\gamma_0|^{-2}k_2$$

and the lim inf is bigger than or equal to $|\gamma_0|^{-2}k_1$.

Next consider the case $\alpha = 1$. The process $(\{X_t\}, P)$ is a 1-semi-stable process on $\mathbb{R}$ which is not strictly 1-semi-stable. We have $\mathrm{Re}\,\psi(z) = -|z|\eta_1(z)$ and $\mathrm{Im}\,\psi(z) = -|z|\eta_2(z) + \gamma z$ with $\eta_1(bz) = \eta_1(z)$ and $\eta_2(bz) = \eta_2(z) + \theta\,\mathrm{sgn}\,z$, $\theta = \int_{1<|x|\leq b} x\nu(dx) \neq 0$. The constants k_1 and k_2 do exist, but k_3 does not. We have $\eta_2(b^n z) = \eta_2(z) + n\theta\,\mathrm{sgn}\,z$ for $n = 1, 2, \ldots$. By continuity of $\eta_2(z)$ for $z \neq 0$, there is k_4 such that $|\eta_2(z)| \leq k_4$ for $1 \leq |z| \leq b$. We have

$$|z|\mathrm{Re}\left(\frac{1}{q-\psi(z)}\right) = \frac{|z|^{-1}q + \eta_1(z)}{(|z|^{-1}q + \eta_1(z))^2 + (\eta_2(z) - \gamma\,\mathrm{sgn}\,z)^2}.$$

Given $z \neq 0$, choose n such that $b^n < |z| \leq b^{n+1}$, that is, $n < \frac{\log|z|}{\log b} \leq n+1$. Since $\eta_2(z) = \eta_2(b^{-n}z) + n\theta\,\mathrm{sgn}\,z$, we have

$$-k_4 + \tfrac{|\theta|}{\log b}\log|z| - |\theta| \leq |\eta_2(z)| \leq k_4 + \tfrac{|\theta|}{\log b}\log|z|.$$

Thus $|\eta_2(z)| \sim \frac{|\theta|}{\log b}\log|z|$ as $|z| \to \infty$. Hence

$$|z|\mathrm{Re}\left(\frac{1}{q-\psi(z)}\right) \leq \frac{|z|^{-1}q + k_2}{(\eta_2(z) - |\gamma|)^2} \sim \frac{k_2(\log b)^2}{\theta^2(\log|z|)^2},$$

$$|z|\mathrm{Re}\left(\frac{1}{q-\psi(z)}\right) \geq \frac{k_1}{(|z|^{-1}q + k_2)^2 + (|\eta_2(z)| + |\gamma|)^2} \sim \frac{k_1(\log b)^2}{\theta^2(\log|z|)^2}.$$

Again the two processes $(\{X_t\}, P_1)$ and $(\{X_t\}, P_2)$ satisfy (42.37) and also the inequality with ψ_1 and ψ_2 interchanged. This finishes the proof for case (3).

Our processes satisfy (ACP) (see Example 41.21). Thus polarity and essential polarity are equivalent by Theorem 41.15. Hence the last statement of the theorem follows. □

REMARK 42.31. A way of expressing the smallness of a set B in $\mathbb{R}^d$ is its Hausdorff dimension, $\dim_{\mathrm{H}} B$ (see [**34**]). If B has positive d-dimensional Lebesgue measure, then $\dim_{\mathrm{H}} B = d$. If B is a countable set, then $\dim_{\mathrm{H}} B = 0$. The converses are not true. Let $\{X_t\}$ be a nondegenerate Lévy process on $\mathbb{R}^d$. Suppose that it is α-semi-stable with $1 < \alpha \le 2$ or strictly α-semi-stable with $0 < \alpha \le 1$. Then, an F_σ set B is polar if $\dim_{\mathrm{H}} B < d - \alpha$. It is not polar if $\dim_{\mathrm{H}} B > d - \alpha$. The results follow from the rotation invariant case of Taylor [**525**] by Theorem 42.30.

43. Hitting probability and regularity of a point

When the process starts at a point x, let us study whether it hits a point y with positive probability and whether $T_{\{x\}} = 0$ with probability 1. We derive criteria for these properties and consider relations with generating triplets.

DEFINITION 43.1. Let $c^q = C^q(\{x\})$, the q-capacity of a one-point set $\{x\}$. Note that $C^q(\{x\})$ is independent of x by Proposition 42.12(vi). Write $T_x = T_{\{x\}}$ and

$$h^q(x) = E^0[\mathrm{e}^{-qT_x}], \qquad q \ge 0.$$

We understand that $h^0(x) = P^0[T_x < \infty]$. Write

$$\Sigma_0 = \{x\colon P^0[T_x < \infty] > 0\} = \{x\colon P^{-x}[T_0 < \infty] > 0\} = \{x\colon h^q(x) > 0\},$$

the set of points which can be hit from the origin with positive probability. Given an F_σ set B, we call a point x *regular for* B if $P^x[T_B = 0] = 1$; *irregular for* B if $P^x[T_B = 0] = 0$. Note that $P^x[T_B = 0] = 0$ or 1 by Blumenthal's 0–1 law (Proposition 40.4). Write $B^{\mathrm{reg}} = \{x\colon \text{regular for } B\}$ and $B^{\mathrm{irreg}} = \{x\colon \text{irregular for } B\} = \mathbb{R}^d \setminus B^{\mathrm{reg}}$. We say that x is *regular for itself* if x is regular for the set $\{x\}$.

Notice that x is regular for itself if and only if 0 is regular for itself, since $P^x[T_x = 0] = P^0[T_0 = 0]$. In terms of the function h^q, x is regular for itself if and only if $h^q(0) = 1$ for some (equivalently, for all) $q > 0$.

The set Σ_0 is universally measurable by Proposition 40.2 and by the expression $\Sigma_0 = \{x\colon P^{-x}[T_0 < \infty] > 0\}$. Similarly, B^{reg} and B^{irreg} are universally measurable sets.

PROPOSITION 43.2. *The following are equivalent.*

(1) *Some one-point set is essentially polar.*
(2) *Any one-point set is essentially polar.*
(3) $\operatorname{Leb}\Sigma_0 = 0$.

Proof. A set $\{x\}$ is essentially polar if and only if $P^y[T_x = \infty] = 1$ for a. e. y. Since $P^y[T_x = \infty] = P^{y-x}[T_0 = \infty]$, $\{x\}$ is essentially polar if and only if $\{0\}$ is essentially polar. From the definition of Σ_0, $\operatorname{Leb}\Sigma_0 = 0$ if and only if $\{0\}$ is essentially polar. □

Similarly we see that the following are equivalent. (1) Some one-point set is polar. (2) Any one-point set is polar. (3) Some one-point set is co-polar. (4) Any one-point set is co-polar. (5) Σ_0 is empty.

THEOREM 43.3. *Let $q > 0$. The following are equivalent.*

(1) *A one-point set is not essentially polar.*

(2) $c^q > 0$.

(3) *$V^q(\mathrm{d}x)$ has a bounded density.*

(4) *Condition* (ACP) *is satisfied, u^q is bounded, and*

$$c^q u^q(x) = E^{-x}[\mathrm{e}^{-qT_0}] = E^0[\mathrm{e}^{-qT_x}] = h^q(x), \qquad x \in \mathbb{R}^d. \tag{43.1}$$

In the transient case the equivalence holds also for $q = 0$.

Proof. Equivalence of (1) and (2) is proved in Theorem 42.19.

(2) $\Rightarrow$ (3), (4). Since the q-capacitary measure of $\{0\}$ is $c^q\delta_0$,

$$c^q\widetilde{V}^q(B) = \int_B E^x[\mathrm{e}^{-qT_0}]\mathrm{d}x, \qquad B \in \mathcal{B}(\mathbb{R}^d),$$

by Theorem 42.5. Hence $\widetilde{V}^q$ is absolutely continuous and it has density $E^x[\mathrm{e}^{-qT_0}]/c^q$. Thus V^q has density $E^{-x}[\mathrm{e}^{-qT_0}]/c^q$ and $E^{-x}[\mathrm{e}^{-qT_0}] = E^0[\mathrm{e}^{-qT_x}] = h^q(x)$. Thus (43.1) holds for a. e. x. By q-co-excessiveness it holds everywhere. The u^q is bounded, since $u^q \le 1/c^q$.

Obviously (4) implies (3). Let us show that (3) implies (2). We assume that V^q is absolutely continuous with density bounded by a constant k. There is the function u^q of Theorem 41.16. We have $u^q \le k$ a. e., hence everywhere. Let $B_{1/n} = \{x\colon |x| \le 1/n\}$. Then $T_{B_{1/n}} \uparrow T_0$ P^x-a. s. for any $x \ne 0$ by quasi-left-continuity. Using (42.3), we have

$$\int P^q_{B_{1/n}}(x, \mathrm{d}z)u^q(-z) = \int u^q(z-x)\widetilde{P}^q_{B_{1/n}}(0, \mathrm{d}z) = u^q(-x). \tag{43.2}$$

Hence $u^q(-x) \le kE^x[\mathrm{e}^{-qT(B_{1/n})}]$. Hence $u^q(-x) \le kE^x[\mathrm{e}^{-qT_0}]$ for $x \ne 0$. It follows that $E^x[\mathrm{e}^{-qT_0}] > 0$ for x of positive Lebesgue measure. Hence $\{0\}$ is not essentially polar.

Let us consider the transient case. Equivalence of (1) and $c^0 > 0$ is known by Theorem 42.19. Equivalence of (2), (3), and (4) for $q = 0$ is proved in the same way as above. □

PROPOSITION 43.4. *For any x and y*

$$h^q(x+y) \ge h^q(x)h^q(y), \qquad q \ge 0. \tag{43.3}$$

If Σ_0 is nonempty, then it is closed under addition. If $c^q > 0$, then Σ_0 is a nonempty open set and $\Sigma_0 = \{x\colon u^q(x) > 0\}$. If $c^q > 0$ and $\Sigma = \mathbb{R}^d$, then $\Sigma_0 = \mathbb{R}^d$.

Proof. Let $T = \inf\{t > T_x\colon X_t = x+y\}$. Then

$$h^q(x+y) \ge E^0[\mathrm{e}^{-qT}] = E^0[\mathrm{e}^{-qT_x}E^{X(T_x)}[\mathrm{e}^{-qT_{x+y}}]] = h^q(x)h^q(y).$$

Hence we have (43.3) and $\Sigma_0 = \{x\colon h^q(x) > 0\}$ is closed under addition. If $c^q > 0$, then (43.1) shows that $\Sigma_0 = \{x\colon u^q(x) > 0\}$, which is open by the lower semi-continuity of u^q. Suppose that $c^q > 0$ and $\Sigma = \mathbb{R}^d$. Then, by Theorem 42.22, $P^x[T_0 < \infty] > 0$ for a. e. x, and hence $h^q(x) > 0$ for a. e. x by (43.1). Given an arbitrary $x \in \mathbb{R}^d$, there is y such that $h^q(y) > 0$ and $h^q(x-y) > 0$. Hence $h^q(x) \geq h^q(y)h^q(x-y) > 0$ by (43.3). Hence $\Sigma_0 = \mathbb{R}^d$. □

THEOREM 43.5. *Let $q > 0$. The following are equivalent:*

(1) *$\{0\}$ is not essentially polar and 0 is regular for itself;*
(2) *$V^q(\mathrm{d}x)$ has a bounded density which is continuous at $x = 0$;*
(3) *(ACP) holds and u^q is bounded, continuous, and positive on $\mathbb{R}^d$.*

In the transient case we can replace "$q > 0$" by "$q \geq 0$".

Proof. Assume (1). Let us show (3). By Theorem 43.3 (ACP) holds, $c^q > 0$, and u^q is bounded and satisfies (43.1). Thus $E^x[\mathrm{e}^{-qT_0}] = c^q u^q(-x)$. Since $E^x[\mathrm{e}^{-qT_0}]$ is q-excessive, it is lower semi-continuous. Hence

$$\liminf_{x\to 0} E^x[\mathrm{e}^{-qT_0}] \geq E^0[\mathrm{e}^{-qT_0}] = 1.$$

It follows that $E^x[\mathrm{e}^{-qT_0}]$ is continuous at $x = 0$. Hence $u^q(x)$ is continuous at $x = 0$ and $u^q(0) = 1/c^q$. We have

$$u^q(x+y) \geq c^q u^q(x)u^q(y) \tag{43.4}$$

from (43.1) and (43.3). Therefore $\liminf_{y\to 0} u^q(x+y) \geq u^q(x)$. It follows from (43.4) that $u^q(x) \geq c^q u^q(x+y)u^q(-y)$, and hence $u^q(x) \geq \limsup_{y\to 0} u^q(x+y)$. Thus u^q is continuous. Since $u^q(0) > 0$, there is $\varepsilon > 0$ such that $u^q(x) > 0$ for $|x| < \varepsilon$. This, combined with (43.4), implies that u^q is positive everywhere.

The implication (3) $\Rightarrow$ (2) is trivial.

(2) $\Rightarrow$ (1). By Theorem 43.3, $c^q > 0$ and a one-point set is not essentially polar. Let $v^q(x)$ be a density of V^q, continuous at $x = 0$ and $|v^q| \leq k$. Then

$$u^q(x) = \lim_{r\to\infty} r \int \widetilde{U}^{q+r}(x,\mathrm{d}y)v^q(y) = \lim_{r\to\infty} r \int \widetilde{U}^{q+r}(0,\mathrm{d}y)v^q(x+y)$$

by (41.12). For any $\varepsilon > 0$, there is $\eta > 0$ such that $|v^q(x) - v^q(0)| < \varepsilon$ for $|x| < \eta$. If $|x| < \eta/2$, then

$$\begin{aligned} u^q(x) &\leq \lim_{r\to\infty} rk \int_{|y|\geq\eta/2} \widetilde{U}^{q+r}(0,\mathrm{d}y) + \lim_{r\to\infty} r \int_{|y|<\eta/2} \widetilde{U}^{q+r}(0,\mathrm{d}y)(v^q(0)+\varepsilon) \\ &= v^q(0) + \varepsilon \end{aligned}$$

and, similarly, $u^q(x) \geq v^q(0) - \varepsilon$. In particular, $v^q(0) - \varepsilon \leq u^q(0) \leq v^q(0) + \varepsilon$. Hence $\lim_{x\to 0} u^q(x) = v^q(0) = u^q(0)$. Therefore we have

$$u^q(-x) = E^x[\mathrm{e}^{-qT(B_{1/n})}u^q(-X_{T(B_{1/n})})] \to E^x[\mathrm{e}^{-qT_0}]u^q(0), \qquad n\to\infty,$$

for $x \neq 0$, using (43.2) and quasi-left-continuity. Hence, combined with (43.1), $E^x[\mathrm{e}^{-qT_0}]/c^q = u^q(-x) = E^x[\mathrm{e}^{-qT_0}]u^q(0)$ for $x \neq 0$. Since $E^x[\mathrm{e}^{-qT_0}] >$

0 for some $x \neq 0$, we get $u^q(0) = 1/c^q$. On the other hand, $u^q(0) = E^0[e^{-qT_0}]/c^q$ by (43.1). Thus $E^0[e^{-qT_0}] = 1$, that is, 0 is regular for itself.

In the transient case the same argument works even for $q = 0$. □

REMARK 43.6. A one-point set is not essentially polar (that is, $c^q > 0$) if and only if

$$\int_{\mathbb{R}^d} \mathrm{Re}\left(\tfrac{1}{q-\psi(z)}\right)dz < \infty \tag{43.5}$$

for some (equivalently, for all) $q > 0$. Indeed $c^q > 0$, $e^q(\{0\}) < \infty$, and (43.5) are equivalent by Theorem 42.27 and by (42.32). If

$$\int_{\mathbb{R}^d} \left|\tfrac{1}{q-\psi(z)}\right| dz < \infty \quad \text{for some } q > 0, \tag{43.6}$$

then, in addition, any point is regular for itself. Since $\widehat{(qV^q)}(z) = q/(q - \psi(z))$, V^q has a bounded continuous density in this case and we can use Theorem 43.5.

EXAMPLES 43.7. (i) Let $(\{X_t\}, P^0)$ be the Brownian motion on $\mathbb{R}^2$. Then $\mathrm{Re}\,\frac{1}{q-\psi} = (q + \frac{1}{2}|z|^2)^{-1}$. Thus $\int \mathrm{Re}\,\frac{1}{q-\psi}dz = \infty$ and $\mathrm{Leb}\,\Sigma_0 = 0$. Since (ACP) is satisfied, $\Sigma_0 = \emptyset$ by Theorem 41.15(6).

(ii) Let $(\{X_t\}, P^0)$ be a Cauchy process on $\mathbb{R}$ with γ zero or non-zero. Then $\psi(z) = -c|z| + i\gamma z$ with some $c > 0$. Hence $\mathrm{Re}\,\frac{1}{q-\psi} = (q + c|z|)/((q + c|z|)^2 + \gamma^2 z^2) \sim c(c^2 + \gamma^2)^{-1}|z|^{-1}$ as $|z| \to \infty$. Thus $\int \mathrm{Re}\,\frac{1}{q-\psi}dz = \infty$. Since (ACP) is satisfied, $\Sigma_0 = \emptyset$.

(iii) Let $(\{X_t\}, P^0)$ be a 1-stable, not strictly stable, process on $\mathbb{R}$. Then, by Theorem 14.15, $\psi(z) = -c|z| + ic'z\log|z| + i\tau z$ with $c > 0$, c' a non-zero real, and τ real. Hence $-\mathrm{Re}\,\psi(z) \sim c|z|$ and $\mathrm{Im}\,\psi(z) \sim c'z\log|z|$ as $|z| \to \infty$. It follows that $\mathrm{Re}\,\frac{1}{q-\psi} \sim c|c'|^{-2}|z|^{-1}(\log|z|)^{-2}$. Thus, $\int \mathrm{Re}\,\frac{1}{q-\psi}dz < \infty$ and $\mathrm{Leb}\,\Sigma_0 > 0$. It satisfies (ACP), as Example 41.21 says.

The facts in (ii) and (iii) extend to semi-stable processes on $\mathbb{R}$ [**449**]. That is, if $(\{X_t\}, P^0)$ is non-trivial and strictly 1-semi-stable, then $\Sigma_0 = \emptyset$; if $(\{X_t\}, P^0)$ is non-trivial, 1-semi-stable and not strictly 1-semi-stable, then $\mathrm{Leb}\,\Sigma_0 > 0$. They satisfy (ACT) by Example 28.2.

PROPOSITION 43.8. *Let $B_\varepsilon = \{x\colon |x| \le \varepsilon\}$. The following are equivalent.*

(1) *A one-point set is not essentially polar.*
(2) $\liminf_{\varepsilon\downarrow 0} \varepsilon^{-d}V^q(B_\varepsilon) < \infty$ *for some* $q > 0$.
(3) *For every $q > 0$ we have $c^q > 0$ and*

$$\lim_{\varepsilon\downarrow 0}\sup_{x\in B_\varepsilon} V^q(x + B_\varepsilon)/\,\mathrm{Leb}\,B_\varepsilon = 1/c^q. \tag{43.7}$$

Notice that (43.7) implies that c^q is an increasing function of q.

Proof of proposition. By Theorem 43.3, (1) implies (2). Since $\mathrm{Leb}\,B_\varepsilon = \varepsilon^d\,\mathrm{Leb}\,B_1$, (3) implies (2). Fix q and let $k_\varepsilon = \sup_{x\in B_\varepsilon} V^q(x + B_\varepsilon)/\,\mathrm{Leb}\,B_\varepsilon$.

We claim that, if there are $\varepsilon_n \downarrow 0$ such that k_{ε_n} tends to some $k < \infty$, then (1) holds. We have

$$V^q(x + B_\varepsilon) = E^0[e^{-qT}U^q(X_T, x + B_\varepsilon)] \le E^0[e^{-qT}]\sup_{z\in B_\varepsilon} V^q(z + B_\varepsilon),$$

where $T = T_{x+B_\varepsilon}$. Let G be bounded and open. Then

$$\int_G (V^q(x+B_\varepsilon)/\operatorname{Leb} B_\varepsilon)\mathrm{d}x \le k_\varepsilon \int_G E^0[\mathrm{e}^{-qT(x+B_\varepsilon)}]\mathrm{d}x.$$

The left-hand side equals

$$\int V^q(\mathrm{d}y)\int(1_{B_\varepsilon}(y-x)/\operatorname{Leb} B_\varepsilon)1_G(x)\mathrm{d}x = \int V^q(\mathrm{d}y)\int(1_{B_\varepsilon}(x)/\operatorname{Leb} B_\varepsilon)1_G(y-x)\mathrm{d}x.$$

Let $\varepsilon = \varepsilon_n$ and let $n \to \infty$. If $x \ne 0$, then $T_{x+B(\varepsilon_n)} \uparrow T_x$ P^0-almost surely. Hence

$$\begin{aligned} k\int_G E^0[\mathrm{e}^{-qT_x}]\mathrm{d}x &\ge \int V^q(\mathrm{d}y)\liminf_{n\to\infty}\int(1_{B(\varepsilon_n)}(x)/\operatorname{Leb} B_{\varepsilon_n})1_G(y-x)\mathrm{d}x \\ &\ge \int V^q(\mathrm{d}y)1_G(y) = V^q(G). \end{aligned}$$

Hence, for every Borel set B, $k\int_B E^0[\mathrm{e}^{-qT_x}]\mathrm{d}x \ge V^q(B)$. It follows that V^q is absolutely continuous with density $\le kE^0[\mathrm{e}^{-qT_x}]$. Our claim is thus proved. Since

$$k_\varepsilon \le V^q(B_{2\varepsilon})/\operatorname{Leb} B_\varepsilon = 2^d V^q(B_{2\varepsilon})/\operatorname{Leb} B_{2\varepsilon}, \tag{43.8}$$

we see that (2) implies (1).

Assume (1). Let us prove (3). We see from (43.8) that k_ε is bounded, since V^q has a bounded density. Suppose that $\varepsilon_n \downarrow 0$ is such that k_{ε_n} tends to some k. Let G be bounded and open and K be a compact subset of G. Since

$$V^q(x+B) \ge E^0\left[\int_{T_x}^\infty \mathrm{e}^{-qt}1_{x+B}(X_t)\mathrm{d}t\right] = E^0[\mathrm{e}^{-qT_x}]V^q(B)$$

for any x and B, we have

$$\begin{aligned} (V^q(z+B_\varepsilon)/\operatorname{Leb} B_\varepsilon)\int_K E^0[\mathrm{e}^{-qT_x}]\mathrm{d}x &\le (1/\operatorname{Leb} B_\varepsilon)\int_K V^q(x+z+B_\varepsilon)\mathrm{d}x \\ &= (1/\operatorname{Leb} B_\varepsilon)\int V^q(\mathrm{d}y)\int_K 1_{z+B_\varepsilon}(y-x)\mathrm{d}x, \end{aligned}$$

which is $\le \int V^q(\mathrm{d}y)1_G(y)$ if ε is small enough and if $z \in B_\varepsilon$, because then $1_{z+B_\varepsilon}(y-x) = 0$ for $x \in K$ and $u \notin G$. It follows that $k_\varepsilon\int_K E^0[\mathrm{e}^{-qT_x}]\mathrm{d}x \le V^q(G)$ if ε is small enough. Let $\varepsilon = \varepsilon_n$ and $n \to \infty$. We get

$$k\int_K E^0[\mathrm{e}^{-qT_x}]\mathrm{d}x \le V^q(G).$$

Let K approach G. Combining this with the preceding argument, we get $k\int_B E^0[\mathrm{e}^{-qT_x}]\mathrm{d}x = V^q(B)$ for all Borel set B. Since we know, by Theorem 43.3, that $u^q(x) = E^0[\mathrm{e}^{-qT_x}]/c^q$, we see that $k = 1/c^q$, independently of the sequence ε_n. Hence (43.7) is true. □

THEOREM 43.9. *If $d \ge 2$, then any one-point set is essentially polar, that is, $c^q = 0$.*

REMARK 43.10. Combined with Theorem 41.15(6), any one-point set is polar if $d \ge 2$ and if (ACP) holds. As a matter of fact, Kesten [**272**] proves that, in the genuinely d-dimensional case with $d \ge 2$, if no one-dimensional projection of the process is compound Poisson, then any one-point set is polar. He also proves that, for $d = 1$, if the process is non-zero and not compound Poisson, then any essentially polar one-point set is polar. See Bretagnolle [**68**] for an improved proof. We will not give the proof of this result. To do this, we have to develop the potential theory further. In particular we need measurability of hitting times of sets of a larger class, deeper properties of excessive functions, and introduction of fine topology.

LEMMA 43.11. *Let (A, ν, γ) be the generating triplet of the Lévy process we are considering and let $\psi(z) = \log \widehat{\mu}(z)$ for $z \in \mathbb{R}^d$.*

(i) *If $A = 0$, then $|\psi(z)| = o(|z|^2)$ as $|z| \to \infty$.*

(ii) *If the process is of type A or B and if the drift $\gamma_0 = 0$, then $|\psi(z)| = o(|z|)$ as $|z| \to \infty$.*

Proof. (i) We have $\psi(z) = \int(\mathrm{e}^{\mathrm{i}\langle z,x\rangle} - 1 - \mathrm{i}\langle z,x\rangle 1_{\{|x|\leq 1\}}(x))\nu(\mathrm{d}x) + \mathrm{i}\langle \gamma, z\rangle$. Recall that $\int(1 \wedge |x|^2)\nu(\mathrm{d}x) < \infty$. Since $|\mathrm{e}^{\mathrm{i}\langle z,x\rangle} - 1| \leq 2$ and $|\mathrm{e}^{\mathrm{i}\langle z,x\rangle} - 1 - \mathrm{i}\langle z,x\rangle| \leq 2^{-1}|\langle z,x\rangle|^2 \leq 2^{-1}|z|^2|x|^2$ by Lemma 8.6, we can use the dominated convergence theorem to show that $\psi(z)/|z|^2 \to 0$ as $|z| \to \infty$.

(ii) We have $\psi(z) = \int(\mathrm{e}^{\mathrm{i}\langle z,x\rangle} - 1)\nu(\mathrm{d}x)$ with $\int(1 \wedge |x|)\nu(\mathrm{d}x) < \infty$. Hence, the proof is similar. □

Proof of Theorem 43.9. Use the generating triplet (A, ν, γ). By (i) of the lemma above, we have $\psi(z) = 2^{-1}\langle z, Az\rangle + \psi_0(z)$ with $\psi_0(z) = o(|z|^2)$ as $|z| \to \infty$.

Step 1. Case $d = 2$. Suppose that A is of rank 2. Let us show that the integral in (43.5) is infinite. We have $\mathrm{Re}\,\frac{1}{q-\psi} = (q - \mathrm{Re}\,\psi)/|q - \psi|^2$ and, for large $|z|$,

$$q - \mathrm{Re}\,\psi(z) \geq 2^{-1}\langle z, Az\rangle + o(|z|^2) \geq k_1|z|^2,$$
$$|q - \psi(z)|^2 = |q + 2^{-1}\langle z, Az\rangle + o(|z|^2)|^2 \leq k_2|z|^4$$

with positive constants k_1 and k_2. Hence $\int_{|z|>a} \mathrm{Re}\left(\frac{1}{q-\psi(z)}\right)\mathrm{d}z \geq \mathrm{const}\int_a^\infty \frac{\mathrm{d}r}{r} = \infty$ for some $a > 0$, which implies that $c^q = 0$.

Next suppose that A is of rank 1 or 0. Let ρ be the Gaussian distribution with $\widehat{\rho}(z) = \mathrm{e}^{-\langle z,Az\rangle/2}$. As $t \downarrow 0$, the distribution of $t^{-1/2}X_t$ under P^0 tends to ρ, because

$$E^0[\mathrm{e}^{\mathrm{i}\langle z,t^{-1/2}X_t\rangle}] = \mathrm{e}^{t\psi(t^{-1/2}z)} = \mathrm{e}^{-\langle z,Az\rangle/2 + t\psi_0(t^{-1/2}z)}$$

and $t\psi_0(t^{-1/2}z) = t\,o(t^{-1}|z|^2) = o(1)$ as $t \downarrow 0$. Hence, for $B_\varepsilon = \{x\colon |x| \leq \varepsilon\}$ and $a > 0$,

$$\varepsilon^{-2}V^q(B_\varepsilon) \geq \varepsilon^{-2}\int_0^{\varepsilon^2/a^2} \mathrm{e}^{-qt}P^0[|X_t| \leq at^{1/2}]\mathrm{d}t$$
$$= \int_0^{1/a^2} \mathrm{e}^{-q\varepsilon^2 t}P^0[|X_{\varepsilon^2 t}| \leq a\varepsilon t^{1/2}]\mathrm{d}t \to a^{-2}\rho(B_a)$$

as $\varepsilon \downarrow 0$. Hence $\liminf_{\varepsilon\downarrow 0} \varepsilon^{-2}V^q(B_\varepsilon) \geq a^{-2}\rho(B_a)$. Since ρ is a degenerate Gaussian, $a^{-2}\rho(B_a) \to \infty$ as $a \downarrow 0$. Hence, by Proposition 43.8, $c^q = 0$.

Step 2. Case $d \geq 3$. We shall show that $\Sigma_0 = \{x \in \mathbb{R}^d\colon P^0[T_x < \infty] > 0\}$ has Lebesgue measure 0. Let $X_t' = (X_j(t))_{j=1,2}$ be the projection of $X_t = (X_j(t))_{1\leq j\leq d}$ onto the first two components. Then $\{X_t'\}$ under P^0 is a Lévy process on $\mathbb{R}^2$. Define $T_y' = \inf\{t > 0\colon X_t' = y\}$ for $y \in \mathbb{R}^2$ and

$\Sigma_0' = \{y \in \mathbb{R}^2 \colon P^0[T_y' < \infty] > 0\}$. The set Σ_0' has two-dimensional Lebesgue measure 0 by Step 1. Since $\Sigma_0 \subset \Sigma_0' \times \mathbb{R}^{d-2} = \{x = (x_j)_{1\le j\le d} \colon (x_j)_{j=1,2} \in \Sigma_0'\}$, Σ_0 has d-dimensional Lebesgue measure 0. □

REMARK 43.12. The notion of point recurrence is introduced in Remark 37.21. For any point recurrent process, the set Σ_0 contains 0. Thus, if $d \ge 2$ and (ACP) holds, then the process is not point recurrent, by Remark 43.10. Let $d = 1$ in the following. A Lévy process which is not compound Poisson is point recurrent if and only if it is recurrent and $0 \in \Sigma_0$. It is also true that it is point recurrent if and only if it is recurrent and $\Sigma_0 = \mathbb{R}$. Hence, by Theorem 37.16 combined with Examples 43.22 and 43.23 below, a non-trivial α-semi-stable process is point recurrent if and only if it is strictly α-semi-stable with $1 < \alpha \le 2$. A compound Poisson process is point recurrent if and only if $V^0(\{0\}) = \infty$. See Fristedt [**150**] for the proof.

THEOREM 43.13. *Let $d = 1$. Suppose that the process we are considering is of type A or B and let γ_0 be its drift. Then, any one-point set is essentially polar (that is, $c^q = 0$) if and only if $\gamma_0 = 0$.*

Proof. Since $\psi(z) = \int (e^{izx} - 1)\nu(dx) + i\gamma_0 z$, we have $\psi(z)/z \to i\gamma_0$ as $|z| \to \infty$ by Lemma 43.11. Thus $E^0[e^{izX_t/t}] = e^{t\psi(z/t)} \to e^{i\gamma_0 z}$ as $t \downarrow 0$, that is, $X_t/t \to \gamma_0$ in probability under P^0.

Assume that $\gamma_0 = 0$. Then, for any $a > 0$,

$$\varepsilon^{-1} V^q(B_\varepsilon) \ge \varepsilon^{-1} \int_0^{\varepsilon/a} e^{-qt} P^0[|X_t| \le at] dt$$
$$= \int_0^{1/a} e^{-q\varepsilon t} P^0[|X_{\varepsilon t}| \le a\varepsilon t] dt \to a^{-1}, \quad \varepsilon \downarrow 0.$$

Hence $\liminf_{\varepsilon\downarrow 0} \varepsilon^{-1} V^q(B_\varepsilon) = \infty$, which shows that $c^q = 0$ by Proposition 43.8.

Assume that $\gamma_0 \neq 0$. Consider $\operatorname{Re} \frac{1}{q-\psi} = (q - \operatorname{Re}\psi)/|q-\psi|^2$. We have $|q - \psi(z)|^2 \sim |\gamma_0|^2 |z|^2$ as $|z| \to \infty$. Notice that

$$\int_{|z|>1} \frac{-\operatorname{Re}\psi(z)}{z^2} dz = 2 \int_{\mathbb{R}} \nu(dx) \int_1^\infty \frac{1 - \cos zx}{z^2} dz$$

and $\int_1^\infty \frac{1-\cos zx}{z^2} dz \le \text{const}(1 \wedge |x|)$ since it is bounded by $\int_1^\infty \frac{2}{z^2} dz$ and $|x| \int_0^\infty \frac{1-\cos z}{z^2} dz$. Hence $\int_{|z|>1} \operatorname{Re} \frac{1}{q-\psi} dz < \infty$. This implies that $c^q > 0$ by Remark 43.6. □

In order to clarify how the properties in this section are related to the generating triplet, we need to prepare some propositions.

PROPOSITION 43.14. *Let f be a q-excessive function for some $q \ge 0$. If B is an F_σ set and $x_0 \in B^{\text{reg}}$, then*

$$\inf_{x\in B} f(x) \le f(x_0) \le \sup_{x\in B} f(x).$$

Proof. By Proposition 41.5(iii) we may assume $q > 0$. Choose compact sets K_n, $n = 1, 2, \ldots$, increasing to B. Then $T_{K_n} \downarrow T_B = 0$ under P^0. Since $X(T_{K_n}) \in K_n$ if $T_{K_n} < \infty$, we have

$$f(x_0) \geq E^{x_0}[\mathrm{e}^{-qT(K_n)} f(X(T_{K_n}))] \geq E^{x_0}[\mathrm{e}^{-qT(K_n)}] \inf_{x \in B} f(x)$$

by Proposition 41.5(vi). Hence the lower bound of $f(x_0)$ follows. Proposition 41.5(v) ensures the existence of nonnegative bounded universally measurable functions g_k such that $U^q g_k \uparrow f$. We have

$$U^q g_k(x_0) = U^q_{K_n} g_k(x_0) + P^q_{K_n} U^q g_k(x_0)$$

by (41.3). As $n \to \infty$, the first term of the right-hand side tends to 0, while the second term is bounded by $\sup_{x \in B} f(x)$. Hence we get the upper bound of $f(x_0)$, letting $k \to \infty$. □

PROPOSITION 43.15. *Let B be an F_σ set. For any probability measure ρ on $\mathbb{R}^d$, $X_{T_B} \in B \cup B^{\mathrm{reg}}$ P^ρ-almost surely on $\{T_B < \infty\}$.*

Proof. Recall the definition of T_B and use the strong Markov property. Then

$$P^\rho[T_B < \infty,\ X_{T_B} \notin B] = E^\rho[P^{X(T_B)}[T_B = 0];\ T_B < \infty,\ X_{T_B} \notin B].$$

Hence $P^{X(T_B)}[T_B = 0] = 1$, that is, $X_{T_B} \in B^{\mathrm{reg}}$, P^ρ-a. s. on $\{T_B < \infty, X_{T_B} \notin B\}$. □

PROPOSITION 43.16. *Assume* (ACP). *Let f be q-excessive for some $q \geq 0$. If B is an F_σ set and $x_0 \in B^{\mathrm{reg}}$, then there is a sequence $y_n \in B$, $n = 1, 2, \ldots$, such that $y_n \to x_0$ and $f(y_n) \to f(x_0)$.*

Proof. Suppose that we cannot find a sequence $y_n \in B$ having the properties above. Then, for some $a > 0$ and $\varepsilon > 0$,

$$B \cap \{x\colon |x - x_0| < \varepsilon\} \cap \{x\colon f(x_0) - a < f(x) \leq f(x_0) + a\} = \emptyset.$$

Write $D_1 = \{x\colon f(x) \leq f(x_0) - a\}$, $D_2 = \{x\colon f(x) > f(x_0) + a\}$, and $D = D_1 \cup D_2$. Since f is lower semi-continuous by (ACP) (Theorem 41.15), D_1 is closed and D_2 is open. Hence we can apply Proposition 43.14 to D_1 and D_2. If $x_0 \in {D_1}^{\mathrm{reg}}$, then $f(x_0) \leq \sup_{x \in D_1} f(x) \leq f(x_0) - a$, which is absurd. Thus $x_0 \in {D_1}^{\mathrm{irreg}}$. If $x_0 \in {D_2}^{\mathrm{reg}}$, then $f(x_0) \geq \inf_{x \in D_2} f(x) \geq f(x_0) + a$, which is absurd. Hence $x_0 \in {D_2}^{\mathrm{irreg}}$. It follows that $x_0 \in D^{\mathrm{irreg}}$. Hence, almost surely, paths starting at x_0 stay in $\{x\colon |x - x_0| < \varepsilon\} \cap \{x\colon f(x_0) - a < f(x) \leq f(x_0) + a\}$ for some positive time, which means that they stay in $\mathbb{R}^d \setminus B$ for some positive time. This contradicts $x \in B^{\mathrm{reg}}$. □

PROPOSITION 43.17. *Let $q \geq 0$. Let $y_n \neq 0$, $n = 1, 2, \ldots$, be a sequence such that $y_n \to 0$ and $\inf_n h^q(y_n) > 0$. Let $B = \bigcup_{n=1}^{\infty} \{y_n\}$. If $0 \in B^{\mathrm{irreg}}$, then $h^q(-y_n) \to 1$ as $n \to \infty$ and $0 \in (-B)^{\mathrm{reg}}$.*

Proof. Write $B_n = \bigcup_{k=n}^{\infty} \{y_k\}$ and $S_n = T_{B_n}$. Then $X_{S_n} \in \overline{B_n} = B_n \cup \{0\}$ on $\{S_n < \infty\}$ and S_n increases to some S. By quasi-left-continuity, $X_S = \lim_{n \to \infty} X_{S_n} = 0$ a. s. on $\{S < \infty\}$. The assumption $0 \in B^{\mathrm{irreg}}$ implies that $0 \in {B_n}^{\mathrm{irreg}}$. Thus $X_{S_n} \in B_n$ a. s. on $\{S_n < \infty\}$ by Proposition 43.15. Therefore

$S_n < S$ a.s. on $\{S < \infty\}$. Let $S'_n = \inf\{t > S_n \colon X_t = 0\}$. Then $S'_n \le S$ a.s. on $\{S < \infty\}$ and

$$E^0[\mathrm{e}^{-qS}] \le E^0[\mathrm{e}^{-qS'_n}] = E^0[\mathrm{e}^{-qS_n} E^{X(S_n)}[\mathrm{e}^{-qT_0}]]$$
$$\le E^0[\mathrm{e}^{-qS_n}]\sup_{y\in B_n} h^q(-y) \to E^0[\mathrm{e}^{-qS}]\lim_{n\to\infty}\sup_{y\in B_n} h^q(-y).$$

Let $a = \inf_n h^q(y_n)$. We have $a > 0$. Since $E^0[\mathrm{e}^{-qS_n}] \ge E^0[\mathrm{e}^{-qT(y_n)}] = h^q(y_n) \ge a$, we have $E^0[\mathrm{e}^{-qS}] \ge a$. Therefore we get $\limsup_{n\to\infty} h^q(-y_n) = 1$. As we can replace $\{y_n\}$ by any subsequence of it, we conclude that $\lim_{n\to\infty} h^q(-y_n) = 1$. Now $P^0[T_{-B} = 0] = 1$, since $E^0[\mathrm{e}^{-qT_{-B}}] \ge E^0[\mathrm{e}^{-qT(-y_n)}] = h^q(-y_n) \to 1$. □

PROPOSITION 43.18. *Assume* (ACP). *Let* $q \ge 0$. *Then any limit point of* $h^q(x)$ *as* $x \to 0$ *belongs to the set* $\{0, h^q(0), 1\}$.

Proof. First consider the case that $h^q(0) > 0$. Let $x_n \ne 0$ be a sequence such that $x_n \to 0$ and $h^q(x_n)$ tends to some a. Choosing a subsequence if necessary, we may assume that $h^q(-x_n)$ tends to some b. Let $B = \bigcup_n \{x_n\}$.

Case 1 ($0 \in (-B)^{\mathrm{reg}}$). Apply Proposition 43.16 to $-B$ and $f(x) = E^x[\mathrm{e}^{-qT_0}] = h^q(-x)$. There are $y_k \in -B$ such that $y_k \to 0$ and $h(-y_k) \to h^q(0)$. A subsequence of $\{y_k\}$ is a subsequence of $\{-x_n\}$. Hence $a = h^q(0)$.

Case 2 ($0 \in (-B)^{\mathrm{irreg}} \cap B^{\mathrm{reg}}$). Apply the same proposition to B and $h^q(-x)$. Then there are $y_k \in B$ with $y_k \to 0$ and $h^q(-y_k) \to h^q(0)$. Now a subsequence of $\{y_k\}$ is a subsequence of $\{x_n\}$, and we have $b = h^q(0) > 0$. Thus we can assume $\inf_n h^q(-x_n) > 0$, deleting some initial terms if necessary. Then, by Proposition 43.17, $h^q(x_n) \to 1$. Hence $a = 1$.

Case 3 ($0 \in (-B)^{\mathrm{irreg}} \cap B^{\mathrm{irreg}}$). In this case, $a = 0$. In fact, if $a > 0$, then we may assume $\inf_n h^q(x_n) > 0$, deleting some terms if necessary, and we can apply Proposition 43.17 to obtain $0 \in (-B)^{\mathrm{reg}}$, a contradiction.

Next consider the case that $h^q(0) = 0$. The assertion is trivial if h^q identically vanishes. Assume that $h^q(x_0) > 0$ for some $x_0 \ne 0$. Let $c > 0$. Enlarging the probability space $(\Omega, \mathcal{F}, P^0)$, consider a compound Poisson process $\{Y_t\}$ with Lévy measure $c\delta_{-x_0}$, independent of the process $\{X_t\}$. Let $Z_t = X_t + Y_t$. Then $\{Z_t\}$ is a Lévy process, which satisfies (ACP) by Exercise 44.4. Distinguish the quantities related to $\{X_t\}$ and $\{Z_t\}$ by attaching X and Z, respectively, as a sub- or superscript. We can show that $h^q_Z(0) > 0$. In fact, letting J_1 and J_2 be the first and second jumping times of $\{Y_t\}$, we have

$$P^0[T^Z_0 < \infty] \ge P^0[J_1 < T^X_{x_0} < J_2] = \textstyle\int P^0[J_1 < t < J_2] P^0[T^X_{x_0} \in \mathrm{d}t] > 0.$$

Therefore our proposition is true for the process $\{Z_t\}$. For any x,

$$|h^q_X(x) - h^q_Z(x)| = |E^0[\mathrm{e}^{-qT^X_x} - \mathrm{e}^{-qT^Z_x}]| \le E^0[\mathrm{e}^{-qT^X_x} + \mathrm{e}^{-qT^Z_x};\, J_1 \le T^X_x]$$
$$\le E^0[2\mathrm{e}^{-qJ_1};\, J_1 \le T^X_x] \le 2E^0[\mathrm{e}^{-qJ_1}] = 2\textstyle\int_0^\infty \mathrm{e}^{-qt} c\mathrm{e}^{-ct}\mathrm{d}t = \frac{2c}{c+q}.$$

In the above the first inequality is because $T^X_x = T^Z_x$ on $\{J_1 > T^X_x\}$ and the second one is because $J_1 \le T^Z_x$ on $\{J_1 \le T^X_x\}$. Now let $x_n \ne 0$ be such that $x_n \to 0$ and $h^q_X(x_n)$ tends to some a. Choose c small enough. Then we have $\liminf_{n\to\infty} h^q_Z(x_n) > a/2$ and $h^q_Z(0) < a/2$. Here we have used that $h^q_X(0) = 0$.

It follows that $\liminf_{n\to\infty} h_Z^q(x_n) = 1$. Hence $\lim_{n\to\infty} h_Z^q(x_n) = 1$. Hence $a \geq 1 - \frac{2c}{c+q}$. As c is arbitrarily small, we have $a = 1$. □

Propositions 43.16 and 43.18 were proved by Bretagnolle [**68**] without the assumption of (ACP). See Remark 43.10 and Notes at the end of the chapter.

THEOREM 43.19. *Let $d = 1$. Assume that a one-point set is not essentially polar (that is, $c^q > 0$). Let $q > 0$. Then the following are true.*

(i) *The function $h^q(x)$ is continuous on $\mathbb{R} \setminus \{0\}$.*

(ii) *The set of limit points of $h^q(x)$ as $x \to 0$ is exactly $\{1, h^q(0)\}$.*

(iii) *One of the following, (a) and (b), holds: (a) $h^q(0+) = 1$ and $h^q(0-) = h^q(0)$; (b) $h^q(0+) = h^q(0)$ and $h^q(0-) = 1$.*

(iv) *Assume that (a) holds. If $h^q(0) = 1$, then $\Sigma_0 = \mathbb{R}$ and $0 \in (0,\infty)^{\mathrm{reg}} \cap (-\infty, 0)^{\mathrm{reg}}$. If $0 < h^q(0) < 1$, then $\Sigma_0 = \mathbb{R}$ and $0 \in (0,\infty)^{\mathrm{reg}} \cap (-\infty,0)^{\mathrm{irreg}}$. If $h^q(0) = 0$, then $\Sigma_0 = (0,\infty)$ and $0 \in (0,\infty)^{\mathrm{reg}} \cap (-\infty,0)^{\mathrm{irreg}}$, and the process is a subordinator.*

Proof. (i) Let $\varepsilon > 0$ and $B = \{x\colon |x| \geq \varepsilon\}$. For any $x \in B$, $h^r(x) = E^0[\mathrm{e}^{-rT_x}] \leq E^0[\mathrm{e}^{-rT_B}] \to 0$ as $r \to \infty$. Thus $h^r(x) \to 0$ uniformly on B as $r \to \infty$. Our process satisfies (ACP) and $u^r(x) = h^r(x)/c^r$ by Theorem 43.3. As r increases, $1/c^r$ decreases by Proposition 43.8. Hence $u^r(x) \to 0$ uniformly on B. We have, for $0 < q < r$,

$$u^q(x) = u^r(x) + (r-q)\int_{\mathbb{R}^d} u^r(x-y)u^q(y)\mathrm{d}y, \quad x \in \mathbb{R}^d. \tag{43.9}$$

This is true for a. e. x by the resolvent equation (41.2) and, hence, for all x by r-co-excessiveness of both sides. Since $\int u^r(x-y)u^q(y)\mathrm{d}y$ is continuous in x by Theorem 41.15 applied to the dual process, (43.9) shows that $u^q(x)$ is the uniform limit of continuous functions on B. Hence u^q and h^q are continuous on $\mathbb{R} \setminus \{0\}$.

(ii) Since $h^q(x)$ is lower semi-continuous, $h^q(0) \leq \liminf_{x\to 0} h^q(x)$. By (ACP), the process is not compound Poisson. Hence 0 is regular for $\mathbb{R} \setminus \{0\}$. By Proposition 43.16 for $f(x) = h^q(-x)$, there are $y_n \neq 0$ such that $y_n \to 0$ and $h^q(-y_n) \to h^q(0)$. It follows that

$$\liminf_{x\to 0} h^q(x) = h^q(0). \tag{43.10}$$

Hence by virtue of Proposition 43.18 limit points of $h^q(x)$ as $x \to 0$ belong to $\{1, h^q(0)\}$. If $h^q(x)$ is continuous at $x = 0$, then so is $u^q(x)$ and $u^q(0) = 1/c^q$ by (43.7), which implies that $h^q(0) = 1$. If $h^q(x)$ is discontinuous at $x = 0$, then limsup and liminf must be different. In either case we have

$$\limsup_{x\to 0} h^q(x) = 1. \tag{43.11}$$

(iii) Write $h^q(0) = a$. If $a = 1$, then there is nothing to prove. Assume that $a < 1$. Suppose that $h^q(0+)$ does not exist. Then, by (ii), there are sequences $x_n \downarrow 0$ and $y_n \downarrow 0$ such that $h^q(x_n) \to 1$ and $h^q(y_n) \to a$. For large n we have $h^q(x_n) > \frac{1+a}{2} > h^q(y_n)$. Hence, by (i), there is z_n between x_n and y_n such that $h^q(z_n) = \frac{1+a}{2}$. Since $z_n \downarrow 0$, this contradicts (ii). Hence $h^q(0+)$ exists. Similarly $h^q(0-)$ exists. Thus (iii) follows.

(iv) We have $h^q(0+) = 1$ and $h^q(0-) = h^q(0) = a$. If $x_n \downarrow 0$, then $E^0[\mathrm{e}^{-qT_{(0,\infty)}}] \geq E^0[\mathrm{e}^{-qT(x_n)}] = h^q(x_n) \to 1$. Hence $0 \in (0,\infty)^{\mathrm{reg}}$. If $a = 1$, then $0 \in (-\infty, 0)^{\mathrm{reg}}$ by the same reasoning. If $a > 0$, then $\Sigma_0 = \mathbb{R}$, because h^q is positive on a neighborhood of 0 and, by (43.3), positive on $\mathbb{R}$. If $a < 1$, then $0 \in (-\infty,0)^{\mathrm{irreg}}$, since otherwise Proposition 43.16 says that, for some $y_n < 0$ with $y_n \to 0$, $h^q(-y_n) \to h^q(0) = a < 1$, contradicting $h^q(0+) = 1$.

Let $a = 0$. Since h^q is positive on some $(0, \varepsilon)$, we have $h^q(x) > 0$ for $x > 0$ by (43.3). For every $x \in \mathbb{R}$, $0 = h^q(0) \geq h^q(x)h^q(-x)$ by (43.3). Hence $h^q(x) = 0$ for $x < 0$. Hence $\Sigma_0 = (0,\infty)$. It follows that $u^q(x) = h^q(x)/c^q = 0$ on $(-\infty, 0]$. Hence the process is a subordinator. □

Let us prove a result of Shtatland [**486**] on short time behavior of processes of type A or B.

THEOREM 43.20. *Assume that the process we are considering is of type A or B with drift γ_0. Then*

$$P^0\left[\lim_{t\downarrow 0} t^{-1}X_t = \gamma_0\right] = 1. \tag{43.12}$$

Proof. Since the component processes are also of type A or B, it is enough to give a proof for $d = 1$. Further, we may and do assume that $\gamma_0 = 0$. We use the Lévy–Itô decomposition

$$X_t(\omega) = \int_{(0,t]\times(\mathbb{R}\setminus\{0\})} xJ(\mathrm{d}(s,x),\omega) \quad P^0\text{-a. s.}$$

in Theorem 19.3, where $J(C,\omega) = \#\{s\colon (s, X_s - X_{s-}) \in C\}$ is a Poisson random measure on $(0,\infty)\times(\mathbb{R}\setminus\{0\})$ with intensity measure $\mathrm{d}t\,\nu(\mathrm{d}x)$. Let $B_n = \{x\colon |x| \geq 2^{-n}\}$. Then

$$\sum_n P^0[J((0,2^{-n}]\times B_n) \geq 1] \leq \sum_n E[J((0,2^{-n}]\times B_n)] = \sum_n 2^{-n}\nu(B_n) < \infty,$$

since

$$\infty > \int_{|x|<1} |x|\nu(\mathrm{d}x) \geq \sum_{n=1}^{\infty} 2^{-n}\nu(B_n \setminus B_{n-1}) \geq \sum_{n=1}^{\infty} 2^{-n-1}\nu(B_n) - 2^{-1}\nu(B_0).$$

Hence, by the Borel–Cantelli lemma, almost surely there is $n_0(\omega)$ such that $J((0,2^{-n}]\times B_n) = 0$ for $n \geq n_0(\omega)$. If $n \geq n_0(\omega)$ and $2^{-n-1} < t \leq 2^{-n}$, then $|t^{-1}X_t| \leq Y_n$, where

$$Y_n = 2^{n+1}\int_{(0,2^{-n}]\times\{|x|<2^{-n}\}} |x|J(\mathrm{d}(s,x),\omega).$$

We can prove from Proposition 19.5 that

$$EY_n = 2\int_{|x|<2^{-n}} |x|\nu(\mathrm{d}x) \text{ and } \operatorname{Var} Y_n = 2^{n+2}\int_{|x|<2^{-n}} |x|^2\nu(\mathrm{d}x).$$

Hence $EY_n \to 0$ as $n \to \infty$ and

$$\begin{aligned}\sum_{n=0}^{\infty} \operatorname{Var} Y_n &= \int_{|x|<1} \sum_{n=0}^{\infty} 2^{n+2}|x|^2 1_{\{|x|<2^{-n}\}}(x)\nu(\mathrm{d}x)\\ &= \int_{|x|<1} |x|^2 \sum_{2^n<1/|x|} 2^{n+2}\nu(\mathrm{d}x) \le 8\int_{|x|<1} |x|\nu(\mathrm{d}x) < \infty.\end{aligned}$$

It follows that $\sum_{n=0}^{\infty}(Y_n - EY_n)^2 < \infty$ a. s. and hence $Y_n \to 0$ a. s. □

For an arbitrary Lévy process on $\mathbb{R}$, either it belongs to one of the following cases, or its dual belongs to one of them.

THEOREM 43.21. *Let $d = 1$. Let (A, ν, γ) be the generating triplet. Let γ_0 be the drift in case of type* A *or* B. *Let $q > 0$.*

Case 1 (type A, $\gamma_0 = 0$, that is, compound Poisson or zero process). $0 \in \{0\}^{\mathrm{reg}} \cap (\mathbb{R} \setminus \{0\})^{\mathrm{irreg}}$. (ACP) *does not hold. If ν is continuous, then* $\Sigma_0 = \{0\}$. *If ν is not continuous, then $\Sigma_0 = \{0\} \cup D$, where D is the smallest set which contains $D_0 = \{x\colon \nu\{x\} > 0\}$ and which is closed under addition.*

Case 2 (type A or B, $\gamma_0 > 0$, $\nu(-\infty, 0) = 0$, that is, subordinator with positive drift). $\Sigma_0 = (0, \infty)$, $0 \in \{0\}^{\mathrm{irreg}} \cap (0, \infty)^{\mathrm{reg}} \cap (-\infty, 0)^{\mathrm{irreg}}$, (ACP) *holds, u^q is bounded,* $0 = h^q(0) = h^q(0-) < 1 = h^q(0+)$.

Case 3 (type A or B, $\gamma_0 > 0$, $\nu(-\infty, 0) > 0$). $\Sigma_0 = \mathbb{R}$, $0 \in \{0\}^{\mathrm{irreg}} \cap (0, \infty)^{\mathrm{reg}} \cap (-\infty, 0)^{\mathrm{irreg}}$, (ACP) *holds, u^q is bounded,* $0 < h^q(0) = h^q(0-) < 1 = h^q(0+)$.

Case 4 (type B, $\gamma_0 = 0$). There are two subcases: Subcase 1 ($\operatorname{Leb} \Sigma_0 = 0^{(*)}$, (ACP) *does not hold). Subcase 2 (*$\Sigma_0 = \emptyset$, (ACP) *holds, u^q is unbounded). Each subcase is nonvoid.*

Case 5 (type C, $A = 0$). Divided into three subcases: Subcase 1 ($\operatorname{Leb} \Sigma_0 = 0^{(*)}$, (ACP) *does not hold). Subcase 2 (*$\Sigma_0 = \emptyset$, (ACP) *holds, u^q is unbounded). Subcase 3 (*$\operatorname{Leb} \Sigma_0 > 0$, (ACP) *holds, u^q is bounded). Each subcase is nonvoid.*

Case 6 (type C, $A > 0$). $\Sigma_0 = \mathbb{R}$, $0 \in \{0\}^{\mathrm{reg}} \cap (0, \infty)^{\mathrm{reg}} \cap (-\infty, 0)^{\mathrm{reg}}$, (ACP) *holds, u^q is bounded and continuous.*

In the above, $\operatorname{Leb} \Sigma_0 = 0^{(*)}$ means that we prove here $\operatorname{Leb} \Sigma_0 = 0$ but it is known that Σ_0 is the empty set (see Remark 43.10). Emptiness of Σ_0 implies, in particular, that $0 \in \{0\}^{\mathrm{irreg}}$. Regularity of 0 for $(0, \infty)$ or $(-\infty, 0)$ in Cases 4 and 5 will be analyzed in Theorems 47.1 and 47.5. In particular, it will be shown that, in Case 5, $0 \in (0, \infty)^{\mathrm{reg}} \cap (-\infty, 0)^{\mathrm{reg}}$. Hence, in Subcase 3 of Case 5, $0 \in \{0\}^{\mathrm{reg}}$ and $\Sigma_0 = \mathbb{R}$ by Theorem 43.19.

Proof of theorem. Case 1. (ACP) does not hold, as we see in Example 41.21. Since sample functions are step functions, $0 \in \{0\}^{\mathrm{reg}} \cap (\mathbb{R} \setminus \{0\})^{\mathrm{irreg}}$. Let $x_1, x_2, \ldots$ be an enumeration of D_0. The set D consists of points $n_1 x_1 + n_2 x_2 + \cdots + n_k x_k$, where $k \in \mathbb{N}$ and $n_1, n_2, \ldots \in \mathbb{Z}_+$. Let $c_j = \nu\{x_j\}$. Then

$$P^0[X_t = n_1 x_1 + \cdots + n_k x_k] \geq \mathrm{e}^{-t(\nu(\mathbb{R}) - c_1 - \cdots - c_k)} \prod_{j=1}^{k} \mathrm{e}^{-tc_j} \frac{(tc_j)^{n_j}}{n_j!} > 0$$

and $P^0[X_t = 0] \geq \mathrm{e}^{-t\nu(\mathbb{R})} > 0$. Hence $\{0\} \cup D \subset \Sigma_0$. Let $Y_l(\omega)$ be the amount of the lth jump of $X_t(\omega)$. For any $x \neq 0$, $P^0[T_x < \infty] \leq \sum_{l=1}^{\infty} P^0[Y_1 + \cdots + Y_l = x]$. If $x \notin \{0\} \cup D$, then $P^0[Y_1 + \cdots + Y_l = x] \leq \sum_{j=1}^{l} P^0[Y_1 + \cdots + Y_l = x,\ Y_j \notin D_0]$. We have $P^0[Y_1 + \cdots + Y_l = x,\ Y_1 \notin D_0] = \int P^0[Y_1 + y = x,\ Y_1 \notin D_0] P^0[Y_2 + \cdots + Y_l \in \mathrm{d}y] = 0$. Similarly the other terms are 0. Hence $\Sigma_0 = \{0\} \cup D$.

Cases 2 and 3. By Theorem 43.20, $0 \in \{0\}^{\mathrm{irreg}} \cap (0, \infty)^{\mathrm{reg}} \cap (-\infty, 0)^{\mathrm{irreg}}$. We have $c^q > 0$ by Theorem 43.13. Hence the other assertion follows from Theorem 43.19.

Case 4. In this case $c^q = 0$ by Theorem 43.13. Hence $\mathrm{Leb}\,\Sigma_0 = 0$. By Theorem 43.3 V^q cannot have bounded density. So there is no case other than Subcases 1 and 2. If (ACP) holds, then $\Sigma_0 = \emptyset$ by Theorem 41.15(6). Examples 41.22 and 41.23 give processes of Subcase 1. Strictly stable processes of index $0 < \alpha < 1$ are examples of Subcase 2.

Case 5. If $\mathrm{Leb}\,\Sigma_0 = 0$, then Subcase 1 or 2 occurs, since V^q cannot have bounded density, by Theorem 43.3, and $\Sigma_0 = \emptyset$ in the case of (ACP). If $\mathrm{Leb}\,\Sigma_0 > 0$, then V^q has bounded density by the same theorem. Example 41.23 furnishes processes of Subcase 1. Example 43.7 gives processes of Subcases 2 and 3.

Case 6. We have $\psi(z) = -\frac{1}{2}Az^2 + o(z^2)$ as $|z| \to \infty$ by Lemma 43.11(i). It follows that $|1/(q - \psi)| \leq 1/(|\psi| - q) \leq \mathrm{const}|z|^{-2}$ for large $|z|$. Thus $\int |1/(q - \psi)| \mathrm{d}z < \infty$. Hence, by Remark 43.6 and Theorem 43.5, $c^q > 0$, $0 \in \{0\}^{\mathrm{reg}}$, and V^q has bounded continuous density. Hence, by Theorem 43.19, $\Sigma_0 = \mathbb{R}$ and $0 \in (0, \infty)^{\mathrm{reg}} \cap (-\infty, 0)^{\mathrm{reg}}$. □

EXAMPLE 43.22. Let $(\{X_t\}, P^0)$ be a non-trivial α-stable process on $\mathbb{R}$ with $0 < \alpha \leq 2$. If $0 < \alpha < 2$, let (α, β, τ, c) be its parameters in Definition 14.16. Then, the set Σ_0 and the regularity of 0 for itself are described as follows.

(1) If $1 < \alpha \leq 2$, then $\Sigma_0 = \mathbb{R}$ and $0 \in \{0\}^{\mathrm{reg}}$.
(2) If $\alpha = 1$ and $\beta = 0$, then $\Sigma_0 = \emptyset$ and $0 \in \{0\}^{\mathrm{irreg}}$.
(3) If $\alpha = 1$ and $\beta \neq 0$, then $\Sigma_0 = \mathbb{R}$ and $0 \in \{0\}^{\mathrm{reg}}$.
(4) If $0 < \alpha < 1$ and $\tau = 0$, then $\Sigma_0 = \emptyset$ and $0 \in \{0\}^{\mathrm{irreg}}$.
(5) If $0 < \alpha < 1$, $\tau > 0$, and $\beta = 1$, then $\Sigma_0 = (0, \infty)$ and $0 \in \{0\}^{\mathrm{irreg}}$.
(6) If $0 < \alpha < 1$, $\tau > 0$, and $\beta \neq 1$, then $\Sigma_0 = \mathbb{R}$ and $0 \in \{0\}^{\mathrm{irreg}}$.

If $0 < \alpha < 1$ and $\tau < 0$, then we have the situation dual to (5) and (6).

Let us check these facts. First, in general, $0 \notin \Sigma_0$ implies $0 \in \{0\}^{\mathrm{irreg}}$. Second, note that $(\{X_t\}, P^0)$ satisfies (ACT) by Example 28.2. If $\alpha = 2$, then Case 6 of Theorem 43.21 applies. If $1 < \alpha < 2$, then $\mathrm{Re}\,(1/(q-\psi)) \leq 1/(q - \mathrm{Re}\,\psi) \leq \mathrm{const}|z|^{-\alpha}$ and hence $\int \mathrm{Re}\,(1/(q-\psi))\mathrm{d}z < \infty$. Thus (1) is true by Remark 43.6 and by Theorem 43.21 and the remark after it; (2) and (3) are shown by Examples 43.7 combined with the remark after Theorem 43.21; (4), (5), and (6) are respectively in Cases 4, 2, and 3 of Theorem 43.21.

EXAMPLE 43.23. Let $(\{X_t\}, P^0)$ be a non-trivial α-semi-stable process on $\mathbb{R}$, $0 < \alpha \leq 2$. Then, (1)–(6) of Example 43.22 remain valid with the following replacements [**449**]: $\beta = 0$ in (2) and $\beta \neq 0$ in (3) by strictly 1-semi-stable and not strictly 1-semi-stable, respectively; τ in (4), (5), and (6) by the drift γ_0; $\beta = 1$ in (5) and $\beta \neq 1$ in (6) by $\nu(-\infty, 0) = 0$ and $\nu(-\infty, 0) > 0$, respectively.

Let us examine Case 5 of Theorem 43.21.

THEOREM 43.24. *Let $d = 1$ and consider the case of type C with $A = 0$. If $\int_{(0,1]} x\nu(\mathrm{d}x) < \infty$ or $\int_{[-1,0)} |x|\nu(\mathrm{d}x) < \infty$, then* $\mathrm{Leb}\,\Sigma_0 > 0$, *that is, the process is in Subcase 3 of Case 5 of Theorem 43.21.*

Examples 41.23 and 43.7 show that the case of type C satisfying $A = 0$, $\int_{(0,1]} x\nu(\mathrm{d}x) = \infty$, and $\int_{[-1,0)} |x|\nu(\mathrm{d}x) = \infty$ contains processes of each of Subcases 1, 2, and 3 of Case 5.

To prove this theorem, we analyze the sample functions, following Kesten [**272**].

LEMMA 43.25. *Let $\Theta_t(\omega)$ be the set $\{X_s(\omega)\colon 0 \leq s \leq t\}$, the range of the path in the time interval $[0,t]$, and let $\overline{\Theta}_t$ be its closure. Then* $\mathrm{Leb}\,\overline{\Theta}_t(\omega)$ *is $\mathcal{F}^0$-measurable. If $P^0[\mathrm{Leb}\,\overline{\Theta}_t > 0] > 0$ for some $t > 0$, then* $\mathrm{Leb}\,\Sigma_0 > 0$.

Proof. Fix $t > 0$. Define $a(x,\omega)$ as follows. For $x \neq X_0(\omega)$, $a(x,\omega) = 1$ if $X_s(\omega) = x$ or $X_{s-}(\omega) = x$ for some $s \in (0,t]$, and $a(x,\omega) = 0$ otherwise. For $x = X_0(\omega)$, $a(x,\omega) = 0$. We claim that $a(t,\omega)$ is $(\mathcal{B}_{\mathbb{R}} \times \mathcal{F}^0)$-measurable in (x,ω). Indeed, define, for $x \neq X_0(\omega)$, $a_n(x,\omega) = 1$ if $|X_s(\omega) - x| < 1/n$ for some $s \in (0,t]$, $a_n(x,\omega) = 0$ otherwise, and, for $x = X_0(\omega)$, $a_n(x,\omega) = 0$. Then $a_n(x,\omega) \downarrow a(x,\omega)$ as $n \to \infty$. Since, for $x \neq X_0(\omega)$, $a_n(x,\omega) = 1$ if and only if $|X_s(\omega) - x| < 1/n$ for some $s \in (\mathbb{Q} \cap (0,t]) \cup \{t\}$, $a_n(x,\omega)$ is $(\mathcal{B}_{\mathbb{R}} \times \mathcal{F}^0)$-measurable. Hence, so is $a(x,\omega)$, as claimed. Since $\overline{\Theta}_t(\omega) \setminus \{X_0(\omega)\} = \{x\colon a(x,\omega) = 1\}$, $\mathrm{Leb}\,\overline{\Theta}_t(\omega) = \int a(x,\omega)\mathrm{d}x$ is $\mathcal{F}^0$-measurable. Suppose that $P^0[\mathrm{Leb}\,\overline{\Theta}_t > 0] > 0$. Then $0 < E^0[\mathrm{Leb}\,\overline{\Theta}_t] = \int E^0[a(x,\omega)]\mathrm{d}x$ by Fubini's theorem. Let $h(x) = P^0[T_x < \infty]$, $h^*(x) = P^0[X_s = x \text{ or } X_{s-} = x \text{ for some } s > 0]$, $S_x = \inf\{s > 0\colon X_s = x \text{ or } X_{s-} = x\}$, $T_x^{\varepsilon} = \inf\{s \geq \varepsilon\colon X_s = x\}$, and $S_x^{\varepsilon} = \inf\{s \geq \varepsilon\colon X_s = x \text{ or } X_{s-} = x\}$. Given x, let $G_n = \{y\colon |y - x| < 1/n\}$ and $T_{G_n}^{\varepsilon} = \inf\{s \geq \varepsilon\colon X_s \in G_n\}$. Then $T_{G_n}^{\varepsilon} \uparrow T_x^{\varepsilon}$ a.s. as $n \to \infty$ by quasi-left-continuity. Since $T_{G_n}^{\varepsilon/2} \leq S_x^{\varepsilon} \leq T_x^{\varepsilon}$ and since $T_x^{\varepsilon} \downarrow T_x$ and $S_x^{\varepsilon} \downarrow S_x$ as $\varepsilon \downarrow 0$, we have $T_x = S_x$ a.s. Hence $h^*(x) = h(x)$. Now $0 < \int E^0[a(x,\omega)]\mathrm{d}x \leq \int h^*\mathrm{d}x = \int h(x)\mathrm{d}x$, and we have $\mathrm{Leb}\,\Sigma_0 > 0$. $\square$

LEMMA 43.26. *Let $d = 1$. Assume that $A = 0$ and $\int_{(0,1]} x\nu(\mathrm{d}x) < \infty$. Let $X_t^+ = \sum_{s\le t}(X_s - X_{s-}) \vee 0$, the sum of positive jumps up to time t. Suppose that $P^0[X_t > X_t^+] > 0$ for some $t > 0$. Then* $\mathrm{Leb}\,\Sigma_0 > 0$.

Proof. By the previous lemma, it suffices to show that $P^0[\mathrm{Leb}\,\overline{\Theta}_t > 0] > 0$ for some $t > 0$. Fix $t > 0$. Denote $\Lambda = \Lambda(\omega) = \bigcup[X_{s-}, X_s]$, where the union is taken over all $s \in (0, t]$ such that $X_{s-} < X_s$. Then $\mathrm{Leb}\,\Lambda \le X_t^+$. Assuming $X_t(\omega) > 0$, suppose that $x \in (0, X_t(\omega)] \setminus \Lambda(\omega)$. Then $0 < T_{[x,\infty)}(\omega) \le t$ and $X_{T_{[x,\infty)}-}(\omega) \le x \le X_{T_{[x,\infty)}}(\omega)$. Since $x \notin \Lambda(\omega)$, $X_{T_{[x,\infty)}-}(\omega) = X_{T_{[x,\infty)}}(\omega)$. Hence $X_{T_{[x,\infty)}}(\omega) = x$. It follows that $a(x, \omega) = 1$, where $a(x, \omega)$ is the function in the proof above. Hence $(0, X_t(\omega)] \setminus \Lambda(\omega) \subset \overline{\Theta}_t(\omega)$. Now

$$P^0[\mathrm{Leb}\,\overline{\Theta}_t > 0] \ge P^0[X_t > 0 \text{ and } \mathrm{Leb}((0, X_t] \setminus \Lambda) > 0] \ge P^0[X_t > X_t^+],$$

since $\mathrm{Leb}((0, X_t] \cap \Lambda) \le \mathrm{Leb}\,\Lambda \le X_t^+$. Hence $P^0[\mathrm{Leb}\,\overline{\Theta}_t > 0] > 0$. □

Proof of Theorem 43.24. We assume that $\int_{(0,1]} x\nu(\mathrm{d}x) < \infty$. Define X_t^+ as in Lemma 43.26 and $Y_t = X_t - X_t^+$. Then $(\{Y_t\}, P^0)$ is a Lévy process with generating triplet $(0, 1_{(-\infty,0)}(x)\nu(\mathrm{d}x), \gamma)$. Since $\int_{[-1,0)} |x|\nu(\mathrm{d}x) = \infty$, we have $P^0[Y_t > 0] > 0$ for any $t > 0$ by Theorem 24.10. Thus Lemma 43.26 applies. □

REMARK 43.27. Fix a point $x \in \mathbb{R}^d$. Regularity of x for itself is related to the existence of local time at x. $\{L_t(\omega) : t \ge 0\}$ is called the *local time at x* of a Lévy process $(\{X_t\}, P)$ if L_t is $\mathcal{F}_t$-measurable for any t and $E^y[L_t] > 0$ for some t and y and if there is $\Omega_0 \in \mathcal{F}$ with $P^y[\Omega_0] = 1$ for every y such that, for all $\omega \in \Omega_0$, the following are satisfied:

(1) as a function of t, $L_t(\omega)$ is continuous and increasing, and $L_0(\omega) = 0$;
(2) $L_{s+t}(\omega) = L_s(\omega) + L_t(\theta_s\omega)$ for all s and t, where the *shift* θ_s of sample functions is defined by $(\theta_s\omega)(t) = \omega(s + t)$;
(3) $\int_0^\infty 1_{\mathbb{R}^d\setminus\{x\}}(X_t(\omega))\mathrm{d}L_t(\omega) = 0$, where the integral is Stieltjes in t.

The local time at x exists if and only if x is regular for itself. If $\{L_t\}$ and $\{L'_t\}$ are both local times at x, then there is a constant $k > 0$ such that, for all y, $P^y[L'_t = kL_t \text{ for all } t] = 1$. See Blumenthal and Getoor [**45**]. Local times are generalization of the local time of the one-dimensional Brownian motion, which was a great discovery by Lévy (see [**318**]). Consult [**147**], [**228**], [**258**], [**289**] for its importance.

Suppose that $(\{X_t\}, P)$ is non-zero and not compound Poisson. Then, for any x, $\int_0^\infty 1_{\{x\}}(X_t)\mathrm{d}t = 0$ almost surely, because $V^q\{x\} = 0$. If its local time $\{L_t\}$ at x exists, then for any $\omega \in \Omega_0$ the Stieltjes measure on $[0, \infty)$ induced by the function $L_t(\omega)$ of t is continuous singular. This follows from (1) and (3).

Let $d = 1$. Assume that $\Sigma_0 = \mathbb{R}$ and that 0 is regular for itself. Then, there is a function $L(x, t, \omega)$ measurable in (x, t, ω), called the *occupation density*, such that, for every fixed x, $L(x, t, \omega)$ is the local time at x and, for every nonnegative measurable function f and for every y,

$$P^y\big[\textstyle\int_{\mathbb{R}} f(x)L(x, t, \omega)\mathrm{d}x = \int_0^t f(X_s(\omega))\mathrm{d}s \text{ for all } t\big] = 1.$$

For the Brownian motion Trotter [**537**] proves the existence of $L(x,t,\omega)$ continuous in (x,t). When x is taken as time parameter, Ray [**411**] and Knight [**288**] prove its Markov property and find its infinitesimal generator. Boylan [**61**] proves the continuity in (x,t) for α-stable processes with $\alpha > 1$. But Getoor and Kesten [**161**] shows that for 1-stable, not strictly stable, processes the occupation density $L(x,t,\omega)$ does not have an (x,t)-continuous version; further, it does not have a locally (x,t)-bounded version, as shown by Millar and Tran [**351**]. A necessary and sufficient condition for the existence of an (x,t)-continuous version of $L(x,t,\omega)$, in terms of some metric expressible in $\psi(z)$, is obtained by Barlow [**8, 9**] and Barlow and Hawkes [**10**]. See also Hawkes [**191**]. In the symmetric case Marcus and Rosen [**337, 338**] introduce a new method to study the occupation density, using the associated Gaussian processes based on an isomorphism theorem of Dynkin [**122**] and E 44.26.

The Hilbert transform with respect to x of $L(x,t,\omega)$, denoted by $H(y,t,\omega)$, is studied by Yor [**595**], Yamada [**584**], Biane and Yor [**32**], and Fitzsimmons and Getoor [**143**]. A simple formula involving $H(y,t)$ evaluated at the inverse local time is known.

See Bertoin [**26**] for more detailed accounts.

44. Exercises 8

E 44.1. Show that the support Σ of a Lévy process is identical with the support of V^q for any $q > 0$ (for any $q \geq 0$ in the transient case).

E 44.2 (Hawkes [**190**]). Suppose that $d = 1$ and $(\{X_t\}, P)$ is a Poisson process with drift 1 added. Show that $V(\mathrm{d}x)$ has a continuous density

$$v(x) = 1_{(0,\infty)}(x)\textstyle\sum_{k=0}^{[x]}(\mathrm{e}^{-(x-k)}(x-k)^k/k!).$$

E 44.3. Suppose that $\{X_t\}$ and $\{Y_t\}$ are independent Lévy processes both satisfying Condition (ACP). Show that the Lévy process $\{X_t + Y_t\}$ does not necessarily satisfy Condition (ACP).

E 44.4. Let $\{X_t\}$ and $\{Y_t\}$ be independent Lévy processes on $\mathbb{R}^d$. Let $Z_t = X_t + Y_t$. Prove that, if $\{X_t\}$ satisfies (ACP) and if $\{Y_t\}$ is a compound Poisson process, then the Lévy process $\{Z_t\}$ satisfies (ACP).

E 44.5. Show the equivalence of the following statements for a Lévy process $(\{X_t\}, P)$ on $\mathbb{R}^d$.

(1) Condition (ACT) holds.
(2) If f is bounded and Borel-measurable with compact support, then, for any $t > 0$, $P_t f$ is continuous.
(3) If f is bounded and universally measurable, then, for any $t > 0$, $P_t f$ is continuous.

E 44.6 (Hawkes [**190**]). Show that if a Lévy process $(\{X_t\}, P)$ on $\mathbb{R}^d$ satisfies (ACT), then there are functions $p_t(x)$, $t > 0$, satisfying the following conditions:

(1) $p_t(x)$ is nonnegative and Borel-measurable in (t,x);
(2) $p_t(x)$ is lower semi-continuous in x;

(3) $\int p_t(x-y)p_s(y)\mathrm{d}y = p_{t+s}(x)$ for $t>0$, $s>0$, $x\in\mathbb{R}^d$;

(4) $P[X_t\in B] = \int_B p_t(x)\mathrm{d}x$ for $t>0$, $B\in\mathcal{B}(\mathbb{R}^d)$.

Further, show that the function u^q in Theorem 41.16 is represented as $u^q(x) = \int_0^\infty \mathrm{e}^{-qt}p_t(x)\mathrm{d}t$.

E 44.7. Prove that, if B is an F_σ set with $C^q(B)=0$, then $\mathrm{Leb}\,B=0$.

E 44.8 (Kesten [**272**]). Let $(\{X_t\},P^\sharp)$ be the symmetrization (see E 39.7) of a Lévy process $(\{X_t\},P)$. Show that, if $\{0\}$ is essentially polar for $(\{X_t\},P)$, then $\{0\}$ is essentially polar for $(\{X_t\},P^\sharp)$.

E 44.9. Let $(\{X_t\},P)$ be a Lévy process on $\mathbb{R}$ satisfying $\int_{\mathbb{R}}|\widehat{\mu}(z)|^t\mathrm{d}z<\infty$ for all $t>0$. Denote by $p_t(x)$ the continuous density of μ^t for $t>0$. Show that $\{0\}$ is polar if and only if $\int_0^\infty \mathrm{e}^{-qt}p_t(0)\mathrm{d}t=\infty$ for some (equivalently, for all) $q>0$.

E 44.10. A Lévy process is recurrent in the sense of Blumenthal and Getoor [**45**], p. 89 (call it BG-recurrent), if and only if there are no 0-excessive functions other than constant functions. Show that a Lévy process is BG-recurrent if and only if it is recurrent and satisfies (ACP). Notice that there is a recurrent Lévy process which is not BG-recurrent.

E 44.11. Assume that $(\{X_t\},P)$ is a BG-recurrent Lévy process on $\mathbb{R}^d$ in the words of the preceding exercise. Show that, if B is a Borel set with positive Lebesgue measure, then $P^x\left[\int_0^\infty 1_B(X_t)\mathrm{d}t=\infty\right]=1$ for every x.

E 44.12. Consider the Brownian motion on $\mathbb{R}^d$, $d\geq 3$. Then $u(x)=c_d|x|^{2-d}$ with $c_d=2^{-1}\pi^{-d/2}\Gamma((d/2)-1)$ by (35.19). For the unit closed ball $B=\{x\colon |x|\leq 1\}$ show that $C(B)={c_d}^{-1}$.

E 44.13. Let $(\{X_t\},P)$ be the Brownian motion on $\mathbb{R}^d$, $d\geq 1$. Let $S_a=\{x\colon |x|=a\}$, $a>0$. Show that, for $a\leq |x|\leq b$,

$$P^x[T_{S_a}<T_{S_b}] = \begin{cases} (|x|^{2-d}-b^{2-d})/(a^{2-d}-b^{2-d}) & \text{if } d\geq 3,\\ (\log b-\log|x|)/(\log b-\log a) & \text{if } d=2,\\ (b-x)/(b-a) & \text{if } d=1 \text{ and } a\leq x\leq b. \end{cases}$$

Show that, if $d\geq 3$, then $P^x[T_{S_a}<\infty]=(a/|x|)^{d-2}$ for $|x|\geq a$.

E 44.14. Let $(\{X_t\},P)$ be the Brownian motion on $\mathbb{R}^d$, $d\geq 2$. The measure $P_B(x,\mathrm{d}y)=P^x[X_{T_B}\in \mathrm{d}y]$ is called the *harmonic measure*. Show that, for $|x|<a$ and for any Borel set C in $S_a=\{x\colon |x|=a\}$,

$$P^x[X(T_{S_a})\in C] = (a\sigma_1(S_1))^{-1}\int_C (a^2-|x|^2)|x-y|^{-d}\sigma_a(dy),$$

where σ_a is the area measure on S_a and hence $\sigma_1(S_1)=2\pi^{d/2}/\Gamma(d/2)$. Show that, for $B=\{y=(y_j)_{1\leq j\leq d}\colon y_d=0\}$ and for $x=(x_j)_{1\leq j\leq d}$ with $x_d>0$,

$$P^x[X(T_B)\in C] = \Gamma(d/2)\pi^{-d/2}\int_{C'} x_d(|x'-y|^2+{x_d}^2)^{-d/2}\,\mathrm{Leb}_{d-1}(\mathrm{d}y),$$

where $x'=(x_j)_{1\leq j\leq d-1}$ for $x=(x_j)_{1\leq j\leq d}$, Leb_{d-1} is the Lebesgue measure on $\mathbb{R}^{d-1}$, C is any Borel set in B, and $C'\subset\mathbb{R}^{d-1}$ is such that $C=\{y=(y_j)_{1\leq j\leq d}\colon (y_j)_{1\leq j\leq d-1}\in C' \text{ and } y_d=0\}$. This is the $(d-1)$-dimensional Cauchy distribution as in Example 2.12.

E 44.15. Let $(\{X_t\}, P)$ be the Brownian motion on $\mathbb{R}^d$, $d \geq 3$. Let $B_a = \{x : |x| \leq a\}$. Show that

$$U1_{B_a}(x) = \begin{cases} \frac{2a^d}{d(d-2)}|x|^{2-d} & \text{for } |x| > a, \\ \frac{a^2}{d-2} - \frac{1}{d}|x|^2 & \text{for } |x| \leq a. \end{cases}$$

E 44.16. Let $(\{X_t\}, P)$ be the Brownian motion on $\mathbb{R}^d$, $d \geq 3$. Show that, if B is a bounded F_σ set, then, for any Borel set C, $P_B(x, C)/u^0(x)$ tends to $m_B(C)$ as $|x| \to \infty$.

E 44.17 (Hunt [**209**]). Let $B \subset \mathbb{R}^4$ be the set of points whose coordinates are rational. Let ν be a symmetric probability measure concentrated on B with positive mass at each point in B. Suppose that $(\{X_t\}, P)$ is the Lévy process on $\mathbb{R}^4$ with generating triplet $(I, \nu, 0)$, where I is the unit matrix. Show that it satisfies (ACP) and that $u^q(x) = \infty$ for $q \geq 0$ at every $x \in B$. Thus the potential density u^q is infinite at every point of a dense set.

E 44.18. Assume (ACP). Prove that, for $q > 0$, the q-capacity $C^q(B)$ of a closed set B is the supremum of the total masses $\rho(B)$ of measures ρ supported on B satisfying $\int u^q(y - x)\rho(\mathrm{d}y) \leq 1$, $x \in \mathbb{R}^d$. If the process is transient and B is bounded, the same is true also for $q = 0$.

E 44.19 ([**519**]). Let $(\{X_t\}, P)$ be a rotation invariant α-stable process on $\mathbb{R}^d$, $d \geq 1$, $0 < \alpha \leq 2$. Show that $B = \{x \in \mathbb{R}^d : |x| = 1\}$ is a polar set if and only if $\alpha \leq 1$, regardless of the dimension. Generalization to rotation invariant Lévy processes is given by Millar [**349**].

E 44.20. Let $\{X_t\}$ be a subordinator of type B with drift 0. Suppose that $\log(\nu(x, \infty))$ is convex in $x \in (0, \infty)$. Show that then the function $u^0(x)$ of Theorem 41.16 is a continuous decreasing function on $(0, \infty)$.

E 44.21. Suppose that $(\{X_t\}, P)$ is a Lévy process on $\mathbb{R}$ which is non-zero and not compound Poisson. Let $a > 0$ and $R_a = T_{(a,\infty)}$, the hitting time of (a, ∞). Show that $P^0[\, X(R_a-) < a = X(R_a)\,] = 0$ and $P^0[\, X(R_a-) = a < X(R_a)\,] = 0$.

E 44.22 (Dynkin's formula). Let $(\{X_t\}, P)$ be a Lévy process on $\mathbb{R}^d$. Let L be the infinitesimal generator of its transition semigroup in $C_0(\mathbb{R}^d)$. Show the following. If T is a stopping time with $E[T] < \infty$ and if $g \in C_0^2$, then

$$E^x\left[\int_0^T Lg(X_t)\mathrm{d}t\right] = E^x[g(X_T)] - g(x), \qquad x \in \mathbb{R}^d.$$

E 44.23. Let $(\{X_t\}, P)$ be a subordinator and let $R'_a = T_{[a,\infty)}$ for $a > 0$. Show that, for any Borel sets $C \subset [0, a)$ and $D \subset [a, \infty)$ and for any $q \geq 0$,

$$E^0[\mathrm{e}^{-qR'_a};\, X(R'_a-) \in C,\, X(R'_a) \in D\,] = \int_C U^q(0, \mathrm{d}y)\nu(D - y).$$

(This problem is related to E 6.16 and E 50.5. If $\{X_t\}$ is non-zero and not compound Poisson, then $E^0[\mathrm{e}^{-qR'_a};\, X(R'_a-) = a\,] = E^0[\mathrm{e}^{-qR'_a};\, X(R'_a-) = a = X(R'_a)\,] = E^0[\mathrm{e}^{-qR'_a};\, X(R'_a) = a\,]$ by E 44.21 and this is evaluated in E 50.6.)

E 44.24. Let $\{X_t\}$ be the strictly α-stable subordinator with $E[e^{-uX_t}] = e^{-tu^\alpha}$, $0 < \alpha < 1$. Using the notation in the preceding exercise, show that

$$P^0[\,X(R'_a-) \in C,\, X(R'_a) \in D\,] = \tfrac{\alpha \sin \pi\alpha}{\pi} \int_C y^{\alpha-1} dy \int_D (z-y)^{-1-\alpha} dz.$$

E 44.25 (Bertoin [**26**]). Let $(\{X_t\}, P)$ be a subordinator with $\gamma_0 > 0$. It satisfies (ACP) by Theorem 43.21. Show the following. (i) On $(0,\infty)$, $u^0(x)$ is continuous and positive. On $(-\infty, 0]$, $u^0(x) = 0$. (ii) $u^0(0+) = 1/\gamma_0$. (iii) $P^0[\,X_{T_{(x,\infty)}} = x\,] = \gamma_0 u^0(x)$ for all $x > 0$. (A related problem is E 50.6.)

E 44.26. Show that, if $\{X_t\}$ is a symmetric Lévy process on $\mathbb{R}^d$ satisfying (ACP), then, for any $q > 0$, the function $u^q(x)$ of Theorem 41.16 is nonnegative-definite, that is, (2.3) holds for all n with u^q replacing $\widehat{\mu}$.

Notes

Classical potential theory reflects behaviors of Brownian motions. Kakutani's works [**249, 250, 251**] in the 1940s brought this to light and Doob's works [**107, 108**] followed. See his book [**109**] for a comprehensive survey and also the book [**387**] of Port and Stone. Hunt's work [**209**] built up foundations of potential-theoretic aspects of Markov process theory. Sections 40 and 41 are based on [**209**] and the book [**45**] of Blumenthal and Getoor. Chung's book [**81**] is a good introduction. Hunt [**209**] proves that the hitting times of Borel sets are stopping times. Those of the sets called nearly Borel or, more generally, nearly analytic are also stopping times. His proof uses Choquet's capacitability theorem. See [**45**], [**345**]. If we prove this fact, then many of the results for F_σ sets in this chapter are shown to be true for nearly analytic sets (see also Remark 43.10). Excessive functions are nearly-Borel-measurable. On time reversal different from Proposition 41.8 see Nagasawa [**359**].

Section 42 contains the capacity theory of Hunt [**209**], Port and Stone [**386**], Kanda [**254**], and Hawkes [**190**]. The presentation is influenced by Bertoin [**26**]. Hunt's capacity theory does not cover those Lévy processes which do not satisfy Condition (ACP). The q-capacity in Section 42 was introduced by Port and Stone [**386**]. But our definition of q-capacitary measure is different from that of [**386**] and [**26**] in two points. First, our q-capacitary measure is the q-co-capacitary measure of [**386**] and [**26**] and our q-co-capacitary measure is their q-capacitary measure. By this change of the definition, our q-capacitary measure now coincides with the q-capacitary measure of [**45, 81, 209, 254**] whenever the latter is definable. The q-capacity of a set is the total mass of its q-capacitary measure. It is not affected by the difference of the definition, since q-capacitary and q-co-capacitary measures have a common total mass. See (42.10). Second, in the definition of q-capacitary and q-co-capacitary measures in [**386**] and [**26**], they use $T'_B = \inf\{t \geq 0 \colon X_t \in B\}$ instead of the hitting time T_B of B. But this does not have any influence, because $\mathrm{Leb}(B \setminus B^{\mathrm{reg}}) = 0$. Take note that the identities (42.14) and (42.15) are not true in general if T_B is replaced by T'_B.

Theorems 42.27 and 42.29 are by Kanda [**254**] with different constants. A predecessor is Orey [**363**]. Theorem 42.30 is essentially a result of [**254**] on stable processes. For one-dimensional stable processes it is by Orey [**363**]. Kanda [**253, 256**] and Kanda and Uehara [**257**] are papers in this line.

Section 43 is a part of Kesten's work [**272**] with some improvement of results and methods by Bretagnolle [**68**]. We have not given known results in their generality (see Remark 43.10). The difference between Cauchy process and 1-stable, not strictly stable, processes on $\mathbb{R}$ in Examples 43.7 was pointed out earlier by Orey [**363**] and Port and Stone [**385**].

There are notions of thin sets and semi-polar sets. A set is thin if no point is regular for the set. A set is semi-polar if it is the union of a countable number of thin sets. The condition that every semi-polar set is polar is called Condition (H) and important in Hunt's theory. Symmetric Lévy processes satisfy Condition (H) and so do all non-trivial stable processes. See Blumenthal and Getoor [**46**], Kanda [**254, 255**], Fitzsimmons and Kanda [**144**], and Rao [**409**].

An expression for $P^x[T_a < T_b]$ for some classes of Lévy processes on $\mathbb{R}$ is given by Getoor [**159**].

Many limit theorems for hitting times, hitting measures, and potential measures are known for stable processes. See Port [**381, 382, 383**] and Pruitt and Taylor [**400**]. For random walks on $\mathbb{Z}^d$ see Spitzer [**496**]. Some of the results are extended to general Lévy processes by Port and Stone [**386**]. DeBlassie [**95**] considers exit time from a wedge for rotation invariant stable processes.

Subordinators often emerge in connection with Markov processes. Indeed, a local time at a point is defined for a temporally homogeneous Markov process by the properties (1)–(3) in Remark 43.27 similarly, and the right-continuous inverse of a local time at a point is a subordinator possibly annihilated at an independent exponentially distributed time (see [**45**]). For the appearance of the inverse of Brownian local time in limit theorems, see Kasahara and Kotani [**264**] and Kasahara [**263**].

Zabczyk [**599**] takes a different approach to potential theory for Lévy processes. Berg and Forst [**21**], Herz [**198**], and Watanabe [**549**] contain some basic results not covered in this chapter.

CHAPTER 9

Wiener–Hopf factorizations

45. Factorization identities

Let $\{X_t : t \geq 0\}$ be a Lévy process on $\mathbb{R}$ defined on a probability space $(\Omega, \mathcal{F}, P)$ in the sense of Definition 1.6. It induces the following stochastic processes. We use this notation throughout this chapter.

DEFINITION 45.1. Let Ω_0 be the set in Definition 1.6. Define, on Ω_0,

$$M_t = \sup_{0 \leq s \leq t} X_s, \qquad N_t = \inf_{0 \leq s \leq t} X_s, \qquad Y_t = M_t - X_t,$$

$$\widetilde{X}_t = -X_t, \qquad \widetilde{M}_t = \sup_{0 \leq s \leq t} \widetilde{X}_s = -N_t,$$

$$\widetilde{N}_t = \inf_{0 \leq s \leq t} \widetilde{X}_s = -M_t, \qquad \widetilde{Y}_t = \widetilde{M}_t - \widetilde{X}_t = X_t - N_t,$$

for $t \geq 0$. The processes $\{M_t\}$, $\{N_t\}$, and $\{Y_t\}$ are, respectively, the *supremum process*, the *infimum process*, and the *reflecting process* of $\{X_t\}$. The process $\{\widetilde{X}_t\}$ is the *dual process* of $\{X_t\}$. The processes $\{\widetilde{M}_t\}$, $\{\widetilde{N}_t\}$, and $\{\widetilde{Y}_t\}$ are, respectively, the *dual supremum process*, the *dual infimum process*, and the *dual reflecting process* of $\{X_t\}$. Let

$$R_x(\omega) = T_{(x,\infty)}(\omega) = \inf\{t > 0 : X_t(\omega) > x\}$$

for $\omega \in \Omega_0$ and $x \geq 0$ with the convention that the empty set has infimum ∞. The process $\{R_x : x \geq 0\}$ is the *first passage time process* of $\{X_t\}$. Further, let

$$\Lambda_t(\omega) = \inf\{s \in [0, t] : X_s(\omega) \vee X_{s-}(\omega) = M_t(\omega)\}$$

for $\omega \in \Omega_0$ and $t \geq 0$. Here we understand $X_{0-} = 0$. All of these are defined to be 0 on $\Omega \setminus \Omega_0$.

We factorize the Laplace transform (in t) of the distribution of X_t by using the Laplace transforms (in t) of the distributions of the processes defined above. Various factorization identities appear in this connection. These identities and their probabilistic interpretations are called, in general, Wiener–Hopf factorizations of one-dimensional Lévy processes. Spitzer proved some basic identities for random walks on $\mathbb{Z}$ by combinatorial methods in [**494**] and later extended them to factorization identities by Fourier-analytic methods in [**496**], using a technique similar to that which Wiener and Hopf [**574**] introduced in solving some integral equations. They were generalized to Lévy processes on $\mathbb{R}$ by Rogozin [**419**], Pecherskii and Rogozin [**368**], Borovkov [**59**], and others. These results are presented in this chapter.

As in the other chapters, the generating triplet of $\{X_t\}$ is denoted by (A, ν, γ). Since $d = 1$, A is a nonnegative real and γ is a real. When the process is of type A or B, the drift is denoted by γ_0. The distribution of X_1 is μ,

$$\psi(z) = \log \widehat{\mu}(z) = \log E[\mathrm{e}^{\mathrm{i}zX_1}], \qquad z \in \mathbb{R},$$

and the q-potential measure is V^q for $q > 0$,

$$V^q(B) = E\left[\int_0^\infty \mathrm{e}^{-qt} 1_B(X_t)\mathrm{d}t\right], \qquad B \in \mathcal{B}(\mathbb{R});$$

qV^q is a probability measure with characteristic function $q(q - \psi(z))^{-1}$, see Proposition 37.4.

THEOREM 45.2 (Factorization of qV^q). (i) *Let $q > 0$. There exists a unique pair of characteristic functions $\varphi_q^+(z)$ and $\varphi_q^-(z)$ of infinitely divisible distributions having drift* 0 *supported on* $[0, \infty)$ *and* $(-\infty, 0]$, *respectively, such that*

$$q(q - \psi(z))^{-1} = \varphi_q^+(z)\varphi_q^-(z), \quad z \in \mathbb{R}. \tag{45.1}$$

(ii) *The functions $\varphi_q^+(z)$ and $\varphi_q^-(z)$ have the following representations:*

$$\varphi_q^+(z) = \exp\left[\int_0^\infty t^{-1}\mathrm{e}^{-qt}\mathrm{d}t \int_{(0,\infty)} (\mathrm{e}^{\mathrm{i}zx} - 1)\mu^t(\mathrm{d}x)\right], \tag{45.2}$$

$$\varphi_q^-(z) = \exp\left[\int_0^\infty t^{-1}\mathrm{e}^{-qt}\mathrm{d}t \int_{(-\infty,0)} (\mathrm{e}^{\mathrm{i}zx} - 1)\mu^t(\mathrm{d}x)\right]. \tag{45.3}$$

Proof. This theorem is essentially contained in Theorem 30.10. That is, we already know that qV^q is infinitely divisible with Lévy measure

$$\nu_q(B) = \int_0^\infty t^{-1}\mathrm{e}^{-qt}\mu^t(B)\mathrm{d}t \quad \text{for } B \in \mathcal{B}(\mathbb{R}),\ B \not\ni 0, \tag{45.4}$$

and

$$q(q - \psi(z))^{-1} = \exp\left[\int_{-\infty}^\infty (\mathrm{e}^{\mathrm{i}zx} - 1)\nu_q(\mathrm{d}x)\right]. \tag{45.5}$$

Hence, if we define φ_q^+ and φ_q^- by (45.2) and (45.3), then (45.1) holds. To see the uniqueness, let φ_q^+ and φ_q^- be functions with the desired properties and denote their Lévy measures by ρ_q^+ and ρ_q^-. Then ρ_q^+ and ρ_q^- are concentrated on $(0, \infty)$ and $(-\infty, 0)$, respectively, by Theorem 24.7. We have $\nu_q = \rho_q^+ + \rho_q^-$ by the uniqueness of the Lévy–Khintchine representation. Hence, by (45.4), φ_q^+ and φ_q^- must satisfy (45.2) and (45.3). □

REMARK 45.3. Throughout this chapter, φ_q^+ and φ_q^- are meant to be the functions in the theorem above. The function φ_q^+ continuously extends to a bounded analytic function on the upper half plane without zero points. Similarly the function φ_q^- on the lower half plane. To see this, consider the function in the right-hand side of (45.2) or (45.3), with $z \in \mathbb{R}$ replaced by a complex variable z with $\operatorname{Im} z \geq 0$ or $\operatorname{Im} z \leq 0$, respectively, and argue as in the proof of Proposition 2.6. The extensions are unique, as is proved like the proof of Theorem 24.11. They are expressed by the same symbols φ_q^+ and φ_q^-.

EXAMPLE 45.4. Suppose that the Lévy measure ν of $\{X_t\}$ is 0. Then $\psi(z) = -\frac{1}{2}Az^2 + \mathrm{i}\gamma z$. Let $A > 0$. We have

$$q(q-\psi(z))^{-1} = \{c_+(c_+ - \mathrm{i}z)^{-1}\}\{c_-(c_- + \mathrm{i}z)^{-1}\} \tag{45.6}$$

with $c_\pm = (A^{-2}\gamma^2 + 2qA^{-1})^{1/2} \mp A^{-1}\gamma > 0$. The first factor in the right-hand side of (45.6) is the characteristic function of the exponential distribution with parameter c_+ and the second one is the characteristic function of the dual of the exponential distribution with parameter c_-. Hence (45.6) is the factorization in Theorem 45.2. Thus qV^q is the two-sided exponential distribution of Example 15.14. Hence the q-potential density $u^q(x)$ of $\{X_t\}$ is

$$u^q(x) = c\{1_{[0,\infty)}(x)\exp(-c_+x) + 1_{(-\infty,0)}(x)\exp(c_-x)\}$$

with $c = (\gamma^2 + 2qA)^{-1/2}$. In particular, if $\{X_t\}$ is the Brownian motion, then

$$u^q(x) = (2q)^{-1/2}\exp(-(2q)^{1/2}|x|).$$

In the case where $A = 0$ and $\gamma > 0$, we have

$$q(q-\psi(z))^{-1} = c_+(c_+ - \mathrm{i}z)^{-1}$$

with $c_+ = q\gamma^{-1}$, that is, φ_q^+ and φ_q^- are the characteristic functions of the exponential distribution and δ_0, respectively.

In this section let us prove several identities concerning joint distributions of some of $\{M_t\}$, $\{N_t\}$, $\{Y_t\}$, $\{\widetilde{Y}_t\}$, and $\{\Lambda_t\}$ for compound Poisson processes on $\mathbb{R}$, and then extend some of them to general Lévy processes on $\mathbb{R}$.

THEOREM 45.5 (Compound Poisson process). *Suppose that $\{X_t\}$ is a compound Poisson process. Let $q > 0$. Then, for $z \in \mathbb{R}$, $w \in \mathbb{R}$, and $p \geq 0$,*

$$q\int_0^\infty \mathrm{e}^{-qt}E[\mathrm{e}^{\mathrm{i}zM_t}]\mathrm{d}t = q\int_0^\infty \mathrm{e}^{-qt}E[\mathrm{e}^{\mathrm{i}z(X_t-N_t)}]\mathrm{d}t = \varphi_q^+(z), \tag{45.7}$$

$$q\int_0^\infty \mathrm{e}^{-qt}E[\mathrm{e}^{\mathrm{i}zN_t}]\mathrm{d}t = q\int_0^\infty \mathrm{e}^{-qt}E[\mathrm{e}^{\mathrm{i}z(X_t-M_t)}]\mathrm{d}t = \varphi_q^-(z), \tag{45.8}$$

$$q\int_0^\infty e^{-qt}E[e^{izM_t+iw(X_t-M_t)}]dt=\varphi_q^+(z)\varphi_q^-(w), \tag{45.9}$$

$$\begin{aligned} q\int_0^\infty & e^{-qt}E[e^{izM_t+iw(X_t-M_t)-p\Lambda_t}]dt \\ &=\varphi_{q+p}^+(z)\varphi_q^-(w)\exp\left[\int_0^\infty t^{-1}e^{-qt}(e^{-pt}-1)P[X_t>0]dt\right]. \end{aligned} \tag{45.10}$$

Later we will extend these identities to general Lévy processes on $\mathbb{R}$; the equalities (45.7), (45.8), and (45.9) will be proved in Theorem 45.7 and Corollary 45.8 and the equality (45.10) in Theorem 49.1. First, we study random walks.

LEMMA 45.6. *Let $\{S_n\}$ be a random walk on $\mathbb{R}$ and define*

$$L_n=\max_{0\le k\le n}S_k, \qquad H_n=\min\{m\colon 0\le m\le n, S_m=L_n\},$$
$$T=\min\{n>0\colon S_n>0\}, \qquad D=\min\{n>0\colon S_n\ge 0\},$$
$$\widetilde{T}=\min\{n>0\colon S_n<0\}, \qquad \widetilde{D}=\min\{n>0\colon S_n\le 0\}.$$

Let ξ, η, u, $v\in\mathbb{C}$ with $|\xi|<1$, $|\eta|\le 1$, $\operatorname{Re}u\le 0$, and $\operatorname{Re}v\ge 0$ and let $r\in\mathbb{R}$. Define

$$f_\xi^+(u)=\exp\left[\sum_{n=1}^\infty n^{-1}\xi^nE[e^{uS_n};\,S_n>0]\right],$$
$$f_\xi^-(v)=\exp\left[\sum_{n=1}^\infty n^{-1}\xi^nE[e^{vS_n};\,S_n<0]\right],$$
$$c_\xi=\exp\left[\sum_{n=1}^\infty n^{-1}\xi^nP[S_n=0]\right].$$

Then,

$$(1-\xi E[e^{irS_1}])^{-1}=c_\xi f_\xi^+(ir)f_\xi^-(ir), \tag{45.11}$$

$$\sum_{n=0}^\infty \xi^nE[e^{uS_n};\,\widetilde{D}>n]=f_\xi^+(u), \tag{45.12}$$

$$\sum_{n=0}^\infty \xi^nE[e^{vS_n};\,D>n]=f_\xi^-(v), \tag{45.13}$$

$$\sum_{n=0}^\infty \xi^nE[e^{uS_n};\,\widetilde{T}>n]=c_\xi f_\xi^+(u), \tag{45.14}$$

$$\sum_{n=0}^\infty \xi^nE[e^{vS_n};\,T>n]=c_\xi f_\xi^-(v), \tag{45.15}$$

$$\sum_{n=0}^{\infty} \xi^n E[\mathrm{e}^{uL_n+v(S_n-L_n)}\eta^{H_n}] = c_\xi f^+_{\xi\eta}(u) f^-_\xi(v) \tag{45.16}$$
$$= (1-\xi)^{-1} \exp\Bigg[\sum_{n=1}^{\infty} n^{-1}(\xi\eta)^n E[\mathrm{e}^{uS_n}-1;\ S_n>0]$$
$$+\sum_{n=1}^{\infty} n^{-1}\xi^n E[\mathrm{e}^{vS_n}-1;\ S_n<0]$$
$$+\sum_{n=1}^{\infty} n^{-1}\xi^n(\eta^n-1)P[S_n>0]\Bigg].$$

Proof. We have

$$(1-\xi E[\mathrm{e}^{\mathrm{i}rS_1}])^{-1} = \exp(-\log(1-\xi E[\mathrm{e}^{\mathrm{i}rS_1}]))$$
$$= \exp\left[\sum_{n=1}^{\infty} n^{-1}\xi^n E[\mathrm{e}^{\mathrm{i}rS_n}]\right] = c_\xi f^+_\xi(\mathrm{i}r) f^-_\xi(\mathrm{i}r),$$

that is, (45.11). Let $S_n - S_{n-1} = Z_n$. Since $(Z_1,\ldots,Z_m)$ and $(Z_m,\ldots,Z_1)$ have the same distribution by reversal of time,

$$E[\mathrm{e}^{uS_m};\ \widetilde{D}>m] = E[\mathrm{e}^{uS_m};\ S_1>0,\ S_2>0,\ldots,S_m>0] \tag{45.17}$$
$$= E[\mathrm{e}^{uS_m};\ Z_m>0,\ Z_m+Z_{m-1}>0,\ldots,Z_m+\cdots+Z_1>0]$$
$$= E[\mathrm{e}^{uS_m};\ S_m>S_k \text{ for } k=0,1,\ldots,m-1].$$

Further, for $0\le m\le n$, since $(Z_1\ldots,Z_{n-m}) \stackrel{\mathrm{d}}{=} (Z_{m+1},\ldots,Z_n)$ by shift of time, we have

$$E[\mathrm{e}^{vS_{n-m}};\ T>n-m] = E[\mathrm{e}^{vS_{n-m}};\ S_1\le 0,\ldots,S_{n-m}\le 0] \tag{45.18}$$
$$= E[\mathrm{e}^{v(Z_{m+1}+\cdots+Z_n)};\ Z_{m+1}\le 0,\ Z_{m+1}+Z_{m+2}\le 0,$$
$$\ldots,Z_{m+1}+\cdots+Z_n\le 0]$$
$$= E[\mathrm{e}^{v(S_n-S_m)};\ S_j\le S_m \text{ for } j=m+1,\ldots,n].$$

Write the left-hand sides of (45.12) and (45.15) as $g_\xi(u)$ and $h_\xi(v)$, respectively. Then, using (45.17) and (45.18), we have

$$g_{\xi\eta}(u)h_\xi(v) = \sum_{m=0}^{\infty}(\xi\eta)^m E[\mathrm{e}^{uS_m};\ \widetilde{D}>m]\sum_{n=m}^{\infty}\xi^{n-m}E[\mathrm{e}^{vS_{n-m}};\ T>n-m]$$
$$= \sum_{n=0}^{\infty}\sum_{m=0}^{n}\xi^n\eta^m E[\mathrm{e}^{uS_m};\ \widetilde{D}>m]E[\mathrm{e}^{vS_{n-m}};\ T>n-m]$$
$$= \sum_{n=0}^{\infty}\sum_{m=0}^{n}\xi^n\eta^m E[\mathrm{e}^{uS_m+v(S_n-S_m)};\ B_{n,m}]$$

$$= \sum_{n=0}^{\infty} \xi^n \sum_{m=0}^{n} \eta^m E[\mathrm{e}^{uL_n+v(S_n-L_n)};\ H_n = m]$$

$$= \sum_{n=0}^{\infty} \xi^n E[\mathrm{e}^{uL_n+v(S_n-L_n)} \eta^{H_n}],$$

where $B_{n,m} = \{S_m > S_k$ for $k = 0, \ldots, m-1$ and $S_m \geq S_j$ for $j = m+1, \ldots, n\}$. Therefore, if (45.12) and (45.15) are proved, then the first equality of (45.16) follows. Letting $\eta = 1$ and $u = v = \mathrm{i}r$, $r \in \mathbb{R}$, and using (45.11), we get

$$(45.19) \quad g_\xi(\mathrm{i}r)h_\xi(\mathrm{i}r) = \sum_{n=0}^{\infty} \xi^n E[\mathrm{e}^{\mathrm{i}rS_n}] = \sum_{n=0}^{\infty} (\xi E[\mathrm{e}^{\mathrm{i}rS_1}])^n = c_\xi f_\xi^+(\mathrm{i}r) f_\xi^-(\mathrm{i}r).$$

Notice that, when we fix ξ with $|\xi| < 1$, the functions $f_\xi^+(u)$ and $g_\xi(u)$ are continuous on $\{u\colon \mathrm{Re}\, u \leq 0\}$ and bounded and analytic on $\{u\colon \mathrm{Re}\, u < 0\}$, while the functions $f_\xi^-(v)$ and $h_\xi(v)$ are continuous on $\{v\colon \mathrm{Re}\, v \geq 0\}$ and bounded and analytic on $\{v\colon \mathrm{Re}\, v > 0\}$. To see the analyticity of $f_\xi^+(u)$ for example, use the analyticity of $E[\mathrm{e}^{uS_n};\ S_n > 0]$ and the fact that the limit of a uniformly bounded sequence of analytic functions is analytic. If $|\xi| < 1$, then

$$|f_\xi^+(u)| \geq \exp\left[-\sum_{n=1}^{\infty} n^{-1}|\xi|^n\right] = 1 - |\xi| > 0.$$

If $|\xi| < 1/2$, then

$$|h_\xi(v)| \geq 1 - \sum_{n=1}^{\infty} |\xi|^n = (1 - 2|\xi|)(1 - |\xi|)^{-1} > 0.$$

Hence, if $|\xi| < 1/2$, then, from (45.19)

$$g_\xi(\mathrm{i}r)/f_\xi^+(\mathrm{i}r) = c_\xi f_\xi^-(\mathrm{i}r)/h_\xi(\mathrm{i}r)$$

for $r \in \mathbb{R}$, which shows that the bounded analytic functions $g_\xi(u)/f_\xi^+(u)$ on the left half plane and $c_\xi f_\xi^-(v)/h_\xi(v)$ on the right half plane have a common boundary value on the imaginary axis. Therefore, by Morera's theorem, the function pasted together is bounded and analytic on the whole plane and hence constant. The constant is 1, because, by the definitions of $g_\xi(u)$ and $f_\xi^+(u)$, $g_\xi(u)/f_\xi^+(u) \to 1$ when u goes to $-\infty$ along the real axis. It follows that

$$g_\xi(u) = f_\xi^+(u) \quad \text{and} \quad h_\xi(v) = c_\xi f_\xi^-(v)$$

for $|\xi| < 1/2$. Thus (45.12), (45.15), and consequently the first equality in (45.16) are true for $|\xi| < 1/2$. But, since the functions that appear are analytic in ξ on $\{\xi\colon |\xi|<1\}$, they are true for $|\xi|<1$. We can show the second equality in (45.16), rewriting

$$\begin{aligned}1-\xi &= \exp\left[-\sum_{n=1}^{\infty} n^{-1}\xi^n\right]\\ &= c_\xi^{-1}\exp\Bigg[-\sum_{n=1}^{\infty} n^{-1}(\xi\eta)^n P[S_n>0]\\ &\qquad -\sum_{n=1}^{\infty} n^{-1}\xi^n P[S_n<0]-\sum_{n=1}^{\infty} n^{-1}(\xi^n-(\xi\eta)^n)P[S_n>0]\Bigg]\end{aligned}$$

and using the definitions of $f^+_{\xi\eta}(u)$ and $f^-_\xi(v)$. The remaining identities (45.13) and (45.14) are obtained from (45.12) and (45.15), when we consider $-S_n$ in place of S_n. □

To verify factorization through splitting the complex plane into two parts as in the proof above is the technique of Wiener and Hopf.

Proof of Theorem 45.5. Since $\{X_t\}$ is compound Poisson,

$$\psi(z)=\int_{|x|>0}(\mathrm{e}^{\mathrm{i}zx}-1)\nu(\mathrm{d}x) \quad \text{with } 0<c=\nu(\mathbb{R})<\infty.$$

Write $\sigma=c^{-1}\nu$. There are a random walk $\{S_n\}$ having σ as the distribution of S_1 and a Poisson process $\{Z_t\}$ with parameter c such that $\{S_n\}$ and $\{Z_t\}$ are independent and $X_t=S_{Z_t}$. Let $0<J_1<J_2<\cdots$ be the jumping times of $\{Z_t\}$ and hence of $\{X_t\}$. Let $J_0=0$. Then $\{J_n\}$ is a random walk with J_1 having exponential distribution with parameter c.

The identities (45.7)–(45.9) follow from (45.10). Indeed, (45.9) is a special case of (45.10) with $p=0$, and (45.9) contains

$$\varphi_q^+(z)=q\int_0^\infty \mathrm{e}^{-qt}E[\mathrm{e}^{\mathrm{i}zM_t}]\mathrm{d}t, \qquad z\in\mathbb{R}, \tag{45.20}$$

$$\varphi_q^-(z)=q\int_0^\infty \mathrm{e}^{-qt}E[\mathrm{e}^{\mathrm{i}z(X_t-M_t)}]\mathrm{d}t, \qquad z\in\mathbb{R}, \tag{45.21}$$

as special cases. The identities (45.20), (45.21) for the dual process $\{\widetilde{X}_t\}$ in place of $\{X_t\}$ are respectively

$$\varphi_q^-(-z)=q\int_0^\infty \mathrm{e}^{-qt}E[\mathrm{e}^{-\mathrm{i}zN_t}]\mathrm{d}t,$$

$$\varphi_q^+(-z)=q\int_0^\infty \mathrm{e}^{-qt}E[\mathrm{e}^{-\mathrm{i}z(X_t-N_t)}]\mathrm{d}t.$$

Thus we get (45.7) and (45.8).

Let us deduce (45.10) from Lemma 45.6. Let $z \in \mathbb{R}$, $w \in \mathbb{R}$, $p \geq 0$. If $J_n \leq t < J_{n+1}$, then $X_t = S_n$, $M_t = L_n$, and $\Lambda_t = J_{H_n}$. Incidentally, this shows the measurability of Λ_t in the compound Poisson case. Hence

$$\begin{aligned} q \int_0^\infty & \mathrm{e}^{-qt} E[\mathrm{e}^{\mathrm{i}zM_t + \mathrm{i}w(X_t - M_t) - p\Lambda_t}]\mathrm{d}t \\ &= \sum_{n=0}^\infty q \int_0^\infty \mathrm{e}^{-qt} E[\mathrm{e}^{\mathrm{i}zL_n + \mathrm{i}w(S_n - L_n) - pJ(H_n)};\ J_n \leq t < J_{n+1}]\mathrm{d}t \\ &= I_1, \quad \text{say.} \end{aligned}$$

Use $q \int_{J_n}^{J_{n+1}} \mathrm{e}^{-qt}\mathrm{d}t = \mathrm{e}^{-qJ_n} - \mathrm{e}^{-qJ_{n+1}}$. Then

$$\begin{aligned} I_1 &= \sum_{n=0}^\infty \sum_{m=0}^n E[\mathrm{e}^{\mathrm{i}zL_n + \mathrm{i}w(S_n - L_n) - pJ_m}(\mathrm{e}^{-qJ_n} - \mathrm{e}^{-qJ_{n+1}});\ H_n = m] \\ &= \sum_{n=0}^\infty \sum_{m=0}^n E[\mathrm{e}^{\mathrm{i}zL_n + \mathrm{i}w(S_n - L_n)};\ H_n = m]\, E[\mathrm{e}^{-pJ_m}(\mathrm{e}^{-qJ_n} - \mathrm{e}^{-qJ_{n+1}})] \\ &= I_2, \quad \text{say.} \end{aligned}$$

We have, for $m \leq n$,

$$\begin{aligned} &E[\mathrm{e}^{-pJ_m}(\mathrm{e}^{-qJ_n} - \mathrm{e}^{-qJ_{n+1}})] \\ &\quad = (c/(q+p+c))^m \xi^{n-m} - (c/(q+p+c))^m \xi^{n+1-m} = (q/(q+c))\xi^n \eta^m, \end{aligned}$$

where $\xi = c/(q+c)$ and $\eta = (q+c)/(q+p+c)$. Thus

$$I_2 = (q/(q+c)) \sum_{n=0}^\infty \xi^n E[\mathrm{e}^{\mathrm{i}zL_n + \mathrm{i}w(S_n - L_n)} \eta^{H_n}] = I_3, \quad \text{say.}$$

Now employ the identitity (45.16) in Lemma 45.6. Then

$$\begin{aligned} I_3 = \exp\Bigg[& \sum_{n=1}^\infty n^{-1}(\xi\eta)^n E[\mathrm{e}^{\mathrm{i}zS_n} - 1;\ S_n > 0] \\ & + \sum_{n=1}^\infty n^{-1}\xi^n E[\mathrm{e}^{\mathrm{i}wS_n} - 1;\ S_n < 0] + \sum_{n=1}^\infty n^{-1}\xi^n(\eta^n - 1)P[S_n > 0]\Bigg] \\ = I_4, \quad & \text{say.} \end{aligned}$$

On the other hand, using ν_q of (45.4), we have

$$\begin{aligned} \text{right-hand side of (45.10)} &= \exp\Bigg[\int_{(0,\infty)} (\mathrm{e}^{\mathrm{i}zx} - 1)\nu_{q+p}(\mathrm{d}x) \\ & + \int_{(-\infty,0)} (\mathrm{e}^{\mathrm{i}zx} - 1)\nu_q(\mathrm{d}x) + \int_0^\infty t^{-1}\mathrm{e}^{-qt}(\mathrm{e}^{-pt} - 1)P[X_t > 0]\mathrm{d}t\Bigg]. \end{aligned}$$

This equals I_4, because, using

$$\nu_q(B) = \sum_{n=1}^{\infty} \int_0^{\infty} t^{-1} \mathrm{e}^{-qt-ct} (n!)^{-1} (ct)^n \sigma^n(B) \mathrm{d}t = \sum_{n=1}^{\infty} n^{-1} \xi^n \sigma^n(B)$$

for any Borel set B not containing 0, we see that

$$\int_{(-\infty,0)} (\mathrm{e}^{\mathrm{i}zx} - 1) \nu_q(\mathrm{d}x) = \sum_{n=1}^{\infty} n^{-1} \xi^n \int_{(-\infty,0)} (\mathrm{e}^{\mathrm{i}zx} - 1) \sigma^n(\mathrm{d}x),$$

$$\int_{(0,\infty)} (\mathrm{e}^{\mathrm{i}zx} - 1) \nu_{q+p}(\mathrm{d}x) = \sum_{n=1}^{\infty} n^{-1} (\xi\eta)^n \int_{(0,\infty)} (\mathrm{e}^{\mathrm{i}zx} - 1) \sigma^n(\mathrm{d}x),$$

and further

$$\begin{aligned} &\int_0^{\infty} t^{-1} \mathrm{e}^{-qt} (\mathrm{e}^{-pt} - 1) P[X_t > 0] \mathrm{d}t \\ &\quad = \int_0^{\infty} t^{-1} \mathrm{e}^{-qt} (\mathrm{e}^{-pt} - 1) \sum_{n=1}^{\infty} \mathrm{e}^{-ct} (n!)^{-1} (ct)^n P[S_n > 0] \mathrm{d}t \\ &\quad = \sum_{n=1}^{\infty} n^{-1} \xi^n (\eta^n - 1) P[S_n > 0]. \end{aligned}$$

Thus (45.10) is proved. □

Let us prove (45.7), (45.8), and (45.9) for general Lévy processes on $\mathbb{R}$.

THEOREM 45.7 (General Lévy process). *For any $q > 0$, $z \in \mathbb{R}$, and $w \in \mathbb{R}$,*

$$q \int_0^{\infty} \mathrm{e}^{-qt} E[\mathrm{e}^{\mathrm{i}zM_t + \mathrm{i}w(X_t - M_t)}] \mathrm{d}t = \varphi_q^+(z) \varphi_q^-(w). \tag{45.22}$$

COROLLARY 45.8. *For any $q > 0$, $z \in \mathbb{R}$, and $w \in \mathbb{R}$,*

$$q \int_0^{\infty} \mathrm{e}^{-qt} E[\mathrm{e}^{\mathrm{i}wN_t + \mathrm{i}z(X_t - N_t)}] \mathrm{d}t = \varphi_q^+(z) \varphi_q^-(w). \tag{45.23}$$

For any $q > 0$ and $z \in \mathbb{R}$, (45.7) and (45.8) hold.

Proof. Consider the dual process $\{\widetilde{X}_t\}$ instead of $\{X_t\}$. Then (45.22) turns into (45.23). Letting $z = 0$ or $w = 0$ in (45.22) or (45.23), we get (45.7) and (45.8). □

REMARK 45.9. We have, for each $t \geq 0$,

$$M_t \stackrel{\mathrm{d}}{=} X_t - N_t, \tag{45.24}$$

$$N_t \stackrel{\mathrm{d}}{=} X_t - M_t. \tag{45.25}$$

Indeed, it follows from (45.7) and (45.8) and from the right-continuity in t that $E[\mathrm{e}^{\mathrm{i}zM_t}] = E[\mathrm{e}^{\mathrm{i}z(X_t - N_t)}]$ and $E[\mathrm{e}^{\mathrm{i}zN_t}] = E[\mathrm{e}^{\mathrm{i}z(X_t - M_t)}]$ for $z \in \mathbb{R}$. We can also see these relations as consequences of Proposition 41.8 as follows. Let $t > 0$. For $Z_s = X_{(t-s)-} - X_{t-}$, $0 \le s < t$, and $Z_t = -X_{t-}$ we have $\{-Z_s : 0 \le s \le t\} \stackrel{\mathrm{d}}{=} \{X_s : 0 \le s \le t\}$ and hence $M_t^{-Z} \stackrel{\mathrm{d}}{=} M_t$, where we define $M_t^{-Z} = \sup_{0 \le s \le t}(-Z_s)$. Since

$$M_t^{-Z} = \sup_{0 \le s < t} (-X_{(t-s)-} + X_{t-}) = -\inf_{0 < s \le t} X_{s-} + X_t = -N_t + X_t \quad \text{a. s.},$$

we get (45.24). Similarly (45.25). Notice that, however, the processes $\{M_t\}$ and $\{N_t\}$ are not identical in law with the processes $\{X_t - N_t\}$ and $\{X_t - M_t\}$, respectively, in general. While $\{M_t\}$ is an increasing process, $\{X_t - N_t\}$ is not in general.

We show Theorem 45.7, approximating a Lévy process pathwise by compound Poisson processes as in Skorohod [**493**].

LEMMA 45.10. *If $\{Z_t\}$ is a Poisson process with parameter* 1, *then*

$$P\Big[\lim_{n\to\infty} \sup_{t \le u} |t - n^{-2} Z_{n^2 t}| = 0,\ \forall u \ge 0\Big] = 1.$$

Proof. Fix u. Partition the interval $[0, u]$ and use the inequality in Lemma 20.5. Letting the partition become finer and finer, we get, for $\varepsilon > 0$,

$$P\Big[\sup_{t \le u} |t - n^{-2} Z_{n^2 t}| > \varepsilon\Big] \le 3^3 \varepsilon^{-2} \operatorname{Var}(n^{-2} Z_{n^2 u}) = 3^3 \varepsilon^{-2} n^{-2} u.$$

Choose $\varepsilon_n \downarrow 0$ such that $\sum \varepsilon_n{}^{-2} n^{-2} < \infty$ ($\varepsilon_n = n^{-1/4}$ for example). Then, by the Borel–Cantelli lemma,

$$P\Big[\sup_{t \le u} |t - n^{-2} Z_{n^2 t}| > \varepsilon_n \text{ for infinitely many } n\Big] = 0.$$

This proves the assertion for fixed u. Now let $u = u_k \uparrow \infty$, to finish the proof. □

LEMMA 45.11. *Suppose that $\{X_t\}$ is not a zero process and has continuous sample functions. Let $\{Z_t\}$ be a Poisson process with parameter* 1, *independent of $\{X_t\}$. Let $X_t^n = X(n^{-2} Z_{n^2 t})$ for $n = 1, 2, \ldots$. Then $\{X_t^n\}$ are compound Poisson processes satisfying*

$$P\Big[\lim_{n\to\infty} \sup_{t \le u} |X_t - X_t^n| = 0,\ \forall u \ge 0\Big] = 1.$$

Proof. The process $\{X_t^n\}$ is a Lévy process subordinate to $\{X_t\}$ by the subordinator $\{n^{-2} Z_{n^2 t}\}$ (see Theorem 30.1). Since $\{X_t^n\}$ has piecewise constant sample functions, it is compound Poisson by Theorem 21.2. Since the sample functions of $\{X_t\}$ are continuous, they are approximated by the sample functions of $\{X_t^n\}$ uniformly on any bounded time interval by virtue of Lemma 45.10. □

LEMMA 45.12. *Suppose that $\{X_t\}$ is not a zero process and $\{Z_t\}$ is a Poisson process with parameter 1, independent of $\{X_t\}$. Then there exists a sequence of compound Poisson processes $\{X_t^n\}$, $n = 1, 2, \ldots$, expressible by $\{X_t\}$ and $\{Z_t\}$ such that*

$$P\Big[\lim_{n\to\infty} \sup_{t\le u} |X_t - X_t^n| = 0,\ \forall u \ge 0\Big] = 1.$$

Proof. Use the Lévy–Itô decomposition of sample functions of $\{X_t\}$ in Theorem 19.2. That is, with probability 1,

$$\begin{aligned}(45.26)\qquad X_t =& X_t^\sharp + \lim_{\varepsilon\downarrow 0}\Big(\int_{(0,t]\times\{\varepsilon<|x|\le 1\}} xJ(\mathrm{d}(s,x)) - t\int_{\varepsilon<|x|\le 1} x\nu(\mathrm{d}x)\Big)\\ &+ \int_{(0,t]\times\{|x|>1\}} xJ(\mathrm{d}(s,x)),\end{aligned}$$

where $J(\mathrm{d}(s,x))$ is the Poisson random measure determined by the jumps of $\{X_t\}$, and $\{X_t^\sharp\}$, written as $\{X_t^2\}$ there, is a Lévy process with continuous sample functions. $\{J(\mathrm{d}(s,x))\}$ and $\{X_t^\sharp\}$ are independent and the convergence in (45.26) is uniform in t on any bounded interval. Let $\{k_n\}$ be a sequence of positive integers increasing to ∞, which will be specified later. Define

$$\begin{aligned}(45.27)\qquad X_t^n =& X^\sharp({k_n}^{-2} Z_{{k_n}^2 t}) - {k_n}^{-2} Z_{{k_n}^2 t}\int_{1/n<|x|\le 1} x\,\nu(\mathrm{d}x)\\ &+ \int_{(0,t]\times\{1/n<|x|<\infty\}} xJ(\mathrm{d}(s,x)).\end{aligned}$$

The difference of the first and second terms on the right-hand side is compound Poisson by Lemma 45.11 and it is independent of the third term. Since the third term is also compound Poisson, $\{X_t^n\}$ is compound Poisson. We have

$$(45.28)\qquad X_t - X_t^n = W_1 + W_2 + W_3$$

with

$$W_1 = X_t^\sharp - X^\sharp({k_n}^{-2} Z_{{k_n}^2 t}),\quad W_2 = -(t - {k_n}^{-2} Z_{{k_n}^2 t})\int_{1/n<|x|\le 1} x\,\nu(\mathrm{d}x),$$

$$W_3 = \lim_{\varepsilon\downarrow 0}\Big(\int_{(0,t]\times\{\varepsilon<|x|\le 1/n\}} xJ(\mathrm{d}(s,x)) - t\int_{\varepsilon<|x|\le 1/n} x\,\nu(\mathrm{d}x)\Big).$$

Let $n \to \infty$. Lemma 45.11 tells us that, with probability 1, W_1 tends to 0 uniformly in t on any bounded interval. By the uniformity of the convergence in (45.26) on bounded intervals, W_3 goes to 0 uniformly on any bounded interval with probability 1. We have, for $\varepsilon_n > 0$,

$$P\Big[\sup_{t\le u}|W_2| > \varepsilon_n\Big] \le 3^3{\varepsilon_n}^{-2}\Big(\int_{1/n<|x|\le 1} x\,\nu(\mathrm{d}x)\Big)^2 {k_n}^{-2}u$$

from Lemma 20.5. Now choose k_n so that ${k_n}^{-2}\big(\int_{1/n<|x|\le 1} x\nu(\mathrm{d}x)\big)^2 \le n^{-2}$ and let $\varepsilon_n = n^{-1/4}$. Then we can use the Borel–Cantelli lemma and obtain that $\sup_{t\le u}|W_2| \to 0$ a. s. It follows that $X_t - X_t^n$ tends to 0 uniformly on any bounded interval with probability 1. □

Proof of Theorem 45.7. If $\{X_t\}$ is a zero process, then (45.22) is trivial. If $\{X_t\}$ is compound Poisson, then it is proved in Theorem 45.5. Suppose that $\{X_t\}$ is not a zero process, nor a compound Poisson process. By Lemma 45.12 it is approximated by a sequence of compound Poisson processes $\{X_t^n\}$. The lemma assumes the existence of a Poisson process $\{Z_t\}$ independent of $\{X_t\}$, but this is accomplished by enlarging the probability space. Denote the supremum process of $\{X_t^n\}$ by $\{M_t^n\}$ and the distribution of X_1^n by μ_n. It follows from the convergence of sample functions uniform on bounded time intervals that

$$P\Big[\lim_{n\to\infty} M_t^n = M_t,\ \forall t\ge 0\Big] = 1.$$

Hence, for $z,\ w\in\mathbb{R}$,

$$q\int_0^\infty \mathrm{e}^{-qt}E[\mathrm{e}^{\mathrm{i}zM_t^n+\mathrm{i}w(X_t^n-M_t^n)}]\mathrm{d}t \to q\int_0^\infty \mathrm{e}^{-qt}E[\mathrm{e}^{\mathrm{i}zM_t+\mathrm{i}w(X_t-M_t)}]\mathrm{d}t.$$

This is the left-hand side of (45.22). Since the functions $\varphi_q^+(z)$ and $\varphi_q^-(w)$ are expressed by (45.2) and (45.3), the proof is complete if

$$\int_0^\infty t^{-1}\mathrm{e}^{-qt}\mathrm{d}t\int_{(0,\infty)}(\mathrm{e}^{\mathrm{i}zx}-1){\mu_n}^t(\mathrm{d}x) \to \int_0^\infty t^{-1}\mathrm{e}^{-qt}\mathrm{d}t\int_{(0,\infty)}(\mathrm{e}^{\mathrm{i}zx}-1)\mu^t(\mathrm{d}x) \tag{45.29}$$

and if this is true with $(-\infty,0)$ in place of $(0,\infty)$. Note that (45.29) is equivalent to

$$\int_0^\infty t^{-1}\mathrm{e}^{-qt}E[(\mathrm{e}^{\mathrm{i}zX_t^n}-1)1_{(0,\infty)}(X_t^n)-(\mathrm{e}^{\mathrm{i}zX_t}-1)1_{(0,\infty)}(X_t)]\mathrm{d}t \to 0.$$

In order to prove this, it is enough to show

$$\int_0^\infty t^{-1}\mathrm{e}^{-qt}E|X_t-X_t^n|\mathrm{d}t \to 0, \tag{45.30}$$

because

$$|(\mathrm{e}^{\mathrm{i}zx}-1)1_{(0,\infty)}(x)-(\mathrm{e}^{\mathrm{i}zy}-1)1_{(0,\infty)}(y)| \le |z||x-y|.$$

Now let $\{X_t^n\}$ be the sequence constructed in the proof of Lemma 45.12. Then we can estimate $E|X_t - X_t^n|$ from (45.28) as follows. The process $\{X_t^\sharp\}$ is expressed by a Brownian motion $\{B_t\}$ as $X_t^\sharp = \sqrt{A}B_t + \gamma t$. We may suppose that $\{B_t\}$ and $\{Z_t\}$ are independent. Thus

$$E|t - k_n{}^{-2} Z_{k_n{}^2 t}| \le (\mathrm{Var}(k_n{}^{-2} Z_{k_n{}^2 t}))^{1/2} = (k_n{}^{-2} t)^{1/2},$$

$$E|B_t - B(k_n{}^{-2} Z_{k_n{}^2 t})| = \sum_{j=0}^{\infty} E|B_t - B_{k_n{}^{-2} j}|\, P[Z_{k_n{}^2 t} = j]$$

$$\le \sum_{j=0}^{\infty} |t - k_n{}^{-2} j|^{1/2} P[Z_{k_n{}^2 t} = j] = E[|t - k_n{}^{-2} Z_{k_n{}^2 t}|^{1/2}]$$

$$\le [\mathrm{Var}(k_n{}^{-2} Z_{k_n{}^2 t})]^{1/4} = (k_n{}^{-2} t)^{1/4}.$$

These give bounds for $E|W_1|$ and $E|W_2|$. Since W_3 has characteristic function $\exp\left[t \int_{|x|\le 1/n} (\mathrm{e}^{izx} - 1 - izx)\, \nu(\mathrm{d}x)\right]$,

$$E|W_3| \le (\mathrm{Var}\, W_3)^{1/2} = \left(t \int_{|x|\le 1/n} x^2 \nu(\mathrm{d}x) \right)^{1/2}.$$

Hence,

$$E|X_t - X_t^n| \le \sqrt{A}(k_n{}^{-2} t)^{1/4} + |\gamma| (k_n{}^{-2} t)^{1/2}$$
$$+ (k_n{}^{-2} t)^{1/2} \left| \int_{1/n<|x|\le 1} x\, \nu(\mathrm{d}x) \right| + \left(t \int_{|x|\le 1/n} x^2 \nu(\mathrm{d}x) \right)^{1/2}.$$

As we have chosen k_n to satisfy $k_n \to \infty$ and $k_n{}^{-1} \left| \int_{1/n<|x|\le 1} x\, \nu(\mathrm{d}x) \right| \to 0$, (45.30) follows. This completes the proof of (45.29). The proof of the convergence with $(-\infty, 0)$ in place of $(0, \infty)$ is similar. □

46. Lévy processes without positive jumps

In order to use the strong Markov property, we assume in this section that the probability space $(\Omega, \mathcal{F}, P)$ and $\{X_t : t \ge 0\}$ are as in Section 40 with $d = 1$. Thus, $\Omega = D([0,\infty), \mathbb{R})$, $X_t(\omega) = \omega(t)$ for $\omega \in \Omega$, $\mathcal{F}$ is the σ-algebra constructed from $\mathcal{F}^0$ using P, and $\{X_t : t \ge 0\}$ is a Lévy process under P. We make special assumptions in this section that $\{X_t\}$ has the following two properties:

$$P[\, X_t \le X_{t-} \text{ for every } t > 0\,] = 1; \tag{46.1}$$

$$P\left[\limsup_{t\to\infty} X_t = \infty \right] = 1. \tag{46.2}$$

In this situation, let us determine the first passage time process $\{R_x : x \ge 0\}$ given in Definition 45.1. Let (A, ν, γ) be the generating triplet of $\{X_t\}$.

REMARK 46.1. A condition equivalent to (46.1) is that

$$\nu(0,\infty) = 0. \tag{46.3}$$

This is clear since (46.1) means that $\{X_t\}$ does not have positive jumps and since, by Theorem 19.2(i), the expected number of jumping times $s \in (0,t]$ satisfying $X_s - X_{s-} \in B$ equals $t\nu(B)$ for any Borel set B not containing 0. Suppose that (46.1) holds. Then (46.2) holds if and only if

$$0 < E|X_t| < \infty \text{ and } EX_t \geq 0 \text{ for } t > 0. \tag{46.4}$$

The proof is as follows. By (46.3), $E[X_t \vee 0] < \infty$ as an application of Theorem 25.3. If (46.2) holds, then, obviously, $E|X_t| > 0$. If $0 > EX_t \geq -\infty$, then, by Theorems 36.5 and 36.6, $\lim_{t\to\infty} X_t = -\infty$ a. s., contrary to (46.2). Conversely, if (46.4) holds, then $\{X_t\}$ is either transient and going to $+\infty$ by Theorem 36.5, or recurrent and oscillating by Theorems 35.8 and 36.7.

THEOREM 46.2. *The process $\{R_x \colon x \geq 0\}$ is a Lévy process, their sample functions are strictly increasing a. s., and*

$$P[\, X_{R_x} = x \text{ for every } x \geq 0\,] = 1. \tag{46.5}$$

Proof. By Example 40.7 R_x is a stopping time for each x. Let us prove (46.5). Let $x > 0$. Then $R_x > 0$ and $X_{R_x} \geq x$ by the right-continuity of $\omega(t)$. On the other hand, $X_{R_x} \leq X_{R_x-} \leq x$ by (46.1). In the case $x = 0$, $X_{R_0} \geq 0$ is obvious and, if $X_{R_0} > 0$ with positive probability, then $R_0 > 0$ and $X_{R_0-} \leq 0$, contrary to (46.1). Hence (46.5) is true. It follows that R_x is strictly increasing in x a. s. The right-continuity of R_x in x is a direct consequence of the definition. To see that $R_0 = 0$ a. s., note that $X_{R_0+r} = X_{R_0+r} - X_{R_0}$ and that, by the strong Markov property,

$$\begin{aligned} 1 &= P[\, X_{R_0+r_n} > 0 \text{ for some } r_n \downarrow 0 \text{ in } \mathbb{Q}\,] \\ &= P[\, X_{r_n} > 0 \text{ for some } r_n \downarrow 0 \text{ in } \mathbb{Q}\,] = P[\, R_0 = 0\,]. \end{aligned}$$

Let us show that $\{R_x \colon x \geq 0\}$ has stationary independent increments. Let $X_t^{(x)} = X_{R_x+t} - X_{R_x}$. Then, by Theorem 40.10, $\{X_t^{(x)} \colon t \geq 0\}$ is a Lévy process identical in law with $\{X_t\}$ and independent of $\mathcal{F}_{R_x}$. Let $R_y^{(x)}$ be the hitting time of (y,∞) for $\{X_t^{(x)}\}$. Then, $R_x + R_y^{(x)} = R_{x+y}$, since $X_t^{(x)} = X_{R_x+t} - x$. Hence, for $y > 0$, $R_{x+y} - R_x = R_y^{(x)} \stackrel{\mathrm{d}}{=} R_y$ and $R_{x+y} - R_x$ is independent of $\mathcal{F}_{R_x}$. □

We can give an expression for the Laplace transform of the distribution of R_x, based on the factorization identities in the preceding section.

THEOREM 46.3. *Define $\Psi(w)$ for $w \geq 0$ by*

$$\Psi(w) = \frac{1}{2}Aw^2 + \gamma w + \int_{(-\infty,0)} (\mathrm{e}^{wx} - 1 - wx1_{[-1,0)}(x))\,\nu(\mathrm{d}x). \tag{46.6}$$

Then $\Psi(w)$ *is strictly increasing and continuous,* $\Psi(0)=0$, *and* $\Psi(w)\to\infty$ *as* $w\to\infty$. *For* $x\geq 0$ *and* $0\leq u<\infty$ *we have*

$$E[\mathrm{e}^{-uR_x}]=\mathrm{e}^{-x\Psi^{-1}(u)}, \tag{46.7}$$

where $w=\Psi^{-1}(u)$ *is the inverse function of* $u=\Psi(w)$.

Proof. If $\{X_t\}$ is trivial, then $X_t=\gamma t$ with $\gamma>0$ and (46.7) is satisfied, since $R_x=x/\gamma$, $\Psi(w)=\gamma w$, and $\Psi^{-1}(u)=u/\gamma$. Suppose that $\{X_t\}$ is non-trivial. We define $\Psi(w)$ for complex w with $\mathrm{Re}\,w\geq 0$ by (46.6). Then $\Psi(w)$ is continuous on $\{w\colon \mathrm{Re}\,w\geq 0\}$ and analytic on $\{w\colon \mathrm{Re}\,w>0\}$. We have

$$E[\mathrm{e}^{wX_t}]=\mathrm{e}^{t\Psi(w)} \tag{46.8}$$

by (46.3) and Theorem 25.17. When w is real and positive, we have

$$\Psi'(w)=Aw+\gamma+\int_{(-\infty,0)}(x\mathrm{e}^{wx}-x1_{[-1,0)}(x))\,\nu(\mathrm{d}x),$$

$$\Psi''(w)=A+\int_{(-\infty,0)}x^2\mathrm{e}^{wx}\nu(\mathrm{d}x)>0.$$

Noting (46.4), we have

$$0\leq EX_t=t\left(\gamma+\int_{(-\infty,-1)}x\,\nu(\mathrm{d}x)\right)=t\Psi'(0+). \tag{46.9}$$

Hence $\Psi'(w)>0$ for $w>0$. Thus $\Psi(w)$ strictly increases from 0 to ∞ with w.

Since $\{R_x\}$ is a subordinator, $E[\mathrm{e}^{-uR_x}]=\mathrm{e}^{-xB(u)}$, $u\geq 0$, with some function $B(u)$. Since $\{X_t\}$ does not have positive jumps, $R_x<t$ and $M_t>x$ are equivalent. Hence, for $x>0$ and $u>0$,

$$\begin{aligned}\mathrm{e}^{-xB(u)}&=\int_{(0,\infty)}\mathrm{e}^{-ut}\mathrm{d}_tP[R_x\leq t]=u\int_0^\infty \mathrm{e}^{-ut}P[R_x\leq t]\mathrm{d}t\\&=u\int_0^\infty \mathrm{e}^{-ut}P[R_x<t]\mathrm{d}t=u\int_0^\infty \mathrm{e}^{-ut}P[M_t>x]\mathrm{d}t\\&=1-u\int_0^\infty \mathrm{e}^{-ut}P[M_t\leq x]\mathrm{d}t.\end{aligned}$$

Now use (45.7) proved in Corollary 45.8. Then, for $u>0$ and $z\in\mathbb{R}$,

$$\begin{aligned}\varphi_u^+(z)&=u\int_0^\infty \mathrm{e}^{-ut}E[\mathrm{e}^{\mathrm{i}zM_t}]\mathrm{d}t=u\int_0^\infty \mathrm{e}^{-ut}\mathrm{d}t\int_{(0,\infty)}\mathrm{e}^{\mathrm{i}zx}\mathrm{d}_xP[M_t\leq x]\\&=\int_{(0,\infty)}\mathrm{e}^{\mathrm{i}zx}\mathrm{d}_x\left(u\int_0^\infty \mathrm{e}^{-ut}P[M_t\leq x]\mathrm{d}t\right)\\&=-\int_{(0,\infty)}\mathrm{e}^{\mathrm{i}zx}\mathrm{d}_x(\mathrm{e}^{-xB(u)})=B(u)\int_0^\infty \mathrm{e}^{\mathrm{i}zx-xB(u)}\mathrm{d}x\end{aligned}$$

$$= B(u)(B(u) - \mathrm{i}z)^{-1}.$$

Here we have used that $B(u) > 0$ for $u > 0$. Since $\psi(z) = \Psi(\mathrm{i}z)$, we have

$$\varphi_u^-(z)(u - \Psi(\mathrm{i}z)) = u(B(u) - \mathrm{i}z)/B(u)$$

by (45.1) of Theorem 45.2. The function $\varphi_u^-(z)$ can be analytically extended to the lower half plane by Remark 45.3. Hence, for $w \geq 0$,

$$\varphi_u^-(-\mathrm{i}w)(u - \Psi(w)) = u(B(u) - w)/B(u). \tag{46.10}$$

Now we see that $B(u) = \Psi^{-1}(u)$ by letting $w = \Psi^{-1}(u)$. □

THEOREM 46.4 (Distribution of R_x). (i) *For any Borel sets B and G in $(0, \infty)$*

$$\int_G P[\, R_x \in B\,]\mathrm{d}x = \int_B t^{-1}\mathrm{d}t \int_G x\mu^t(\mathrm{d}x). \tag{46.11}$$

(ii) *If, for any $t > 0$, $[\mu^t]_{(0,\infty)}$ is absolutely continuous, then, for any $x > 0$, the distribution of R_x is absolutely continuous and there exist nonnegative functions $m(t, x)$ and $h(x, t)$ having the following properties:*

1. *$m(t, x)$ and $h(x, t)$ are measurable in the two variables in $(0, \infty) \times (0, \infty)$;*
2. *for any fixed $t > 0$ $m(t, x)$ is the density of $[\mu^t]_{(0,\infty)}$;*
3. *for any fixed $x > 0$ $h(x, t)$ is the density of the distribution of R_x;*
4. *for almost every (t, x) in $(0, \infty) \times (0, \infty)$*

$$h(x, t) = t^{-1} x\, m(t, x). \tag{46.12}$$

Proof. (i) Define, for $q > 0$ and $x > 0$,

$$g_q(x) = \int_0^\infty t^{-1}\mathrm{e}^{-qt}\mathrm{d}t \int_{(0,x]} y\mu^t(\mathrm{d}y).$$

By Lemma 30.3, this is finite, right-continuous, and increasing. Hence, for $p > 0$,

$$\int_0^\infty x^{-1}(1 - \mathrm{e}^{-px})\mathrm{d}g_q(x)$$
$$= \int_0^\infty t^{-1}\mathrm{e}^{-qt}\mathrm{d}t \int_{(0,\infty)} (1 - \mathrm{e}^{-px})\mu^t(\mathrm{d}x) = -\log \varphi_q^+(\mathrm{i}p),$$

since φ_q^+ is extended to the lower half plane in the same form as (45.2) by Remark 45.3. We obtain $\varphi_q^+(z) = B(q)(B(q) - \mathrm{i}z)^{-1}$, $z \in \mathbb{R}$, as in the proof of Theorem 46.3. Hence $\varphi_q^+(\mathrm{i}p) = B(q)(B(q) + p)^{-1}$. Thus differentiation with respect to p gives

$$\int_{(0,\infty)} \mathrm{e}^{-px}\mathrm{d}g_q(x) = (B(q) + p)^{-1}.$$

As the right-hand side is the Laplace transform of $\mathrm{e}^{-B(q)x}$, we get

$$g_q(x) = \int_0^x \mathrm{e}^{-B(q)y}\mathrm{d}y = \int_0^x \mathrm{d}y \int_{(0,\infty)} \mathrm{e}^{-qt}\mathrm{d}_t P[R_y \le t]$$
$$= \int_{(0,\infty)} \mathrm{e}^{-qt}\mathrm{d}_t\left(\int_0^x P[R_y \le t]\mathrm{d}y\right).$$

Therefore

$$\int_0^s t^{-1}\mathrm{d}t \int_{(0,x]} y\mu^t(\mathrm{d}y) = \int_0^x P[R_y \le s]\mathrm{d}y, \tag{46.13}$$

which shows (46.11).

(ii) We assume that $[\mu^t]_{(0,\infty)}$ is absolutely continuous for each $t > 0$. Define

$$m_n(t,x) = 2^n P[k/2^n < X_t \le (k+1)/2^n]$$

for $k/2^n < x \le (k+1)/2^n$ with nonnegative integer k. We see that $m_n(t,x)$ is measurable in (t,x), observing that $E[f(X_t)]$ is measurable in t for any bounded measurable function f. Let C be the set of points (t,x) such that $\lim_{n\to\infty} m_n(t,x)$ exists and is finite. Then C is a measurable set. Define $m(t,x)$ as this limit for $(t,x) \in C$ and let $m(t,x) = 0$ for $(t,x) \notin C$. Then $m(t,x)$ is measurable in (t,x). By the martingale argument in [**34**], p. 494, or [**80**], p. 353, for each fixed $t > 0$, almost every $x > 0$ satisfies $(t,x) \in C$ and $m(t,x)$ is the density of $[\mu^t]_{(0,\infty)}$. Then, it follows from (46.11) that, for almost every $x > 0$,

$$P[\,R_x \le s\,] = \int_0^s t^{-1}x\, m(t,x)\mathrm{d}t, \quad s \in \mathbb{Q} \cap (0,\infty).$$

Hence, for almost every $x > 0$, R_x has absolutely continuous distribution. Since $\{R_x\}$ is a Lévy process, we now see by Lemma 27.1(iii) that the absolute continuity holds for every $x > 0$. Hence, by a similar argument, we can construct a nonnegative measurable function $h(x,t)$ on $(0,\infty) \times (0,\infty)$ satisfying (3). Thus (46.11) is written as

$$\int_G \mathrm{d}x \int_B h(x,t)\mathrm{d}t = \int_B t^{-1}\mathrm{d}t \int_G x\, m(t,x)\mathrm{d}x,$$

which shows (4). □

EXAMPLE 46.5. Let $\{X_t\}$ be the Brownian motion. Assumptions (46.1) and (46.2) are satisfied. We have $\Psi(w) = \frac{1}{2}w^2$ and $\Psi^{-1}(u) = (2u)^{1/2}$. Hence $\{R_x\}$ is a strictly $\frac{1}{2}$-stable subordinator. This is already proved in Example 40.14 in another way. It follows from Example 25.10 that $ER_x = \infty$ for every $x > 0$. Thus, however small $x > 0$ may be, the hitting time of x has infinite expectation. The relation (46.12) is reconfirmed by the explicit forms of the Gaussian density and the one-sided strictly $\frac{1}{2}$-stable density.

EXAMPLE 46.6. Let $\{X_t\}$ be the non-trivial Lévy process with continuous sample functions. Then $\Psi(w) = \frac{1}{2}Aw^2 + \gamma w$ with $A > 0$. It satisfies (46.2) if and only if $\gamma \geq 0$. In this case,

$$\Psi^{-1}(u) = A^{-1}((\gamma^2 + 2Au)^{1/2} - \gamma).$$

This shows that, if $\gamma > 0$, then $\{R_x\}$ is identical in law with some density transformation, as discussed in Example 33.15, of a strictly $\frac{1}{2}$-stable subordinator. Thus we find there an explicit form of the distribution density of R_x, which can be obtained also from (46.12). It follows from this form that $ER_x < \infty$ for any $x > 0$ if $\gamma > 0$.

EXAMPLE 46.7. Let $\{X_t\}$ be a stable process with parameters (α, β, τ, c) satisfying $1 < \alpha < 2$, $\beta = -1$, $\tau \geq 0$, and $c > 0$. Then $\{X_t\}$ satisfies (46.1) and (46.2). See Example 25.10 to check (46.4). We can show that

$$\Psi(w) = \lambda w^\alpha + \tau w, \qquad w \geq 0, \tag{46.14}$$

with $\lambda = c|\cos\frac{1}{2}\pi\alpha|^{-1}$. Indeed, by Theorem 14.19 and Remark 14.20,

$$E[\mathrm{e}^{\mathrm{i}zX_t}] = \exp[t(-\lambda|z|^\alpha \mathrm{e}^{-\mathrm{i}(\pi/2)(2-\alpha)\operatorname{sgn} z} + \mathrm{i}\tau z)]$$

for $z \in \mathbb{R}$. The analytic extension of the right-hand side to $\{z \in \mathbb{C}\colon \operatorname{Im} z \leq 0,\ z \neq 0\}$ equals

$$\exp[t(-\lambda \mathrm{e}^{\alpha \log z - \mathrm{i}(\pi/2)(2-\alpha)} + \mathrm{i}bz)],$$

where the branch of the logarithm is chosen as $\log z = \log|z| + \mathrm{i}\arg z$, $-\pi \leq \arg z \leq 0$. Letting $z = -\mathrm{i}w$, $w \geq 0$, and using (46.8), we get (46.14). Now the distribution of R_x has Laplace transform written by the inverse function of the Ψ of (46.14). If $\tau = 0$, then $\{R_x\}$ is a strictly $\frac{1}{\alpha}$-stable subordinator and the relation (46.12) is a special case of Zolotarev's formula in Remark 14.21.

The following fact is useful in the next section.

PROPOSITION 46.8. *$P[\,X_t > 0\,] \geq 1/16$ for $t > 0$.*

Proof. For the function $\Psi(w)$ of (46.6) we claim that

$$\Psi(2w) \leq 4\Psi(w), \qquad w \geq 0. \tag{46.15}$$

We have

$$\Psi(w) = \frac{1}{2}Aw^2 + \gamma_1 w + \int_{(-\infty,0)} (\mathrm{e}^{wx} - 1 - wx)\nu(\mathrm{d}x)$$

with $\gamma_1 = \gamma + \int_{(-\infty,-1)} x\,\nu(\mathrm{d}x) \geq 0$ by (46.9). Notice that $\mathrm{e}^{-2z} - 1 + 2z \leq 4(\mathrm{e}^{-z} - 1 + z)$ for $z \geq 0$, which is shown by differentiation twice. Then, $\mathrm{e}^{2wx} - 1 - 2wx \leq 4(\mathrm{e}^{wx} - 1 - wx)$ for $w \geq 0$ and $x \leq 0$ and we get (46.15). By Schwarz's inequality

$$(E[\mathrm{e}^{wX_t};\, X_t > 0])^2 \leq E[\mathrm{e}^{2wX_t}]\, P[\,X_t > 0\,].$$

Hence, using (46.8), (46.15), and $\Psi(w) \geq 0$, we get, for $w \geq 0$,

$$P[\,X_t > 0\,] \geq \left(E[\mathrm{e}^{wX_t}] - E[\mathrm{e}^{wX_t};\, X_t \leq 0]\right)^2 / E[\mathrm{e}^{2wX_t}]$$
$$\geq (\mathrm{e}^{t\Psi(w)} - 1)^2 \mathrm{e}^{-t\Psi(2w)} \geq (\mathrm{e}^{t\Psi(w)} - 1)^2\, \mathrm{e}^{-4t\Psi(w)}.$$

Recall that $\Psi(w)$ continuously increases from 0 to ∞ and choose w such that $t\Psi(w) = \log 2$ to get the bound from below. □

47. Short time behavior

Let $\{X_t \colon t \geq 0\}$ be a Lévy process on $\mathbb{R}$ defined on a probability space $(\Omega, \mathcal{F}, P)$, as in Section 45. If $\{X_t\}$ is of type A or B, then

$$P\left[\lim_{t \downarrow 0} t^{-1} X_t = \gamma_0\right] = 1, \tag{47.1}$$

where γ_0 is the drift. This is deduced from Theorem 43.20. Indeed, let $(\{X_t^\sharp\}, \Omega^\sharp, \mathcal{F}^\sharp, P^\sharp)$ be the process in the set-up of Chapter 8, identical in law with $\{X_t\}$. Then, using the set Ω_0 in Definition 1.6, we have $P[(\lim_{t\downarrow 0} t^{-1} X_t = \gamma_0) \cap \Omega_0] = P^\sharp[\,\lim_{t\downarrow 0} t^{-1} X_t^\sharp = \gamma_0\,]$, which equals 1 by Theorem 43.20. Let us now consider the case of type C.

THEOREM 47.1. *If $\{X_t\}$ is of type C, then*

$$P\left[\limsup_{t \downarrow 0} t^{-1} X_t = \infty \text{ and } \liminf_{t \downarrow 0} t^{-1} X_t = -\infty\right] = 1. \tag{47.2}$$

Proof. Let $\{X_t\}$ be of type C. We have two steps.

Step 1. Assume that $\{X_t\}$ does not have positive jumps, that is, (46.1) is satisfied. As is explained above, it is enough to prove (47.2) under the assumption that $\{X_t\}$ is the process in the set-up of Chapter 8. Moreover we may and do assume that

$$E[\mathrm{e}^{\mathrm{i}zX_t}] = \exp\left[t\left(-\frac{1}{2}Az^2 + \mathrm{i}\gamma z + \int_{[-1,0)} (\mathrm{e}^{\mathrm{i}zx} - 1 - \mathrm{i}zx)\nu(\mathrm{d}x)\right)\right]$$

with $\gamma \geq 0$, because the remaining part satisfies (47.1). Then $EX_t \geq 0$ and we can apply Theorem 46.2. Let

$$\Psi(w) = \frac{1}{2}Aw^2 + \gamma w + \int_{[-1,0)} (\mathrm{e}^{wx} - 1 - wx)\nu(\mathrm{d}x), \qquad w \geq 0.$$

The first passage time process $\{R_x \colon x \geq 0\}$ is a subordinator and $E[\mathrm{e}^{-uR_x}] = \mathrm{e}^{-xB(u)}$, $u \geq 0$, with $B(u) = \Psi^{-1}(u)$. Let ρ and β_0 be the Lévy measure and the drift of $\{R_x\}$. Then

$$B(u) = \beta_0 u + \int_{(0,\infty)} (1 - \mathrm{e}^{-uy})\rho(\mathrm{d}y), \quad B'(u) = \beta_0 + \int_{(0,\infty)} y\mathrm{e}^{-uy}\rho(\mathrm{d}y)$$

for $u > 0$. Since $\{R_x\}$ is not compound Poisson, by Theorem 46.2, $B(u)$ tends to ∞ as $u \to \infty$. We have

$$\beta_0 = \lim_{u\to\infty} B'(u) = \lim_{u\to\infty} (\Psi'(B(u)))^{-1} = \lim_{w\to\infty} (\Psi'(w))^{-1} = 0,$$

because

$$\Psi'(w) = Aw + \gamma + \int_{[-1,0)} (-x)(1 - \mathrm{e}^{wx})\nu(\mathrm{d}x) \to \infty, \qquad w \to \infty,$$

as $\{X_t\}$ is of type C. Hence we have

$$\lim_{x\downarrow 0} x^{-1} R_x = 0 \qquad \text{a. s.},$$

applying the result (47.1) to $\{R_x\}$. Therefore, $X_{R_x}/R_x = x/R_x \to \infty$ as $x \downarrow 0$ a. s. and hence $\limsup_{t\downarrow 0} t^{-1} X_t = \infty$ a. s.

Next let us show that $\liminf_{t\downarrow 0} t^{-1} X_t = -\infty$ a. s. It is enough to show that $\liminf_{t\downarrow 0} t^{-1}(X_t + \beta t) \le 0$ a. s. for any fixed $\beta > 0$. Thus it is enough to show that $\inf_{t<\delta}(X_t + \beta t) < 0$ a. s. for any fixed $\beta > 0$ and $\delta > 0$. We can replace $\{X_t\}$ by $\{X_t + \beta t\}$. So we have only to prove that

$$\text{(47.3)} \qquad N_t < 0 \text{ for any } t > 0 \qquad \text{a. s.}$$

Let $u > 0$. We have (45.8) by Corollary 45.8 and

$$\varphi_u^-(-\mathrm{i}w) = u \int_0^\infty \mathrm{e}^{-ut} E[\mathrm{e}^{wN_t}]\mathrm{d}t \to u \int_0^\infty \mathrm{e}^{-ut} P[N_t = 0]\mathrm{d}t$$

as $w \to \infty$. On the other hand, by (46.10),

$$\begin{aligned}\lim_{w\to\infty} \varphi_u^-(-\mathrm{i}w) &= \lim_{w\to\infty} uB(u)^{-1}(B(u) - w)(u - \Psi(w))^{-1} \\ &= uB(u)^{-1} \lim_{w\to\infty} w(\Psi(w) - u)^{-1}.\end{aligned}$$

If $A > 0$, then clearly $w^{-1}\Psi(w) \to \infty$. If $A = 0$, then, for every $\varepsilon > 0$,

$$w^{-1}\Psi(w) \ge \gamma + \int_{[-1,-\varepsilon)} (w^{-1}(\mathrm{e}^{wx} - 1) - x)\nu(\mathrm{d}x) \to \gamma + \int_{[-1,-\varepsilon)} (-x)\nu(\mathrm{d}x),$$

which again implies that $w^{-1}\Psi(w) \to \infty$, since $\{X_t\}$ is of type C. Therefore $\varphi_u^-(-\mathrm{i}w) \to 0$. It follows that $P[N_t = 0] = 0$ for any t, that is, (47.3).

Step 2. Consider the general case of type C. Let $I_1 = \int_{(0,1]} x\nu(\mathrm{d}x)$ and $I_2 = \int_{[-1,0)}(-x)\nu(\mathrm{d}x)$. If $I_1 < \infty$, then either $A > 0$ or $I_2 = \infty$. Hence, if $I_1 < \infty$, then $X_t = X_t^1 + X_t^2$ with $\{X_t^1\}$ being of type C without positive jumps and $\{X_t^2\}$ being of type A or B, and $\{X_t\}$ satisfies (47.2) by Step 1 and (47.1). If $I_2 < \infty$, then we get (47.2) again, considering $\{-X_t\}$ instead of $\{X_t\}$. There remains the case that $I_1 = \infty$ and $I_2 = \infty$. In this case $\{X_t\}$ is the sum of three independent processes $\{X_t^1\}$, $\{X_t^2\}$, and $\{X_t^3\}$, where $\{X_t^1\}$ is of type C without positive jumps and $EX_t^1 \ge 0$, $\{X_t^2\}$ is of type C without negative jumps, and $\{X_t^3\}$ is of type A or B. Applying Step 1 to $\{-X_t^2\}$, we have

$$\limsup_{t\downarrow 0} t^{-1}X_t^2 = \infty \qquad \text{a. s.} \tag{47.4}$$

If

$$P\left[\limsup_{n\to\infty} {t_n}^{-1}X_{t_n}^1 \geq 0\right] = 1 \tag{47.5}$$

for any sequence $t_n \downarrow 0$ which does not depend on ω, then we have

$$\limsup_{t\downarrow 0} t^{-1}X_t = \infty \qquad \text{a. s.} \tag{47.6}$$

To see this, choose, for P-almost all ω, functions $V_n(\omega)$ of $X_t^2(\omega)$, $t \geq 0$, such that $V_n(\omega) > 0$, $V_n(\omega) \downarrow 0$, and $V_n(\omega)^{-1}X_{V_n(\omega)}^2(\omega) \geq n$. This is possible by (47.4), letting $V_0(\omega) = 1$ and

$$W_n = \sup\{t\colon t \leq V_{n-1} \wedge n^{-1} \text{ and } t^{-1}X_t^2 \geq n\},$$
$$V_n = \inf\{t\colon 2^{-1}W_n \leq t \leq W_n \text{ and } t^{-1}X_t^2 \geq n\}$$

for $n = 1, 2, \ldots$. Use Proposition 1.16 extended to independent stochastic processes from independent random variables. Then

$$P\left[\limsup_{n\to\infty} {V_n}^{-1}X_{V_n}^1 \geq 0\right] = E[f(V_1, V_2, \ldots)],$$

where $f(t_1, t_2, \ldots) = P[\limsup_{n\to\infty} {t_n}^{-1}X_{t_n}^1 \geq 0]$, which equals 1 by (47.5). Hence $\limsup_{n\to\infty} {V_n}^{-1}X_{V_n} = \infty$ a. s., and (47.6) follows.

Now let us prove (47.5). Choose a subsequence $\{s_k\}$ of $\{t_n\}$ as follows: let $s_1 = t_1$ and, using s_k, select s_{k+1} such that

$$P[\,{s_k}^{-1}|X_{s_{k+1}}^1| \geq k^{-1}\,] \leq k^{-2}.$$

This is possible since $X_t^1 \to 0$, $t \downarrow 0$, a. s. Then ${s_k}^{-1}X_{s_{k+1}}^1 \to 0$ a. s. by the Borel–Cantelli lemma. Note that

$${s_k}^{-1}X_{s_k}^1 = {s_k}^{-1}(X_{s_k}^1 - X_{s_{k+1}}^1) + {s_k}^{-1}X_{s_{k+1}}^1.$$

Proposition 46.8 implies that $\sum_k P[\,X_{s_k}^1 - X_{s_{k+1}}^1 > 0\,] = \infty$. Hence, by the Borel–Cantelli lemma, $P[\,X_{s_k}^1 - X_{s_{k+1}}^1 > 0$ for infinitely many $k\,] = 1$. It follows that $\limsup_{k\to\infty} {s_k}^{-1}X_{s_k}^1 \geq 0$ a. s., hence (47.5).

The second property in (47.2) follows from the first, if we consider $\{-X_t\}$ in place of $\{X_t\}$. □

A criterion for whether $R_0 = 0$ a. s. or $R_0 > 0$ a. s. is obtained directly from Corollary 45.8. In the terminology of Chapter 8, this is a criterion whether the point 0 is regular, or irregular, for the set $(0, \infty)$.

THEOREM 47.2. *$R_0 = 0$ a. s. if and only if*

$$\int_0^1 t^{-1} P[\,X_t > 0\,]\mathrm{d}t = \infty. \tag{47.7}$$

$R_0 > 0$ a. s. if and only if

$$\int_0^1 t^{-1} P[\,X_t > 0\,]\mathrm{d}t < \infty. \tag{47.8}$$

Proof. The identity (45.7) is proved by Corollary 45.8. It can be extended to the upper half plane. Thus, for $q > 0$ and $u \geq 0$,

$$q \int_0^\infty \mathrm{e}^{-qt} E[\mathrm{e}^{-uM_t}]\mathrm{d}t = \exp\left[\int_0^\infty t^{-1}\mathrm{e}^{-qt}\mathrm{d}t \int_{(0,\infty)} (\mathrm{e}^{-ux} - 1)\mu^t(\mathrm{d}x)\right]. \tag{47.9}$$

Let $u \to \infty$. Then

$$q \int_0^\infty \mathrm{e}^{-qt} P[\,M_t = 0\,]\mathrm{d}t = \exp\left[-\int_0^\infty t^{-1}\mathrm{e}^{-qt} P[\,X_t > 0\,]\mathrm{d}t\right], \tag{47.10}$$

where the integral in the right-hand side may possibly be infinite. If (47.7) holds, then $P[\,M_t = 0\,] = 0$ for almost every t and we have $P[\,M_t > 0 \text{ for all } t > 0\,] = 1$, using the increasingness of $\{M_t\}$. Hence, (47.7) implies that $R_0 = 0$ a. s. If (47.8) holds, then the right-hand side of (47.10) tends to 1 as $q \to \infty$. On the other hand,

$$\begin{aligned} \text{left-hand side of (47.10)} &= \int_0^\infty \mathrm{e}^{-t} P[\,M_{t/q} = 0\,]\mathrm{d}t \\ &\to P[\,M_s = 0 \text{ for some } s > 0\,] \end{aligned}$$

as $q \to \infty$. Hence (47.8) implies that $R_0 > 0$ a. s. □

REMARK 47.3. As $t \to \infty$, $P[\,M_t = 0\,]$ tends to $P[\,M_s = 0 \text{ for all } s\,] = P[\,R_0 = \infty\,]$. Hence, letting $q \downarrow 0$ in (47.10), we get

$$P[\,R_0 = \infty\,] = \exp\left[-\textstyle\int_0^\infty t^{-1} P[\,X_t > 0\,]\mathrm{d}t\right]. \tag{47.11}$$

The integral in the right is possibly infinite. Later, in Corollary 49.7, we give a representation of the Laplace transform of the distribution of R_0; (47.11) follows also from it.

EXAMPLE 47.4. Let $\{X_t\}$ and $\{X'_t\}$ be independent strictly stable subordinators with indices α and α', respectively. Suppose that $0 < \alpha < \alpha' < 1$. Let us show that

$$P[\,X_t < X'_t \text{ for all sufficiently small } t > 0\,] = 1. \tag{47.12}$$

Let $Z_t = X_t - X'_t$. Then $\{Z_t\}$ is a Lévy process. Let R_0 be the hitting time of $(0, \infty)$ for $\{Z_t\}$. We claim that $R_0 > 0$ a. s. If this is true, then $X_t \leq X'_t$ for all sufficiently small $t > 0$ a. s., which implies (47.12) since X_t and X'_t are positive for $t > 0$ and since, for any $c > 0$, $\{cX_t\}$ is a strictly α-stable subordinator. As $X_t \overset{\mathrm{d}}{=} t^{1/\alpha} X_1$ and $X'_t \overset{\mathrm{d}}{=} t^{1/\alpha'} X'_1$, we have

$$P[\, Z_t > 0\,] = P[\, t^{1/\alpha} X_1 - t^{1/\alpha'} X'_1 > 0\,] = \int_{[0,\infty)} P[\, X_1 > t^{1/\alpha' - 1/\alpha} x\,]\, P[\, X'_1 \in \mathrm{d}x\,]$$
$$\leq P[\, X'_1 < t^{\eta}\,] + P[\, X_1 > t^{1/\alpha' - 1/\alpha + \eta}\,]$$

for any η. Choose η so that $0 < \eta < 1/\alpha - 1/\alpha'$. We have $P[\, X'_1 < t^{\eta}\,] \leq \mathrm{const}\, t^{\eta}$ since X'_1 has a continuous density by Example 28.2. Since $E[X_1{}^{\theta}] < \infty$ for $0 < \theta < \alpha$ by Example 25.10, we see $x^{\theta} P[\, X_1 \geq x\,] \to 0$ as $x \to \infty$ by Lemma 26.7. Hence $P[\, X_1 > t^{1/\alpha' - 1/\alpha + \eta}\,] \leq \mathrm{const}\, t^{\theta(1/\alpha - 1/\alpha' - \eta)}$. It follows that $\int_0^1 t^{-1} P[\, Z_t > 0\,] \mathrm{d}t < \infty$. Hence, by Theorem 47.2, $R_0 > 0$ a. s. □

Now we consider R_0 in each type.

THEOREM 47.5. (i) *If type A and $\gamma_0 > 0$, then $R_0 = 0$ a. s.*

(ii) *If type A and $\gamma_0 \leq 0$, then $R_0 > 0$ a. s.*

(iii) *If type B and $\gamma_0 > 0$, then $R_0 = 0$ a. s.*

(iv) *If type B and $\gamma_0 < 0$, then $R_0 > 0$ a. s.*

(v) *If type B, $\gamma_0 = 0$, and $\nu(-\infty, 0) < \infty$, then $R_0 = 0$ a. s.*

(vi) *If type B, $\gamma_0 = 0$, and $\nu(0, \infty) < \infty$, then $R_0 > 0$ a. s.*

(vii) *Among processes of type B with $\gamma_0 = 0$, $\nu(-\infty, 0) = \infty$, and $\nu(0, \infty) = \infty$, there are the case that $R_0 = 0$ a. s. and the case that $R_0 > 0$ a. s.*

(viii) *If type C, then $R_0 = 0$ a. s.*

Proof. Sample functions of a process of type A equal $\gamma_0 t$ until the first jumping time. Hence we have (i) and (ii). The property (47.1) proves (iii) and (iv). Theorem 47.1 shows (viii).

Let $\{X_t\}$ be of type B with $\gamma_0 = 0$. Then sample functions are of bounded variation in any finite time interval and their continuous parts are 0. If $\nu(-\infty, 0) < \infty$, then positive jumps immediately occur but negative jumps do not occur for a while, and hence $R_0 = 0$ a. s. If $\nu(0, \infty) < \infty$, then the situation is opposite and hence $R_0 > 0$ a. s. Finally, to show (vii), we give examples. The process $\{Z_t\}$ in Example 47.4 furnishes a process in (vii) with $R_0 > 0$ a. s. and, by virtue of (47.12), $\{-Z_t\}$ is a process in (vii) with $R_0 = 0$ a. s. □

THEOREM 47.6. *Let $\{X_t\}$ be a non-trivial α-stable process on $\mathbb{R}$ with $0 < \alpha \leq 2$. In the case $\alpha \neq 2$, let (α, β, τ, c) be its parameters as in Definition 14.16.*

(i) *If $0 < \alpha < 1$, $-1 < \beta \leq 1$, and $\tau < 0$, then $R_0 > 0$ a. s.*

(ii) *If $0 < \alpha < 1$, $\beta = -1$, and $\tau \leq 0$, then $R_0 > 0$ a. s.*

(iii) *$R_0 = 0$ a. s. in all other cases.*

Proof. We can apply Theorem 47.5 unless $0 < \alpha < 1$, $|\beta| < 1$, and $\gamma_0 = 0$. In the latter case the process falls into the category (vii) of Theorem

47.5. Here we give a proof using Theorem 47.2 in all cases. Let $I = \int_0^1 t^{-1}P[\,X_t > 0\,]\mathrm{d}t$.

Suppose that $\alpha \neq 1$. In the case $\alpha = 2$, define $\tau = EX_1$. Let $X_t^0 = X_t - \tau t$. Then $\{X_t^0\}$ is strictly α-stable process by Theorem 14.8. Thus

$$
\begin{aligned}
(47.13) \qquad t^{-1/\alpha}X_t &= t^{-1/\alpha}X_t^0 + \tau t^{1-1/\alpha} \\
&\overset{\mathrm{d}}{=} X_1^0 + \tau t^{1-1/\alpha} = X_1 + \tau(t^{1-1/\alpha} - 1).
\end{aligned}
$$

Hence

$$I = \int_0^1 t^{-1}P[\,X_1 > \tau(1 - t^{1-1/\alpha})]\mathrm{d}t.$$

If $1 < \alpha \leq 2$, then $I \geq \int_0^1 t^{-1}P[\,X_1 > 0 \vee \tau\,]\mathrm{d}t = \infty$, which gives $R_0 = 0$ a. s. If $0 < \alpha < 1$ and $\tau < 0$, then

$$
\begin{aligned}
I &= \int_0^1 t^{-1}P[\,X_1 > |\tau|(t^{1-1/\alpha} - 1)]\mathrm{d}t \\
&= \alpha(1-\alpha)^{-1}\int_1^\infty u^{-1}P[\,|\tau|^{-1}X_1 + 1 > u\,]du \\
&= -\alpha(1-\alpha)^{-1}\int_1^\infty (\log u)d_uP[\,|\tau|^{-1}X_1 + 1 > u\,] < \infty,
\end{aligned}
$$

which gives $R_0 > 0$ a. s. Here we have used Example 25.10 for the finiteness of the integral. If $0 < \alpha < 1$ and $\tau > 0$, or if $0 < \alpha < 1$, $\beta \neq -1$, and $\tau = 0$, then $I \geq \int_0^1 t^{-1}P[\,X_1 > 0\,]\mathrm{d}t = \infty$ and $R_0 = 0$ a. s. If $0 < \alpha < 1$, $\beta = -1$, and $\tau = 0$, then the process is decreasing and we have $R_0 = \infty$ a. s.

Consider the case $\alpha = 1$. Thus, for $t > 0$, we have

$$(47.14) \qquad t^{-1}X_t \overset{\mathrm{d}}{=} X_1 + b\log t$$

with $b = 2\pi^{-1}c\beta$, directly from (14.25) or from Exercise 18.6. Hence

$$I = \int_0^1 t^{-1}P[\,X_1 > -b\log t\,]\mathrm{d}t = \int_0^\infty P[\,X_1 > bu\,]\mathrm{d}u.$$

If $b \leq 0$, then $I \geq \int_0^\infty P[\,X_1 > 0\,]\mathrm{d}u = \infty$. If $b > 0$, then $\beta > 0$ and it follows that $I = b^{-1}E[X_1 \vee 0] = \infty$. Hence $R_0 = 0$ a. s. □

REMARK 47.7. The assertions (i), (ii), and (iii) of Theorem 47.6 remain true for non-trivial α-semi-stable processes with $0 < \alpha \leq 2$ with replacement of $-1 < \beta \leq 1$ in (i) and $\beta = -1$ in (ii) by $\nu(0,\infty) > 0$ and $\nu(0,\infty) = 0$, respectively, and of τ in (i) and (ii) by the drift γ_0 [**449**].

Let us give an overview of short time fluctuation results for Lévy processes, apart from applications of Wiener–Hopf factorizations. We state them in a series of propositions without proof.

One of the fundamental results in probability theory is the following *law of the iterated logarithm.*

PROPOSITION 47.8. *Let $\{S_n\}$ be a random walk on $\mathbb{R}$ with $ES_1 = 0$ and* $\operatorname{Var} S_1 = 1$. *Then*

$$\limsup_{n\to\infty} \frac{S_n}{(2n\log\log n)^{1/2}} = 1 \quad a.\,s. \tag{47.15}$$

It follows that, under the same assumption,

$$\liminf_{n\to\infty} \frac{S_n}{(2n\log\log n)^{1/2}} = -1 \quad \text{a. s.} \tag{47.16}$$

and hence

$$\limsup_{n\to\infty} \frac{|S_n|}{(2n\log\log n)^{1/2}} = 1 \quad \text{a. s.} \tag{47.17}$$

The corresponding fact for the Brownian motion is as follows.

PROPOSITION 47.9. *The Brownian motion $\{X_t\}$ on $\mathbb{R}$ satisfies*

$$\limsup_{t\to\infty} \frac{X_t}{(2t\log\log t)^{1/2}} = 1 \quad a.\,s. \tag{47.18}$$

and

$$\limsup_{t\downarrow 0} \frac{X_t}{(2t\log\log(1/t))^{1/2}} = 1 \quad a.\,s. \tag{47.19}$$

By symmetry, (47.18) and (47.19) remain true if we replace limsup by liminf in the left-hand sides and 1 by -1 in the right-hand sides. Also, (47.18) and (47.19) remain true if X_t is replaced by $|X_t|$. In this form they hold for the d-dimensional Brownian motion.

In the case of the simplest random walk where S_1 takes only two values, Proposition 47.8 was proved by Khintchine [**275**] following the work of Hausdorff, Hardy and Littlewood, and Steinhaus. See Feller [**135**] for the history. Kolmogorov [**291**] weakened the assumption. In the form of Proposition 47.8 it is proved by Hartman and Wintner [**183**]. Proposition 47.9 for the Brownian motion $\{X_t\}$ was proved by Khintchine [**276**]. Since $\{tX_{1/t}\}$ is again the Brownian motion by Theorem 5.4, the two assertions (47.18) and (47.19) are equivalent.

The result can be made more precise in the following form, called an integral test. We state it for the behavior as $t\downarrow 0$.

PROPOSITION 47.10. *Suppose that $g(t)$ belongs to the class of*

(47.20) *functions positive, continuous, and decreasing on some* $(0,\delta]$.

Let $\{X_t\}$ be the Brownian motion on $\mathbb{R}$. Then

$$P[\, X_t \le t^{1/2} g(t) \text{ for all sufficiently small } t\,] = 1 \quad \textit{or} \quad 0, \tag{47.21}$$

according as

$$\textstyle\int_0^\delta t^{-1} g(t) \mathrm{e}^{-g(t)^2/2} \mathrm{d}t < \infty \quad \textit{or} \quad = \infty. \tag{47.22}$$

For example, if $g(t) = \{2(1+\varepsilon)\log\log\frac{1}{t}\}^{1/2}$, then (47.22) holds according as $\varepsilon > 0$ or $\varepsilon \leq 0$. If we denote the n-fold iteration of the logarithmic function by $\log_{(n)}$ and if $g(t) = \{2(\log_{(2)}\frac{1}{t} + (\frac{3}{2}+\varepsilon)\log_{(3)}\frac{1}{t})\}^{1/2}$, then (47.22) holds according as $\varepsilon > 0$ or $\varepsilon \leq 0$. More generally, if $g(t) = \{2(\log_{(2)}\frac{1}{t} + \frac{3}{2}\log_{(3)}\frac{1}{t} + \log_{(4)}\frac{1}{t} + \cdots + \log_{(n-1)}\frac{1}{t} + (1+\varepsilon)\log_{(n)}\frac{1}{t})\}^{1/2}$ with some $n \geq 4$, then (47.22) holds according as $\varepsilon > 0$ or $\varepsilon \leq 0$.

Proposition 47.10 is given by Petrowsky [**372**] and Kolmogorov (see Lévy [**318**], p. 88). Petrowsky finds it as a criterion of regularity of a point in the boundary-value problem for the heat equation. For random walks Erdős [**128**] and Feller [**135, 136**] studied conditions for validity of the criterion (47.22).

Results of the type of (47.19) for more general Lévy processes are obtained in various cases. The following two facts are due to Khintchine [**282**].

PROPOSITION 47.11. *Let $\{X_t\}$ be a Lévy process on $\mathbb{R}$ and let $A \geq 0$ be its Gaussian variance. Then*

$$\limsup_{t\downarrow 0} \frac{|X_t|}{(2t\log\log(1/t))^{1/2}} = \sqrt{A} \quad a.\,s. \tag{47.23}$$

PROPOSITION 47.12. *Suppose that $h(t)$ belongs to the class of*

(47.24) *functions positive, continuous, and increasing on some* $(0, \delta]$

and that

$$\lim_{t\downarrow 0} \frac{h(t)}{(t\log\log(1/t))^{1/2}} = 0.$$

Then there is a purely non-Gaussian Lévy process $\{X_t\}$ on $\mathbb{R}$ such that

$$\limsup_{t\downarrow 0} \frac{|X_t|}{h(t)} = \infty \quad a.\,s.$$

That is, fluctuation of a purely non-Gaussian Lévy process is smaller than that of the Brownian motion, but it can be arbitrarily close.

When $\{X_t\}$ is a stable process on $\mathbb{R}$ with one-sided jumps, the following results are known. Proposition 47.13 is by Breiman [**67**] and Propositions 47.14 and 47.15 are by Mijnheer [**346**]. They extended Motoo's method [**357**] for a new proof of Proposition 47.10. These results deal with bounding of sample functions in small time only from the direction without jumps.

PROPOSITION 47.13. *Let $\{X_t\}$ be a stable process on $\mathbb{R}$ with parameters (α, β, τ, c) in Definition 14.16 equal to $(\alpha, 1, 0, 1)$ with $0 < \alpha < 1$ (hence a subordinator). Let $k(t)$ be a function in the class (47.24) and let*

$$g(t) = (2B_\alpha)^{1/2} k(t)^{-\alpha/(2(1-\alpha))} \quad \textit{with } B_\alpha = (1-\alpha)\alpha^{\alpha/(1-\alpha)}(\cos 2^{-1}\pi\alpha)^{-1/(1-\alpha)}.$$

Then

$$P[\,X_t \geq t^{1/\alpha}k(t) \text{ for all sufficiently small } t\,] = 1 \quad or \quad 0$$

according to the condition (47.22) for $g(t)$. In particular,

$$\liminf_{t\downarrow 0} \frac{X_t}{t^{1/\alpha}(2\log\log(1/t))^{-(1-\alpha)/\alpha}} = (2B_\alpha)^{(1-\alpha)/\alpha} \quad a.\,s. \tag{47.25}$$

PROPOSITION 47.14. *Let $\{X_t\}$ be stable with $(\alpha, \beta, \tau, c) = (1, -1, 0, 1)$ (hence no positive jumps). Let $k(t)$ be a function in the class (47.20) and let*

$$g(t) = 2(\pi \mathrm{e})^{-1/2} \mathrm{e}^{\pi k(t)/4}.$$

Then

$$P[\, X_t - \tfrac{\pi}{2} t \log \tfrac{1}{t} \le t k(t) \text{ for all sufficiently small } t \,] = 1 \quad \textit{or} \quad 0$$

according to the condition (47.22) for $g(t)$. In particular,

$$\limsup_{t \downarrow 0} \left(\frac{X_t - 2\pi^{-1} t \log(1/t)}{t} - \frac{2}{\pi} \log \log \log \frac{1}{t} \right) = \frac{2}{\pi} \left(1 + \log \frac{\pi}{2} \right) \quad \textit{a. s.}$$

It follows from this that

$$\limsup_{t \downarrow 0} \frac{X_t}{2\pi^{-1} t \log(1/t)} = 1 \quad \textit{a. s.} \tag{47.26}$$

PROPOSITION 47.15. *Let $\{X_t\}$ be stable with $(\alpha, \beta, \tau, c) = (\alpha, -1, 0, 1)$ with $1 < \alpha < 2$ (hence no positive jumps). Let $k(t)$ be a function in the class (47.20) and let*

$$g(t) = (2B_\alpha)^{1/2} k(t)^{\alpha/(2(\alpha-1))} \quad \textit{with } B_\alpha = (\alpha - 1)\alpha^{\alpha/(\alpha-1)} |\cos 2^{-1}\pi\alpha|^{1/(\alpha-1)}.$$

Then

$$P[\, X_t \le t^{1/\alpha} k(t) \text{ for all sufficiently small } t \,] = 1 \quad \textit{or} \quad 0$$

according to the condition (47.22) for $g(t)$. In particular,

$$\limsup_{t \downarrow 0} \frac{X_t}{t^{1/\alpha} (2 \log \log(1/t))^{(\alpha-1)/\alpha}} = (2B_\alpha)^{-(\alpha-1)/\alpha} \quad \textit{a. s.} \tag{47.27}$$

In this way we have a group of laws of the iterated logarithm. The result (47.25) was obtained by Fristedt [**148**] prior to [**67**]. Zolotarev [**605**] announced (47.27) and some of Proposition 47.14 for $t \to \infty$ instead of $t \downarrow 0$. The following result on bounding of sample functions in small time from the other direction is by Fristedt [**150**]. The same statement is true with X_t replaced by $|X_t|$ in (47.28), which was earlier obtained by Khintchine [**280**].

PROPOSITION 47.16. *Let $\{X_t\}$ be a strictly α-stable process on $\mathbb{R}$ with $0 < \alpha < 2$ satisfying $\nu(0, \infty) > 0$ for the Lévy measure ν. Let $h(t)$ be a function in the class (47.24). Then*

$$\limsup_{t \downarrow 0} \frac{X_t}{h(t)} = 0 \ \textit{a. s.} \quad \textit{or} \quad = \infty \ \textit{a. s.} \tag{47.28}$$

according as

$$\textstyle\int_0^\delta h(t)^{-\alpha} \mathrm{d}t < \infty \quad \textit{or} \quad = \infty. \tag{47.29}$$

This is a typical result on short time behavior of Lévy processes. It implies that no function $h(t)$ in the class (47.24) satisfies $\limsup_{t \downarrow 0}(X_t/h(t)) = C$ a. s. with a finite positive constant C. The following fact on subordinators found by Fristedt [**149**] is closely connected.

PROPOSITION 47.17. *Let $\{X_t\}$ be a subordinator with drift $\gamma_0 = 0$. Let $h(t)$ be a function such that $t^{-1}h(t)$ is in the class (47.24). Then, we have (47.28) according as*

$$\int_0^\delta \nu[h(t),\infty)\mathrm{d}t < \infty \quad \textit{or} \quad = \infty. \tag{47.30}$$

Fristedt and Pruitt [**152**] extended the liminf result (47.25) as follows.

PROPOSITION 47.18. *Let $\{X_t\}$ be a subordinator of type B with drift $\gamma_0 = 0$. Let ν be its Lévy measure, write $F(u) = \int_{(0,\infty)}(1 - \mathrm{e}^{-ux})\,\nu(\mathrm{d}x)$, and let G be the inverse function of F. (Since F is continuous and strictly increasing from 0 to ∞ on $[0,\infty)$, it has an inverse function.) Let*

$$h(t) = \frac{\log\log(1/t)}{G(t^{-1}\log\log(1/t))}.$$

If $\int_{(0,1]} x^{\varepsilon}\nu(\mathrm{d}x) = \infty$ for some $\varepsilon > 0$, then

$$\liminf_{t\downarrow 0} \frac{X_t}{h(t)} = C \text{ a. s. with } 0 < C = \text{const} < \infty.$$

They mention that there is $\{X_t\}$ satisfying $\int_{(0,1]} x^{\varepsilon}\nu(\mathrm{d}x) < \infty$ for all $\varepsilon > 0$ such that, for any $h(t)$ in the class (47.24), $\liminf_{t\downarrow 0}\frac{X_t}{h(t)}$ is either 0 a. s. or ∞ a. s. Pruitt [**398**] studied extension of the integral test in Proposition 47.13 to subordinators. This is connected with Jain and Pruitt [**234**].

Let $\{X_t\}$ be the Brownian motion on $\mathbb{R}^d$. If $d \geq 3$, then it is transient; Dvoretzky and Erdős [**115**] studied the speed of $|X_t|$ going to ∞ as $t \to \infty$. If $d = 2$, then $\{X_t\}$ is recurrent but it never hits the starting point 0 (Example 43.7(i)); Spitzer [**495**] studied the speed of $|X_t|$ approaching 0 in the lim inf sense as $t \to \infty$. The following two propositions state the results equivalent to theirs in the form of short time behavior.

PROPOSITION 47.19. *Let $\{X_t\}$ be the Brownian motion on $\mathbb{R}^d$ with $d \geq 3$. Let $g(t)$ be a function in the class (47.24). Then*

$$P[\,|X_t| \geq t^{1/2}g(t) \text{ for all sufficiently small } t\,] = 1 \quad \textit{or} \quad 0 \tag{47.31}$$

according as

$$\int_0^\delta t^{-1}g(t)^{d-2}\mathrm{d}t < \infty \quad \textit{or} \quad = \infty. \tag{47.32}$$

It follows that

$$\liminf_{t\downarrow 0} \frac{|X_t|}{t^{1/2}(\log(1/t))^{-(1+\varepsilon)/(d-2)}} = \infty \text{ a. s.} \quad \textit{or} \quad = 0 \text{ a. s.} \tag{47.33}$$

according as $\varepsilon > 0$ or $\varepsilon \leq 0$. There is no choice of $g(t)$ in the class (47.24) such that

$$\liminf_{t\downarrow 0} \frac{|X_t|}{t^{1/2}g(t)} = C \text{ a. s. with } 0 < C = \text{const} < \infty. \tag{47.34}$$

PROPOSITION 47.20. *Let $\{X_t\}$ be the Brownian motion on $\mathbb{R}^2$. Let $g(t)$ be a function in the class (47.24) satisfying $g(\delta) < 1$. Then we have (47.31) according as*

$$\int_0^{\delta} t^{-1}\{\log(1/g(t))\}^{-1} dt < \infty \quad or \quad = \infty.$$

In particular,

$$\liminf_{t \downarrow 0} \frac{|X_t|}{\exp\{(\log t)(\log\log(1/t))^{1+\varepsilon}\}} = \infty \text{ a. s.} \quad or \quad = 0 \text{ a. s.}$$

according as $\varepsilon > 0$ or $\varepsilon \le 0$. It is impossible to find $g(t)$ in the class (47.24) satisfying (47.34).

This liminf problem for $|X_t|$ is meaningless for the one-dimensional Brownian motion, since $T_{\{0\}} = 0$ a. s. Takeuchi [**517**] extends Proposition 47.19 to transient rotation invariant α-stable processes on $\mathbb{R}^d$ (that is, $\alpha < d$) and shows that the same statement is true if $d - 2$ in (47.32) and (47.33) is replaced by $d - \alpha$ and if $t^{1/2}$ in (47.31), (47.33), and (47.34) is replaced by $t^{1/\alpha}$. Takeuchi and Watanabe [**518**] considers the one-dimensional Cauchy process and proves formally the same statement as Proposition 47.20 with $t^{1/2}$ replaced by t in (47.31) and (47.34).

Short time behavior of

$$X_t^*(\omega) = \sup_{0 \le s \le t} |X_s(\omega)| \tag{47.35}$$

has also been studied. Again some laws of the iterated logarithm appear. Taylor [**526**] and Pruitt and Taylor [**399, 400**] prove the following results for strictly stable processes.

PROPOSITION 47.21. *Let $d \ge 1$. Let $\{X_t\}$ be a strictly α-stable process on $\mathbb{R}^d$ with $0 < \alpha \le 2$ such that none of its one-dimensional projections is a subordinator. When $\alpha = 1$, we assume that $\tau = 0$ in the representation (14.16) of the distribution of X_1. Then*

$$\liminf_{t \downarrow 0} \frac{X_t^*}{t^{1/\alpha}(\log\log(1/t))^{-1/\alpha}} = C \text{ a. s. with } 0 < C = \text{const} < \infty.$$

PROPOSITION 47.22. *Let $d \ge 1$ and $0 < \alpha < 1$. Let $\{X_t\}$ be a nondegenerate strictly α-stable process on $\mathbb{R}^d$ such that one of its one-dimensional projections is a subordinator. Then*

$$\liminf_{t \downarrow 0} \frac{X_t^*}{t^{1/\alpha}(\log\log(1/t))^{-(1-\alpha)/\alpha}} = C \text{ a. s. with } 0 < C = \text{const} < \infty.$$

For the limsup behavior we remark that $\limsup_{t\downarrow 0} \frac{X_t^*}{h(t)} = \limsup_{t\downarrow 0} \frac{|X_t|}{h(t)}$ whenever $h(t)$ is in the class (47.24).

It is a hard problem to determine $h(t)$ satisfying

$$\liminf_{t \downarrow 0} \frac{X_t^*}{h(t)} = C \text{ a. s. with } 0 < C = \text{const} < \infty$$

for a more general Lévy process $\{X_t\}$ on $\mathbb{R}$. Dupuis [**112**] and Wee [**570, 571**] have some results. The following fact is proved by Jain and Pruitt [**232**].

PROPOSITION 47.23. *Let $\{X_t\}$ be a 1-stable, not strictly 1-stable, process on $\mathbb{R}$. Then*

$$\liminf_{t\downarrow 0} \frac{X_t^*}{t\log(1/t)} = C \text{ a. s. with } 0 < C = \text{const} < \infty.$$

What is the analogue of the index of a stable process in the short time behavior of a Lévy process on $\mathbb{R}$? This question has been studied since Blumenthal and Getoor [**43**]. Pruitt [**395**] succeeded in handling $\liminf_{t\downarrow 0} t^{-1/\eta} X_t^*$ and $\limsup_{t\downarrow 0} t^{-1/\eta} X_t^*$ in a dual way, using some analogues of the index. To give his result, let $\{X_t\}$ be a Lévy process on $\mathbb{R}^d$ generated by (A, ν, γ) with $A = 0$. Let

$$h(r) = \int_{|x|>r} \nu(\mathrm{d}x) + r^{-2} \int_{|x|\le r} |x|^2 \nu(\mathrm{d}x) + r^{-1} \Big|\gamma - \int_{r<|x|\le 1} x\nu(\mathrm{d}x)\Big|,$$

and define

$$\beta_L = \inf\Big\{\eta > 0\colon \limsup_{r\downarrow 0} r^\eta h(r) = 0\Big\}, \tag{47.36}$$

$$\delta_L = \inf\Big\{\eta > 0\colon \liminf_{r\downarrow 0} r^\eta h(r) = 0\Big\}, \tag{47.37}$$

where the subscript L alludes Lévy processes. Then $0 \le \delta_L \le \beta_L \le 2$. There is a case with $\delta_L < \beta_L$. If $\{X_t\}$ is α-stable, then $\delta_L = \beta_L = \alpha$. Except in the case where $\int_{|x|\le 1} |x|\nu(\mathrm{d}x) < \infty$ and $\gamma_0 \ne 0$, the β_L coincides with β'_L defined by

$$\beta'_L = \inf\Big\{\eta > 0\colon \int_{|x|\le 1} |x|^\eta \nu(\mathrm{d}x) < \infty\Big\}, \tag{47.38}$$

which is introduced in [**43**] under the name β. It is shown in [**43**] that

$$\begin{aligned}\beta'_L &= \inf\Big\{\eta \ge 0\colon \lim_{|z|\to\infty} |z|^{-\eta}|\psi(z)| = 0\Big\}\\ &= \inf\Big\{\eta \ge 0\colon \lim_{|z|\to\infty} |z|^{-\eta}\mathrm{Re}\,\psi(z) = 0\Big\}.\end{aligned}$$

If $\{X_t\}$ is a subordinator with $\gamma_0 = 0$, then, as in Horowitz [**207**],

$$\delta_L = \sup\Big\{\eta > 0\colon \lim_{r\downarrow 0} r^{\eta-1} \int_0^r \nu(y,\infty)\mathrm{d}y = \infty\Big\}.$$

PROPOSITION 47.24. *Let $\eta > 0$. We have*

$$\limsup_{t\downarrow 0} t^{-1/\eta} X_t^* = 0 \text{ a. s.} \quad \text{or} \quad = \infty \text{ a. s.}$$

according as $\eta > \beta_L$ or $\eta < \beta_L$. We have

$$\liminf_{t\downarrow 0} t^{-1/\eta} X_t^* = 0 \text{ a. s.} \quad \text{or} \quad = \infty \text{ a. s.}$$

according as $\eta > \delta_L$ or $\eta < \delta_L$.

This is from [**395**] but part of it is obtained in [**43**].

48. Long time behavior

As in the first half of the preceding section, let $\{X_t : t \geq 0\}$ be a Lévy process on $\mathbb{R}$ as in Section 45. The following theorem on the criterion for drifting to ∞, to $-\infty$, and oscillating (Definition 37.11) is a major result of this section. It was discovered by Spitzer [**494**] for random walks and extended to Lévy processes by Rogozin [**419**].

THEOREM 48.1. *Let $\{X_t\}$ be a non-zero Lévy process on $\mathbb{R}$.*
(i) *Let*

$$I^+ = \int_1^\infty t^{-1} P[X_t > 0] \mathrm{d}t \quad \text{and} \quad I^- = \int_1^\infty t^{-1} P[X_t < 0] \mathrm{d}t. \tag{48.1}$$

Then, $\{X_t\}$ is drifting to ∞, if and only if $I^- < \infty$; drifting to $-\infty$ if and only if $I^+ < \infty$; and oscillating if and only if $I^+ = \infty$ and $I^- = \infty$.
(ii) *Let $M_\infty = \sup_t X_t$ and $N_\infty = \inf_t X_t$. If $\{X_t\}$ is drifting to ∞, then $N_\infty > -\infty$ a. s., P_{N_∞} is infinitely divisible, and*

$$E[\mathrm{e}^{uN_\infty}] = \exp\left[\int_0^\infty t^{-1} \mathrm{d}t \int_{(-\infty,0)} (\mathrm{e}^{ux} - 1) \mu^t(\mathrm{d}x)\right], \qquad u \geq 0. \tag{48.2}$$

If $\{X_t\}$ is drifting to $-\infty$, then $M_\infty < \infty$ a. s., P_{M_∞} is infinitely divisible, and

$$E[\mathrm{e}^{-uM_\infty}] = \exp\left[\int_0^\infty t^{-1} \mathrm{d}t \int_{(0,\infty)} (\mathrm{e}^{-ux} - 1) \mu^t(\mathrm{d}x)\right], \qquad u \geq 0. \tag{48.3}$$

REMARK 48.2. We have $I^+ + I^- = \infty$ for any non-zero Lévy process, since $\int_1^\infty t^{-1} P[\, X_t = 0\,] \mathrm{d}t < \infty$ by Lemma 48.3 below. Hence the theorem above ensures again that any non-zero Lévy process on $\mathbb{R}$ is drifting to ∞, drifting to $-\infty$, or oscillating. Another criterion for those three cases is given in terms of the Lévy measure of $\{X_t\}$. See Remark 37.13.

LEMMA 48.3. *Let $\{X_t\}$ be a non-zero Lévy process on $\mathbb{R}$. Then, for any finite interval K,*

$$P[\, X_t \in K\,] = O(t^{-1/2}) \quad \text{as } t \to \infty. \tag{48.4}$$

Proof. We use the generating triplet (A, ν, γ). Choose independent Lévy processes $\{X_t^1\}$ and $\{X_t^2\}$ satisfying $X_t = X_t^1 + X_t^2$ as follows. If $A > 0$, then let $\{X_t^1\}$ and $\{X_t^2\}$ be generated by $(A, 0, 0)$ and $(0, \nu, \gamma)$, respectively, and see that

$$P[\, X_t \in K\,] \leq \sup_x P[\, X_t^1 \in K - x\,], \tag{48.5}$$

which is bounded by a constant multiple of $t^{-1/2}$. Consider the case that $\nu(0, \infty) > 0$. Using $\varepsilon > 0$ satisfying $\nu(\varepsilon, \infty) = c > 0$, let $\{X_t^1\}$ be the compound Poisson process with Lévy measure $[\nu]_{(\varepsilon,\infty)}$. Represent $\{X_t^1\}$ as $X_t^1 = S_{Z_t}$, where $\{S_n\}$ is a random walk with $P_{S_1} = c^{-1}[\nu]_{(\varepsilon,\infty)}$ and $\{Z_t\}$ is a Poisson process with parameter c, independent of $\{S_n\}$. We have (48.5) and

$$P[\,X_t^1 \in K - x\,] \le \mathrm{e}^{-ct}\left(\max_{n\ge 0}\frac{(ct)^n}{n!}\right)\sum_{n=0}^{\infty} P[\,S_n \in K - x\,].$$

Let b be the length of the interval K. Since $S_n - S_{n-1} > \varepsilon$ a. s., we have

$$\sum_{n=0}^{\infty} P[\,S_n \in K - x\,] = E\left[\sum_{n=0}^{\infty} 1_{K-x}(S_n)\right] \le \varepsilon^{-1}b + 1.$$

The sequence $\{(ct)^n/n!\}$ has the maximum element at $n = [ct]$. We have

$$(ct)^{[ct]}/[ct]! \le ([ct]+1)^{[ct]}/[ct]! \sim (2\pi[ct])^{-1/2}\mathrm{e}^{[ct]+1}$$

by Stirling's formula. Hence (48.4) follows. The case $\nu(-\infty,0) > 0$ is similar. The remaining case is that $A = 0$, $\nu = 0$, and $\gamma \ne 0$. In this case, $P[\,X_t \in K\,] = 0$ for large t. □

REMARK 48.4. Suppose that $\{X_t\}$ is non-trivial. Fix $0 < b < \infty$. Then (48.4) holds uniformly for all intervals K with length b. This is shown in the proof above.

Proof of Theorem 48.1. Our basic formula is (47.9). Since its left-hand side equals $\int_0^\infty \mathrm{e}^{-t} E[\mathrm{e}^{-uM_{t/q}}]\mathrm{d}t$, we get (48.3) for $u > 0$ by letting $q \downarrow 0$. The double integral in the right-hand side of (48.3) is possibly $-\infty$. We claim that,

$$\text{if } I^+ = \infty, \text{ then } M_\infty = \infty \text{ a. s.,} \tag{48.6}$$

$$\text{if } I^+ < \infty, \text{ then } M_\infty < \infty \text{ a. s.} \tag{48.7}$$

Fix $u > 0$ and choose c such that $1 - \mathrm{e}^{-uc} \ge 1/2$. Then

$$\int_1^\infty t^{-1}\mathrm{d}t \int_{(0,\infty)} (1 - \mathrm{e}^{-ux})\mu^t(\mathrm{d}x) \ge \frac{1}{2}\int_1^\infty t^{-1}\mu^t(c,\infty)\mathrm{d}t. \tag{48.8}$$

Note that $\int_1^\infty t^{-1}\mu^t(0,c]\,\mathrm{d}t$ is finite by Lemma 48.3. Thus, if $I^+ = \infty$, then the right-hand side of (48.8) is infinite and hence $M_\infty = \infty$ a. s. by (48.3). Since $1 - \mathrm{e}^{-ux} \le 1 - \mathrm{e}^{-x} \le x$ for $0 < u \le 1$ and $x > 0$, we have

$$\int_0^\infty t^{-1}\mathrm{d}t \int_{(0,\infty)} (1 - \mathrm{e}^{-x})\mu^t(\mathrm{d}x)$$
$$\le I^+ + \int_0^1 t^{-1}\mathrm{d}t \int_{(0,1]} x\mu^t(\mathrm{d}x) + \int_0^1 t^{-1}\mathrm{d}t \int_{(1,\infty)} \mu^t(\mathrm{d}x).$$

Lemma 30.3 tells us that the second and third terms in the right-hand side are finite. Hence, if $I^+ < \infty$, then the dominated convergence theorem applies and

$$\int_0^\infty t^{-1}\mathrm{d}t \int_{(0,\infty)} (1 - \mathrm{e}^{-ux})\mu^t(\mathrm{d}x) \to 0 \quad \text{as } u \downarrow 0,$$

which means $M_\infty < \infty$ a. s. by (48.3). Thus (48.6) and (48.7) are shown. We next claim that,

$$\text{(48.9)} \qquad \text{if } N_\infty > -\infty \text{ a. s., then } \lim_{t\to\infty} X_t = \infty \text{ a. s.}$$

We may and do assume the set-up of Chapter 8. Suppose that $N_\infty > -\infty$ a. s. For $a_n \uparrow \infty$ let $B_n = \{X_t < a_n \text{ for some } t > R_{2a_n}\}$. Since

$$P[B_n] \le P[X_t - X(R_{2a_n}) < -a_n \text{ for some } t > R_{2a_n}],$$

we have $P[B_n] \le P[N_\infty < -a_n]$ by the strong Markov property. Choosing $\{a_n\}$ such that $P[N_\infty < -a_n] < n^{-2}$ and using the Borel–Cantelli lemma, we get $P[\limsup_{n\to\infty} B_n] = 0$. Hence $P[\liminf_{t\to\infty} X_t < \infty] = 0$, that is, $\{X_t\}$ is drifting to ∞. By using the dual process, we get similarly

$$\text{(48.10)} \qquad \text{if } I^- = \infty, \text{ then } N_\infty = -\infty \text{ a. s.,}$$

$$\text{(48.11)} \qquad \text{if } I^- < \infty, \text{ then } N_\infty > -\infty \text{ a. s.,}$$

$$\text{(48.12)} \qquad \text{if } M_\infty < \infty \text{ a. s., then } \lim_{t\to\infty} X_t = -\infty \text{ a. s.}$$

Since sample functions are bounded in any finite time interval, $M_\infty = \infty$ is equivalent to $\limsup_{t\to\infty} X_t = \infty$, and $N_\infty = -\infty$ is equivalent to $\liminf_{t\to\infty} X_t = -\infty$. Therefore, we get the 'if' parts of the three 'if and only if' assertions in (i). Then the 'only if' parts are automatic.

Let us show (ii). Suppose that $\{X_t\}$ is drifting to $-\infty$. Then, obviously, $M_\infty < \infty$ a. s. The identity (48.3) is already shown. It follows from $I^+ < \infty$ and from Lemmas 30.3 and 48.3 that $\int_0^\infty t^{-1}dt \int_{(0,\infty)} (1 \wedge x)\mu^t(dx) < \infty$. Hence the right-hand side of (48.3) is the Laplace transform of an infinitely divisible distribution. Thus we get the latter half of (ii). The former half is similar. □

EXAMPLE 48.5. As in Example 47.4, let $\{X_t\}$ and $\{X'_t\}$ be independent strictly stable subordinators with indices α and α', respectively, and let $0 < \alpha < \alpha' < 1$. Then

$$\text{(48.13)} \qquad P[X_t - X'_t \to \infty \text{ as } t \to \infty] = 1.$$

The proof is to show that $\int_1^\infty t^{-1} P[Z_t < 0]dt < \infty$ for $Z_t = X_t - X'_t$. We obtain this since, as in the argument in Example 47.4,

$$\begin{aligned} P[Z_t < 0] &= \textstyle\int_{[0,\infty)} P[X_1 < t^{1/\alpha'-1/\alpha}x]\, P[X'_1 \in dx] \\ &\le P[X_1 < t^{1/\alpha'-1/\alpha+\eta}] + P[X'_1 > t^\eta] \le \text{const}\, t^{1/\alpha'-1/\alpha+\eta} + \text{const}\, t^{-\theta\eta} \end{aligned}$$

for $0 < \eta < 1/\alpha - 1/\alpha'$ and $0 < \theta < \alpha'$. Now we see the interesting almost sure behavior that $X_t < X'_t$ for small t and $X_t > X'_t$ for large t.

Let us apply Theorem 48.1 to stable processes.

THEOREM 48.6. *Let $\{X_t\}$ be non-trivial and α-stable on $\mathbb{R}$ with $0 < \alpha \le 2$. In the case $\alpha \ne 2$, let (α, β, τ, c) be its parameters as in Definition 14.16. In the case $\alpha = 2$, let $\tau = EX_1$.*

(i) *If* $1 < \alpha \le 2$, *then* $\{X_t\}$ *is drifting to* ∞, *drifting to* $-\infty$, *or oscillating according as* τ *is positive, negative, or* 0, *respectively.*

(ii) *If* $0 < \alpha \le 1$, *then* $\{X_t\}$ *is drifting to* ∞, *drifting to* $-\infty$, *or oscillating according as* $\beta = 1$, $\beta = -1$, *or* $|\beta| < 1$, *respectively.*

REMARK 48.7. An α-stable process on $\mathbb{R}$ is transient and oscillating if and only if it is in one of the following two cases:

(1) $\alpha = 1$ and $0 < |\beta| < 1$;
(2) $0 < \alpha < 1$ and $|\beta| < 1$.

This is a consequence of Theorem 48.6 combined with Corollary 37.17.

Proof of Theorem 48.6. (i) Let $1 < \alpha \le 2$. Then $\tau = EX_1$. Hence, if τ is positive or negative, then $\{X_t\}$ is drifting to ∞ or to $-\infty$, respectively, by the analogue of the strong law of large numbers in Theorem 36.5. If $\tau = 0$, then it is recurrent by Theorem 36.7 and hence oscillating.

(ii) If $0 < \alpha \le 1$, then $\beta = 1$ or -1 respectively implies $EX_1 = \infty$ or $-\infty$, and hence drifting to ∞ or to $-\infty$ by Theorem 36.6. If $0 < \alpha < 1$ and $|\beta| < 1$, then, using (47.13), we have, for $t \ge 1$,

$$P[\,X_t > 0\,] = P[\,X_1 > \tau(1 - t^{1-1/\alpha})\,] \ge P[\,X_1 > |\tau|\,] > 0$$

and hence $I^+ = \infty$. In this case, we have also $I^- = \infty$ for the same reason, and hence the process is oscillating. If $\alpha = 1$ and $|\beta| < 1$, then, using (47.14) and $b = 2\pi^{-1}c\beta$, we get

$$P[\,X_t > 0\,] = P[\,X_1 > -b\log t\,] \ge P[\,X_1 > |b|\log t\,]$$

and hence

$$I^+ \ge \int_0^\infty P[\,X_1 > |b|u\,]du.$$

This integral equals $|b|^{-1}E[X_1 \vee 0] = \infty$ if $|b| \ne 0$. It equals $\int_0^\infty P[\,X_1 > 0\,]du = \infty$ if $|b| = 0$. Hence $I^+ = \infty$ and, similarly, $I^- = \infty$ in this case and the process is oscillating. □

REMARK 48.8. Analogous results for a non-trivial α-semi-stable process, $0 < \alpha \le 2$, are as follows [**449**].

(i) If $1 < \alpha \le 2$, then it is drifting to ∞, to $-\infty$, or oscillating according as EX_1 is positive, negative, or 0, respectively.

(ii) Assume that $0 < \alpha \le 1$. Then, it is drifting to ∞ if $\nu(-\infty, 0) = 0$; drifting to $-\infty$ if $\nu(0, \infty) = 0$; oscillating if $\nu(0, \infty) > 0$ and $\nu(-\infty, 0) > 0$.

A natural gneralization to Lévy processes of the law of the iterated logarithm of random walks is in long time behavior. Thus Gnedenko [**166**] shows the following.

PROPOSITION 48.9. *Let $\{X_t\}$ be a Lévy process on $\mathbb{R}$. If $EX_1 = 0$ and $E[X_1{}^2] < \infty$, then*

$$\limsup_{t\to\infty} \frac{|X_t|}{(2t\log\log t)^{1/2}} = (E[X_1{}^2])^{1/2} \quad a.\,s. \tag{48.14}$$

If $E[X_1{}^2] = \infty$, then (48.14) remains true.

In the Brownian motion $\{X_t\}$, long time behavior and short time behavior are derived from each other, as $\{tX_{1/t}\} \stackrel{\mathrm{d}}{=} \{X_t\}$. If $\{X_t\}$ is a strictly α-stable process with $0 < \alpha < 2$, then, letting $Z_t = t^{2/\alpha}X_{1/t}$, we see that $Z_t \stackrel{\mathrm{d}}{=} X_t$ for any fixed $t > 0$, but $\{Z_t\}$ is not a Lévy process (Exercise 18.18). Thus, in this case, we cannot derive long time behavior from short time behavior. But, nevertheless, Propositions 47.13, 47.14, 47.15, 47.16, 47.21, and 47.22 have their counterparts in long time behaviors. They are given in the same papers, as similar proofs work.

In many cases, a technique similar to that employed in proving an assertion on short time behavior of a Lévy process works in giving an assertion on long time behavior. That is the case for Propositions 47.18 and 47.23. Pruitt [**395**] proves the following analogue of Proposition 47.24. Let $\{X_t\}$ be a Lévy process on $\mathbb{R}^d$ generated by (A, ν, γ) with $A = 0$. Letting

$$\overline{h}(r) = \textstyle\int_{|x|>r}\nu(\mathrm{d}x) + r^{-2}\int_{|x|\le r}|x|^2\nu(\mathrm{d}x) + r^{-1}\left|\gamma + \int_{1<|x|\le r}x\nu(\mathrm{d}x)\right|,$$

define the analogues of the index of a stable process, relevant to long time behavior:

$$\overline{\beta}_L = \sup\Big\{\eta\colon\ \limsup_{r\to\infty} r^{\eta}\overline{h}(r) = 0\Big\},$$

$$\overline{\delta}_L = \sup\Big\{\eta\colon\ \liminf_{r\to\infty} r^{\eta}\overline{h}(r) = 0\Big\}.$$

If $\{X_t\}$ is α-stable, then $\overline{\beta}_L = \overline{\delta}_L = \alpha$. We have $0 \le \overline{\beta}_L \le \overline{\delta}_L \le 2$. There is a case where $\overline{\beta}_L < \overline{\delta}_L$. Let $X_t^* = \sup_{0\le s\le t}|X_s|$ as in the previous section.

PROPOSITION 48.10. *Let $\eta > 0$. Then*

$$\limsup_{t\to\infty} t^{-1/\eta}X_t^* = 0 \ a.\,s. \quad or \quad = \infty \ a.\,s.$$

according as $\eta < \overline{\beta}_L$ or $\eta > \overline{\beta}_L$. We have

$$\liminf_{t\to\infty} t^{-1/\eta}X_t^* = 0 \ a.\,s. \quad or \quad = \infty \ a.\,s.$$

according as $\eta < \overline{\delta}_L$ or $\eta > \overline{\delta}_L$. We have

$$\overline{\beta}_L = \sup\Big\{\eta \in [0,2]\colon \textstyle\int_{|x|>1}|x|^{\eta}\nu(\mathrm{d}x) < \infty\Big\}$$

except in the case where $E|X_t| < \infty$ and $EX_t \ne 0$.

It is the case that $\limsup_{t\to\infty}(X_t^*/h(t)) = \limsup_{t\to\infty}(|X_t|/h(t))$ whenever $h(t)$ increases to ∞ as $t \to \infty$.

The following is the first law of the iterated logarithm of the type of Proposition 47.21, given by Chung [**78**] in the case of random walks.

PROPOSITION 48.11. *Let $\{X_t\}$ be the Brownian motion on $\mathbb{R}$. Let $k(t)$ be one of the*

$$\text{functions positive, continuous, and decreasing on some } [c,\infty), \tag{48.15}$$

and let $g(t) = 2^{-1}\pi k(t)^{-1}$. Then

$$P[\,X_t^* \geq t^{1/2}k(t) \text{ for all sufficiently large } t\,] = 1 \quad \textit{or} \quad 0$$

according as

$$\int_c^\infty t^{-1}g(t)^2 \mathrm{e}^{-g(t)^2/2}\mathrm{d}t < \infty \quad \textit{or} \quad = \infty. \tag{48.16}$$

As a consequence,

$$\liminf_{t\to\infty} \frac{X_t^*}{t^{1/2}(\log\log t)^{-1/2}} = \frac{\pi}{2\sqrt{2}} \quad \textit{a. s.}$$

On the other hand, the counterpart of Proposition 47.10 is as follows.

PROPOSITION 48.12. *Let $\{X_t\}$ be the Brownian motion on $\mathbb{R}$ and let $g(t)$ be a function positive, continuous, and increasing on some $[c,\infty)$. Then*

$$P[\,X_t \leq t^{1/2}g(t) \text{ for all sufficiently large } t\,] = 1 \quad \textit{or} \quad 0$$

according as

$$\int_c^\infty t^{-1}g(t) \mathrm{e}^{-g(t)^2/2}\mathrm{d}t < \infty \quad \textit{or} \quad = \infty. \tag{48.17}$$

Notice that the integral tests (48.16) and (48.17) are different. Thus, when we consider functions more delicate than $g(t) = (2(1+\varepsilon)\log\log t)^{1/2}$, they do not necessarily give the same results.

The behavior of $M_t = \sup_{0\leq s\leq t} X_s$ is not similar to that of X_t^*.

PROPOSITION 48.13. *Let $\{X_t\}$ be the Brownian motion on $\mathbb{R}$ and let $g(t)$ be in the class (48.15). Then*

$$P[\,M_t \geq t^{1/2}g(t) \text{ for all sufficiently large } t\,] = 1 \quad \textit{or} \quad 0$$

according as

$$\int_c^\infty t^{-1}g(t)\mathrm{d}t < \infty \quad \textit{or} \quad = \infty.$$

Strassen [**506**] made a beginning to a new development of the law of the iterated logarithm for random walks and the Brownian motion, bringing it to distributions on the path space. See Freedman's book [**147**], Stroock's book [**507**], de Acosta [**2**], and Bingham's review [**37**]. Proofs of Propositions 48.11 and 48.13 and related studies, including the random walk case and its connection with the result of Strassen, were made by Chung and Erdős [**82**], Chung [**78**], Sirao [**488**], Hirsch [**204**], Jain and Pruitt [**232, 233**], Jain and Taylor [**235**], and Csáki [**88, 89**]. For random walks $\{S_n\}$ on $\mathbb{R}$, Stone [**504**] studies $\limsup_{n\to\infty} n^{-1/2}S_n$, Kesten [**273**] works on limit points of $n^{-\alpha}S_n$ for $0 < \alpha \leq 1$, Pruitt [**397**] studies the choice of β_n to consider $\liminf_{n\to\infty} \beta_n{}^{-1}|S_n|$, and Kesten [**274**] and Pruitt [**394**] discuss many generalizations of the law of the iterated logarithm. Some other delicate long time behaviors of the Brownian motions on $\mathbb{R}^d$, $d \geq 2$, are studied by Adelman, Burdzy, and Pemantle [**5**] and Erdős and Révész [**129**].

49. Further factorization identities

Of the two factorization identities (45.9) and (45.10) for compound Poisson processes on $\mathbb{R}$, the first is extended to general Lévy processes on $\mathbb{R}$ in Theorem 45.7. In this section we prove that (45.10) also holds in general. This concerns the Laplace transform (in t) of the 3-variate joint distribution of M_t, $M_t - X_t$, and Λ_t. Then we will give another beautiful factorization identity dealing with the Laplace transform (in x) of the joint distribution of R_x and Γ_x, where $R_x = T_{(x,\infty)}$ is the first passage time process and Γ_x is the *overshoot* defined for $x \geq 0$ by

$$\Gamma_x = X_{R_x} - x \quad \text{whenever } R_x < \infty.$$

That is, we will prove the following two theorems.

THEOREM 49.1 (General Lévy process on $\mathbb{R}$). *For any $q > 0$, $p \geq 0$, $z \in \mathbb{R}$, and $w \in \mathbb{R}$,*

$$(49.1) \qquad q\int_0^\infty \mathrm{e}^{-qt} E[\mathrm{e}^{\mathrm{i}zM_t + \mathrm{i}w(X_t - M_t) - p\Lambda_t}]\mathrm{d}t$$
$$= \varphi_{q+p}^+(z)\varphi_q^-(w)\exp\left[\int_0^\infty t^{-1}\mathrm{e}^{-qt}(\mathrm{e}^{-pt} - 1)P[X_t > 0]\mathrm{d}t\right].$$

THEOREM 49.2 (General Lévy process on $\mathbb{R}$). *For any $q > 0$, $u > 0$, and $v > 0$ with $u \neq v$,*

$$(49.2) \qquad u\int_0^\infty \mathrm{e}^{-ux} E[\mathrm{e}^{-qR_x - v\Gamma_x}]\mathrm{d}x = \frac{u}{u-v}\left(1 - \frac{\varphi_q^+(\mathrm{i}u)}{\varphi_q^+(\mathrm{i}v)}\right).$$

The functions φ_{q+p}^+, φ_q^-, and φ_q^+ are those of Theorem 45.2 and Remark 45.3. Since the formulas (49.1) and (49.2) are invariant under the transfer to Lévy processes equivalent in law, we may and do assume in this section that $(\Omega, \mathcal{F}, P)$ and $\{X_t\}$ are as in Section 40 with $d = 1$. So we use the strong Markov property and the quasi-left-continuity. Define the following random quantities:

$$(49.3) \qquad R'_x = T_{[x,\infty)} = \inf\{t > 0\colon X_t \geq x\},$$
$$(49.4) \qquad R''_x = \inf\{t > 0\colon X_t \vee X_{t-} \geq x\},$$
$$(49.5) \qquad \Lambda'_t = \sup\{s \in [0,t]\colon X_s \vee X_{s-} = M_t\}$$

with the understanding that the empty set has infimum ∞ and that $X_{0-} = X_0$. The $\mathcal{F}$-measurability of R'_x follows from Theorem 40.13. For that of R''_x, Λ_t, and Λ'_t, see Exercise 50.3.

LEMMA 49.3. *Suppose that $\{X_t\}$ is either of type B with $R_0 = 0$ a. s. or of type C. Then, for every $t > 0$, M_t has a continuous distribution.*

Proof. Note that, if $\{X_t\}$ is of type C, then $R_0 = 0$ a.s. by Theorem 47.1. We have $P[\, M_t > 0 \text{ for } t > 0\,] = 1$. Hence $P[\, M_t = 0\,] = 0$ for $t > 0$. Let $t > 0$ and $x > 0$. We will prove that

$$P[\, M_t = x\,] = 0. \tag{49.6}$$

Let $I_1 = P[\, M_t = x,\, R''_x = t\,]$ and $I_2 = P[\, M_t = x,\, R''_x < t\,]$. We have $P[\, M_t = x\,] = I_1 + I_2$. Since

$$I_1 \le P[\, X_t = x \text{ or } X_{t-} = x\,] \le P[\, X_t = x\,] + P[\, X_t \ne X_{t-}\,],$$

$I_1 = 0$ by Theorem 27.4 and (1.10). Suppose that

$$P[\, R'_x = R''_x\,] = 1 \quad \text{for } x > 0. \tag{49.7}$$

Then $I_2 = P[\, R'_x < t,\, M_t = x\,] = P[\, R'_x < t,\, M_t = x,\, X_{R'_x} = x\,]$, which is bounded by the probability that $X_{R'_x} = x$ and $X_{R'_x + s} \le x$ for all sufficiently small $s \ge 0$. Hence $I_2 = 0$ by the strong Markov property and by $R_0 = 0$ a.s. The proof of (49.7) is as follows. By definition, $R'_x \ge R''_x$. We have

$$\begin{aligned} P[\, R''_x < R'_x\,] &\le P[\, t < R'_x \text{ and } X_{t-} \ge x,\, \exists t > 0\,] \\ &\le P[\, t < R'_x \text{ and } R'_{x-1/n} \le t,\, n = 1, 2, \ldots,\, \exists t > 0\,], \end{aligned}$$

which is 0, since, by the quasi-left-continuity and by $x > 0$,

$$\lim_{n\to\infty} R'_{x-1/n} = R'_x \quad \text{a.s. on } \left\{ \lim_{n\to\infty} R'_{x-1/n} < \infty \right\}. \tag{49.8}$$

Thus (49.7) is proved. □

LEMMA 49.4. *Suppose that $\{X_t\}$ is non-zero and not compound Poisson. Then, for any $t > 0$,*

$$P[\, \Lambda_t = \Lambda'_t\,] = 1. \tag{49.9}$$

Proof. Let $0 < s < t$. It is enough to prove that

$$P[\, \Lambda_t < s < \Lambda'_t\,] = 0, \tag{49.10}$$

because $P[\, \Lambda_t < \Lambda'_t\,] = P[\, \Lambda_t < s < \Lambda'_t,\, \exists s \in (0,t) \cap \mathbb{Q}\,]$. We divide the proof into three cases: 1. Either of type B with $R_0 = 0$ a.s. or of type C. 2. Of type A with drift $\gamma_0 > 0$. 3. Either of type A with $\gamma_0 < 0$ or of type B with $R_0 > 0$ a.s.

Case 1. We can use Lemma 49.3. Thus

$$\begin{aligned} P[\, \Lambda_t < s < \Lambda'_t\,] &\le P\Big[\, M_s - X_s = \sup_{s \le u \le t} (X_u - X_s)\Big] \\ &= \int P[(X_s, M_s) \in \mathrm{d}(x,y)]\, P[\, M_{t-s} = y - x\,], \end{aligned}$$

which is 0.

Case 2. The sample functions are right-continuous step functions with a linear function $\gamma_0 t$ added. Hence, for $x > 0$, almost surely on $\{R''_x <$

∞, $X_{R''_x-} = x\}$, we have $R'_{x-1/n} < R''_x$, and hence $R'_x \le R''_x$ by (49.8), which implies $R'_x = R''_x$ and $X_{R''_x-} = X_{R''_x}$. Here $X_{R''_x-}$ denotes $\lim_{u\downarrow 0} X_{R''_x-u}$. Thus

$$P[\, R''_x < \infty,\, X_{R''_x-} = x,\, X_{R''_x} \ne x\,] = 0. \tag{49.11}$$

Denote M_{t-s} and Λ'_{t-s} for the sample function $\{X_{s+u}(\omega) - X_s(\omega) \colon u \ge 0\}$ by $M^{(s)}_{t-s}(\omega)$ and $\Lambda'^{(s)}_{t-s}(\omega)$, respectively. Then

$$\begin{aligned} P[\,\Lambda_t < s < \Lambda'_t < t\,] &\le P[\,M_s - X_s = M^{(s)}_{t-s},\, \Lambda'^{(s)}_{t-s} < t-s\,] \\ &= \int P[(X_s, M_s) \in \mathrm{d}(x,y)]\, P[\,M_{t-s} = y-x,\, \Lambda'^{(s)}_{t-s} < t-s\,]. \end{aligned}$$

Recall that $P[\,M_{t-s} = 0\,] = 0$ by $\gamma_0 > 0$ and that, for $z > 0$, $P[\,M_{t-s} = z,\, \Lambda'^{(s)}_{t-s} < t-s\,] \le P[\,X_{R''_z-} = z,\, X_{R''_z} < z\,] = 0$ by (49.11). Hence $P[\,\Lambda_t < s < \Lambda'_t < t\,] = 0$. Next we claim that

$$P[\,\Lambda_t < s < \Lambda'_t = t\,] = 0. \tag{49.12}$$

Let $J_0 = 0$ and let $J_1 < J_2 < \cdots$ be the jumping times of $\{X_t\}$. Denote by $\{Z_u \colon 0 \le u < t\}$ the process obtained by time reversal of Proposition 41.8. That is, $Z_u = X_{(t-u)-} - X_{t-}$. Then, in our case, we get

$$\begin{aligned} P[\,\Lambda_t < s < \Lambda'_t = t\,] &\le P[\,Z_u = 0 > Z_{u-},\, \exists u \in (0,t)\,] \\ &= P[\,X_u = 0 < X_{u-},\, \exists u \in (0,t)\,] \le P[\,X_{J_k} = 0,\, \exists k \ge 1\,]. \end{aligned}$$

We have $P[\,X_{J_1} = x\,] = \int_0^\infty \mathrm{e}^{-\nu(\mathbb{R})u} \nu\{x - \gamma_0 u\}\mathrm{d}u = 0$ for $x \in \mathbb{R}$, because $\nu\{x - \gamma_0 u\} = 0$ except for a countable number of u. Since $\{X_{J_k}\}$ is a random walk, we have, for $k \ge 2$,

$$P[X_{J_k} = 0] = P[X_{J_k} - X_{J_{k-1}} = -X_{J_{k-1}}] = E\big[\big(P[X_{J_1} = x]\big)_{x = -X(J_{k-1})}\big] = 0.$$

Therefore (49.12) and hence (49.10) are true.

Case 3. We have $T_{(-\infty,0)} = 0$ a. s. in this case. We use $\{Z_u \colon 0 \le u < t\}$ obtained by time reversal again. Define $Z_t = -X_{t-}$. Then $\{Z_u \colon 0 \le u \le t\}$ is identical in law with the restriction to $[0,t]$ of a Lévy process either of type A with drift positive or of type B with $R_0 = 0$. Denote the quantities for $\{Z_u \colon 0 \le u \le t\}$ by putting the superscript Z. Then, using (1.10),

$$P[\,\Lambda_t < \Lambda'_t\,] = P[\,\Lambda_t < \Lambda'_t,\, X_t = X_{t-}\,] = P[\,\Lambda^Z_t < \Lambda'^Z_t\,],$$

which is 0 by Cases 1 and 2. □

Proof of Theorem 49.1. If $\{X_t\}$ is a zero process, (49.1) is evident. If $\{X_t\}$ is a compound Poisson process, then it is proved in Theorem 45.5. Let us consider the remaining case. We approximate $\{X_t\}$ by a sequence of compound Poisson processes $\{X^n_t\}$, $n = 1, 2, \ldots$, using Lemma 45.12. Denote the quantities M_t, Λ_t, φ^+_{q+p}, and φ^-_q related to $\{X^n_t\}$ by M^n_t, Λ^n_t, $\varphi^+_{q+p}(n,\cdot)$, and $\varphi^-_q(n,\cdot)$, respectively. Since X^n_t tends to X_t uniformly in t in any bounded interval almost surely, we have $P[\,\lim_{n\to\infty} M^n_t = M_t$ for any $t\,] = 1$

and, further, $P[\lim_{n\to\infty}\Lambda_t^n=\Lambda_t]=1$ for any t. Indeed, since $\Lambda_t=\Lambda'_t$ a. s. by Lemma 49.4, for any $\delta>0$, we have $P[\Lambda_t<\delta$ or $\sup_{s\in[0,\Lambda_t-\delta]}X_s<M_t]=1$, $P[\Lambda_t>t-\delta$ or $\sup_{s\in[\Lambda_t+\delta,t]}X_s<M_t]=1$, and hence $P[\Lambda_t-\delta\leq\Lambda_t^n\leq\Lambda_t+\delta$ for all large $n]=1$. Consequently,

$$q\int_0^\infty e^{-qt}E\left[e^{izM_t^n+iw(X_t^n-M_t^n)-p\Lambda_t^n}\right]dt$$

tends to the left-hand side of (49.1) as $n\to\infty$. Now suppose that the approximate sequence $\{X_t^n\}$ is the one constructed in the proof of Lemma 45.12. Then, $\varphi_{q+p}^+(n,z)$ and $\varphi_q^-(n,w)$ tend to $\varphi_{q+p}^+(z)$ and $\varphi_q^-(w)$, respectively, as is shown in the proof of Theorem 45.7 (at end of Section 45). To finish the proof of (49.1), it suffices to verify that $I_n\to I$, where

$$I=\int_0^\infty t^{-1}e^{-qt}(1-e^{-pt})P[X_t>0]dt$$

and I_n is defined by replacing X_t by X_t^n. We have

$$\begin{aligned}P[X_t>0]&\leq P\left[\liminf_{n\to\infty}\{X_t^n>0\}\right]\leq\liminf_{n\to\infty}P[X_t^n>0]\\&\leq\limsup_{n\to\infty}P[X_t^n>0]\leq P\left[\limsup_{n\to\infty}\{X_t^n>0\}\right]\leq P[X_t\geq 0].\end{aligned}$$

Hence

$$I\leq\liminf_{n\to\infty}I_n\leq\limsup_{n\to\infty}I_n\leq\int_0^\infty t^{-1}e^{-qt}(1-e^{-pt})P[X_t\geq 0]dt. \tag{49.13}$$

We have $\int_0^\infty e^{-rt}P[X_t=0]dt=0$ for $r>0$, noting that V^r is continuous by Theorem 27.4, since rV^r is an infinitely divisible distribution with infinite Lévy measure by Theorem 30.10(ii). Thus the extreme right member of (49.13) equals I. Hence $I_n\to I$. □

REMARK 49.5. A probability measure $\mu_{1/2}$ on $[0,1]$ is the *arcsine distribution* if

$$\mu_{1/2}[0,x]=\tfrac{2}{\pi}\arcsin\sqrt{x}=\tfrac{1}{\pi}\int_0^x y^{-1/2}(1-y)^{-1/2}dy\quad\text{for }0\leq x\leq 1.$$

For $0<a<1$, a probability measure μ_a on $[0,1]$ is called the *generalized arcsine distribution* with parameter a if

$$\mu_a[0,x]=\tfrac{\sin a\pi}{\pi}\int_0^x y^{a-1}(1-y)^{-a}dy\quad\text{for }0\leq x\leq 1.$$

For a Lévy process $\{X_t\}$ on $\mathbb{R}$ let $D_t=\int_0^t 1_{(0,\infty)}(X_s)ds$, the time spent on the positive axis during the time interval $[0,t]$. If $\{X_t\}$ is the Brownian motion, it is proved by Lévy [**313**] that $t^{-1}D_t$ has distribution $\mu_{1/2}$ for all $t>0$. The same result is shown by Kac [**247**] when $\{X_t\}$ is symmetric and stable. If

$$P[X_t>0]=a\qquad\text{for }t>0, \tag{49.14}$$

for some $a\in(0,1)$ which is independent of t, then $t^{-1}D_t$ has distribution μ_a for all $t>0$. In fact, if $\{X_t\}$ is not compound Poisson, we can prove that $t^{-1}D_t\overset{d}{=}t^{-1}\Lambda_t$ and then apply Theorem 49.1 with $z=w=0$ to get this result.

The condition (49.14) holds if $\{X_t\}$ is either symmetric with $P[X_t = 0] = 0$ or strictly stable. See Exercise 18.10 for evaluation of a in the strictly stable case. Getoor and Sharpe [**163**] and Bertoin [**26**] show the following. As $t \to \infty$, the following conditions are equivalent:

(1) $t^{-1} \int_0^t P[X_s > 0]ds$ tends to $a \in [0, 1]$;
(2) the distribution of $t^{-1}D_t$ converges;
(3) the distribution of $t^{-1}\Lambda_t$ converges.

If (2) or (3) holds, the limit distribution is μ_a for $0 < a < 1$, δ_0 for $a = 0$, and δ_1 for $a = 1$. The same assertion is true when $t \to 0$ in place of $t \to \infty$. The condition (1) is an analogue of Spitzer's condition in [**494**] for random walks. Bertoin and Doney [**29**] prove that (1) is equivalent to $P[X_t > 0] \to a$. Other necessary and sufficient conditions are known. Not all Lévy processes satisfy the condition (1); see [**26**].

Let us prove Theorem 49.2. We need the following fact.

LEMMA 49.6. *Suppose that $\{X_t\}$ is non-zero and not compound Poisson. Then, for any $x > 0$, $P[\,R_x = R'_x = R''_x\,] = 1$.*

Proof. It follows from the definition that $R''_x \le R'_x \le R_x$. Let $x > 0$. The proof of $P[\,R'_x = R''_x\,] = 1$ in (49.7) works for any Lévy process on $\mathbb{R}$. So we have only to show that

$$P[\,R'_x < R_x\,] = 0. \tag{49.15}$$

Let $R_x^{(R'_x)}$ be the R_x for $\{X_{R'_x+t} - X_{R'_x} : t \ge 0\}$. If $R_0 = 0$ a. s., then, by the strong Markov property,

$$P[\,R'_x < R_x\,] = P[\,R'_x < \infty,\ X_{R'_x} = x,\ R_0^{(R'_x)} > 0\,] = 0.$$

If $\{X_t\}$ is of type A and $\gamma_0 < 0$, then, using the jumping times $J_1 < J_2 < \cdots$ of $\{X_t\}$, we have

$$P[\,R'_x < R_x\,] \le P[\,R'_x < \infty,\ X_{R'_x-} < X_{R'_x} = x\,] \le P[\,X_{J_k} = x,\ \exists k \ge 1\,],$$

which is 0 by the argument in Case 2 in the proof of Lemma 49.4. Next, assume that $\{X_t\}$ is of type B or C. We will show (49.15) in this case, which will finish the proof. Notice that

$$P[\,R'_x < R_x\,] \le P[\,M_t = x \text{ and } \Lambda_t < t,\ \exists t \in \mathbb{Q} \cap (0, \infty)\,].$$

As in the proof of Lemma 49.4 we use the process $\{Z_s,\ 0 \le s < t\}$ obtained by time reversal in Proposition 41.8. That is, $Z_s = X_{(t-s)-} - X_{t-}$. Define $Z_t = -X_{t-}$. Then, by Lemma 49.4 and (1.10),

$$\begin{aligned} P[\,M_t = x,\ \Lambda_t < t\,] &= P[\,M_t = x,\ 0 < \Lambda_t = \Lambda'_t < t,\ X_t = X_{t-}\,] \\ &= P[\,M_t^Z = x + Z_t,\ 0 < \Lambda_t^Z = \Lambda_t'^Z < t\,] \end{aligned}$$

for $t > 0$. For any s with $0 < s < t$, we have

$$P[\,0 < \Lambda_t^Z < s,\ M_t^Z = x + Z_t\,] \le P[\,Z_t = M_s^Z - x\,]$$

$$= P[\,Z_t - Z_s = M_s^Z - Z_s - x\,]$$
$$= \int P[(Z_s, M_s^Z) \in d(y,z)]\, P[\,Z_{t-s} = z - y - x\,],$$

which is 0, since processes of type B or C have continuous distributions at any positive time. Hence $P[\,M_t = x,\, \Lambda_t < t\,] = 0$. Thus we have (49.15). □

Proof of Theorem 49.2. Let $x > 0$ and $z \geq 0$. If $R_x < \infty$ and $\Gamma_x > z$, then $R_{x+z} = R_x$. If $R_x < \infty$ and $\Gamma_x \leq z$, then $R_{x+z} = R_x + R^{(R_x)}_{x+z-X(R_x)} = R_x + R^{(R_x)}_{z-\Gamma_x}$, where the superscript (R_x) indicates quantities for the sample function $\{X_{R_x+s} - X_{R_x},\, s \geq 0\}$. The strong Markov property shows that

$$(49.16)\qquad E[\mathrm{e}^{-qR_{x+z}}] = I_1 + I_2,$$
$$I_1 = E[\mathrm{e}^{-qR_{x+z}};\, \Gamma_x > z\,] = E[\mathrm{e}^{-qR_x};\, \Gamma_x > z\,],$$
$$I_2 = E[\mathrm{e}^{-qR_{x+z}};\, \Gamma_x \leq z\,] = E\big[\mathrm{e}^{-qR_x}\big(E[\mathrm{e}^{-qR_{z-y}}]\big)_{y=\Gamma_x};\, \Gamma_x \leq z\,\big].$$

We define $R_y = 0$ for $-\infty < y < 0$. Then

$$(49.17)\qquad I_1 + I_2 = I_1 + \int_{[0,z]} E[\mathrm{e}^{-qR_{z-y}}]\, E[\mathrm{e}^{-qR_x};\, \Gamma_x \in \mathrm{d}y]$$
$$= \int_{[0,\infty)} E[\mathrm{e}^{-qR_{z-y}}]\, E[\mathrm{e}^{-qR_x};\, \Gamma_x \in \mathrm{d}y].$$

Assume that $\{X_t\}$ is non-trivial and not compound Poisson. It is convenient to use a random variable τ_q exponentially distributed with parameter q such that $\{X_t\}$ and τ_q are independent. Such τ_q exists if the probability space is appropriately enlarged. We can prove, for $x > 0$,

$$(49.18)\qquad E[\mathrm{e}^{-qR_x}] = P[\,M_{\tau_q} > x\,].$$

Indeed, we have $R_x = R''_x$ by Lemma 49.6 and, in general, $R''_x \leq t$ if and only if $M_t \geq x$. Since $R_x > 0$, we have

$$E[\mathrm{e}^{-qR_x}] = \int_{(0,\infty)} \mathrm{e}^{-qt}\mathrm{d}_t P[\,R''_x \leq t\,] = \int_{(0,\infty)} \mathrm{e}^{-qt}\mathrm{d}_t P[\,M_t \geq x\,]$$
$$= q\int_0^\infty \mathrm{e}^{-qt} P[\,M_t \geq x\,]\mathrm{d}t = P[\,M_{\tau_q} \geq x\,]$$

and, noting that R_x is right-continuous in x, we get (49.18). Actually we see that $E[\mathrm{e}^{-qR_x}]$ is continuous in $x > 0$. Consider the Laplace transforms with respect to z of the Stieltjes signed measures defined by (49.16) and (49.17). First,

$$(49.19)\qquad \int_{[0,\infty)} \mathrm{e}^{-vz}\mathrm{d}_z E[\mathrm{e}^{-qR_{x+z}}] = \int_{(0,\infty)} \mathrm{e}^{-vz}\mathrm{d}_z P[\,M_{\tau_q} > x + z\,]$$

$$= -\int_{(0,\infty)} \mathrm{e}^{-vz} \mathrm{d}_z P[\, M_{\tau_q} \le x + z\,] = -E[\mathrm{e}^{-v(M(\tau_q)-x)};\ M_{\tau_q} > x\,].$$

Second, since the Laplace transform of the convolution of two bounded signed measures is the product of their Laplace transforms,

$$\int_{[0,\infty)} \mathrm{e}^{-vz} \mathrm{d}_z (I_1 + I_2) = J_1 J_2 \tag{49.20}$$

with

$$J_1 = \int_{[0,\infty)} \mathrm{e}^{-vz} \mathrm{d}_z E[\mathrm{e}^{-qR_z}], \qquad J_2 = \int_{[0,\infty)} \mathrm{e}^{-vz} E[\mathrm{e}^{-qR_x};\ \Gamma_x \in \mathrm{d}z].$$

We have

$$\begin{aligned} J_1 &= E[\mathrm{e}^{-qR_0}] - 1 + \int_{(0,\infty)} \mathrm{e}^{-vz} \mathrm{d}_z E[\mathrm{e}^{-qR_z}] \\ &= E[\mathrm{e}^{-qR_0}] - 1 - E[\mathrm{e}^{-vM(\tau_q)};\ M_{\tau_q} > 0\,] = -E[\mathrm{e}^{-vM(\tau_q)}]. \end{aligned} \tag{49.21}$$

The second equality here is obtained similarly to (49.19). The last equality is checked as follows. Notice that $E[\mathrm{e}^{-vM_t};\ M_t = 0] = P[\, M_t = 0\,]$ is bounded by $P[\, R_0 \ge t\,]$ and $P[\, R_0 > t\,]$ from above and below, and that $\int_0^\infty \mathrm{e}^{-qt} P[\, R_0 > t]\mathrm{d}t = \int_0^\infty \mathrm{e}^{-qt} P[\, R_0 \ge t]\mathrm{d}t$, since $P[\, R_0 = t\,] = 0$ except at a countable number of times t. Then

$$\begin{aligned} q\int_0^\infty \mathrm{e}^{-qt} E[\mathrm{e}^{-vM_t};\ M_t = 0]\mathrm{d}t &= q\int_0^\infty \mathrm{e}^{-qt} P[\, R_0 > t\,]\mathrm{d}t \\ &= 1 - q\int_0^\infty \mathrm{e}^{-qt} P[\, R_0 \le t\,]\mathrm{d}t \\ &= 1 - P[\, R_0 = 0\,] - \int_0^\infty \mathrm{e}^{-qt} \mathrm{d}_t P[\, R_0 \le t\,] = 1 - E[\mathrm{e}^{-qR_0}], \end{aligned}$$

and we get the last equality in (49.21). Now (49.16), (49.19), (49.20), and (49.21) together give

$$E[\mathrm{e}^{-v(M(\tau_q)-x)};\ M_{\tau_q} > x] = E[\mathrm{e}^{-vM(\tau_q)}]\, E[\mathrm{e}^{-qR_x - v\Gamma_x}]. \tag{49.22}$$

Multiply this by $u\mathrm{e}^{-ux}$ and integrate over x. The left-hand side becomes

$$\begin{aligned} u\int_0^\infty & \mathrm{e}^{-ux} E[\mathrm{e}^{-v(M(\tau_q)-x)};\ M_{\tau_q} > x]\mathrm{d}x \\ &= u\int_0^\infty \mathrm{e}^{-ux}\mathrm{d}x \int_{(z,\infty)} \mathrm{e}^{-v(y-x)} P[\, M_{\tau_q} \in \mathrm{d}y\,] \\ &= u\int_{(0,\infty)} \mathrm{e}^{-vy} P[\, M_{\tau_q} \in \mathrm{d}y\,] \int_0^y \mathrm{e}^{(v-u)x}\mathrm{d}x \\ &= \frac{u}{u-v} \int_{(0,\infty)} (\mathrm{e}^{-vy} - \mathrm{e}^{-uy}) P[\, M_{\tau_q} \in \mathrm{d}y\,] \end{aligned}$$

$$= \frac{u}{u-v}(E[\mathrm{e}^{-vM(\tau_q)}] - E[\mathrm{e}^{-uM(\tau_q)}])$$

for $u \neq v$. The extension of (45.22) with $w = 0$ to the upper half plane gives

$$E[\mathrm{e}^{-uM(\tau_q)}] = q\int_0^\infty \mathrm{e}^{-qt}E[\mathrm{e}^{-uM_t}]\mathrm{d}t = \varphi_q^+(\mathrm{i}u) \tag{49.23}$$

for $q > 0$ and $u > 0$. Thus we get (49.2) in the case where $\{X_t\}$ is non-trivial and not compound Poisson.

If $\{X_t\}$ is trivial, then (49.2) is obvious. Consider the case where $\{X_t\}$ is compound Poisson. Let $X_t^n(\omega) = X_t(\omega) + n^{-1}t$ for $n = 1, 2, \ldots$. Then we have

$$u\int_0^\infty \mathrm{e}^{-ux}E[\mathrm{e}^{-qR_x^n - v\Gamma_x^n}]\mathrm{d}x = \frac{u}{u-v}\left(1 - \frac{\varphi_q^+(n,\mathrm{i}u)}{\varphi_q^+(n,\mathrm{i}v)}\right), \tag{49.24}$$

where R_x^n, Γ_x^n, and $\varphi_q^+(n,\cdot)$ are the quantities related to $\{X_t^n(\omega)\}$. We have $M_t(\omega) \leq M_t^n(\omega) \leq M_t(\omega) + n^{-1}t$. Hence it follows from (49.23) that $\varphi_q^+(n,\mathrm{i}u) \to \varphi_q^+(\mathrm{i}u)$ for $u > 0$ as $n \to \infty$. Recall that $X_t(\omega)$ is a right-continuous step function for each ω. If x does not belong to the range $\{X_t(\omega)\colon t \in [0,\infty)\}$, then $\sup_{t<R_x(\omega)} X_t(\omega) < x$ and hence $R_x^n(\omega) = R_x(\omega)$ for sufficiently large n and $\Gamma_x^n(\omega) \to \Gamma_x(\omega)$ as $n \to \infty$. Therefore

$$\int_0^\infty \mathrm{e}^{-ux - qR_x^n(\omega) - v\Gamma_x^n(\omega)}\mathrm{d}x \to \int_0^\infty \mathrm{e}^{-ux - qR_x(\omega) - v\Gamma_x(\omega)}\mathrm{d}x$$

and the left-hand side of (49.24) tends to $u\int_0^\infty \mathrm{e}^{-ux}E[\mathrm{e}^{-qR_x - v\Gamma_x}]\mathrm{d}x$. Hence we get (49.2). □

COROLLARY 49.7. *Suppose that $R_0 > 0$ a. s. Then, for $q > 0$ and $v > 0$,*

$$E[\mathrm{e}^{-qR_0 - v\Gamma_0}] = 1 - \exp\left[-\int_0^\infty t^{-1}\mathrm{e}^{-qt}\mathrm{d}t\int_{(0,\infty)} \mathrm{e}^{-vx}\mu^t(\mathrm{d}x)\right], \tag{49.25}$$

$$E[\mathrm{e}^{-qR_0}] = 1 - \exp\left[-\int_0^\infty t^{-1}\mathrm{e}^{-qt}P[\,X_t > 0\,]\mathrm{d}t\right], \tag{49.26}$$

$$\Gamma_0 > 0 \quad \text{a. s. on } \{R_0 < \infty\}. \tag{49.27}$$

Proof. Since $R_x \to R_0$ and $\Gamma_x \to \Gamma_0$ as $x \downarrow 0$,

$$u\int_0^\infty \mathrm{e}^{-ux}E[\mathrm{e}^{-qR_x - v\Gamma_x}]\mathrm{d}x = \int_0^\infty \mathrm{e}^{-x}E[\mathrm{e}^{-qR_{x/u} - v\Gamma_{x/u}}]\mathrm{d}x \to E[\mathrm{e}^{-qR_0 - v\Gamma_0}]$$

as $u \to \infty$. We have

$$\frac{\varphi_q^+(\mathrm{i}u)}{\varphi_q^+(\mathrm{i}v)} = \exp\left[\int_0^\infty t^{-1}\mathrm{e}^{-qt}\mathrm{d}t\int_{(0,\infty)} (\mathrm{e}^{-ux} - \mathrm{e}^{-vx})\mu^t(\mathrm{d}x)\right].$$

Since $\int_0^1 t^{-1}P[\,X_t > 0\,]dt < \infty$ by Theorem 47.2, we can use the dominated convergence theorem to obtain

$$\frac{\varphi_q^+(\mathrm{i}u)}{\varphi_q^+(\mathrm{i}v)} \to \exp\Big[-\int_0^\infty t^{-1}\mathrm{e}^{-qt}\mathrm{d}t\int_{(0,\infty)}\mathrm{e}^{-vx}\mu^t(\mathrm{d}x)\Big]$$

as $u\to\infty$. Hence (49.25). Letting $v\downarrow 0$ in (49.25), we get (49.26). The limit of (49.25) as $v\uparrow\infty$ gives $E[\mathrm{e}^{-qR_0};\,\Gamma_0=0]=0$, that is, (49.27). □

Greenwood and Pitman [**171**] noticed that Theorem 49.1 shows the fact that we will state in Theorem 49.8. They proved this theorem directly, using the local time and the excursions of the reflecting process $\{Y_t\}$, and derived Theorems 45.7 and 49.1 in the converse direction, thus avoiding the Wiener–Hopf argument in analytic functions.

THEOREM 49.8. *Let τ_q be a random variable exponentially distributed with parameter $q>0$ and independent of $\{X_t\}$. Then, the two $\mathbb{R}^2$-valued random variables $(\Lambda_{\tau_q}, M_{\tau_q})$ and $(\tau_q-\Lambda_{\tau_q}, X_{\tau_q}-M_{\tau_q})$ are independent.*

Proof. Write $\tau=\tau_q$. Let $\xi,\eta,z,w\in\mathbb{R}$ satisfy $q+\eta>0$ and $\xi-\eta\geq 0$. By Theorem 49.1,

$$\begin{aligned}
&q\int_0^\infty \mathrm{e}^{-qt}E[\mathrm{e}^{-\xi\Lambda_t+\mathrm{i}zM_t-\eta(t-\Lambda_t)+\mathrm{i}w(X_t-M_t)}]dt\\
&=q\int_0^\infty \mathrm{e}^{-(q+\eta)t}E[\mathrm{e}^{\mathrm{i}zM_t+\mathrm{i}w(X_t-M_t)-(\xi-\eta)\Lambda_t}]dt\\
&=\frac{q}{q+\eta}\varphi_{q+\xi}^+(z)\varphi_{q+\eta}^-(w)\exp\Big[\int_0^\infty t^{-1}\mathrm{e}^{-(q+\eta)t}(\mathrm{e}^{-(\xi-\eta)t}-1)P[X_t>0]\mathrm{d}t\Big].
\end{aligned}$$

Since the integral under the exponential sign is split into

$$\begin{aligned}
&\int_0^1 t^{-1}(1-\mathrm{e}^{-(q+\eta)t})P[X_t>0]\mathrm{d}t-\int_0^1 t^{-1}(1-\mathrm{e}^{-(q+\xi)t})P[X_t>0]\mathrm{d}t\\
&\quad+\int_1^\infty t^{-1}\mathrm{e}^{-(q+\xi)t}P[X_t>0]\mathrm{d}t-\int_1^\infty t^{-1}\mathrm{e}^{-(q+\eta)t}P[X_t>0]\mathrm{d}t,
\end{aligned}$$

there exist two functions $f(z,\xi)$ and $g(w,\eta)$ such that

$$E[\mathrm{e}^{-\xi\Lambda_\tau+\mathrm{i}zM_\tau-\eta(\tau-\Lambda_\tau)+\mathrm{i}w(X_\tau-M_\tau)}]=f(z,\xi)g(w,\eta).$$

Consequently,

$$\begin{aligned}
E[\mathrm{e}^{-\xi\Lambda_\tau+\mathrm{i}zM_\tau}]&=f(z,\xi)g(0,0)\quad\text{if }\xi\geq 0,\ z\in\mathbb{R},\\
E[\mathrm{e}^{-\eta(\tau-\Lambda_\tau)+\mathrm{i}w(X_\tau-M_\tau)}]&=f(0,0)g(w,\eta)\quad\text{if }-q<\eta\leq 0,\ w\in\mathbb{R},
\end{aligned}$$

and $1=f(0,0)g(0,0)$. Hence

$$\begin{aligned}
&E[\mathrm{e}^{-\xi\Lambda_\tau+\mathrm{i}zM_\tau-\eta(\tau-\Lambda_\tau)+\mathrm{i}w(X_\tau-M_\tau)}]\\
&\qquad=E[\mathrm{e}^{-\xi\Lambda_\tau+\mathrm{i}zM_\tau}]\,E[\mathrm{e}^{-\eta(\tau-\Lambda_\tau)+\mathrm{i}w(X_\tau-M_\tau)}]
\end{aligned}\tag{49.28}$$

for $\xi \geq 0$, $-q < \eta \leq 0$, $z \in \mathbb{R}$, and $w \in \mathbb{R}$. If we fix $-q < \eta \leq 0$, $z \in \mathbb{R}$, and $w \in \mathbb{R}$, then (49.28) holds for $\xi \in \mathbb{C}$ with $\mathrm{Re}\,\xi \geq 0$, since both sides are continuous there and analytic on $\{\mathrm{Re}\,\xi > 0\}$. Next, fixing z, w, and ξ, we see that (49.28) holds for $\eta \in \mathbb{C}$ with $\mathrm{Re}\,\eta > -q$, since both sides are analytic there. Thus, with $\xi = -\mathrm{i}\xi'$ and $\eta = \mathrm{i}\eta'$, (49.28) is true whenever ξ', η', z, and w are real. That is, (Λ_τ, M_τ) and $(\tau - \Lambda_\tau, X_\tau - M_\tau)$ have the joint distribution whose characteristic function is the product of their characteristic functions. Hence they are independent. □

REMARK 49.9. In the following let $\{X_t\}$ be of type B or C such that the set Σ_0 in Definition 43.1 contains $(0, \infty)$; Theorem 43.21 tells us in what cases this condition is satisfied. Millar [**348**] finds an analytic condition that $P^0[\,\Gamma_x = 0\,] > 0$. Let

$$K(y) = \lim_{q \downarrow 0} \int_{-\infty}^{\infty} (1 - \cos yz) \mathrm{Re}\left(\tfrac{1}{q - \psi(z)}\right) \mathrm{d}z,$$
$$I_+ = \int_0^1 K(y)\, \nu(y, 1) \mathrm{d}y, \qquad I_- = \int_{-1}^0 K(y)\, \nu(-1, y) \mathrm{d}y.$$

Millar's results are as follows. $P^0[\,\Gamma_x = 0\,] > 0$ for all $x > 0$ if $I_+ < \infty$; $P^0[\,\Gamma_x = 0\,] = 0$ for all $x > 0$ if $I_+ = \infty$. In particular, the latter case occurs if the process is symmetric with $A = 0$ or if it is an α-stable process, $0 < \alpha < 2$, with $\nu(0, \infty) > 0$. Bertoin [**26**], p. 174, calls $P^0[\,\Gamma_x = 0\,]$ the probability that the process creeps across x.

When $\{X_t\}$ hits a point x, the behavior of the path immediately before the hitting time T_x is described as follows. For $x > 0$ let $\Omega_x = \{T_x < \infty\}$ and

$$\Omega_x^+ = \Omega_x \cap \{\exists \varepsilon > 0 \text{ such that } X_t < x \ \forall t \in [T_x - \varepsilon, T_x)\},$$
$$\Omega_x^- = \Omega_x \cap \{\exists \varepsilon > 0 \text{ such that } X_t > x \ \forall t \in [T_x - \varepsilon, T_x)\},$$
$$\Omega_x^\pm = \Omega_x \cap \{\exists t_n \uparrow T_x \text{ and } \exists s_n \uparrow T_x \text{ such that } X_{t_n} < x < X_{s_n}\}.$$

Then Ω_x^+, Ω_x^-, and $\Omega_x^\pm$ are disjoint and their union is Ω_x. Ikeda and Watanabe [**214**] show the following. If $A > 0$, then $P^0[\Omega_x^+ \cup \Omega_x^- \mid \Omega_x] = 1$ for $x > 0$. If $A > 0$ and $\nu(0, \infty) = 0$, then $P^0[\Omega_x^+ \mid \Omega_x] = 1$ for $x > 0$. If $A > 0$ and $\nu(0, \infty) > 0$, then $P^0[\Omega_x^+ \mid \Omega_x] > 0$ and $P^0[\Omega_x^- \mid \Omega_x] > 0$ for $x > 0$.

The following facts are obtained from Millar's results.

$$I_+ < \infty \text{ and } I_- < \infty \text{ if and only if } A > 0,$$
$$I_+ < \infty \text{ and } I_- = \infty \text{ if and only if } P^0[\Omega_x^+ \mid \Omega_x] = 1 \text{ for } x > 0,$$
$$I_+ = \infty \text{ and } I_- < \infty \text{ if and only if } P^0[\Omega_x^- \mid \Omega_x] = 1 \text{ for } x > 0,$$
$$I_+ = \infty \text{ and } I_- = \infty \text{ if and only if } P^0[\Omega_x^\pm \mid \Omega_x] = 1 \text{ for } x > 0.$$

Consult Ikeda and Watanabe [**214**] and Bertoin [**26**], p. 175. A related paper is Takada [**511**].

A more delicate problem is whether a Lévy process $\{X_t\}$ on $\mathbb{R}$ has increase times as defined in Remark 5.10. The probability that the sample function has increase times is 0 or 1. The problem was studied by Bertoin [**23**] first for processes without positive jumps. A necessary and sufficient condition for a. s. existence of increase times is obtained by Doney [**102**]. It is known that a strictly stable process $\{X_t\}$ has increase times a. s. if and only if $P^0[X_t > 0] > 1/2$ for $t > 0$. See Bertoin [**25, 26**]. Hence, by Remark 14.20 and Exercise 18.10, a strictly stable process has increase times a. s. if and only if the parameters in Definition 14.16 satisfy one of the following three conditions: (a) $1 < \alpha < 2$ and $\beta < 0$; (b) $\alpha = 1$ and $\tau > 0$; (c) $0 < \alpha < 1$ and $\beta > 0$.

At this point, as additional remarks, we mention results on some sample function properties that we have not discussed so far.

REMARK 49.10. For $t > 0$ let $\Theta_t(\omega)$ be the range of the path in the time interval $[0, t]$, for a Lévy process $\{X_t\}$ on $\mathbb{R}^d$, defined as in Lemma 43.25. Let $\Theta(\omega)$ be the range of the path, that is, $\Theta(\omega) = \bigcup_{t>0} \Theta_t(\omega)$. Then $\mathrm{Leb}\,\Theta = 0$ a. s. if and only if $\mathrm{Leb}\,\Sigma_0 = 0$, where Σ_0 is defined in Definition 43.1. Indeed, $E[\mathrm{Leb}\,\Theta] = \int_{\mathbb{R}^d} P[T_{\{x\}} < \infty]dx$. Blumenthal and Getoor [**42**] shows that $\dim_{\mathrm{H}} \Theta(\omega)$, the Hausdorff dimension of $\Theta(\omega)$, equals $\alpha \wedge d$ a. s. for any nondegenerate α-stable process on $\mathbb{R}^d$. For $d = 1 < \alpha$ one can prove that, moreover, $\Theta(\omega) = \mathbb{R}$ a. s. The $\dim_{\mathrm{H}} \Theta(\omega)$ is studied by Horowitz [**207**] for subordinators and Pruitt [**392**] for general Lévy processes. The latter shows that $\dim_{\mathrm{H}} \Theta(\omega) = \gamma_L$ a. s., where

$$\gamma_L = \sup\left\{\eta \geq 0\colon \limsup_{r\downarrow 0} r^{-\eta} \textstyle\int_0^1 P[|X_t| \leq r]dt < \infty\right\},$$

another analogue of the index of a stable process. Its relation to the parameters defined by (47.36) and (47.37) is that $\gamma_L \leq \delta_L$ in general and that $\gamma_L = \delta_L$ in the case of subordinators. A refinement of results on $\dim_{\mathrm{H}} \Theta(\omega)$ is to find the correct measure function for $\Theta_t(\omega)$, that is, the measure function that makes the corresponding Hausdorff measure (see Rogers [**417**] for definition) of $\Theta_t(\omega)$ finite and non-zero. The correct measure function in the case of the Brownian motion on $\mathbb{R}^d$ is found by Lévy [**316**] and Ciesielski and Taylor [**86**] for $d \geq 3$ and by Taylor [**524**] for $d = 2$. Generalization to stable processes is made by Ray [**412**], Pruitt and Taylor [**399, 402**], and Fristedt [**150**]. Further, Fristedt and Pruitt [**152**] show how to determine the correct measure function for general subordinators.

A notion akin to the Hausdorff measure is packing measure. See Taylor and Tricot [**530**] for its definition and for the correct measure function, in the sense of the packing measure, for the Brownian paths for $d \geq 3$. Their result is closely connected with the evaluation of the sojourn time $S_\varepsilon = \int_0^\infty 1_{B(\varepsilon)}(X_t)dt$ for small ε, where $B(\varepsilon) = \{x\colon |x| < \varepsilon\}$. They show that

$$\liminf_{\varepsilon\downarrow 0} \frac{\log\log(1/\varepsilon)}{\varepsilon^2} S_\varepsilon = \frac{1}{2} \quad \text{a. s.}$$

for the Brownian motion for $d \geq 3$. Gruet and Shi [**176**] refine this result in the form of integral test. The packing dimension $\dim_{\mathrm{p}}$ is defined from packing measures; it is greater than or equal to the Hausdorff dimension. It is known that, for any Lévy process on $\mathbb{R}^d$, $\dim_{\mathrm{p}} \Theta(\omega) = \gamma'_L$ a. s., where

$$\gamma'_L = \sup\left\{\eta \geq 0\colon \liminf_{r\downarrow 0} r^{-\eta} \textstyle\int_0^1 P[|X_t| \leq r]dt < \infty\right\}.$$

See Pruitt and Taylor [**405**] for the study of γ'_L.

REMARK 49.11. A sample function $X_t(\omega)$ is said to have a *multiple point of multiplicity* n, or n-multiple point, if there are $t_1 < \cdots < t_n$ such that $X_{t_1}(\omega) = \cdots = X_{t_n}(\omega)$. Almost surely the Brownian motion on $\mathbb{R}^d$ does not have a 2-multiple point if $d \geq 4$, has 2-multiple points but no 3-multiple point if $d = 3$, and has a multiple point with continuum multiplicity if $d = 1$ or 2. These are proved by Kakutani [**249**], Dvoretzky, Erdős, and Kakutani [**116**], and others. Further properties of the planar Brownian motion are surveyed in Le Gall [**307**]. It is known that a nondegenerate strictly α-stable process on $\mathbb{R}^d$ has an n-multiple point a. s. if and only if $(d-\alpha)n < d$. In this case the Hausdorff dimension of the set of n-multiple points is known. See Takeuchi [**516**], Taylor [**525**], and others. For general Lévy processes Evans [**134**], Fitzsimmons and Salisbury [**145**], and Le Gall, Rosen, and Shieh [**308**] obtain a necessary and sufficient condition for a. s. existence of n-multiple points.

REMARK 49.12. Given x, let $Z_t^x(\omega) = \{s \in [0,t] \colon X_s(\omega) = x\}$ for a Lévy process $\{X_t\}$ on $\mathbb{R}$. Suppose that it is non-zero and not compound Poisson. It is known that $\dim_{\mathrm{H}} Z_t^x = 1 - \alpha^{-1}$ for an α-stable process with $\alpha > 1$ (Blumenthal and Getoor [**44**]). Hawkes [**188**] proves that $\dim_{\mathrm{H}} Z_t^x = 1 - {b_L}^{-1}$ a. s. in general, where b_L is another analogue of the index, defined as

$$ {b_L}^{-1} = \inf \{\eta \in [0,1] \colon \textstyle\int_{\mathbb{R}} \{1 + \mathrm{Re}\,[(-\psi(z))^{\eta}]\}^{-1} \mathrm{d}z < \infty\} $$

with the understanding that the infimum of the empty set is 1. The relation to β_L of (47.36) is that $b_L \leq \beta_L$ if $\beta_L > 1$ and that $b_L = 1$ if $\beta_L \leq 1$. The β_L, δ_L, γ_L, γ'_L, and b_L are different in general. It is possible to find a measure function such that the corresponding Hausdorff measure of $Z_t^x(\omega)$ is equal to the local time at x for all t a. s. In the case of the Brownian motion on $\mathbb{R}$, it is a constant multiple of $(t \log\log \frac{1}{t})^{1/2}$ as is proved by Taylor and Wendel [**531**]. Further, for a fairly general class of Lévy processes on $\mathbb{R}$, Barlow, Perkins, and Taylor [**11**] show that it holds simultaneously for all x a. s.

REMARK 49.13. Let $\{X_t\}$ be a Lévy process on $\mathbb{R}^d$. Write $X(B,\omega) = \{x \in \mathbb{R}^d \colon X_t(\omega) = x$ for some $t \in B\}$, the image set of B. For any strictly α-stable process, $0 < \alpha \leq 2$, and any $B \in \mathcal{B}_{[0,\infty)}$,

$$ P[\dim_{\mathrm{H}} X(B,\omega) = (\alpha \dim_{\mathrm{H}} B) \wedge d\,] = 1, $$

as is shown by McKean [**342**] and Blumenthal and Getoor [**41, 42**]. A much stronger result proved by Kaufman [**267**] for the Brownian motion on $\mathbb{R}^d$ with $d \geq 2$ is that

$$ P[\dim_{\mathrm{H}} X(B,\omega) = (\alpha \dim_{\mathrm{H}} B) \wedge d \text{ for all } B \in \mathcal{B}_{[0,\infty)}\,] = 1 \tag{49.29} $$

with $\alpha = 2$. In the case of strictly α-stable processes, Hawkes [**186**] and Hawkes and Pruitt [**193**] show that (49.29) is true for $\alpha \leq d$, but not true for $\alpha > d = 1$. See Perkins and Taylor [**370**] and Hawkes [**192**] for further development. Extension to general Lévy processes has the form $\dim_{\mathrm{H}} X(B,\omega) \leq \beta'_L \dim_{\mathrm{H}} B$ with β'_L of (47.38), as is shown in Blumenthal and Getoor [**43**], Millar [**347**], and [**193**]. In the case of type B subordinators the extension has the form

$$ \sigma \dim_{\mathrm{H}} B \leq \dim_{\mathrm{H}} X(B,\omega) \leq \beta'_L \dim_{\mathrm{H}} B, $$

where σ is defined by

$$\sigma = \sup\Big\{\eta \geq 0\colon \lim_{u\to\infty} u^{-\eta} \textstyle\int_{(0,\infty)} (1 - \mathrm{e}^{-ux})\nu(\mathrm{d}x) = \infty\Big\}. \tag{49.30}$$

We have $\sigma \leq \beta'_L$, because

$$\beta'_L = \inf\Big\{\eta \geq 0\colon \lim_{u\to\infty} u^{-\eta} \textstyle\int_{(0,\infty)} (1 - \mathrm{e}^{-ux})\nu(\mathrm{d}x) = 0\Big\}$$

for subordinators; there are examples for $\sigma < \beta'_L$.

REMARK 49.14. Let $t > 0$ and $\eta > 0$. The *η-variation* of a function $f(s)$ from $[0,t]$ to $\mathbb{R}^d$ is defined to be the supremum of $\sum_{k=1}^n |f(t_k) - f(t_{k-1})|^\eta$ over all finite partitions $0 = t_0 < t_1 < \cdots < t_n = t$ of $[0,t]$. For any rotation invariant α-stable process $\{X_t\}$ on $\mathbb{R}^d$, $0 < \alpha \leq 2$, Blumenthal and Getoor [**41**] prove that the η-variation of the sample function $X_s(\omega)$, $0 \leq s \leq t$, is finite a. s. if $\eta > \alpha$, and infinite a. s. if $\eta \leq \alpha$. In [**43**] they give some results which suggest that extension of this result is related to the parameter β'_L of (47.38). In the case of the Brownian motion $\{X_t\}$ on $\mathbb{R}^d$, letting $g(u) = u^2(2\log\log(1/u))^{-1}$ for $0 < u \leq \mathrm{e}^{-\mathrm{e}}$, $g(0) = 0$, $g(u) = g(\mathrm{e}^{-\mathrm{e}})$ for $u > \mathrm{e}^{-\mathrm{e}}$, Taylor [**527**] proves that, for each $t > 0$,

$$\lim_{\varepsilon\downarrow 0} \sup_{\mathrm{mesh}(\Delta)<\varepsilon} \textstyle\sum_{k=1}^n g(|X_{t_k} - X_{t_{k-1}}|) = 1 \quad \text{a. s.},$$

where Δ is a finite partition $0 = t_0 < t_1 < \cdots < t_n = t$ of $[0,t]$. His work was continued in Fristedt and Taylor [**153**].

See Fristedt [**150**], Pruitt [**393**], and Taylor [**528, 529**] for other facts related to the five remarks above.

REMARK 49.15. Let $\{X_t\}$ be the Brownian motion on $\mathbb{R}$. Lévy [**317**] shows the order of uniform continuity for its sample function:

$$\lim_{\varepsilon\downarrow 0} \sup_{\substack{0\leq t\leq 1\\ 0<s<\varepsilon}} \frac{|X_{t+s} - X_t|}{(2s\log(1/s))^{1/2}} = 1 \qquad \text{a. s.}$$

Chung, Erdős, and Sirao [**83**] refine this result in the form of integral test. On the other hand, the law of the iterated logarithm (Proposition 47.9) says that, for fixed t, the order of smallness of $|X_{t+s} - X_t|$ as $s \downarrow 0$ is $(2s\log\log(1/s))^{1/2}$ a. s. Keeping this in mind, we say that t is a *fast point* or a *slow point* of $X(\cdot,\omega)$ according as

$$\limsup_{s\downarrow 0} \frac{|X(t+s,\omega) - X(t,\omega)|}{(2s\log\log(1/s))^{1/2}}$$

is bigger than 1 or smaller than 1. Lévy's result above tells us that, almost surely, fast points of order $(2s\log(1/s))^{1/2}$ exist densely in the time interval $[0,\infty)$. Orey and Taylor [**365**] study problems on the size of the set of fast points. Jain and Taylor [**235**] and Kôno [**294**] take up closely related problems, the so-called two-sided growth. Concerning slow points, Dvoretzky [**114**], Davis [**93**], and Greenwood and Perkins [**170**] prove that

$$\inf_t \limsup_{s\downarrow 0} \frac{|X_{t+s} - X_t|}{s^{1/2}} = 1 \quad \text{a. s.}$$

and analyze the set of slow points. If $\{X_t\}$ is a strictly α-stable subordinator, then Hawkes [**187**] and Davis [**94**] prove that

$$\lim_{\varepsilon\downarrow 0}\inf_{\substack{0\le t\le 1\\ 0<s<\varepsilon}}\frac{X_{t+s}-X_t}{s^{1/\alpha}(\log(1/s))^{-(1-\alpha)/\alpha}}=C \qquad \text{a. s.}$$

with a positive finite constant C, and study fast points also. Other properties of the increments of the Brownian path and its local time are studied in Révész's book [**414**] and papers cited there.

REMARK 49.16. There are many properties that are known for stable processes but unknown for general Lévy processes. In this connection we have mentioned several analogues of the index of stable processes. In the multi-dimensional case, processes whose components are independent stable processes (see Notes of Chapter 3) are studied as the next simplest case in Pruit and Taylor [**399**], Hendricks [**194, 195, 196**], Khoshnevisan [**283**], and Shieh [**482**].

REMARK 49.17. Consider a 1-stable process on $\mathbb{R}$ with parameters $(\alpha,\beta,\tau,c) = (1,\beta,0,1)$. Its behavior is quite different between the case $\beta=0$ and the case $\beta\neq 0$. This is already pointed out in Sections 37 and 43. Many intriguing fine structures of sample functions in the case $\beta\neq 0$ are shown by Pruitt and Taylor [**401, 403, 404**]. For example, the range $\Theta(\omega)$ is dense in $\mathbb{R}$ with Lebesgue measure 0 a. s. if $\beta=0$; $\Theta(\omega)$ is a countable union of intervals and has infinite Lebesgue measure a. s. if $|\beta|=1$; $\Theta(\omega)$ is nowhere dense in $\mathbb{R}$ and has infinite Lebesgue measure a. s. if $0<|\beta|<1$. Another example is a limit theorem for $\mathrm{Leb}(\Theta(\omega)\cap[0,a])$, as $a\to\infty$, involving the geometric distribution.

50. Exercises 9

All processes below are $\mathbb{R}$-valued.

E 50.1. Show that, if $\{X_t\}$ is a symmetric Lévy process of type B, then $R_0=0$ a. s.

E 50.2. (i) Let $\{X_t\}$ be the Lévy process generated by $(A,0,\gamma)$ with $A>0$ and $\gamma<0$. Show that then $M_\infty<\infty$ a. s. and M_∞ has exponential distribution with mean $A/(2|\gamma|)$. (ii) Let $\{X_t\}$ be a Lévy process satisfying (46.1), $0>EX_1\ge-\infty$, and $P[\,R_0<\infty\,]>0$. Show that $M_\infty<\infty$ a. s. and M_∞ has exponential distribution.

E 50.3. In the set-up of Section 40 with $d=1$, show that R''_x and Λ_t are $\mathcal{F}^0$-measurable on $\{X_0=0\}$ and that Λ'_t is $\mathcal{F}^0$-measurable on $\{X_0=0 \text{ and } X_t=X_{t-}\}$.

E 50.4. Suppose that $\{X_t\}$ is a non-trivial Lévy process and is not a compound Poisson process. Show that $P[\,R_0=R'_0=R''_0\,]=1$. This complements Lemma 49.6.

E 50.5. Let $\{X_t\}$ be a Lévy process. Let $a > 0$. Show that, for any $q \geq 0$ and any Borel sets $C \subset (-\infty, a)$ and $D \subset [a, \infty)$,

$$E[\mathrm{e}^{-qR'_a};\ X(R'_a-) \in C,\ X(R'_a) \in D] = \textstyle\int_C U^q_B(0, \mathrm{d}y)\, \nu(D - y),$$

where $B = [a, \infty)$.

E 50.6. Let $\{X_t\}$ be a non-zero subordinator. Show that

$$\textstyle\int_0^\infty \mathrm{e}^{-ux} E[\mathrm{e}^{-qR_x};\ X_{R_x} = x]\mathrm{d}x = \gamma_0 \{q + \gamma_0 u + \int_{(0,\infty)} (1 - \mathrm{e}^{-ux})\nu(\mathrm{d}x)\}^{-1}$$

for $q > 0$ and $u > 0$. If $\gamma_0 = 0$, then does $P[\, X_{R_x} = x\,] = 0$ for all positive x?

E 50.7 (Bingham [**35**]). Let $\{X_t\}$ be a strictly α-stable subordinator, $0 < \alpha < 1$, with $E[\mathrm{e}^{-uX_t}] = \mathrm{e}^{-tu^\alpha}$. Show that then, for any $x > 0$, $x^{-\alpha} R_x$ has Mittag–Leffler distribution with parameter α. Thus it follows from E 29.19 that R_x does not have an infinitely divisible distribution.

E 50.8. Let $\{X_t\}$ be a Poisson process. Show that then R_x has infinitely divisible distribution for any $x \geq 0$ but the first passage time process $\{R_x \colon x \geq 0\}$ is not a Lévy process.

E 50.9. Let $\{X_t\}$ be strictly α-stable, $1 < \alpha < 2$, and without positive jumps. Prove that M_t has a scaled Mittag–Leffler distribution for each $t > 0$ and that

$$P[M_t > x] = P[X_t > x]/P[X_t > 0]$$

for $t > 0$ and $x > 0$.

E 50.10 (Prabhu [**389**]). Let $\{X_t\}$ be an arbitrary Lévy process and let L be its infinitesimal generator in the Banach space $C_0(\mathbb{R})$. Prove that, for each $q > 0$, there uniquely exist $q_1 > 0$, $q_2 > 0$, L_q^+, and L_q^- such that

(1) $qI - L = (q_1 I - L_q^+)(q_2 I - L_q^-)$ on C_0^∞ defined in Section 31,
(2) L_q^+ and L_q^- are the infinitesimal generators in $C_0(\mathbb{R})$ of a subordinator and the negative of a subordinator, respectively.

We have $q_1 q_2 = q$. What is the decomposition in the case where $\{X_t\}$ is the Brownian motion?

Notes

As is mentioned in Section 45, this chapter is based on Rogozin [**419**], Pecherskii and Rogozin [**368**], and Borovkov [**59**]. Here proofs are made precise and sometimes different from the original ones. We have used the method of Skorohod [**493**]. Lemma 45.6 is from [**368**], but, when $\{S_n\}$ is integer-valued, (45.11)–(45.15) and (45.16) with $\eta = 1$ were found by Spitzer [**496**]. Feller's book [**139**] contains proofs of (45.11)–(45.15). Theorem 45.7 was obtained also by Gusak and Korolyuk [**177**] in a different form. Earlier Baxter and Donsker [**18**] got essentially the identity (45.7).

Theorems 46.3 and 46.4 are by Keilson [**269**], Zolotarev [**606**], and Borovkov [**57**].

Theorem 47.1 is by Shtatland [**486**] and Rogozin [**420**]. Theorems 47.2 and 47.6 are by Rogozin [**419**]. Example 47.4 is taken from [**420**].

Lemma 48.3 follows from the Kolmogorov–Rogozin inequality in Hengartner and Theodorescu [**197**]. The present proof is from [**428**]. Theorem 48.6 is from [**419**].

The factorization identity in Theorem 49.2 is from [**368**] and [**59**]. Our proof of it is based on the book of Bratijchuk and Gusak [**62**]. Lemmas 49.3, 49.4, and 49.6 are given by [**368**].

Suppose that 0 is regular for $(0,\infty)$ for $\{X_t\}$. Then the reflecting process $\{Y_t\}$ is a temporally homogeneous Markov process having local time $L(t)$ at the point 0. Let $L^{-1}(t)$ be the right-continuous inverse function of $L(t)$. Fristedt [**150**] shows that $\{(L^{-1}(t), M(L^{-1}(t)))\}$ is a Lévy process on $\mathbb{R}^2$ and proves the formula

$$E[\mathrm{e}^{-uL^{-1}(t)-vM(L^{-1}(t))}] = \exp\big[-c\exp\big[\int_0^\infty t^{-1}\mathrm{d}t\int_{(0,\infty)}(\mathrm{e}^{-t}-\mathrm{e}^{-ut-vx})\mu^t(\mathrm{d}x)\big]\big]$$

for $u \geq 0$ and $v \geq 0$, where c is a positive constant determined by the normalization of the local time $L(t)$. This is intimately connected with (49.1). See Bertoin and Doney [**28**] for application of this combined with the renewal theorem E 39.14. The reflecting processes for symmetric stable processes are studied by Watanabe [**547**].

This chapter is a part of the area called fluctuation theory of Lévy processes on $\mathbb{R}$. It is a wide area with a large number of results. It is a continuous analogue of the fluctuation theory of random walks on $\mathbb{R}$ expounded in the books [**139, 496**]. Application to queueing and storage processes has been an impetus to the theory; see the books of Takács [**510**], Borovkov [**60**], and Prabhu [**390**] and the survey by Bingham [**36**].

Based on the theory of excursions and local times for temporally homogeneous Markov processes developed by Motoo [**358**], Itô [**226**], and Maisonneuve [**336**], the whole theory can be directly developed without approximation by random walks or compound Poisson processes in Sections 45 and 49. Bertoin's book [**26**] takes this road, presents a wide view of this field, and contains various beautiful applications to limit theorems. Among many related works, we mention Bertoin [**24**], Bratijchuk and Gusak [**62**], Greenwood and Pitman [**171**], Isozaki [**219**], Millar [**350**], Prabhu and Rubinovitch [**391**], Silverstein [**487**], Tanaka [**521, 522, 523**], and in the case of stable processes Bingham [**35**], Darling [**92**], Doney [**101**], Heyde [**199**], and Monrad and Silverstein [**355**].

The formula in E 50.5 involving the Lévy measure can be extended to a more general B in place of $[a,\infty)$. See Ikeda and Watanabe [**213**].

CHAPTER 10

More distributional properties

51. Infinite divisibility on the half line

In this chapter we discuss more properties of infinitely divisible distributions on the half line $\mathbb{R}_+ = [0, \infty)$ or on the line $\mathbb{R}$. We give a characterization of infinite divisibility on the half line and introduce some prominent classes. Then we are mainly concerned with unimodality and multimodality. Special attention will be paid to selfdecomposable distributions and processes.

Our starting point is that a probability measure μ with support S_μ in $\mathbb{R}_+$ is infinitely divisible if and only if its Laplace transform $L_\mu(u) = \int_{[0,\infty)} \mathrm{e}^{-ux}\mu(\mathrm{d}x)$ has the form

$$L_\mu(u) = \exp\left[-\gamma_0 u - \int_{(0,\infty)} (1 - \mathrm{e}^{-ux})\nu(\mathrm{d}x)\right] \tag{51.1}$$

with $\gamma_0 \geq 0$ and with a measure ν on $(0, \infty)$ satisfying

$$\int_{(0,\infty)} (1 \wedge x)\nu(\mathrm{d}x) < \infty. \tag{51.2}$$

This is what we have mentioned repeatedly, in Remark 21.6, Theorem 24.7, Corollary 24.8, and Theorem 24.11. The measure ν and the nonnegative real γ_0 are uniquely determined by μ; they are the Lévy measure and the drift of μ, respectively. This fact is reformulated as follows.

THEOREM 51.1. *Let μ be a probability measure with support in $\mathbb{R}_+$. Then μ is infinitely divisible if and only if there exist $\gamma_0 \geq 0$ and a measure ν on $(0, \infty)$ satisfying (51.2) such that*

$$\int_{[0,x]} y\mu(\mathrm{d}y) = \int_{(0,x]} \mu([0, x - y])y\nu(\mathrm{d}y) + \gamma_0\mu([0, x]) \quad \textit{for } x > 0. \tag{51.3}$$

Proof. Suppose that μ is infinitely divisible. Let γ_0 and ν be those in (51.1). Then,

$$\frac{\mathrm{d}}{\mathrm{d}u}\int_{(0,\infty)} (1 - \mathrm{e}^{-ux})\nu(\mathrm{d}x) = \int_{(0,\infty)} x\mathrm{e}^{-ux}\nu(\mathrm{d}x) \quad \text{for } u > 0$$

by Lebesgue's dominated convergence theorem, since

$$|(\mathrm{e}^{-(u+\varepsilon)x} - \mathrm{e}^{-ux})/\varepsilon| \leq \sup\nolimits_{\theta\in(0,1)}|x\mathrm{e}^{-(u+\theta\varepsilon)x}| \leq x\mathrm{e}^{-ux/2}$$

for $0 < |\varepsilon| < u/2$. Hence, differentiating (51.1), we get

$$-\int_{[0,\infty)} x\mathrm{e}^{-ux}\mu(\mathrm{d}x) = L_\mu(u)\Big[-\gamma_0 - \int_{(0,\infty)} x\mathrm{e}^{-ux}\nu(\mathrm{d}x)\Big]. \tag{51.4}$$

Let $\widetilde{\nu}(\mathrm{d}x) = \gamma_0\delta_0(\mathrm{d}x) + x\nu(\mathrm{d}x)$. Then, for $v > 0$,

$$\int_{[0,\infty)} x\mathrm{e}^{-ux-vx}\mu(\mathrm{d}x) = \int_{[0,\infty)} \mathrm{e}^{-ux-vx}\mu(\mathrm{d}x)\cdot\int_{[0,\infty)} \mathrm{e}^{-ux-vx}\widetilde{\nu}(\mathrm{d}x), \quad u \geq 0.$$

Hence, by the extension of Proposition 2.6(ii) to finite measures, $x\mathrm{e}^{-vx}\mu(\mathrm{d}x)$ is the convolution of $\mathrm{e}^{-vx}\mu(\mathrm{d}x)$ and $\mathrm{e}^{-vx}\widetilde{\nu}(\mathrm{d}x)$. That is,

$$\int_{[0,x]} y\mathrm{e}^{-vy}\mu(\mathrm{d}y) = \int_{[0,x]}\Big(\int_{[0,x-y]} \mathrm{e}^{-vz}\mu(\mathrm{d}z)\Big)\mathrm{e}^{-vy}\widetilde{\nu}(\mathrm{d}y).$$

Letting $v \downarrow 0$, we get (51.3).

Conversely, if (51.3) holds, then

$$\int 1_{[0,x]}(y)y\mu(\mathrm{d}y) = \iint 1_{[0,x]}(y+z)\widetilde{\nu}(\mathrm{d}y)\mu(\mathrm{d}z),$$

which implies

$$\int f(y)y\mu(\mathrm{d}y) = \iint f(y+z)\widetilde{\nu}(\mathrm{d}y)\mu(\mathrm{d}z)$$

for any nonnegative measurable function f. Set $f(x) = \mathrm{e}^{-ux}$. Then we get (51.4), that is, the function $G(u) = \log L_\mu(u)$ satisfies $G'(u) = -\gamma_0 - \int_{(0,\infty)} x\mathrm{e}^{-ux}\nu(\mathrm{d}x)$ on $(0,\infty)$. Moreover $G(u)$ is continuous on $[0,\infty)$ and $G(0) = 0$. Hence $G(u) = -\gamma_0 u - \int_{(0,\infty)}(1-\mathrm{e}^{-ux})\nu(\mathrm{d}x)$ and we obtain (51.1). $\square$

COROLLARY 51.2. *Let μ be a probability measure with $S_\mu \subset \mathbb{Z}_+$, $\mu = \sum_{n\in\mathbb{Z}_+} p_n\delta_n$, with $p_0 > 0$. Then, μ is infinitely divisible if and only if there are $q_n \geq 0$, $n = 1, 2, \ldots$, such that*

$$np_n = \sum_{k=1}^{n} kq_k p_{n-k}, \qquad n = 1, 2, \ldots. \tag{51.5}$$

A sufficient condition for the infinite divisibility of μ is the existence of $r_n \geq 0$, $n = 1, 2, \ldots$, such that

$$p_n = \sum_{k=1}^{n} r_k p_{n-k}, \qquad n = 1, 2, \ldots. \tag{51.6}$$

Proof. Assume that μ is infinitely divisible. Then $S_\nu \subset \mathbb{Z}_+ \setminus \{0\}$ and $\gamma_0 = 0$ by Corollaries 24.6 and 24.8, and the identity (51.3) takes the form

$$\sum_{k=0}^{n} kp_k = \sum_{k=1}^{n} kq_k \sum_{j=0}^{n-k} p_j \quad \text{for } n \geq 1, \tag{51.7}$$

where $\nu = \sum_{n=1}^{\infty} q_n \delta_n$. The increments of both sides of (51.7) as n goes to $n+1$ give (51.5) for $n \geq 2$. For $n = 1$, (51.5) is identical with (51.7). Conversely, suppose that there are nonnegative q_n, $n = 1, 2, \ldots$, satisfying (51.5). Then they satisfy (51.7) and $\sum_{n=1}^{\infty} q_n < \infty$ since $p_n \geq q_n p_0$. Setting $\nu = \sum_{n=1}^{\infty} q_n \delta_n$ and $\gamma_0 = 0$, we get (51.3), and hence μ is infinitely divisible.

In order to prove the latter half of the assertion, first note that the unique existence of $\{q_n\}$ and $\{r_n\}$ satisfying (51.5) and (51.6), respectively, is obvious, since we can give them inductively, beginning with $q_1 = r_1 = p_1/p_0$. We need to examine their nonnegativity. Let us use generating functions $P(z) = \sum_{n \geq 0} p_n z^n$, $Q(z) = \sum_{n \geq 1} n q_n z^n$, $R(z) = \sum_{n \geq 1} r_n z^n$ for $|z| < 1$. See [**138, 151, 413**] for general accounts of generating functions. The identities (51.5) and (51.6) are written as $zP'(z) = Q(z)P(z)$ and $P(z) - p_0 = R(z)P(z)$, respectively. It follows that $P(z) = p_0(1 - R(z))^{-1}$ and $P'(z) = p_0 R'(z)(1 - R(z))^{-2}$. Hence $zR'(z)(1 - R(z))^{-1} = Q(z)$. That is, $Q(z) = zR'(z) + Q(z)R(z)$. This means that

$$nq_n = nr_n + \sum_{k=1}^{n-1} k q_k r_{n-k}.$$

If $r_k \geq 0$ for all k, then we get $q_n \geq 0$ inductively, and hence μ is infinitely divisible by the first half of this corollary. □

THEOREM 51.3. *Let μ be a probability measure on $\mathbb{Z}_+$ with $\mu\{n\} = p_n > 0$ for all n. If it is log-convex in the sense that*

$$\log p_n - \log p_{n-1} \leq \log p_{n+1} - \log p_n, \quad n = 1, 2, \ldots, \tag{51.8}$$

then it is infinitely divisible.

In this case, $\sum_{n \geq 0} p_n \delta_{an+b}$ with $a > 0$ and $b \in \mathbb{R}$ is also infinitely divisible.

Proof of theorem. Assume the log-convexity. Then $p_n/p_{n-1} \leq p_{n+1}/p_n$. Consider the equation (51.6). Beginning with $r_1 = p_1/p_0$, it is solved inductively. If $r_n \geq 0$ for $n = 1, 2, \ldots$, then μ is infinitely divisible by Corollary 51.2. Evidently $r_1 > 0$. Now suppose that $r_1, \ldots, r_n$ are nonnegative. Then

$$\begin{aligned} p_{n+1} &= \frac{p_{n+1}}{p_n}(p_{n-1}r_1 + p_{n-2}r_2 + \cdots + p_0 r_n) \\ &\geq \frac{p_n}{p_{n-1}} p_{n-1} r_1 + \frac{p_{n-1}}{p_{n-2}} p_{n-2} r_2 + \cdots + \frac{p_1}{p_0} p_0 r_n \\ &= p_n r_1 + p_{n-1} r_2 + \cdots + p_1 r_n = p_{n+1} - p_0 r_{n+1}. \end{aligned}$$

Hence $r_{n+1} \geq 0$. □

THEOREM 51.4. *Let $\mu(\mathrm{d}x) = c\delta_0(\mathrm{d}x) + f(x) 1_{(0,\infty)}(x)\mathrm{d}x$ be a probability measure such that $0 \leq c < 1$ and that $f(x)$ is a positive function on $(0, \infty)$,* *log-convex in the sense that* $\log f(x)$ *is convex on* $(0,\infty)$. *Then* μ *is infinitely divisible.*

Proof. The function $f(x)$ is continuous and decreasing. Define $f_n(x)$ on $(0,\infty)$ by $f_n(x) = f(2^{-n}k)$ for $2^{-n}(k-1) < x \le 2^{-n}k$ and c_n by $c_n = c + \int_0^\infty f_n(x)\mathrm{d}x$. Then $f_n(x) \uparrow f(x)$ and $c_n \to c + \int_0^\infty f(x)\mathrm{d}x = 1$. Define $\mu_n = c_n{}^{-1}\big(c\delta_{2^{-n}} + \sum_{k=1}^\infty f(2^{-n}k)2^{-n}\delta_{2^{-n}k}\big)$. Then $\mu_n \to \mu$, since $\mu_n[0,x] \to \mu[0,x]$ for $x > 0$. It follows from the log-convexity of f that

$$\log f(2^{-n}k) - \log f(2^{-n}(k-1)) \le \log f(2^{-n}(k+1)) - \log f(2^{-n}k)$$

for $k = 2,3,\ldots$. By Theorem 51.3 μ_n is infinitely divisible and so is μ. □

DEFINITION 51.5. Let $\{\mu_\theta\colon \theta \in \Theta\}$ be a family of probability measures on $\mathbb{R}^d$. Assume that Θ is equipped with a σ-algebra $\mathcal{B}$ and that, for any $B \in \mathcal{B}(\mathbb{R}^d)$, $\mu_\theta(B)$ is $\mathcal{B}$-measurable in θ. Let ρ be a probability measure on $(\Theta,\mathcal{B})$. Then $\mu(B) = \int_\Theta \mu_\theta(B)\rho(\mathrm{d}\theta)$, $B \in \mathcal{B}(\mathbb{R}^d)$, is a probability measure on $\mathbb{R}^d$. It is called the *mixture* of $\{\mu_\theta\colon \theta\in\Theta\}$ by the *mixing measure* ρ.

Special mixtures have already appeared in subordination (Theorem 30.1) and in Markov processes (Remark 10.8).

A function $f(x)$ on $(0,\infty)$ is called *completely monotone* if it is of class C^∞ and if $(-1)^n(\mathrm{d}^n/\mathrm{d}x^n)f(x) \ge 0$ on $(0,\infty)$ for $n = 0,1,\ldots$. If

$$f(x) = \int_{[0,\infty)} \mathrm{e}^{-xy}\rho(\mathrm{d}y), \quad x > 0, \tag{51.9}$$

with some measure ρ such that this integral is finite, then

$$\frac{\mathrm{d}^n f}{\mathrm{d}x^n}(x) = \int_{[0,\infty)} (-y)^n \mathrm{e}^{-xy}\rho(\mathrm{d}y), \quad x > 0,\ n = 1,2,\ldots,$$

and f is completely monotone. Bernstein's theorem tells that, conversely, any completely monotone function can be expressed as in (51.9) [**139, 151, 572**]. The measure ρ is uniquely determined by f, because, for $x_0 > 0$, the measure $\mathrm{e}^{-x_0 y}\rho(\mathrm{d}y)$ is uniquely determined (Proposition 2.6). Note that $\rho(\{0\}) = \lim_{x\to\infty} f(x)$.

THEOREM 51.6. *Consider a probability measure* μ *on* $\mathbb{R}_+$ *such that* $\mu = c\delta_0 + f(x)1_{(0,\infty)}(x)\mathrm{d}x$ *with* $0 \le c < 1$ *and* $f(x)$ *being completely monotone on* $(0,\infty)$. *Then* μ *is infinitely divisible.*

Notice that the function $f(x)$ above is not necessarily bounded (see Exercise 55.1).

Proof of theorem. By Theorem 51.4 it is enough to show that $f(x)$ is positive and log-convex on $(0,\infty)$. Since we have (51.9), differentiating it and using the Schwarz inequality, we get

$$f'(x)^2 = \left(\int y\mathrm{e}^{-xy}\rho(\mathrm{d}y)\right)^2 \le \left(\int \mathrm{e}^{-xy}\rho(\mathrm{d}y)\right)\left(\int y^2\mathrm{e}^{-xy}\rho(\mathrm{d}y)\right)$$

$$= f(x)f''(x).$$

Hence $(\log f)'' = (ff'' - (f')^2)/f^2 \geq 0$ and f is log-convex. ☐

DEFINITION 51.7. Consider the family which is the union of $\{\delta_0\}$ and the class of all exponential distributions. The class of mixtures of this family is called *the class ME*.

PROPOSITION 51.8. *The class ME coincides with the class of μ considered in Theorem 51.6.*

Proof. If μ is in ME, then, by Definition 51.5, of mixtures, either $\mu = \delta_0$ or there are $0 \leq c < 1$ and a measure ρ on $(0, \infty)$ with total mass $1 - c$ such that $\mu(B) = c\delta_0(B) + \int_{(0,\infty)} \rho(\mathrm{d}\alpha) \int_{B\cap(0,\infty)} \alpha \mathrm{e}^{-\alpha x}\mathrm{d}x$. In the case $\mu \neq \delta_0$, $\mu = c\delta_0 + f(x)1_{(0,\infty)}\mathrm{d}x$ with $f(x) = \int_{(0,\infty)} \alpha \mathrm{e}^{-\alpha x}\rho(\mathrm{d}\alpha)$. This $f(x)$ is completely monotone. The converse follows from Bernstein's theorem. ☐

DEFINITION 51.9. Consider the smallest class that contains ME and that is closed under convergence and convolution. We call it *the class B*. Sometimes it is called the Bondesson class or the Goldie–Steutel–Bondesson class, because it is related to Goldie [**168**], Steutel [**499**], and Bondesson [**53**].

THEOREM 51.10. *Let μ be a probability measure on $\mathbb{R}_+$. Then, $\mu \in B$ if and only if*

$$L_\mu(u) = \exp\left[-\gamma_0 u - \int_0^\infty (1 - \mathrm{e}^{-ux})l(x)\mathrm{d}x\right], \quad u \geq 0, \tag{51.10}$$

where $\gamma_0 \geq 0$, $\int_0^\infty (1 \wedge x)l(x)\mathrm{d}x < \infty$, and $l(x)$ is completely monotone.

REMARK 51.11. By Bernstein's theorem, the function $l(x)$ in the theorem above is uniquely expressed as

$$l(x) = \int_{(0,\infty)} \mathrm{e}^{-xy}Q(\mathrm{d}y), \quad x > 0, \tag{51.11}$$

by a measure Q on $(0, \infty)$. The condition on Q appearing here is

$$\int_{(0,\infty)} \frac{Q(\mathrm{d}y)}{y(y+1)} < \infty. \tag{51.12}$$

The expression (51.10) is written as

$$L_\mu(u) = \exp\left[-\gamma_0 u - \int_{(0,\infty)} \left(\frac{1}{y} - \frac{1}{y+u}\right)Q(\mathrm{d}y)\right]. \tag{51.13}$$

In fact,

$$\int_0^\infty (1 - \mathrm{e}^{-ux})l(x)\mathrm{d}x = \int_0^\infty (1 - \mathrm{e}^{-ux})\mathrm{d}x \int_{(0,\infty)} \mathrm{e}^{-xy}Q(\mathrm{d}y)$$

$$= \int_{(0,\infty)} Q(\mathrm{d}y) \int_0^\infty (\mathrm{e}^{-xy} - \mathrm{e}^{-(y+u)x})\mathrm{d}x$$

$$= \int_{(0,\infty)} \Big(\frac{1}{y} - \frac{1}{y+u}\Big) Q(\mathrm{d}y) = \int_{(0,\infty)} \frac{u}{y(y+u)} Q(\mathrm{d}y).$$

The equivalence of the condition $\int (1\wedge x) l(x)\mathrm{d}x < \infty$ to (51.12) is seen from the equality above by letting $u = 1$.

THEOREM 51.12. *Let μ be a probability measure on $\mathbb{R}_+$. Then, $\mu \in ME$ if and only if $L_\mu(u)$ is expressed by (51.10) with $\gamma_0 = 0$ and*

$$l(x) = \int_0^\infty \mathrm{e}^{-xy} q(y)\mathrm{d}y, \tag{51.14}$$

where $q(y)$ is measurable, $\int_0^1 (q(y)/y)\mathrm{d}y < \infty$, and

$$0 \le q(y) \le 1. \tag{51.15}$$

REMARK 51.13. Theorem 51.12 shows that if $\mu \in ME$, then $\mu^t \in ME$ for any $t \in [0, 1]$. Thus, if $\mu_1, \dots, \mu_n$ are Γ-distributions with a common parameter c (see Example 2.15) in $(0, 1]$, then any mixture of $\mu_1, \dots, \mu_n$ is infinitely divisible. Steutel [**499**] conjectured that this statement is true for $c \in (0, 2]$. The proof was given by Kristiansen [**295**] many years later.

We prove Theorems 51.10 and 51.12 after two lemmas.

LEMMA 51.14. *For $0 < a < \infty$ let $\mu_a = a\mathrm{e}^{-ax} 1_{(0,\infty)}\mathrm{d}x$, the exponential distribution with parameter a. Let $\mu_\infty = \delta_0$. Fix $m \ge 2$ and $0 < a_1 < \dots < a_{m-1} < a_m \le \infty$. Let μ be a probability measure on $\mathbb{R}_+$. Then the following are equivalent:*

(1) $\mu = \sum_{k=1}^m q_k \mu_{a_k}$ *with some* $q_k > 0$, $\sum_{k=1}^m q_k = 1$*;*

(2) *there are b_k, $k = 1, \dots, m$, with*

$$0 < a_1 < b_1 < a_2 < b_2 < \dots < a_{m-1} < b_{m-1} < a_m \le b_m = \infty$$

such that

$$L_\mu(u) = \exp\Big[-\int_0^\infty \frac{1-\mathrm{e}^{-ux}}{x} \sum_{k=1}^m (\mathrm{e}^{-a_k x} - \mathrm{e}^{-b_k x})\mathrm{d}x\Big] \tag{51.16}$$

$$= \exp\Big[-\int_0^\infty \Big(\frac{1}{y} - \frac{1}{y+u}\Big) \sum_{k=1}^m 1_{(a_k,b_k)}(y)\mathrm{d}y\Big].$$

Proof. Case 1. $a_m < \infty$. Assume that (1) holds. Then

$$L_\mu(u) = \sum_{k=1}^m \frac{q_k a_k}{u + a_k} = P(u) \Big/ \prod_{k=1}^m (u + a_k),$$

where $P(u)$ is a polynomial of degree $m-1$. The function $L_\mu(u)$ is the restriction to $\mathbb{R}_+$ of a rational function $g(z)$ on $\mathbb{C}$. Consider $g(u)$ on $\mathbb{R}$. Then, $g'(u) < 0$ for $u \neq -a_1, \ldots, -a_k$ and $g(u) \uparrow \infty$ or $\downarrow -\infty$ as $u \downarrow -a_k$ or $u \uparrow -a_k$, respectively. Hence $g(u)$ has one and only one zero point in each interval $(-a_{k+1}, -a_k)$. Denote it by $-b_k$. Thus

$$P(u) = c\prod_{k=1}^{m-1}(u+b_k)$$

with $c > 0$. Since $1 = L_\mu(0) = c\prod_{j=1}^{m-1} b_j \,/\, \prod_{k=1}^{m} a_k$, we get

$$L_\mu(u) = \prod_{k=1}^{m}\frac{a_k}{u+a_k} \Big/ \prod_{j=1}^{m-1}\frac{b_j}{u+b_j}. \tag{51.17}$$

Since

$$\int_0^\infty \mathrm{e}^{-ux} a\mathrm{e}^{-ax}\mathrm{d}x = \frac{a}{u+a} = \exp\left[-\int_0^\infty (1-\mathrm{e}^{-ux})\frac{\mathrm{e}^{-ax}}{x}\mathrm{d}x\right]$$

by Example 8.10 and since $x^{-1}(\mathrm{e}^{-ax} - \mathrm{e}^{-bx}) = \int_0^\infty \mathrm{e}^{-xy} 1_{(a,b)}(y)\mathrm{d}y$ for $0 < a < b \leq \infty$, we obtain (2).

Conversely, assume that (2) holds. Then we get (51.17) from (51.16). The partial fraction expansion gives

$$\prod_{k=1}^{m}\frac{a_k}{u+a_k}\prod_{j=1}^{m-1}\frac{u+b_j}{b_j} = \sum_{k=1}^{m} q_k\frac{a_k}{u+a_k},$$

where

$$q_k = \prod_{l\neq k}\frac{a_l}{a_l-a_k}\prod_j \frac{b_j-a_k}{b_j} > 0.$$

Letting $u=0$, we have $\sum_k q_k = 1$. Hence we have (1).

Case 2. $a_m = \infty$. Assume (1). We have

$$L_\mu(u) = \sum_{k=1}^{m-1}\frac{q_k a_k}{u+a_k} + q_m = P(u) \Big/ \prod_{k=1}^{m-1}(u+a_k)$$

with a polynomial $P(u)$ of degree $m-1$. This time the extension $g(u)$ increases to q_m as $u \downarrow -\infty$ and has a zero point $-b_{m-1}$ in $(-\infty, -a_{m-1})$. Now we have

$$L_\mu(u) = \prod_{k=1}^{m-1}\frac{a_k}{u+a_k} \Big/ \prod_{j=1}^{m-1}\frac{b_j}{u+b_j}$$

instead of (51.17) and we get (2) with $b_m = \infty$ similarly. Conversely, if (2) holds, then

$$\prod_{k=1}^{m-1}\frac{a_k}{u+a_k}\frac{u+b_k}{b_k} = \sum_{k=1}^{m-1} q_k\frac{a_k}{u+a_k} + q_m,$$

where

$$q_k = \prod_{l \neq k} \frac{a_l}{a_l - a_k} \prod_j \frac{b_j - a_k}{b_j} > 0 \quad \text{for } 1 \leq k \leq m-1$$

and $q_m = \prod_l \frac{a_l}{b_l} > 0$. □

EXAMPLE 51.15. If $\mu = q_1\mu_{a_1} + q_2\mu_{a_2}$ with $0 < a_1 < a_2 < \infty$, $q_1 > 0$, $q_2 > 0$, and $q_1 + q_2 = 1$, then

$$L_\mu(u) = q_1 \frac{a_1}{u + a_1} + q_2 \frac{a_2}{u + a_2} = \frac{a_1}{u + a_1} \frac{a_2}{u + a_2} \frac{u + b_1}{b_1},$$

where b_1 is determined by $1/b_1 = (q_2/a_1) + (q_1/a_2)$. Thus $a_1 < b_1 < a_2$ and

$$\begin{aligned} L_\mu(u) &= \exp\left[-\int_0^\infty (1 - \mathrm{e}^{-ux}) x^{-1} (\mathrm{e}^{-a_1 x} - \mathrm{e}^{-b_1 x} + \mathrm{e}^{-a_2 x}) \mathrm{d}x\right] \\ &= \exp\left[-\int_0^\infty \left(\tfrac{1}{y} - \tfrac{1}{y+u}\right)(1_{(a_1,b_1)}(y) + 1_{(a_2,\infty)}(y)) \mathrm{d}y\right]. \end{aligned}$$

If $\mu = q\delta_0 + (1-q)\mu_a$ with $0 < q < 1$ and $0 < a < \infty$, then

$$\begin{aligned} L_\mu(u) &= q + (1-q)\tfrac{a}{u+a} = \tfrac{a}{u+a} \tfrac{u+a/q}{a/q} \\ &= \exp\left[-\int_0^\infty (1 - \mathrm{e}^{-ux}) x^{-1} (\mathrm{e}^{-ax} - \mathrm{e}^{-(a/q)x}) \mathrm{d}x\right] \\ &= \exp\left[-\int_0^\infty \left(\tfrac{1}{y} - \tfrac{1}{y+u}\right) 1_{(a,a/q)}(y) \mathrm{d}y\right]. \end{aligned}$$

LEMMA 51.16. *Let μ_n, $n = 1, 2, \ldots$, be probability measures on $\mathbb{R}_+$ such that $L_{\mu_n}(u)$ has the representation (51.13) with $\gamma_{0,n}$ and Q_n satisfying $\gamma_{0,n} \geq 0$ and $\int \frac{Q_n(\mathrm{d}y)}{y(y+1)} < \infty$ in place of γ_0 and Q. Let μ be a probability measure on $\mathbb{R}_+$. Then, $\mu_n \to \mu$ as $n \to \infty$ if and only if $L_\mu(u)$ has the representation (51.13) with some γ_0 and Q satisfying $\gamma_0 \geq 0$, $\int \frac{Q(\mathrm{d}y)}{y(y+1)} < \infty$, and the following conditions:*

(1) *if g is a bounded continuous function on $[0, \infty)$ such that $g(x) = 0$ on some $[x_0, \infty)$, then*

$$\int \frac{g(y)}{y} Q_n(\mathrm{d}y) \to \int \frac{g(y)}{y} Q(\mathrm{d}y), \quad n \to \infty;$$

(2) $\lim_{c \to \infty} \limsup_{n \to \infty} \left|\gamma_{0,n} + \int_{(c,\infty)} y^{-2} Q_n(\mathrm{d}y) - \gamma_0\right| = 0.$

Proof. Assume that $\mu_n \to \mu$. Define finite measures R_n on $[0, \infty]$ by $R_n\{0\} = 0$, $R_n\{\infty\} = \gamma_{0,n}$, and $[R_n]_{(0,\infty)} = \frac{1}{y(y+1)} Q_n(\mathrm{d}y)$. Then

$$\log L_{\mu_n}(u) = -\gamma_{0,n} u - \int_{(0,\infty)} \frac{u}{y(y+u)} Q_n(\mathrm{d}y) = -\int_{[0,\infty]} \frac{u(y+1)}{y+u} R_n(\mathrm{d}y).$$

We see that $R_n[0, \infty]$ is bounded in n, since $\log L_{\mu_n}(1)$ is bounded below. By the selection theorem we can find a convergent subsequence $\{R_{n_k}\}$ of $\{R_n\}$. Let $R_{n_k} \to R$, a finite measure on $[0, \infty]$. Then

$$\log L_\mu(u) = -R\{0\} - R\{\infty\}u - \int_{(0,\infty)} \frac{u(y+1)}{y+u} R(\mathrm{d}y) \quad \text{for } u > 0.$$

We get $R\{0\} = 0$, letting $u \downarrow 0$. Denoting $\gamma_0 = R\{\infty\}$ and $Q(\mathrm{d}y) = y(y+1)R(\mathrm{d}y)$ on $(0, \infty)$, we get (51.13). By the uniqueness of the representation we have $R_n \to R$ on $[0, \infty]$ for the full sequence. Thus (1) follows, since

$$\int g(y)(y+1)R_n(\mathrm{d}y) \to \int g(y)(y+1)R(\mathrm{d}y).$$

To see (2), let $g_c(y)$ be a bounded continuous function such that $g_c(y) = 0$ on $[0, c-1]$ and $g_c(y) = 1$ on $[c, \infty]$. Then

$$\int g_c(y)\frac{y+1}{y}R_n(\mathrm{d}y) \to \int g_c(y)\frac{y+1}{y}R(\mathrm{d}y),$$

that is

$$\gamma_{0,n} + \int g_c(y)y^{-2}Q_n(\mathrm{d}y) \to \gamma_0 + \int g_c(y)y^{-2}Q(\mathrm{d}y),$$

which goes to γ_0 as $c \to \infty$. This proves (2).

Conversely, if μ has the representation (51.13) with γ_0 and Q and if (1) and (2) are satisfied, then $L_{\mu_n}(u) \to L_\mu(u)$, which means that $\mu_n \to \mu$. □

Proof of Theorem 51.12. Assume that $\mu \in ME$. Let μ_a, $0 < a \leq \infty$, be defined as in Lemma 51.14. Then there is a probability measure ρ on $(0, \infty]$ such that $\mu(B) = \int_{(0,\infty]} \mu_a(B)\rho(\mathrm{d}a)$, $B \in \mathcal{B}(\mathbb{R})$. Since $\mu_a[0, x]$ is continuous in a if $x > 0$, we see that μ is the limit of some $\mu_{(n)}$, $n = 1, 2, \ldots$, where each $\mu_{(n)}$ has a mixing measure supported on a finite number of points. By Lemma 51.14, each $\mu_{(n)}$ has Laplace transform of the form (51.13) with $\gamma_{0,n} = 0$ and with Q_n absolutely continuous having density $0 \leq q_n(y) \leq 1$. Hence, by Lemma 51.16, μ has Laplace transform (51.13) with $\gamma_0 = 0$ and with some Q. Using the property (1) of Lemma 51.16, we have $Q(a, b] \leq b-a$ for $0 < a < b < \infty$. Hence Q has density $q(y)$ with $0 \leq q(y) \leq 1$. We have $\int_0^1 \frac{q(y)}{y}\mathrm{d}y < \infty$ by (51.12).

Conversely, suppose that $L_\mu(u)$ is expressed by (51.10) with $\gamma_0 = 0$ and that (51.14) and (51.15) are satisfied together with $\int_0^1 \frac{q(y)}{y}\mathrm{d}y < \infty$. We can find piecewise constant functions $q_n(y)$ with $0 \leq q_n(y) \leq 1$ and $\int_0^1 \frac{q_n(y)}{y}\mathrm{d}y < \infty$ such that (1) of Lemma 51.16 holds with $Q_n(\mathrm{d}y) = q_n(y)\mathrm{d}y$. Let $\mu_{(n)}$ be the distribution corresponding to Q_n and $\gamma_0 = 0$. Then $\mu_{(n)} \to \mu$. Further, for each n, we can construct $q_{n,k}(y)$ such that, as $k \to \infty$, $Q_{n,k}(\mathrm{d}y) = q_{n,k}(y)\mathrm{d}y$ tends to Q_n in the sense of (1) of Lemma 51.16 and that each $q_{n,k}(y)$ is of the form $\sum_{k=1}^m 1_{(a_k,b_k)}(y)$ with $0 < a_1 < b_1 < \cdots < a_m \leq b_m = \infty$. Let $\mu_{(n,k)}$ be the distribution corresponding to $Q_{n,k}$ and $\gamma_0 = 0$. Then $\mu_{(n,k)} \to \mu_{(n)}$. By Lemma 51.14, $\mu_{(n,k)}$ is a mixture of a finite number of exponential distributions and, possibly, of δ_0. Since ME is closed under convergence (Exercise 55.6), we have $\mu_{(n)} \in ME$ and then $\mu \in ME$. □

In the proof above we have not used Theorem 51.6 on the infinite divisibility of distributions of class ME. We have actually given an alternative proof of Theorem 51.6.

Proof of Theorem 51.10. Suppose that $\mu \in B$. Then we can find $\mu_{(n)}$, $n = 1, 2, \ldots$, such that $\mu_{(n)} \to \mu$ and that each $\mu_{(n)}$ is the convolution of a finite number of distributions of class ME. Hence, by Theorem 51.12 and Lemma 51.16, $L_\mu(u)$ has the representation as asserted.

Conversely, suppose that μ is a probability measure on $\mathbb{R}_+$ such that $L_\mu(u)$ admits the representation (51.13) with $\gamma_0 \geq 0$ and Q satisfying (51.12). Let $Q_n = Q + \gamma_0 n^2 \delta_n$ and let $\mu_{(n)}$ be the distribution corresponding to the pair of $\gamma_{0,n} = 0$ and Q_n. Then $\mu_{(n)} \to \mu$ by Lemma 51.16. Next, we can find, for each n, distributions $\mu_{(n,k)}$ such that $\mu_{(n,k)} \to \mu_{(n)}$ as $k \to \infty$ and that $L_{\mu_{(n,k)}}(u)$ has the representation (51.13) with $\gamma_0 = 0$ and $Q_{n,k}$ supported on a finite number of points. This is again by Lemma 51.16. Hence, in order to show that $\mu \in B$, we have only to prove it in the case that $\gamma_0 = 0$ and $Q = c\delta_a$ with some $c > 0$ and $a > 0$. Such a case is approximated by the case that $Q(\mathrm{d}y) = n1_{(a,a+c/n)}(y)\mathrm{d}y$, and this case is the convolution of n identical mixtures of δ_0 and an exponential distribution, as is seen from Lemma 51.14 or Example 51.15. Thus it belongs to B. $\square$

52. Unimodality and strong unimodality

Unimodality of a measure ρ on $\mathbb{R}$ is defined in Definition 23.2. To wit, ρ is said to be unimodal with mode $a \in \mathbb{R}$ if $\rho = c\delta_a + f(x)\mathrm{d}x$ with $0 \leq c \leq \infty$ and with $f(x)$ increasing on $(-\infty, a)$, decreasing on (a, ∞), and $\int_{|x-a|>\varepsilon} f(x)\mathrm{d}x < \infty$ for $\varepsilon > 0$. Recall that we are using the words *increase* and *decrease* in the wide sense allowing flatness. Thus, if ρ is a finite measure, $\rho = f(x)\mathrm{d}x$, such that $f(x)$ is equal to a constant c on $[a_1, a_2]$, increasing on $(-\infty, a_1)$, decreasing on (a_2, ∞), and $f(a_1-)$, $f(a_2+) \leq c$, then, for any $a \in [a_1, a_2]$, ρ is unimodal with mode a. Now, we give a general definition.

DEFINITION 52.1. Let ρ be a measure on $\mathbb{R}$, finite outside of some compact set. Let F be a finite set, possibly empty. Assume that

$$\rho(\mathrm{d}x) = \sum_{r \in F} c_r \delta_r(\mathrm{d}x) + f(x)\mathrm{d}x, \tag{52.1}$$

where $0 < c_r \leq \infty$ for $r \in F$ and f is a piecewise monotone function taking values in $[0, +\infty]$, but $f(x) < \infty$ except at finitely many points. Here *piecewise monotone*, of f, means that $\mathbb{R}$ is partitioned into finitely many disjoint intervals I_j, $j = 1, 2, \ldots, k$, such that, on each I_j, f is either increasing or decreasing. Further assume that $f(x) = f(x-) \vee f(x+)$ for every x. We call a point $a \in \mathbb{R}$ a *mode* of ρ if one of the following conditions holds.

(1) $a \in F$.

(2) There are a', a'', and ε such that $\varepsilon > 0$, $a' \leq a \leq a''$, $[a', a''] \cap F = \emptyset$, $f(x) = f(a)$ on $[a', a'']$, and $f(x) < f(a)$ on $(a' - \varepsilon, a') \cup (a'', a'' + \varepsilon)$.

In the case where (1) holds or in the case where (2) holds with $a' = a = a''$, the one-point set $\{a\}$ is called a *modal interval*. In the case where (2) holds with $a' < a''$, the interval $[a', a'']$ is called a modal interval. Two modal intervals are either disjoint or identical. The measure ρ is called *n-modal* if ρ has exactly n disjoint modal intervals; ρ is called *n-modal with modes* $a_1, \ldots, a_n$ if ρ is n-modal and the modal intervals containing $a_1, \ldots, a_n$ are disjoint. The word *bimodal* is used for 2-modal. The "unimodal" in Definition 23.2 coincides with 1-modal. In the following the function $f(x)$ for an n-modal measure is always taken to be piecewise monotone and to satisfy $f(x) = f(x+) \vee f(x-)$.

In Example 23.4 we have given a Lévy process $\{Y_t\}$ on $\mathbb{R}$ such that the distribution of Y_t is unimodal for $0 \leq t \leq 2$ but not unimodal for $t > 2$. Hence, if $0 < t \leq 2$, $0 < s \leq 2$, and $t + s > 2$, then $P_{Y_t} * P_{Y_s}$ is not unimodal although P_{Y_t} and P_{Y_s} are both unimodal. Thus convolution does not preserve unimodality. This phenomenon was first noticed by Chung [**79**]. A simple example of a unimodal probability measure whose convolution with itself is n-modal is given in [**442**]. In this connection, we give the following definition.

DEFINITION 52.2. A probability measure μ is *strongly unimodal* if the convolution of μ with any unimodal probability measure ρ is unimodal.

The following characterization of strong unimodality is known. We say that a function $f(x)$ is *log-concave* on an interval (b, c) if, on (b, c), $f(x) > 0$ and $\log f(x)$ is concave.

THEOREM 52.3. *Let μ be a non-trivial probability measure on $\mathbb{R}$. It is strongly unimodal if and only if*

(52.2) *there is an interval (b, c) with $\mu(b, c) = 1$ such that μ is absolutely continuous with density $f(x)$ log-concave on (b, c).*

The meaning of non-trivial is given in Definition 13.6. We prepare lemmas.

LEMMA 52.4. *Let μ_n be a unimodal probability measure with mode a_n for $n = 1, 2, \ldots$. If μ_n tends to a probability measure μ as $n \to \infty$, then μ is unimodal and $[\liminf_{n\to\infty} a_n, \limsup_{n\to\infty} a_n]$ is a subset of the modal interval of μ.*

Proof. See Exercise 29.20. □

LEMMA 52.5. *If μ is a unimodal probability measure, then there are unimodal probability measures μ_n, $n = 1, 2, \ldots$, such that $\mu_n \to \mu$ and such that each μ_n has C^∞ density on $\mathbb{R}$ and has compact support.*

Proof. We use the metric r of the convergence of probability measures in Exercise 6.9. Let μ be unimodal with mode a. Given $\varepsilon > 0$, choose a continuous ρ_1 unimodal with mode a such that $r(\mu, \rho_1) < \varepsilon$. For this it suffices to make $\rho_1 = [\mu]_{\mathbb{R}\setminus\{a\}} + \mu\{a\}\rho_{[a-\eta,a+\eta]}$ with $\eta > 0$ small enough. Here ρ_J denotes the uniform distribution on an interval J. Next, choose ρ_2, unimodal with mode a, such that $r(\rho_1, \rho_2) < \varepsilon$, ρ_2 has a piecewise constant density and has a compact support. For this, let

$$\rho_2 = (\rho_1[b_{-n}, b_n])^{-1} \sum_{j=-n}^{n-1} \rho_1[b_k, b_{k+1}]\rho_{[b_k,b_{k+1}]},$$

choosing $b_{-n} < \cdots < b_{-1} < b_0 = a < b_1 < \cdots < b_n$ with b_n and $-b_{-n}$ large enough and $\max_{-n\le k\le n-1}(b_{k+1} - b_k)$ small enough. Then choose ρ_3, unimodal with mode a, $r(\rho_2, \rho_3) < \varepsilon$, such that ρ_3 has C^∞ density and compact support. For this, let f_2 be the density of ρ_2 and choose a C^∞ function f_3 so that $f_3(x)$ vanishes on $(-\infty, b_{-n}] \cup [b_n, \infty)$, increases on $[b_{-n}, a]$, decreases on $[a, b_n]$, and $\int |f_2(x) - f_3(x)|dx$ small enough. Then $\rho_3 = cf_3(x)dx$ with a normalizing constant c will do. Since ε is arbitrary, the assertion is proved. □

LEMMA 52.6. *Suppose that, for $n = 1, 2, \ldots$, μ_n are probability measures satisfying (52.2) with (b, c) replaced by $(-\infty, \infty)$. If μ_n tends to a non-trivial probability measure μ, then μ satisfies (52.2) with some (b, c).*

Proof. Let f_n be the log-concave density on $(-\infty, \infty)$ of μ_n. Then there is $a_n \in \mathbb{R}$ such that $\log f_n$ is increasing on $(-\infty, a_n)$ and decreasing on (a_n, ∞). Hence μ_n is unimodal with mode a_n. Hence, by Lemma 52.4, there is a subsequence $\{\mu_{n_k}\}$ such that $a_{n_k} \to a$ with some a and μ is unimodal with mode a. We identify this subsequence with the full sequence $\{\mu_n\}$. Given any $\varepsilon > 0$, we have $f_n(x) \le 1/\varepsilon$ for $|x - a_n| > \varepsilon$. Hence, if n is so large that $|a_n - a| < \varepsilon/2$, then $f_n(x) \le 2/\varepsilon$ for $|x - a| > \varepsilon$. Hence, using Helly's selection theorem and choosing a further subsequence by the diagonal argument, we can find a function $h(x)$ on $\mathbb{R} \setminus \{a\}$ and a subsequence $\{f_{n_k}\}$ of $\{f_n\}$ such that $f_{n_k}(x) \to h(x)$ at any continuity point of $h(x)$. The function $h(x)$ can be chosen right-continuous and increasing on $(-\infty, a)$ and left-continuous and decreasing on (a, ∞). Then $h(x) = f(x)$. Let b and c be the left and right endpoints, respectively, of the support of μ. Let $g_n = \log f_n$ and $g = \log f$. From the assumption of log-concavity of f_n, we have

$$\frac{g_n(x_2) - g_n(x_1)}{x_2 - x_1} \ge \frac{g_n(x_3) - g_n(x_2)}{x_3 - x_2} \quad \text{for } x_1 < x_2 < x_3. \tag{52.3}$$

Hence

$$\frac{g(x_2)-g(x_1)}{x_2-x_1} \geq \frac{g(x_3)-g(x_2)}{x_3-x_2} \quad \text{for } b < x_1 < x_2 < x_3 < c \tag{52.4}$$

as long as x_1, x_2, and x_3 are continuity points of $g(x)$ in $(b,c)\setminus\{a\}$. By the right- or left-continuity we get (52.4) as long as x_1, x_2, and x_3 are not equal to a. Moreover, we have $g_n(x) \leq K$ with some K independent of n. To see this, use (52.3) for $x_1 < x_2 < x < a$ or $a < x < x_2 < x_3$. It follows that $\mu_n(a-\varepsilon, a+\varepsilon) \leq 2\varepsilon \mathrm{e}^K$ for any $\varepsilon > 0$. Hence $\mu(a-\varepsilon, a+\varepsilon) \leq 2\varepsilon \mathrm{e}^K$ and thus $\mu\{a\} = 0$. It also follows that $f(a+) < \infty$ if $a < c$ and that $f(a-) < \infty$ if $b < a$. Now let $f(a) = f(a-) \vee f(a+)$. Then we get (52.4) without any exceptional point. Thus $g(x)$ is convex, and hence continuous, on (b,c). □

Proof of Theorem 52.3. Step 1. Assume that μ has a log-concave density $f(x)$ on $(-\infty,\infty)$. Let us show that μ is strongly unimodal. By Lemma 52.5 and by Lemma 52.4, it suffices to prove the unimodality of $\mu*\rho$ under the assumption that ρ is unimodal, has C^∞ density $g(x)$ on $\mathbb{R}$ and has compact support. Convolution with a δ-distribution translates the mode. Hence we may assume that ρ has mode 0. The distribution $\mu*\rho$ is absolutely continuous with density

$$h(x) = \int_{-\infty}^{\infty} g(x-y)f(y)\mathrm{d}y.$$

It is of class C^∞ and

$$h'(x) = \int_{-\infty}^{\infty} g'(x-y)f(y)\mathrm{d}y = \int_{-\infty}^{\infty} g'(y)f(x-y)\mathrm{d}y.$$

If $x_1 < x_2$, then

$$f(x_2-y)/f(x_1-y) \geq f(x_2)/f(x_1) \quad \text{for } y > 0,$$
$$f(x_2-y)/f(x_1-y) \leq f(x_2)/f(x_1) \quad \text{for } y < 0.$$

Hence, using that $g'(y) \leq 0$ for $y > 0$ and that $g'(y) \geq 0$ for $y < 0$, we get

$$\begin{aligned} h'(x_2) &= \int_{-\infty}^{\infty} g'(y)\frac{f(x_2-y)}{f(x_1-y)}f(x_1-y)\mathrm{d}y \\ &\leq \int_{-\infty}^{\infty} g'(y)\frac{f(x_2)}{f(x_1)}f(x_1-y)\mathrm{d}y = \frac{f(x_2)}{f(x_1)}h'(x_1). \end{aligned}$$

It follows that

$$h'(x_2) \leq 0 \quad \text{if } h'(x_1) = 0 \text{ and } x_1 < x_2. \tag{52.5}$$

Obviously $h'(x) > 0$ at some x and $h'(y) < 0$ at some y. Let $a = \sup\{x\colon h'(x) > 0\}$. Then, by (52.5), $a < \infty$ and $h'(x) > 0$ for $x < a$ while $h'(x) \leq 0$ for $x > a$. Hence $\mu*\rho$ is unimodal with mode a.

Step 2. Assume that μ satisfies (52.2) with a log-concave density $f(x)$ on (b,c). We prove that μ is strongly unimodal. If $b = -\infty$ and $c = \infty$, then this is shown in Step 1. Suppose that $b > -\infty$ and $c = \infty$. For $\eta > 0$ and $\alpha > 0$, modify the function $f(x)$ and define

$$f_1(x) = \begin{cases} \beta f(x) & \text{for } x \geq b + \eta, \\ \beta f(b+\eta)\mathrm{e}^{\alpha(x-b-\eta)} & \text{for } x < b + \eta, \end{cases}$$

where β is chosen so that $\int f_1(x)\mathrm{d}x = 1$. Given $\varepsilon > 0$, we can choose $\eta > 0$ small and $\alpha > 0$ large so that $\mu_1 = f_1(x)\mathrm{d}x$ satisfies $r(\mu, \mu_1) < \varepsilon$ and that $f_1(x)$ is log-concave on $\mathbb{R}$. Here r is the metric in Exercise 6.9. By Step 1, μ_1 is strongly unimodal. If $b = -\infty$ and $c < \infty$, or if $b > -\infty$ and $c < \infty$, then make similar modification of f for large x or for large and small x. In each case μ is the limit of a sequence of strongly unimodal distributions. Hence we see strong unimodality of μ by using Lemma 52.4.

Step 3. Suppose that μ is strongly unimodal and has a continuous positive density $f(x)$ on $(-\infty, \infty)$. Let us show that $f(x)$ is log-concave. Suppose that it is not. Then there are $x_0 \in \mathbb{R}$ and $a > 0$ such that

$$f(x_0)^2 < f(x_0 - a)f(x_0 + a).$$

We claim that there exists a unimodal distribution ρ such that $\mu{*}\rho$ is not unimodal. Let $c > a$, $0 < \alpha < 1$, and $\beta = 1 - \alpha$. Let $\rho = g(x)\mathrm{d}x$ with

$$g(x) = \tfrac{\alpha}{c}1_{[a-c,0)}(x) + \tfrac{1}{c}1_{[0,a)}(x) + \tfrac{\beta}{c}1_{[a,c)}(x).$$

Then ρ is a unimodal probability measure. The density $h(x)$ of $\mu{*}\rho$ is of class C^1 and

$$\begin{aligned} h(x) &= \int_{-\infty}^{\infty} f(y)g(x-y)\mathrm{d}y \\ &= \frac{\alpha}{c}\int_x^{x+c-a} f(y)\mathrm{d}y + \frac{1}{c}\int_{x-a}^{x} f(y)\mathrm{d}y + \frac{\beta}{c}\int_{x-c}^{x-a} f(y)\mathrm{d}y, \\ h'(x) &= c^{-1}[\beta f(x) - \alpha f(x-a) + \alpha f(x+c-a) - \beta f(x-c)]. \end{aligned}$$

Now fix α so that $f(x_0)/f(x_0 - a) < \alpha/\beta < f(x_0 + a)/f(x_0)$. The strong unimodality of μ implies unimodality of it, because $\mu = \mu{*}\delta_0$. Hence $f(x) \to 0$ as $|x| \to \infty$. Therefore, as $c \to \infty$,

$$\begin{aligned} \frac{ch'(x_0)}{\beta f(x_0 - a)} &\to \frac{f(x_0)}{f(x_0 - a)} - \frac{\alpha}{\beta} < 0, \\ \frac{ch'(x_0 + a)}{\beta f(x_0)} &\to \frac{f(x_0 + a)}{f(x_0)} - \frac{\alpha}{\beta} > 0. \end{aligned}$$

This shows that $\mu{*}\rho$ is not unimodal if we choose c sufficiently large.

Step 4. Let μ be a general strongly unimodal distribution. Let ρ_n be Gaussian with mean 0 and variance $1/n$. Since the density $g_n(x)$ of ρ_n is log-concave on $(-\infty, \infty)$, ρ_n is strongly unimodal by Step 1. Hence $\mu{*}\rho_n$ is strongly unimodal. The density $\int g_n(x-y)\mu(\mathrm{d}y)$ of $\mu{*}\rho_n$ is positive and continuous, and hence log-concave on $(-\infty, \infty)$ by Step 3. We have $\mu*\rho_n \to \mu$ as $n \to \infty$. Now, by Lemma 52.6, μ satisfies (52.2), completing the proof. □

EXAMPLE 52.7. Theorem 52.3 enables us to check whether a given distribution with an explicit density is strongly unimodal or not. Gaussian and exponential distributions are thus strongly unimodal. The Cauchy distribution (Example 2.11) and the one-sided strictly stable distribution of index 1/2 (Example 2.13) are unimodal, but not strongly unimodal. We also see that a Γ-distribution as in Example 2.15 is strongly unimodal if and only if the parameter c is greater than or equal to 1, because the second derivative of the logarithm of the density is $(1-c)x^{-2}$ on $(0, \infty)$. Distributions of class ME are unimodal with mode 0, but they are not strongly unimodal except exponential distributions themselves. To see this, recall that the density $f(x)$ satisfies $ff'' - (f')^2 \geq 0$, as is shown in the proof of Theorem 51.6.

REMARK 52.8. If μ is strongly unimodal, then, by Theorem 52.3, its density $f(x)$ satisfies $f(x) \leq \mathrm{e}^{-c|x|}$ for $|x| > b$ with some $b > 0$ and $c > 0$. Combined with Example 25.10, it follows that any semi-stable distribution of index $\alpha \neq 2$ is not strongly unimodal.

The following two lemmas are useful.

LEMMA 52.9. *Let μ and ρ be probability measures on $\mathbb{R}$. Let $a, b \in \mathbb{R}$. Suppose that μ is supported on $[a, \infty)$ and $[\rho]_{(-\infty,b)}$ is absolutely continuous with increasing density. Then, $[\mu*\rho]_{(-\infty,a+b)}$ is absolutely continuous with increasing density.*

Proof. Considering translations we may and do assume that $a = b = 0$. Thus $\rho = f(x)\mathrm{d}x + \rho_0$, where $f(x)$ is increasing on $(-\infty, 0)$ and 0 on $(0, \infty)$ and ρ_0 is supported on $[0, \infty)$. We have

$$\mu*\rho = \left(\int_{[0,\infty)} f(x-y)\mu(\mathrm{d}y)\right)\mathrm{d}x + \mu*\rho_0.$$

Notice that $\mu*\rho_0$ is supported on $[0, \infty)$ and that $\int_{[0,\infty)} f(x-y)\mu(\mathrm{d}y)$ is increasing on $(-\infty, 0)$. □

LEMMA 52.10. *Let μ and ρ be probability measures on $\mathbb{R}$. Suppose that μ is unimodal with mode a and supported on $[a, \infty)$ and ρ is unimodal with mode b and supported on $(-\infty, b]$. Then, $\mu*\rho$ is unimodal with mode $a + b$.*

Proof. By the definition of unimodality, $[\rho]_{(-\infty,b)}$ is absolutely continuous with increasing density. Hence, by the preceding lemma, $[\mu*\rho]_{(-\infty,a+b)}$ is absolutely continuous with increasing density. Using the dual version of the lemma, we conclude that $[\mu*\rho]_{(a+b,\infty)}$ is absolutely continuous with decreasing density. □

Recall that, whenever μ is unimodal with mode a, we choose the density $f(x)$ of μ on $\mathbb{R}\setminus\{a\}$ right-continuous increasing on $(-\infty,a)$ and left-continuous decreasing on (a,∞), and define $f(a) = f(a-)\vee f(a+)$. Yamazato [**588**] introduces the following property and exhibits its power.

DEFINITION 52.11. Let μ be a probability measure unimodal with mode a. Let $b \geq -\infty$ be the left end of its support. We say that μ has *the Yamazato property* if $a > b$, $\mu\{a\} = 0$, $\log f(x)$ is concave on (b,a), and $f(a-) \geq f(a+)$, or if $a = b$.

THEOREM 52.12 (Yamazato [**588**]). *Let μ and ρ be unimodal with modes a and b, respectively. Let $\widetilde{\rho}$ be the dual of ρ, $\widetilde{\rho}(B) = \rho(-B)$ for $B \in \mathcal{B}(\mathbb{R})$. Suppose that μ and $\widetilde{\rho}$ have the Yamazato property. Then $\mu*\rho$ is unimodal. If, in addition, the left end of the support of μ and the right end of the support of ρ are both* 0*, then $\mu*\rho$ has a mode in $[b,a]$.*

Proof. Step 1. Let f and g be the densities of μ and ρ on $\mathbb{R}\setminus\{a\}$ and $\mathbb{R}\setminus\{b\}$, respectively. We prove the theorem under the assumptions that $S_\mu = [0,\infty)$, $S_\rho = (-\infty,0]$, that $f(0+) = 0$ if $a > 0$, that $g(0-) = 0$ if $b < 0$, that $\mu\{0\} = 0$ if $a = 0$, and that $\rho\{0\} = 0$ if $b = 0$. Let $h(x) = \int_0^\infty g(x-y)f(y)\mathrm{d}y = \int_{-\infty}^0 f(x-y)g(y)\mathrm{d}y$. Then h is the density of $\mu*\rho$ and

$$h \text{ is strictly increasing on } (-\infty,b) \text{ and strictly decreasing on } (a,\infty). \tag{52.6}$$

Indeed, if $x_1 < x_2 < b$, then

$$h(x_1) = \int_0^\infty g(x_1-y)f(y)\mathrm{d}y \leq \int_0^\infty g(x_2-y)f(y)\mathrm{d}y = h(x_2);$$

thus h is increasing on $(-\infty,b)$. If $h(x_1) = h(x_2)$ for some $x_1 < x_2 < b$, then $g(x_1-y) = g(x_2-y)$ for every y, which contradicts the unboundedness of S_ρ. Hence h is strictly increasing on $(-\infty,b)$. With similar discussion on (a,∞), we have (52.6).

If $a > 0$, then $f(a-) < \infty$ by the log-concavity. If $b < 0$, then $g(b+) < \infty$. If $a > 0$, then define, for $\varepsilon > 0$,

$$A_\varepsilon(x) = f(x+\varepsilon)/f(x), \quad x > 0.$$

We have

$$A_\varepsilon(x) \geq 1 \quad \text{on } (0,a-\varepsilon], \tag{52.7}$$

$$A_\varepsilon(x) \leq 1 \quad \text{on } (a,\infty), \tag{52.8}$$

$$A_\varepsilon(x) \text{ is decreasing on } (0,a]. \tag{52.9}$$

To see (52.7) and (52.8), use that f increases on $(0,a]$ and decreases on (a,∞). If $0 < x_1 < x_2 \leq a-\varepsilon$, then $A_\varepsilon(x_1) \geq A_\varepsilon(x_2)$ from the log-concavity. If $a-\varepsilon \leq x_1 < x_2 \leq a$, then $f(x_1+\varepsilon) \geq f(x_2+\varepsilon)$ (recall that

$f(a) \geq f(a+)$) and $f(x_1) \leq f(x_2)$, and hence $A_\varepsilon(x_1) \geq A_\varepsilon(x_2)$ again. This proves (52.9).

Let us prove the unimodality of $\mu*\rho$, considering three cases.

Case 1. $b = a = 0$. The unimodality with mode 0 follows from (52.6).

Case 2. Either $b = 0 < a$ or $b < 0 = a$. Assume the former; the latter case is similar. Assume, for the time being, that

$$f \text{ is continuous on } \mathbb{R} \text{ and } g(0-) < \infty. \tag{52.10}$$

Let m be the finite measure on $(-\infty, 0)$ induced by g by $m(x_1, x_2] = g(x_2) - g(x_1)$. Then

$$\begin{aligned} h(x) &= \int_{-\infty}^{0} f(x-y)g(y)\mathrm{d}y = \int_{-\infty}^{0} f(x-y)\left(g(0-) - \int_{(y,0)} m(\mathrm{d}z)\right)\mathrm{d}y \\ &= g(0-)\int_{x}^{\infty} f(y)\mathrm{d}y - \int_{(-\infty,0)} m(\mathrm{d}z)\int_{x-z}^{\infty} f(y)\mathrm{d}y. \end{aligned}$$

Hence h is of class C^1 on $\mathbb{R}$ and

$$h'(x) = \int_{(-\infty,0)} f(x-y)m(\mathrm{d}y) - g(0-)f(x).$$

We claim that

$$\text{if } h'(x_0) \leq 0 \text{ for some } x_0 > 0, \text{ then } h'(x) \leq 0 \text{ for every } x > x_0. \tag{52.11}$$

This fact will tell us that $\mu*\rho$ is unimodal with mode $\sup\{x \colon h'(x) > 0\}$. If $x_0 \geq a$ or if $x_0 < a \leq x$, then (52.11) follows from (52.6). Let $h'(x_0) \leq 0$ and $x_0 < x < a$. Let $x - x_0 = \varepsilon$. Then

$$h'(x) = \int_{(-\infty,0)} f(x_0-y)A_\varepsilon(x_0-y)m(\mathrm{d}y) - g(0-)f(x_0)A_\varepsilon(x_0).$$

Since $A_\varepsilon(x-y)$ is continuous in $y \leq 0$, we can find $y_1 < 0$ such that

$$h'(x) = A_\varepsilon(x_0 - y_1)\int_{(-\infty,0)} f(x_0-y)m(\mathrm{d}y) - g(0-)f(x_0)A_\varepsilon(x_0).$$

If $x_0 - y_1 \leq a$, then $A_\varepsilon(x_0-y_1) \leq A_\varepsilon(x_0)$ by (52.9). If $x_0 - y_1 > a$, then $A_\varepsilon(x_0 - y_1) \leq 1 \leq A_\varepsilon(x_0)$ by (52.7) and (52.8). Thus

$$h'(x) \leq A_\varepsilon(x_0)\left(\int_{(-\infty,0)} f(x_0-y)m(\mathrm{d}y) - g(0-)f(x_0)\right) = A_\varepsilon(x_0)h'(x_0).$$

Hence we get (52.11), and $\mu*\rho$ is unimodal. If (52.10) is not satisfied, then choose $\mu_n \to \mu$ and $\rho_n \to \rho$ such that μ_n is unimodal with mode a, continuous, and has support $[0,\infty)$, that ρ_n is unimodal with mode 0, continuous, and has support $(-\infty, 0]$, that the density f_n of μ_n is continuous on $\mathbb{R}$ and equal to f on $[0,a]$, and that the density g_n of ρ_n satisfies $g_n(0-) < \infty$. It is possible to choose such μ_n and ρ_n. Then $\mu_n*\rho_n$ is unimodal with mode in $[0,a]$. Hence, by Lemma 52.4, $\mu*\rho$ is unimodal with mode in $[0,a]$.

Case 3. $b < 0 < a$. Let us assume that

$$f \text{ and } g \text{ are continuous on } \mathbb{R}. \tag{52.12}$$

Let m be the signed measure induced by g by $m(x_1, x_2] = g(x_2) - g(x_1)$. It is a finite measure on $(-\infty, b]$ and $-m$ is a finite measure on (b, ∞). Then

$$h(x) = -\int_{-\infty}^{0} f(x-y)\mathrm{d}y \int_{(y,0)} m(\mathrm{d}z) = -\int_{(-\infty,0)} m(\mathrm{d}z) \int_{x-z}^{\infty} f(y)\mathrm{d}y,$$

which shows that h is of class C^1 on $\mathbb{R}$ and

$$h'(x) = \int_{(-\infty,0)} f(x-y)m(\mathrm{d}y) = \int_{(-\infty,0\wedge x)} f(x-y)m(\mathrm{d}y).$$

We claim that,

$$\text{if } h'(x_0) \geq 0 \text{ for some } x_0 \in (b, a+b], \text{ then } h'(x) \geq 0 \text{ for every } x \in (b, x_0). \tag{52.13}$$

To prove this, let $b < x = x_0 - \varepsilon < x_0 \leq a + b$ and see that

$$h'(x) = \int_{(-\infty,0\wedge x)} f(x_0 - y)A_\varepsilon(x-y)^{-1}m(\mathrm{d}y),$$

and then, from the continuity of $A_\varepsilon(x-y)$ in y, there are $-\infty < y_1 < b < y_2 < 0 \wedge x$ satisfying

$$\begin{aligned} h'(x) =& A_\varepsilon(x-y_1)^{-1} \int_{(-\infty,b]} f(x_0-y)m(\mathrm{d}y) \\ &+ A_\varepsilon(x-y_2)^{-1} \int_{(b,0\wedge x)} f(x_0-y)m(\mathrm{d}y). \end{aligned}$$

If $x - y_1 \leq a$, then $A_\varepsilon(x-y_1)^{-1} \geq A_\varepsilon(x-y_2)^{-1}$ by (52.9). If $x - y_1 > a$, then $A_\varepsilon(x-y_1)^{-1} \geq 1 \geq A_\varepsilon(x-y_2)^{-1}$ by (52.7) and (52.8). It follows that

$$h'(x) \geq A_\varepsilon(x-y_2)^{-1} \int_{(-\infty,0\wedge x)} f(x_0-y)m(\mathrm{d}y) = A_\varepsilon(x-y_2)^{-1}h'(x_0),$$

which proves (52.13). The dual assertion is that

$$\text{if } h'(x_0) \leq 0 \text{ for some } x_0 \in [a+b, a), \text{ then } h'(x) \leq 0 \text{ for every } x \in (x_0, a). \tag{52.14}$$

Now pick $c = \sup\{x \colon h'(x) > 0\}$. We have $b \leq c \leq a$ by (52.6). Let us see that $\mu{*}\rho$ is unimodal with mode c. Indeed, $h'(c) = 0$, $h'(x) \leq 0$ for every $x > c$, and points x satisfying $h'(x) > 0$ cluster at c from the left. We have $h' \geq 0$ on $(-\infty, b]$. We also have $h' \geq 0$ on (b, c), because, if $h'(x) < 0$ at some $x \in (b, c)$, then there are $b < x_1 - \varepsilon_1 < x_1 < x_2 < c$ such that $h' < 0$ on $(x_1 - \varepsilon_1, x_1)$, $h'(x_1) = 0$, and $h'(x_2) > 0$, which contradicts (52.13) if $x_1 \leq a + b$, and (52.14) if $x_1 > a + b$. We have shown that $\mu{*}\rho$ is unimodal with mode in $[b, a]$ under the assumption (52.12). In the case where (52.12) is violated, approximation by probability measures satisfying (52.12) gives unimodality with mode in $[b, a]$.

Step 2. Let μ and ρ be the probability measures given in the theorem. Denote by $a_0 \geq -\infty$ and $b_0 \leq \infty$, respectively, the left end of S_μ and the right end of S_ρ. Consider the case that $a_0 < a$ and $b < b_0$. To see the unimodality of $\mu * \rho$, we may assume that $a_0 > -\infty$ and $b_0 < \infty$, because approximation works. Further, by translation, we may assume that $a_0 = b_0 = 0$. For any $\varepsilon \in (0, a)$ we can choose f_1 such that $f_1 = f$ on $[\varepsilon, \infty)$, f_1 is log-concave on $(0, a)$, and $\lim_{x \downarrow 0} \log f_1 = -\infty$, and make a probability measure by multiplying a normalizing constant close to 1. For any $\varepsilon \in (b, 0)$ we can choose g_1 similarly for g. Thus μ and ρ are approximated by distributions satisfying the conditions in Step 1 and the result follows. In the case that $a_0 < a$ and $b = b_0$, approximate μ as above and, if $\rho\{b\} > 0$, approximate ρ by a distribution without point mass. The case that $a_0 = a$ and $b < b_0$ is similar. The case that $a_0 = a$ and $b_0 = b$ is treated by Lemma 52.10. □

EXAMPLE 52.13. Let μ_c be the one-sided strictly stable distribution of index 1/2 in Example 2.13. Then, for the density $f_c(x)$ on $(0, \infty)$, $(\log f_c)' = \frac{1}{2x^2}(c^2 - 3x)$ and $(\log f_c)'' = -\frac{1}{2x^3}(2c^2 - 3x)$. Thus μ_c is unimodal with mode $c^2/3$ and f_c is log-concave on $(0, 2c^2/3)$. By Theorem 52.12, $\mu_{c_1} * \widetilde{\mu_{c_2}}$ is unimodal for any $c_1 > 0$ and $c_2 > 0$. This example is a special case of the results given in the next section.

53. Selfdecomposable processes

A selfdecomposable process $\{X_t\}$ is defined, in Definition 15.6, to be a Lévy process with selfdecomposable distribution at $t = 1$. It has a selfdecomposable distribution at any t. In one dimension it is characterized in its generating triplet (A, ν, γ) by

$$\nu(\mathrm{d}x) = \frac{k(x)}{|x|}\mathrm{d}x \tag{53.1}$$

with $k(x)$ increasing on $(-\infty, 0)$ and decreasing on $(0, \infty)$ (Corollary 15.11). This ν is a special case of Lévy measures unimodal with mode 0. Hence we choose $k(x)$ right-continuous on $(-\infty, 0)$ and left-continuous on $(0, \infty)$. This we call the *k-function* of the selfdecomposable process $\{X_t\}$ or the selfdecomposable distribution $\mu = P_{X_1}$. Stable processes belong to the class of selfdecomposable processes. The k-function of an α-stable process with $\alpha \neq 2$ is a constant multiple of $|x|^{-\alpha}$ on each of $(-\infty, 0)$ and $(0, \infty)$.

The unimodality of selfdecomposable distributions on $\mathbb{R}$ had been an open problem for many years, since the 1940s. The affirmative answer was given in the symmetric case by Wintner [**576**] in 1956, in the one-sided case by Wolfe [**577**] in 1971, and finally in the general case by Yamazato [**588**] in 1978. As a consequence, the unimodality of stable distributions on $\mathbb{R}$ was given a complete proof for the first time.

THEOREM 53.1 (Yamazato [**588**]). *If $\{X_t\}$ is a selfdecomposable process on $\mathbb{R}$, then its distribution at any $t \geq 0$ is unimodal.*

This is equivalent to saying that any selfdecomposable distribution μ on $\mathbb{R}$ is unimodal. We will examine the one-sided case carefully and show that Theorem 52.12 is applicable in the two-sided case.

LEMMA 53.2. *Let μ be a selfdecomposable distribution on $\mathbb{R}_+$ with drift 0 and with k-function $k(x)$ being piecewise constant and, for some $0 < a_1 < \cdots < a_n < \infty$ and positive reals $c_1, \ldots, c_n$ with $c = \sum_{j=1}^n c_j$,*

$$k(x) = \begin{cases} c & \text{for } 0 < x \leq a_1, \\ c - \sum_{l=1}^{j} c_l & \text{for } a_j < x \leq a_{j+1} \\ 0 & \text{for } x > a_n. \end{cases}$$

Then the following are true.

(i) *The distribution μ is absolutely continuous and has a density $f(x)$ which is positive and continuous on $(0,\infty)$, equal to Kx^{c-1} on $(0, a_1]$ with $K = \text{const} > 0$, and of class C^1 on $(0,\infty) \setminus \{a_1, \ldots, a_n\}$.*

(ii) *If $c \leq 1$, then $f'(x) < 0$ on $(a_1, \infty) \setminus \{a_2, \ldots, a_n\}$.*

(iii) *If $c > 1$, then f is of class C^1 on $(0,\infty)$ and there is a point $a > b$ with $b = \sup\{x : k(x) \geq 1\}$ such that f' is positive on $(0, a)$, 0 at a, and negative on (a, ∞) and f is log-concave on $(0, a)$.*

Notice that, in the case $c = 1$, μ provides an example of a unimodal infinitely divisible distribution with modal interval not being a one-point set.

Proof of lemma. (i) It is shown in Theorem 28.4 that μ is absolutely continuous with density $f(x)$ continuous on $(0,\infty)$. It has the Lévy measure $\nu = 1_{(0,\infty)}(x)\frac{k(x)}{x}\mathrm{d}x$ and the drift $\gamma_0 = 0$. By Theorem 51.1, it satisfies the equation

$$\int_{(0,x]} y\mu(\mathrm{d}y) = \int_0^x \mu[0, x-y]k(y)\mathrm{d}y, \quad x > 0. \tag{53.2}$$

Let $F(x) = \mu(-\infty, x]$, the distribution function of μ, and let $\rho = \sum_{j=1}^n c_j \delta_{a_j}$. Since $k(y) = \rho[y,\infty)$, we have

$$\begin{aligned}\int_0^x \mu[0, x-y]k(y)\mathrm{d}y &= \int_0^x F(x-y)\mathrm{d}y \int_{[y,\infty)} \rho(\mathrm{d}z) \\ &= \int_{(0,\infty)} \rho(\mathrm{d}z) \int_0^{x\wedge z} F(x-y)\mathrm{d}y = \int_{(0,\infty)} \rho(\mathrm{d}z) \int_{x-(x\wedge z)}^x F(y)\mathrm{d}y\end{aligned}$$

$$= \int_{(0,\infty)} \rho(\mathrm{d}z) \left(\int_0^x F(y)\mathrm{d}y - \int_0^{x-(x\wedge z)} F(y)\mathrm{d}y \right)$$

and

$$\int_0^{x-(x\wedge z)} F(y)\mathrm{d}y = \int_0^{0\vee(x-z)} F(y)\mathrm{d}y = \int_z^{z\vee x} F(y-z)\mathrm{d}y = \int_0^x F(y-z)\mathrm{d}y.$$

Hence, (53.2) is written as

$$\int_0^x yf(y)\mathrm{d}y = \int_0^x \mathrm{d}y \int_{(0,\infty)} (F(y) - F(y-z))\rho(\mathrm{d}z).$$

It follows that

$$\int_{(0,\infty)} (F(x-y) - F(x))\rho(\mathrm{d}y) + xf(x) = 0, \quad x \neq 0. \tag{53.3}$$

In our case

$$\sum_{j=1}^{n} c_j(F(x-a_j) - F(x)) + xf(x) = 0, \quad x \neq 0. \tag{53.4}$$

Consequently, $f(x)$ is of class C^1 for $x \neq 0, a_1, \ldots, a_n$, and

$$\sum_{j=1}^{n} c_j(f(x-a_j) - f(x)) + f(x) + xf'(x) = 0, \quad x \neq 0, \tag{53.5}$$

that is,

$$xf'(x) = (c-1)f(x) - \int_{(0,x)} f(x-y)\rho(\mathrm{d}y). \tag{53.6}$$

If $0 < x < a_1$, then $xf'(x) = (c-1)f(x)$ by (53.6) and hence $f(x) = Kx^{c-1}$. Since $S_\mu = [0,\infty)$, the constant K is positive. It follows that $F(x)$ is strictly increasing on $[0, a_1]$. Hence, by (53.4), $f(x) > 0$ on $(0, a_1 + a_n)$. Repeating this, we conclude that

$$f(x) > 0 \quad \text{for } x > 0. \tag{53.7}$$

(ii) If $c \leq 1$, then we consider (53.6) on $(a_1, \infty) \setminus \{a_2 \ldots, a_n\}$ and find that the right-hand side is not greater than $-c_1 f(x-a_1)$, which means $f'(x) < 0$ by (53.7).

(iii) Assume $c > 1$. Now f is continuous on the whole line and hence f is of class C^1 for $x \neq 0$ and satisfies (53.5) and (53.6) even at $x = a_1, \ldots, a_n$. We claim that

$$f'(x) > 0 \quad \text{for } 0 < x \leq b. \tag{53.8}$$

Suppose that (53.8) is not true. Then, there is $0 < x_1 \le b$ such that $f' > 0$ on $(0, x_1)$ and $f'(x_1) = 0$. By (53.5),

$$f(x_1) = \int (f(x_1) - f(x_1 - y))\rho(\mathrm{d}y)$$
$$= \int_{(0,x_1)} (f(x_1) - f(x_1 - y))\rho(\mathrm{d}y) + k(x_1)f(x_1).$$

Since $k(x_1) \ge 1$ and since $f(x_1) - f(x_1 - y) > 0$ for $y \in (0, x_1)$, we get $k(x_1) = 1$ and $\int_{(0,x_1)} \rho(\mathrm{d}y) = k(0+) - k(x_1) = 0$, which contradicts $c > 1$. Hence (53.8) is proved. Next we claim that

(53.9) two points $0 < x_1 < x_2$ with the following properties do not exist: $f'(x_1) = f'(x_2) = 0$ and f is strictly increasing on $(0, x_1)$, and decreasing on (x_1, x_2).

Suppose that there are such two points. Let $x_2 - x_1 = \varepsilon$. It follows from (53.5) that

$$f(x_2) = \int (f(x_2) - f(x_2 - y))\rho(\mathrm{d}y) \le \int_{(\varepsilon,\infty)} (f(x_2) - f(x_2 - y))\rho(\mathrm{d}y)$$
$$= k(\varepsilon+)f(x_2) - \int_{(\varepsilon,x_2)} f(x_2 - y)\rho(\mathrm{d}y).$$

Since $f(x_2) > 0$ by (53.7), we get $k(\varepsilon+) \ge 1$. On the other hand (53.5) says that

$$f(x_1) = \int_{(0,\varepsilon]} (f(x_1) - f(x_1 - y))\rho(\mathrm{d}y) + k(\varepsilon+)f(x_1)$$
$$- \int_{(\varepsilon,x_2)} f(x_1 - y)\rho(\mathrm{d}y).$$

We have $x_1 > b$ by (53.8) and hence $k(x_2) < 1 < c$. Besides, $f(x_1) - f(x_1 - y) > 0$ and, for $y \in (\varepsilon, x_2)$, $f(x_1 - y) < f(x_2 - y)$. Consequently,

$$f(x_1) > k(\varepsilon+)f(x_1) - \int_{(\varepsilon,x_2)} f(x_2 - y)\rho(\mathrm{d}y),$$

which implies

$$(1 - k(\varepsilon+))f(x_2) > - \int_{(\varepsilon,x_2)} f(x_2 - y)\rho(\mathrm{d}y).$$

This is in contradiction with the equality above. Hence (53.9) is proved.

There exists a point $x > 0$ such that $f'(x) = 0$. Let a be the infimum of all such points. Then $a > b \ge a_1$, $f'(a) = 0$, and $f' > 0$ on $(0, a)$. Now, let us show that

(53.10) $$f' < 0 \quad \text{on } (a, \infty).$$

We have only to prove that $f'(a+\varepsilon) < 0$ for all sufficiently small $\varepsilon > 0$, since we have (53.9). Fix $0 < \varepsilon_0 < a_1$ and $0 < \alpha < a-b$, and let $0 < \varepsilon < \varepsilon_0$. Using (53.6) at $x = a$ and at $x = a+\varepsilon$, we have

$$(a+\varepsilon)f'(a+\varepsilon) =(c-1)(f(a+\varepsilon)-f(a)) \\ - \int_{(0,a+\varepsilon)} (f(a+\varepsilon-y)-f(a-y))\rho(\mathrm{d}y).$$

Noting that $\rho(0,a_1) = 0$ and $\rho(a,a+\varepsilon) = 0$ for small ε, we see that, as $\varepsilon \downarrow 0$,

$$(a+\varepsilon)f'(a+\varepsilon) = -\int_{[a_1,a]} (f(a+\varepsilon-y)-f(a-y))\rho(\mathrm{d}y) + o(\varepsilon) \\ \leq -\int_{[b,b+\alpha]} (f(a+\varepsilon-y)-f(a-y))\rho(\mathrm{d}y) + o(\varepsilon).$$

Write $M = \min\{f'(y) \colon y \in [a-b-\alpha, a+\varepsilon_0-b]\}$. Then $M > 0$ and

$$f(a+\varepsilon-y)-f(a-y) \geq \varepsilon \min_{z\in[a-y,a+\varepsilon-y]} f'(z) \geq \varepsilon M$$

for $y \in [b, b+\alpha]$. Since $\rho\{b\} > 0$, we get $(a+\varepsilon)f'(a+\varepsilon) < 0$ when $\varepsilon > 0$ is small enough. Hence (53.10).

It remains to prove that f is log-concave on $(0,a)$. Differentiating (53.5), we have, for $x \neq 0, a_1, \ldots, a_n$,

$$\sum_{j=1}^{n} c_j(f'(x-a_j)-f'(x)) + 2f'(x) + xf''(x) = 0, \tag{53.11}$$

that is,

$$xf''(x) = (c-2)f'(x) - \int_{(0,x)} f'(x-y)\rho(\mathrm{d}y). \tag{53.12}$$

If $1 < c \leq 2$, then the right-hand side of (53.12) is nonpositive on $(0,a) \setminus \{a_1, \ldots, a_n\}$ and hence, on $(0,a)$, f' is decreasing, $(\log f)' = f'/f$ is decreasing, and f is thus log-concave. Let us consider the case $c > 2$. We need more delicate argument. We have $f'(x) = K(c-1)x^{c-2}$ and hence f is of class C^1 on the whole line and, from (53.11), f is of class C^2 for $x \neq 0$. Write $g(x) = (\log f)'' = (f''f-(f')^2)/f^2$. Then, $g = (1-c)/x^2 < 0$ on $(0,a_1]$. Using (53.6) and (53.12) on $\mathbb{R}$, we have

$$x(f''(x)f(x)-f'(x)^2) \\ = -f'(x)f(x) - \int_{(0,x)} (f'(x-y)f(x)-f(x-y)f'(x))\rho(\mathrm{d}y). \tag{53.13}$$

Suppose that g has a zero in $(0,a)$. Then we can choose $x_0 \in (0,a)$ such that $g < 0$ on $(0,x_0)$ and $g(x_0) = 0$. The left-hand side of (53.13) vanishes at $x = x_0$, but the right-hand side is negative since

$$f'(x_0-y)f(x_0)-f(x_0-y)f'(x_0) = f(x_0)f(x_0-y)\left(\frac{f'(x_0-y)}{f(x_0-y)}-\frac{f'(x_0)}{f(x_0)}\right) > 0.$$

This is absurd. Hence $g > 0$ on $(0, a)$, that is, f is log-concave on $(0, a)$. □

REMARK 53.3. In the lemma above, let m be the mean of μ. If $c > 1$, then we can prove that $a < m$. Indeed, we have $m = \int_0^\infty k(x)\mathrm{d}x$ and, from (53.3),

$$\begin{aligned} af(a) &= \textstyle\int_{(0,\infty)}(F(a) - F(a-y))\rho(\mathrm{d}y) = \int_{(0,\infty)}\rho(\mathrm{d}y)\int_{a-y}^{a} f(x)\mathrm{d}x \\ &= \textstyle\int_0^a f(x)\mathrm{d}x \int_{[a-x,\infty)}\rho(\mathrm{d}y) = \int_0^a f(x)k(a-x)\mathrm{d}x \\ &< f(a)\textstyle\int_0^a k(a-x)\mathrm{d}x \le f(a)m. \end{aligned}$$

Proof of Theorem 53.1. Let μ be selfdecomposable. In order to show the unimodality of μ, it is enough to consider the purely non-Gaussian case, because addition of the Gaussian part preserves the unimodality by Theorem 52.3. The Lévy measure ν of μ has expression (53.1). Thus μ is the limit of μ_n defined by

$$\widehat{\mu}_n(z) = \exp\left[\int_{-\infty}^{\infty}(\mathrm{e}^{\mathrm{i}zx} - 1 - \mathrm{i}zx 1_{[-1,1]}(x))(k(x)\wedge n)\frac{\mathrm{d}x}{|x|} + \mathrm{i}\gamma z\right].$$

Here $\gamma \in \mathbb{R}$ and $k(x)$ is increasing on $(-\infty, 0)$ and decreasing on $(0, \infty)$. By Lemma 52.4, we have only to show the unimodality of μ_n. By translation it is enough to consider the case of μ with drift 0 and with $0 < k(0-)+k(0+) < \infty$. This μ is the limit of a sequence of distributions, each of which has piecewise constant k-function with a finite number of steps. Thus, again by Lemma 52.4, we have only to consider $\mu = \mu_1 * \widetilde{\mu}_2$, where μ_1 and μ_2 are of the type treated in Lemma 53.2 and $\widetilde{\mu}_2$ is the dual of μ_2. The lemma says that μ_1 and μ_2 are unimodal and have the Yamazato property. Hence, by Theorem 52.12, μ is unimodal. □

The preceding arguments give consequences on the location of modes.

THEOREM 53.4. *Let μ be a selfdecomposable distribution on $\mathbb{R}$ with k-function $k(x)$ and*

$$\widehat{\mu}(z) = \exp\left[\int_{-\infty}^{\infty}(\mathrm{e}^{\mathrm{i}zx} - 1)\frac{k(x)}{|x|}\mathrm{d}x\right], \tag{53.14}$$

$$\int_{|x|<1} k(x)\mathrm{d}x < \infty \quad \textit{and} \quad 0 < k(0-) + k(0+) \le \infty. \tag{53.15}$$

If $k(0-) = 0$ and $k(x) \ge 1$ for some $x > 0$, then denote by b the supremum of such x.

(i) *If $k(0-) \le 1$, then μ has a mode $a \ge 0$.*

(ii) *If $k(0-) \le 1$ and $k(0+) \le 1$, then μ has a mode 0.*

(iii) *If* $k(0-) = 0$, $k(0+) = 1$, *and* $k(x) = 1$ *for some* $x > 0$, *then the modal interval of* μ *contains* $[0, b]$.

(iv) *If* $k(0-) = 0$ *and* $k(0+) > 1$, *then* μ *has a mode* a *such that* $b \le a \le m$, *where* m *is the mean of* μ.

Proof. If μ has a piecewise constant k-function $k(x)$ with $k(0-) = 0$ and $k(0+) \le 1$, then μ is unimodal with mode 0 by Lemma 53.2. By approximation by such distributions, we see that,

$$\text{if } k(0-) = 0 \text{ and } k(0+) \le 1, \text{ then } \mu \text{ is unimodal with mode } 0. \tag{53.16}$$

In order to show (iii), we have only to make this approximation with the interval $\{x: k(x) = 1\}$ unaltered and to use Lemmas 52.4 and 53.2. The proof of (iv) is similar, if we use Remark 53.3 and approximate $m = \int_0^\infty k(x)\mathrm{d}x$ simultaneously. To show (i), represent μ as $\mu = \mu_1 * \mu_2$, where μ_1 and μ_2 are selfdecomposable distributions supported on $[0, \infty)$ and $(-\infty, 0]$, respectively. Applying (53.16) to the dual $\widetilde{\mu}_2$ of μ_2 and using Lemma 52.9, we get (i). Then (ii) follows from (i) and its dual. □

REMARK 53.5. In (iii) of the theorem above, the conclusion can be strengthened to saying that the modal interval is exactly $[0, b]$. Any selfdecomposable distribution whose modal interval is not a one-point set falls into this case by translation and, possibly, by going to the dual. The statement of (iv) can be strengthened to $b < a < m$. See [**466**].

Let us study the asymptotics of the density $f(x)$ as $x \to 0$, when the drift is 0. First we examine the one-sided case, using the Tauberian theorem of Laplace transforms.

THEOREM 53.6 ([**466**]). *Let* μ *be a selfdecomposable distribution on* $\mathbb{R}_+$ *with k-function* $k(x)$ *satisfying* $c = k(0+) < \infty$, *and let*

$$L_\mu(u) = \exp\left[\int_0^\infty (\mathrm{e}^{-ux} - 1)\frac{k(x)}{x}\mathrm{d}x\right], \quad u \ge 0. \tag{53.17}$$

Define

$$K(x) = \exp\left[\int_x^1 (c - k(y))\frac{\mathrm{d}y}{y}\right], \quad x > 0. \tag{53.18}$$

Then, the density $f(x)$ *of* μ *satisfies*

$$f(x) \sim \frac{\kappa}{\Gamma(c)} x^{c-1} K(x) \quad \text{as } x \downarrow 0, \tag{53.19}$$

where κ *is the constant given by*

$$\kappa = \exp\left[c\int_0^1 (\mathrm{e}^{-x} - 1)\frac{\mathrm{d}x}{x} + c\int_1^\infty \mathrm{e}^{-x}\frac{\mathrm{d}x}{x} - \int_1^\infty k(x)\frac{\mathrm{d}x}{x}\right]. \tag{53.20}$$

Proof. The function $K(x)$ is slowly varying at 0, that is, $K(ux)/K(x)$ tends to 1 as $x \downarrow 0$ for any $0 < u < 1$. In fact,

$$\frac{K(ux)}{K(x)} = \exp\left[\int_{ux}^{x} (c - k(y))\frac{\mathrm{d}y}{y}\right] = \exp\left[\int_{u}^{1} (c - k(xy))\frac{\mathrm{d}y}{y}\right] \to 1.$$

Let us show that

$$L_\mu(u) \sim \kappa u^{-c} K(1/u) \quad \text{as } u \to \infty. \tag{53.21}$$

Split $\int_0^\infty (\mathrm{e}^{-ux} - 1)\frac{k(x)}{x}\mathrm{d}x$ into three and observe that, as $u \to \infty$,

$$\int_0^{1/u} (\mathrm{e}^{-ux} - 1)\frac{k(x)}{x}\mathrm{d}x = \int_0^1 (\mathrm{e}^{-x} - 1)\frac{k(x/u)}{x}\mathrm{d}x \to c\int_0^1 (\mathrm{e}^{-x} - 1)\frac{\mathrm{d}x}{x},$$

$$\int_1^\infty (\mathrm{e}^{-ux} - 1)\frac{k(x)}{x}\mathrm{d}x \to -\int_1^\infty \frac{k(x)}{x}\mathrm{d}x,$$

$$\begin{aligned}\int_{1/u}^1 &(\mathrm{e}^{-ux} - 1)\frac{k(x)}{x}\mathrm{d}x + c\log u - \log K(1/u)\\ &= \int_{1/u}^1 (\mathrm{e}^{-ux} - 1)\frac{k(x)}{x}\mathrm{d}x + c\int_{1/u}^1 \frac{\mathrm{d}x}{x} - \int_{1/u}^1 (c - k(x))\frac{\mathrm{d}x}{x}\\ &= \int_{1/u}^1 \mathrm{e}^{-ux}\frac{k(x)}{x}\mathrm{d}x = \int_1^u \mathrm{e}^{-x}\frac{k(x/u)}{x}\mathrm{d}x \to c\int_1^\infty \mathrm{e}^{-x}\frac{\mathrm{d}x}{x}.\end{aligned}$$

Then we see that

$$\frac{L_\mu(u)}{u^{-c}K(1/u)} \to \exp\left[c\int_0^1 (\mathrm{e}^{-x} - 1)\frac{\mathrm{d}x}{x} - \int_1^\infty k(x)\frac{\mathrm{d}x}{x} + c\int_1^\infty \mathrm{e}^{-x}\frac{\mathrm{d}x}{x}\right] = \kappa,$$

that is, (53.21). We obtain from (53.21) that

$$\mu[0, x] \sim \frac{\kappa}{\Gamma(c+1)} x^c K(x) \quad \text{as } x \downarrow 0 \tag{53.22}$$

by Karamata's Tauberian theorem (Feller [**139**], p. 445). Since the density $f(x)$ is monotone in a right neighborhood of the origin by the unimodality, (53.22) leads to (53.19) by the dual version of a theorem of Feller [**139**], p. 446. □

REMARK 53.7. In Theorem 53.6 above, define $N \in \mathbb{Z}_+$ by $N < c \le N + 1$. Then, by Theorem 28.4, the density $f(x)$ of μ is of class C^N on $(0, \infty)$. In [**466**] it is proved that, for $j = 1, \ldots, N$, the jth derivative $f^{(j)}(x)$ satisfies

$$f^{(j)}(x) \sim \frac{\kappa}{\Gamma(c - j)} x^{c-j-1} K(x) \quad \text{as } x \downarrow 0. \tag{53.23}$$

In the two-sided case, we have to manipulate the characteristic function.

THEOREM 53.8 ([**466**]). *Let μ be a selfdecomposable distribution on $\mathbb{R}$ satisfying (53.14) with k-function $k(x)$ satisfying*

$$c = k(0+) + k(0-) < \infty, \quad k(0+) > 0, \quad \textit{and} \quad k(0-) > 0. \tag{53.24}$$

Let $c' = k(0+) - k(0-)$. *Define*

$$K(x) = \exp\left[\int_{|x|}^{1} (c - k(y) - k(-y))\frac{\mathrm{d}y}{y}\right], \quad x \neq 0, \tag{53.25}$$

$$L(x) = \int_{|x|}^{1} K(y)\frac{\mathrm{d}y}{y}, \quad x \neq 0, \tag{53.26}$$

$$\kappa = \exp\biggl[c\int_0^1 (\mathrm{e}^{-x} - 1)\frac{\mathrm{d}x}{x} + c\int_1^\infty \mathrm{e}^{-x}\frac{\mathrm{d}x}{x} - \int_1^\infty (k(x) + k(-x))\frac{\mathrm{d}x}{x}\biggr]. \tag{53.27}$$

Then the density $f(x)$ *of* μ *satisfies the following.*

(i) *If* $c < 1$, *then*

$$f(x) \sim \frac{\kappa \sin k(0+)\pi}{\Gamma(c)\sin c\pi} x^{c-1} K(x) \quad \textit{as } x \downarrow 0 \tag{53.28}$$

and

$$f(x) \sim \frac{\kappa \sin k(0-)\pi}{\Gamma(c)\sin c\pi} |x|^{c-1} K(x) \quad \textit{as } x \uparrow 0. \tag{53.29}$$

(ii) *If* $c = 1$, *then*

$$f(x) \sim \frac{\kappa}{\pi}\Bigl(\cos\frac{c'\pi}{2}\Bigr) L(x) \quad \textit{as } x \to 0 \tag{53.30}$$

and

$$\lim_{x\downarrow 0} \frac{f(x) - f(-x)}{K(x)} = \kappa \sin\frac{c'\pi}{2} \quad \textit{as } x \downarrow 0. \tag{53.31}$$

LEMMA 53.9. *Let* μ *be as in Theorem 53.8. Then, there is a constant* M *such that, for* $|z| \geq 1$,

$$|\widehat{\mu}(z)| \leq M|z|^{-c} K(|z|^{-1}). \tag{53.32}$$

Proof. Since $\widehat{\mu}(-z) = \overline{\widehat{\mu}(z)}$, we may assume $z \geq 1$. Let $r(x) = k(x) + k(-x)$. Then we have, from (53.14), that

$$\begin{aligned}
|\widehat{\mu}(z)| &= \exp\bigl[\textstyle\int_0^\infty (\cos zx - 1)\frac{r(x)}{x}\mathrm{d}x\bigr] \leq \exp\bigl[\int_{1/z}^1 (\cos zx - 1)\frac{r(x)}{x}\mathrm{d}x\bigr] \\
&= \exp\bigl[-c\log z + \textstyle\int_{1/z}^1 (c - r(x))\frac{\mathrm{d}x}{x} + \int_{1/z}^1 (\cos zx)\frac{r(x)}{x}\mathrm{d}x\bigr] \\
&= z^{-c} K(z^{-1}) \exp\bigl[\textstyle\int_1^z (\cos x) r(\frac{x}{z})\frac{\mathrm{d}x}{x}\bigr].
\end{aligned}$$

Using the bounded signed measure ρ defined by $\rho(x_1, x_2] = r(x_2+) - r(x_1+)$ for $0 < x_1 < x_2$, we have

$$\begin{aligned}
\textstyle\int_1^z (\cos x) r(\frac{x}{z})\frac{\mathrm{d}x}{x} &= r(z^{-1}+)\textstyle\int_1^z \cos x\frac{\mathrm{d}x}{x} + \int_1^z \cos x\frac{\mathrm{d}x}{x}\int_{(1/z, x/z]}\rho(\mathrm{d}y) \\
&= r(z^{-1}+)\textstyle\int_1^z \cos x\frac{\mathrm{d}x}{x} + \int_{(1/z,1]}\rho(\mathrm{d}y)\int_{zy}^z \cos x\frac{\mathrm{d}x}{x},
\end{aligned}$$

which tends to $c\lim_{\eta\to\infty}\int_1^\eta \cos x\frac{\mathrm{d}x}{x}$ as $z \to \infty$. □

Proof of Theorem 53.8. The function $K(x)$ is slowly varying at 0. So is the function $L(x)$, as is seen by l'Hospital's rule, since $L(0+) = \infty$. Let $r(x) = k(x) + k(-x)$ and $s(x) = k(x) - k(-x)$.

Step 1. We prove that, for any $u \neq 0$,

$$\widehat{\mu}(\tfrac{u}{x}) \sim \kappa |u|^{-c} \mathrm{e}^{\mathrm{i}(\operatorname{sgn} u)c'\pi/2} x^c K(x) \quad \text{as } x \downarrow 0 \tag{53.33}$$

and that, for any $u \neq 0$ and $0 < |x| \le 1$,

$$|\widehat{\mu}(\tfrac{u}{x})| \le M|u|^{-c} K(|u|^{-1} \wedge 1)|x|^c K(x) \tag{53.34}$$

with the constant M in Lemma 53.9.

It follows from (53.14) that

$$\widehat{\mu}(\tfrac{u}{x}) = \exp\big[\textstyle\int_0^\infty (\cos \frac{uy}{x} - 1)\frac{r(y)}{y}\mathrm{d}y + \mathrm{i}\int_0^\infty (\sin \frac{uy}{x})\frac{s(y)}{y}\mathrm{d}y\big].$$

Let $u > 0$ and let $x \downarrow 0$. Split the integrals into those over $(0, x/u)$, $(x/u, 1)$, and $(1, \infty)$. Then

$$\textstyle\int_1^\infty (\cos \frac{uy}{x} - 1)\frac{r(y)}{y}\mathrm{d}y \to -\int_1^\infty \frac{r(y)}{y}\mathrm{d}y \quad \text{and} \quad \int_1^\infty (\sin \frac{uy}{x})\frac{s(y)}{y}\mathrm{d}y \to 0$$

by the Riemann–Lebesgue theorem, and

$$\textstyle\int_0^{x/u} (\cos \frac{uy}{x} - 1)\frac{r(y)}{y}\mathrm{d}y = \int_0^1 (\cos y - 1) r(\frac{xy}{u})\frac{\mathrm{d}y}{y} \to c\int_0^1 (\cos y - 1)\frac{\mathrm{d}y}{y},$$

$$\textstyle\int_0^{x/u} (\sin \frac{uy}{x})\frac{s(y)}{y}\mathrm{d}y = \int_0^1 (\sin y) s(\frac{xy}{u})\frac{\mathrm{d}y}{y} \to c' \int_0^1 \sin y \frac{\mathrm{d}y}{y}.$$

Writing improper Riemann integrals on $[0, \infty)$ as $\int_0^{\to\infty}$, we have

$$\textstyle\int_{x/u}^1 (\cos \frac{uy}{x} - 1)\frac{r(y)}{y}\mathrm{d}y + c\log \frac{u}{x} - \log K(x) \to c\int_1^{\to\infty} \cos y \frac{\mathrm{d}y}{y},$$

because

$$\begin{aligned}
&\textstyle\int_{x/u}^1 (\cos \frac{uy}{x} - 1)\frac{r(y)}{y}\mathrm{d}y + c\log \frac{u}{x} - \log K(x) \\
&\quad = \textstyle\int_{x/u}^1 (\cos \frac{uy}{x})\frac{r(y)}{y}\mathrm{d}y + \int_{x/u}^x (c - r(y))\frac{\mathrm{d}y}{y} \\
&\quad = \textstyle\int_1^{u/x} (\cos y) r(\frac{xy}{u})\frac{\mathrm{d}y}{y} + \int_{1/u}^1 (c - r(xy))\frac{\mathrm{d}y}{y},
\end{aligned}$$

where the second term tends to 0 and the first term tends to $c\int_1^{\to\infty} \cos y \frac{\mathrm{d}y}{y}$ as in the proof of Lemma 53.9. Similarly,

$$\begin{aligned}
\textstyle\int_{x/u}^1 (\sin \frac{uy}{x})\frac{s(y)}{y}\mathrm{d}y &= \textstyle\int_1^{u/x} (\sin y) s(\frac{xy}{u})\frac{\mathrm{d}y}{y} \\
&= \textstyle s(\frac{x}{u}+)\int_1^{u/x} \sin y \frac{\mathrm{d}y}{y} + \int_1^{u/x} (\sin y)(s(\frac{xy}{u}+) - s(\frac{x}{u}+))\frac{\mathrm{d}y}{y} \\
&\to \textstyle c' \int_1^{\to\infty} \sin y \frac{\mathrm{d}y}{y}.
\end{aligned}$$

Together we have

$$\begin{aligned}
\widehat{\mu}(\tfrac{u}{x})\frac{1}{x^c K(x)} \to \frac{1}{u^c}\exp\big[&\textstyle -\int_1^\infty r(y)\frac{\mathrm{d}y}{y} + c\int_0^1 (\cos y - 1)\frac{\mathrm{d}y}{y} \\
&\textstyle + c\int_1^{\to\infty} \cos y \frac{\mathrm{d}y}{y} + \mathrm{i}c' \int_0^{\to\infty} \sin y \frac{\mathrm{d}y}{y}\big].
\end{aligned}$$

Since $\int_0^{\to\infty} \sin y \frac{\mathrm{d}y}{y} = \frac{\pi}{2}$ and $\int_0^{\to\infty} (\cos y - \mathrm{e}^{-y})\frac{\mathrm{d}y}{y} = 0$, the limit equals $\kappa u^{-c} \mathrm{e}^{\mathrm{i}c'\pi/2}$. In the case where $u < 0$ and $x \downarrow 0$, we have

$$\widehat{\mu}(\tfrac{u}{x})\frac{1}{x^c K(x)} = \overline{\widehat{\mu}(|u|/x)}\frac{1}{x^c K(x)} \to \kappa |u|^{-c} \mathrm{e}^{-\mathrm{i}c'\pi/2}.$$

Hence (53.33).

To see (53.34), let $u > 0$ and $0 < x \le 1$. If $u/x \le 1$, then (53.34) is trivial with 1 in place of M since $K(y) \ge 1$ for $0 < y \le 1$. If $u/x > 1$, then, by Lemma 53.9,

$$|\widehat{\mu}(\tfrac{u}{x})| \le M u^{-c} x^{c} K(\tfrac{x}{u})$$

and $K(\frac{x}{u}) \le K(x)$ for $u \le 1$ while

$$K(\tfrac{x}{u}) = K(x)\exp\bigl[\textstyle\int_{x/u}^{x}(c - r(y))\frac{\mathrm{d}y}{y}\bigr] = K(x)\exp\bigl[\int_{1/u}^{1}(c - r(xy))\frac{\mathrm{d}y}{y}\bigr] \le K(x)K(\tfrac{1}{u})$$

for $u > 1$.

Step 2. Define a finite measure ρ by $\rho[x,\infty) = k(x)$ on $(0,\infty)$, $\rho(-\infty,x] = k(x)$ on $(-\infty,0)$, and $\rho\{0\} = 0$. In the proof of Theorem 28.4 we have proved that

$$F(x) - F(0) = \tfrac{\operatorname{sgn} x}{2\pi}\textstyle\int_{-\infty}^{\infty}\frac{\mathrm{e}^{-\mathrm{i}u}-1}{-\mathrm{i}u}\widehat{\mu}(\frac{u}{x})\mathrm{d}u \quad \text{for } x \ne 0, \tag{53.35}$$

$$\tfrac{\mathrm{d}}{\mathrm{d}z}\widehat{\mu}(z) = \widehat{\mu}(z)\tfrac{1}{z}\textstyle\int_{\mathbb{R}}(\mathrm{e}^{\mathrm{i}zx} - 1)\rho(\mathrm{d}x) \quad \text{for } z \ne 0, \tag{53.36}$$

$$f(x) = \tfrac{\operatorname{sgn} x}{2\pi}\textstyle\int_{-\infty}^{\infty}\frac{\mathrm{e}^{-\mathrm{i}u}-1}{-\mathrm{i}u}\frac{\mathrm{d}}{\mathrm{d}x}(\widehat{\mu}(\frac{u}{x}))\mathrm{d}u \quad \text{for } x \ne 0. \tag{53.37}$$

Here $F(x) = \mu(-\infty,x]$. Let us prove that, for every $\varepsilon > 0$, there is $\eta_0 > 0$ such that, for any $\eta \ge \eta_0$ and $0 < |x| \le 1$,

$$\Bigl|\tfrac{f(x)}{|x|^{c-1}K(x)} - \tfrac{1}{2\pi|x|^{c}K(x)}\textstyle\int_{-\eta}^{\eta}\mathrm{e}^{-\mathrm{i}u}\widehat{\mu}(\frac{u}{x})\mathrm{d}u\Bigr| < \varepsilon. \tag{53.38}$$

First, notice that, for any $\alpha > 0$, $K(x) = o(x^{-\alpha})$ as $x \downarrow 0$, since $K(x)$ is slowly varying at 0. Thus, it follows from (53.34) that, for any $0 < \alpha < c$, there is a constant M_1 such that, for any $0 < |x| \le 1$ and $|u| \ge 1$,

$$|\widehat{\mu}(\tfrac{u}{x})|\tfrac{1}{|x|^{c}K(x)} \le M_1|u|^{-\alpha}. \tag{53.39}$$

It follows from (53.36) that

$$|\tfrac{\mathrm{d}}{\mathrm{d}z}\widehat{\mu}(z)| \le |\widehat{\mu}(z)|\tfrac{2c}{|z|} \quad \text{for } z \ne 0$$

and hence

$$|\tfrac{\mathrm{d}}{\mathrm{d}x}(\widehat{\mu}(\tfrac{u}{x}))| \le |\widehat{\mu}(\tfrac{u}{x})|\tfrac{2c}{|x|} \quad \text{for } x \ne 0,\ u \ne 0. \tag{53.40}$$

Thus we can find η_0 such that

$$\tfrac{1}{2\pi|x|^{c-1}K(x)}\Bigl|\textstyle\int_{|u|>\eta}\frac{\mathrm{e}^{-\mathrm{i}u}-1}{-\mathrm{i}u}\frac{\mathrm{d}}{\mathrm{d}x}(\widehat{\mu}(\frac{u}{x}))\mathrm{d}u\Bigr| < \tfrac{\varepsilon}{2}$$

for $\eta \ge \eta_0$ and $0 < |x| \le 1$. Now use that $\frac{1}{-u}\frac{\mathrm{d}}{\mathrm{d}x}(\widehat{\mu}(\frac{u}{x})) = \frac{1}{x}\frac{\mathrm{d}}{\mathrm{d}u}(\widehat{\mu}(\frac{u}{x}))$. Then

$$\begin{aligned}\tfrac{\operatorname{sgn} x}{2\pi|x|^{c-1}K(x)}\textstyle\int_{-\eta}^{\eta}\frac{\mathrm{e}^{-\mathrm{i}u}-1}{-\mathrm{i}u}\frac{\mathrm{d}}{\mathrm{d}x}(\widehat{\mu}(\frac{u}{x}))\mathrm{d}u &= \tfrac{\operatorname{sgn} x}{2\pi|x|^{c-1}K(x)}\tfrac{1}{\mathrm{i}x}\textstyle\int_{-\eta}^{\eta}(\mathrm{e}^{-\mathrm{i}u} - 1)\frac{\mathrm{d}}{\mathrm{d}u}(\widehat{\mu}(\frac{u}{x}))\mathrm{d}u\\ &= \tfrac{1}{2\pi|x|^{c}K(x)}\textstyle\int_{-\eta}^{\eta}\mathrm{e}^{-\mathrm{i}u}\widehat{\mu}(\frac{u}{x})\mathrm{d}u + R\end{aligned}$$

with $|R| \le 2M_1\eta^{-\alpha}$ by (53.39). Hence we can find η_0 such that $|R| < \varepsilon/2$ for $\eta \ge \eta_0$.

Step 3. Proof of (i). Assume that $c < 1$. We can use Lebesgue's dominated convergence theorem by (53.34) and get, by (53.33),

$$\tfrac{1}{2\pi}\textstyle\int_{-\eta}^{\eta}\frac{\mathrm{e}^{-\mathrm{i}u}}{|x|^{c}K(x)}\widehat{\mu}(\frac{u}{x})\mathrm{d}u \to \tfrac{\kappa}{2\pi}\int_{-\eta}^{\eta}\frac{\mathrm{e}^{-\mathrm{i}u}}{|u|^{c}}\mathrm{e}^{\mathrm{i}(\operatorname{sgn} u)c'\pi/2}\mathrm{d}u$$

as $x \downarrow 0$. The limit equals

$$\tfrac{\kappa}{2\pi}\int_0^\eta \tfrac{1}{u^c}\cos(u - \tfrac{c'\pi}{2})\mathrm{d}u = \tfrac{\kappa}{\pi}\big(\cos\tfrac{c'\pi}{2}\int_0^\eta u^{-c}\cos u\mathrm{d}u + \sin\tfrac{c'u}{2}\int_0^\eta u^{-c}\sin u\mathrm{d}u\big),$$

which tends to

$$\tfrac{\kappa}{\pi}\Gamma(1-c)(\cos\tfrac{c'\pi}{2}\cos\tfrac{(1-c)\pi}{2} + \sin\tfrac{c'\pi}{2}\sin\tfrac{(1-c)\pi}{2}) = \tfrac{\kappa}{\pi}\Gamma(1-c)\sin k(0+)\pi$$

as $\eta \to \infty$. Combining this with the result in Step 2 and using $\Gamma(c)\Gamma(1-c) = \pi/\sin c\pi$, we obtain (53.28). Its dual is (53.29).

Step 4. Proof of (ii). Assume that $c = 1$. Let us show that

$$\lim_{x\downarrow 0}\tfrac{x}{K(x)}\tfrac{\mathrm{d}}{\mathrm{d}x}(\tfrac{F(x)-F(0)}{x}) = -\tfrac{\kappa}{\pi}\cos\tfrac{c'\pi}{2}. \tag{53.41}$$

Let $x > 0$. We have, by (53.35) and (53.37),

$$\begin{aligned}\tfrac{x}{K(x)}\tfrac{\mathrm{d}}{\mathrm{d}x}(\tfrac{F(x)-F(0)}{x}) &= \tfrac{1}{K(x)}(f(x) - \tfrac{F(x)-F(0)}{x})\\ &= \tfrac{1}{2\pi K(x)}\int_{-\infty}^{\infty}\tfrac{\mathrm{e}^{-\mathrm{i}u}-1}{-\mathrm{i}u}[\tfrac{\mathrm{d}}{\mathrm{d}x}(\widehat{\mu}(\tfrac{u}{x})) - \tfrac{1}{x}\widehat{\mu}(\tfrac{u}{x})]\mathrm{d}u.\end{aligned}$$

For any $\varepsilon > 0$, by (53.39) and (53.40),

$$\big|\tfrac{1}{2\pi K(x)}\int_{|u|>\eta}\tfrac{\mathrm{e}^{-\mathrm{i}u}-1}{-\mathrm{i}u}[\tfrac{\mathrm{d}}{\mathrm{d}x}(\widehat{\mu}(\tfrac{u}{x})) - \tfrac{1}{x}\widehat{\mu}(\tfrac{u}{x})]\mathrm{d}u\big| < \varepsilon$$

for large η uniformly in $0 < |x| \le 1$. Now

$$\begin{aligned}\tfrac{1}{2\pi K(x)}\int_{-\eta}^{\eta}\tfrac{\mathrm{e}^{-\mathrm{i}u}-1}{-\mathrm{i}u}[\tfrac{\mathrm{d}}{\mathrm{d}x}(\widehat{\mu}(\tfrac{u}{x})) - \tfrac{1}{x}\widehat{\mu}(\tfrac{u}{x})]\mathrm{d}u &= \tfrac{1}{2\pi xK(x)}\int_{-\eta}^{\eta}\tfrac{\mathrm{e}^{-\mathrm{i}u}-1}{u}\tfrac{\mathrm{d}}{\mathrm{d}u}(-\mathrm{i}u\widehat{\mu}(\tfrac{u}{x}))\mathrm{d}u\\ &= \tfrac{-1}{2\pi xK(x)}\int_{-\eta}^{\eta}(-\mathrm{i}u)\widehat{\mu}(\tfrac{u}{x})h(u)\mathrm{d}u + R_1,\end{aligned}$$

where $h(u) = \tfrac{\mathrm{d}}{\mathrm{d}u}(\tfrac{\mathrm{e}^{-\mathrm{i}u}-1}{u})$ and $|R_1| < \varepsilon$ for large η uniformly in $0 < |x| \le 1$, again by (53.39). Let $x \downarrow 0$ and use (53.33) and (53.34). Then

$$\begin{aligned}\tfrac{-1}{2\pi xK(x)}\int_{-\eta}^{\eta}(-\mathrm{i}u)\widehat{\mu}(\tfrac{u}{x})h(u)\mathrm{d}u &\to \tfrac{-\kappa}{2\pi}\int_{-\eta}^{\eta}(-\mathrm{i}\,\mathrm{sgn}\,u)\mathrm{e}^{\mathrm{i}(\mathrm{sgn}\,u)c'\pi/2}h(u)\mathrm{d}u\\ &= \tfrac{-\kappa}{2\pi}\big(\mathrm{e}^{\mathrm{i}(c'-1)\pi/2}\int_0^\eta h(u)\mathrm{d}u + \mathrm{e}^{\mathrm{i}(1-c')\pi/2}\int_{-\eta}^0 h(u)\mathrm{d}u\big) = \tfrac{-\kappa}{\pi}\cos\tfrac{c'\pi}{2} + R_2\end{aligned}$$

with $|R_2| < \varepsilon$ for large η. This proves (53.41). As $x \downarrow 0$,

$$\tfrac{F(x)-F(0)}{xL(x)} = \tfrac{F(1)-F(0)}{L(x)} - \tfrac{1}{L(x)}\int_x^1(\tfrac{F(y)-F(0)}{y})'\mathrm{d}y \to \tfrac{\kappa}{\pi}\cos\tfrac{c'\pi}{2}$$

by $L(0+) = \infty$, (53.41), and l'Hospital's rule. Now, note that

$$K(x)/L(x) \to 0 \quad \text{as } x \downarrow 0,$$

which is evident if $K(0+) < \infty$ and by l'Hospital's rule if $K(0+) = \infty$. We see that

$$\tfrac{f(x)}{L(x)} = \tfrac{K(x)}{L(x)}\tfrac{x}{K(x)}\tfrac{\mathrm{d}}{\mathrm{d}x}\big(\tfrac{F(x)-F(0)}{x}\big) + \tfrac{F(x)-F(0)}{xL(x)} \to \tfrac{\kappa}{\pi}\cos\tfrac{c'\pi}{2}.$$

Combined with the dual result, this gives (53.30). The asymptotic (53.31) is proved similarly. That is, by (53.38),

$$\begin{aligned}\tfrac{f(x)-f(-x)}{K(x)} &= \tfrac{1}{2\pi}\int_{-\eta}^{\eta}\tfrac{\mathrm{e}^{-\mathrm{i}u}}{xK(x)}(\widehat{\mu}(\tfrac{u}{x}) - \widehat{\mu}(-\tfrac{u}{x}))\mathrm{d}u + R_3\\ &= \tfrac{1}{\pi}\int_{-\eta}^{\eta}\tfrac{\sin u}{xK(x)}\mathrm{Im}\,\widehat{\mu}(\tfrac{u}{x})\mathrm{d}u + R_3,\end{aligned}$$

where $|R_3| < \varepsilon$ for large η uniformly in $0 < x \le 1$. By (53.33) and (53.34), as $x \downarrow 0$,

$$\tfrac{1}{\pi}\int_{-\eta}^{\eta} \tfrac{\sin u}{xK(x)} \operatorname{Im}\widehat{\mu}(\tfrac{u}{x})du \to \tfrac{\kappa}{\pi}\sin\tfrac{c'u}{2}\int_{-\eta}^{\eta}\tfrac{\sin u}{u}du = \kappa\sin\tfrac{c'\pi}{2} + R_4$$

with $|R_4| < \varepsilon$ for large η. The proof is complete. □

REMARK 53.10. Some consequences of Theorems 53.6 and 53.8 are as follows. We assume (53.14).

(i) If $k(0-) = 0$ and $k(0+) < 1$, then $f(0+) = \infty$.

(ii) If $k(0-) = 0$ and $k(0+) = 1$, then $f(0+)$ is finite or infinite according as $\int_0^1 (1 - k(x))\frac{dx}{x}$ is finite or infinite.

(iii) If $k(0-) > 0$, $k(0+) > 0$, and $k(0-) + k(0+) \le 1$, then $f(0-) = \infty$ and $f(0+) = \infty$.

Notice that, when $c = 1$, there is an essential difference between the one-sided and two-sided cases, as is seen in (ii) and (iii).

The remaining part of the proof of Theorem 28.4 is now obtained.

Our Theorem 53.8 on the two-sided case deals only with the case $c \le 1$. When $c > 1$, we can obtain similar asymptotics of $f^{(N)}(x)$ as $x \to 0$, provided that neither $k(0+)$ nor $k(0-)$ is an integer. Some of the results in [**466**] are as follows. Assume (53.14).

(iv) If $1 < c < 2$, $0 < k(0-) < 1$, and $0 < k(0+) < 1$, then $f'(0-) = \infty$ and $f'(0+) = -\infty$.

(v) If $1 < c \le 2$ and $0 < k(0-) < 1 < k(0+)$, then the mode a is positive and $f'(0-) = \infty$ and $f'(0+) = \infty$.

Some analysis in the case that $k(0-)$ or $k(0+)$ is an integer is made in [**467**], but the situation is complicated.

REMARK 53.11. Let $\{X_t\}$ be a selfdecomposable subordinator with k-function $k(x)$. Then, by Lemma 52.9, its mode is increasing with time. If $k(0+) < \infty$, then Theorem 53.4, Remark 53.5, and Theorem 53.6 describe how the shape of the density function varies as time goes on, since the Lévy measure of the distribution at time t is $t\frac{k(x)}{x}dx$. Thus, except at at most one moment $t_0 = 1/k(0+)$, the modal interval of the distribution P_{X_t} consists of only one point, which we denote by $a(t)$. As $t \to \infty$, we have

$$a(t)/t \to E[X_1]. \tag{53.42}$$

In fact, if $E[X_1] < \infty$, then (53.42) follows from $X_t/t \to E[X_1]$ a.s. in Theorem 36.5 and from Lemma 52.4, since X_t/t has mode $a(t)/t$. If $E[X_1] = \infty$, then consider a selfdecomposable subordinator $\{X_n(t)\}$ with k-function $k(x)1_{(0,n]}(x)$ and with drift in common with $\{X_t\}$. Then the mode $a_n(t)$ of $X_n(t)$ does not exceed $a(t)$ and thus $a(t)/t \to \infty$. See [**435**], [**436**] for more results.

REMARK 53.12. Let $\{X_t\}$ be an α-stable process on $\mathbb{R}$. The distribution of X_t is unimodal by Theorem 53.1. The modal interval is a one-point set for each t by Remark 53.5. Denote it by $\{a(t)\}$. Use the parameters (α, β, τ, c) for $\alpha \ne 2$ in Definition 14.16 or the variance A and the mean τ for $\alpha = 2$. Let $X_t^0 = X_t - t\tau$

and denote the mode of X_t^0 by $a_0(t)$. Then, it follows from Exercise 18.6 that

$$a(t) = \begin{cases} t^{1/\alpha}a_0(1) + t\tau & \text{for } \alpha \neq 1, \\ ta_0(1) + t\tau + \frac{2}{\pi}c\beta t \log t & \text{for } \alpha = 1. \end{cases} \tag{53.43}$$

Thus we see that $a(t)$ is not always monotone in t. If $\alpha = 2$, then $a_0(1) = 0$, as $\{X_t^0\}$ is symmetric. Zolotarev [**608**], p. 140, proves that

$$\operatorname{sgn} a_0(1) = \begin{cases} \operatorname{sgn}\beta & \text{for } 0 < \alpha < 1, \\ -\operatorname{sgn}\beta & \text{for } 1 < \alpha < 2, \end{cases}$$

and suggests numerically that $\operatorname{sgn} a_0(1) = -\operatorname{sgn}\beta$ for $\alpha = 1$. Behavior of $a_0(1)$ as $\beta \to 0$ is studied by Hall [**179**], Sato [**435**], and Gawronski and Wießner [**157**]. Recall that the meaning of τ for $\{X_t\}$ is its drift for $0 < \alpha < 1$, and its center for $1 < \alpha \leq 2$. As is explained in Definition 14.16, $\beta > 0$, or $\beta < 0$, means, respectively, that the positive axis has more, or less, Lévy measure than the negative axis. The graphs of the densities are contained in [**608**], pp. 144–146.

54. Unimodality and multimodality in Lévy processes

In this section let $\{X_t\}$ be a Lévy process on $\mathbb{R}$ generated by (A, ν, γ). If it is of type A or B, the drift is denoted by γ_0. If the Lévy measure ν is absolutely continuous, then we write $\nu = l(x)\mathrm{d}x$, and $l(x) = k(x)/|x|$, although we do not assume selfdecomposability of $\{X_t\}$. Write $\mu = P_{X_1}$. Thus $\mu^t = P_{X_t}$.

THEOREM 54.1 (Wolfe [**581**]). *If there is a sequence of times $t_n > 0$, $t_n \to 0$, such that μ^{t_n} is unimodal, then ν is unimodal with mode* 0.

Proof. Let μ^{t_n} be unimodal with mode a_n. Since $\mu^{t_n} \to \delta_0$, we have $a_n \to 0$ by Lemma 52.4. The proof of Theorem 8.1(i) (before Corollary 8.8) shows that $\exp(t_n{}^{-1}(\widehat{\mu}(z)^{t_n} - 1)) \to \widehat{\mu}(z)$. We regard this as convergence of a sequence of infinitely divisible probability measures. Thus, by Theorem 8.7,

$$\frac{1}{t_n}\int h(x)\mu^{t_n}(\mathrm{d}x) \to \int h(x)\nu(\mathrm{d}x)$$

for any bounded continuous function $h(x)$ which vanishes on a neighborhood of 0. For $x > 0$, let $G(x) = \nu(x, \infty)$ and $G_n(x) = t_n{}^{-1}\mu^{t_n}(x, \infty)$. Given $0 < x_1 < x_2 < x_3$, we have $a_n < x_1$ for large n and hence

$$\frac{G_n(x_2) - G_n(x_1)}{x_2 - x_1} \leq \frac{G_n(x_3) - G_n(x_2)}{x_3 - x_2}.$$

If x_1, x_2, and x_3 are continuity points of $G(x)$, then, letting $n \to \infty$, we get

$$\frac{G(x_2) - G(x_1)}{x_2 - x_1} \leq \frac{G(x_3) - G(x_2)}{x_3 - x_2}.$$

By the right-continuity of $G(x)$, this holds even at points of discontinuity of $G(x)$. Hence $G(x)$ is convex on $(0,\infty)$. Similarly $\nu(-\infty, x]$ is convex on $(-\infty, 0)$. Hence ν is unimodal with mode 0. □

THEOREM 54.2 (Medgyessy [**343**]). *If $\{X_t\}$ is symmetric and if ν is unimodal with mode 0, then μ^t is unimodal with mode 0 for every $t \geq 0$.*

Proof. From the symmetry, ν is symmetric and $\gamma = 0$ (Exercise 18.1). In order to show the unimodality of μ^t, we can assume that $A = 0$ by the strong unimodality of Gaussian distributions (Example 52.7). Thus we consider

$$\widehat{\mu}(z)^t = \exp\left[t\int(\mathrm{e}^{\mathrm{i}zx} - 1 - \mathrm{i}zx1_{[-1,1]}(x))l(x)\mathrm{d}x\right],$$

where $l(x) = l(-x)$ and $l(x)$ is decreasing on $(0,\infty)$. Define μ_n by

$$\widehat{\mu}_n(z) = \exp\left[\int(\mathrm{e}^{\mathrm{i}zx} - 1)l_n(x)\mathrm{d}x\right]$$

with $l_n(x) = l(x \vee n^{-1})$ for $x > 0$ and $l_n(x) = l_n(-x)$ for $x < 0$. Then ${\mu_n}^t \to \mu^t$ as $n \to \infty$. Since μ_n is compound Poisson, we have

$$\widehat{\mu}_n(z)^t = \mathrm{e}^{-c_n t}\sum_{j=0}^{\infty}\frac{1}{j!}(c_n t)^j\widehat{\sigma}_n(z)^j,$$

where $c_n = \int l_n(x)\mathrm{d}x$ and $\sigma_n = {c_n}^{-1}l_n(x)\mathrm{d}x$. Since σ_n is symmetric and unimodal, so is ${\sigma_n}^j$ for $j = 1, 2, \ldots$ by Wintner's result (Exercise 29.22). Note that any symmetric unimodal measure is unimodal with mode 0. Since ${\mu_n}^t$ is a mixture of such measures, it is symmetric and unimodal with mode 0. Hence μ^t is symmetric and unimodal with mode 0 by Lemma 52.4. □

The two theorems above give the condition on the Lévy measure in order that a symmetric Lévy process on $\mathbb{R}$ be unimodal for every t. In the non-symmetric case, however, the unimodality of ν with mode 0 does not guarantee the unimodality of μ^t for every t. A process in Example 23.4 shows this fact. We do not know a necessary and sufficient condition in terms of Lévy measures. A sufficient condition is the selfdecomposability, which means that $k(x)$ is increasing on $(-\infty, 0)$ and decreasing on $(0,\infty)$. This is shown by Theorem 53.1.

REMARK 54.3. By Theorem 54.2 there are many symmetric Lévy processes which are not selfdecomposable and for which μ^t is unimodal for all t. Watanabe [**551, 555**] constructs non-symmetric Lévy processes which are not selfdecomposable but whose distributions are unimodal for every t. Among them, the subordinator $\{X_t\}$ having

$$k(x) = 2 \cdot 1_{(0,1]}(x) + (1 + ax)1_{(1,1+\varepsilon]}(x)$$

with $0 < a < 1/6$ and $0 < \varepsilon < 1/2$ has this property.

REMARK 54.4. There is a Lévy process such that, for some $0 < t_1 < t_2$, μ^t is non-unimodal for $t \in (0, t_1)$ and unimodal for $t \in [t_2, \infty)$. A method to give such a process is to use Remark 54.21 given later, namely, consider a subordinator $\{X_t\}$ for which $k(x)$ is log-concave on $(0, \infty)$, $k(0+) > 0$, and $k(x)/x$ is strictly increasing on some interval in $(0, \infty)$. For example, let $k(x) = \mathrm{e}^{-(x-c)^2}$ with $c > \sqrt{2}$, which satisfies, at $x = c/2$, $(\log \frac{k(x)}{x})' = \frac{c^2-2}{c} > 0$. Then, for $t \geq 1/k(0+)$, μ^t is strongly unimodal, and hence unimodal. But, by Theorem 54.1, μ^t is non-unimodal for t small enough. If we consider a subordinator $\{Y_t\}$ equivalent in law with this $\{X_t\}$ such that $\{X_t\}$ and $\{Y_t\}$ are independent, then the symmetric process $\{X_t - Y_t\}$ has the same character. Thus, being symmetric and unimodal is a time dependent distributional property in the class of Lévy processes. Another example for showing this time dependence is given in Exercise 55.11. Notice that there are symmetric unimodal infinitely divisible distributions whose Lévy measures are not unimodal with mode 0, as these examples show.

REMARK 54.5. Let $\{X_t\}$ and $\{Y_t\}$ be independent and $\{Y_t\}$ be a constant multiple of the Brownian motion. If $\{X_t\}$ has unimodal distribution at every t, then $\{X_t + Y_t\}$ has unimodal distribution at every t. But the converse is not true, as an example of Watanabe [**557**] shows.

REMARK 54.6. A property of selfdecomposable subordinators can be extended to more general subordinators. *If $\{X_t\}$ is a subordinator with μ^t unimodal for every t, then μ^t has the Yamazato property for every t.* This is proved by Watanabe [**554**]. It follows from this result and Theorem 52.12 that if $\{X_t\}$ and $\{Y_t\}$ are independent subordinators whose distributions are unimodal at every t, then $\{X_t - Y_t\}$ has a unimodal distribution at every t. But the converse is not true. An example is the case that $l(x) = \mathrm{e}^{-x}$ on $(0, \infty)$ and $\{Y_t\} \stackrel{\mathrm{d}}{=} \{X_t\}$. In fact, in this case, X_t is non-unimodal for $t > 2$ (Example 23.4), but $X_t - Y_t$ is unimodal for every t (Theorem 54.2). A non-symmetric example is given in [**557**]. It is also to be noted that a unimodal infinitely divisible distribution on $\mathbb{R}_+$ does not always have the Yamazato property [**554**].

Let us consider semi-stable processes in Theorems 54.7 and 54.9 and Remark 54.8.

THEOREM 54.7 ([**445**]). *There is a semi-stable process $\{X_t\}$ such that μ^t is non-unimodal for every $t > 0$. In particular, if $A = 0$ and $\nu(\mathrm{d}x) = h(x)|x|^{-\alpha-1}\mathrm{d}x$ with $0 < \alpha < 2$ and $h(x)$ nonnegative, bounded, measurable, and satisfying $h(x) = h(bx)$ for some $b > 1$ and if $h(x)x^{-\alpha-1}$ on $(0, \infty)$ does not equal a. e. a decreasing function, then $\{X_t\}$ is α-semi-stable having b as a span and μ^t is non-unimodal for every $t > 0$.*

An example of $h(x)$ on $(0, \infty)$ is $h(x) = 1 + \sin(2\pi \frac{\log x}{\log b})$.

Proof of theorem. Let $A = 0$ and $\nu = h(x)|x|^{-\alpha-1}\mathrm{d}x$ with $h(x)$ satisfying the conditions above. Then, $\nu(B) = b^{-\alpha}\nu(b^{-1}B)$ for $B \in \mathcal{B}(\mathbb{R})$. Hence, by Theorem 14.3, $\{X_t\}$ is α-semi-stable and b is a span. That is, $\{X_{b^\alpha t}\} \stackrel{\mathrm{d}}{=}$

$\{bX_t + \gamma' t\}$ with some $\gamma' \in \mathbb{R}$. If X_{t_0} has a unimodal distribution for some $t_0 > 0$, then so does $X_{b^{n\alpha}t_0}$ for any $n \in \mathbb{Z}$. Since $b^{n\alpha}t_0 \to 0$ as $n \to -\infty$, Theorem 54.1 tells that ν is unimodal with mode 0. This contradicts the assumption. □

REMARK 54.8. Watanabe [**557**] shows the following fact. Let $0 < \alpha < 1$. For some $b > 1$ there is an α-semi-stable process $\{X_t\}$ having b as a span such that, for some $t_0 > 0$ and $t_1 > 0$, μ^t is unimodal for $t = t_0 b^{\alpha n}$, $n \in \mathbb{Z}$, and non-unimodal for $t = t_1 b^{\alpha n}$, $n \in \mathbb{Z}$. Notice that, in this process, times of unimodality and times of non-unimodality both cluster at 0 and at ∞. He shows that in particular, for $0 < c < 1/2$, a subordinator $\{X_t\}$ with $l(x) = b^{-\alpha n} x^{-1}(1 + cb^{-n}x)$ on $(b^{n-1}, b^n]$, $n \in \mathbb{Z}$, is a process of this kind with $t_0 = 1/2$ and some $t_1 \neq 1/2$, provided that b is sufficiently large.

THEOREM 54.9 (Watanabe [**556**]). *Let $0 < \alpha < 2$ and $b > 1$. There exists a process $\{X_t\}$ selfdecomposable and α-semi-stable having b as a span such that, for some $t_0 > 0$, the mode $a(t)$ of μ^t is not 0 at $t_n = t_0 b^{\alpha n/2}$, $n \in \mathbb{Z}$, and satisfies $a(t_n) = (-1)^n b^{n/2} a(t_0)$. For this process, automatically,*

$$\limsup_{t\to\infty} t^{-c} a(t) = \infty \quad \textit{and} \quad \liminf_{t\to\infty} t^{-c} a(t) = -\infty$$

for any c with $0 < c < 1/\alpha$.

Outline of proof. Let $\{X_t\}$ have $A = 0$ and $\nu = h(x)|x|^{-\alpha-1}\mathrm{d}x$ with $h(x)$ nonnegative, bounded and measurable. Then it is α-semi-stable with b as a span if and only if $h(x) = h(bx)$ for $x \in \mathbb{R}$. Choose $h(x)$ satisfying $h(x) = h(-\sqrt{b}x)$ for $x \in \mathbb{R}$. Assume that the drift is 0 in the case $0 < \alpha < 1$, that the center is 0 in the the case $1 < \alpha < 2$, and that

$$\widehat{\mu}(z) = \exp\left[\int_{-\infty}^{\infty}\{\mathrm{e}^{\mathrm{i}zx} - 1 - \mathrm{i}zx\tfrac{1}{2}(1_{[-1,1]}(x) + 1_{[-\sqrt{b},\sqrt{b}]}(x))\}h(x)x^{-2}\mathrm{d}x\right]$$

in the case $\alpha = 1$. Then we have $\{X_{b^{\alpha/2}t}\} \overset{\mathrm{d}}{=} \{-\sqrt{b}X_t\}$. Further assume that $h(x)x^{-\alpha}$ is decreasing on $(0, \infty)$. Then $\{X_t\}$ is selfdecomposable. Hence μ^t is unimodal and the modal interval is a one-point set for each t (Theorem 53.1 and Remark 53.5). Denote the mode by $a(t)$. By Lemma 52.4, $a(t)$ is continuous in $t \in [0, \infty)$. We have $a(b^{\alpha/2}t) = -\sqrt{b}a(t)$. Hence $a(b^{\alpha n/2}t) = (-1)^n b^{n/2} a(t)$ for any $t \in [0, \infty)$ and $n \in \mathbb{Z}$. Assume, moreover, that ν is non-symmetric. Then we can prove that $a(t_0) \neq 0$ for some $t_0 > 0$, but we omit the details. Then, letting $t_n = t_0 b^{\alpha n/2}$, we have $a(t_n)/t_n{}^c = (-1)^n b^{(1-\alpha c)n/2} a(t_0)/t_0{}^c$. An example of $h(x)$ satisfying all requirements is

$$h(x) = 1 + (\operatorname{sgn} x) a \sin(2\pi \tfrac{\log|x|}{\log b})$$

with a constant a such that $0 < |a| < (1 + (\frac{2\pi}{\alpha \log b})^2)^{-1/2}$.

REMARK 54.10. Let $\mu = Ce^{h(x)}1_{(0,\infty)}(x)\mathrm{d}x$, where C is a positive constant and $h(x)$ is a convex function on $(0, \infty)$. Theorem 51.4 guarantees infinite divisibility of μ. Consider a subordinator $\{X_t\}$ with distribution μ at $t = 1$. Let n be an integer ≥ 2. Suppose that $h(x)$ is expressed by some

$0 = a_0 < a_1 < \cdots < a_{n-1} < a_n = \infty$ and $c_1 > c_2 > \cdots > c_n > 0$ as a linear function with slope $-c_j$ on each interval $[a_{j-1}, a_j)$. It is proved in Sato [**443**] that there is a choice of $a_1, \ldots, a_{n-1}$ and $c_1, \ldots, c_n$ such that the distribution μ^2 of X_t at $t = 2$ is n-modal. Since μ is unimodal, this shows that time evolution from unimodal to n-modal is possible. In this process we can show that μ^2 has a density of class C^1 on $(0, \infty)$ and that μ^t is non-unimodal for $t \in (0, 1)$. Another way to show the possibility of time evolution from unimodal to n-modal is to use Remarks 54.19, (i) and (ii).

The following lemma is sometimes useful in showing unimodality.

LEMMA 54.11. *Let $f(x) = \sum_{n=0}^{\infty} b_n x^n$, a power series with the radius of convergence being ∞. Suppose that there is $n_0 \geq 1$ such that $b_n > 0$ for $n < n_0$ and $b_n < 0$ for $n > n_0$. Then there is a point x_0 in $(0, \infty)$ such that $f(x)$ is positive on $[0, x_0)$, 0 at x_0, and negative on (x_0, ∞).*

Proof. Since $f(0) > 0$ and $f(x) \to -\infty$ as $x \to \infty$, $f(x)$ has a zero in $(0, \infty)$. If x_0 is a zero of $f(x)$, then

$$x_0 f'(x_0) = \sum_{n=1}^{\infty} n b_n x_0^n < n_0 \sum_{n=0}^{\infty} b_n x_0^n = 0.$$

Hence there is only one zero of $f(x)$ in $(0, \infty)$. □

PROPOSITION 54.12 ([**445**]). *Let $\{X_t\}$ be a subordinator with drift 0 and $\nu(\mathrm{d}x) = 1_{(0,\infty)}(x)\mathrm{e}^{-x}\mathrm{d}x$. Then μ^t is unimodal with mode 0 for $0 \leq t \leq 2$ and bimodal with leftmost mode 0 for $t > 2$.*

Proof. We have shown in Example 23.4 that $\mu^t = \mathrm{e}^{-t}\delta_0 + 1_{(0,\infty)} f_t(x)\mathrm{d}x$ with

$$f_t'(x) = \mathrm{e}^{-t-x} \sum_{n=0}^{\infty} \frac{t^{n+1} x^n}{(n+1)!n!} \Big(\frac{t}{n+2} - 1 \Big).$$

If $0 < t \leq 2$, then $f_t'(x) < 0$ on $(0, \infty)$. If $t > 2$, then the power series in x in the expression for $f_t'(x)$ satisfies the condition in Lemma 54.11. Hence, if $t > 2$, then the measure $1_{(0,\infty)}(x) f_t(x)\mathrm{d}x$ is unimodal with positive mode. □

THEOREM 54.13 ([**445**]). *Let $\{X_t\}$ be a subordinator with $\mu = p\delta_0 + (1-p)1_{(0,\infty)}(x)\mathrm{e}^{-x}\mathrm{d}x$, $0 < p < 1$. Then μ^t is unimodal with mode 0 for $0 \leq t \leq \frac{1+p}{1-p}$ and bimodal with leftmost mode 0 for $t > \frac{1+p}{1-p}$.*

Outline of proof. We use the confluent hypergeometric function [**1**]

$$M(\alpha, \beta, z) = 1 + \tfrac{\alpha z}{\beta} + \tfrac{\alpha(\alpha+1)z^2}{\beta(\beta+1)2!} + \cdots = \textstyle\sum_{n=0}^{\infty} \tfrac{\alpha(\alpha+1)\ldots(\alpha+n-1)z^n}{\beta(\beta+1)\ldots(\beta+n-1)n!}.$$

It is shown in [**445**] and reproved by Watanabe [**561**] that

$$\mu^t = p^t \delta_0 + 1_{(0,\infty)}(x) f_t(x)\mathrm{d}x,$$

$$f_t(x) = tp^t b e^{-x} M(1-t, 2, -bx) = tp^t b e^{-x/p} M(1+t, 2, bx).$$

It follows that

$$f_t'(x) = tp^{t-1} b e^{-x/p} \sum_{n=0}^{\infty} (-pn - 1 - p + (1-p)t) \tfrac{(t+1)(t+2)\dots(t+n)}{(n+2)!n!} (bx)^n.$$

Hence, using Lemma 54.11, the same reasoning as in the proof of Proposition 54.12 works. □

Watanabe [**561**] introduces a transform from subordinators to subordinators, which he calls the Bessel transform. This is iteration of two subordinations of special kinds. It makes analysis of unimodality and multimodality of a class of subordinators possible.

DEFINITION 54.14. Let $\{X_t\}$, $\{N_t\}$, and $\{G_t\}$ be independent subordinators. Assume that $\{N_t\}$ is the Poisson process with parameter 1, $E[e^{-uN_t}] = e^{t(e^{-u}-1)}$, and that $\{G_t\}$ is the Γ-process with parameter 1, $E[e^{-uG_t}] = \exp\bigl[\int_0^\infty (e^{-ux} - 1) e^{-x} \frac{dx}{x}\bigr]$. Given constants $a \geq 0$ and $s > 0$, define $Y_t = G(at + N(sX_t))$. The subordinator $\{Y_t\}$ is called the *Bessel transform* of $\{X_t\}$ with parameters a and s. In the case of a probability measure μ on $\mathbb{R}_+$, let X be a random variable with distribution μ and suppose that X, $\{N_t\}$, and $\{G_t\}$ are independent. Then the distribution ρ of $G(a + N_{sX})$ is called the *Bessel transform* of μ with parameters a and s.

LEMMA 54.15. *Let ρ be the Bessel transform, with parameters $a \geq 0$ and $s > 0$, of a probability measure μ on $\mathbb{R}_+$. We have*

$$\rho(dx) = 1_{(0,\infty)}(x) \left[e^{-x} \int_{[0,\infty)} e^{-su} \left(\tfrac{x}{su}\right)^{(a-1)/2} I_{a-1}(2\sqrt{sxu}) \mu(du) \right] dx$$

if $a > 0$, and

$$\rho(dx) = \left[\int_{[0,\infty)} e^{-su} \mu(du) \right] \delta_0(dx)$$
$$+ 1_{(0,\infty)}(x) \left[e^{-x} \int_{[0,\infty)} e^{-su} \sqrt{\tfrac{su}{x}} I_1(2\sqrt{sxu}) \mu(du) \right] dx$$

if $a = 0$. If $\{Y_t\}$ is the Bessel transform, with parameters $a \geq 0$ and $s > 0$, of a subordinator $\{X_t\}$, then $\{Y_t\}$ has drift $\gamma_Y = 0$ and Lévy measure

$$\nu_Y = 1_{(0,\infty)}(x) e^{-x} \left[\tfrac{a}{x} + s\gamma + \int_0^\infty e^{-su} \sqrt{\tfrac{su}{x}} I_1(2\sqrt{sxu}) \nu(du) \right] dx,$$

where ν and γ are the Lévy measure and the drift of $\{X_t\}$.

Here $I_\alpha(x)$ is the modified Bessel function (4.11). The proof is given by direct calculation. A special case ($s = 1$ and $X_t = t$) is given in Exercise 34.1. For the expression for ν_Y use Theorem 30.1. Now we state one of Watanabe's results.

THEOREM 54.16 ([**561**]). *Let ρ be the Bessel transform of μ. If μ is n-modal, then ρ is either $(n+1)$-modal with leftmost mode 0 or at most n-modal. If μ is n-modal with leftmost mode 0, then ρ is at most n-modal.*

We do not give his proof here. The tools are the relationship between multimodality and total positivity of Karlin [**259**], which is a generalization of Theorem 52.3, and the power series version ([**379**], p. 41) of Descartes' rule of signs, which is an extension of Lemma 54.11.

REMARK 54.17. Using the result above and making careful additional observations, Watanabe [**561**] shows the following. *Let $\{Y_t\}$ be the Bessel transform with parameters $a \geq 0$ and $s > 0$ of a non-zero selfdecomposable subordinator $\{X_t\}$. Let ν and γ be the Lévy measure and the drift of $\{X_t\}$. Define*

$$C = s\gamma + \int_0^\infty \mathrm{e}^{-sx} sx\nu(\mathrm{d}x).$$

(i) *Suppose that $a \geq C$. Then $\{Y_t\}$ is a selfdecomposable process and hence has a unimodal distribution at each t. If $0 \leq t \leq \frac{1}{a}$, then Y_t has mode 0. If $t > \frac{1}{a}$, then Y_t has a positive mode.*

(ii) *Suppose that $a = 0$. Define*

$$t_0 = \tfrac{1}{C^2}\big(2C - \int_0^\infty \mathrm{e}^{-sx}(sx)^2\nu(\mathrm{d}x)\big).$$

Then $t_0 \geq \frac{1}{C}$. The process $\{Y_t\}$ is a compound Poisson process and the distribution of Y_t is unimodal with mode 0 for $0 \leq t \leq t_0$ and bimodal with leftmost mode 0 for $t > t_0$.

(iii) *Suppose that $0 < a < C$. Then there exists $t_1 \in (\frac{1}{C}, \frac{1}{a})$ such that the distribution of Y_t is unimodal with mode 0 for $0 \leq t \leq t_1$, bimodal with leftmost mode 0 for $t_1 < t < \frac{1}{a}$, and unimodal with positive mode for $t \geq \frac{1}{a}$.*

Actually he proves a large part of the assertions above for more general processes, that is, for the Bessel transforms of subordinators $\{X_t\}$ such that P_{X_t} is unimodal for every t.

REMARK 54.18. Application of Remark 54.17 gives the following [**561**]. Let $\{Y_t\}$ be a subordinator satisfying one of the following:

(1) $\quad l_Y(x) = \mathrm{e}^{-x}(s + \frac{a}{x})$,

(2) $\quad l_Y(x) = \frac{1}{x}((a-1)\mathrm{e}^{-x} + \mathrm{e}^{-x/(s+1)})$,

where s and a are positive constants. Write $C = s$ in case (1) and $C = \frac{s}{s+1}$ in case (2). If $a \geq C$, then $\{Y_t\}$ is selfdecomposable, and hence Y_t has unimodal distribution for every t. If $a < C$, then there is $t_1 \in (\frac{1}{C}, \frac{1}{a})$ such that the distribution of Y_t is unimodal with mode 0 for $0 \leq t \leq t_1$, bimodal with leftmost mode 0 for $t_1 < t < \frac{1}{a}$, and unimodal with positive mode for $t \geq \frac{1}{a}$.

The process $\{Y_t\}$ in case (1) or (2), respectively, is the Bessel transform of the trivial process $X_t = t$ or the Γ-process $\{X_t\}$ with parameter 1.

Part of the special case $s = 1$ of case (1) was obtained by Wolfe [**581**] and Sato [**445**]. In the above $a = 0$ is not allowed. But if we make $a = 0$, then (1) is the process in Proposition 54.12 with a time change and (2) is that of Theorem 54.13 with $p = \frac{1}{s+1}$.

REMARK 54.19. Here are some applications of Theorem 54.16 [**561**].

(i) Suppose that $\{X_t\}$ is the subordinator with

$$\mu(\mathrm{d}x) = 1_{(0,\infty)}(x)\textstyle\sum_{j=1}^{n} q_j a_j \mathrm{e}^{-a_j x}\mathrm{d}x,$$

where $q_j > 0$ for $j = 1,\ldots,n$, $\sum_{j=1}^{n} q_j = 1$, and $a_1,\ldots,a_n$ are distinct positive reals. Then, μ^t is at most n-modal for every t, and unimodal for all t large enough. If the constants q_j and a_j are chosen appropriately, then, for some $t > 0$, μ^t is n-modal.

(ii) Suppose that $\{X_t\}$ is the subordinator with

$$\mu(\mathrm{d}x) = q_0\delta_0(\mathrm{d}x) + 1_{(0,\infty)}(x)\textstyle\sum_{j=1}^{n} q_j a_j \mathrm{e}^{-a_j x}\mathrm{d}x,$$

where $q_j > 0$ for $j = 0, 1,\ldots,n$, $\sum_{j=0}^{n} q_j = 1$, and $a_1,\ldots,a_n$ are distinct positive reals. Then, the conclusions in (i) are true with "n-modal" and "unimodal" replaced by "$(n+1)$-modal" and "bimodal", respectively.

(iii) Suppose that $\{X_t\}$ is a subordinator with Lévy measure

$$\nu(\mathrm{d}x) = 1_{(0,\infty)}(x)\textstyle\sum_{j=1}^{n}(b_j + c_j x^{-1})\mathrm{e}^{-a_j x}\mathrm{d}x,$$

where $b_j \geq 0$, $c_j \in \mathbb{R}$, $b_j + |c_j| > 0$, and $\sum_{k=1}^{j} c_k \geq 0$ for $j = 1,\ldots,n$, and $0 < a_1 < \cdots < a_n$. Then, μ^t is at most $(n+1)$-modal for any $t > 0$.

The processes are obtained by iteration of Bessel transforms from the Γ-process (in the case of (i) and (ii)) or from the trivial process (in the case of (iii)). Part of the result for $n = 2$ of (i) is obtained by Yamamuro [**585**].

Here we add some results related to strong unimodality.

THEOREM 54.20 (Watanabe [**552**]). *A Lévy process $\{X_t\}$ on $\mathbb{R}$ has a strongly unimodal distribution at every $t \geq 0$ if and only if $\nu = 0$.*

Proof. If $\nu = 0$, then the distribution is Gaussian, hence log-concave, which means strong unimodality by Theorem 52.3. Conversely, assume that μ^t is strongly unimodal for every $t \geq 0$. Suppose that $\nu \neq 0$. We assume $\nu(0,\infty) > 0$. In general, as in the proof of Theorem 54.1, $t^{-1}\int h(x)\mu^t(\mathrm{d}x) \to \int h(x)\nu(\mathrm{d}x)$ as $t \downarrow 0$ for any bounded continuous function h which vanishes on a neighborhood of 0. Fix a continuity point $b > 0$ of ν such that $\nu[b,\infty) > 0$. Then $t^{-1}\mu^t[b,\infty) \to \nu[b,\infty)$ as $t \downarrow 0$. Write $\rho_t = (\mu^t[b,\infty))^{-1}[\mu^t]_{[b,\infty)}$. Then $\rho_t \to (\nu[b,\infty))^{-1}[\nu]_{[b,\infty)}$ as $t \downarrow 0$. By the characterization of Theorem 52.3, strong unimodality is inherited by ρ_t from μ^t. In general, by Lemma 52.4, strong unimodality is preserved in passing to a limit. Hence $(\nu[b,\infty))^{-1}[\nu]_{[b,\infty)}$ is strongly unimodal. Again by Theorem 52.3, it follows that ν is absolutely continuous and there is an interval $(b_1, b_2) \subset (0,\infty)$ such that $\nu((0,\infty) \setminus (b_1,b_2)) = 0$ and ν has a density $l(x)$ log-concave on (b_1,b_2). We have $l(b_1+) < \infty$. We can have a similar conclusion on $(-\infty,0)$, if $\nu(-\infty,0) > 0$. Hence $\nu(\mathbb{R})$ is finite. Hence

$$\widehat{\mu}(z)^t = \exp\bigl[t\bigl(-\tfrac{1}{2}Az^2 + \mathrm{i}\gamma_0 z + \textstyle\int_{\mathbb{R}}(\mathrm{e}^{\mathrm{i}zx} - 1)l(x)\mathrm{d}x\bigr)\bigr].$$

We assume $\gamma_0 = 0$ (otherwise consider $\{X_t - \gamma_0 t\}$ in place of $\{X_t\}$). There are two cases: (1) $A = 0$ and (2) $A > 0$. In case (1), μ is compound Poisson and hence has a point mass at 0, which contradicts the strong unimodality by Theorem

52.3. Consider case (2). Let $f_t(x)$ be the log-concave density of μ^t on $(-\infty, \infty)$. Let $g_t(x) = (2\pi At)^{-1/2}e^{-x^2/(2At)}$ and let σ be the compound Poisson distribution with Lévy measure $l(x)dx$. Then $f_t(x) = \int_{\mathbb{R}} g_t(x-y)\sigma^t(dy)$, $x \in \mathbb{R}$. Thus

$$f_t(0) = g_t(0)\sigma^t\{0\} + \int_{\mathbb{R}\setminus\{0\}} g_t(-y)\sigma^t(dy),$$

$$f_t(0)/g_t(0) \to 1, \text{ that is, } f_t(0) \sim (2\pi At)^{-1/2} \text{ as } t \downarrow 0$$

since $\sigma^t\{0\} \to 1$. Choose $x_0 \in (b_1, b_2)$ and recall that $f_t(0)f_t(x_0) \le f_t(\frac{x_0}{2})^2$, by the log-concavity. Let t be so small that $a(t) < \frac{x_0}{4}$ for the mode $a(t)$ of μ^t. Then $\frac{x_0}{4} f_t(\frac{x_0}{2}) \le \mu^t(\frac{x_0}{4}, \frac{x_0}{2})$ and $f_t(x_0) \ge \mu^t(x_0, x_0+1)$. Thus

$$t^{1/2} f_t(0) \cdot t^{-1}\mu^t(x_0, x_0+1) \le t^{3/2}\big(t^{-1}\tfrac{4}{x_0}\mu^t(\tfrac{x_0}{4}, \tfrac{x_0}{2})\big)^2.$$

As $t \downarrow 0$, the left-hand side tends to $(2\pi A)^{-1/2}\nu(x_0, x_0+1) > 0$, while the right-hand side tends to 0. This is absurd. Hence $\nu = 0$. □

REMARK 54.21. Yamazato [**589**] proves the following. Let $\{X_t\}$ be a subordinator with $\nu(dx) = x^{-1}k(x)dx$. Assume that there is $0 < b \le \infty$ such that $k(x)$ is log-concave on $(0, b)$ and 0 on $[b, \infty)$. If $t \ge 1/k(0+)$, then μ^t is strongly unimodal. If $0 < t < 1/k(0+)$, then μ^t is not strongly unimodal.

For example, if $k(x) = e^{-x}$ on $(0, \infty)$, then $\{X_t\}$ is a Γ-process and the result is already shown in Example 52.7. Another example is $k(x) = e^{-(x-c)^2}$ with $c > 0$ employed in Example 54.4.

REMARK 54.22. Here is another result of Yamazato [**589**]. Let μ be an infinitely divisible distribution on $\mathbb{R}_+$, of class B, defined in Definition 51.9. Suppose that $l(x)$ satisfies (51.14) with $q(y)$ nonnegative, measurable, and $\int_0^1 (q(y)/y)dy < \infty$. If there is b, $0 < b < \infty$, such that $1 \le q(y) \le 2$ on $[b, \infty)$ and $q(y) = 0$ on $(0, b)$, then μ is strongly unimodal. If there is b, $0 < b < \infty$, such that $1 \le q(y) \le 2$ on $[b, \infty)$, $0 \le q(y) \le 1$ on $(0, b)$, and $\mathrm{Leb}\{y \in (0, b) \colon 0 < q(y) < 1\} > 0$, then μ is not strongly unimodal.

For example if $q(y) = 1_{[b,\infty)}(y)$, then $k(x) = e^{-bx}$ and μ is exponential, for which case strong unimodality is known. Another strongly unimodal example is $q(y) = 1_{[b,\infty)}(y) + 1_{[\alpha,\beta)}(y)$ with $0 < b < \alpha < \beta$, that is, $k(x) = e^{-bx} + e^{-\alpha x} - e^{-\beta x}$. Since $k''k - (k')^2 \sim (\alpha - b)^2 e^{-bx-\alpha x}$ as $x \to \infty$, $(\log k)'' > 0$ for large x, which shows that k is not log-concave on $(0, \infty)$ and Remark 54.21 is not applicable.

Yamazato [**594**] considers strong unimodality in another class.

55. Exercises 10

E 55.1. Show that the following are examples of completely monotone integrable functions on $(0, \infty)$: $(1+x)^{-\alpha-1}$ with $\alpha > 0$; $x^{\beta-1}e^{-x^\alpha}$ with $0 < \alpha \le 1$ and $0 < \beta \le 1$; $\exp(e^{-x} - x)$.

E 55.2. Give examples of the infinitely divisible distributions μ in Theorem 51.4 with $f(x)$ not completely monotone.

E 55.3 (Steutel [**499**]). Let λ be infinitely divisible on $\mathbb{R}^d$. For $a > 0$, let μ_a be the infinitely divisible distribution with $\widehat{\mu}_a(z) = a(a - \log \widehat{\lambda}(z))^{-1}$ as in

(30.23) with $t = 1$. Define $\mu_\infty = \delta_0$. Show that any mixture of the family $\{\mu_a : a \in (0, \infty]\}$ is infinitely divisible.

E 55.4. Let $LCV_{\mathbb{Z}_+}$ be the class of probability measures μ on $\mathbb{Z}_+$ such that $\mu = \sum_n p_n \delta_n$ with $p_n > 0$ for $n \in \mathbb{Z}_+$ and $\{p_n\}$ is log-convex. Show the following. (i) If $\mu_1, \ldots, \mu_m \in LCV_{\mathbb{Z}_+}$, then any mixture of $\{\mu_1, \ldots, \mu_m\}$ belongs to $LCV_{\mathbb{Z}_+}$. (ii) If $\mu_k \in LCV_{\mathbb{Z}_+}$ for $k = 1, 2, \ldots$ and μ_k tends to a probability measure μ as $k \to \infty$, then $\mu \in LCV_{\mathbb{Z}_+}$. (iii) Let $\gamma > 0$. If $\mu = \sum_n p_n \delta_n \in LCV_{\mathbb{Z}_+}$ and $c = \sum_n p_n{}^\gamma < \infty$, then $\rho = c^{-1} \sum_n p_n{}^\gamma \delta_n$ belongs to $LCV_{\mathbb{Z}_+}$.

E 55.5. Let LCV be the class of probability measures μ on $\mathbb{R}_+$ such that μ is either δ_0 or $\mu = c\delta_0 + f(x)\mathrm{d}x$ with $0 \le c < 1$ and with $f(x)$ positive and log-convex on $(0, \infty)$. Show the following. (i) $\mu \in LCV$ if and only if the distribution function $F(x) = \mu(\infty, x]$ satisfies

$$\text{(lcv)} \quad F(\alpha x + \beta y + \varepsilon) - F(\alpha x + \beta y) \le (F(x + \varepsilon) - F(x))^\alpha (F(y + \varepsilon) - F(y))^\beta,$$

for any positive x, y, α, β, and ε with $\alpha + \beta = 1$. (ii) If $\mu_1, \ldots, \mu_m \in LCV$, then any mixture of them belongs to LCV. (iii) If $\mu_k \in LCV$ and μ_k tends to a probability measure μ as $k \to \infty$, then $\mu \in LCV$. (iv) Let $\alpha > 0$. If $\mu = f(x)\mathrm{d}x \in LCV$ and $\int f(x)^\alpha \mathrm{d}x < \infty$, then, for any $c \ge 0$, $\rho = a(c\delta_0 + f(x)^\alpha \mathrm{d}x)$ with a normalizing constant a belongs to LCV.

E 55.6. Show that, if $\mu_{(n)} \in ME$, $n = 1, 2, \ldots$, and if $\mu_{(n)}$ tends to a probability measure μ, then $\mu \in ME$.

E 55.7. The smallest class that contains all stable distribution on $\mathbb{R}_+$ and that is closed under convergence and convolution is called *the class* L_∞. Show that $\mu \in L_\infty$ if and only if, in (51.1), $\gamma_0 \ge 0$ and $\nu(\mathrm{d}x) = x^{-1} h(\log x)\mathrm{d}x$ with $h(y)$ satisfying $(-1)^n (\mathrm{d}^n/\mathrm{d}y^n) h(y) \ge 0$ for $n = 0, 1, \ldots$ on $\mathbb{R}$.

E 55.8. Let T be the smallest class that contains all Γ-distributions and that is closed under convergence and convolution. Show that $\mu \in T$ if and only if, in (51.1), $\gamma_0 \ge 0$ and $\nu(\mathrm{d}x) = x^{-1} k(x)\mathrm{d}x$ with $k(x)$ completely monotone. (A distribution in T is called generalized gamma convolution or GGC.)

E 55.9. Let L be the class of selfdecomposable distributions on $\mathbb{R}_+$. Show that $L \supset T \supset L_\infty$ and $B \supset ME \cup T$.

E 55.10 (Keilson and Steutel [**270**]). Let μ_j, $j = 1, \ldots, n$, be symmetric α_j-stable distributions on $\mathbb{R}$ with $0 < \alpha_j \le 1$. Show that any mixture of $\mu_1, \ldots, \mu_n$ is infinitely divisible.

E 55.11 (Wolfe [**581**]). Let $\nu = (c_1 1_{[-1,1]}(x) + c_2 1_{[-2,-1) \cup (1,2]}(x))\mathrm{d}x$ with $0 < c_1 < c_2$. Let μ be an infinitely divisible distribution generated by $(0, \nu, 0)$. Let μ_0 be Gaussian with mean 0 and variance 1. Show that, when c_1 and c_2 are suitably chosen, then μ is non-unimodal, $\mu * \mu$ is unimodal, and $\mu * \mu_0$ is unimodal.

E 55.12 (Yamazato [**589**]). Let μ be of class B with Lévy measure density $l(x)$ satisfying $l(x) = \int_0^\infty \mathrm{e}^{-xy} q(y)\mathrm{d}y$, where $q(y)$ is nonnegative and increasing, and $\int_0^\infty \frac{q(y)}{y(y+1)}\mathrm{d}y < \infty$. Show that μ is unimodal.

E 55.13 (Yamazato [**589**]). Prove the unimodality of α-stable distributions on $\mathbb{R}$, $0 < \alpha < 2$, with one-sided Lévy measures, using the preceding exercise.

E 55.14. Let μ be of class B with Lévy measure density $l(x)$ satisfying $l(x) = \int_0^\infty \mathrm{e}^{-xy} q(y)\mathrm{d}y$ with $q(y) = a1_{[\alpha,\infty)}(y) + b1_{[\beta,\infty)}(y)$ with $0 < \alpha < \beta$, $a > 0$, and $b \geq 0$. Show the following.

(i) If $a \geq 1$, then μ is strongly unimodal.

(ii) If $a < 1$ and $a + b \leq 2$, then μ is not strongly unimodal.

Notes

Theorem 51.1 is a reformulation by Steutel [**499**], p. 86, of Feller's characterization in [**139**]. The first half of Corollary 51.2 was originally pointed out by Katti [**266**]. This, combined with an argument of Goldie [**168**], proves the second half of the corollary and Theorem 51.3; see [**499**]. Theorem 51.4 is the continuous analogue, in [**499**], of Theorem 51.3. The fact stated in Theorem 51.6 was discovered by Goldie [**168**]. An alternative proof is given by Steutel [**498**]. Theorems 51.10 and 51.12 are by Bondesson [**53**] and Steutel [**499**], respectively.

The analogues of unimodality, log-concavity, stability, and selfdecomposability for distributions on $\mathbb{Z}_+$ are studied. Mixing of Poisson distributions by a probability measure ρ on $\mathbb{R}_+$ is a transformation from ρ to a distribution on $\mathbb{Z}_+$, called the Poisson transform. It has proved useful in analyzing properties of distributions on $\mathbb{R}_+$. It is embedded in the Bessel transform of Definition 54.14. See Holgate [**206**], Puri and Goldie [**406**], Forst [**146**], Steutel and van Harn [**502**], and Watanabe [**553, 554**].

Facts analogous to Theorems 51.4 and 51.6 are known in characteristic functions of symmetric distributions. See Keilson and Steutel [**270**]. The class B is related to characterization of hitting time distributions of diffusion and birth-death processes and, further, to their spectral decomposition and Krein's correspondence. See Itô and McKean [**228**], Kent [**271**], Knight [**290**], and Yamazato [**591, 592**]. For some other classes of infinitely divisible distributions see Pruitt [**396**], Bondesson's monograph [**55**], and the papers [**508, 509, 550**] by Sugitani and Watanabe.

Theorem 52.3 on characterization of strong unimodality is given by Ibragimov [**210**]. See also Dharmadhikari and Joag-dev [**97**]. Related more general results are found in Karlin [**259**]. Bounds on the location of the mode of a unimodal distribution by the absolute pth moments around 0 or around the mean are discussed in Johnson and Rogers [**242**], Sato [**437**], and [**97**].

Equation (53.3) can be looked upon as expressing that a selfdecomposable distribution is the invariant distribution of a process of Ornstein–Uhlenbeck type, discussed in Section 17. From this viewpoint the equation can be generalized to distributions on $\mathbb{R}^d$ [**469**].

We are far from complete knowledge on time evolution of unimodality and multimodality in Lévy processes. See [**441, 445, 448, 553, 561**] for additional information.

Supplement

Some of the following sections treat subjects closely connected with the contents of the ten chapters. Some others are selected from the development in the areas familiar to the author.

56. Forms of Lévy-Khintchine representation

The Lévy-Khintchine representation of characteristic functions of infinitely divisible distributions on $\mathbb{R}^d$ in Theorem 8.1 is the basis of the whole theory. As in Remark 8.4, if $c(x)$ is a bounded measurable function from $\mathbb{R}^d$ into $\mathbb{R}$ satisfying (8.3) and (8.4), then the representation (8.5) for a general infinitely divisible distribution on $\mathbb{R}^d$ is obtained; the generating triplet in (8.5) is written as $(A, \nu, \gamma_c)_c$. Examples of $c(x)$ used in the literature are mentioned there. For the purpose of obtaining this representation (8.5), we can weaken (8.3) to

$$c(x) = 1 + O(|x|), \quad \text{as } |x| \to 0, \tag{56.1}$$

but the condition (8.3) is needed in Theorem 8.7.

Maruyama [**340**] introduces a 'big' generating triplet $(\widetilde{A}, \widetilde{\nu}, \widetilde{\gamma})$ in the infinite product space for every infinitely divisible process, by which he means a stochastic process all of whose finite-dimensional marginal distributions are infinitely divisible. Using this triplet, he characterizes the mixing property of a stationary infinitely divisible process. In order to construct the 'big' triplet, he uses a new form of the representation, which replaces $\langle z, x\rangle c(x)$ in (8.5) by $\sum_{j=1}^{d} z_j((-1) \vee (x_j \wedge 1))$. Let us generalize this new form and (8.5) simultaneously. Let $a_j(x)$, $j = 1, \ldots, d$, be bounded measurable functions from $\mathbb{R}^d$ into $\mathbb{R}$ satisfying

$$a_j(x) = x_j + O(|x|^2) \quad \text{as } |x| \to 0, \tag{56.2}$$

$$a_j(x) = O(1) \quad \text{as } |x| \to \infty. \tag{56.3}$$

Let $a(x) = (a_j(x))_{j=1,\ldots,d}$. Then, the characteristic function of any infinitely divisible distribution μ on $\mathbb{R}^d$ is written as

$$\widehat{\mu}(z) = \exp\left[-\tfrac{1}{2}\langle z, Az\rangle + \mathrm{i}\langle \gamma_{(a)}, z\rangle + \textstyle\int_{\mathbb{R}^d}(\mathrm{e}^{\mathrm{i}\langle z,x\rangle} - 1 - \mathrm{i}\langle z, a(x)\rangle)\nu(\mathrm{d}x)\right] \tag{56.4}$$

for $z \in \mathbb{R}^d$, where A and ν are the Gaussian covariance matrix and the Lévy measure of μ, respectively, and $\gamma_{(a)} \in \mathbb{R}^d$. Note that, for each z, $\mathrm{e}^{\mathrm{i}\langle z,x\rangle} - 1 - \mathrm{i}\langle z, a(x)\rangle = O(|x|^2)$ as $|x| \to 0$. Let us denote the triplet appearing in (56.4) as $(A, \nu, \gamma_{(a)})_{(a)}$. It is easy to see that its relation with the usual triplet (A, ν, γ) of μ in (8.1) is that

$$\gamma_{(a)} = \gamma + \textstyle\int_{\mathbb{R}^d}(a(x) - x1_{\{|x|\le 1\}}(x))\nu(\mathrm{d}x), \tag{56.5}$$

noting that $a(x)-x=O(|x|^2)$ as $|x|\to 0$. As examples of $a_j(x)$ satisfying (56.2) and (56.3) we have $x_j 1_{\{|x_j|\le 1\}}(x_j)$, $x_j/(1+x_j^2)$, $\sin x_j$, and $a_j(x)=(-1)\vee(x_j\wedge 1)$. These examples satisfy

$$a_j(x)=x_j+o(|x|^2) \quad \text{as } |x|\to 0, \tag{56.6}$$

which is stronger than (56.2).

The extension of Theorem 8.7 for the triplet $(A,\nu,\gamma)_{(a)}$ is as follows.

THEOREM 56.1. *Let $a(x)=(a_j(x))_{j=1,\dots,d}$ be a bounded continuous function from $\mathbb{R}^d$ into $\mathbb{R}^d$ satisfying (56.3) and (56.6). Suppose that μ_n $(n=1,2,\dots)$ are infinitely divisible distributions on $\mathbb{R}^d$ and that each $\widehat{\mu}_n(z)$ has the Lévy–Khintchine representation by triplet $(A_n,\nu_n,\beta_n)_{(a)}$. Let μ be a probability measure on $\mathbb{R}^d$. Then $\mu_n\to\mu$ if and only if μ is infinitely divisible and $\widehat{\mu}(z)$ has the Lévy–Khintchine representation by the triplet $(A,\nu,\beta)_{(a)}$ with A,ν,β satisfying the three conditions* (1), (2), *and* (3) *of Theorem 8.7 word for word.*

Proof. Let $c(x)$ be a bounded continuous function from $\mathbb{R}^d$ into $\mathbb{R}$ satisfying (8.3) and (8.4). Using this c, μ_n has triplet $(A_n,\nu_n,\gamma_n)_c$. The relation between β_n and γ_n is that $\beta_n=\gamma_n+\int_{\mathbb{R}^d}(a(x)-xc(x))\nu_n(dx)$. Then, by Theorem 8.7, $\mu_n\to\mu$ if and only if μ is infinitely divisible and $\widehat{\mu}(z)$ has the Lévy–Khintchine representation by the triplet $(A,\nu,\gamma)_c$ with A,ν,γ satisfying (1) and (2) of Theorem 8.7 and $\gamma_n\to\gamma$. An infinitely divisible distribution μ has triplet $(A,\nu,\gamma)_c$ if and only if it has triplet $(A,\nu,\beta)_{(a)}$ with $\beta=\gamma+\int_{\mathbb{R}^d}(a(x)-xc(x))\nu(dx)$. So, it is enough to show that, under the assumption that μ is infinitely divisible and that (1) and (2) are satisfied, we have $\beta_n\to\beta$ if and only if $\gamma_n\to\gamma$. This assumption implies (8.12) as we saw in the proof of 'if' part of Theorem 8.7. Choose $\varepsilon>0$ such that $\nu\{|x|=\varepsilon\}=0$. Then $\int_{|x|>\varepsilon}(a(x)-xc(x))\nu_n(dx)$ tends to $\int_{|x|>\varepsilon}(a(x)-xc(x))\nu(dx)$, since $a(x)-xc(x)$ is bounded continuous. Since $|a(x)-xc(x)|=o(|x|^2)$ as $|x|\to 0$ from (8.3) and (56.6), we have $\int_{|x|\le\varepsilon}(a(x)-xc(x))\nu_n(dx)\to 0$ uniformly in n as $\varepsilon\downarrow 0$, using (8.12). This finishes the proof. □

Some counterexamples. If the continuity of $c(x)$ is dropped, then Theorem 8.7 is not true. For example, if $d=1$, $c(x)=1_{\{|x|\le 1\}}(x)$ and if μ_n has triplet $(0,\delta_{1+1/n},0)_c$ and μ has triplet $(0,\delta_1,1)_c$, then $\mu_n\to\mu$ but (3) of Theorem 8.7 does not hold. Similarly, if the continuity of $a(x)$ is dropped, then Theorem 56.1 is not true.

In Theorem 8.7, the assumption (8.3) on $c(x)$ cannot be weakened to (56.1). To show this, let $c(x)=(1+|x|)1_{\{|x|\le 1\}}(x)+(2/|x|)1_{\{|x|>1\}}(x)$, which satisfies (56.1), but not (8.3). Let $d=1$ and let μ_n and μ be infinitely divisible with triplets $(0,n^2\delta_{1/n},0)_c$ and $(1,0,-1)_c$, respectively. Then $\widehat{\mu}_n(z)\to\exp(-z^2/2-iz)$, that is, $\mu_n\to\mu$, but (3) of Theorem 8.7 is not true. Likewise, in Theorem 56.1, the assumption (56.6) cannot be weakened to (56.2).

The representation (56.4) with $a_j(x)=(-1)\vee(x_j\wedge 1)$ is transferable in coordinate projection. This property is noticed by Maruyama and is essential in the proof of Theorem 1 of [**340**] for the existence of the 'big' generating triplet for an infinitely divisible process. Let us generalize this fact.

Let d and n be integers with $1 \le n \le d$. Let $j_1, \dots, j_n$ be n distinct elements in the set $\{1, 2, \dots, d\}$. Let U be an $n \times d$ matrix such that $y = Ux$ for $x = (x_j)_{j=1,\dots,d} \in \mathbb{R}^d$ and $y = (y_k)_{k=1,\dots,n} \in \mathbb{R}^n$ if and only if $y_k = x_{j_k}$ for $k = 1, \dots, n$. Let us call U the coordinate projection from $\{1, \dots, d\}$ to $\{j_1, \dots, j_n\}$. The following theorem shows that a special form of generating triplet of infinitely divisible distributions is transferable in coordinate projection.

THEOREM 56.2. *Let U be the coordinate projection from $\{1, \dots, d\}$ to $\{j_1, \dots, j_n\}$. Let $b(u)$ be a bounded measurable function from $\mathbb{R}$ into $\mathbb{R}$ such that $b(0) = 0$, $b(u) = u + O(u^2)$ as $u \to 0$, and $b(u) = O(1)$ as $|u| \to \infty$. Define $a(x) = (b(x_j))_{j=1,\dots,d}$ for $x = (x_j)_{j=1,\dots,d} \in \mathbb{R}^d$ and $a_U(y) = (b(y_k))_{k=1,\dots,n}$ for $y = (y_k)_{k=1,\dots,n} \in \mathbb{R}^n$. Let X be an $\mathbb{R}^d$-valued random variable having infinitely divisible distribution μ with triplet $(A, \nu, \beta)_{(a)}$. Then UX has infinitely divisible distribution on $\mathbb{R}^n$ with triplet $(A_U, \nu_U, \beta_U)_{(a_U)}$ given by*

$$A_U = UAU', \tag{56.7}$$

$$\nu_U = [\nu U^{-1}]_{\mathbb{R}^n \setminus \{0\}}, \tag{56.8}$$

$$\beta_U = U\beta. \tag{56.9}$$

Here, U' is the transpose of U; $[\nu U^{-1}]_{\mathbb{R}^n \setminus \{0\}}$ is the restriction of the measure $(\nu U^{-1})(B) = \nu(\{x : Ux \in B\})$ to $\mathbb{R}^n \setminus \{0\}$. The identity (56.9) comes from our choice of $a(x)$ and $a_U(x)$.

Proof of Theorem 56.2. Let $c(x) = 1_{\{|x| \le 1\}}(x)$ and let μ have triplet $(A, \nu, \gamma)_c$. Then it follows from Proposition 11.10 that UX has infinitely divisible distribution with triplet $(A_U, \nu_U, \gamma_U)_c$ satisfying (56.7), (56.8), and

$$\gamma_U = U\gamma + \int_{\mathbb{R}^d} Ux(1_{\{|Ux| \le 1\}} - 1_{\{|x| \le 1\}})\nu(dx).$$

We have

$$\beta = \gamma + \int_{\mathbb{R}^d} (a(x) - x1_{\{|x| \le 1\}})\nu(dx),$$
$$\beta_U = \gamma_U + \int_{\mathbb{R}^n} (a_U(y) - y1_{\{|y| \le 1\}})\nu_U(dy).$$

Hence

$$\beta_U = U\beta + \int_{\mathbb{R}^d} (Ux1_{\{|x| \le 1\}} - U(a(x)))\nu(dx) + \int_{\mathbb{R}^d} Ux(1_{\{|Ux| \le 1\}} - 1_{\{|x| \le 1\}})\nu(dx)$$
$$+ \int_{\{x \in \mathbb{R}^d :\, Ux \ne 0\}} (a_U(Ux) - Ux1_{\{|Ux| \le 1\}})\nu(dx).$$

Since the last integral is equal to the integral over $\mathbb{R}^d$, we have

$$\beta_U = U\beta + \int_{\mathbb{R}^d} (a_U(Ux) - U(a(x)))\, \nu(dx).$$

Noting that

$$a_U(Ux) = (b((Ux)_k))_{k=1,\dots,n} = (b(x_{j_k}))_{k=1,\dots,n} = ((a(x))_{j_k})_{k=1,\dots,n} = U(a(x)),$$

we obtain (56.9). □

We remark that, under the condition of existence of mean (center) or drift, the appropriate form of generating triplet is transferable in linear transformation.

THEOREM 56.3. *Let X be an $\mathbb{R}^d$-valued random variable having infinitely divisible distribution μ with Gaussian covariance matrix A and Lévy measure ν. Let n be a positive integer and let U be an $n \times d$ matrix. Then UX has infinitely divisible distribution μ_U with Gaussian covariance matrix A_U and Lévy measure ν_U given by (56.7) and (56.8). Further the following are true.*

(i) *Suppose that $\int_{|x|\le 1} |x|\nu(\mathrm{d}x) < \infty$ and that μ has triplet $(A, \nu, \beta)_0$ of (8.7), that is, β is the drift of μ. Then $\int_{|y|\le 1} |y|\nu_U(\mathrm{d}y) < \infty$ and μ_U has triplet $(A_U, \nu_U, \beta_U)_0$ of (8.7) satisfying (56.9); β_U is the drift of μ_U.*

(ii) *Suppose that $\int_{|x|>1} |x|\nu(\mathrm{d}x) < \infty$ and that μ has triplet $(A, \nu, \beta)_1$ of (8.8), that is, β is the mean of μ. Then $\int_{|y|>1} |y|\nu_U(\mathrm{d}y) < \infty$ and μ_U has triplet $(A_U, \nu_U, \beta_U)_1$ of (8.8) satisfying (56.9); β_U is the mean of μ_U.*

Proof. It follows from Proposition 11.10 that μ_U is infinitely divisible with A_U and ν_U given by (56.7) and (56.8). To show (i), note that

$$\int_{|y|\le 1}|y|\nu_U(\mathrm{d}y) = \int_{|Ux|\le 1}|Ux|\nu(\mathrm{d}x) \le \|U\|\int_{|x|\le 1}|x|\nu(\mathrm{d}x) + \int_{|x|>1}\nu(\mathrm{d}x) < \infty$$

and that, for $z \in \mathbb{R}^n$,

$$\begin{aligned} E[\mathrm{e}^{\mathrm{i}\langle z,UX\rangle}] &= E[\mathrm{e}^{\mathrm{i}\langle U'z,X\rangle}] \\ &= \exp\big[-\tfrac{1}{2}\langle U'z, AU'z\rangle + \mathrm{i}\langle \beta, U'z\rangle + \textstyle\int_{\mathbb{R}^d}(\mathrm{e}^{\mathrm{i}\langle U'z,x\rangle} - 1)\nu(\mathrm{d}x)\big] \\ &= \exp\big[-\tfrac{1}{2}\langle z, UAU'z\rangle + \mathrm{i}\langle U\beta, z\rangle + \textstyle\int_{\mathbb{R}^n}(\mathrm{e}^{\mathrm{i}\langle z,y\rangle} - 1)\nu_U(\mathrm{d}y)\big]. \end{aligned}$$

To see (ii), note that

$$\int_{|y|>1}|y|\nu_U(\mathrm{d}y) = \int_{|Ux|>1}|Ux|\nu(\mathrm{d}x) \le \|U\|\int_{|x|>1/\|U\|}|x|\nu(\mathrm{d}x) < \infty$$

and make a similar manipulation of $E[\mathrm{e}^{\mathrm{i}\langle z,UX\rangle}]$ for $z \in \mathbb{R}^n$. An alternative proof of (ii) is to use Example 25.12. □

Another example of $c(x)$. The representation (8.5) on $\mathbb{R}^d$ with

$$c(x) = 1_{\{|x|\le 1\}}(x) + |x|^{-1}1_{\{|x|>1\}}(x) \tag{56.10}$$

is used in Rajput and Rosinski [**407**], Kwapień and Woyczyński [**302**], and Sato and Ueda [**460**].

57. Independently scattered random measures

The following concept is closely related with additive and Lévy processes. Let $\Theta = \mathbb{R}_+$ or $\mathbb{R}$ in this section. Let $\mathcal{B}_\Theta$ be the class of Borel sets on Θ and $\mathcal{B}_\Theta^0$ the class of bounded Borel sets on Θ.

DEFINITION 57.1. A collection $\{M(B)\colon B \in \mathcal{B}_\Theta^0\}$ of $\mathbb{R}^d$-valued random variables is called $\mathbb{R}^d$-valued *independently scattered random measure* on Θ if the following three conditions are satisfied.

(1) If $\{B_1, B_2, \ldots\}$ is a sequence of disjoint sets in $\mathcal{B}_\Theta^0$ with $\bigcup_{n=1}^\infty B_n \in \mathcal{B}_\Theta^0$, then $\sum_{n=1}^\infty M(B_n)$ converges a. s. and equals $M(\bigcup_{n=1}^\infty B_n)$ a. s. (countable additivity)

(2) For any finite sequence $B_1, \ldots, B_n$ of disjoint sets in $\mathcal{B}_\Theta^0$, $M(B_1), \ldots, M(B_n)$ are independent. (independent increments property)

(3) $M(\{a\}) = 0$ a. s. for every one-point set $\{a\}$. (atomless property)

An independently scattered random measure $\{M(B)\colon B \in \mathcal{B}^0_\Theta\}$ is called *homogeneous* if

(4) $M(B) \stackrel{\mathrm{d}}{=} M(B+a)$ for any $B \in \mathcal{B}^0_\Theta$ and $a > 0$.

Here we adopt the usual usage, though 'random signed measure' is appropriate rather than 'random measure' if $d = 1$. This notion was introduced by Urbanik and Woyczyński [**546**]. After that it is extensively studied by Rajput and Rosinski [**407**], Kwapień and Woyczyński [**302**], Sato [**452, 453**] and others. Note that property (1) implies $M(\emptyset) = 0$ a. s. Note also that the exceptional set in the a. s. assertion in (1) depends on the sequence $\{B_n\}$ in general.

PROPOSITION 57.2. *If $\{M(B)\colon B \in \mathcal{B}^0_\Theta\}$ is an $\mathbb{R}^d$-valued independently scattered random measure on Θ, then, for each $B \in \mathcal{B}^0_\Theta$, the distribution $\mu_{M(B)}$ of $M(B)$ is infinitely divisible.*

Proof. Let $\{M(B)\colon B \in \mathcal{B}^0_{\mathbb{R}_+}\}$ be an independently scattered random measure on $\mathbb{R}_+$. Then, for each $B \in \mathcal{B}^0_{\mathbb{R}_+}$, $\{M(B \cap [0,t])\colon t \in \mathbb{R}_+\}$ is an additive process in law. Hence the infinite divisibility of $\mu_{M(B)}$ follows from Theorem 9.1, choosing t so large that $[0,t] \supset B$. The case $\Theta = \mathbb{R}$ is proved by using the result for $\Theta = \mathbb{R}_+$. □

In this section let us formulate basic results on independently scattered random measures. We omit the proofs, which can be given by using the discussion in [**452, 453**]. The Nikodým theorem in Dunford and Schwartz [**111**] is used in the proof of Propositions 57.9 and 57.10. We begin with the introduction of the notion of a natural additive process in law in order to clarify the relation between independently scattered random measures and additive processes.

DEFINITION 57.3. An additive process in law $\{X_t\}$ on $\mathbb{R}^d$ with distribution μ_t at time t is called *natural* if, for each $z \in \mathbb{R}^d$, $\widehat{\mu}_t(z)$ is locally of finite variation in t, that is, $\{\widehat{\mu}_t(z)\colon t \in [0,t_0]\}$ has finite variation for any $t_0 \in (0,\infty)$.

PROPOSITION 57.4. *Let c be a bounded measurable function on $\mathbb{R}^d$ satisfying (8.3) and (8.4). An additive process in law $\{X_t\}$ on $\mathbb{R}^d$ with triplet $(A_t, \nu_t, \gamma_t)_c$ is natural if and only if γ_t is locally of finite variation in t.*

It is also known that an additive process in law on $\mathbb{R}^d$ is natural if and only if it is a semimartingale, but semimartingale theory is beyond the scope of this book.

THEOREM 57.5. *If $\{M(B)\colon B \in \mathcal{B}^0_{\mathbb{R}_+}\}$ is an $\mathbb{R}^d$-valued independently scattered random measure, then $\{X_t\colon t \in \mathbb{R}_+\}$ defined by $X_t = M((0,t])$ is a natural additive process in law. Conversely, if $\{X_t\colon t \in \mathbb{R}_+\}$ is a natural additive process in law on $\mathbb{R}^d$ defined on a probability space $(\Omega, \mathcal{F}, P)$, then there exists a unique $\mathbb{R}^d$-valued independently scattered random measure $\{M(B)\colon B \in \mathcal{B}^0_{\mathbb{R}_+}\}$ defined on this space $(\Omega, \mathcal{F}, P)$ such that $M((0,t]) = X_t$ a. s. for each $t \in \mathbb{R}_+$.*

In saying the uniqueness, two independently scattered random measures $\{M^j(B)\colon B \in \mathcal{B}^0_{\mathbb{R}_+}\}$, $j = 1,2$, satisfying $M^1(B) = M^2(B)$ a. s. for each $B \in \mathcal{B}^0_{\mathbb{R}_+}$ are identified.

COROLLARY 57.6. *Fix* $t_0 \in \mathbb{R}$. *If* $\{M(B)\colon B \in \mathcal{B}^0_{\mathbb{R}}\}$ *is an* $\mathbb{R}^d$*-valued independently scattered random measure, then* $\{X^j_t\colon t \in \mathbb{R}_+\}$, $j = 1, 2$, *defined by* $X^1_t = M((t_0, t_0 + t])$ *and* $X^2_t = M((t_0 - t, t_0])$ *are independent natural additive processes in law. Conversely, if* $\{X^j_t\colon t \in \mathbb{R}_+\}$, $j = 1, 2$, *are independent natural additive processes in law on* $\mathbb{R}^d$ *defined on a probability space* $(\Omega, \mathcal{F}, P)$, *then there exists a unique* $\mathbb{R}^d$*-valued independently scattered random measure* $\{M(B)\colon B \in \mathcal{B}^0_{\mathbb{R}}\}$ *defined on* $(\Omega, \mathcal{F}, P)$ *such that* $M((t_0, t_0 + t]) = X^1_t$ *and* $M((t_0 - t, t_0]) = X^2_t$ *a. s. for each* $t \in \mathbb{R}_+$.

Lévy processes in law are natural additive processes in law. Thus any Lévy process in law is associated to a unique homogeneous independently scattered random measure on $\mathbb{R}_+$. Selfsimilar additive processes in law are also natural, but semi-selfsimilar additive processes in law are not necessarily natural.

From now on in this section, let $\{X_t\colon t \in \mathbb{R}_+\}$ be a Lévy process on $\mathbb{R}^d$ and let $\{M(B)\colon B \in \mathcal{B}^0_{\mathbb{R}_+}\}$ be the $\mathbb{R}^d$-valued homogeneous independently scattered random measure on $\mathbb{R}_+$ associated to $\{X_t\}$ as in Theorem 57.5. Let ρ denote the distributions of X_1. Let $c(x)$ be either a bounded continuous function on $\mathbb{R}^d$ satisfying (8.3) and (8.4) or the function $1_{\{|x|\le 1\}}(x)$. For the proof in the case $c(x) = 1_{\{|x|\le 1\}}(x)$, see also [**457**]. Let $(A^\rho, \nu^\rho, \gamma^\rho)_c$ be the triplet of ρ.

PROPOSITION 57.7. *We have, for* $B \in \mathcal{B}^0_{\mathbb{R}_+}$,

$$\log \widehat{\mu}_{M(B)}(z) = \mathrm{Leb}(B) \log \widehat{\rho}(z), \qquad z \in \mathbb{R}^d,$$

where $\log \widehat{\mu}_{M(B)}(z)$ *and* $\log \widehat{\rho}(z)$ *are the distinguished logarithms of* $\widehat{\mu}_{M(B)}$ *and* $\widehat{\rho}$, *respectively.*

The definition of the integral with respect to $\{M(B)\}$ is given as follows.

DEFINITION 57.8. (i) Call a function $f(s)$ on $\mathbb{R}_+$ a *simple function* if $f(s) = \sum_{j=1}^n a_j 1_{B_j}(s)$ for some n, where $B_1, \ldots, B_n$ are disjoint sets in $\mathcal{B}_{\mathbb{R}_+}$ and $a_1, \ldots, a_n \in \mathbb{R}$. If $f(s)$ is a simple function of this form, we define the integral of $f(s)$ over $B \in \mathcal{B}^0_{\mathbb{R}_+}$ with respect to $\{M(B)\}$ as

$$\textstyle\int_B f(s) M(\mathrm{d}s) = \sum_{j=1}^n a_j M(B \cap B_j).$$

(ii) Suppose that $f(s)$ is a measurable function on $\mathbb{R}_+$ and that there is a sequence of simple functions $f_n(s)$, $n = 1, 2, \ldots$, such that

(1) $f_n(s) \to f(s)$ a. e. on $\mathbb{R}_+$ as $n \to \infty$,
(2) for any $B \in \mathcal{B}^0_{\mathbb{R}_+}$, the sequence $\int_B f_n(s) M(\mathrm{d}s)$ is convergent in probability as $n \to \infty$.

Then the limit in probability in (2) is denoted by $\int_B f(s) M(\mathrm{d}s)$ and we say that $f(s)$ is *locally M-integrable* on $\mathbb{R}_+$. We also write $\int_{t_1}^{t_2} f(s) \mathrm{d}X_s$ or $\int_{t_1}^{t_2} f(s) M(\mathrm{d}s)$ for $\int_{[t_1, t_2]} f(s) M(\mathrm{d}s)$.

PROPOSITION 57.9. *The definition of* $\int_B f(s) M(\mathrm{d}s)$ *does not depend on the choice of simple functions* $f_n(s)$ *satisfying (1) and (2) in (ii) of Definition 57.8, up to on a set of probability* 0 *for each* B.

PROPOSITION 57.10. *Let $f(s)$ be a measurable function on $\mathbb{R}_+$.*
(i) *The following are equivalent.*

(1) *$f(s)$ is locally M-integrable on $\mathbb{R}_+$.*
(2) *$\int_0^t |\log\widehat{\rho}(f(s)z)|\mathrm{d}s < \infty$ for $t \in \mathbb{R}_+$ and $z \in \mathbb{R}^d$. Here $\log\widehat{\rho}(f(s)z)$ means $\log\widehat{\rho}(w)$ evaluated at $w = f(s)z$.*
(3) *For every $t \in \mathbb{R}_+$, $\int_0^t f(s)^2 \operatorname{tr} A^\rho \mathrm{d}s$, $\int_0^t \mathrm{d}s \int_{\mathbb{R}^d}(|f(s)x|^2 \wedge 1)\nu^\rho(\mathrm{d}x)$, and $\int_0^t \left|f(s)\gamma^\rho + \int_{\mathbb{R}^d} f(s)x(c(f(s)x) - c(x))\nu^\rho(\mathrm{d}x)\right| \mathrm{d}s$ are finite. Here $\operatorname{tr} A^\rho$ is the trace of the matrix A^ρ.*

(ii) *Let $f(s)$ be locally M-integrable on $\mathbb{R}_+$ and $N(B) = \int_B f(s)M(\mathrm{d}s)$. Then $\{N(B)\colon B \in \mathcal{B}^0_{\mathbb{R}_+}\}$ is an $\mathbb{R}^d$-valued independently scattered random measure on $\mathbb{R}_+$ and*

$$\log\widehat{\mu}_{N(B)}(z) = \int_B \log\widehat{\rho}(f(s)z)\mathrm{d}s, \qquad B \in \mathcal{B}^0_{\mathbb{R}_+}.$$

The triplet $(A^{N(B)}, \nu^{N(B)}, \gamma^{N(B)})_c$ of $\mu_{N(B)}$ are

$$A^{N(B)} = \int_B f(s)^2 A^\rho \mathrm{d}s,$$
$$\nu^{N(B)}(C) = \int_B \mathrm{d}s \int_{\mathbb{R}^d} 1_C(f(s)x)\nu^\rho(\mathrm{d}x), \quad C \in \mathcal{B}(\mathbb{R}^d \setminus \{0\}),$$
$$\gamma^{N(B)} = \int_B \left(f(s)\gamma^\rho + \int_{\mathbb{R}^d} f(s)x(c(f(s)x) - c(x))\nu^\rho(\mathrm{d}x)\right) \mathrm{d}s.$$

COROLLARY 57.11. *If $f(s)$ is locally square-integrable on $\mathbb{R}_+$, then $f(s)$ is locally M-integrable.*

DEFINITION 57.12. We say that the improper integral of $f(s)$ with respect to $\{M(B)\}$ is *definable* if $f(s)$ is locally M-integrable and $\int_0^t f(s)M(\mathrm{d}s)$ is convergent in probability as $t \to \infty$. The limit is written as $\int_0^{\infty-} f(s)M(\mathrm{d}s)$ or $\int_0^{\infty-} f(s)\mathrm{d}X_s$.

PROPOSITION 57.13. *Let $f(s)$ be a locally M-integrable function on $\mathbb{R}_+$ and let $N(B) = \int_B f(s)M(\mathrm{d}s)$.*
(i) *The following are equivalent.*

(1) *$\int_0^{\infty-} f(s)M(\mathrm{d}s)$ is definable.*
(2) *$\int_0^t \log\widehat{\rho}(f(s)z)\mathrm{d}s$ is convergent in $\mathbb{C}$ as $t \to \infty$ for each $z \in \mathbb{R}^d$.*
(3) *$(3)_1$ $\int_0^\infty f(s)^2 \operatorname{tr} A^\rho \mathrm{d}s < \infty$, $(3)_2$ $\int_0^\infty \mathrm{d}s \int_{\mathbb{R}^d}(|f(s)x|^2 \wedge 1)\nu^\rho(\mathrm{d}x) < \infty$, and $(3)_3$ $\gamma^{N([0,t])}$ is convergent in $\mathbb{R}^d$ as $t \to \infty$.*

(ii) *Suppose that $\int_0^{\infty-} f(s)M(\mathrm{d}s)$ is definable. Let μ be its distribution and $(A^\mu, \nu^\mu, \gamma^\mu)_c$ the triplet of μ. Then,*

$$\log\widehat{\mu}(z) = \lim_{t\to\infty} \int_0^t \log\widehat{\rho}(f(s)z)\mathrm{d}s,$$
$$A^\mu = \int_0^\infty f(s)^2 A^\rho \mathrm{d}s,$$
$$\nu^\mu(C) = \int_0^\infty \mathrm{d}s \int_{\mathbb{R}^d} 1_C(f(s)x)\nu^\rho(\mathrm{d}x), \quad C \in \mathcal{B}(\mathbb{R}^d \setminus \{0\}),$$
$$\gamma^\mu = \lim_{t\to\infty} \gamma^{N([0,t])}.$$

REMARK 57.14. The definability of $\int_0^{\infty-} f(s)M(\mathrm{d}s)$ is equivalent to almost sure convergence of the additive process modification $\{\widetilde{Y}_t\}$ of $Y_t = \int_0^t f(s)M(\mathrm{d}s)$ as $t \to \infty$. It is also equivalent to the convergence in distribution.

The condition (2) in Proposition 57.13 leads to the following stronger definition of the definability of the improper integral.

DEFINITION 57.15. We say that $\int_0^{\infty-} f(s)M(\mathrm{d}s)$ is *absolutely definable* if $f(s)$ is locally M-integrable and $\int_0^{\infty} |\log \widehat{\rho}(f(s)z)|\mathrm{d}s < \infty$ for all $z \in \mathbb{R}^d$.

PROPOSITION 57.16. *Let $f(s)$ be a locally M-integrable function on $\mathbb{R}_+$. The following are equivalent.*

(1) *$\int_0^{\infty-} f(s)M(\mathrm{d}s)$ is absolutely definable.*
(2) *$(3)_1$ and $(3)_2$ hold and $\int_0^{\infty} \left|f(s)\gamma^{\rho} + \int_{\mathbb{R}^d} f(s)x(c(f(s)x) - c(x))\nu^{\rho}(\mathrm{d}x)\right| \mathrm{d}s < \infty$.*

EXAMPLE 57.17. Fix $H > 0$. The following are equivalent:

(1) $\int_0^{\infty-} \mathrm{e}^{-Hs}M(\mathrm{d}s)$ is definable.
(2) $\int_0^{\infty-} \mathrm{e}^{-Hs}M(\mathrm{d}s)$ is absolutely definable.
(3) $E(\log^+ |M(B)|) < \infty$ for $B \in \mathcal{B}^0_{\mathbb{R}_+}$.
(4) $E(\log^+ |X_1|) < \infty$.

Here $\log^+ u = (\log u) \vee 0$ for $u \geq 0$. We say that $\{X_t\}$ (or $\{M(B)\}$) has finite log-moment if (2) or (3) is satisfied. This example is a reformulation of a result in Section 17.

EXAMPLE 57.18. Consider $g(t) = \int_t^{\infty} u^{-2}\mathrm{e}^{-u}\mathrm{d}u$, $t \in (0,\infty)$, or $g_p(t) = (1/\Gamma(p))\int_t^1 (1-u)^{p-1}u^{-2}\mathrm{d}u$, $t \in (0,1]$, with $p > 0$. Let $t = f(s)$ for $s \in (0,\infty)$ or $t = f_p(s)$ for $s \in [0,\infty)$ be the inverse function of $s = g(t)$ or $s = g_p(t)$, respectively. Let $f(0) = \infty$. Then there is a homogeneous independently scattered random measure $\{M(B)\}$ on $\mathbb{R}_+$ such that $\int_0^{\infty-} f(s)M(\mathrm{d}s)$ or $\int_0^{\infty-} f_p(s)M(\mathrm{d}s)$, respectively, is definable but is not absolutely definable. This is shown in [**457**]. Thus absolute definability of improper integrals is a strictly stronger concept than definability. Here we have $f(s) \sim s^{-1}$ and $f_p(s) \sim s^{-1}/\Gamma(p)$ as $s \to \infty$.

All results and definitions ranging from Proposition 57.7 to Proposition 57.16 are extended in [**452, 453**] to natural additive processes in law and (not necessarily homogeneous) independently scattered random measures with due modification using the factoring of distributions of a natural additive process in law. A notion called essential definability of improper integrals, which is weaker than the definability, is also discussed. For $t_0 < \infty$ the improper integral $\int_{0+}^{t_0} f(s)M(\mathrm{d}s)$ defined by the limit in probability of $\int_t^{t_0} f(s)M(\mathrm{d}s)$ as $t \downarrow 0$ can be treated in parallel. See Sato [**452**]–[**457**]. Other development was made by Barndorff-Nielsen, Rosiński, and Thorbjørnsen [**16**].

58. Relations of representations of selfdecomposable distributions

An $\mathbb{R}^d$-valued stochastic process $\{Z_s : s \in \mathbb{R}\}$ with time parameter in $\mathbb{R}$ is called *stationary* if, for every $n \in \mathbb{N}$ and $s_1 < \cdots < s_n$ in $\mathbb{R}$, the distribution of $(Z_{s_j})_{j=1,\dots,n}$ is invariant under shift of time, that is, $(Z_{s_j})_{j=1,\dots,n} \stackrel{\mathrm{d}}{=} (Z_{s_j+u})_{j=1,\dots,n}$ for any $u \in \mathbb{R}$. A transformation between stationary processes and selfsimilar processes is known. This is called *Lamperti transformation* and described by the following proposition.

PROPOSITION 58.1. *Fix $H > 0$. If $\{Z_s : s \in \mathbb{R}\}$ is a stationary process on $\mathbb{R}^d$, then the process $\{Y_t : t \in \mathbb{R}_+\}$ defined by $Y_t = t^H Z_{\log t}$ for $t > 0$ and by $Y_0 = 0$ is an $\mathbb{R}^d$-valued process satisfying*

$$\{Y_{at} : t \in \mathbb{R}_+\} \stackrel{\mathrm{d}}{=} \{a^H Y_t : t \in \mathbb{R}_+\} \quad \text{for every } a > 0. \tag{58.1}$$

Conversely, if $\{Y_t : t \in \mathbb{R}_+\}$ is an $\mathbb{R}^d$-valued process satisfying (58.1), then the $\mathbb{R}^d$-valued process $\{Z_s : s \in \mathbb{R}\}$ defined by $Z_s = \mathrm{e}^{-Hs} Y(\mathrm{e}^s)$ for $s \in \mathbb{R}$ is a stationary process.

Proof is straightforward and omitted. In the case $d = 1$ it is given in Embrechts and Maejima [**125**], p. 11. Recall that (58.1) means that $\{Y_t : t \in \mathbb{R}_+\}$ is an H-selfsimilar process, if it is stochastically continuous and $Y_0 = 0$ a. s. (see Theorem 13.11, Definition 13.12, and Remark 13.13). The result above is due to Lamperti [**305**], but Doob [**105**] already recognized it in the case of symmetric stable process.

The class of selfdecomposable distributions is represented by distributions of selfsimilar additive processes in Section 16, and by limit distributions of processes of Ornstein–Uhlenbeck type in Section 17, where the processes of Ornstein–Uhlenbeck type are generated by Lévy processes with finite log-moment. Let us formulate those connections more clearly. For the proofs of the following three propositions consult Jeanblanc, Pitman, and Yor [**239**] and Maejima and Sato [**334**].

PROPOSITION 58.2. *Let $\{M(B) : B \in \mathcal{B}^0_{\mathbb{R}}\}$ be an $\mathbb{R}^d$-valued homogeneous independently scattered random measure on $\mathbb{R}$. Let $H > 0$ and $s_0 \in \mathbb{R}$ and let Ξ be an $\mathbb{R}^d$-valued random variable. Then there exists a unique (in the a. s. sense) $\mathbb{R}^d$-valued process $\{Z_s : s \in \mathbb{R}\}$ right-continuous with left limits such that*

$$Z_{s_2} - Z_{s_1} = M((s_1, s_2]) - H \textstyle\int_{s_1}^{s_2} Z_u \mathrm{d}u, \qquad s_1 < s_2 \tag{58.2}$$

and $Z_{s_0} = \Xi$ a. s. This process $\{Z_s\}$ is represented as

$$Z_s = \mathrm{e}^{(s_0 - s)H} \Xi + \mathrm{e}^{-sH} \textstyle\int_{s_0}^{s} \mathrm{e}^{uH} M(\mathrm{d}s), \quad s \in \mathbb{R}, \quad a.\,s., \tag{58.3}$$

where we understand $\int_{s_0}^{s} \mathrm{e}^{uH} M(\mathrm{d}s) = -\int_{s}^{s_0} \mathrm{e}^{uH} M(\mathrm{d}s)$ if $s < s_0$.

DEFINITION 58.3. An $\mathbb{R}^d$-valued process $\{Z_s : s \in \mathbb{R}\}$ is called a *process of Ornstein–Uhlenbeck type generated by $\{M(B)\}$ and H* if it is right-continuous with left limits a. s. and satisfies (58.2).

PROPOSITION 58.4. *Let $\{M(B)\colon B \in \mathcal{B}^0_{\mathbb{R}}\}$ be an $\mathbb{R}^d$-valued homogeneous independently scattered random measure on $\mathbb{R}$ and let $H > 0$. Then the following are equivalent.*

(1) *$\{M(B)\}$ has finite log-moment.*

(2) *$\int_{(-\infty)+}^{0} \mathrm{e}^{sH} M(\mathrm{d}s)$ is definable, that is, the limit in probability of $\int_t^0 \mathrm{e}^{sH} M(\mathrm{d}s)$ exists as $t \to -\infty$.*

(3) *There exists a stationary process of Ornstein–Uhlenbeck type generated by $\{M(B)\}$ and H.*

If (3) and hence also (1) and (2) are satisfied, then a stationary process of Ornstein–Uhlenbeck type $\{Z_s\colon s \in \mathbb{R}\}$ generated by $\{M(B)\}$ and H is unique and expressed as

$$Z_s = \mathrm{e}^{-sH} \int_{(-\infty)+}^{s} \mathrm{e}^{uH} M(\mathrm{d}u), \quad s \in \mathbb{R}, \quad a.\,s., \tag{58.4}$$

and the H-selfsimilar process $\{Y_t\colon t \in \mathbb{R}_+\}$ on $\mathbb{R}^d$ obtained from $\{Z_s\}$ through Lamperti transformation is an additive process in law expressed as

$$Y_t = \int_{(-\infty)+}^{\log t} \mathrm{e}^{uH} M(\mathrm{d}u), \quad t > 0, \quad a.\,s. \tag{58.5}$$

REMARK 58.5. In Proposition 58.4, $\{M(B)\}$ is recovered from $\{Z_s\}$ by (58.2). If we define the integral with respect to an independently scattered random measure on $\mathbb{R}_+$ as a generalization of Section 57, then $\{M(B)\}$ is recovered from $\{Y_t\}$ by

$$M(B) = \int_{\exp B} t^{-H} N(\mathrm{d}t), \quad B \in \mathcal{B}^0_{\mathbb{R}}, \tag{58.6}$$

where $\exp B = \{t = \mathrm{e}^s\colon s \in B\}$ and $\{N(C)\colon C \in \mathcal{B}^0_{\mathbb{R}_+}\}$ is the independently scattered random measure on $\mathbb{R}_+$ associated to the selfsimilar additive process $\{Y_t\}$.

PROPOSITION 58.6. *Fix $H > 0$. A distribution μ on $\mathbb{R}^d$ given by $\mu = P_{Y_1} = P_{Z_0}$ in Proposition 58.4 is selfdecomposable. Conversely, for any selfdecomposable distribution μ on $\mathbb{R}^d$, there is, uniquely in law, an $\mathbb{R}^d$-valued homogeneous independently scattered random measure $\{M(B)\}$ on $\mathbb{R}$ with finite log-moment such that $\mu = P_{Y_1} = P_{Z_0}$ in Proposition 58.4.*

Extension to semi-selfdecomposable distributions is made in [**334**].

59. Remarkable classes of infinitely divisible distributions

Let $d \geq 1$. Let $ID(\mathbb{R}^d)$, $\mathfrak{S}(\mathbb{R}^d)$, and $L(\mathbb{R}^d)$ denote the classes of infinitely divisible distributions, stable distributions, and selfdecomposable distributions, respectively, on $\mathbb{R}^d$. Let us consider some other subclasses of $ID(\mathbb{R}^d)$. In this section, for any subclass C of $ID(\mathbb{R}^d)$, $\widetilde{C}$ denotes the smallest class that is closed under convergence and convolution and contains C. If $C = ID(\mathbb{R}^d)$, then $\widetilde{C} = C$ (Lemmas 7.4 and 7.8). The Lévy measure of μ is denoted by ν^{μ}.

PROPOSITION 59.1. *If $C = L(\mathbb{R}^d)$, then $\widetilde{C} = C$.*

Proof. Use Definition 15.1 of selfdecomposability. Closedness under convolution is clear. To show closedness under convergence, suppose that $\mu_n \in L(\mathbb{R}^d)$ for $n = 1, 2, \ldots$ and that $\mu_n \to \mu$. Note that $\mu_n \in ID(\mathbb{R}^d)$ and hence $\mu \in ID(\mathbb{R}^d)$ and $\widehat{\mu}_n(z)$ and $\widehat{\mu}(z)$ have no zeros. For every $b > 1$ there is a distribution ρ_n such that $\widehat{\rho}_n(z) = \widehat{\mu}_n(z)/\widehat{\mu}_n(b^{-1}z)$, which tends to a continuous function $\widehat{\mu}(z)/\widehat{\mu}(b^{-1}z)$ as $n \to \infty$. Hence there is a distribution ρ such that $\rho_n \to \rho$ (Proposition 2.5(viii)). Thus $\widehat{\mu}(z) = \widehat{\mu}(b^{-1}z)\widehat{\rho}(z)$ and $\mu \in L(\mathbb{R}^d)$. □

DEFINITION 59.2. Define $L_\infty(\mathbb{R}^d) = \widetilde{C}$ for $C = \mathfrak{S}(\mathbb{R}^d)$.

We use the following lemma.

LEMMA 59.3. *Let ν be a σ-finite measure on $\mathbb{R}^d$ satisfying $\nu(\{0\}) = 0$. Then there are a finite measure λ on $S = \{\xi\colon |\xi| = 1\}$ with $\lambda(S) \geq 0$ and a family $\{\nu_\xi\colon \xi \in S\}$ of σ-finite measures on $(0, \infty)$ with $\nu_\xi((0, \infty)) > 0$ such that $\nu_\xi(B)$ is measurable in ξ for each $B \in \mathcal{B}(S)$ and*

$$\nu(B) = \int_S \lambda(\mathrm{d}\xi) \int_{(0,\infty)} 1_B(r\xi)\nu_\xi(\mathrm{d}r), \qquad B \in \mathcal{B}(\mathbb{R}^d). \tag{59.1}$$

Here $\lambda(\mathrm{d}\xi)$ and ν_ξ are uniquely determined in the following sense: if $(\lambda(\mathrm{d}\xi), \nu_\xi)$ and $(\lambda^\sharp(\mathrm{d}\xi), \nu_\xi^\sharp)$ both have these properties, then there is a measurable function $c(\xi)$ on S such that $0 < c(\xi) < \infty$, $c(\xi)\lambda^\sharp(\mathrm{d}\xi) = \lambda(\mathrm{d}\xi)$, and $\nu_\xi^\sharp(\mathrm{d}r) = c(\xi)\nu_\xi(\mathrm{d}r)$ for λ-a. e. $\xi \in S$.

We do not give the proof of the lemma, but it is similar to a part of the proofs of Theorems 14.3 and 15.10; see [**15**] or [**457**]. If $\nu = 0$, then $\lambda = 0$ and $\{\nu_\xi\colon \xi \in S\}$ is arbitrary with $\nu_\xi((0, \infty)) > 0$.

DEFINITION 59.4. The pair $(\lambda(\mathrm{d}\xi), \nu_\xi)$ in Lemma 59.3 is called the *polar decomposition* of ν; $\lambda(\mathrm{d}\xi)$ and ν_ξ are called the *spherical part* and the *radial part* of ν, respectively.

PROPOSITION 59.5. *Let $\mu \in ID(\mathbb{R}^d)$. Then the following are equivalent.*

(1) $\mu \in L_\infty(\mathbb{R}^d)$.
(2) *ν^μ has polar decomposition $(\lambda(\mathrm{d}\xi), r^{-1}h_\xi(\log r)\mathrm{d}r)$, where $h_\xi(u)$ is measurable in $\xi \in S$ and, for λ-a. e. ξ, $(-\mathrm{d}/\mathrm{d}u)^n h_\xi(u) \geq 0$, $u \in (-\infty, \infty)$, for $n = 0, 1, \ldots$.*
(3) *ν^μ has representation*

$$\nu^\mu(B) = \int_{(0,2)} \Gamma(\mathrm{d}\alpha) \int_S \lambda_\alpha(\mathrm{d}\xi) \int_0^\infty 1_B(r\xi) r^{-\alpha-1}\mathrm{d}r, \qquad B \in \mathcal{B}(\mathbb{R}^d), \tag{59.2}$$

where Γ is a measure on $(0, 2)$ satisfying $\int_{(0,2)}(\alpha^{-1} + (2-\alpha)^{-1})\Gamma(\mathrm{d}\alpha) < \infty$, λ_α is a probability measure on S for each α, and $\lambda_\alpha(B)$ is measurable in α for each $B \in \mathcal{B}(S)$.

We omit the proof; see [**416, 431, 545**]. The class $L_\infty(\mathbb{R})$ was first introduced by Urbanik [**544**].

DEFINITION 59.6. Let us call Vx an elementary Γ-variable, an elementary mixed-exponential variable, or an elementary compound Poisson variable, respectively, on $\mathbb{R}^d$ if x is a non-random, non-zero element of $\mathbb{R}^d$ and V is a real random variable having a Γ-distribution, a mixture of a finite number of exponential distributions, or a compound Poisson distribution in Definition 4.1 with $c > 0$ and σ (jump size distribution) being uniform on the interval $[0, a]$ for some $a > 0$, respectively. Let $T(\mathbb{R}^d)$, $B(\mathbb{R}^d)$, or $U(\mathbb{R}^d)$, respectively, denote the class $\widetilde{C}$ for the class C of distributions of elementary Γ-variables, elementary mixed-exponential variables, or elementary compound Poisson variables, respectively, on $\mathbb{R}^d$. We call these three classes the *Thorin class*, the *Goldie–Steutel–Bondesson class*, and the *Jurek class*.

If $d = 1$ and if we confine our attention to infinitely divisible distributions with support in $[0, \infty)$, then the classes $L_\infty(\mathbb{R}^d)$, $T(\mathbb{R}^d)$, and $B(\mathbb{R}^d)$ are exactly the classes L_∞, T, and B in Exercises 55.7, 55.8, and Definition 51.9. Concerning T and B, see Thorin [**532**]–[**534**], Bondesson [**55**], and James, Roynette, and Yor [**237**] for more information.

Let us state the representations of the three classes in Definition 59.6. We use stochastic integrals of Section 57.

DEFINITION 59.7. For $\rho \in ID(\mathbb{R}^d)$ let $\{X_t^{(\rho)}\colon t \geq 0\}$ be a Lévy process in law on $\mathbb{R}^d$ such that $X_t^{(\rho)}$ has distribution ρ at $t = 1$. For a locally square-integrable function $f(s)$ on $[0, \infty)$, let $\mathfrak{D}(\Phi_f)$ be the class of $\rho \in ID(\mathbb{R}^d)$ such that $\int_0^{\infty-} f(s) \mathrm{d}X_s^{(\rho)}$ is definable. Let $\Phi_f(\rho)$ be the distribution of $\int_0^{\infty-} f(s) \mathrm{d}X_s^{(\rho)}$ for $\rho \in \mathfrak{D}(\Phi_f)$. We call Φ_f the *stochastic integral mapping* for f. Its domain equals $\mathfrak{D}(\Phi_f)$; its range is denoted by $\mathfrak{R}(\Phi_f)$.

PROPOSITION 59.8. *Let $\mu \in ID(\mathbb{R}^d)$. Then the following are equivalent.*

(1) $\mu \in T(\mathbb{R}^d)$.
(2) ν^μ *has polar decomposition* $(\lambda(\mathrm{d}\xi), r^{-1}k_\xi(r)\mathrm{d}r)$, *where* $k_\xi(r)$ *is measurable in* $\xi \in S$ *and, for* λ*-a. e.* ξ, *completely monotone in* $r \in (0, \infty)$.
(3) $\mu \in \mathfrak{R}(\Phi_f)$, *where* f *is defined by* $s = \int_{f(s)}^\infty u^{-1}\mathrm{e}^{-u}\mathrm{d}u$ *for* $s > 0$. *(This* f *satisfies* $f(s) \sim -\log s$, $s \downarrow 0$, *and* $f(s) \sim a\mathrm{e}^{-s}$, $s \to \infty$, *with some* $a > 0$.*)*

PROPOSITION 59.9. *Let $\mu \in ID(\mathbb{R}^d)$. Then the following are equivalent.*

(1) $\mu \in B(\mathbb{R}^d)$.
(2) ν^μ *has polar decomposition* $(\lambda(\mathrm{d}\xi), k_\xi(r)\mathrm{d}r)$, *where* $k_\xi(r)$ *is measurable in* $\xi \in S$ *and, for* λ*-a. e.* ξ, *completely monotone in* $r \in (0, \infty)$.
(3) $\mu \in \mathfrak{R}(\Phi_f)$, *where* $f(s) = (-\log s)1_{(0,1]}(s)$.

PROPOSITION 59.10. *Let $\mu \in ID(\mathbb{R}^d)$. Then the following are equivalent.*

(1) $\mu \in U(\mathbb{R}^d)$.
(2) ν^μ *has polar decomposition* $(\lambda(\mathrm{d}\xi), k_\xi(r)\mathrm{d}r)$, *where* $k_\xi(r)$ *is measurable in* $\xi \in S$ *and, for* λ*-a. e.* ξ, *nonnegative, decreasing, and right-continuous in* $r \in (0, \infty)$.
(3) $\mu \in \mathfrak{R}(\Phi_f)$, *where* $f(s) = (1 - s)1_{(0,1]}(s)$.
(4) *For any* $b > 1$ *there is* $\rho_b \in ID(\mathbb{R}^d)$ *such that* $\widehat{\mu}(z) = \widehat{\mu}(b^{-1}z)^{b^{-1}}\widehat{\rho}_b(z)$, $z \in \mathbb{R}^d$.

For the proof of Propositions 59.8–59.10 we refer to [**15, 457**] and Jurek [**244**]. In Proposition 59.8, $\mathfrak{D}(\Phi_f)$ equals the class of $\rho \in ID(\mathbb{R}^d)$ with finite log-moment, and it coincides with the class of ρ for which $\int_0^{\infty-} f(s)\mathrm{d}X_s^{(\rho)}$ is absolutely definable. In Propositions 59.9 and 59.10, $\mathfrak{D}(\Phi_f)$ equals $ID(\mathbb{R}^d)$.

A decreasing sequence of classes $L_m(\mathbb{R}^d)$, $m = 0, 1, \ldots$, is defined by $L_0(\mathbb{R}^d) = L(\mathbb{R}^d)$ and $L_m(\mathbb{R}^d) = \{\mu \in L(\mathbb{R}^d)\colon \rho_b \text{ in (15.1) belongs to } L_{m-1}(\mathbb{R}^d) \text{ for any } b > 1\}$. Then $\bigcap_{m=0}^{\infty} L_m(\mathbb{R}^d) = L_\infty(\mathbb{R}^d)$. This sequence was introduced by Urbanik [**544**]. The class $L_m(\mathbb{R}^d)$ equals $\mathfrak{R}((\Phi_f)^{m+1})$ with $f(s) = \mathrm{e}^{-s}$ and its members are called $(m+1)$-times selfdecomposable. In this connection distributions in $L_\infty(\mathbb{R}^d)$ are sometimes called *completely selfdecomposable.* Continuous-parameter interpolation and extrapolation of these classes are known in Thu [**535**] and Sato [**457**]. The limit class $\bigcap_{n=1}^{\infty} \mathfrak{R}((\Phi_f)^n)$ is studied and shown to coincide with $L_\infty(\mathbb{R}^d)$ in the cases treated in Maejima and Sato [**335**], which include Φ_f of Propositions 59.8–59.10.

Sato [**454, 457**] studied $\mathfrak{R}(\Phi_f)$ for some explicitly given functions f having one or two continuous parameters, and made two schemes of $\mathfrak{R}(\Phi_f)$ each with two continuous parameters. The classes L, L_m, T, B, and U on $\mathbb{R}^d$ have their locations in the schemes. The functions f and f_p in Example 57.18 are also related to the schemes. Rosiński's paper [**422**] deals with properties of Lévy processes associated to some classes in these schemes. Given a function f, not only the class $\mathfrak{R}(\Phi_f)$ but also the class of $\Phi_f(\rho)$ such that $\rho \in ID(\mathbb{R}^d)$ and $\int_0^{\infty-} f(s)\mathrm{d}X_s^{(\rho)}$ is absolutely definable is studied. A similar study is made also for another kind of improper integrals, $\int_{0+}^{t_0} f(s)\mathrm{d}X_s(\rho)$, mentioned at the end of Section 57. In characterization of those many classes we use the new concepts of weak mean, absolute weak mean, weak drift, absolute weak drift, and inversion in Sato [**457, 459**] and Sato and Ueda [**460**].

60. Lebesgue decomposition for path space measures

For σ-finite measures ρ, $\rho^\sharp$ on a measurable space $(\Theta, \mathcal{B})$, we use the following notation: $\rho^\sharp \ll \rho$ means that $\rho^\sharp$ is absolutely continuous with respect to ρ (that is, $\rho(B) = 0$ implies $\rho^\sharp(B) = 0$); $\rho^\sharp \perp \rho$ means that $\rho^\sharp$ is singular with respect to ρ (that is, there is $B \in \mathcal{B}$ such that $\rho(B) = 0$ and $\rho^\sharp(\Theta \setminus B) = 0$), equivalently, ρ is singular with respect to $\rho^\sharp$; $\rho^\sharp \approx \rho$ means that $\rho^\sharp \ll \rho$ and $\rho \ll \rho^\sharp$ (that is, ρ and $\rho^\sharp$ are mutually absolutely continuous). We write $\int_B f \mathrm{d}\rho$ for $\int_B f(\theta)\rho(\mathrm{d}\theta)$ for short. With a nonnegative measurable f, we write $f\rho$ for the measure defined as $(f\rho)(B) = \int_B f \mathrm{d}\rho$.

Let $\mathbf{D} = D([0,\infty), \mathbb{R}^d)$, $x_t(\xi) = \xi(t)$ for $\xi \in \mathbf{D}$, and σ-algebras $\mathcal{F}_{\mathbf{D}}$, $\mathcal{F}_t$ be as in Section 33. Let $(\{x_t\}, P)$ and $(\{x_t\}, P^\sharp)$ be Lévy processes on $\mathbb{R}^d$, where P and $P^\sharp$ are probability measures on $(\mathbf{D}, \mathcal{F}_{\mathbf{D}})$. The generating triplets of $(\{x_t\}, P)$ and $(\{x_t\}, P^\sharp)$ are denoted by (A, ν, γ) and $(A^\sharp, \nu^\sharp, \gamma^\sharp)$, respectively. Let us write $P_t = [P]_{\mathcal{F}_t}$ and $P_t^\sharp = [P^\sharp]_{\mathcal{F}_t}$. The condition for $P_t^\sharp \approx P_t$ and the description of the Radon–Nikodým density $\mathrm{d}P_t^\sharp/\mathrm{d}P_t$ in the case $P_t^\sharp \approx P_t$ are given in Section 33. Now let us describe the Lebesgue decomposition of $P_t^\sharp$ with respect to P_t. The study in this area was made by Skorohod [**490**]–[**492**], Kunita and S. Watanabe [**300**], Newman [**360, 361**], and Sato [**450**].

We begin with the preparation of the Hellinger–Kakutani inner product and distance. They are powerful tools in studying absolute continuity and singularity, as Kakutani [**252**], Brody [**70**], Newman [**360, 361**], and Memin and Shiryayev [**344**] have shown. Measures ρ, $\rho^\sharp$, $\widetilde{\rho}$ below are assumed to be σ-finite.

DEFINITION 60.1. Let $0 < \alpha < 1$. The *Hellinger–Kakutani inner product* of ρ and $\rho^\sharp$ of order α is the measure $H_\alpha(\rho, \rho^\sharp)$ defined by

$$H_\alpha(\rho, \rho^\sharp) = (\mathrm{d}\rho/\mathrm{d}\widetilde{\rho})^\alpha (\mathrm{d}\rho^\sharp/\mathrm{d}\widetilde{\rho})^{1-\alpha}\widetilde{\rho},$$

where ρ, $\rho^\sharp \ll \widetilde{\rho}$. It is independent of the choice of $\widetilde{\rho}$ (to prove this, for ρ, $\rho^\sharp \ll \widetilde{\rho}_j$, $j = 1, 2$, consider $\widetilde{\rho}_3$ such that $\widetilde{\rho}_1$, $\widetilde{\rho}_2 \ll \widetilde{\rho}_3$). Sometimes we write $\mathrm{d}H_\alpha(\rho, \rho^\sharp) = (\mathrm{d}\rho)^\alpha(\mathrm{d}\rho^\sharp)^{1-\alpha}$. The total mass is written as $h_\alpha(\rho, \rho^\sharp) = \int_\Theta \mathrm{d}H_\alpha(\rho, \rho^\sharp)$.

LEMMA 60.2. (i) $H_\alpha(\rho, \rho^\sharp) \leq \alpha\rho + (1-\alpha)\rho^\sharp$. (ii) $\rho \perp \rho^\sharp$ *if and only if* $h_\alpha(\rho, \rho^\sharp) = 0$.

Proof. To show (i), note that

$$\textstyle\int_B f_1^\alpha f_2^{1-\alpha}\mathrm{d}\rho \leq \left(\int_B f_1 \mathrm{d}\rho\right)^\alpha \left(\int_B f_2 \mathrm{d}\rho\right)^{1-\alpha} \leq \alpha \int_B f_1 \mathrm{d}\rho + (1-\alpha)\int_B f_2 \mathrm{d}\rho$$

for f_1, $f_2 \geq 0$ by Hölder's inequality and concavity of $\log x$. To show the 'only if' part of (ii), assume $\rho \perp \rho^\sharp$, choose $\widetilde{\rho} = \rho + \rho^\sharp$ and B with $\rho(B) = 0 = \rho^\sharp(B^c)$, and see that $\mathrm{d}\rho/\mathrm{d}\widetilde{\rho} = 1_{B^c}$ and $\mathrm{d}\rho^\sharp/\mathrm{d}\widetilde{\rho} = 1_B$. The 'if' part of (ii) is clear. □

DEFINITION 60.3. Given two σ-finite measures ρ and $\rho^\sharp$, write

$$C(\rho) = \{\theta \in \Theta \colon \mathrm{d}\rho/\mathrm{d}\widetilde{\rho} > 0\}, \quad C(\rho^\sharp) = \{\theta \in \Theta \colon \mathrm{d}\rho^\sharp/\mathrm{d}\widetilde{\rho} > 0\},$$

where ρ, $\rho^\sharp \ll \widetilde{\rho}$. We call $C(\rho)$ and $C(\rho^\sharp)$ the *carriers* of ρ and $\rho^\sharp$, respectively, relative to $\{\rho, \rho^\sharp\}$. (They depend not only on ρ and $\rho^\sharp$, but also on the choice of $\widetilde{\rho}$ and versions of the Radon–Nikodým densities. However, $1_{C(\rho)}\rho^\sharp$ and $1_{C(\rho^\sharp)}\rho$ are determined by ρ and $\rho^\sharp$ as in the next lemma.)

LEMMA 60.4. *Let* $\rho^\sharp = \rho^\sharp_{\mathrm{ac}} + \rho^\sharp_{\mathrm{s}}$ *be the Lebesgue decomposition of* $\rho^\sharp$ *with respect to* ρ, *where* $\rho^\sharp_{\mathrm{ac}}$ *is absolutely continuous and* $\rho^\sharp_{\mathrm{s}}$ *is singular with respect to* ρ. *Then,* $\rho^\sharp_{\mathrm{ac}} = 1_{C(\rho)}\rho^\sharp$ *and* $\rho^\sharp_{\mathrm{s}} = 1_{C(\rho)^c}\rho^\sharp$, *where* $C(\rho)^c = \Theta \setminus C(\rho)$. *Hence* $\rho^\sharp_{\mathrm{ac}}(\Theta) = \rho^\sharp(C(\rho))$ *and* $\rho^\sharp_{\mathrm{s}}(\Theta) = \rho^\sharp(C(\rho)^c)$. *If* ρ *and* $\rho^\sharp$ *are finite, then* $\lim_{\alpha\downarrow 0} h_\alpha(\rho, \rho^\sharp) = \rho^\sharp(C(\rho))$ *and* $\lim_{\alpha\uparrow 1} h_\alpha(\rho, \rho^\sharp) = \rho(C(\rho^\sharp))$.

Proof. If $G \in \mathcal{B}$ satisfies $\widetilde{\rho}(G \cap C(\rho)) > 0$, then $\rho(G) = \int_G (\mathrm{d}\rho/\mathrm{d}\widetilde{\rho})\mathrm{d}\widetilde{\rho} = \int_{G\cap C(\rho)}(\mathrm{d}\rho/\mathrm{d}\widetilde{\rho})\mathrm{d}\widetilde{\rho} > 0$. Hence if $\rho(G) = 0$, then $\widetilde{\rho}(G \cap C(\rho)) = 0$ and thus $\rho^\sharp(G \cap C(\rho)) = 0$. It follows that $1_{C(\rho)}\rho^\sharp$ is absolutely continuous with respect to ρ. On the other hand, $1_{C(\rho)^c}\rho^\sharp$ is singular with respect to ρ, since $\rho(C(\rho)^c) = \rho(\{\mathrm{d}\rho/\mathrm{d}\widetilde{\rho} = 0\}) = 0$. Therefore, $\rho^\sharp_{\mathrm{ac}} = 1_{C(\rho)}\rho^\sharp$ and $\rho^\sharp_{\mathrm{s}} = 1_{C(\rho)^c}\rho^\sharp$. Assuming that ρ and $\rho^\sharp$ are finite, we choose $\widetilde{\rho} = \rho + \rho^\sharp$ and see that $h_\alpha(\rho, \rho^\sharp) \to \rho^\sharp(C(\rho))$, $\alpha \downarrow 0$, since $h_\alpha(\rho, \rho^\sharp) = \int_{C(\rho)}(\mathrm{d}\rho/\mathrm{d}\widetilde{\rho})^\alpha(\mathrm{d}\rho^\sharp/\mathrm{d}\widetilde{\rho})^{1-\alpha}\mathrm{d}\widetilde{\rho} \to \int_{C(\rho)}(\mathrm{d}\rho^\sharp/\mathrm{d}\widetilde{\rho})\mathrm{d}\widetilde{\rho} = \rho^\sharp(C(\rho))$ by the dominated convergence. This implies $h_\alpha(\rho, \rho^\sharp) \to \rho(C(\rho^\sharp))$, $\alpha \uparrow 1$, since $h_\alpha(\rho, \rho^\sharp) = h_{1-\alpha}(\rho^\sharp, \rho)$. □

DEFINITION 60.5. Let $0 < \alpha < 1$. Choose $\widetilde{\rho}$ with $\rho, \rho^\sharp \ll \widetilde{\rho}$ and let $f = \mathrm{d}\rho/\mathrm{d}\widetilde{\rho}$ and $f^\sharp = \mathrm{d}\rho^\sharp/\mathrm{d}\widetilde{\rho}$. Define

$$K_\alpha(\rho, \rho^\sharp)(B) = \int_B (\alpha f + (1-\alpha) f^\sharp - f^\alpha f^{\sharp\, 1-\alpha}) \mathrm{d}\widetilde{\rho}, \quad B \in \mathcal{B}. \tag{60.1}$$

Then $K_\alpha(\rho, \rho^\sharp)$ is a σ-finite measure from Lemma 60.2(i). Similarly to $H_\alpha(\rho, \rho^\sharp)$, we can prove that $K_\alpha(\rho, \rho^\sharp)$ does not depend on the choice of $\widetilde{\rho}$. The total mass $k_\alpha(\rho, \rho^\sharp) = \int_\Theta \mathrm{d}K_\alpha(\rho, \rho^\sharp)$ is called the *Hellinger–Kakutani distance* of order α between ρ and $\rho^\sharp$.

For $\alpha = 1/2$ we sometimes write $\mathrm{d}K_{1/2}(\rho, \rho^\sharp) = (1/2)(\sqrt{\mathrm{d}\rho} - \sqrt{\mathrm{d}\rho^\sharp})^2$. Let $\|\rho - \rho^\sharp\|$ be the total variation norm of $\rho - \rho^\sharp$, admitting infinity.

LEMMA 60.6. (i) $\|\rho - \rho^\sharp\| \ge 2\, k_{1/2}(\rho, \rho^\sharp)$. (ii) *If ρ and $\rho^\sharp$ are finite measures, then $\|\rho - \rho^\sharp\| \le c\, k_{1/2}(\rho, \rho^\sharp)^{1/2}$, where $c = 2(\rho(\Theta) + \rho^\sharp(\Theta))^{1/2}$.* (iii) *If $k_\alpha(\rho, \rho^\sharp) < \infty$ for some $\alpha \in (0,1)$, then $k_\alpha(\rho, \rho^\sharp) < \infty$ for all $\alpha \in (0,1)$, $\lim_{\alpha\downarrow 0} k_\alpha(\rho, \rho^\sharp) = \rho^\sharp(C(\rho)^c) = \rho^\sharp_s(\Theta) < \infty$, and $\lim_{\alpha\uparrow 1} k_\alpha(\rho, \rho^\sharp) = \rho(C(\rho^\sharp)^c) < \infty$.*

Proof. Let $\widetilde{\rho}$, f, and $f^\sharp$ be as in Definition 60.5. To show (i), note that $2k_{1/2}(\rho, \rho^\sharp) = \int (f + f^\sharp - 2\sqrt{f f^\sharp}) \mathrm{d}\widetilde{\rho} = \int (\sqrt{f} - \sqrt{f^\sharp})^2 \mathrm{d}\widetilde{\rho} \le \int |f - f^\sharp| \mathrm{d}\widetilde{\rho} = \|\rho - \rho^\sharp\|$. To see (ii), note that

$$\begin{aligned} \int |f - f^\sharp| \mathrm{d}\widetilde{\rho} &\le \Big(\int (\sqrt{f} - \sqrt{f^\sharp})^2 \mathrm{d}\widetilde{\rho}\Big)^{1/2} \Big(\int (\sqrt{f} + \sqrt{f^\sharp})^2 \mathrm{d}\widetilde{\rho}\Big)^{1/2} \\ &\le (2k_{1/2}(\rho, \rho^\sharp))^{1/2} \Big(2\int (f + f^\sharp) \mathrm{d}\widetilde{\rho}\Big)^{1/2} = c k_{1/2}(\rho, \rho^\sharp)^{1/2}. \end{aligned}$$

Let us prove (iii). Suppose that $k_\alpha(\rho, \rho^\sharp) < \infty$ for a given α. Let $C = \{f > 0,\ f^\sharp > 0\}$. We have $[\rho]_C \approx [\rho^\sharp]_C$. Letting $\mathrm{d}\rho^\sharp/\mathrm{d}\rho = \mathrm{e}^g$ on C, we have $K_\alpha(\rho, \rho^\sharp)(C) = \int_C (\alpha + (1-\alpha)\mathrm{e}^g - \mathrm{e}^{(1-\alpha)g}) \mathrm{d}\rho = \int_C \varphi_\alpha(g) \mathrm{d}\rho$ from (60.1), where $\varphi_\alpha(u) = \alpha + (1-\alpha)\mathrm{e}^u - \mathrm{e}^{(1-\alpha)u}$. For nonnegative functions $\varphi(u)$, $\psi(u)$, we say that $\varphi(u) \asymp \psi(u)$ on a set B if there are positive constants c_1, c_2 such that $c_1\psi(u) \le \varphi(u) \le c_2\psi(u)$ on B. We have $\varphi_\alpha(0) = 0$ and $\varphi_\alpha(u) > 0$ for $u \ne 0$ since e^u is strictly convex. Hence $\varphi_\alpha(u) \asymp u^2 1_{\{|u|\le 1\}} + \mathrm{e}^u 1_{\{u>1\}} + 1_{\{u<-1\}}$ on $\mathbb{R}$. Since

$$k_\alpha(\rho, \rho^\sharp) = \alpha \int_{\{f^\sharp=0\}} \mathrm{d}\rho + (1-\alpha)\int_{\{f=0\}} \mathrm{d}\rho^\sharp + \int_C \mathrm{d}K_\alpha(\rho, \rho^\sharp), \tag{60.2}$$

we have

$$\rho(C(\rho^\sharp)^c) + \rho^\sharp(C(\rho)^c) + \int_C (g^2 1_{\{|g|\le 1\}} + \mathrm{e}^g 1_{\{g>1\}} + 1_{\{g<-1\}}) \mathrm{d}\rho < \infty. \tag{60.3}$$

Conversely, if (60.3) is satisfied, then $k_\alpha(\rho, \rho^\sharp) < \infty$. As the condition (60.3) does not involve α, we see that $k_\alpha(\rho, \rho^\sharp) < \infty$ for all $\alpha \in (0,1)$. Since there is c_3 such that, for all $\alpha \in (0,1)$ and $u \in \mathbb{R}$, $0 \le \varphi_\alpha(u) \le c_3(u^2 1_{\{|u|\le 1\}} + \mathrm{e}^u 1_{\{u>1\}} + 1_{\{u<-1\}})$ and since $\varphi_\alpha(u) \to 0$ as $\alpha \downarrow 0$, it follows from (60.2) that $\lim_{\alpha\downarrow 0} k_\alpha(\rho, \rho^\sharp) = \rho^\sharp(C(\rho)^c)$ by the dominated convergence. Now $\lim_{\alpha\uparrow 1} k_\alpha(\rho, \rho^\sharp) = \rho(C(\rho^\sharp)^c) < \infty$ since $k_\alpha(\rho, \rho^\sharp) = k_{1-\alpha}(\rho^\sharp, \rho)$. □

We say that $\{B_n\}$ is a measurable partition of $B \in \mathcal{B}$, if $\{B_n\}$ is a finite or countably infinite family of disjoint sets in $\mathcal{B}$ such that $\bigcup_n B_n = B$. Let $\mathcal{P}_B$ denote the collection of all measurable partitions of B.

LEMMA 60.7. *For any $B \in \mathcal{B}$ we have*

$$H_\alpha(\rho, \rho^\sharp)(B) = \inf_{\{B_n\} \in \mathcal{P}_B} \textstyle\sum_n \rho(B_n)^\alpha \rho^\sharp(B_n)^{1-\alpha}. \tag{60.4}$$

Proof. To show (60.4) with $\leq$ in place of $=$, use Hölder's inequality as in the proof of Lemma 60.2(i). To show the reverse inequality use, for $b > 1$, $B_{l,m} = \{\theta \in B \colon b^l \leq f(\theta)^\alpha < b^{l+1}$ and $b^m \leq f^\sharp(\theta)^{1-\alpha} < b^{m+1}\}$, $l, m \in \mathbb{Z}$, and let $b \downarrow 1$. □

LEMMA 60.8. *Let $(\Theta', \mathcal{B}')$ be another measurable space and let $\varphi \colon \Theta \to \Theta'$ be a measurable mapping. Then $h_\alpha(\rho\varphi^{-1}, \rho^\sharp\varphi^{-1}) \geq h_\alpha(\rho, \rho^\sharp)$, where $(\rho\varphi^{-1})(B') = \rho(\varphi^{-1}(B'))$ and $(\rho^\sharp\varphi^{-1})(B') = \rho^\sharp(\varphi^{-1}(B'))$ for $B' \in \mathcal{B}'$.*

Proof. Use Lemma 60.7. □

LEMMA 60.9. *Let Θ be a metric space and $\mathcal{B}$ the Borel σ-algebra. Let ρ_n, ρ, $\rho_n^\sharp$, $\rho^\sharp$, and σ be finite measures on $\mathcal{B}$. Fix $\alpha \in (0,1)$. If $\rho_n \to \rho$, $\rho_n^\sharp \to \rho^\sharp$, and $H_\alpha(\rho_n, \rho_n^\sharp) \to \sigma$ as $n \to \infty$ and $\inf_n h_\alpha(\rho_n, \rho_n^\sharp) \geq h_\alpha(\rho, \rho^\sharp)$, then $H_\alpha(\rho, \rho^\sharp) = \sigma$.*

Proof. Since $\int \varphi \, \mathrm{d}H_\alpha(\rho_n, \rho_n^\sharp) \leq (\int \varphi \, \mathrm{d}\rho_n)^\alpha (\int \varphi \, \mathrm{d}\rho_n^\sharp)^{1-\alpha}$ for any bounded continuous $\varphi \geq 0$, we have $\int \varphi \, \mathrm{d}\eta \leq (\int \varphi \, \mathrm{d}\rho)^\alpha (\int \varphi \, \mathrm{d}\rho^\sharp)^{1-\alpha}$. It follows that $\sigma(B) \leq \rho(B)^\alpha \rho^\sharp(B)^{1-\alpha}$. Hence $\sigma \leq H_\alpha(\rho, \rho^\sharp)$ by Lemma 60.7. On the other hand we have $\sigma(\Theta) \geq h_\alpha(\rho, \rho^\sharp)$ from $h_\alpha(\rho_n, \rho_n^\sharp) \geq h_\alpha(\rho, \rho^\sharp)$. Hence $\sigma = H_\alpha(\rho, \rho^\sharp)$. □

LEMMA 60.10. *Suppose that $k_\alpha(\nu, \nu^\sharp) < \infty$. Then $\int_{|x| \leq 1} |x| \, \mathrm{d}|\nu - \nu^\sharp|$, $\int_{|x| \leq 1} |x| \, \mathrm{d}|\nu - H_\alpha(\nu, \nu^\sharp)|$, and $\int_{|x| \leq 1} |x| \, \mathrm{d}|\nu^\sharp - H_\alpha(\nu, \nu^\sharp)|$ are finite. Here, for a signed measure σ, $|\sigma|$ denotes the total variation measure of σ.*

Proof. Let ν, $\nu^\sharp \ll \widetilde{\nu}$, $\mathrm{d}\nu/\mathrm{d}\widetilde{\nu} = f$, and $\mathrm{d}\nu^\sharp/\mathrm{d}\widetilde{\nu} = f^\sharp$. Then $|\nu - \nu^\sharp| = |f - f^\sharp| \widetilde{\nu}$ and, similarly to the proof of Lemma 60.6(ii), $\int_{|x| \leq 1} |x| \, \mathrm{d}|\nu - \nu^\sharp| \leq (2k_{1/2}(\nu, \nu^\sharp))^{1/2} \left(2 \int_{|x| \leq 1} |x|^2 (f + f^\sharp) \mathrm{d}\widetilde{\nu}\right)^{1/2} < \infty$. Since $\nu - H_\alpha(\nu, \nu^\sharp) = K_\alpha(\nu, \nu^\sharp) + (1-\alpha)(\nu - \nu^\sharp)$ and $\nu^\sharp - H_\alpha(\nu, \nu^\sharp) = K_\alpha(\nu, \nu^\sharp) - \alpha(\nu - \nu^\sharp)$, the other assertions are clear. □

Fully using the lemmas above, Newman [**361**] showed the following Theorem 60.11. We omit the proof. See Sato [**450**] for detail.

THEOREM 60.11. (i) *Suppose that*

$$k_\alpha(\nu, \nu^\sharp) < \infty, \quad A = A^\sharp, \quad \text{and} \quad \widetilde{\gamma} \in \mathfrak{R}(A), \tag{NS}$$

where $\widetilde{\gamma} = \gamma^\sharp - \gamma - \int_{|x| \leq 1} x \, \mathrm{d}(\nu^\sharp - \nu)$ and $\mathfrak{R}(A) = \{Ax \colon x \in \mathbb{R}^d\}$. Then

$$H_\alpha(P_t, P_t^\sharp) = \exp(-t l_\alpha) P_t^\alpha \quad \text{for } t > 0,\ 0 < \alpha < 1, \tag{60.5}$$

where $l_\alpha = (1/2)\alpha(1-\alpha)\langle \eta, A\eta \rangle + k_\alpha(\nu, \nu^\sharp)$ with η satisfying $A\eta = \widetilde{\gamma}$, P^α is the probability measure for which $(\{x_t\}, P^\alpha)$ is the Lévy process with generating triplet $(A, H_\alpha(\nu, \nu^\sharp), \gamma^\alpha)$ with

$$\gamma^\alpha = \alpha\gamma + (1-\alpha)\gamma^\sharp - \int_{|x| \leq 1} x \, \mathrm{d}K_\alpha(\nu, \nu^\sharp), \tag{60.6}$$

and $P_t^\alpha = [P^\alpha]_{\mathcal{F}_t}$.

(ii) *Suppose that* (NS) *is not satisfied. Then*

$$H_\alpha(P_t, P_t^\sharp) = 0 \quad \text{for } t > 0,\ 0 < \alpha < 1. \tag{60.7}$$

Condition (NS) is referred to as the nonsingularity condition. The quantity l_α does not depend on the choice of η satisfying $A\eta = \widetilde{\gamma}$, since $A\eta = A\eta_1$ implies $\langle \eta, A\eta \rangle = \langle \eta, A\eta_1 \rangle = \langle A\eta, \eta_1 \rangle = \langle A\eta_1, \eta_1 \rangle$. When we say that (NS) is not satisfied, we mean that one of the following holds: (1) $k_\alpha(\nu, \nu^\sharp) = \infty$; (2) $k_\alpha(\nu, \nu^\sharp) < \infty$ and $A \neq A^\sharp$; (3) $k_\alpha(\nu, \nu^\sharp) < \infty$, $A = A^\sharp$, and $\widetilde{\gamma} \notin \mathfrak{R}(A)$. Note that, if $k_\alpha(\nu, \nu^\sharp) = \infty$, then $\widetilde{\gamma}$ may not be defined. The following six consequences are from Theorem 60.11 and the preceding lemmas. Corollary 60.15 is a restatement of Theorem 33.1.

COROLLARY 60.12. *The following three conditions are equivalent:* (1) $P_t^\sharp \not\perp P_t$ *for some* $t > 0$; (2) $P_t^\sharp \not\perp P_t$ *for all* $t > 0$; (3) *Condition* (NS) *is satisfied.*

COROLLARY 60.13. *If* $P_t^\sharp \not\perp P_t$, *then* $\nu^\sharp(C(\nu)^c) < \infty$, $\nu(C(\nu^\sharp)^c) < \infty$, $P_t^\sharp(C(P_t)) = \exp(-t\nu^\sharp(C(\nu)^c))$, *and* $P_t(C(P_t^\sharp)) = \exp(-t\nu(C(\nu^\sharp)^c))$. *Here* $C(\nu)$ *and* $C(\nu^\sharp)$ *are the carriers relative to* $\{\nu, \nu^\sharp\}$, *and* $C(P_t)$ *and* $C(P_t^\sharp)$ *are the carriers relative to* $\{P_t, P_t^\sharp\}$.

COROLLARY 60.14. *The following three conditions are equivalent:* (1) $P_t^\sharp \ll P_t$ *for some* $t > 0$; (2) $P_t^\sharp \ll P_t$ *for all* $t > 0$; (3) $\nu^\sharp \ll \nu$ *and Condition* (NS) *is satisfied.*

COROLLARY 60.15. *The following three conditions are equivalent:* (1) $P_t^\sharp \approx P_t$ *for some* $t > 0$; (2) $P_t^\sharp \approx P_t$ *for all* $t > 0$; (3) $\nu^\sharp \approx \nu$ *and Condition* (NS) *is satisfied.*

COROLLARY 60.16 (dichotomy). *If* $\nu^\sharp \approx \nu$, *then either* $P_t^\sharp \approx P_t$ *for all* $t > 0$ *or* $P_t^\sharp \perp P_t$ *for all* $t > 0$.

COROLLARY 60.17 (on the whole $\mathcal{F}_{\mathbf{D}}$). *If* $P^\sharp \neq P$, *then* $P^\sharp \perp P$.

Now we state in Lemma 60.18 and Theorem 60.19 the result on the Lebesgue decomposition of $P_t^\sharp$ with respect to P_t in the formulation of Sato [**450**]. See [**450**] for their full proofs. Let $P_t^\sharp = (P_t^\sharp)_{ac} + (P_t^\sharp)_s$ be the Lebesgue decomposition of $P_t^\sharp$ with respect to P_t, and $\nu^\sharp = \nu_{ac}^\sharp + \nu_s^\sharp$ the Lebesgue decomposition of $\nu^\sharp$ with respect to ν. Let $\widetilde{\nu} = \nu + \nu^\sharp$. Choose the versions f and $f^\sharp$ of $d\nu/d\widetilde{\nu}$ and $d\nu^\sharp/d\widetilde{\nu}$, respectively, satisfying $f \geq 0$, $f^\sharp \geq 0$, and $f + f^\sharp = 1$ everywhere on $\mathbb{R}^d$. Denote $D = \{f = 1 \text{ and } f^\sharp = 0\}$, $D^\sharp = \{f = 0 \text{ and } f^\sharp = 1\}$, and $C = \{f > 0 \text{ and } f^\sharp > 0\} = (D \cup D^\sharp)^c$. Thus $\nu_{ac}^\sharp = 1_C \nu^\sharp$, $\nu_s^\sharp = 1_{D^\sharp}\nu^\sharp = 1_{D \cup D^\sharp}\nu^\sharp$, and $(f^\sharp/f)1_C$ is a version of $d\nu_{ac}^\sharp/d\nu$. Define

$$g(x) = \begin{cases} \log(f^\sharp/f) & \text{on } C \\ -\infty & \text{on } D \cup D^\sharp, \end{cases} \qquad \widetilde{g}(x) = \begin{cases} g(x) & \text{on } C \\ 0 & \text{on } D \cup D^\sharp. \end{cases}$$

Let $x_t^\nu(\xi)$ and $J(B, \xi)$ be defined by (33.5) and (33.18). Thus $(\{x_t^\nu\}, P)$ and $(\{x_t - x_t^\nu\}, P)$ are the jump part and the continuous part, respectively, of $(\{x_t\}, P)$ defined in Theorem 19.2.

LEMMA 60.18. *Suppose that $P_t^\sharp \not\perp P_t$ for $0 < t < \infty$. Then the following are true.*

(i) *Using g and $\widetilde{g}$, we can define*

$$V_t = \lim_{\varepsilon \downarrow 0} \Big(\textstyle\sum_{(s, x_s - x_{s-}) \in (0,t] \times \{|x| > \varepsilon\}} \widetilde{g}(x_s - x_{s-}) - t \int_{|x|>\varepsilon} (e^{g(x)} - 1) \nu(dx) \Big);$$

the right-hand side exists P-a. s. and the convergence is uniform on any bounded time interval P-a. s.

(ii) *Let $\eta \in \mathbb{R}^d$ and define*

$$U_t^{(\eta)} = \langle \eta, x_t - x_t^\nu \rangle - (t/2)\langle \eta, A\eta \rangle - t\langle \gamma, \eta \rangle + V_t.$$

Then $\{U_t^{(\eta)} \colon t \geq 0\}$ is, under P, a Lévy process on $\mathbb{R}$ with generating triplet $(A_{U^{(\eta)}}, \nu_U, \gamma_{U^{(\eta)}})$ given by

$$A_{U^{(\eta)}} = \langle \eta, A\eta \rangle,$$

$$\nu_U(B) = \textstyle\int_{\mathbb{R}^d} 1_B(g(x)) \nu(dx) \quad \text{for } B \in \mathcal{B}_{\mathbb{R}\setminus\{0\}},$$

$$\gamma_{U^{(\eta)}} = -(1/2)\langle \eta, A\eta \rangle - \textstyle\int_{\mathbb{R}^d} (e^{g(x)} - 1 - g(x) 1_{\{|g(x)| \leq 1\}}(x)) \nu(dx).$$

The processes $\{U_t^{(\eta)} \colon t \geq 0\}$ and $\{J((0,t] \times (D \cup D^\sharp)) \colon t \geq 0\}$ are independent under P.

THEOREM 60.19. *Define $\Lambda_t \in \mathcal{F}_t$ by*

$$\Lambda_t = \{J((0,t] \times (D \cup D^\sharp)) = 0\} = \{x_s - x_{s-} \notin D \cup D^\sharp,\ \forall s \in (0,t]\}.$$

Suppose that $P_t^\sharp \not\perp P_t$ for $0 < t < \infty$. Then the following are true.

(i) *The Lebesgue decomposition of $P_t^\sharp$ with respect to P_t is given by*

$$(P_t^\sharp)_{ac} = 1_{\Lambda_t} P_t^\sharp, \qquad (P_t^\sharp)_s = 1_{\mathbf{D}\setminus\Lambda_t} P_t^\sharp. \tag{60.8}$$

We have $P(\Lambda_t) = \exp(-t\nu(D))$ and $P^\sharp(\Lambda_t) = \exp(-t\nu^\sharp(D^\sharp))$.

(ii) *The Radon–Nikodým density of $(P_t^\sharp)_{ac}$ is given by*

$$d(P_t^\sharp)_{ac}/dP_t = 1_{\Lambda_t} \exp(-t\nu^\sharp(D^\sharp) + U_t), \tag{60.9}$$

where $U_t = U_t^{(\eta)}$ with η satisfying $A\eta = \widetilde{\gamma}$.

(iii) *Let Q be the probability measure on $(\mathbf{D}, \mathcal{F}_{\mathbf{D}})$ such that $(\{x_t\}, Q)$ is the Lévy process with generating triplet $(A, \nu_{ac}^\sharp, \gamma^\sharp - \int_{|x|\leq 1} x\, d\nu_s^\sharp)$, and $Q_t = [Q]_{\mathcal{F}_t}$. Then*

$$(P_t^\sharp)_{ac} = \exp(-t\nu^\sharp(D^\sharp))\, Q_t. \tag{60.10}$$

Recall that $\nu_s^\sharp$ is a finite measure under Condition (NS) by Lemma 60.6(iii). If $P_t^\sharp \approx P_t$, then Theorem 60.19 coincides with Theorem 33.2.

EXAMPLE 60.20. *Scaled Poisson processes with drift.* Suppose that both $(\{x_t\}, P)$ and $(\{x_t\}, P^\sharp)$ are scaled Poisson processes with drift. That is,

$$E^P[e^{izx_t}] = \exp[t(b(e^{iaz} - 1) + i\gamma_0 z)], \quad z \in \mathbb{R}$$

and $E^{P^\sharp}[e^{izx_t}]$ has the same expression with b, a, γ_0 replaced by $b^\sharp$, $a^\sharp$, $\gamma_0^\sharp$. Here b, $b^\sharp > 0$, a, $a^\sharp \in \mathbb{R} \setminus \{0\}$, and γ_0, $\gamma_0^\sharp \in \mathbb{R}$. Thus $\nu = b\delta_a$ and $\nu^\sharp = b^\sharp \delta_{a^\sharp}$. This is the case studied by Dvoretzky, Kiefer, and Wolfowitz [**118**]. We have $P_t^\sharp \not\perp P_t$ if and only if $\gamma_0^\sharp = \gamma_0$. Under the condition that $\gamma_0^\sharp = \gamma_0$, there are two cases.

Case 1: $a^\sharp = a$. In this case we have $P_t^\sharp \approx P_t$ and $P_t^\sharp = (b^\sharp/b)^{N_t} \mathrm{e}^{-t(b^\sharp - b)} P_t$, where $N_t = N_t(\xi)$ is the number of jumps of $x_s(\xi)$ for $s \le t$.

Case 2: $a^\sharp \neq a$. In this case we have $(P_t^\sharp)_{ac} = \mathrm{e}^{-t(b^\sharp - b)} 1_{\Lambda_t} P_t$, where $\Lambda_t = \{x_s - x_{s-} \neq a, a^\sharp \text{ for } s \in (0,t]\}$. Further we have $(P_t^\sharp)_{ac}(\mathbf{D}) = \mathrm{e}^{-tb^\sharp}$ and $(P_t^\sharp)_{ac} = \mathrm{e}^{-tb^\sharp} Q_t$, where $(\{x_t\}, Q)$ is a deterministic motion, $Q(x_t = t\gamma_0^\sharp \text{ for } t \ge 0) = 1$.

EXAMPLE 60.21. Suppose that $(\{x_t\}, P)$ and $(\{x_t\}, P^\sharp)$ satisfy $\nu(\mathbb{R}^d) < \infty$, $\nu^\sharp(\mathbb{R}^d) < \infty$, $A = A^\sharp$, and $\widetilde{\gamma} \in \mathfrak{R}(A)$. Then $(P_t^\sharp)_{ac}(\mathbf{D}) = \exp(-t\nu_s^\sharp(\mathbb{R}^d))$. Thus, $P_t^\sharp \ll P_t$ if and only if $\nu^\sharp \ll \nu$.

EXAMPLE 60.22. *Absolutely continuous change of Lévy measures.* We start from one Lévy process $(\{x_t\}, P)$ on $\mathbb{R}^d$ with generating triplet (A, ν, γ). Suppose that we are given a measurable function $g(x)$ with values $-\infty \le g(x) < \infty$ and a vector $\eta \in \mathbb{R}^d$. Assume that

$$\int_{\mathbb{R}^d} (\mathrm{e}^{g(x)/2} - 1)^2 \nu(\mathrm{d}x) < \infty. \tag{60.11}$$

Define $(A^\sharp, \nu^\sharp, \gamma^\sharp)$ by $A^\sharp = A$, $\nu^\sharp(\mathrm{d}x) = \mathrm{e}^{g(x)}\nu(\mathrm{d}x)$, $\gamma^\sharp = \gamma + \int_{|x|\le 1} x\mathrm{d}(\nu^\sharp - \nu) + A\eta$. Notice that (60.11) means that $k_{1/2}(\nu, \nu^\sharp) < \infty$. Hence $\gamma^\sharp$ is definable by Lemma 60.10. The condition (60.11) is equivalent to the property that $\int_{|g|\le 1} g^2 \mathrm{d}\nu + \int_{g>1} \mathrm{e}^g \mathrm{d}\nu + \int_{g<-1} \mathrm{d}\nu < \infty$. It follows that $\int (|x|^2 \wedge 1)\nu^\sharp(\mathrm{d}x) < \infty$. Hence a new Lévy process $(\{x_t\}, P^\sharp)$ with generating triplet $(A^\sharp, \nu^\sharp, \gamma^\sharp)$ exists. We have $P_t^\sharp \ll P_t$ by Corollary 60.14 and $P_t^\sharp = \mathrm{e}^{U_t} 1_{\Lambda_t} P_t$ in the notation of Theorem 60.19. Let us consider two special cases.

(1) If $G \in \mathcal{B}(\mathbb{R}^d)$ satisfies $\nu(G) < \infty$, then, letting $g(x) = -\infty$ on G and $g(x) = 0$ on G^c, we obtain (60.11) and the resulting $(\{x_t\}, P^\sharp)$ satisfies $P_t^\sharp \ll P_t$, $A^\sharp = A$, and $\nu^\sharp = 1_{G^c}\, \nu$ for any $\eta \in \mathbb{R}^d$; this is truncation of the support of Lévy measure. If, on the contrary, $\nu(G) = \infty$, then we cannot make a Lévy process $(\{x_t\}, P^\sharp)$ satisfying $P_t^\sharp \ll P_t$ and $\nu^\sharp = 1_{G^c}\, \nu$.

(2) If $g(x)$ is finite-valued and satisfies (60.11), then we have $\nu^\sharp \approx \nu$ and $P_t^\sharp \approx P_t$ and this procedure to get $(\{x_t\}, P^\sharp)$ is the density transformation in Definition 33.4. In particular, if $\eta \neq 0$ and $\int_{\langle \eta, x\rangle > 1} \mathrm{e}^{\langle \eta, x\rangle} \nu(\mathrm{d}x) < \infty$, then $g(x) = \langle \eta, x \rangle$ satisfies (60.11) and we have, using this η also in the expression of $\gamma^\sharp$, $P^\sharp[B] = \mathrm{e}^{-t\Psi(\eta)} E^P[\mathrm{e}^{\langle \eta, x_t\rangle} 1_B]$ for $B \in \mathcal{F}_t$ as in Example 33.14, where Ψ is of (25.11). This particular case is called *Esscher transformation*, as it was introduced by Esscher [**132**] in compound Poisson processes on $\mathbb{R}$. The Lévy process on $\mathbb{R}$ corresponding to the distribution of stochastic area (Example 15.15) and its Esscher transformation are called Meixner processes; see Schoutens and Teugels [**473**] and Grigelionis [**172**].

61. Supports of Lévy processes

The support Σ of a Lévy process $\{X_t\}$ on $\mathbb{R}^d$, $d \geq 1$, is defined in Definition 24.13 as the smallest closed set that contains almost all paths. Equivalent descriptions of Σ are given in Proposition 24.14. One of them says that Σ is the closure of $\bigcup_{t\geq 0} S(X_t)$, where $S(X_t)$ is the support of the distribution of X_t. In this section we state six results on Σ without proofs.

A set M in $\mathbb{R}^d$ is called *symmetric* if $M = -M$. A set M in $\mathbb{R}^d$ is called *one-sided* if $M \subset \{x \in \mathbb{R}^d \colon \langle a, x\rangle \geq 0\}$ for some $a \in \mathbb{R}^d \setminus \{0\}$. The first result is in Sato and Watanabe [**463**].

THEOREM 61.1. *Let $\{X_t\}$ be a Lévy process on $\mathbb{R}^d$ with support Σ. Then Σ is either one-sided or symmetric.*

The dependence on t of $S(X_t)$ for a Lévy process $\{X_t\}$ on $\mathbb{R}$ ($d = 1$) is described in Theorem 24.10. The result is extended to a general dimension by Theorems 61.2 and 61.3 below, which are due to Tortrat [**536**] and Sharpe [**479**] and amended by Sato and Watanabe [**464**]. A set H in $\mathbb{R}^d$ is called a closed additive semigroup if it is a closed set satisfying $H + H \subset H$.

THEOREM 61.2. *Let $\{X_t\}$ be a Lévy process on $\mathbb{R}^d$. Then there are $a \in \mathbb{R}^d$ and a closed additive semigroup H containing 0 such that $S(X_t) = ta + H$ for all $t > 0$.*

Note that in this theorem H is uniquely determined by $\{X_t\}$, because if a_1 and H_1 also satisfy the condition on a and H, then $t(a - a_1) + h \in H_1$ for any $h \in H$, which shows $H \subset H_1$ by letting $t \downarrow 0$. On the other hand, a is not necessarily unique (consider, for example, the case $H = \mathbb{R}^d$). Following Sharpe [**479**], we call the set H the *invariant semigroup* of the process $\{X_t\}$.

THEOREM 61.3. *Let $\{X_t\}$ be a Lévy process on $\mathbb{R}^d$ with generating triplet (A, ν, γ). Let M be the set of all $y \in \mathbb{R}^d$ such that $\int_{|x|\leq 1} |\langle y, x\rangle| \nu(\mathrm{d}x) < \infty$. Then M is a linear subspace of $\mathbb{R}^d$. Let Π_M be the orthogonal projection from $\mathbb{R}^d$ onto M and let $\{X_t^M\}$ be the Lévy process defined by $X_t^M = \Pi_M X_t$. Then the Lévy measure ν_M of $\{X_t^M\}$ satisfies $\int_{|x|\leq 1} |x| \nu_M(\mathrm{d}x) < \infty$. The invariant semigroup H of $\{X_t\}$ is equal to the closure of $\Pi_M^{-1}\operatorname{Sgp}(\nu_M) + A(\mathbb{R}^d)$, where $\operatorname{Sgp}(\nu_M)$ is the smallest closed additive semigroup containing 0 and S_{ν_M}, the support of ν_M. The drift γ_0^M of $\{X_t^M\}$ can be taken as the vector a in Theorem 61.2.*

If $\{X_t\}$ is of type A or B in Definition 11.9, then $A = 0$, $M = \mathbb{R}^d$, and $H = \operatorname{Sgp}(\nu)$. If $M = \{0\}$, then $S(X_t) = H = \mathbb{R}^d$ for $t > 0$.

Sharpe [**479**] further described the positivity of the density under Condition (ACT) of Definition 41.11.

THEOREM 61.4. *Let $\{X_t\}$ be a Lévy process on $\mathbb{R}^d$ satisfyig Condition* (ACT) *and let $p_t(x)$ be the density of the distribution of X_t in Exercise 44.6 for $t > 0$. Then the set $G(X_t) = \{x \in \mathbb{R}^d \colon p_t(x) > 0\}$ for $t > 0$ satisfies*

$$G(X_t) = \operatorname{int} S(X_t) = t\gamma_0^M + \operatorname{int} H,$$

where int *means "the interior of".*

COROLLARY 61.5. *Let $\{X_t\}$ be a Lévy process on $\mathbb{R}^d$ in Theorem 61.4.*

(i) *Suppose that $\{X_t\}$ is of type A or B. Then $G(X_t) = \mathbb{R}^d$ for all $t > 0$ if and only if S_ν is not one-sided.*

(ii) *Suppose that $\{X_t\}$ is of type C and purely non-Gaussian. Then $G(X_t) = \mathbb{R}^d$ for all $t > 0$ if and only if either $M = \{0\}$ or S_{ν_M} is not one-sided in M.*

Application to supports of stable or semi-stable processes gives

THEOREM 61.6. *Let $\{X_t\}$ be a nondegenerate α-semi-stable process on $\mathbb{R}^d$ with $0 < \alpha \le 2$.*

(i) *Suppose that $0 < \alpha < 1$ and S_ν is one-sided. Then, H is one-sided, convex, closed under multiplication by nonnegative reals, and equal to* $\mathrm{Sgp}(\nu)$*; $S(X_t) = t\gamma_0 + H$ and $G(X_t) = t\gamma_0 + \operatorname{int} H$ for $t > 0$, where γ_0 is the drift of $\{X_t\}$.*

(ii) *Suppose that $1 \le \alpha \le 2$, or suppose that $0 < \alpha < 1$ and S_ν is not one-sided. Then $S(X_t) = G(X_t) = \mathbb{R}^d$ for $t > 0$.*

See [**464**] for the proof. In Theorem 61.6 $p_t(x)$ is continuous in (t, x), as is shown from Propositions 2.5(xii) and 24.20.

62. Densities of multivariate stable distributions

Let μ be a non-trivial α-stable distribution on $\mathbb{R}$ with $0 < \alpha < 2$. Let $p(x)$ be the continuous density of μ. Let ν be the Lévy measure of μ. For $\xi \in \{1, -1\}$ satisfying $S_\nu \supset \{r\xi \colon r \in (0,\infty)\}$, we have $p(r\xi) \sim c_\xi r^{-(1+\alpha)}$ as $r \to \infty$ with some $c_\xi > 0$ which depends on ξ. If $S_\nu \not\supset \{r\xi \colon r \in (0,\infty)\}$, then $\nu(\{r\xi \colon r \in (0,\infty)\}) = 0$ and there are two cases: (1) if $0 < \alpha < 1$, then $p(r\xi) = 0$ for all large r; (2) if $1 \le \alpha < 2$, then $p(x) > 0$ for all x and $p(r\xi) = o(\mathrm{e}^{-r^c})$ for some $c > 0$ as $r \to \infty$. See (14.37) of Remark 14.18.

In the case of a nondegenerate stable distribution μ on $\mathbb{R}^d$ with $d \ge 2$, the behavior of the continuous density $p(x)$ of μ for large $|x|$ is much more complicated.

EXAMPLE 62.1. If μ is a non-trivial rotation invariant 1-stable distribution on $\mathbb{R}^d$, that is, d-dimensional Cauchy distribution in Example 2.12 with $\gamma = 0$, then $p(x) = \text{const}\,(|x|^2 + c^2)^{-(d+1)/2}$ with some $c > 0$, and hence $p(x) \sim c_1|x|^{-(d+1)}$ with some $c_1 > 0$ as $|x| \to \infty$. More generally, let μ be a non-trivial rotation invariant α-stable distribution on $\mathbb{R}^d$ with $d \ge 2$ and $0 < \alpha < 2$. Then $\widehat{\mu}(z) = \mathrm{e}^{-c|z|^\alpha}$, $z \in \mathbb{R}^d$, with some $c > 0$ (Theorem 14.14). It is known that the continuous density $p(x)$ of μ satisfies $p(x) \sim c_1|x|^{-(d+\alpha)}$ with some $c_1 > 0$ as $|x| \to \infty$, that is, $p(r\xi) \sim c_1 r^{-(d+\alpha)}$ uniformly in $\xi \in S$ as $r \to \infty$. See Blumenthal and Getoor [**41**], where c_1 is given explicitly. As in the proof of Theorem 14.14, the Lévy measure ν of μ is expressed as

$$\nu(B) = c_2\textstyle\int_S \lambda_1(\mathrm{d}\xi)\int_0^\infty 1_B(r\xi) r^{-1-\alpha}\mathrm{d}r, \qquad B \in \mathcal{B}(\mathbb{R}^d),$$

where $c_2 > 0$ and λ_1 is the surface area measure on the unit sphere S. Let λ_r be the surface area measure on $S_r^{d-1} = \{r\xi \in \mathbb{R}^d \colon |\xi| = 1\} = rS$. Then

$$\nu(B) = c_2\textstyle\int_0^\infty r^{-1-\alpha}\mathrm{d}r\int_S 1_B(r\xi)\lambda_1(\mathrm{d}\xi)$$

$$= c_2 \textstyle\int_0^\infty r^{-1-\alpha} \mathrm{d}r \int_{S_r^{d-1}} 1_B(\eta) r^{-(d-1)} \lambda_r(\mathrm{d}\eta) = c_2 \int_{\mathbb{R}^d} 1_B(x) |x|^{-(d+\alpha)} \mathrm{d}x.$$

Thus $p(x)$ and the density of ν are of the same order.

EXAMPLE 62.2. Let $\{X_j : j = 1, \ldots, d\}$ be independent identically distributed random variables on $\mathbb{R}$ such that each X_j has non-trivial α-stable distribution μ_0 with $0 < \alpha < 2$. Then the distribution μ of $X = (X_j)_{j=1,\ldots,d}$ is nondegenerate α-stable on $\mathbb{R}^d$ and the Lévy measure ν of μ is supported on the union of the coordinate axes (Exercise 12.10). Hence the measure λ on S in the representation (14.4) of ν is concentrated on d or $2d$ points (according as the support of the Lévy measure of μ_0 is one-sided or not) at which S is intersected with the coordinate axes. Let $p_0(x_j)$ be the density of the distribution of X_j. Then μ has density $p(x) = p_0(x_1) \cdots p_0(x_d)$, where $x = (x_j)_{j=1,\ldots,d}$. Consider two special cases.

(1) Suppose that μ_0 is symmetric. Then $p_0(x_j) > 0$ for all $x_j \in \mathbb{R}$, and $p_0(x_j) \sim c_0 |x_j|^{-(1+\alpha)}$ with some $c_0 > 0$ as $|x_j| \to \infty$. If $\xi = (\xi_j)_{j=1,\ldots,d} \in S$ is such that $\xi_j \neq 0$ for all j, then

$$p(r\xi) = p_0(r\xi_1) \cdots p_0(r\xi_d) \sim c_0^d \, |\xi_1 \cdots \xi_d|^{-(1+\alpha)} r^{-d(1+\alpha)}, \qquad r \to \infty.$$

If $\xi \in S$ is such that $\xi_j = 0$ for some j, then, letting $n \in \{1, \ldots, d-1\}$ be the number of j such that $\xi_j \neq 0$, we have

$$p(r\xi) = p_0(0)^{d-n} \textstyle\prod_{\xi_j \neq 0} p_0(r\xi_j) \sim c_0^n \, p_0(0)^{d-n} \left| \prod_{\xi_j \neq 0} \xi_j \right|^{-(1+\alpha)} r^{-n(1+\alpha)}, \quad r \to \infty.$$

(2) Suppose that $0 < \alpha < 1$ and μ_0 has support $S_{\mu_0} = [0, \infty)$. Then $p_0(0) = 0$, $p_0(x_j) \sim c_0 x_j^{-(1+\alpha)}$ with some $c_0 > 0$ as $x_j \to \infty$, and $S_\mu = [0, \infty)^d$. If $\xi \in S$ satisfies $\xi_j = 0$ for some j, then $p(r\xi) = 0$ for all r. If $\xi \in S$ satisfies $\xi_j > 0$ for all j, then

$$p(r\xi) \sim c_0^d \, (\xi_1 \cdots \xi_d)^{-(1+\alpha)} r^{-d(1+\alpha)}, \qquad r \to \infty.$$

In the two theorems below we formulate some of Watanabe's results in [**563**] without proofs. Let $d \geq 2$ and $0 < \alpha < 2$ and let μ be a nondegenerate α-stable distribution on $\mathbb{R}^d$. As in (14.4), the Lévy measure ν of μ is expressed as

$$\nu(B) = c \textstyle\int_S \sigma(\mathrm{d}\xi) \int_0^\infty 1_B(r\xi) r^{-1-\alpha} \mathrm{d}r, \qquad B \in \mathcal{B}(\mathbb{R}^d), \tag{62.1}$$

where c is a positive constant and σ is a probability measure on the unit sphere S. This σ is considered as a measure on $\mathbb{R}^d$ with $\sigma(S^c) = 0$. A function $\varphi(r)$ on $(0, \infty)$ is called of dominated variation if it is positive and decreasing and if there is $c_1 > 0$ such that $\varphi(r) \leq c_1 \varphi(2r)$, $\forall r > 0$. Let $B_a = \{x \in \mathbb{R}^d : |x| < a\}$. Define $\sigma_\xi(r) = \sigma(\xi + B_{1/r})$ for $\xi \in S$ and $r > 0$, and $\sigma^*(r) = \sup_{\xi \in S} \sigma_\xi(r)$. Then it is easy to see that $\sigma^*(r)$ is of dominated variation. Let S_ν and S_σ be the supports of ν and σ, respectively. For $n = 1, \ldots, d$, let $C_\sigma^0(n)$ be the set of $\xi \in S$ expressible as $\xi = \sum_{j=1}^n c_j \xi_j$ with $c_j > 0$ such that $\xi_1, \ldots, \xi_n$ are in S_σ and linearly independent. Define $T_\sigma(1) = C_\sigma^0(1) = S_\sigma$ and $T_\sigma(n) = \overline{C_\sigma^0(n)} \setminus \overline{C_\sigma^0(n-1)}$ for $n = 2, \ldots, d$, where the overline denotes the closure. Then $\overline{C_\sigma^0(n)} \supset \overline{C_\sigma^0(n-1)}$. Write $C_\sigma^0 = C_\sigma^0(d)$ and $C_\sigma = \overline{C_\sigma^0}$ and let $\operatorname{int} C_\sigma$ be the interior of C_σ in the relative topology on S. We have $\operatorname{int} C_\sigma \supset C_\sigma^0$. Let $S^* = \operatorname{int} C_\sigma$ if $0 < \alpha < 1$ and S_ν is one-sided; let $S^* = S$ otherwise. Let $p(x)$ be the continuous density of μ; $c_1, c_2, \ldots$ below are positive constants that depend on μ. If $0 < \alpha < 1$ and S_ν is one-sided, then in the estimate of $p(x)$ from below in (ii) and (iii) of Theorem 62.3 we implicitly assume that $|x|$ is sufficiently large for the argument x of $p(x)$, noting that $p(x) = 0$ for some x.

THEOREM 62.3. (i) *There is* c_1 *such that*

$$p(x) \le c_1(1+|x|)^{-(1+\alpha)}\sigma^*(1+|x|) \qquad \text{for } x \in \mathbb{R}^d,$$

hence

$$p(x) \le c_1(1+|x|)^{-(1+\alpha)} \qquad \text{for } x \in \mathbb{R}^d.$$

(ii) *Let* $\xi^0 \in S^*$. *Then, for any* $\delta_1 > 0$, *there is* c_2, *which is independent of* ξ^0, *such that*

$$p(r\xi^0 + y) \ge c_2(1+r)^{-(1+\alpha)}\sigma_{\xi^0}(1+r) \qquad \text{for } r > 0,\ |y| \le \delta_1.$$

(iii) *For any compact set* K_1 *in* C_σ^0 *and for any* $\delta_2 > 0$, *there is* c_3 *such that*

$$p(r\xi + y) \ge c_3(1+r)^{-(1+\alpha)d} \qquad \text{for } r > 0,\ \xi \in K_1,\ |y| \le \delta_2.$$

(iv) *For any compact set* K_2 *in* S *with* $K_2 \cap C_\sigma = \emptyset$, *for any* $\delta_3 > 0$, *and for any* $c > 0$, *there is* c_4 *such that*

$$p(r\xi + y) \le c_4 \exp(-cr\log r) \qquad \text{for } r > 0,\ \xi \in K_2,\ |y| \le \delta_3.$$

THEOREM 62.4. *Let* $d \ge 2$. *Fix* $\alpha \in (0,2)$. *Let* β *be an arbitrary real in* $(1+\alpha, d+\alpha)$. *Then we can construct a nondegenerate* α*-stable distribution* μ *on* $\mathbb{R}^d$ *with some* σ *in* (62.1) *such that, for any* $\delta_4 > 0$, *there are* c_5 *and* c_6 *satisfying*

$$c_5(1+r)^{-\beta} \le p(r\xi + y) \le c_6(1+r)^{-\beta} \qquad \text{for } r > 0,\ \xi \in S_\sigma,\ |y| \le \delta_4.$$

In the proof of Theorem 62.4 Watanabe chooses, as the measure σ, the so-called Hausdorff s-measure on S with $0 < s < d-1$ and $\beta = 1 + s + \alpha$. This suggests how a diversity of spherical parts of Lévy measures of stable distributions induces a diversity of tail behaviors of their densities.

The second estimate of $p(x)$ in (i) of Theorem 62.3 was proved in Pruitt and Taylor [**400**] by Fourier-theoretic method except in the case where μ is 1-stable and not strictly 1-stable. Their argument could not treat this case. The method of Watanabe [**563**] is more probabilistic, based on the decomposition $\mu = \mu_1 * \mu_2$, where μ_1 has Lévy measure with bounded support and μ_2 is a compound Poisson distribution whose Lévy measure has support not containing the origin; it can handle all stable distributions. In [**563**] many other estimates of $p(x)$ are given; for example the estimates of $p(r\xi^0 + y)$ for $\xi^0 \in T_\sigma(n)$ in relation to the Hausdorff dimension of the measure σ are found; the case where σ is absolutely continuous with respect to the surface measure on S with bounded density, the case where σ is discrete with S_σ being a finite set, and mixture of these two cases are treated. The case of S_σ being a finite set was studied earlier by Hiraba [**202, 203**]; the introduction of the sets $C_\sigma^0(n)$ and $T_\sigma(n)$ is due to him.

63. Conditions stronger than subexponentiality

In the study of the relation of tail behaviors of an infinitely divisible distribution on $\mathbb{R}_+$ and its Lévy measure, the subexponentiality in Definition 25.13 is an important concept. A basic result is the theorem of Embrechts, Goldie, and Veraverbeke [**123**] in Remark 25.14. In this section we review results in Asmussen, Foss, and Korshunov [**7**] and Watanabe and Yamamuro [**565, 566**] on some conditions stronger than subexponentiality with application to Lévy measure densities of some distributions including log-normal. Throughout this section (except in the paragraph next to the last) we always assume that μ is a distribution on $\mathbb{R}_+$. Following [**566**], we say that a function $f(x)$ is of class **L** if it is nonnegative, measurable, $f(x) > 0$ for all large x, and $f(x+a) \sim f(x)$, $x \to \infty$, for every $a > 0$. We begin with some definitions in [**7**].

DEFINITION 63.1. (i) We say that μ is *long-tailed* (or of *class* $\mathcal{L}$) if $\mu(x,\infty)$ is of class **L**. (ii) Let $c > 0$ ($c = \infty$ is always excluded). We say that μ is $(0,c]$-*long-tailed* (or of class $\mathcal{L}_{(0,c]}$) if $\mu(x, x+c]$ is of class **L**, that is, if $\mu(x,x+c] > 0$ for all large x and, for every $a > 0$, $\mu(x+a, x+a+c] \sim \mu(x,x+c]$, $x \to \infty$. (iii) Let $c > 0$. We say that μ is $(0,c]$-*subexponential* (or of class $\mathcal{S}_{(0,c]}$) if μ is $(0,c]$-long-tailed and if $\mu * \mu(x,x+c] \sim 2\mu(x,x+c]$, $x \to \infty$.

As in Definition 25.13 and a remark following it, μ is subexponential if and only if $\mu(x,\infty) > 0$ for all x and $\mu * \mu(x,\infty) \sim 2\mu(x,\infty)$ as $x \to \infty$. Let $\mathcal{S}$ be the class of subexponential distributions on $[0,\infty)$. It is known that $\mathcal{S} \subset \mathcal{L}$ (Remark 25.14). Hence the definition of $(0,c]$-subexponentiality is obtained from that of subexponentiality by replacing $(0,\infty)$ with $(0,c]$.

PROPOSITION 63.2. *We have the following properties of* $\mathcal{L}_{(0,c]}$.
(i) $\mathcal{L}_{(0,c]} \subset \mathcal{L}_{(0,nc]}$ *for* $c > 0$ *and* $n \in \mathbb{N}$.
(ii) $\mathcal{L}_{(0,c_1]} \not\supset \mathcal{L}_{(0,c_2]}$ *for any* c_1 *and* c_2 *satisfying* $0 < c_1 < c_2$.
(iii) $\mathcal{L}_{(0,c_1]} \not\subset \mathcal{L}_{(0,c_2]}$ *for some* c_1 *and* c_2 *satisfying* $0 < c_1 < c_2$.
(iv) $\mathcal{L}_{(0,c]} \subset \mathcal{L}$ *for any* $c > 0$.

Proof. (i) Let $\mu \in \mathcal{L}_{(0,c]}$. Given $a > 0$ and $\varepsilon > 0$, we have

$$\begin{aligned}\mu(x+a, x+a+nc] &= \textstyle\sum_{j=0}^{n-1} \mu(x+a+jc, x+a+(j+1)c] \\ &\le (1+\varepsilon)\textstyle\sum_{j=0}^{n-1} \mu(x+jc, x+(j+1)c] = (1+\varepsilon)\mu(x,x+nc]\end{aligned}$$

for all large x, and similarly $\mu(x+a, x+a+nc] \ge (1-\varepsilon)\mu(x,x+nc]$. Hence $\mu \in \mathcal{L}_{(0,nc]}$. (ii) Let $\mu = b\sum_{j=2}^{\infty} j^{-1}(\log j)^{-2}\delta_{jc_2}$ with a constant $b > 0$. Then $\mu \in \mathcal{L}_{(0,c_2]}$. But $\mu \notin \mathcal{L}_{(0,c_1)}$ for $c_1 < c_2$, since $\mu(jc_1, (j+1)c_1] = 0$ for infinitely many $j \in \mathbb{N}$. (iii) Let $c_2 = (3/2)c_1$. Let $\mu = b\sum_{j=2}^{\infty} j^{-1}(\log j)^{-2}\delta_{jc_1}$ with a constant $b > 0$. Then $\mu \in \mathcal{L}_{(0,c_1]}$. But $\mu \notin \mathcal{L}_{(0,c_2]}$, since $\mu((2n+1)c_2, (2n+2)c_2] = \mu\{(3n+2)c_1\} + \mu\{(3n+3)c_1\} \sim 2\mu\{(3n+1)c_1\} = 2\mu(2nc_2, (2n+1)c_2]$ as $n \to \infty$. (iv) is proved similarly to (i). $\square$

PROPOSITION 63.3. *The statements in* (i)–(iv) *of Proposition 63.2 remain true when* $\mathcal{L}$ *is replaced by* $\mathcal{S}$. *In particular,* $\mathcal{S}_{(0,c]} \subset \mathcal{S}$ *for any* $c > 0$.

Proof. The analogues of (i) and (iv) are shown by a similar argument. The analogues of (ii) and (iii) are obtained by using the same μ. □

Now we give more definitions and their simple consequences.

DEFINITION 63.4. (i) We say that μ is *locally of class* $\mathcal{L}$ (or of class $\mathcal{L}_{\mathrm{loc}}$) if it is of class $\mathcal{L}_{(0,c]}$ for all $c > 0$. (ii) We say that μ is *locally subexponential* (or of class $\mathcal{S}_{\mathrm{loc}}$) if it is of class $\mathcal{S}_{(0,c]}$ for all $c > 0$.

DEFINITION 63.5. (i) We say that μ has a *long-tailed density* (or μ is of class $\mathcal{L}_{\mathrm{ac}}$) if μ is absolutely continuous and a version $p(x)$ of its density is of class $\mathbf{L}$. (ii) We say that μ has a *subexponential density* (or μ is of class $\mathcal{S}_{\mathrm{ac}}$) if μ is absolutely continuous and a version $p(x)$ of its density is of class $\mathbf{L}$ and $p * p(x) \sim 2p(x)$ as $x \to \infty$. (Here, for two distribution density functions $p(x)$, $q(x)$ on $\mathbb{R}_+$, we define $p * q(x) = \int_0^x p(x-y)q(y)\mathrm{d}y$; this is a version of the density function of $(p(x)\mathrm{d}x) * (q(x)\mathrm{d}x)$.)

PROPOSITION 63.6. (i) $\mathcal{L}_{\mathrm{ac}} \subset \mathcal{L}_{\mathrm{loc}} \subset \mathcal{L}$. (ii) $\mathcal{S}_{\mathrm{ac}} \subset \mathcal{S}_{\mathrm{loc}} \subset \mathcal{S}$.

Proof. Let $\mu \in \mathcal{L}_{\mathrm{ac}}$, $\mu(\mathrm{d}x) = p(x)\mathrm{d}x$, and $p \in \mathbf{L}$. Then $\mu(x, x+c] = \int_x^{x+c} p(y)\mathrm{d}y > 0$ for all large x and $\mu(x+a, x+a+c] = \int_x^{x+c} p(y+a)\mathrm{d}y \sim \int_x^{x+c} p(y)\mathrm{d}y = \mu(x, x+c]$. Hence $\mu \in \mathcal{L}_{\mathrm{loc}}$. We have $\mathcal{L}_{\mathrm{loc}} \subset \mathcal{L}$ from Proposition 63.2(iv). The proof of (ii) is similar. □

The following three theorems are known. Theorem 63.7 is from Theorem 7 of [**7**] and Theorem 1.2 of [**565**]. Theorems 63.8 and 63.9 are Theorems 1.1 and 1.2 of [**566**] obtained by an involved argument. We omit the proofs. For an infinitely divisible distribution μ, its Lévy measure is denoted by ν and $[\nu]_{(1,\infty)}/\nu(1,\infty)$ is denoted by $\nu_{(1)}$ (that is, $\nu_{(1)}(B) = \nu(B \cap (1,\infty))/\nu(1,\infty))$ if $\nu(1,\infty) > 0$.

THEOREM 63.7. *Fix $c > 0$. Let μ be an infinitely divisible distribution on $\mathbb{R}_+$. Assume that $\nu(x, x+c] > 0$ for all large x and that $\nu_{(1)} \in \mathcal{L}_{(0,c]}$. Then the following are equivalent:* (1) $\mu \in \mathcal{S}_{(0,c]}$. (2) $\nu_{(1)} \in \mathcal{S}_{(0,c]}$. (3) $\mu(x, x+c] \sim \nu(x, x+c]$, $x \to \infty$.

THEOREM 63.8. *Let μ be an infinitely divisible distribution on $\mathbb{R}_+$. Then the following are equivalent:*

(1) $\mu \in \mathcal{S}_{\mathrm{loc}}$.
(2) $\nu(1,\infty) > 0$ *and* $\nu_{(1)} \in \mathcal{S}_{\mathrm{loc}}$.
(3) $\nu(1,\infty) > 0$, $\nu_{(1)} \in \mathcal{L}_{\mathrm{loc}}$, *and* $\mu(x, x+c] \sim \nu(x, x+c]$, $x \to \infty$, *for all* $c > 0$.
(4) $\nu(1,\infty) > 0$, $\nu_{(1)} \in \mathcal{L}_{\mathrm{loc}}$, *and there is* $C \in (0,\infty)$ *such that, for all* $c > 0$, $\mu(x, x+c] \sim C\nu(x, x+c]$, $x \to \infty$.

THEOREM 63.9. *Let μ be an infinitely divisible distribution on $\mathbb{R}_+$.*

(i) *If $\mu^t \in \mathcal{S}_{\mathrm{loc}}$ for some $t > 0$, then, for all $t > 0$, $\mu^t \in \mathcal{S}_{\mathrm{loc}}$ and $\mu^t(x, x+c] \sim t\mu(x, x+c]$, $x \to \infty$, for all $c > 0$.*

(ii) *If $\mu \in \mathcal{L}_{\mathrm{loc}}$ and if, for some $t \in (0,1) \cup (1,\infty)$, there is $C \in (0,\infty)$ such that, for all $c > 0$, $\mu^t(x, x+c] \sim C\mu(x, x+c]$, $x \to \infty$, then $C = t$ and $\mu \in \mathcal{S}_{\mathrm{loc}}$.*

If μ is a non-trivial selfdecomposable distribution on $\mathbb{R}_+$, then μ is absolutely continuous and unimodal with some mode $a \geq 0$ and $\nu(\mathrm{d}x) = x^{-1}k(x)1_{(0,\infty)}(x)\mathrm{d}x$

with an decreasing function $k(x)$ (Theorem 53.1 and Corollary 15.11). In this case we choose the density function $p(x)$ of μ increasing on $(0, a)$ (if $a > 0$) and decreasing on (a, ∞). Since μ^t is non-trivial selfdecomposable for any $t > 0$ in this case, the similarly chosen density of μ^t is denoted by $p_t(x)$. The following two theorems are Theorems 1.3 and 1.4 of [**566**].

THEOREM 63.10. *Let μ be a selfdecomposable distribution on $\mathbb{R}_+$. Then the following are equivalent:* (1) $\mu \in \mathcal{S}_{ac}$. (2) $\nu(1,\infty) > 0$ *and* $\nu_{(1)} \in \mathcal{S}_{ac}$. (3) $k(x)$ *is of class* **L** *and* $k(x) \sim xp(x)$, $x \to \infty$. (4) $k(x)$ *is of class* **L** *and there is* $C \in (0,\infty)$ *such that* $k(x) \sim Cxp(x)$, $x \to \infty$.

THEOREM 63.11. *Let μ be a selfdecomposable distribution on $\mathbb{R}_+$.* (i) *If* $\mu^t \in \mathcal{S}_{ac}$ *for some* $t > 0$, *then, for all* $t > 0$, $\mu^t \in \mathcal{S}_{ac}$ *and* $p_t(x) \sim tp(x)$, $x \to \infty$. (ii) *If* $\mu \in \mathcal{L}_{ac}$ *and if, for some* $t \in (0,1) \cup (1,\infty)$, *there is* $C \in (0,\infty)$ *such that* $p_t(x) \sim Cp(x)$, $x \to \infty$, *then* $C = t$ *and* $\mu \in \mathcal{S}_{ac}$.

We omit the proofs of Watanabe and Yamamuro. They apply the results to log-normal and Weibull with parameter $\alpha \in (0,1)$. For unimodal absolutely continuous distribution μ they notice that the three properties (1) $\mu \in \mathcal{L}_{(0,c]}$ for some $c > 0$, (2) $\mu \in \mathcal{L}_{loc}$, (3) $\mu \in \mathcal{L}_{ac}$, are equivalent and that the same statement is true with $\mathcal{S}$ in place of $\mathcal{L}$. For μ in Theorems 63.10 and 63.11, they make the decomposition $\mu = \mu_1 * \mu_2$ where μ_1 and μ_2 have Lévy measures $x^{-1}(k(x) \wedge k(1))\mathrm{d}x$ and $x^{-1}(k(x) - k(1))1_{(0,1]}(x)\mathrm{d}x$, respectively, and apply Theorems 63.8 and 63.9 to μ_1 and μ_2. Then they derive properties of μ from those of μ_1 and μ_2.

EXAMPLE 63.12. Let μ_{p}, μ_{ln}, μ_{w}, and $\mu_{\mathrm{i\Gamma}}$ denote Pareto, log-normal, Weibull with $0 < \alpha < 1$, and inverse-Γ distributions, respectively. The last one is the distribution of X^{-1}, where X has Γ-distribution with parameters $c > 0$ and 1, which is mentioned in Exercise 34.13 as a special case of generalized inverse Gaussian distributions. It is known that those four distributions are of class T of Exercise 55.8 and hence selfdecomposable (see Steutel and van Harn [**503**] p. 414 or Bondesson [**55**]; see also Example 15.13 for μ_{p} and μ_{ln}). We have, for $x > 0$,

$$\begin{aligned}
\mu_{\mathrm{p}}(\mathrm{d}x) &= \alpha(1+x)^{-\alpha-1}\mathrm{d}x \qquad (\alpha > 0),\\
\mu_{\mathrm{ln}}(\mathrm{d}x) &= (\alpha/\pi)^{1/2}x^{-1}\mathrm{e}^{-\alpha(\log x)^2}\mathrm{d}x \qquad (\alpha > 0),\\
\mu_{\mathrm{w}}(\mathrm{d}x) &= \alpha x^{\alpha-1}\mathrm{e}^{-x^\alpha}\mathrm{d}x, \qquad (1 > \alpha > 0),\\
\mu_{\mathrm{i\Gamma}}(\mathrm{d}x) &= (1/\Gamma(c))x^{-c-1}\mathrm{e}^{-1/x}\mathrm{d}x \qquad (c > 0).
\end{aligned}$$

Proposition 11 of [**7**] gives sufficient conditions for $\mu \in \mathcal{S}_{ac}$, using explicit expression of the density of μ. Thus it is shown in [**7**] that $\mu_{\mathrm{p}}, \mu_{\mathrm{ln}}, \mu_{\mathrm{w}} \in \mathcal{S}_{ac}$. We can also show $\mu_{\mathrm{i\Gamma}} \in \mathcal{S}_{ac}$ by the same method. Now it follows from Theorems 63.10 and 63.11 that those four distributions satisfy $k(x) \sim xp(x)$ and $p_t(x) \sim tp(x)$ as $x \to \infty$, although we do not know $k(x)$ and $p_t(x)$ for $t \neq 1$ explicitly.

REMARK 63.13. Even if μ is a selfdecomposable subexponential distribution on $\mathbb{R}_+$, it is not necessarily true that $k(x) \sim xp(x)$ as $x \to \infty$. This fact is shown in Remark 4.1 of [**566**] by an example of $k(x)$ of the following form: $k(x) = 2$ for $0 < x < 1$ and $k(x) = 1/n!$ for $\mathrm{e}^{n(n-1)/2} \leq x < \mathrm{e}^{n(n+1)/2}$ for $n \in \mathbb{N}$.

For a distribution μ on $\mathbb{R}$ not supported on the half line, behavior of the right tail $\mu(0,\infty)$ is studied; classes $\mathcal{L} \supset \mathcal{L}_{(0,c]} \supset \mathcal{L}_{\rm loc} \supset \mathcal{L}_{\rm ac}$ and $\mathcal{S} \supset \mathcal{S}_{(0,c]} \supset \mathcal{S}_{\rm loc} \supset \mathcal{S}_{\rm ac}$ of distributions on $\mathbb{R}$ are defined formally in the same manner. The result corresponding to Theorem 63.7 is shown for distributions on $\mathbb{R}$ by [**565**] under the condition that $\int_{-\infty}^{\infty} \mathrm{e}^{-\gamma x}\mu(\mathrm{d}x) < \infty$ for some $\gamma > 0$. Theorems 63.10 and 63.11 are proved by [**566**] for selfdecomposable distributions μ on $\mathbb{R}$.

In the direction opposite to this section, conditions weaker than subexponentiality are studied in several papers; among them we mention Shimura and Watanabe [**485**] and Watanabe and Yamamuro [**567**].

64. Class of c-decomposable distributions

Let $c \in (0,1)$ in this section.

DEFINITION 64.1. A distribution μ on $\mathbb{R}^d$ is called *c-decomposable* if there is a distribution ρ on $\mathbb{R}^d$ such that

$$(64.1) \qquad \widehat{\mu}(z) = \widehat{\mu}(cz)\widehat{\rho}(z), \qquad z \in \mathbb{R}^d.$$

In this case ρ is called a *c-factor* of μ.

The name c-decomposable follows Loève [**327**], Vol. 1, p. 352. Without the name, c-decomposable distributions are already treated in Exercises 18.14 and 29.13. Moreover a semi-selfdecomposable distribution in Section 15 is, for some c, c-decomposable with an infinitely divisible c-factor.

PROPOSITION 64.2. (i) *Suppose that μ is a c-decomposable distribution on $\mathbb{R}^d$ with a c-factor ρ. Then ρ has finite log-moment and*

$$(64.2) \qquad \widehat{\mu}(z) = \textstyle\prod_{n=0}^{\infty} \widehat{\rho}(c^n z),$$

that is, μ is the limit of the distribution of $\sum_{k=0}^{n} c^k Z_k$ as $n \to \infty$, where Z_n, $n = 0, 1, \cdots$, are independent random variables each with distribution ρ.

(ii) *If ρ is a distribution on $\mathbb{R}^d$ with finite log-moment, then there is a distribution μ which is c-decomposable with a c-factor ρ.*

Proof. If (64.1) holds, then $\widehat{\mu}(z) = \widehat{\mu}(c^{n+1}z)\prod_{k=0}^{n}\widehat{\rho}(c^k z)$ and hence (64.2) holds. The series $\sum_{k=0}^{\infty} c^k Z_k$ is convergent in distribution for independent random variables $\{Z_k\}$ each with distribution ρ if and only if ρ has finite log-moment. This is a special case of Theorem 1 of Grintsevichyus [**173**], and similar to our Theorems 17.5 and 17.11. On the other hand, if (64.2) holds, then $\widehat{\mu}(cz) = \prod_{k=1}^{\infty}\widehat{\rho}(c^k z)$ and (64.1) holds. □

The following fact is due to Loève [**326**][1].

PROPOSITION 64.3. *There exist distributions μ, ρ_1, and ρ_2 on $\mathbb{R}$ such that $\rho_1 \neq \rho_2$ and μ is c-decomposable with c-factors ρ_1 and ρ_2.*

[1]The latter half of Theorem 7 of [**326**] is incorrect. But our proof of Proposition 64.3 was suggested by his argument.

Proof. Let $a = 2^{-1}(1+c^{-1}) > 1$ and let ρ_1 and ρ_2 be such that $\widehat{\rho}_1(z) = (1-|z|) \vee 0$ for $z \in \mathbb{R}$, $\widehat{\rho}_2(z) = \widehat{\rho}_1(z)$ for $|z| \leq 2a-1$, $\widehat{\rho}_2(z) = \widehat{\rho}_2(-z)$ and $\widehat{\rho}_2(z) = \widehat{\rho}_2(z+2a)$ for $z \in \mathbb{R}$. Then ρ_1 is absolutely continuous with density $(1-\cos x)/(\pi x)^2$ and ρ_2 is discrete and concentrated on $\{\pi a^{-1} n \colon n \in \mathbb{Z}\}$ with $\rho_2(\{0\}) = (2a)^{-1}$ and $\rho_2(\{\pi a^{-1} n\}) = a(1-\cos \pi a^{-1} n)/(\pi n)^2$ for $n \neq 0$. They are distinct distributions, whose characteristic functions coincide on a neighborhood of 0 (see Loève [**327**], Vol. 1, p. 231 or Feller [**139**], pp. 506–507). Since they have finite log-moment, we can find c-decomposable distributions μ_1 and μ_2 with c-factors ρ_1 and ρ_2, respectively. (In fact we do not have to use the explicit form of $\mu_j(dx)$, $j=1,2$. Note that $\prod_{k=0}^{n} \widehat{\rho}_j(c^k z) = \exp \sum_{k=0}^{n} \log(1-c^k|z|)$ is convergent uniformly for $|z| \leq 1-\varepsilon$ ($\varepsilon > 0$) as $n \to \infty$ and, for all z, $\prod_{k=0}^{n} \widehat{\rho}_j(c^k z)$ equals $\prod_{k=n_0}^{n} \widehat{\rho}_j(c^k z) \prod_{l=0}^{n_0-1} \widehat{\rho}_j(c^l z)$ and is convergent as $n \to \infty$, where n_0 is such that $c^{n_0}|z| < 1-\varepsilon$.)

We claim that $\mu_1 = \mu_2$, which will finish the proof. Clearly $\widehat{\mu}_1(z) = 0$ for $|z| \geq 1$. We have $\widehat{\mu}_2(z) = \widehat{\mu}_1(z)$ for $|z| \leq 2a-1$ from (64.2). Then, by induction on $n = 0, 1, \ldots$, we see that $\widehat{\mu}_2(z) = 0$ for $(2a-1)^n \leq |z| \leq (2a-1)^{n+1}$ using (64.1) and $2a-1 = c^{-1}$. Now $\widehat{\mu}_2(z) = 0 = \widehat{\mu}_1(z)$ for $|z| \geq 1$. Hence $\mu_2 = \mu_1$. □

PROPOSITION 64.4. *Let μ be c-decomposable on $\mathbb{R}^d$ with a c-factor ρ. If $\widehat{\mu}(z) \neq 0$ for z in a dense subset of $\mathbb{R}^d$, then the c-factor is unique. If ρ is infinitely divisible, then μ is infinitely divisible.*

Proof. For z satisfying $\widehat{\mu}(cz) \neq 0$, $\widehat{\rho}(z)$ is determined by μ as $\widehat{\rho}(z) = \widehat{\mu}(z)/\widehat{\mu}(cz)$. Hence the first assertion follows from the continuity of characteristic functions. The second assertion is seen from (64.2). □

REMARK 64.5. The converse of the second assertion in Proposition 64.4 is not true. That is, even if μ is c-decomposable and infinitely divisible, the c-factor ρ is not necessarily infinitely divisible. This was shown by Mišeikis [**352**] and Niedbalska-Rajba [**362**]; more examples for this will be given in Example 64.11.

REMARK 64.6. Let μ be non-trivial and c-decomposable with a c-factor ρ. If ρ has a bounded support, then μ is not infinitely divisible. See the solution to Exercise 18.14.

If μ is a non-trivial c-decomposable distribution on $\mathbb{R}^d$, then μ is either absolutely continuous or continuous singular (Exercise 29.13). To find a criterion in this dichotomy is an open problem hard to solve. But the following four results are remarkable. Pisot–Vijayaraghavan (PV) numbers in $(1,\infty)$ and Peres–Solomyak (PS) numbers in $(0,1)$ are introduced in Remark 27.22,

PROPOSITION 64.7. *If c^{-1} is a PV number, then there exists a nondegenerate, continuous singular, c-decomposable distribution on $\mathbb{R}^d$.*

PROPOSITION 64.8. *If c is a PS number and if μ is nondegenerate, c-decomposable on $\mathbb{R}^d$, then, for all sufficiently large integers n, μ^n is absolutely continuous with a continuous density $f_n(x)$ which tends to 0 as $|x| \to \infty$. Here μ^n is the n-fold convolution of μ.*

PROPOSITION 64.9. *(No restriction on c.) If μ is c-decomposable on $\mathbb{R}^d$ with a c-factor ρ being discrete, then, $\dim_{\mathrm{H}} \mu \leq H(\rho)/(-\log c)$.*

Here $\dim_{\mathrm{H}} \mu$, the Hausdorff dimension of μ, is defined as the infimum of the Hausdorff dimensions of sets B over all Borel sets B with $\mu(B) = 1$, and $H(\rho)$ is the entropy of ρ in Exercise 29.23. Note that if $\dim_{\mathrm{H}} \mu < d$, then μ is singular.

PROPOSITION 64.10. *(No restriction on c.) If μ is nondegenerate, c-decomposable with a c-factor ρ on $\mathbb{R}^d$ and ρ is also c-decomposable, then μ is absolutely continuous with continuous density $f(x)$ which tends to 0 as $|x| \to \infty$.*

Concerning the proofs see Erdős [**127**] and Theorem 1.1 of Watanabe [**564**] for the first proposition, Peres and Solomyak [**369**] and Theorem 1.2 of [**564**] for the second, Theorem 2.2 of Watanabe [**560**] for the third, and Theorems 2 and 4 of Watanabe [**562**], who extended an idea of Zakusilo [**600**], for the fourth.

EXAMPLE 64.11. This is from Lindner and Sato [**321, 322**]; see [**458**] for survey. Recall that elements of $\mathbb{R}^d$ are column vectors and that a prime denotes going to the transpose. Let u, v, w be nonnegative and $u + w > 0$ and $v + w > 0$. For each $k \in \mathbb{Z}$ let $\{(N_t^{(k)}, Y_t^{(k)})' \colon t \geq 0\}$ be a compound Poisson process on $\mathbb{R}^2$ with Lévy measure supported on the three points $(1,0)'$, $(0,1)'$, $(1, c^k)'$ with mass u, v, w, respectively. Then it follows from Theorem 56.3 that $\{N_t^{(k)}\}$ is a Poisson process with parameter $u + w$ and $\{Y_t^{(k)}\}$ is a compound Poisson process with Lévy measure supported on two points at most. Let

$$\mu_k = \text{distribution of } \textstyle\int_0^{\infty-} c^{N_{s-}^{(k)}} \mathrm{d}Y_s^{(k)}. \tag{64.3}$$

The convergence of the improper integral is shown by the strong law of large numbers (Theorem 36.3). This μ_k is c-decomposable with a c-factor ρ_k being the distribution of $Y_T^{(k)}$, where T is the first jump time for $\{N_t^{(k)}\}$. Indeed, we have

$$\textstyle\int_0^{\infty-} c^{N_{s-}^{(k)}} \mathrm{d}Y_s^{(k)} = \int_0^T c^{N_{s-}^{(k)}} \mathrm{d}Y_s^{(k)} + c \int_T^{\infty-} c^{N_{s-}^{(k)} - N_T^{(k)}} \mathrm{d}Y_s^{(k)},$$

the two terms in the right-hand side being independent, and the last improper integral has distribution μ_k (Theorem 40.10). Let p, q, r be the normalization of u, v, w (hence $p + q + r = 1$). The characteristic functions of μ_k and ρ_k are written explicitly and we can see that they are determined by p, q, r, c, and k. We can completely classify μ_k and ρ_k by infinite divisibility. For instance, letting $ID = ID(\mathbb{R})$, we have (1) $\mu_k \in ID$ implies $\mu_{k+1} \in ID$; (2) $0 < p \leq r$ implies $\mu_k, \rho_k \notin ID$; (3) $0 < r \leq pq$ implies $\mu_0, \rho_0 \in ID$; (4) $pq < r < p$ implies $\mu_0, \rho_0 \notin ID$; and

$$0 < r \leq pq,\ k > 0,\ c^{-k} \neq 2 \quad \Longrightarrow \quad \mu_k \in ID,\ \rho_k \notin ID.$$

The classification of μ_k by the continuity properties (that is, whether absolutely continuous or continuous singular) is made only partially. But the following are known: (1) The continuity properties and $\dim_{\mathrm{H}} \mu_k$ do not depend on k. (2) If c^{-1} is a PV number, then μ_k is continuous singular. (3) If p, q, r are fixed, μ_k is continuous singular for all sufficiently small c. For example, if $p = 1/2$ and $c < 1/4$, then μ_k is continuous singular.

If a Lévy process $\{Y_t \colon t \geq 0\}$ on $\mathbb{R}$ and a random variable U are independent, then the process $\{V_t \colon t \geq 0\}$ of Ornstein–Uhlenbeck type generated by $\{Y_t\}$ (or the Ornstein–Uhlenbeck process driven by $\{Y_t\}$) with $V_0 = U$ is expressed as $V_t = \mathrm{e}^{-t}(U + \int_0^t \mathrm{e}^s \mathrm{d}Y_s)$ as in Definition 58.3 with $H = 1$. We extend this notion, to explain the background of Example 64.11. Let $\{(X_t, Y_t)' \colon t \geq 0\}$ be a Lévy process on $\mathbb{R}^2$ and let U be a random variable independent of $\{(X_t, Y_t)'\}$. Define a process $\{V_t \colon t \geq 0\}$ by

$$(64.4) \qquad V_t = \mathrm{e}^{-X_t}\left(U + \int_0^t \mathrm{e}^{X_{s-}} \mathrm{d}Y_s\right).$$

This process is called the *generalized Ornstein–Uhlenbeck process* (driven by $\{(X_t, Y_t)'\}$ with $V_0 = U$) by de Haan and Karandikar [**96**]; see also Carmona, Petit, and Yor [**73, 74**]. This is a temporally homogeneous Markov process. Here we do not give the definition of the integral in (64.4), but it is well-defined pathwise if $\{Y_t\}$ has sample functions of bounded variation on any finite time interval a. s. Lindner and Maller [**320**] proved the following fact. Define

$$(64.5) \qquad L_t = Y_t + \sum_{0<s\leq t} (\mathrm{e}^{-(X_s - X_{s-})} - 1)(Y_s - Y_{s-}).$$

Then

$$(64.6) \qquad Y_t = L_t + \sum_{0<s\leq t} (\mathrm{e}^{X_s - X_{s-}} - 1)(L_s - L_{s-}).$$

If $\{X_t\}$ and $\{Y_t\}$ are independent, then $L_t = Y_t$ a. s. Assume that there is no constant a satisfying (64.4) with $U = a$ and $V_t = a$. If

$$(64.7) \qquad \int_0^{\infty-} \mathrm{e}^{-X_{s-}} \mathrm{d}L_s \text{ is convergent a. s.,}$$

then its distribution is a stationary distribution of the generalized Ornstein–Uhlenbeck process $\{V_t\}$ driven by $\{(X_t, Y_t)'\}$. If (64.7) does not hold, then the process $\{V_t\}$ does not have a stationary distribution. If $\int_0^{\infty-} \mathrm{e}^{-X_{s-}} \mathrm{d}Y_s$ is convergent a. s., then (64.7) holds, but the converse is not true. If $E|X_1| < \infty$ and $EX_1 > 0$, then (64.7) is equivalent to $E[\log^+ |L_1|] < \infty$, and also to $E[\log^+ |Y_1|] < \infty$. Erickson and Maller [**131**] found a condition for (64.7) in terms of the triplet of $\{(X_t, L_t)'\}$.

In Example 64.11 we can consider a sequence of generalized Ornstein–Uhlenbeck processes $\{(X_t^{(k)}, Y_t^{(k)})' \colon t \geq 0\}$, $k \in \mathbb{Z}$, letting $X_t^{(k)} = -(\log c) N_t^{(k)}$. Define $L_t^{(k)}$ by (64.5) with $\{L_t^{(k)}\}, \{X_t^{(k-1)}\}, \{Y_t^{(k-1)}\}$ replacing $\{L_t\}, \{X_t\}, \{Y_t\}$. Then (64.7) is true with $\{X_t^{(k-1)}\}, \{L_t^{(k)}\}$ replacing $\{X_t\}, \{L_t\}$. We can see that $\{(X_t^{(k)}, Y_t^{(k)})'\} \stackrel{\mathrm{d}}{=} \{(X_t^{(k-1)}, L_t^{(k)})'\}$. Hence μ_k is the law of $\int_0^{\infty-} \mathrm{e}^{-X_s^{(k)}} \mathrm{d}Y_s^{(k)}$ on the one hand, and is the stationary distribution of the generalized Ornstein–Uhlenbeck process $\{(X_t^{(k-1)}, Y_t^{(k-1)})'\}$ on the other. Now we notice how delicate the properties of stationary distributions of the generalized Ornstein–Uhlenbeck processes are.

The class of semi-stable distributions is another important class of distributions c-decomposable for some c. We mention Sato and Watanabe [**464**] and Watanabe and Yamamuro [**568**] for new results on semi-stable distributions.

Solutions to exercises

Chapter 1

E 6.1. Use Taylor expansion of the exponential function.

E 6.2. By orthogonal transformation, calculation of the characteristic function reduces to the case that A is diagonal. Thus it is enough to prove the one-dimensional case, that is, (2.7) for μ of Example 2.8. Change of variables gives $\widehat{\mu}(z) = \mathrm{e}^{\mathrm{i}\gamma z}\varphi(a^{1/2}z)$ with

$$\varphi(z) = (2\pi)^{-1/2}\int_{-\infty}^{\infty} \mathrm{e}^{\mathrm{i}zx-x^2/2}\mathrm{d}x = (2/\pi)^{1/2}\int_0^{\infty} \mathrm{e}^{-x^2/2}\cos zx\,\mathrm{d}x.$$

By differentiation under the integral sign and integration by parts, $\varphi'(z) = -z\varphi(z)$. Since $\varphi(0) = 1$, this gives $\varphi(z) = \mathrm{e}^{-z^2/2}$.

E 6.3. By change of variables it is enough to see (2.9) for $\gamma = 0$ and $c = 1$. Let μ_1 be the distribution on $\mathbb{R}$ with density $2^{-1}\mathrm{e}^{-|x|}$. Then $\widehat{\mu}_1(z) = (z^2+1)^{-1}$. Hence Proposition 2.5(xii) gives the characteristic function of μ.

E 6.4. We get (2.15) for $L_\mu(u)$ from the definition by change of variables, using the definition of the Γ-function by the integral. Analytic extension of (2.15) to the half plane as in the proof of Proposition 2.6 leads to (2.16).

E 6.5. See Lukacs [**329**], p. 24.

E 6.6. The 'only if' part follows from Proposition 1.12(ii). To show the 'if' part, suppose that X_n does not converge to X in prob. Then, for some $\varepsilon > 0$, there is a subsequence $\{X_{n_k}\}$ of $\{X_n\}$ such that $P[\,|X_{n_k} - X| > \varepsilon\,]$ tends to a positive real. Choose a further subsequence $\{X_{n'_k}\}$ that converges to X a.s. Then $X_{n'_k} \to X$ in prob. by Proposition 1.12(i). This is a contradiction.

E 6.7. Use that the function $\theta/(1+\theta)$ of $\theta \geq 0$ is increasing and concave.

E 6.8. $P[\,|X_n - \gamma| > \varepsilon\,]$ tends to 0 for every $\varepsilon > 0$ if and only if $E[f(X_n)]$ tends to $f(\gamma)$ for every bounded continuous f.

E 6.9. Choose a countable family $\{f_k \colon k \in \mathbb{N}\}$ of continuous functions on $\mathbb{R}^d$ with $|f_k(x)| \leq 1$ such that $\mu_n \to \mu$ if and only if $\int f_k(x)\mu_n(\mathrm{d}x) \to \int f_k(x)\mu(\mathrm{d}x)$ for every k. Let $r(\mu_1,\mu_2) = \sum_{k\in\mathbb{N}} 2^{-k}|\int f_k(x)\mu_1(\mathrm{d}x) - \int f_k(x)\mu_2(\mathrm{d}x)|$.

E 6.10. Similar to the check of $\mu(\mathbb{R}) = 1$ at the beginning of Example 2.13.

E 6.11. $\mathrm{Re}\,(1-\widehat{\mu}(2z)) = 2\int \sin^2\langle z,x\rangle\mu(\mathrm{d}x)$, which is bounded by $4\,\mathrm{Re}\,(1-\widehat{\mu}(z)) = 4\int(1-\cos\langle z,x\rangle)\mu(\mathrm{d}x)$.

E 6.12. We get $\int_0^\infty((1-\cos z)/z^2)\mathrm{d}z = \pi/2$ from $\lim_{a\to\infty}\int_0^a(\sin z/z)\mathrm{d}z = \pi/2$ by integration by parts. Hence

$$\pi\int_{-\infty}^{\infty}|x|\mu(\mathrm{d}x) = \int_{-\infty}^{\infty}\mu(\mathrm{d}x)\int_{-\infty}^{\infty}((1-\cos|x|z)/z^2)\mathrm{d}z = \int_{-\infty}^{\infty}(\mathrm{Re}\,(1-\widehat{\mu}(z))/z^2)\mathrm{d}z.$$

E 6.13. Check the assertion in the case where $X(\omega) = \sum_{k=1}^{n} x_k 1_{A_k}(\omega)$ with disjoint A_k, $k = 1\ldots, n$. Then extend it to the general case.

E 6.14. Use Proposition 1.16 and E 6.13. Then $E|X+Y| = \int E|x+Y|P_X(\mathrm{d}x) \geq \int |E(x+Y)|P_X(\mathrm{d}x) = \int |x|P_X(\mathrm{d}x) = E|X|$.

E 6.15. Suppose that $\{X_t\}$ is associated also with c' and σ'. It follows from $\mathrm{e}^{c(\widehat{\sigma}(z)-1)} = \mathrm{e}^{c'(\widehat{\sigma}'(z)-1)}$ that $c(\widehat{\sigma}(z)-1) = c'(\widehat{\sigma}'(z)-1)$ by Lemma 7.6. Hence $c(\sigma - \delta_0) = c'(\sigma' - \delta_0)$, because finite signed measures are determined by their Fourier transforms. Using $\sigma\{0\} = \sigma'\{0\} = 0$, we get $c = c'$. Hence $\sigma = \sigma'$.

E 6.16. Suppose $q > 0$. Using the notation in Theorems 3.2 and 4.3, observe that

$$\begin{aligned}
&E[\mathrm{e}^{-qT_B};\, 0 < T_B < \infty,\, X_{T_B-} \in C,\, X_{T_B} \in D] \\
&\quad = \textstyle\sum_{n=1}^{\infty} E[\mathrm{e}^{-qJ_n};\, S_1, \ldots, S_{n-2} \notin B,\, S_{n-1} \in C,\, S_n \in D] \\
&\quad = \textstyle\sum E[\mathrm{e}^{-qJ_n}] \int \cdots \int 1_{\{x_1, x_1+x_2, \ldots, x_1+\cdots+x_{n-2} \notin B\}} \sigma(\mathrm{d}x_1) \cdots \sigma(\mathrm{d}x_{n-2}) \int \sigma(\mathrm{d}x_{n-1}) \\
&\qquad\quad \times 1_C(x_1 + \cdots + x_{n-1}) \sigma(D - x_1 - \cdots - x_{n-1}) \\
&\quad = \textstyle\sum E[\mathrm{e}^{-qJ_n}]\, E[\sigma(D - S_{n-1});\, S_1, \ldots, S_{n-2} \notin B,\, S_{n-1} \in C]
\end{aligned}$$

and that

$$\begin{aligned}
&E\big[\textstyle\int_0^{T_B} \mathrm{e}^{-qt} 1_C(X_t) c\sigma(D - X_t) \mathrm{d}t\big] \\
&\quad = E\big[\textstyle\sum_{n=1}^{\infty} \int_{J_{n-1}}^{J_n} \mathrm{e}^{-qt} 1_C(S_{n-1}) c\sigma(D - S_{n-1}) 1_{\{J_n \leq T_B\}} \mathrm{d}t\big] \\
&\quad = \textstyle\sum E[q^{-1}(\mathrm{e}^{-qJ_{n-1}} - \mathrm{e}^{-qJ_n})]\, E[c\sigma(D - S_{n-1});\, S_{n-1} \in C,\, S_1, \ldots, S_{n-2} \notin B].
\end{aligned}$$

Since $E[\mathrm{e}^{-qJ_n}] = \left(\frac{c}{c+q}\right)^n$, we get the equality to be shown. The case $q = 0$ is handled by letting $q \downarrow 0$.

E 6.17. (i) For any $\varepsilon > 0$, $P[|Z_n/a_n| > \varepsilon] = P[|Z_1| > a_n\varepsilon] \to 0$. (ii) Choose $a_n \uparrow \infty$ so that $\sum_n P[|Z_1| > a_n] < \infty$. Then $P[|Z_n| > a_n$ infinitely often$] = 0$, that is, $P[|Z_n|/(na_n) > 1/n$ infinitely often$] = 0$. Hence $Z_n/(na_n) \to 0$ a. s. (iii) Choose $a_n \uparrow \infty$ so that $\sum_n P[|Z_1| > a_n] = \infty$.

E 6.18. No. If the answer is yes, (iii) of the preceding problem is not true, because $Z_n/a_n = (Z_n/\sqrt{a_n})/\sqrt{a_n}$.

Chapter 2

E 12.1. The proof of Lemma 7.5 works and $\widehat{\mu}(z) \neq 0$ for any z. Hence we can apply Lemma 7.6. Therefore μ is the limit of compound Poisson distributions by the argument in the proof of Theorem 8.1(i) (before Corollary 8.8).

E 12.2. Let $\nu_k = \nu_{k1} - \nu_{k2}$ be the Jordan decomposition of ν_k, where ν_{k1} and ν_{k2} are measures. Let $A_k = A_{k1} - A_{k2}$, where A_{k1} and A_{k2} are symmetric and nonnegative-definite. Then $\varphi_1(z) = \varphi_2(z)$ shows that two triplets $(A_{11} + A_{22}, \nu_{11} + \nu_{22}, \gamma_1)$ and $(A_{12} + A_{21}, \nu_{12} + \nu_{21}, \gamma_2)$ represent an identical infinitely divisible distribution. Hence the uniqueness in Theorem 8.1 applies.

E 12.3. Apply E 12.2 and Theorem 8.1.

E 12.4. Ibragimov [**211**] and Linnik and Ostrovskii [**325**], Chap. 6, §7, show this by using the fact of E 12.3. Dwass and Teicher [**119**] shows it by using Lévy's result [**314**].

E 12.5. Using the function $c(x)$ in Theorem 8.7, consider the generating triplets $(A_\mu, \nu_\mu, \beta_\mu)_c$ of $\mu \in M$. Consider the condition

(4a) $\sup_{\mu\in M} |\beta_\mu| < \infty$.

Then, (1), (2), (3), and (4) are satisfied if and only if (1), (2), (3), and (4a) are satisfied. Now use the arguments in the proof of Theorem 8.7.

E 12.6. The Laplace transforms of probability measures are positive. By Proposition 2.6, $L_{\mu_k{}^n}(u) = L_{\mu_k}(u)^n$. Hence, if $L_{\mu_1{}^n}(u) = L_{\mu_2{}^n}(u)$, then $L_{\mu_1}(u) = L_{\mu_2}(u)$.

E 12.7. Notice that uniform continuity on any $[0, t_0]$ of $\{X_t\}$ in the metric r follows from the stochastic continuity. Then use $P[\,|X_t - X_s| > \varepsilon] \le (1 + \varepsilon)\varepsilon^{-1} r(X_s, X_t)$.

E 12.8. To see the 'only if' part, let $X^{(1)}$ and $X^{(2)}$ be independent. Their distributions are infinitely divisible by Proposition 11.10. Let $(A^{(l)}, \nu^{(l)}, \gamma^{(l)})$ be the generating triplet of $X^{(l)}$. Write, for $z = (z_1, \ldots, z_{n+m})$, $z^{(1)} = (z_1, \ldots, z_n)$ and $z^{(2)} = (z_{n+1}, \ldots, z_{n+m})$. Then

$$\widehat{P}_X(z) = \widehat{P}_{X^{(1)}}(z^{(1)})\widehat{P}_{X^{(2)}}(z^{(2)}),$$

and hence $\langle z, Az\rangle = \langle z^{(1)}, A^{(1)}z^{(1)}\rangle + \langle z^{(2)}, A^{(2)}z^{(2)}\rangle$ and

$$\int_{\mathbb{R}^{n+m}} f(x_1, \ldots, x_{n+m})\nu(dx) = \int_{\mathbb{R}^n} f(x_1^{(1)}, \ldots, x_n^{(1)}, 0, \ldots, 0)\nu^{(1)}(dx^{(1)}) + \int_{\mathbb{R}^m} f(0, \ldots, 0, x_1^{(2)}, \ldots, x_m^{(2)})\nu^{(2)}(dx^{(2)})$$

for any nonnegative measurable f.

E 12.9. The 'only if' part is obvious. The 'if' part is proved by double induction in n and m. If $n = m = 1$, then there is nothing to prove. Let $n = 1$ and $m \ge 2$. Suppose that the assertion is true with $m - 1$ in place of m. Assume that, for $2 \le q \le 1 + m$, X_1 and X_q are independent. Consider

$$B = \{x \in \mathbb{R}^{1+m} : x_1 \neq 0 \text{ and } (x_2, \ldots, x_{1+m}) \neq (0, \ldots, 0)\},$$
$$B_1 = \{x \in \mathbb{R}^{1+m} : x_1 \neq 0 \text{ and } (x_2, \ldots, x_m) \neq (0, \ldots, 0)\},$$
$$B_2 = \{x \in \mathbb{R}^{1+m} : x_1 \neq 0 \text{ and } (x_3, \ldots, x_{1+m}) \neq (0, \ldots, 0)\},$$
$$C = \{x \in \mathbb{R}^m : x_1 \neq 0 \text{ and } (x_2, \ldots, x_m) \neq (0, \ldots, 0)\}.$$

Then $B = B_1 \cup B_2$. By the induction hypothesis, X_1 and $(X_2, \ldots, X_m)$ are independent. Hence, using Proposition 11.10 for the projection to the first m coordinates, and then using E 12.8, we get

$$\nu(B_1) = \int_{\mathbb{R}^{1+m}} 1_C(x_1, \ldots, x_m)\nu(dx) = 0$$

and $A_{1q} = 0$ for $2 \le q \le m$. Similarly $\nu(B_2) = 0$ and $A_{1q} = 0$ for $3 \le q \le m + 1$. It follows that $\nu(B) = 0$. Hence X_1 and $(X_2, \ldots, X_{1+m})$ are independent by E 12.8. This shows that the assertion is true for $n = 1$ and any m. Let $n \ge 2$. By a similar argument it is shown that, if the assertion is true for $n - 1$ in place of n, then it is true for n.

E 12.10. (i) To see the 'if' part, it is easy to check the condition in Proposition 2.5(iv). Use E 12.7 and Proposition 11.10 for the 'only if' part. (ii) If $d = 2$, there is nothing to prove. If the assertion is true for $d-1$ in place of d, then $X_1, \ldots, X_{d-1}$

are independent and, moreover, $(X_1,\ldots,X_{d-1})$ and X_d are independent by E 12.8.

E 12.11. (i) Use that $\widehat{\mu}(-z)=\overline{\widehat{\mu}(z)}$. The examples below in (iii) show that $\widehat{\mu}(z)$ of a symmetric μ may take negative values. This does not happen for the infinitely divisible case, by Lemma 7.5. (ii) Note that $\operatorname{Re}\widehat{\mu}(z)=\frac{1}{2}(\widehat{\mu}(z)+\widehat{\mu}_1(z))=\widehat{\mu}_0(z)$, where μ_1 is the dual of μ. (iii) Consider a non-symmetric infinitely divisible distribution μ such that $\operatorname{Re}\widehat{\mu}(z)$ vanishes for some z, which shows that μ_0 is not infinitely divisible. A simple example is $\mu=\delta_\gamma$ with $\gamma\neq 0$, for which $\operatorname{Re}\widehat{\mu}(z)=\cos\langle\gamma,z\rangle$. A similar example is a Gaussian with mean $\neq 0$. (iv) Let $\mu=p\delta_0+q\delta_1$ on $\mathbb{R}$ with $p>0$, $q>0$, and $p+q=1$. Let μ_2 be the symmetrization of μ. Then $\mu_2=(p^2+q^2)\delta_0+pq\delta_1+pq\delta_{-1}$. If there is a probability measure ρ such that $\widehat{\rho}(z)=|\widehat{\mu}(z)|$, then ρ is symmetric, $\mu_2=\rho*\rho$, and the support of ρ consists of two points (if the support of ρ has at least three points, then the support of μ_2 would have at least five points.) Hence it is necessary that $\rho=\frac{1}{2}\delta_{1/2}+\frac{1}{2}\delta_{-1/2}$, which is a contradiction if $p\neq\frac{1}{2}$.

E 12.12 ([**91**]). If $X=(X_j)_{j=1,\ldots,d}$ is a random variable on $\mathbb{R}^d$, then $P[\,|X|>\varepsilon\,]\leq\sum_{j=1}^d P[\,|X_j|>\varepsilon/\sqrt{d}\,]$ and $\widehat{P}_{X_j}(z)$ is the value of $\widehat{P}_X$ on the jth axis. Hence it is enough to deal with the one-dimensional case. For any probability measure μ on $\mathbb{R}$, we have

$$\begin{aligned}a^{-1}\textstyle\int_{-a}^{a}(1-\widehat{\mu}(z))\mathrm{d}z&=a^{-1}\textstyle\int\mu(\mathrm{d}x)\int_{-a}^{a}(1-\mathrm{e}^{\mathrm{i}zx})\mathrm{d}z\\&=2\textstyle\int(1-(\sin ax)/(ax))\mu(\mathrm{d}x)\geq 2\int_{|x|\geq 2/a}(1-(a|x|)^{-1})\mu(\mathrm{d}x)\\&\geq\mu(\{x\colon|x|\geq 2/a\})\end{aligned}$$

for any $a>0$.

E 12.13. Repeated integration by parts gives

$$\begin{aligned}P[\,X_t>x\,]&=\tfrac{1}{(t-1)!}\textstyle\int_x^\infty y^{t-1}\mathrm{e}^{-y}\mathrm{d}y=\tfrac{1}{(t-1)!}x^{t-1}\mathrm{e}^{-x}+\tfrac{1}{(t-2)!}\int_x^\infty y^{t-2}\mathrm{e}^{-y}\mathrm{d}y\\&=\tfrac{1}{(t-1)!}x^{t-1}\mathrm{e}^{-x}+\tfrac{1}{(t-2)!}x^{t-2}\mathrm{e}^{-x}+\cdots+x\mathrm{e}^{-x}+\mathrm{e}^{-x}.\end{aligned}$$

E 12.14. Let $\{X_t^0\}$ be the Brownian motion on $\mathbb{R}$ and A_t be a continuous increasing function such that $A_0=0$ and $A_t\sim(-\log t)^{-1}$ as $t\downarrow 0$. Define $X_t=X^0(A_t)$. Then $\{X_t\}$ is an additive process with continuous paths. Using $1-3x^{-4}$ in place of $1+x^{-2}$ in (11.7), we have

$$\textstyle\int_c^\infty\mathrm{e}^{-x^2/2}\mathrm{d}x\geq\mathrm{e}^{-c^2/2}(c^{-1}-c^{-3})\quad\text{for } c>0.$$

Using this, prove that, with some $a>0$, $P[|X_t|>\varepsilon]\geq at^{\varepsilon^2}(-\log t)^{-1/2}$ for small t. Hence (11.5) does not hold for $0<\varepsilon<1$.

E 12.15. No. For example, choose $\widehat{\mu}(z)=\mathrm{e}^{-z^2/2}$ and $X_t=\sqrt{t}X_1$. See E 18.18 for another example.

Chapter 3

E 18.1. (i) Symmetry of μ is equivalent to $\widehat{\mu}(z)=\widehat{\mu}(-z)$. The infinitely divisible distribution corresponding to $\widehat{\mu}(-z)$ is the dual $\widetilde{\mu}$ of μ and generated by $(A,T_{-1}\nu,-\gamma)$. Now use the uniqueness of the generating triplet. (ii) $\mu^\sharp=\mu*\widetilde{\mu}$.

E 18.2. The statements (2) and (3) are equivalent, because, in general, $\widehat{\mu}(-z) = \overline{\widehat{\mu}(z)}$. By definition, (1) implies (4). If μ satisfies (2), then (1) holds, because, for any orthogonal U, $|U'z| = |z|$ and $\widehat{\mu}(U'z) = \widehat{\mu}(z)$. To complete the proof, it is enough to show that (4) implies (2). Assume (4). Let $z_1, z_2 \in \mathbb{R}^d$ be such that $|z_1| = |z_2| = c > 0$. Let e_1 be the unit vector with first component 1. Choose, for each $j = 1, 2$, an orthogonal matrix U_j with $\det U_j = 1$ such that $U_j e_1 = c^{-1} z_j$. The matrix $U = U_1 U_2'$ satisfies $U' z_1 = z_2$. Hence $\widehat{\mu}(z_2) = \widehat{\mu}(U'z_1) = \widehat{\mu}(z_1)$ by (4). This means (2).

E 18.3. Use Proposition 11.10 for orthogonal matrices U.

E 18.4. (i) It follows that

$$\widehat{\mu}(z) = [\widehat{\mu}(b_n^{-1}z)\exp(-\mathrm{i}\langle c_n, n^{-1}b_n^{-1}z\rangle)]^n.$$

Hence μ is infinitely divisible and the function within the square brackets equals $\widehat{\mu}(z)^{1/n}$. Hence, for any $m \in \mathbb{N}$,

$$\widehat{\mu}(z)^{m/n} = \widehat{\mu}(b_m b_n^{-1} z)\mathrm{e}^{\mathrm{i}\langle c_{n,m}, z\rangle}$$

with some $c_{n,m}$. Assuming that μ is non-trivial, use Lemma 13.10. Then, for any $t > 0$, there are $b_t > 0$ and c_t such that $\widehat{\mu}(z)^t = \widehat{\mu}(b_t z)\mathrm{e}^{c_t, z}$. (ii) By induction, for any $k \in \mathbb{N}$, there is c_k such that $\widehat{\mu}(z)^{n^k} = \widehat{\mu}(b^k z)\mathrm{e}^{\mathrm{i}\langle c_k, z\rangle}$. Hence, there is μ_k such that $\widehat{\mu}(z) = \widehat{\mu}_k(z)^{n^k}$. Thus μ is infinitely divisible by E 12.1.

E 18.5. Let μ be non-trivial and α-stable and $\{X(t)\}$ be the corresponding Lévy process. Given $a_1 > 0$ and $a_2 > 0$, let $b = (a_1{}^\alpha + a_2{}^\alpha)^{1/\alpha}$. Then $X(a_1{}^\alpha + a_2{}^\alpha) \overset{\mathrm{d}}{=} bX(1) + c_0$, $X(a_1{}^\alpha + a_2{}^\alpha) - X(a_2{}^\alpha) \overset{\mathrm{d}}{=} X(a_1{}^\alpha) \overset{\mathrm{d}}{=} a_1 X(1) + c_1$, and $X(a_2{}^\alpha) \overset{\mathrm{d}}{=} a_2 X(1) + c_2$ with some c_0, c_1, and c_2. Note that $X(a_1{}^\alpha + a_2{}^\alpha) - X(a_2{}^\alpha)$ and $X(a_2{}^\alpha)$ are independent. This proves the 'only if' part. To prove the 'if' part, show, by induction, that, for $Z_1, Z_2, \ldots$ independent identically distributed each with μ, $Z_1 + \cdots + Z_n \overset{\mathrm{d}}{=} b_n Z_1 + c_n$ with some $b_n > 0$ and c_n, and then use E 18.4.

E 18.6. (i) Let $\mu = P_{X_1}$. Check from Theorem 14.10 $\widehat{\mu}(z)^a = \widehat{\mu}(a^{1/\alpha}z)\mathrm{e}^{\mathrm{i}\langle c_a, z\rangle}$. (ii) Replace a and t in the result (i) by t and 1, respectively.

E 18.7. Let $\{X_t\}$ be an α-stable process on $\mathbb{R}^d$ with $\alpha \neq 1$. E 18.6 and the proof of Proposition 13.14 show that $k(t)$ can be chosen as $k(t) = (t - t^{1/\alpha})\tau$. On the other hand, the proof of Theorem 14.8 shows that we can choose $k(t) = t\tau$. In general, for any $k(t)$ in Proposition 13.14, $k(t) + t^H \gamma$ with any $\gamma \in \mathbb{R}^d$ is usable in place of $k(t)$.

E 18.8. Look at the proof of Theorem 14.10 to get

$$\lambda_1 = \begin{cases} -\Gamma(-\alpha)(\cos\frac{\pi\alpha}{2})\lambda & \text{if } 0 < \alpha < 1 \text{ or } 1 < \alpha < 2, \\ \frac{\pi}{2}\lambda & \text{if } \alpha = 1. \end{cases}$$

Then we obtain the result, using the formulas $\Gamma(\frac{1+\alpha}{2})\Gamma(\frac{1-\alpha}{2}) = \pi/\cos\frac{\pi\alpha}{2}$ and $\Gamma(1-\alpha) = \pi^{-1/2}2^{-\alpha}\Gamma(\frac{1-\alpha}{2})\Gamma(\frac{2-\alpha}{2})$.

E 18.9. We have, by Theorem 14.10 and E 18.8, $c = c_\alpha \int_S |\langle\zeta, \xi\rangle|^\alpha \lambda(\mathrm{d}\xi)$ for $\zeta \in S$. Choosing ζ to be the vector e_1 in the solution of E 18.2, we get

$$c = c_\alpha \textstyle\int_S |\xi_1|^\alpha \lambda(\mathrm{d}\xi).$$

If $d = 1$, the last integral equals 1. Let $d \geq 2$ and let $b(d)$ be the surface area of S in $\mathbb{R}^d$. Using polar coordinates in $\mathbb{R}^d$, we get

$$\int_S |\xi_1|^\alpha \lambda(\mathrm{d}\xi) = b(d)^{-1} \int |\cos\theta_1|^\alpha \sin^{d-2}\theta_1 \sin^{d-3}\theta_2 \ldots \sin\theta_{d-2} \mathrm{d}\theta_1 \mathrm{d}\theta_2 \ldots \mathrm{d}\theta_{d-1},$$
$$b(d) = \int \sin^{d-2}\theta_1 \sin^{d-3}\theta_2 \ldots \sin\theta_{d-2} \mathrm{d}\theta_1 \mathrm{d}\theta_2 \ldots \mathrm{d}\theta_{d-1},$$

where the integrals in the right-hand sides are over $0 \leq \theta_j \leq \pi$ for $j = 1, \ldots, d-2$ and $0 \leq \theta_{d-1} \leq 2\pi$. Hence

$$\int_S |\xi_1|^\alpha \lambda(\mathrm{d}\xi) = \int_0^\pi |\cos\theta_1|^\alpha \sin^{d-2}\theta_1 \mathrm{d}\theta_1 \ / \ \int_0^\pi \sin^{d-2}\theta_1 \mathrm{d}\theta_1$$
$$= B(\tfrac{\alpha+1}{2}, \tfrac{d-1}{2}) \ / \ B(\tfrac{1}{2}, \tfrac{d-1}{2})$$

and we get the result.

E 18.10. Since $X_t \stackrel{\mathrm{d}}{=} t^{1/\alpha} X_1$, $P[X_t > 0]$ does not depend on t. Its evaluation is done in [**608**], p. 79, and [**65**].

E 18.11 ([**76**]). In the case $\alpha \neq 1$, the proof is direct from Theorems 14.1, 14.2, and 14.7. Consider the case that $\alpha = 1$ and μ is non-trivial. It follows from the strict 1-semi-stability with span b that $\int_{1<|x|\leq b} x\nu(\mathrm{d}x) = 0$ (Theorem 14.7). Suppose that $\widehat{\mu}(z)^{b'} = \widehat{\mu}(b'z)\mathrm{e}^{\mathrm{i}\langle c,z\rangle}$ with $c \neq 0$. Then $b' \int_{1/b'<|x|\leq 1} x\nu(\mathrm{d}x) = c$. Let $C = \limsup_{a\downarrow 0} \left|\int_{a<|x|\leq 1} x\nu(\mathrm{d}x)\right|$. We have $\int_{b^{-n}<|x|\leq 1} x\nu(\mathrm{d}x) = 0$ and $\int_{(b')^{-n}<|x|\leq 1} x\nu(\mathrm{d}x) = nc$ for $n \in \mathbb{N}$. The latter implies $C = \infty$. But the former implies $C < \infty$, since $\int_{a<|x|\leq b^{-n}} x\nu(\mathrm{d}x) = \int_{b^n a<|x|\leq 1} x\nu(\mathrm{d}x)$ for every $a \in (b^{-n-1}, b^{-n})$. This is a contradiction.

E 18.12. We have $\widehat{\mu}(z)^{a^2} = \widehat{\mu}(-a^{1/\alpha}z)^a \mathrm{e}^{\mathrm{i}\langle ac,z\rangle} = \widehat{\mu}(a^{2/\alpha}z)\mathrm{e}^{\mathrm{i}\langle(a-a^{1/\alpha})c,z\rangle}$. Therefore, μ is α-semi-stable and (iii) holds. The assertion (ii) follows from E 18.11, since $a - a^{1/\alpha} \neq 0$ for $\alpha \neq 1$. See [**556**] for (iv).

E 18.13. Let μ be rotation invariant and selfdecomposable. Then we can choose, in the representation of ν in Theorem 15.10, the surface area measure λ on S and a function $k_\xi(r) = k(r)$ independent of ξ. It follows that $\nu(B) = \int_B |x|^{-d} k(|x|)\mathrm{d}x$, since $\mathrm{Leb}(B) = \int_S \lambda(\mathrm{d}\xi) \int_0^\infty r^{d-1} 1_B(r\xi)\mathrm{d}r$. The converse is similar.

E 18.14 (Loève [**327**]). Let $\{Z_n\}$ be independent and identically distributed. Suppose that each Z_n is bounded and non-constant. Then the distribution ρ of Z_n is not infinitely divisible (Corollary 24.4). Let μ be the distribution of $X = \sum_{n=1}^\infty Z_n b^{1-n}$ with $b > 1$. Then $\widehat{\mu}(z) = \widehat{\mu}(b^{-1}z)\widehat{\rho}(z)$.

E 18.15. By integration by parts, the assertion $c = 1 - \gamma$ is equivalent to

$$\int_0^1 r^{-1}(1 - \cos r)\mathrm{d}r - \int_1^\infty r^{-1} \cos r \mathrm{d}r = \gamma,$$

found in [**169**], Formula 3.782.1.

E 18.16. The function $\psi(z)$ for $\{Z_t\}$ satisfies $b\psi(z) = \psi(b^{1/\alpha}z)$ for every $b > 0$. Hence $\int_0^t \psi(\mathrm{e}^{-cs}z)\mathrm{d}s = \int_0^t \mathrm{e}^{-c\alpha s}\psi(z)\mathrm{d}s \to \frac{1}{c\alpha}\psi(z)$.

E 18.17. The transition function $P_t(x, B)$ of the process of Ornstein–Uhlenbeck type satisfies

$$\int P_t(x, \mathrm{d}y)\mathrm{e}^{\mathrm{i}\langle z,y\rangle} = \exp[\mathrm{i}\langle \mathrm{e}^{-t/\alpha}x, z\rangle + (1 - \mathrm{e}^{-t})\psi(z)].$$

We have $X_t \stackrel{\mathrm{d}}{=} X_0 = Z_1$, since $Z_{bt} \stackrel{\mathrm{d}}{=} b^{1/\alpha} Z_t$ for $b > 0$. The Markov property of $\{X_t\}$ follows from that of $\{Z_t\}$. For any bounded measurable f and g and for $s \in \mathbb{R}$ and $t > 0$,

$$\begin{aligned} E[f(X_s)g(X_{s+t})] &= E[f(\mathrm{e}^{-s/\alpha}Z(\mathrm{e}^s))g(\mathrm{e}^{-(s+t)/\alpha}Z(\mathrm{e}^{s+t}))] \\ &= \int \mu(\mathrm{d}x)f(x)E[g(\mathrm{e}^{-t/\alpha}x + \mathrm{e}^{-(s+t)/\alpha}(Z(\mathrm{e}^{s+t}) - Z(\mathrm{e}^s)))] \\ &= \int \mu(\mathrm{d}x)f(x)\int P_t(x,\mathrm{d}y)g(y), \end{aligned}$$

since $\mathrm{e}^{-(s+t)/\alpha}(Z(\mathrm{e}^{s+t}) - Z(\mathrm{e}^s)) \stackrel{\mathrm{d}}{=} \mathrm{e}^{-(s+t)/\alpha}Z(\mathrm{e}^{s+t} - \mathrm{e}^s) \stackrel{\mathrm{d}}{=} Z(1-\mathrm{e}^{-t})$.

E 18.18 Let $t > 0$. Since $X_{1/t} \stackrel{\mathrm{d}}{=} t^{-1/\alpha}X_1$ and $X_t \stackrel{\mathrm{d}}{=} t^{1/\alpha}X_1$, we have $t^{2/\alpha}X_{1/t} \stackrel{\mathrm{d}}{=} X_t$. Let $\mu = P_{X_1}$. If $\{X_t\}$ is symmetric and if $\{Y_t\}$ has stationary increments, then $X_1 = Y_1 \stackrel{\mathrm{d}}{=} Y_2 - Y_1 = -(X_1 - X_{1/2}) + (2^{2/\alpha}-1)X_{1/2} \stackrel{\mathrm{d}}{=} (X_1 - X_{1/2}) + (2^{2/\alpha}-1)X_{1/2}$, which leads to $\widehat{\mu}(z) = \widehat{\mu}((2^{2/\alpha}-1)z)$, a contradiction. If $\{X_t\}$ is not symmetric, consider the symmetrization.

E 18.19. [**325**], Chapter 2, Section 6. The selfdecomposability is evident from the expression of the Lévy measure; see Theorem 2 of [**475**] for another proof. It also follows from E 29.16 if $a^{-1}b = 1$ and $c = 1$. By (17.16) the corresponding measure ρ equals $c\{b\mathrm{e}^{bx} + (a-b)\mathrm{e}^{(a+b)x}\}(1-\mathrm{e}^{ax})^{-2}1_{(-\infty,0)}(x)\mathrm{d}x$.

Chapter 4

E 22.1. By Theorem 19.2, $P[Y_t \geq a] = P[J((0,t]\times[a,\infty)) \geq 1] = 1 - \mathrm{e}^{-t\nu([a,\infty))}$.

E 22.2. Let $S_0 = 0$, $S_n = \sum_{j=1}^n h(U_j)$ for $n = 1, 2, \ldots$, and $X_t = S_{N(t)}$. For $u \geq 0$,

$$\begin{aligned} E[\mathrm{e}^{-uX(t)}] &= \textstyle\sum_{n=0}^\infty E\big[\exp(-u\sum_{j=1}^n h(U_j)) \mid N_t = n\big]P[N_t = n] \\ &= \textstyle\sum_{n=0}^\infty E\big[\exp(-u\sum_{j=1}^n h(V_j))\big]P[N_t = n] \end{aligned}$$

by Proposition 3.4, where $V_1, V_2, \ldots$ are independent random variables, each uniformly distributed on $[0,t]$. Hence

$$E[\mathrm{e}^{-uX(t)}] = \exp\textstyle\int_0^t(\mathrm{e}^{-uh(s)}-1)\mathrm{d}s.$$

Letting $t \to \infty$, we get

$$E[\mathrm{e}^{-uX(\infty)}] = \exp\textstyle\int_0^\infty(\mathrm{e}^{-uh(s)}-1)\mathrm{d}s = \exp\int_0^\infty(\mathrm{e}^{-ux}-1)\nu(\mathrm{d}x) = L_\mu(u).$$

If μ is strictly α-stable and supported on $[0,\infty)$, then $\nu((x,\infty)) = ax^{-\alpha}$ with some $a > 0$.

E 22.3. Define $S_0 = 0$, $S_n = \sum_{j=1}^n Y_j U_j^{-1/\alpha}$ for $n = 1, 2, \ldots$, and $X_t = S_{N(t)}$. The idea in the solution of E 22.2 gives

$$\begin{aligned} E[\mathrm{e}^{\mathrm{i}zX(t)}] &= \exp\textstyle\int_{\mathbb{R}}\lambda(\mathrm{d}y)\int_0^t(\mathrm{e}^{\mathrm{i}zys^{-1/\alpha}}-1)\mathrm{d}s \\ &= \exp\textstyle\int_{\mathbb{R}}\lambda(\mathrm{d}y)\int_{t^{-1/\alpha}}^\infty(\mathrm{e}^{\mathrm{i}xyz}-1)\alpha x^{-\alpha-1}\mathrm{d}x, \end{aligned}$$

where λ is the distribution of Y_n. It follows that

$$\lim_{t\to\infty} E[\mathrm{e}^{\mathrm{i}zX(t)}] = \exp\big[-c|z|^\alpha\textstyle\int_{\mathbb{R}}|y|^\alpha\lambda(\mathrm{d}y)\big]$$

with some $c > 0$. The almost sure convergence of S_n needs an additional argument. See Rosinski [**421**].

E 22.4. The expression of the characteristic function of X_t is obtained similarly to E 22.3. For the proof that $\{X_t\}$ is an additive process, see Lemmas 2.1 and 2.2 of [**421**].

E 22.5. Let $A = \Omega_0 \cap \{\omega\colon X_t(\omega)$ is not continuous in $t\}$, where Ω_0 is that of Definition 1.6. Suppose that $P[A] > 0$. We can choose t_0 such that $P[A_1] > 0$ for $A_1 = \Omega_0 \cap \{\omega\colon X_t(\omega)$ is not continuous in $t \in [0, t_0]\}$. Let $A_n = \Omega_0 \cap \{\omega\colon X_t(\omega)$ is not continuous in $t \in [(n-1)t_0, nt_0]\}$. By Proposition 10.7 the events A_n, $n = 1, 2, \ldots$, are independent and $P[A_n] = P[A_1]$. Since $A = \bigcup_{n=1}^{\infty} A_n$, $P[\Omega \setminus A] = \prod_{n=1}^{\infty} P[\Omega \setminus A_n] = 0$ and $P[A] = 1$. The other assertions are proved similarly.

E 22.6. To get a simple example let $\{X_t^0\}$ be a compound Poisson process on $\mathbb{R}$ with Lévy measure ν satisfying $\nu((0,\infty)) > 0$ and $\nu((-\infty, 0)) > 0$, and let $h(t)$ be a strictly increasing bounded function on $[0, \infty)$ with $h(0) = 0$. Consider $\{X_t\}$ defined by $X_t = X_{h(t)}^0$.

E 22.7. By Theorem 21.3, $\nu(\mathbb{R}^d) < \infty$. Hence $X_t = X_t^3 + X_t^4$, where $\{X_t^3\}$ is a compound Poisson or a zero process and $\{X_t^4\}$ is generated by $(A, 0, \gamma_0)$. If $A \neq 0$, then, by Theorem 21.9, the variation function of X_t immediately becomes ∞. Hence $X_t^4 = t\gamma_0$. It follows from the assumption that $\gamma_0 = 0$.

E 22.8. Use Propositions 10.7 and 14.5 and Theorems 21.1, 21.3, and 21.9.

E 22.9. Use the definitions of V_t and of α-(semi-)stability together with Lemma 21.8(iv). Another proof is to use Theorem 21.9 and the form of the Lévy measure of α-(semi-)stable process.

E 22.10. Let $Z_t = Z_t^1 + Z_t^2$ be the Lévy–Itô decomposition, where $\{Z_t^1\}$ and $\{Z_t^2\}$ are the jump part and the continuous part, respectively. Hence $X_t = X_t^1 + X_t^2$, where X_t^1 and X_t^2 are the first components of Z_t^1 and Z_t^2, respectively. Similarly $Y_t = Y_t^1 + Y_t^2$. $\{X_t^1\}$ and $\{X_t^2\}$ are independent, since $\{Z_t^1\}$ and $\{Z_t^2\}$ are. $\{X_t^2\}$ is purely non-Gaussian by Proposition 11.10. Since $\{X_t\}$ is Gaussian, $\{X_t^2\}$ must be a trivial process. Hence $\{X_t\}$ is a function of $\{Z_t^1\}$. Similarly $\{Y_t\}$ is a function of $\{Z_t^2\}$. Hence $\{X_t\}$ and $\{Y_t\}$ are independent.

E 22.11. This is an extension of Theorem 21.5 in Skorohod's book [**493**].

Chapter 5

E 29.1. If $S_X = \{1, 2, 3, \ldots\}$ and $S_Y = \{-n - n^{-1}\colon n = 2, 3, \ldots\}$, then 0 is in $\overline{S_X + S_Y}$ but not in $S_X + S_Y$. If K is compact and F is closed, then $K + F$ is closed.

E 29.2. Similar to E 29.3.

E 29.3. The 'if' part. If $S_\mu \subset a + V$ for some a and some linear subspace V, then $x_j - a \in V$ for $j = 0, \ldots, d$ and hence $x_j - x_0 \in V$ for $j = 1, \ldots, d$, which implies $\dim V = d$.

The 'only if' part. We assume μ is nondegenerate. First, there are x_0 and x_1 in S_μ such that $x_0 \neq x_1$. Suppose that, for some $1 \leq k \leq d-1$, there are $x_0, \ldots, x_k$ in S_μ such that $x_1 - x_0, \ldots, x_k - x_0$ are linearly independent. Then S_μ is not contained in $x_0 + V_k$, where V_k is the linear subspace spanned by

$x_1 - x_0, \ldots, x_k - x_0$. Hence we can find $x_{k+1} \in S_\mu$ such that $x_{k+1} \notin x_0 + V_k$. Now $x_1 - x_0, \ldots, x_{k+1} - x_0$ are linearly independent.

E 29.4. (i) If there are x_n in $\mathfrak{G}$ such that $x_n \neq 0$ and $x_n \to 0$, then $kx_n \in \mathfrak{G}$ for $k \in \mathbb{Z}$ and $\mathfrak{G}$ is dense, hence $\mathfrak{G} = \mathbb{R}$. If 0 is not a cluster point of $\mathfrak{G}$, then there is an element $a \neq 0$ of $\mathfrak{G}$ nearest to 0, and hence $\mathfrak{G} = a\mathbb{Z} = |a|\mathbb{Z}$.

(ii) To show the 'if' part, apply Corollary 24.6 to $a^{-1}X_t$ to find $S(X_t) \subset a\mathbb{Z}$. The 'only if' part is proved as follows. Since $S(X_t) \subset a\mathbb{Z}$, Corollary 24.6 tells us that $A = 0$ and $S_\nu \subset a\mathbb{Z}$. Hence $\{X_t\}$ is a compound Poisson process with a drift γ_0 added. But γ_0 must be 0, as $\gamma_0 t \in S(X_t) \subset a\mathbb{Z}$ for any $t \geq 0$.

(iii) The 'if' part is proved as follows. By (ii), $\mathfrak{G}$ is a subgroup of $a\mathbb{Z}$. Hence $\mathfrak{G} = ak\mathbb{Z}$ with some $k \in \mathbb{N}$. Since $S_\nu \subset ak\mathbb{Z}$ by (ii), k must be 1. To show the 'only if' part, see that, by (ii), $A = 0$, $\gamma_0 = 0$, and $S_\nu \subset a\mathbb{Z}$, and that, if $S_\nu \subset a'\mathbb{Z}$ with some $a' > a$, then $a\mathbb{Z} = \mathfrak{G} \subset a'\mathbb{Z}$ by (ii), a contradiction.

(iv) Let ρ be the counting measure on $a\mathbb{Z}$. For any $n \in \mathbb{Z}$, $\rho(\{an\}) = 1$. Hence $\sum_{k \in \mathbb{Z}} \rho(\{ak\}) P[ak + X_t = an] = P[X_t \in a\mathbb{Z}] = 1 = \rho(\{an\})$.

E 29.5. Let $1/2 < p < 1$ and $q = 1 - p$. Let ν be a probability measure, $\nu(\{1\}) = p$ and $\nu(\{-1\}) = q$. Let $\{X_t\}$ be the compound Poisson process with Lévy measure ν. Then the measure ρ on $\mathbb{Z}$ defined by $\rho(\{n\}) = (p/q)^n$ is an invariant measure of $\{X_t\}$. To see this, consider the random walk $\{S_n\}$ on $\mathbb{Z}$ with $P_{S_1} = \nu$. Then $\sum_k \rho(\{k\}) P[k+S_1 = n] = \rho(\{n\})$, and hence $\sum_k \rho(\{k\}) P[k+S_r = n] = \rho(\{n\})$ for $r \in \mathbb{N}$.

E 29.6. Suppose that a non-zero Lévy process $\{X_t\}$ has an invariant distribution ρ. Then $\int \rho(\mathrm{d}x)\mu^t(B - x) = \rho(B)$ for $B \in \mathcal{B}(\mathbb{R}^d)$. If B is compact, then $\mu^t(B - x) \to 0$ as $t \to \infty$ by (17.25). Hence $\rho = 0$, a contradiction.

E 29.7. The representation of μ in Theorem 14.19 has $c_1 = 1$ and $\theta = \frac{2-\alpha}{\alpha}$. Hence $\mu[0, \infty) = 1/\alpha$ by E 18.10. The dual of μ in Remark 14.21 has $\alpha' = 1/\alpha \in (1/2, 1)$, $\theta' = 1$, and thus $\beta' = 1$. The density $p_\mu(x)$ of μ satisfies, by (14.41),

$$p_\mu(x) = x^{-1-\alpha} p(x^{-\alpha}, (1/\alpha, 1, 1)_Z), \qquad x > 0.$$

Since $p(\cdot, (\alpha', 1, 1)_Z)$ is the density of the distribution discussed at the end of Example 24.12, the assertion follows.

E 29.8. Let ν be a finite measure on $[1, \infty)$ such that $\int g(x)\nu(\mathrm{d}x) < \infty$ and $\int g(x)\mathrm{e}^{h(x)}\nu(\mathrm{d}x) = \infty$. (For example, let $c_k > 0$ satisfy $\sum_{k=1}^\infty c_k < \infty$ and choose $1 \leq a_1 < a_2 < \cdots$ such that $\mathrm{e}^{h(a_k)} c_k \geq 1$. Then $\nu = \sum_{k=1}^\infty g(a_k)^{-1} c_k \delta_{a_k}$ meets the requirement.) Let $\{X_t\}$ be the compound Poisson process with Lévy measure ν. Then $E[g(X_t)] = \infty$, because

$$\textstyle\int g(x)\nu^2(\mathrm{d}x) = \iint \mathrm{e}^{(x+y)h(x+y)}\nu(\mathrm{d}x)\nu(\mathrm{d}y) \geq \int \nu(\mathrm{d}y) \int g(x)\mathrm{e}^{yh(x)}\nu(\mathrm{d}x) = \infty.$$

E 29.9. Use Theorem 26.1.

E 29.10. If μ is not Gaussian, we can use Theorem 26.1 and Remark 26.3. Gaussian distributions do not satisfy the condition $\int_{|x|>r} \mu(\mathrm{d}x) \sim c\mathrm{e}^{-r^\alpha}$, since the Gaussian on $\mathbb{R}$ with mean 0 and variance 1 satisfies $\mu(x, \infty) \sim cx^{-1}\mathrm{e}^{-x^2/2}$ as $x \to \infty$ (see [**138**]).

E 29.11 See Millar [**347**], pp. 55–57.

E 29.12. We have $\int x^\beta \nu(\mathrm{d}x) = \sum_{n=1}^\infty 2^{-c^n(\beta-\alpha)} < \infty$ for $\beta > \alpha$ and $\int x^\alpha \nu(\mathrm{d}x) = \sum_{n=1}^\infty 1 = \infty$. For each $t > 0$, P_{X_t} is either absolutely continuous or continuous singular by Theorem 27.16. Let $z_k = 2\pi {a_k}^{-1}$, $k = 1, 2, \ldots$. Then $|\widehat{\mu}(z_k)|^t \to 1$ as $k \to \infty$. The proof is given in Example 41.23. Hence P_{X_t} is not absolutely continuous, by the Riemann–Lebesgue theorem.

E 29.13. Decompose $\mu = \mu_d + \mu_c$ where μ_d is discrete and μ_c is continuous. Then we can show that either $\mu_d = 0$ or $\mu_c = 0$. This is done by a method similar to the proof of the alternative of $\mu_{ac} = 0$ and $\mu_{cs} = 0$ in Theorem 27.15. Next, suppose that μ is discrete. Let us show a contradiction. It suffices to consider the univariate case, considering projections. By Proposition 27.28, $|\widehat{\mu}(z_k)| \to 1$ along some $z_k \to \infty$. Choose $k_n \in \mathbb{Z}_+$ with $b^{k_n} \le z_k < b^{k_n+1}$. Then $1 \ge |\widehat{\mu}(b^{-k_n} z_k)| \ge |\widehat{\mu}(z_k)| \to 1$. Hence there is $z_0 \in [1, b]$ with $|\widehat{\mu}(z_0)| = 1$. Hence μ is supported on $\gamma + 2\pi {z_0}^{-1}\mathbb{Z}$ with some $\gamma \in \mathbb{R}$. This is impossible, because, for any $n \in \mathbb{N}$, μ has $T_{b^{-n}}\mu$ as a convolution factor and S_μ contains two points with arbitrarily small distance. Finally, $\mu_c = \mu_{ac} + \mu_{cs}$ implies $\mu_{ac} = 0$ or $\mu_{cs} = 0$ as in the proof of Theorem 27.15.

E 29.14. A nondegenerate Gaussian is a convolution factor of $\mu^t = P_{X_t}$. Thus $|\widehat{\mu}(z)^t| \le \mathrm{e}^{-ct|z|^2}$ with some constant $c > 0$. Apply Proposition 28.1.

E 29.15. It follows from $\beta = 1$ that the Lévy measure is concentrated on $(0, \infty)$. Hence $\int_{\mathbb{R}} \mathrm{e}^{\eta x}\mu(\mathrm{d}x) < \infty$ by Theorem 25.17. To obtain $\Psi(w)$ in (25.12) for $w \in \mathbb{C}$ with $\mathrm{Re}\, w \le 0$, recall the proof of the theorem. We have only to find the analytic function such that $\mathrm{e}^{\Psi(\mathrm{i}z)}$, $z \in \mathbb{R}$, equals the right-hand side of (14.24) or (14.25) with $\beta = 1$. That is

$$\Psi(w) = \begin{cases} -c|w|^\alpha \mathrm{e}^{\mathrm{i}\alpha(\arg w - \pi/2)}(1 - \mathrm{i}\tan\frac{\pi\alpha}{2}) + \tau w, & \text{for } \alpha \ne 1, \\ \mathrm{i}cw(1 + \mathrm{i}\frac{2}{\pi}(\log|w| + \mathrm{i}(\arg w - \frac{\pi}{2})) + \tau w, & \text{for } \alpha = 1, \end{cases}$$

where $-\pi/4 < \arg w \le 7\pi/4$.

E 29.16. $P[(Z/X)^\alpha > u] = P[Z > Xu^{1/\alpha}] = \int_0^\infty \mathrm{e}^{-xu^{1/\alpha}} P_X(\mathrm{d}x) = \mathrm{e}^{-u} = P[Z > u]$ for $u \ge 0$.

E 29.17 ([**475**]). Let X and Z be those of E 29.16. Then $Z^{-\eta}X^\eta \stackrel{\mathrm{d}}{=} Z^{-\eta/\alpha}$ for $\eta \in \mathbb{R}$. If $\eta < \alpha$, then $E[Z^{-\eta}] = \Gamma(1-\eta)$, $E[Z^{-\eta/\alpha}] = \Gamma(1 - \frac{\eta}{\alpha})$, and $E[Z^{-\eta}]E[X^\eta] = E[Z^{-\eta/\alpha}]$.

E 29.18 See Feller [**139**], p. 453. Another proof, in [**56**], is to use (25.5) and to observe that, for $Y \ge 0$, $E[\mathrm{e}^{-uY}] = \sum_{n=0}^\infty \frac{(-1)^n}{n!} E[Y^n]u^n$, $u \ge 0$, whenever the right-hand side converges on $[0, \infty)$ ([**139**], p. 234).

E 29.19 Let μ_α be the Mittag–Leffler distribution with parameter α. Then $\int \mathrm{e}^{wx}\mu_\alpha(\mathrm{d}x) = E_\alpha(w)$ for $w \in \mathbb{C}$ and $E_\alpha(w) \sim (1/\alpha)\mathrm{e}^{w^{1/\alpha}}$ as $\mathbb{R} \ni w \to \infty$ by the theory of entire functions. Then Kasahara's Tauberian theorem of exponential type [**262**] shows that $-\log \mu_\alpha(x, \infty) \sim cx^{1/(1-\alpha)}$ as $x \to \infty$ with a positive finite constant c. See [**38**], pp. 253, 329, 337. Non-Gaussian infinitely divisible distributions cannot have this tail, by Theorem 26.1. Another proof, in [**56**], is to use the results of E 50.7 and Remark 14.18.

E 29.20. Let F and F_n be the distribution functions of μ and μ_n. Let $\{\mu_{n_k}\}$ be a subsequence of $\{\mu_n\}$ such that a_{n_k} tends to some $c \in [-\infty, \infty]$. For every choice of $x_1 < x_2 < x_3 < c$ we have

$$(x_2 - x_1)^{-1}(F_{n_k}(x_2) - F_{n_k}(x_1)) \le (x_3 - x_2)^{-1}(F_{n_k}(x_3) - F_{n_k}(x_2))$$

for every large k. Hence, if x_1, x_2, x_3 are continuity points of F, then

$$(x_2 - x_1)^{-1}(F(x_2) - F(x_1)) \le (x_3 - x_2)^{-1}(F(x_3) - F(x_2))$$

Then, by the right-continuity of F, this holds even if some of x_1, x_2, x_3 are not continuity points. Thus F is convex on $(-\infty, c)$. By the same argument, $\overline{F}(x) = 1 - F(x)$ is convex on (c, ∞). If $b_1 = \infty$, then F is convex on $\mathbb{R}$, which is contradictory to the fact that F is bounded, increasing, and non-constant. Similarly, if $b_0 = -\infty$, then we have a contradiction. Hence, b_0 and b_1 are finite and, for any $a \in [b_0, b_1]$, μ is unimodal with mode a.

E 29.21. If $a > 0$ or $a < 0$, then aU is uniformly distributed on $[0, a]$ or $[a, 0]$, respectively. Suppose that $X \in \{x_{-n}, \dots, x_{-1}, x_0, x_1, \dots, x_m\}$ with probability one, where $x_{-n}, \dots, x_{-1}$ are negative, $x_0 = 0$, $x_1, \dots, x_m$ are positive. Let $P[X = x_j] = p_j$. Then

$$P[UX \in B] = \textstyle\sum_j P[x_j U \in B,\, X = x_j] = \sum_j p_j P[x_j U \in B].$$

Hence

$$P_{UX} = p_0\delta_0 + \textstyle\sum_{k=1}^{m} p_k {x_k}^{-1} 1_{[0,x_k]}(x)\mathrm{d}x + \sum_{l=1}^{n} p_l |x_{-l}|^{-1} 1_{[x_{-l},0]}(x)\mathrm{d}x,$$

which is unimodal with mode 0. A general X can be approximated in distribution by random variables of the type above. Hence, by E 29.20, P_{UX} is unimodal with mode 0. This proof suggests a proof of the converse assertion.

E 29.22. It is enough to consider the case that each of μ_1 and μ_2 does not have a point mass at 0 and has a step function as density. The general case is proved from this by approximation using E 29.20. Denote the uniform distributions on $[-a, a]$ and $[0, a]$ by ρ_a and σ_a, respectively. Then $\rho_a = \sigma_{2a} * \delta_{-a}$, $\mu_1 = \sum_{j=1}^{n} c_j \rho_{a_j}$ with $c_j > 0$ and $a_j > 0$, and μ_2 also has a similar form. The graph of the density of $\rho_a * \rho_b = \sigma_{2a} * \sigma_{2b} * \delta_{-a-b}$ for $a > 0$ and $b > 0$ is a trapezoid or triangle, symmetric about the origin. The distribution $\mu_1 * \mu_2$ is a mixture of such distributions, and hence symmetric and unimodal with mode 0.

E 29.23. Let $f(u) = -u \log u$ for $0 \le u \le 1$ with the understanding that $0 \log 0 = 0$. Then $f(0) = f(1) = 0$, $f(u) > 0$ for $0 < u < 1$, and f is strictly concave. Let $\rho = \sum_{n=0}^{\infty} p_n \rho_n$ and $C = \bigcup_{n=0}^{\infty} C_{\rho_n}$. Then

$$H(\rho) = \textstyle\sum_{a \in C} f(\rho\{a\}) \ge \sum_a \sum_n p_n f(\rho_n\{a\}) = \sum_n p_n H(\rho_n).$$

The equality holds if and only if $\rho_n = \rho_{n'}$ whenever $p_n > 0$ and $p_{n'} > 0$. On the other hand,

$$\textstyle\sum_a f(\rho\{a\}) \le \sum_a \sum_n f(p_n \rho_n\{a\}) = \sum_n p_n H(\rho_n) + H(\sigma).$$

Hence we have (i). To see (ii), enumerate C_{X_2} as $a_0, a_1, \dots$ and let $p_n = P[X_2 = a_n]$. Then $P[X_1 + X_2 \in B] = \sum_n p_n P[X_1 + a_n \in B]$ for any Borel set B and we can apply (i). An example of X with $H(X) = \infty$ is given by $P[X = n] = c_\alpha n^{-1}(\log n)^{-\alpha}$, $n = 2, 3, \dots$, with $1 < \alpha \le 2$, where c_α is a normalizing constant.

E 29.24. Since $h(t) = H(X_t - \gamma_0 t)$, we may and do assume that $\{X_t\}$ is a compound Poisson process. Let $\{N_t\}$ be the Poisson process with parameter c. It follows from E 29.23(i) and (27.1) that

$$e^{-ct}\sum_{n=0}^{\infty}(n!)^{-1}(ct)^n H(\sigma^n) \le h(t) \le e^{-ct}\sum_{n=0}^{\infty}(n!)^{-1}(ct)^n H(\sigma^n) + H(N_t).$$

Thus (i) is obvious. We can directly check that $H(N_t) < \infty$. Suppose that $H(\sigma) < \infty$. Since $H(\sigma^n) \le nH(\sigma)$, we see $h(t) < \infty$. We have

$$(*) \qquad h(t_1) < h(t_2) \le h(t_1) + h(t_2 - t_1) \quad \text{whenever} \quad t_1 < t_2$$

by the stationary independent increments property and by E 29.23(ii). Let C be the carrier of X_t for $t > 0$ described by Proposition 27.6 with $\gamma_0 = 0$, and let $p_a(t) = P[X_t = a]$ for $a \in C$. If $s_n \to t$, then $\liminf_{n\to\infty}\{X_{s_n} = a\} \cap B_0 = \limsup_{n\to\infty}\{X_{s_n} = a\} \cap B_0 = \{X_t = a\} \cap B_0$, where $B_0 = \{X_t = X_{t-}\}$. Since $P[B_0] = 1$, we have $p_a(s_n) \to p_a(t)$ as $s_n \to t$. Now note that $f(u) = -u \log u$ is increasing for $0 \le u \le e^{-1}$. Note that $e^{ct}p_a(t)$ is increasing in t for each $a \in C$ by virtue of (27.1) and that $p_0(t) \ge e^{-ct}$. Choose $\varepsilon > 0$ such that $e^{c\varepsilon}P[X_\varepsilon \ne 0] \le e^{-1}$. If $t < \varepsilon$, then $f(p_a(t)) \le f(e^{ct}p_a(t)) \le f(e^{c\varepsilon}p_a(\varepsilon))$ for $a \ne 0$. Since $\sum_{a\in C\setminus\{0\}} f(e^{c\varepsilon}p_a(\varepsilon)) < \infty$, the dominated convergence theorem tells us that $h(s) \to h(t)$ as $s \to t$ whenever $t < \varepsilon$. In particular, $h(t) \to 0$ as $t \to 0$. Hence, by the inequality $(*)$, $h(t)$ is continuous. To show the existence of $b_1 > 0$ such that $h(t) \ge b_1 \log t$ for large t, first notice that $\sup_{a\in C} p_a(t) < Kt^{-1/2}$ with some constant K by Lemma 48.3 and Remark 48.4. Write $\eta_t = Kt^{-1/2}$. Choose t so large that $2\eta_t < e^{-1}$. There is a finite partition $C^{(1)}, \ldots, C^{(N)}$ of C such that $\eta_t < \sum_{a\in C^{(j)}} p_a(t) < 2\eta_t$ for $j = 1, \ldots, N-1$ and $\sum_{a\in C^{(N)}} p_a(t) < 2\eta_t$. Thus $1 < 2\eta_t N$. By the property $f(u_1 + u_2) \le f(u_1) + f(u_2)$, we have

$$\begin{aligned} h(t) &\ge \textstyle\sum_{j=1}^{N-1} f(\sum_{a\in C^{(j)}} p_a(t)) \ge f(\eta_t)(N-1) \\ &> f(\eta_t)((2\eta_t)^{-1} - 1) \ge 4^{-1}\log t - \text{const.} \end{aligned}$$

The existence of $b_2 > 0$ with $h(t) \le b_2 t$ for large t is clear from the property $(*)$.

Chapter 6

E 34.1. Use (4.11) and (30.5).

E 34.2. By (30.5), Y_t has distribution density on $(0, \infty)$ equal to

$$\sum_{k=0}^{\infty} \frac{e^{-x}x^{t+2k-1}}{\Gamma(t+2k)} \frac{t}{2k+t} \binom{2k+t}{k} 2^{-2k-t} = e^{-x}\frac{t}{x} I_t(x).$$

We have used (4.11). Integrating the identity $\sum_{k=1}^{\infty} \binom{-1/2}{k} x^{k-1} = ((1+x)^{-1/2} - 1)/x$, $|x| < 1$, we get

$$\sum_{k=1}^{\infty} \binom{-1/2}{k} \frac{x^k}{k} = 2\log(2/(1+\sqrt{1+x})), \qquad |x| < 1.$$

Thus

$$\begin{aligned} \sum_{k=1}^{\infty}(e^{-2ku} - 1)\frac{(2k-1)!}{(k!)^2}2^{-2k} &= \sum_{k=1}^{\infty}(e^{-2ku} - 1)\binom{-1/2}{k}\frac{(-1)^k}{2k} \\ &= \log(2/(1+\sqrt{1-e^{-2u}})) - \log 2 = \log\Big[\int_{(0,\infty)} e^{-ux}\lambda_1^0(dx)\Big]. \end{aligned}$$

Hence the assertion on ρ follows. This together with (30.8) and (4.11) proves the assertion on $\nu^\sharp$. The Laplace transform of $\mu_t^\sharp$ is obtained by (30.19).

E 34.3. Use Theorem 30.1. Since $\rho(\mathrm{d}s) = s^{-1}k(s)\mathrm{d}s$ with a nonnegative decreasing function $k(s)$ by Corollary 15.11,

$$\nu^\sharp(B) = \int_0^\infty s^{-1}k(s)\mathrm{d}s \int_B (2\pi s)^{-d/2}\mathrm{e}^{-|x|^2/(2s)}\mathrm{d}x = \int_B |x|^{-d}k^\sharp(|x|)\mathrm{d}x$$

with $k^\sharp(r) = (2\pi)^{-d/2}\int_0^\infty u^{-1-d/2}\mathrm{e}^{-1/(2u)}k(r^2u)\mathrm{d}u$, which shows the selfdecomposability of $\{Y_t\}$ by E 18.13.

E 34.4. By Example 30.8 $E[\mathrm{e}^{-uY_t}] = (1+u^\alpha)^{-t}$. On the other hand, we know that $1 - E_\alpha(-x^\alpha)$ increases from 0 to 1 on $[0,\infty)$ (Example 24.12). We have

$$\int_0^\infty \mathrm{e}^{-ux}\mathrm{d}_x(1 - E_\alpha(-x^\alpha)) = u\int_0^\infty (1 - E_\alpha(-x^\alpha))\mathrm{e}^{-ux}\mathrm{d}x$$
$$= 1 - u\sum_{n=0}^\infty \int_0^\infty \tfrac{(-x^\alpha)^n}{\Gamma(n\alpha+1)}\mathrm{e}^{-ux}\mathrm{d}x = 1 - u\sum_{n=0}^\infty (-1)^n u^{-n\alpha-1} = (1+u^\alpha)^{-1}.$$

The change of the order of integration and summation here is justified if $u > 1$. The resulting identity is true for $u > 0$ by analytic continuation. Similarly, the Laplace–Stieltjes transform of the right-hand side of the asserted expression for $P[Y_t \le x]$ is shown to be equal to $(1+u^\alpha)^{-t}$. Use the result of [**375**], p. 159, for selfdecomposability.

E 34.5. We have $P[Y_t \in B] = \sum_{n=0}^\infty \mu^n(B)\lambda^t\{n\}$, where $\lambda = P_{Z_1}$. The proof that $\{Y_t\}$ is a Lévy process is similar to that of Theorem 30.1 (before Theorem 30.4). Let $\Phi(w) = \sum_{n=1}^\infty (w^n - 1)\rho\{n\}$ for complex w with $|w| \le 1$. We get $\sum_{n=0}^\infty w^n P[Z_t = n] = \mathrm{e}^{t\Phi(w)}$ for complex w with $|w| \le 1$, since both sides are analytic in $\{|w| < 1\}$ and continuous on $\{|w| \le 1\}$, and coincide for $w = \mathrm{e}^{\mathrm{i}z}$, $z \in \mathbb{R}$. Hence $E[\mathrm{e}^{\mathrm{i}\langle z,Y_t\rangle}] = \sum_{n=0}^\infty \widehat{\mu}(z)^n P[Z_t = n] = \exp[t\Phi(\widehat{\mu}(z))] = \exp\left[t\sum_{n=1}^\infty \int(\mathrm{e}^{\mathrm{i}\langle z,x\rangle} - 1)\mu^n(\mathrm{d}x)\rho\{n\}\right]$. This shows that $\{Y_t\}$ is a compound Poisson process and its Lévy measure is as asserted. If $\{Z_t\}$ is a Poisson process, then $\rho = c\delta_1$ and the construction of $\{Y_t\}$ here is exactly that of a compound Poisson process in Theorem 4.3. If $\{Z_t\}$ is such that Z_1 has a geometric distribution, then $\rho\{k\} = k^{-1}q^k$ for $k \in \mathbb{N}$ and $\mathrm{e}^{t\Phi(w)} = p^t(1-qw)^{-t}$ (Example 4.6).

E 34.6. Let $f \in \mathfrak{D}(L)$. For any $\varepsilon > 0$, there is α such that $\|f - \alpha U^\alpha f\| < \varepsilon$ and $\|Lf - \alpha U^\alpha Lf\| < \varepsilon$. Since $\alpha U^\alpha f \in \mathfrak{D}(L^2)$, this shows that $\mathfrak{D}(L^2)$ is a core of L. A similar discussion works for $\mathfrak{D}(L^n)$, using $(\alpha_{n-1}U^{\alpha_{n-1}})\ldots(\alpha_1 U^{\alpha_1})$ in place of αU^α.

E 34.7. See [**111**], p. 621.

E 34.8. It follows from (31.11) that L is unbounded if $\{X_t\}$ is neither a compound Poisson nor the zero process.

E 34.9. Use the fact that $P_tQ_s = Q_sP_t$.

E 34.10. By the convexity of $u(x) = |x|^p$, we have

$$\int\left|\int \mu^t(\mathrm{d}y)f(x+y)\right|^p\mathrm{d}x \le \int \mathrm{d}x \int \mu^t(\mathrm{d}y)|f(x+y)|^p$$
$$= \int \mu^t(\mathrm{d}y)\int |f(x+y)|^p\mathrm{d}x = \int |f(x)|^p\mathrm{d}x.$$

Hence, if $f_1(x) = f_2(x)$ a. e., then $\int \mu^t(\mathrm{d}y)f_1(x+y) = \int \mu^t(\mathrm{d}y)f_2(x+y)$ a. e. We see that $\|P_tf\| \le \|f\|$ on $L^p(\mathbb{R}^d)$. Similarly,

$$\int |P_tf(x) - f(x)|^p\mathrm{d}x = \int\left|\int \mu^t(\mathrm{d}y)(f(x+y) - f(x))\right|^p\mathrm{d}x$$
$$\le \int \mu^t(\mathrm{d}y)\int |f(x+y) - f(x)|^p\mathrm{d}x,$$

which tends to 0 as $t \downarrow 0$, since $\int |f(x+y) - f(x)|^p \mathrm{d}x$ is bounded and continuous in y.

E 34.11. Note that

$$\iint f(x+y)g(x)\mathrm{d}x\mu^t(\mathrm{d}y) = \iint f(x)g(x-y)\mathrm{d}x\mu^t(\mathrm{d}y).$$

E 34.12. The function $\varphi(x)$ satisfies (33.13)–(33.15). Use Corollary 15.11.

E 34.13. Use the formula (30.28) and $K_\lambda = K_{-\lambda}$ to determine c. If $\chi = 0$ or $\psi = 0$, then use $K_\lambda(x) \sim \Gamma(\lambda)2^{\lambda-1}x^{-\lambda}$ for $\lambda > 0$ as $x \downarrow 0$ in 9.6.9 of [**1**]. The Laplace transform is obtained from the expression for the normalizing constant. For infinite divisibility and selfdecomposability, see [**14**], [**178**]. Another proof of infinite divisibility in the case $\lambda \le 0$ is found in [**13**].

E 34.14. Let $\{X_t\}$ be a Brownian motion with drift γ and $\{Z_t\}$ be a subordinator with distribution at time 1 being the generalized inverse Gaussian with $\lambda = 1$. The distribution density at time 1 of the process $\{Y_t\}$ subordinate to $\{X_t\}$ by $\{Z_t\}$ is $\exp(-\sqrt{\psi+\gamma^2}\sqrt{\chi+x^2}+\gamma x)$ multiplied by a normalizing constant. The calculation to see this is reduced to the Laplace transform in Example 2.13. Hence the distribution with density $g(x)$ is infinitely divisible. The selfdecomposability in the case $\gamma = 0$ is a consequence of E 34.3 and E 34.13. See [**178**] for the case $\gamma \neq 0$.

E 34.15. By (30.5), the density of μ^t equals

$$(\Gamma(t))^{-1}r^t \int_0^\infty s^{t-1}\mathrm{e}^{-rs}(2\pi s)^{-d/2}\mathrm{e}^{-|x|^2/(2s)}\mathrm{d}s,$$

which is calculated by (30.28).

Chapter 7

E 39.1. Calculate

$$v(x) = \int_0^\infty (2\pi t)^{-1/2}\mathrm{e}^{-(x-t\gamma)^2/(2t)}\mathrm{d}t = (2\pi)^{-1/2}\mathrm{e}^{\gamma x}\int_0^\infty \mathrm{e}^{-x^2/(2t)-t\gamma^2/2}t^{-1/2}\mathrm{d}t,$$

using the formula

$$\int_0^\infty \mathrm{e}^{-vt-1/(2t)}t^{-1/2}\mathrm{d}t = (2\pi)^{1/2}(2v)^{-1/2}\mathrm{e}^{-(2v)^{1/2}}, \quad v > 0$$

from Example 2.13.

E 39.2. We have $\psi(z) = -c|z|^\alpha(1 - \mathrm{i}b\operatorname{sgn} z)$ with $b = \beta\tan\frac{\pi\alpha}{2}$. Let $\rho(\mathrm{d}x) = \pi^{-1}(x^2+1)^{-1}\mathrm{d}x$, the Cauchy distribution. The characteristic function $\mathrm{e}^{-|z|}r(r - \psi(z))^{-1}$ of $\rho * (rV^r)$ is integrable. Hence

$$\int_{-\infty}^\infty ((y-x)^2+1)^{-1}v^r(y)\mathrm{d}y = \tfrac{1}{2}\int_{-\infty}^\infty \mathrm{e}^{-\mathrm{i}xz-|z|}(r-\psi(z))^{-1}\mathrm{d}z.$$

Letting $r \downarrow 0$, we get

$$\begin{aligned}\int_{-\infty}^\infty ((y-x)^2+1)^{-1}v(y)\mathrm{d}y &= \tfrac{1}{2}\int_{-\infty}^\infty \mathrm{e}^{-\mathrm{i}xz-|z|}(-\psi(z))^{-1}\mathrm{d}z \\ &= c^{-1}\Gamma(1-\alpha)(1+b^2)^{-1}\operatorname{Re}\left[(1-\mathrm{i}x)^{\alpha-1}(1-\mathrm{i}b)\right],\end{aligned}$$

where $(1-\mathrm{i}x)^{\alpha-1} = \mathrm{e}^{(\alpha-1)\log(1-\mathrm{i}x)}$ with log taken as the principal value (Example 2.15). Let

$$w(x) = |x|^{\alpha-1}(A1_{(0,\infty)}(x) + B1_{(-\infty,0)}(x)),$$

where we choose A and B so that $w(x)$ is equal to the asserted expression for $v(x)$ on $\mathbb{R}\setminus\{0\}$. Then we can get

$$\int_{-\infty}^{\infty}((y-x)^2+1)^{-1}w(y)\mathrm{d}y=\tfrac{\pi}{\sin\pi\alpha}\mathrm{Re}\,[(1-\mathrm{i}x)^{\alpha-1}(A\mathrm{e}^{-\mathrm{i}\pi\alpha/2}+B\mathrm{e}^{\mathrm{i}\pi\alpha/2})],$$

using 3.252.12 of [**169**]. It follows that

$$\int_{-\infty}^{\infty}((y-x)^2+1)^{-1}v(y)\mathrm{d}y=\int_{-\infty}^{\infty}((y-x)^2+1)^{-1}w(y)\mathrm{d}y,$$

since $\Gamma(\alpha)\Gamma(1-\alpha)=\pi/\sin\pi\alpha$. We see that $\int_{-\infty}^{\infty}(|y|^{-2}\wedge 1)v(y)\mathrm{d}y<\infty$. The strict α-stability of $\{X_t\}$ implies $v(ax)=a^{\alpha-1}v(x)$ for $a>0$. The function $w(x)$ satisfies the same relation. Hence

$$\pi^{-1}s\int_{-\infty}^{\infty}((y-x)^2+s^2)^{-1}v(y)\mathrm{d}y=\pi^{-1}s\int_{-\infty}^{\infty}((y-x)^2+s^2)^{-1}w(y)\mathrm{d}y$$

for $s>0$. For any continuous function $h(x)$ with compact support, we have

$$\lim_{s\downarrow 0}\pi^{-1}s\iint h(x)((y-x)^2+s^2)^{-1}v(y)\mathrm{d}x\mathrm{d}y=\int h(y)v(y)\mathrm{d}y$$

and the same convergence with $w(y)$ in place of $v(y)$. This is because

$$\lim_{s\downarrow 0}\pi^{-1}s\int h(x)((y-x)^2+s^2)^{-1}\mathrm{d}x=h(y)\quad\text{boundedly}$$

and

$$\sup_{0<s\le 1}\ \pi^{-1}s\int|h(x)|((y-x)^2+s^2)^{-1}\mathrm{d}x\le\mathrm{const}\,(|y|^{-2}\wedge 1).$$

Hence $v(y)=w(y)$ a. e.

E 39.3. Use Theorem 37.16 and Corollary 37.17 together with Theorem 14.7. To obtain a non-symmetric ν of a 1-semi-stable process satisfying $\int_{1<|x|\le b}x\nu(\mathrm{d}x)=0$, choose, for example, $k>k'>0$ such that $k'\int_{-b}^{-b'}x\mathrm{d}x+k\int_1^{b'}x\mathrm{d}x=0$ with $b'=(1+b)/2$ and extend $[\nu]_{\{1<|x|\le b\}}=k'1_{(-b,-b')}(x)\mathrm{d}x+k1_{(1,b')}(x)\mathrm{d}x$ to $\mathbb{R}$ by $b\nu=T_b\nu$.

E 39.4. For every $a>0$ and $t>0$, $P[|X_s|<a$ for some $s>t]=E[(P[|x+X_s|<a$ for some $s>0])_{x=X_t}]=1$. Hence $P[\liminf_{t\to\infty}|X_t|=0]=1$, that is, $\{X_t\}$ is recurrent. If $\mathfrak{G}=\mathbb{R}^d$, then the converse follows from Theorem 35.8(i) and (ii).

E 39.5. By Theorem 37.8 $\{X_t\}$ is not genuinely d-dimensional. This means that Σ is contained in a proper linear subspace V of $\mathbb{R}^d$. If V has dimension ≥ 3, then repeat this procedure.

E 39.6. Notice that $-\psi_X\le-\mathrm{Re}\,\psi_Y\le|\psi_Y|$, and use Corollary 37.6.

E 39.7. Notice that $-\psi_{X-Y}=-2\mathrm{Re}\,\psi_X\le 2|\psi_X|$. Use Corollary 37.6; if $\{X_t\}$ is recurrent, then $\int_{B_a}\frac{\mathrm{d}z}{|\psi_X|}=\infty$, and hence $\int_{B_a}\frac{\mathrm{d}z}{-\psi_{X-Y}}=\infty$, which means recurrence of $\{X_t-Y_t\}$.

E 39.8. (i) The condition (C) says that there are $b>0$ and $\varepsilon>0$ such that $-\psi_X(z)\ge b|z|$ for $|z|<\varepsilon$. Since

$$\mathrm{Re}\,(1/(-\psi_{X+Y}(z)))=[-\psi_X-\mathrm{Re}\,\psi_Y]/[(-\psi_X-\mathrm{Re}\,\psi_Y)^2+(\mathrm{Im}\,\psi_Y)^2],$$

we have, for $|z|<\varepsilon$,

$$\tfrac{1}{-\psi_X}\Big/\mathrm{Re}\,(\tfrac{1}{-\psi_{X+Y}})\le 1+\tfrac{-\mathrm{Re}\,\psi_Y}{-\psi_X}+\tfrac{(\mathrm{Im}\,\psi_Y)^2}{(-\psi_X)^2}\le 1+\tfrac{-\mathrm{Re}\,\psi_Y}{b|z|}+\tfrac{(\mathrm{Im}\,\psi_Y)^2}{b^2|z|^2}.$$

This is bounded on $\{|z|<\varepsilon\}$, since $\psi_Y(z)$ is differentiable and $\psi_Y(0)=0$. Now use Corollary 37.6. (ii) Note that, for any $x_0>0$, $z^{-1}\int_0^{x_0}(1-\cos zx)\nu_X(\mathrm{d}x)=2z^{-1}\int_0^{x_0}(\sin\frac{1}{2}zx)^2\nu_X(\mathrm{d}x)\to 0$ as $z\downarrow 0$. (iii) If $E|X_t|<\infty$, then ψ_X is of class C^1 and $\psi_X{}'(0)=0$, which contradicts $\limsup_{z\downarrow 0}(\psi_X(z)/z)<0$ of the condition (C). (iv) Note that $\psi_X(z)=-c|z|$ with $c>0$ for a symmetric Cauchy process.

E 39.9. For $N(x)=\nu(x,\infty)$ we have $N(x)\sim x^{-1}(\log x)^{\alpha}$ and $\int_0^x yN(y)\mathrm{d}y\sim x(\log x)^{\alpha}$ as $x\to\infty$ (Feller [**139**], p. 281). Since ν is quasi-unimodal, Theorem 38.3 applies.

E 39.10. Let $\{X_t\}$ be the Brownian motion on $\mathbb{R}$ with drift $\gamma_0\neq 0$ added. It is transient by Theorem 36.7. It does not satisfy (37.8), since $|\psi(z)|=|z|({\gamma_0}^2+\frac{1}{4}z^2)^{1/2}$.

E 39.11. It is obvious that the conditions (1), (2), and (3) in Proposition 37.10 respectively imply the properties asserted here. The converse is also true, since these properties are pairwise exclusive.

E 39.12. Transience is obvious from the definition of subordination. The expression for $v_\alpha(x)$ is given in Example 37.19. To calculate $\int v_\alpha(y-x)v_\beta(z-y)\mathrm{d}y$, use

$$\int p_t(y-x)p_s(z-y)\mathrm{d}y=p_{t+s}(z-x) \qquad \text{a. e. } x \text{ and } z$$

and

$$\tfrac{1}{\Gamma(\alpha)\Gamma(\beta)}\int_0^\infty\int_0^\infty f(t+s)t^{\alpha-1}s^{\beta-1}\mathrm{d}t\mathrm{d}s=\tfrac{1}{\Gamma(\alpha+\beta)}\int_0^\infty f(u)u^{\alpha+\beta-1}\mathrm{d}u$$

for nonnegative measurable functions f, which is obtained from the convolution formula for Γ-distributions by letting the parameter α in Example 2.15 tend to 0.

E 39.13. Since $V(B)=E\big[\int_0^\infty 1_B(X_t)\mathrm{d}t\big]$ for $B\in\mathcal{B}_{[0,\infty)}$, we have

$$\int_0^\infty \mathrm{e}^{-ux}V(\mathrm{d}x)=E\big[\int_0^\infty \mathrm{e}^{-uX_t}\mathrm{d}t\big]=\int_0^\infty\exp\big[t\big(-\gamma_0 u-\int(1-\mathrm{e}^{-ux})\,\nu(\mathrm{d}x)\big)\big]\mathrm{d}t.$$

E 39.14 (Bertoin [**26**]). Define ρ by $\rho(B)=\int_0^\infty \mathrm{e}^{-t}\mu^t(B)\mathrm{d}t$ for $B\in\mathcal{B}(\mathbb{R})$. Then the smallest closed additive subgroup that contains the support of ρ is $\mathbb{R}$. We have $\int|x|\rho(\mathrm{d}x)<\infty$ if and only if $E|X_1|<\infty$. If $EX_1=\gamma_1$, then $\int x\rho(\mathrm{d}x)=\gamma_1$. We have $\widehat{\rho}(z)=(1-\psi(z))^{-1}$. For $0<c<1$ let $W_c(B)=\sum_{n=0}^\infty c^n\rho^n(B)$ and $W(B)=\sum_{n=0}^\infty\rho^n(B)$. Since $\sum_{n=0}^\infty c^n\widehat{\rho}(z)^n=1+c(1-c-\psi(z))^{-1}$, we have $W_c=\delta_0+c\int_0^\infty \mathrm{e}^{-(1-c)t}\mu^t\mathrm{d}t$, and hence $W=\delta_0+V$. The transience of the random walk with one-step probability ρ follows from the transience of $\{X_t\}$. Therefore we can apply the renewal theorem of Feller and Orey [**140**] expounded in Feller's book [**139**]. See Port and Stone [**386**] in the case $d\geq 2$.

E 39.15. Denote the distribution function and the density of μ_k by F_k and f_k, and those of ρ_k by G_k and g_k. Let $\overline{F}_k(x)=1-F_k(x)$ and $\overline{G}_k(x)=1-G_k(x)$. If μ is symmetric, then $2\mu(-\infty,x]=1+\mu[-x,x]$ for $x\geq 0$. Hence, it is enough to show that $(\mu_1*\mu_2)(-\infty,x]\geq(\rho_1*\rho_2)(-\infty,x]$ for $x\geq 0$. For $x\geq 0$,

$$\begin{aligned}
&(\mu_1*\mu_2)(-\infty,x]-(\rho_1*\rho_2)(-\infty,x]\\
&\quad=\int_{-\infty}^\infty(F_1(x-y)\mathrm{d}F_2(y)-G_1(x-y)\mathrm{d}G_2(y))\\
&\quad=\int_{-\infty}^\infty((F_1(x-y)-G_1(x-y))\mathrm{d}F_2(y)+G_1(x-y)(\mathrm{d}F_2(y)-\mathrm{d}G_2(y)))\\
&\quad=I_1(x)+I_2(x),
\end{aligned}$$

where $I_1(x) = \int_{-\infty}^{\infty}(F_1(x-y) - G_1(x-y))\mathrm{d}F_2(y)$ and $I_2(x) = \int_{-\infty}^{\infty}(F_2(x-y) - G_2(x-y))\mathrm{d}G_1(y)$. Then, using symmetry,

$$\begin{aligned} I_1(x) &= \int_{-\infty}^{\infty}(F_1(-y) - G_1(-y))\mathrm{d}F_2(x+y) \\ &= \int_0^{\infty}(F_1(-y) - G_1(-y))\mathrm{d}F_2(x+y) + \int_0^{\infty}(G_1(y) - F_1(y))\mathrm{d}F_2(x-y) \\ &= \int_0^{\infty}(\overline{F}_1(y) - \overline{G}_1(y))\mathrm{d}F_2(x+y) + \int_0^{\infty}(\overline{F}_1(y) - \overline{G}_1(y))\mathrm{d}F_2(x-y) \\ &= \int_0^{\infty}(\overline{F}_1(y) - \overline{G}_1(y))(f_2(x+y) - f_2(x-y))\mathrm{d}y. \end{aligned}$$

By unimodality and symmetry, $f_2(x+y) - f_2(x-y) \le 0$ for $0 \le y \le x$ and $f_2(x+y) - f_2(x-y) = f_2(y+x) - f_2(y-x) \le 0$ for $0 \le x \le y$. By the assumption, $\overline{F}_1(y) - \overline{G}_1(y) \le 0$ for $y \ge 0$. It follows that $I_1(x) \ge 0$. A similar argument gives $I_2(x) \ge 0$.

E 39.16. Shepp [**480**], pp. 150–151, shows that

$$\int_0^1 \left(1 - 2\int_{(0,\infty)}(1-\cos zx)\nu(\mathrm{d}x)\right)^{-1}\mathrm{d}z = \infty,$$

and that the left-hand side is finite if ν is replaced by $\Lambda\nu$.

Chapter 8

E 44.1. Let $B_\varepsilon(x) = \{y\colon |y-x| < \varepsilon\}$. By Proposition 24.14, $x \in \Sigma$ if and only if, for any $\varepsilon > 0$, $P^0[X_t \in B_\varepsilon(x)$ for some $t \ge 0] > 0$. Hence, $x \in \Sigma$ if and only if, for any $\varepsilon > 0$, $P^0\left[\int_0^\infty 1_{B_\varepsilon(x)}(X_t)\mathrm{d}t > 0\right] > 0$.

E 44.2. We have $\widehat{\mu}(z) = \exp(\mathrm{e}^{\mathrm{i}z} - 1 + \mathrm{i}z)$. For $a > 0$, $\mu^t(0,x]$ is 0 for $a < t$ and $\sum_{k=0}^n (\mathrm{e}^{-t}t^k/k!)$ for $t + n \le x < t+n+1$. Hence

$$V(0,x] = \sum_{n=0}^{[x]-1}\int_{x-n-1}^{x-n}\mathrm{e}^{-t}\sum_{k=0}^n \tfrac{t^k}{k!}\mathrm{d}t + \int_0^{x-[x]}\mathrm{e}^{-t}\sum_{k=0}^{[x]}\tfrac{t^k}{k!}\mathrm{d}t$$

for $x > 0$. Here $[x]$ is the greatest integer not exceeding x. Differentiation with respect to x gives $v(x)$.

E 44.3. A simple example is given by $X_t = -Y_t = \gamma t$ with $\gamma \ne 0$ for $d = 1$. See Remark 41.13. Other examples can be given by using the Poisson process with drift added in E 44.2.

E 44.4. Let B be a Borel set with $\operatorname{Leb} B = 0$. If $h(t)$ is a measurable function taking values in a countable set C, then $\operatorname{Leb}(B - C) = 0$ and $E^x\left[\int_0^\infty \mathrm{e}^{-qt}1_B(X_t + h(t))\mathrm{d}t\right] \le E^x\left[\int_0^\infty \mathrm{e}^{-qt}1_{B-C}(X_t)\mathrm{d}t\right] = 0$. Hence $E^x\left[\int_0^\infty \mathrm{e}^{-qt}1_B(X_t+Y_t)\mathrm{d}t\right] = E^x\left[\left(E^x\left[\int_0^\infty \mathrm{e}^{-qt}1_B(X_t + h(t))\mathrm{d}t\right]\right)_{h(t)=Y_t}\right] = 0$.

E 44.5. Similar to a part of the proof of Theorem 41.15.

E 44.6. See [**190**], p. 341. The assertion $u^q(x) = \int_0^\infty \mathrm{e}^{-qt}p_t(x)\mathrm{d}t$ follows from the properties of $p_t(x)$ and Theorem 41.16.

E 44.7. By Theorem 42.19, $U^q(x,B) = 0$ for a. e. x, if $C^q(B) = 0$. Use Proposition 41.9.

E 44.8. Use Remark 43.6. Since $\psi^\sharp = \psi + \overline{\psi} = 2\operatorname{Re}\psi$, $\operatorname{Re}(1/(q-\psi)) \le 1/|q-\psi| \le 1/(q - \operatorname{Re}\psi) = 2/(2q - \psi^\sharp)$.

E 44.9. Since $p_t(x) = (2\pi)^{-1}\int \mathrm{e}^{-\mathrm{i}xz}\widehat{\mu}(z)^t\mathrm{d}z$, we have $p_t(x) \le p_t(0)$. The function u^q is represented as $u^q(x) = \int_0^\infty \mathrm{e}^{-qt}p_t(x)\mathrm{d}t$ (Remark 41.20). Apply Theorem 43.3.

E 44.10. The 'only if' part. The function $\varphi_B(x) = P^x[T_B < \infty]$ is 0-excessive for any F_σ set B. If G is nonempty and open, then $\varphi_G(x) = 1$ for $x \in G$ and hence it is identically 1. Hence the process is recurrent by E 39.4. If Condition (ACP) is not satisfied, then there exists an essentially polar F_σ set B which is not polar (Theorem 41.15) and thus $\varphi_B(x)$ is not identically 0 but vanishes a. e., contradicting BG-recurrence.

The 'if' part. Let f be a 0-excessive function. Let $a < \sup_x f(x)$ and $G = \{x, f(x) > a\}$. Then G is a nonempty open set, since f is lower semi-continuous (Theorem 41.15). Let B be a closed ball contained in G. It follows from the recurrence and Condition (ACP) that $\Sigma = \mathbb{R}^d$ (E 44.1 and Theorem 35.8). Hence, by E 39.4, $P^x[T_B < \infty] = 1$ for all x. Hence $f(x) \geq P_B f(x) = E^x[f(X_{T_B})] > a$ for all x. It follows that $f(x) = \sup f$ for all x.

Because of the equivalence just proved, a recurrent process which is not BG-recurrent is given in Examples 41.22 and 41.23.

E 44.11. We have $\Sigma = \mathbb{R}^d$ by Theorem 41.19. If G is open, then Theorem 35.8 tells us that $P^x\left[\limsup_{t\to\infty} 1_G(X_t) = 1\right] = 1$ for every x. Since any set with positive Lebesgue measure contains an F_σ set with positive Lebesgue measure, it is enough to prove the assertion for F_σ sets B. The rest of the proof is done as in pp. 425–426 of [**415**].

E 44.12. Let σ be the uniform probability measure on the unit sphere $\{x\colon |x| = 1\}$. Since $c_d \int m_B(\mathrm{d}y)|x-y|^{2-d} = P^x[T_B < \infty]$ (see (42.15)), m_B must be rotation invariant. Hence $m_B = C(B)\sigma$ by Remark 42.11. Letting $x = 0$, we get $c_d C(B) = 1$.

E 44.13. See [**81**], p. 168, or [**387**], p. 55. If $d \geq 3$, then the formula for $P^x[T_{S_a} < \infty]$ is obtained from $P^x[T_{S_a} < T_{S_b}]$ by letting $b \to \infty$.

E 44.14. See [**17**], pp. 91–92, or [**81**], p. 170. The latter result is given also in Example 30.7.

E 44.15. See [**17**], p. 105, or [**387**], p. 56.

E 44.16. See [**387**], p. 58.

E 44.17. See Hunt [**209**], III, p. 178, or [**45**], p. 278.

E 44.18. Let c be the supremum of the total masses of measures ρ supported on B satisfying the condition. Since m_B^q is such a measure by Proposition 42.13, $C^q(B) \leq c$. Choose open sets $G_n \supset B$ such that $C^q(G_n) \to C^q(B)$ as $n \to \infty$ (Proposition 42.12). Then $C^q(G_n) = \widetilde{m}^q_{G_n}(\overline{G_n}) \geq \iint \widetilde{m}^q_{G_n}(\mathrm{d}x) u^q(y-x)\rho(\mathrm{d}y) = \int_B \widetilde{E}^y[\mathrm{e}^{-qT(G_n)}]\rho(\mathrm{d}y) = \rho(B)$. Hence $C^q(B) \geq c$. In the transient case the same proof works for $q = 0$ and B bounded.

E 44.19. Consider the transient case (that is, $\alpha < d$). Let ρ be the surface area measure on $B = \{|x| = 1\}$. Let $f(x) = \int_B u^0(y-x)\rho(\mathrm{d}y) = \mathrm{const} \int_B |y-x|^{\alpha-d}\rho(\mathrm{d}y)$. The value of $f(x)$ is constant ($\leq \infty$) on B by the rotation invariance. We claim that $f(x) < \infty$ on B if and only if B is non-polar. Indeed, if $f(x) = \infty$ on B, then B is essentially polar, hence polar, since $u^0(y-x) \geq \int P_B(x, \mathrm{d}z) u^0(y-z)$. If $f(x) = c < \infty$ on B, then $\sup_{x\in\mathbb{R}^d} f(x) = c$ by the maximum principle in [**81**], p. 221, and hence $C(B) > 0$ by E 44.18, that is, B is non-polar. Now, use the polar coordinates to check that $f(x) < \infty$ if and only if $\alpha > 1$. This finishes the proof in the transient case. In the case $d = \alpha = 2$, B is non-polar by the recurrence or by Theorem 42.29. In the case where $d = 1$ and $\alpha \geq 1$, $B = \{-1, 1\}$ is non-polar by (43.5).

E 44.20. Since log-convex functions are absolutely continuous, the process satisfies (ACT) by Theorem 27.7. The function $u^0(x)$ is left-continuous by Proposition 43.16, since it is co-excessive. By the argument of Hawkes [**189**], p. 120, a version of the density of V^0 is decreasing. Its left-continuous modification must coincide with $u^0(x)$, which cannot have a downward jump by the lower semi-continuity in Theorem 41.15.

E 44.21. Let $0 < \varepsilon < a$. It is enough to show $P^0[\, X_{R_a-} < a - \varepsilon,\, X_{R_a} = a\,] = 0$ and $P^0[\, X_{R_a-} = a,\, X_{R_a} > a + \varepsilon\,] = 0$. If $\nu(\varepsilon, \infty) = 0$, then the assertion is obvious. Suppose that $\nu(\varepsilon, \infty) = c > 0$. Let $Z_t(\omega) = \int_{(0,t] \times (\varepsilon, \infty)} xJ(\mathrm{d}(s, x), \omega)$ and $Y_t = X_t - Z_t$, as in the Lévy–Itô decomposition. Then $\{Y_t\}$ and $\{Z_t\}$ are independent Lévy processes, $\{Z_t\}$ is a compound Poisson process, and $\{Y_t\}$ is non-zero and not compound Poisson. Let J_n be the nth jumping time for $\{Z_t\}$. It is the time of the nth positive jump bigger than ε for $\{X_t\}$. Then $P^0[Y_{J_1} = x] = c \int_0^\infty \mathrm{e}^{-ct} P^0[Y_t = x]\mathrm{d}t = 0$ for any x by Theorems 27.4 and 30.10. Hence X_{J_1} has also a continuous distribution. Thus X_{J_n} has a continuous distribution by the strong Markov property. Now $P^0[\, X_{R_a-} < a - \varepsilon,\, X_{R_a} = a\,] \leq \sum_{n=1}^\infty P^0[X_{J_n} = a] = 0$

Let $r > 0$ and let $Z'_t(\omega) = \int_{(r,t] \times (\varepsilon, \infty)} xJ(\mathrm{d}(s, x), \omega)$ for $t > r$ and $Z'_t(\omega) = 0$ for $0 \leq t \leq r$. Using $\{Z'_t\}$, define $\{Y'_t\}$ and J'_n similarly. We have

$$\begin{aligned} P^0[\, R_a > r,\, X_{R_a-} = a,\, X_{R_a} > a + \varepsilon\,] &\leq \textstyle\sum_{n=1}^\infty P^0[X_{J'_n-} = a] \\ &= \textstyle\sum_{n=1}^\infty E^0[P^{X_r}[X_{J_n-} = a]]. \end{aligned}$$

If $\{X_t\}$ is of type B or C, then P_{X_r} is continuous and

$$E^0[P^{X_r}[X_{J_n-} = a]] = \textstyle\int P^0[X_{J_n-} = a - b] P^0[X_r \in \mathrm{d}b] = 0$$

by Lemma 27.1(i), hence $P^0[\, X_{R_a-} = a,\, X_{R_a} > a + \varepsilon\,] = 0$. If $\{X_t\}$ is of type A with drift $\gamma_0 \neq 0$, then, letting J_n be the nth jumping time, we have $P^0[X_{J_1-} = a] = P^0[J_1 = a/\gamma_0] = 0$ and $P^0[X_{J_n-} = a] = 0$ for each n by the strong Markov property.

E 44.22. By Theorem 31.5, g is in the domain of L. Apply (41.3) to $f = (q - L)g$ with $q > 0$. Use $U^q f = g$, and let $q \downarrow 0$.

E 44.23 (Fristedt [**150**]). It is enough to show that, for $t > 0$ and $0 \leq x < a \leq z$,

$$P^0[\, R'_a \leq t,\, X(R'_a-) \leq x,\, X(R'_a) \geq z\,] = E^0\left[\textstyle\int_0^t 1_{[0,x]}(X_s)\nu[z - X_s, \infty)\mathrm{d}s\right].$$

Choose y with $0 \leq x < y < a \leq z$, n with $0 < 1/n < z - y$, and a C^2 function g such that $0 \leq g \leq 1$, $g(w) = 1$ on $[z, z + n]$, and $g(w) = 0$ on $(-\infty, z - 1/n] \cup [z + n + 1, \infty)$. Apply Dynkin's formula to $T = t \wedge R'_y$. Then $E^0[g(X_T)] = E^0\left[\int_0^T \mathrm{d}s \int_{(0,\infty)} g(X_s + w)\nu(\mathrm{d}w)\right]$. Letting $n \to \infty$, obtain $P^0[X_T \geq z] = E^0\left[\int_0^T \nu[z - X_s, \infty)\mathrm{d}s\right]$. Notice that

$$\begin{aligned} P^0[X_T \geq z] &= P^0[R'_y \leq t,\, X(R'_y) \geq z] \\ &= P^0[R'_a \leq t,\, X(R'_a) \geq z,\, X_s < y \text{ for all } s < R'_a] \end{aligned}$$

and that $E^0\big[\int_0^T \nu[z-X_s,\infty)\mathrm{d}s\big] = E^0\big[\int_0^t 1_{[0,y)}(X_s)\nu[z-X_s,\infty)\mathrm{d}s\big]$. Then, let $y\downarrow x$.

E 44.24. By Examples 24.12 and 37.19, $U^0(0,\mathrm{d}y) = \frac{1}{\Gamma(\alpha)}y^{\alpha-1}1_{(0,\infty)}(y)dy$ and $\nu(\mathrm{d}z) = \frac{\alpha}{\Gamma(1-\alpha)}z^{-1-\alpha}1_{(0,\infty)}(z)\mathrm{d}z$. Hence, by E 44.23,

$$\begin{aligned} P^0[\,X(R'_a-)\in C,\, X(R'_a)\in D\,] &= \tfrac{\alpha}{\Gamma(\alpha)\Gamma(1-\alpha)}\int_C y^{\alpha-1}\mathrm{d}y\int_{D-y} z^{-1-\alpha}\mathrm{d}z \\ &= \tfrac{\alpha\sin\pi\alpha}{\pi}\int_C y^{\alpha-1}\mathrm{d}y\int_D (z-y)^{-1-\alpha}\mathrm{d}z. \end{aligned}$$

E 44.25. By Theorem 43.3 we have $c^0>0$ and $u^0(x) = \frac{1}{c^0}P^0[\,T_x<\infty\,] = \frac{1}{c^0}h^0(x)$. The process has the properties in Case 2 of Theorem 43.21. If $x>0$, then $x\in\Sigma_0$ and $u^0(x)>0$. If $x\le 0$, then $u^0(x)=0$. By transience the continuity of $h^0(x)$ on $\mathbb{R}\setminus\{0\}$ is proved similarly to the proof of Theorem 43.19(i). Let $R_x = T_{(x,\infty)}$. Let us see that $1_{\{X(R_x)=x\}} = 1_{\{T_x<\infty\}}$ a. s. The inequality "$\le$" is obvious. Since $P^0[T_x<R_x] = P^0[X_{T_x}=x,\, T_x<R_x]=0$ by the strong Markov property, we have

$$P^0[T_x<\infty] = P^0[T_x<\infty,\, X_{T_x}=x,\, T_x=R_x] = P^0[R_x<\infty,\, X_{R_x}=x]$$

and the asserted a. s. equality follows. Hence $P^0[X_{R_x}=x] = c^0u^0(x)$. We have $P^0[R_x=R'_x]=1$ for $x>0$ by the strong Markov property. Hence

$$P^0[X_{R_x}=x] = 1-P^0[X_{R_x}>x] = 1-\int_0^x u^0(y)\nu(x-y,\infty)\mathrm{d}y$$

for $x>0$ by E 44.23. It follows that

$$c^0\int_0^\infty \mathrm{e}^{-qx}u^0(x)\mathrm{d}x = \tfrac{1}{q}-\int_0^\infty \mathrm{e}^{-qx}\mathrm{d}x\int_0^x u^0(y)\nu(x-y,\infty)\mathrm{d}y$$

for $q>0$. The right-hand side equals $\frac{1}{q}-\frac{1}{q}\int_0^\infty \mathrm{e}^{-qy}u^0(y)\mathrm{d}y\int_0^\infty(1-\mathrm{e}^{-qx})\nu(\mathrm{d}x)$ by Fubini's theorem and Lemma 17.6. Now use E 39.13 and obtain $c^0=\gamma_0$. Finally $u^0(0+)=1/\gamma_0$ because $h^0(0+)=1$.

E 44.26. Use Remark 41.13 and E 44.6. Then it suffices to prove that $p_t(x)$ is positive definite for each $t>0$. For $x_1,\ldots,x_n\in\mathbb{R}^d$ and $\xi_1,\ldots,\xi_n\in\mathbb{C}$ we have

$$\begin{aligned} \textstyle\sum_{j,k=1}^n p_t(x_j-x_k)\xi_j\bar{\xi}_k &= \textstyle\sum_{j,k=1}^n \int p_{t/2}(y-x_k)p_{t/2}(x_j-y)\mathrm{d}y\,\xi_j\bar{\xi}_k \\ &= \textstyle\int \big|\sum_{j=1}^n p_{t/2}(x_j-y)\xi_j\big|^2\mathrm{d}y \ge 0 \end{aligned}$$

by E 44.6(3) and by $p_t(x)=p_t(-x)$.

Chapter 9

E 50.1. Suppose that $R_0=T_{(0,\infty)}>0$ a. s. Then, by symmetry, $T_{(-\infty,0)}>0$ a.s. Hence X_t stays at 0 for a while, which contradicts the assumption of type B.

E 50.2. (i) Since $EX_1=\gamma<0$, $\lim_{t\to\infty}X_t=-\infty$ and $M_\infty<\infty$ a. s. By Example 45.4 and (45.2),

$$\exp\big[\int_0^\infty t^{-1}\mathrm{e}^{-qt}\mathrm{d}t\int_{(0,\infty)}(\mathrm{e}^{-ux}-1)\mu^t(\mathrm{d}x)\big] = c_+(c_++u)^{-1}, \qquad u\ge 0.$$

Let $q\downarrow 0$ and use (48.3). Then $E[\mathrm{e}^{-uM_\infty}] = 2A^{-1}|\gamma|(2A^{-1}|\gamma|+u)^{-1}$.

(ii) $M_\infty<\infty$ a. s. as in (i). Similarly to the proof of Theorem 46.2, $P[\,X_{R_x} = x$ for every x satisfying $R_x<\infty\,]=1$, $P[\,R_x$ is right-continuous in x satisfying

$R_x < \infty\,] = 1$, and $R_{x+y} - R_x = R_y^{(x)}$. Thus, by Theorem 40.10, $P[\,R_{x+y} < \infty\,] = P[\,R_x < \infty\,]\,P[\,R_y < \infty\,]$, that is, $P[\,M_\infty > x+y\,] = P[\,M_\infty > x\,]\,P[\,M_\infty > y\,]$.

E 50.3. If $T = \lim_{n\to\infty} T_{(x-1/n,\infty)}$ for $x > 0$, then $X_{T-} = x$ or $X_T \geq x$. Hence $R''_x = \lim_{n\to\infty} T_{(x-1/n,\infty)}$ on $\{X_0 = 0\}$. Since

$$\{(x,\omega)\colon T_{(x,\infty)}(\omega) < t\} = \{(x,\omega)\colon X_s(\omega) > x,\ \exists s \in \mathbb{Q}\cap(0,t)\},$$

which belongs to $\mathcal{B}_{[0,\infty)} \times \mathcal{F}^0$, $R''_x(\omega)$ is $(\mathcal{B}_{[0,\infty)} \times \mathcal{F}^0)$-measurable in (x,ω) on $\{X_0 = 0\}$. Therefore $\Lambda_t(\omega) = R''_{M_t(\omega)}(\omega)$ is $\mathcal{F}^0$-measurable on $\{X_0 = 0\}$. As to Λ'_t, use that $\Lambda'_t = t - \Lambda^Y_t$ on $\{X_0 = 0$ and $X_t = X_{t-}\}$ in the notation of Case 3 in the proof of Lemma 49.4.

E 50.4. It is enough to show that $P[R''_0 < R_0] = 0$, assuming that $R_0 > 0$ a. s. First,

$$\begin{aligned} P[0 < R''_0 < R_0] &\leq \textstyle\sum_{s\in\mathbb{Q}\cap(0,\infty)} P[s < R''_0 < R_0,\ X_s < 0] \\ &\leq \textstyle\sum \int_{(-\infty,0)} P[X_s \in \mathrm{d}x] P[R''_{-x} < R_{-x}] = 0 \end{aligned}$$

by Lemma 49.6. Second,

$$\begin{aligned} &P[0 = R''_0 < R'_0 \leq R_0] \\ &\quad\leq P[R'_0 > 0 \text{ and } X_{t_n-} = 0 \text{ for some } t_n \to 0 \text{ with } t_n > 0] \\ &\quad\leq P[\Lambda_t < \Lambda'_t,\ \exists t \in \mathbb{Q}\cap(0,\infty)] = 0 \end{aligned}$$

by Lemma 49.4. Third, let $P[0 = R''_0 = R'_0 < R_0] = p$. If $\{X_t\}$ is of type A and $\gamma_0 < 0$, then $p \leq P[R'_0 = 0] = 0$. If $\{X_t\}$ is of type B and $R_0 > 0$ a. s., then

$$\begin{aligned} p &\leq P[X_t = 0 \text{ and } t < R_0,\ \exists t > 0] \\ &\leq \textstyle\sum_{s\in\mathbb{Q}\cap(0,\infty)} \int_{(-\infty,0)} P[X_s \in \mathrm{d}x] P[R'_{-x} < R'_{-x}] = 0 \end{aligned}$$

by Lemma 49.6.

E 50.5. We write $R'_a = T$. The assertion is proved in E 6.16 if $\{X_t\}$ is a compound Poisson process. It is trivial for the zero process. So assume that $\{X_t\}$ is non-zero and not compound Poisson. Since $P[\,X_{T-} \in C,\ X_t = a\,] = 0$ by E 44.21 and since $\int_C U^q_B(0,\mathrm{d}y)\nu\{a-y\} = 0$ by Theorems 27.4 and 30.10, we may assume $D \subset (a,\infty)$. It is enough to show

$$E[f(T)g(X_{T-})h(X_T)] = E\big[\textstyle\int_0^T f(t)g(X_t)\mathrm{d}t \int_{\mathbb{R}} h(X_t+y)\nu(\mathrm{d}y)\big]$$

for f continuous on $[0,\infty)$ with compact support S_f and for g and h both bounded and continuous on $\mathbb{R}$ satisfying $S_g \subset (-\infty,a)$ and $S_h \subset (a,\infty)$. By Lemma 45.12 $\{X_t\}$ is approximated by a sequence of compound Poisson processes $\{X^n_t\}$. Denote the hitting time of $[a,\infty)$ and the Lévy measure for $\{X^n_t\}$ by T^n and ν^n. Then

$$E[f(T^n)g(X^n_{T^n-})h(X^n_{T^n})] = E\big[\textstyle\int_0^{T^n} f(t)g(X^n_t)\mathrm{d}t \int_{\mathbb{R}} h(X^n_t+y)\nu^n(\mathrm{d}y)\big].$$

Recall that X^n_t tends to X_t uniformly on any bounded time interval a. s. Using Lemma 49.6, we can prove that $T^n \to T$ a. s. Thus we have $X^n_{T^n} - X_{T^n} \to 0$ and $X^n_{T^n-} - X_{T^n-} \to 0$ a. s. on $\{T < \infty\}$. By E 44.21, we have $X_{T-} = a = X_T$ or $X_{T-} < a < X_T$ a. s. If $X_{T-} < a < X_T$, then $T^n = T$ for all large n. It follows that $X_{T^n} \to X_T$ and $X_{T^n-} \to X_{T-}$ a. s. on $\{T < \infty\}$. Hence $X^n_{T^n} \to X_T$ and $X^n_{T^n-} \to$

X_{T-} a. s. on $\{T<\infty\}$. Thus $E[f(T^n)g(X^n_{T^n-})h(X^n_{T^n})] \to E[f(T)g(X_{T-})h(X_T)]$. We have $\nu^n \to \nu$ in the sense of Theorem 8.7(1). We can prove that

$$E\big[\int_0^{T^n} f(t)g(X_t^n)\mathrm{d}t\int_{\mathbb{R}} h(X_t^n+y)\nu^n(\mathrm{d}y)\big] \to E\big[\int_0^T f(t)g(X_t)\mathrm{d}t\int_{\mathbb{R}} h(X_t+y)\nu(\mathrm{d}y)\big].$$

In fact, choose $\varepsilon>0$ such that $\mathrm{dis}(S_g,S_h)>2\varepsilon$. Then we can restrict the integral over $\mathbb{R}$ to $\{y\colon |y|>\varepsilon\}$ and we see

$$g(X_t^n)\int_{|y|>\varepsilon} h(X_t^n+y)\nu^n(\mathrm{d}y) \to g(X_t)\int_{|y|>\varepsilon} h(X_t+y)\nu(\mathrm{d}y).$$

Hence we get the desired identity in the limit.

E 50.6. Since $M_t=X_t$, (49.23) gives

$$\varphi_q^+(\mathrm{i}u)=q\int_0^\infty \mathrm{e}^{-qt}E[\mathrm{e}^{-uX_t}]\mathrm{d}t=q/(q-\Psi(-u)),$$

where $\Psi(-u)=-\gamma_0 u+\int_{(0,\infty)}(\mathrm{e}^{-ux}-1)\nu(\mathrm{d}x)$. Let $v\to\infty$ in (49.2). Then

$$\int_0^\infty \mathrm{e}^{-ux}E[\mathrm{e}^{-qR_x};\,\Gamma_x=0]\mathrm{d}x=\lim_{v\to\infty}(\varphi_q^+(\mathrm{i}u)/(v\varphi_q^+(\mathrm{i}v)))=\gamma_0/(q-\Psi(-u)),$$

since $\lim_{v\to\infty}((q-\Psi(-v))/v)=\gamma_0$ as in Lemma 43.11. If $\gamma_0=0$ and $\{X_t\}$ satisfies (ACP), then $\Sigma_0=\emptyset$ by Theorem 43.21, which implies $P[\,\Gamma_x=0\,]=0$ for all $x>0$. If $\gamma_0=0$ and $\{X_t\}$ does not satisfy (ACP) and if, moreover, it is not compound Poisson, then again $\Sigma_0=\emptyset$; but see the remark after Theorem 43.21 concerning its proof. Bertoin [**26**], p. 77, contains another proof.

E 50.7. See [**35**]. Another proof, due to Bondesson, Kristiansen, and Steutel [**56**], is as follows. Almost surely R_x equals R'_x of (49.3) by Lemma 49.6. Hence

$$P[R_x>t]=P[X_t<x]=P[t^{1/\alpha}X_1<x]=P[x^\alpha {X_1}^{-\alpha}>t],$$

that is, $R_x \stackrel{\mathrm{d}}{=} x^\alpha {X_1}^{-\alpha}$. Then use E 29.18.

E 50.8. Since sample functions of $\{X_t\}$ are right-continuous step functions with jump height 1, we have $R_x=T_{(x,\infty)}=T_{\{[x]+1\}}$, where $[x]$ is the integer part of x. Hence R_x has distribution $\rho^{[x]+1}$, where ρ is an exponential distribution.

E 50.9. See Bingham [**36**], p. 749, or combine the results of Example 46.7, E 29.7, and E 50.7.

E50.10. We have $\widehat{(qV^q)}(z)=\varphi_q^+(z)\varphi_q^-(z)$ in (45.1). Hence we have only to show that, for some $q_1>0$ and $q_2>0$, $\varphi_q^+(z)=\widehat{(q_1V_1)}(z)$ and $\varphi_q^-(z)=\widehat{(q_2V_2)}(z)$, where V_1 is the q_1-potential measure of a subordinator and V_2 is the q_2-potential measure of the negative of a subordinator. Such subordinators can be given by using the formula of Fristedt [**150**] mentioned in the Notes of Chapter 9. See [**389**]. The uniqueness assertion is obvious from the uniqueness in Theorem 45.2. The decomposition for the Brownian motion is $q-\frac{1}{2}\big(\frac{\mathrm{d}}{\mathrm{d}x}\big)^2=\big(\sqrt{q}-\frac{1}{\sqrt{2}}\frac{\mathrm{d}}{\mathrm{d}x}\big)\big(\sqrt{q}+\frac{1}{\sqrt{2}}\frac{\mathrm{d}}{\mathrm{d}x}\big)$.

Chapter 10

E 55.1 ([**499**]). Complete monotonicity of the functions can be checked by differentiation n times. Another method for e^{-x^α} with $0<\alpha\le 1$ is to use that it is the Laplace transform of an α-stable distribution. We can also use the fact that the products of completely monotone functions are completely monotone.

E 55.2. Consider a log-convex function $f(x)$ on $(0,\infty)$ which is not of class C^∞. For example, $f(x) = c\exp(-x - x1_{(0,1)}(x))$ on $(0,\infty)$. Then $f(x)$ is not completely monotone. For an example of C^∞, consider $g(x) = c(\mathrm{e}^{-x} + b\mathrm{e}^{-ax})^{1/2}$ with $a > 1$, $b > 0$, and $c > 0$. This is log-convex because

$$gg'' - (g')^2 = 2^{-1}c^2b(a-1)^2(\mathrm{e}^{-x} + b\mathrm{e}^{-ax})^{-1}\mathrm{e}^{-(1+a)x} > 0.$$

If g is completely monotone on $(0,\infty)$, then $g(x) = \int_{(0,\infty)} \mathrm{e}^{-xy}\rho(\mathrm{d}y)$ with some ρ and $\rho{*}\rho$ has support $\{1, a\}$, which is impossible. Hence g is not completely monotone. For another example of C^∞, let $h(x) = ((x+1)^2+1)^{-1}$ on $(0,\infty)$. This is log-convex, but $(\mathrm{d}^4/\mathrm{d}x^4)h(x) < 0$ for small $x > 0$. The last example shows that the Cauchy density $\pi^{-1}(x^2+1)^{-1}$ restricted to $[a,\infty)$ with $a \geq 1$ and multiplied by a normalizing constant gives an infinitely divisible distribution.

E 55.3. Let μ be a mixture of $\{\mu_a \colon a \in (0,\infty]\}$ with mixing measure ρ. Let η be the mixture of the exponential distributions and δ_0 with the mixing measure ρ. We have $a(a - \log\widehat{\lambda}(z))^{-1} = a\int_0^\infty \widehat{\lambda}(z)^t e^{-at}dt$ as in Example 30.8 and hence $\widehat{\mu}(z) = \int_{[0,\infty)} \widehat{\lambda}(z)^t \eta(dt)$. Since η is infinitely divisible by Theorem 51.6, μ is infinitely divisible by Theorem 30.1.

E 55.4. Let $\mu_k = \sum_n p_{n,k}\delta_n$. (i) Let $\mu = \sum_{k=1}^m \mu_k q_k$ with $q_k \geq 0$ and $\sum_k q_k = 1$. Then $\mu = \sum_n p_n\delta_n$ with $p_n = \sum_k p_{n,k}q_k$. We have $p_n > 0$ and

$${p_n}^2 \leq (\textstyle\sum_k (p_{n-1,k}p_{n+1,k})^{1/2}q_k)^2 \leq (\sum_k p_{n-1,k}q_k)(\sum_k p_{n+1,k}q_k) = p_{n-1}p_{n+1}.$$

(ii) First observe that μ has support in $\mathbb{Z}_+$ and let $\mu\{n\} = p_n$. Then $p_{n,k} \to p_n$ as $k \to \infty$. It follows from ${p_{n,k}}^2 \leq p_{n-1,k}p_{n+1,k}$ that ${p_n}^2 \leq p_{n-1}p_{n+1}$. If $p_n = 0$ for some n, then, by this inequality, $p_n = 0$ for all n, a contradiction. Hence $p_n > 0$ for all $n \in \mathbb{Z}_+$. (iii) It follows from ${p_n}^2 \leq p_{n-1}p_{n+1}$ that $({p_n}^\gamma)^2 \leq {p_{n-1}}^\gamma {p_{n+1}}^\gamma$.

E 55.5. A function $f(x)$ positive on $(0,\infty)$ is log-convex if and only if $f(\alpha x + \beta y) \leq f(x)^\alpha f(y)^\beta$ for $x, y \in (0,\infty)$, $\alpha > 0$, and $\beta > 0$ with $\alpha + \beta = 1$. (i) If f is positive and log-convex on $(0,\infty)$, then

$$\begin{aligned}\textstyle\int_0^\varepsilon f(\alpha x + \beta y + u)\mathrm{d}u &\leq \textstyle\int_0^\varepsilon f(x+u)^\alpha f(y+u)^\beta \mathrm{d}u \\ &\leq \big(\textstyle\int_0^\varepsilon f(x+u)\mathrm{d}u\big)^\alpha \big(\int_0^\varepsilon f(y+u)\mathrm{d}u\big)^\beta\end{aligned}$$

by Hölder's inequality. Conversely, suppose that $F(x)$ satisfies the condition (lcv). Then, $F(x)$ is either flat on $(0,\infty)$ or strictly increasing on $(0,\infty)$. Moreover, $F(x)$ is continuous on $(0,\infty)$. Suppose that $F(x)$ is not flat. Define $f(x) = \limsup_{\varepsilon\downarrow 0} \varepsilon^{-1}(F(x+\varepsilon) - F(x))$ for $x > 0$. Then $f(\alpha x + \beta y) \leq f(x)^\alpha f(y)^\beta$. It follows that $0 < f(x) < \infty$ and $\log f(x)$ is convex. Hence $f(x)$ is continuous. The usual proof of the mean value theorem shows that, for any $x > 0$ and $\varepsilon > 0$, there is y with $x < y < x + \varepsilon$ such that $\varepsilon^{-1}(F(x+\varepsilon) - F(x)) \geq \eta^{-1}(F(y+\eta) - F(y))$ for any small $\eta > 0$. Hence $\varepsilon^{-1}(F(x+\varepsilon) - F(x)) \geq f(y)$. It follows that $\liminf_{\varepsilon\downarrow 0} \varepsilon^{-1}(F(x+\varepsilon) - F(x)) = f(x)$. Hence $F(x)$ is of class C^1. Now we get $f(\alpha x + \beta y) \leq f(x)^\alpha f(y)^\beta$. (ii) Note that, if f_1 and f_2 are log-convex, then

$$\begin{aligned}(f_1 + f_2)(\alpha x + \beta y) &\leq f_1(x)^\alpha f_1(y)^\beta + f_2(x)^\alpha f_2(y)^\beta \\ &\leq (f_1(x) + f_2(x))^\alpha (f_1(y) + f_2(y))^\beta\end{aligned}$$

by Hölder's inequality. (iii) Use the characterization (i). (iv) We have $f(\alpha x + \beta y)^{\gamma} \leq f(x)^{\gamma\alpha} f(y)^{\gamma\beta}$.

E 55.6. Let μ_a be as in Lemma 51.14. Express $\mu_{(n)}$ in terms of a probability measure ρ_n as $\mu_{(n)}(B) = \int_{(0,\infty]} \mu_a(B)\rho_n(\mathrm{d}a)$ and use the selection theorem for $\{\rho_n\}$ to get a probability measure ρ on $[0,\infty]$ such that $\mu[0,x] = \int_{[0,\infty]} \mu_a[0,x]\rho(\mathrm{d}a)$ for $0 < x < \infty$. Here $\mu_0 = \delta_\infty$. Letting $x \to \infty$, we have $\rho\{0\} = 0$.

E 55.7. See Urbanik [**544**] or Sato [**431**].

E 55.8. See Thorin [**532**].

E 55.9. Use definitions and characterizations.

E 55.10. See Keilson and Steutel [**270**], p. 245.

E 55.11. See Wolfe [**581**], p. 332.

E 55.12. Write $q(y) = \sum_j 1_{[b_j,\infty)}(y) + q_0(y)$, $0 < b_1 \leq b_2 \leq \cdots$, $0 \leq q(y) \leq 1$. Then $\mu = \delta_{\gamma_0} * \mu_0 * \mu_1$, where μ_0 is of class ME and μ_1 is a (possibly infinite) convolution of exponential distributions. Note that μ_0 is unimodal with mode 0 and μ_1 is strongly unimodal.

E 55.13. Let μ be α-stable, $0 < \alpha < 2$, with Lévy measure ν concentrated on $(0,\infty)$. Then $\nu = l(x)\mathrm{d}x$ on $(0,\infty)$ with

$$l(x) = cx^{-\alpha-1} = \tfrac{c}{\Gamma(\alpha+1)} \int_0^\infty \mathrm{e}^{-xy} y^\alpha \mathrm{d}y, \quad c > 0.$$

Hence, if $0 < \alpha < 1$, then μ is a special case of E 55.12. If $1 \leq \alpha < 2$, note that $l(x) = \lim_{n\to\infty} \int_0^\infty \mathrm{e}^{-xy}(y \wedge n)^\alpha \mathrm{d}y$.

E 55.14. (i) Denote the integer part by square brackets. Write $q(y) = q_0(y) + ([a]-1)1_{[\alpha,\infty)} + [b]1_{[\beta,\infty)}$ with $q_0(y) = (a - [a] + 1)1_{[\alpha,\infty)} + (b - [b])1_{[\beta,\infty)}$. Let μ_0 be the distribution corresponding to q_0. Then μ is the convolution of μ_0 with a finite number of exponential distributions. Hence we may and do assume that $1 \leq a < 2$ and $0 \leq b < 1$. If $a + b \leq 2$, then we can apply Remark 54.22 directly. If $a + b > 2$, then note that $q(y) = (a1_{[\alpha,\beta)} + 1_{[\beta,\infty)}) + (a+b-1)1_{[\beta,\infty)}$, where $1 < a + b - 1 < 2$. (ii) If $a + b \geq 1$, then apply Remark 54.22. If $a + b < 1$, then $\mu \in ME$ by Theorem 51.12 and thus μ is not strongly unimodal by Example 52.7.

References and author index

[1] Abramowitz, M. and Stegun, I. A. (ed.) (1965) *Handbook of Mathematical Functions with Formulas, Graphs, and Mathematical Tables,* Dover Pub., New York. *420,470*

[2] Acosta, A. de (1983) A new proof of the Hartman–Wintner law of the iterated logarithm, *Ann. Probab.* **11**, 270–276. *368*

[3] Acosta, A. de (1994) Large deviations for vector-valued Lévy processes, *Stoch. Proc. Appl.* **51**, 75–115. *249,272*

[4] Adelman, O. (1985) Brownian motion never increases: a new proof to a result of Dvoretzky, Erdős and Kakutani, *Israel J. Math.* **50**, 189–192. *28*

[5] Adelman, O., Burdzy, K. and Pemantle, R. (1998) Sets avoided by Brownian motion, *Ann. Probab.* **26**, 429–464. *368*

[6] *Applebaum, D. (2009) *Lévy Processes and Stochastic Calculus*, 2nd ed., Cambridge Univ. Press, Cambridge. *ix*

[7] *Asmussen, S., Foss, S. and Korshunov, D. (2003) Asymptotics for sums of random variables with local subexponential behaviour, *J. Theoretic. Probab.* **16**, 489–518. *450_2,451,452_2*

[8] Barlow, M. T. (1985) Continuity of local times for Lévy processes, *Zeit. Wahrsch. Verw. Gebiete* **69**, 23–35. *328*

[9] Barlow, M. T. (1988) Necessary and sufficient conditions for the continuity of local time of Lévy processes, *Ann. Probab.* **16**, 1389–1427. *328*

[10] Barlow, M. T. and Hawkes, J. (1985) Application de l'entropie métrique à la continuité des temps locaux des processus de Lévy, *C. R. Acad. Sci. Paris* **301**, 237–239. *328*

[11] Barlow, M. T., Perkins, E. A. and Taylor, S. J. (1986) Two uniform intrinsic constructions for the local time of a class of Lévy processes, *Illinois J. Math.* **30**, 19–65. *380*

[12] Barndorff-Nielsen, O. (1978) Hyperbolic distributions and distributions on hyperbolae, *Scand. J. Statist.* **5**, 151–157. *236*

[13] Barndorff-Nielsen, O., Blæsild, P. and Halgreen, C. (1978) First hitting time models for the generalized inverse Gaussian distribution, *Stoch. Proc. Appl.* **7**, 49–54. *470*

[14] Barndorff-Nielsen, O. and Halgreen, C. (1977) Infinite divisibility of the hyperbolic and generalized inverse Gaussian distributions, *Zeit. Wahrsch. Verw. Gebiete* **38**, 309–311. *235,470*

[15] *Barndorff-Nielsen, O. E., Maejima, M. and Sato, K. (2006) Some classes of infinitely divisible distributions admitting stochastic integral representations, *Bernoulli* **12**, 1–33. *437,439*

Starred references are additions in this new printing.

Slanted numbers at the end of each item indicate the pages where it is cited. If it is cited twice or more in the same page, the number of times is attached as a subscript.

[16] *Barndorff-Nielsen, O. E., Rosiński, J. and Thorbjørnsen, S. (2008) General Υ transformations, *ALEA Lat. Am. J. Probab. Math. Statist.* **4**, 131–165. *434*

[17] Bass, R. F. (1995) *Probabilistic Techniques in Analysis,* Springer, New York. *474_2*

[18] Baxter, G. and Donsker, M. D. (1957) On the distribution of the supremum functional for processes with stationary independent increments, *Trans. Amer. Math. Soc.* **85**, 73–87. *383*

[19] Baxter, G. and Shapiro, J. M. (1960) On bounded infinitely divisible random variables, *Sankhyā* **22**, 253–260. *196*

[20] Berg, C. (1979) Hunt convolution kernels which are continuous singular with respect to Haar measure, *Probability Measures on Groups* (ed. H. Heyer, Lect. Notes in Math. No. 706, Springer, Berlin), 10–21. *294*

[21] Berg, C. and Forst, G. (1975) *Potential Theory on Locally Compact Abelian Groups,* Springer, New York. *236,332*

[22] Berman, S. M. (1986) The supremum of a process with stationary independent and symmetric increments, *Stoch. Proc. Appl.* **23**, 281–290. *167*

[23] Bertoin, J. (1991) Increase of a Lévy process with no positive jumps, *Stoch. and Stoch. Rep.* **37**, 247–251. *378*

[24] Bertoin, J. (1993) Splitting at the infimum and excursions in half-lines for random walks and Lévy processes, *Stoch. Proc. Appl.* **47**, 17–35. *384*

[25] Bertoin, J. (1994) Increase of stable processes, *J. Theoretic. Probab.* **7**, 551–563. *379*

[26] Bertoin, J. (1996) *Lévy Processes,* Cambridge Univ. Press, Cambridge. *29,117, 328,331_5,373_2,378_2,379,384,472,478*

[27] *Bertoin, J. (1999) Subordinators: examples and applications, *Lectures on Probability Theory and Statistics. Ecole d'Eté de Probabilités de Saint-Flour XXVII–1997* (Lect. Notes in Math. No. 1717, Springer, Berlin Heidelberg), 1–91. *ix*

[28] Bertoin, J. and Doney, R. A. (1994) Cramér's estimate for Lévy processes, *Statist. Probab. Letters* **21**, 363–365. *384*

[29] Bertoin, J. and Doney, R. A. (1997) Spitzer's condition for random walks and Lévy processes, *Ann. Inst. Henri Poincaré, Probab. Statist.* **33**, 167–178. *373*

[30] *Bertoin, J. and Yor, M. (2001) On subordinators, self-similar Markov processes and some factorizations of the exponential variable, *Electr. Comm. Probab.* **6**, 95–106. *ix*

[31] *Bertoin, J. and Yor, M. (2002) On the entire moments of self-similar Markov processes and exponential functionals of Lévy processes, *Ann. Fac. Sci. Toulouse* **11**, 33–45. *ix*

[32] Biane, Ph. and Yor, M. (1987) Valeurs principales associées aux temps locaux browniens, *Bull. Sci. Math.* **111**, 23–101. *328*

[33] Billingsley, P. (1968) *Convergence of Probability Measures,* Wiley, New York. *117*

[34] Billingsley, P. (1986) *Probability and Measure,* 2nd ed., Wiley, New York. *xi, 4,6_2,9,22,30,42,128,174,179,228,243,245,312,349*

[35] Bingham, N. H. (1973) Maxima of sums of random variables and suprema of stable processes, *Zeit. Wahrsch. Verw. Gebiete* **26**, 273–296. *383,384,478*

[36] Bingham, N. H. (1975) Fluctuation theory in continuous time, *Adv. Appl. Probab.* **7**, 705–766. *384,478*

[37] Bingham, N. H. (1986) Variants on the law of the iterated logarithm, *Bull. London Math. Soc.* **18**, 433–467. *368*

[38] Bingham, N. H., Goldie, C. M. and Teugels, J. L. (1987) *Regular Variation,* Cambridge Univ. Press, Cambridge. *117,466*

[39] Birnbaum, Z. W. (1948) On random variables with comparable peakedness, *Ann. Math. Statist.* **19**, 76–81. *265,272*

Blæsild, P. *see* [**13**]

[40] Blum, J. R. and Rosenblatt, M. (1959) On the structure of infinitely divisible distribution functions, *Pacific J. Math.* **9**, 1–7. *196*

[41] Blumenthal, R. M. and Getoor, R. K. (1960) Some theorems on stable processes, *Trans. Amer. Math. Soc.* **95**, 263–273. *380,381,447*

[42] Blumenthal, R. M. and Getoor, R. K. (1960) A dimension theorem for sample functions of stable processes, *Illinois J. Math.* **4**, 370–375. *379,380*

[43] Blumenthal, R. M. and Getoor, R. K. (1961) Sample functions of stochastic processes with independent increments, *J. Math. Mech.* **10**, 493–516. *362_4,380,381*

[44] Blumenthal, R. M. and Getoor, R. K. (1962) The dimension of the set of zeros and the graph of a symmetric stable process, *Illinois J. Math.* **6**, 308–316. *380*

[45] Blumenthal, R. M. and Getoor, R. K. (1968) *Markov Processes and Potential Theory,* Academic Press, New York. *284,327,329,331_3,332,474*

[46] Blumenthal, R. M. and Getoor, R. K. (1970) Dual processes and potential theory, *Proc. Twelfth Biennial Sem. Canadian Math. Cong.* (ed. R. Pyke, Canadian Math. Cong., Montreal), 137–156. *332*

[47] Blumenthal, R. M., Getoor, R. K. and Ray, D. B. (1961) On the distribution of first hits for the symmetric stable processes, *Trans. Amer. Math. Soc.* **99**, 540–554. *304*

[48] Bochner, S. (1949) Diffusion equation and stochastic processes, *Proc. Nat. Acad. Sci. USA* **35**, 368–370. *197,233*

[49] Bochner, S. (1955) *Harmonic Analysis and the Theory of Probability,* Univ. California Press, Berkeley and Los Angeles. *197*

[50] Bochner, S. (1962) Subordination of non-Gaussian processes, *Proc. Nat. Acad. Sci. USA* **48**, 19–22. *236*

[51] Bochner, S. and Chandrasekharan, K. (1949) *Fourier Transforms,* Annals of Mathematical Studies No. 19, Princeton Univ. Press, Princeton, NJ. *174_2*

[52] Bohr, H. (1947) *Almost Periodic Functions,* Chelsea Pub., New York. [German original 1933] *188,196*

[53] Bondesson, L. (1981,1982) Classes of infinitely divisible distributions and densities, *Zeit. Wahrsch. Verw. Gebiete* **57**, 39–71; Correction and addendum, **59**, 277. *389,426*

[54] Bondesson, L. (1987) On the infinite divisibility of the half-Cauchy and other decreasing densities and probability functions on the nonnegative line, *Scand. Actuarial J.* **1987**, 225–247. *47*

[55] Bondesson, L. (1992) *Generalized Gamma Convolutions and Related Classes of Distribution Densities,* Lect. Notes in Statistics, No. 76, Springer, New York. *426,438,452*

[56] Bondesson, L., Kristiansen, G. K. and Steutel, F. W. (1996) Infinite divisibility of random variables and their integer parts, *Statist. Probab. Letters* **28**, 271–278. *466_2,478*

[57] Borovkov, A. A. (1965) On the first-passage time for one class of processes with independent increments, *Theory Probab. Appl.* **10**, 331–334. *383*

[58] Borovkov, A. A. (1967) Boundary-value problems for random walks and large deviations in function spaces, *Theory Probab. Appl.* **12**, 575–595. *249*

[59] Borovkov, A. A. (1970) Factorization identities and properties of the distribution of the supremum of sequential sums, *Theory Probab. Appl.* **15**, 359–402. *334,383,384*

[60] Borovkov, A. A. (1976) *Stochastic Processes in Queueing Theory,* Springer, New York. [Russian original 1972] *384*

[61] Boylan, E. S. (1964) Local times for a class of Markov processes, *Illinois J. Math.* **8**, 19–39. *328*

[62] Bratijchuk, N. S. and Gusak, D. V. (1990) *Boundary Problems for Processes with Independent Increments,* Naukova Dumka, Kiev (in Russian). 384_2

[63] Braverman, M. (1997) Suprema and sojourn times of Lévy processes with exponential tails, *Stoch. Proc. Appl.* **68**, 265–283. *167*

[64] Braverman, M. and Samorodnitsky, G. (1995) Functionals of infinitely dvisible stochastic processes with exponential tails, *Stoch. Proc. Appl.* **56**, 207–231. *167*

[65] Breiman, L. (1965) On some limit theorems similar to the arc-sin law, *Theory Probab. Appl.* **10**, 323–331. *462*

[66] Breiman, L. (1968) *Probability,* Addison-Wesley, Reading, Mass. (Republished, SIAM, Philadelphia, 1992.) *4,9,22,68,174,245*

[67] Breiman, L. (1968,1970) A delicate law of the iterated logarithm for non-decreasing stable processes, *Ann. Math. Statist.* **39**, 1814–1824; Correction, **41**, 1126. *116, 358,359*

[68] Bretagnolle, J. (1971) Résultats de Kesten sur les processus à accroissements indépendants, *Séminaire de Probabilités V, Université de Strasbourg* (Lect. Notes in Math. No. 191, Springer, Berlin), 21–36. *317,322,332*

[69] Brockett, P. L. and Tucker, H. G. (1977) A conditional dichotomy theorem for stochastic processes with independent increments, *J. Multivar. Anal.* **7**, 13–27. *236*

[70] *Brody, E. J. (1971) An elementary proof of the Gaussian dichotomy theorem, *Zeit. Wahrsch. Verw. Gebiete* **20**, 217–226. *440*

[71] Burdzy, K. (1990) On nonincrease of Brownian motion, *Ann. Probab.* **18**, 978–980. *28*

Burdzy, K. *see also* [**5**]

[72] Cameron, R. H. and Martin, W. T. (1944) Transformations of Wiener integrals under translations, *Ann. Math.* **45**, 386–396. *232,249*

[73] *Carmona, Ph., Petit, F. and Yor, M. (1997) On the distribution and asymptotic results for exponential functionals of Lévy processes, *Exponential Functionals and Principal Values Related to Brownian Motion* (ed. M. Yor, Bibl. Rev. Mat. Iberoamericana, Madrid), 73–126. *456*

[74] *Carmona, Ph., Petit, F. and Yor, M. (2001) Exponential functionals of Lévy processes, *Lévy Processes, Theory and Applications* (ed. O. E. Barndorff-Nielsen, T. Mikosch and S. I. Resnick, Birkhäuser, Boston), 41–55. *456*

Chandrasekharan, K. *see* [**51**]

[75] Chistyakov, V. P. (1964) A theorem on sums of independent positive random variables and its applications to branching random processes, *Theory Probab. Appl.* **9**, 640–648. *163*

[76] Choi, G. S. (1994) Criteria for recurrence and transience of semistable processes, *Nagoya Math. J.* **134**, 91–106. *117,272,462*

[77] Choi, G. S. and Sato, K. (1995) Recurrence and transience of operator semi-stable processes, *Proc. Japan Acad.* **71**, Ser. A, 87–89. *272*

[78] Chung, K. L. (1948) On the maximum partial sums of sequences of independent random variables, *Trans. Amer. Math. Soc.* **64**, 205–233. *367,368*

[79] Chung, K. L. (1953) Sur les lois de probabilités unimodales, *C. R. Acad. Sci. Paris* **236**, 583–584. *395*

[80] Chung, K. L. (1974) *A Course in Probability Theory,* Academic Press, New York. *xi,6$_2$,9,128,228,245$_2$,272$_2$,349*

[81] Chung, K. L. (1982) *Lectures from Markov Processes to Brownian Motion,* Springer, New York. *6,68,108,272,284,303,310,331$_2$,474$_3$*

[82] Chung, K. L. and Erdős, P. (1947) On the lower limit of sums of independent random variables, *Ann. Math.* **48**, 1003–1013. *368*

[83] Chung, K. L., Erdős, P. and Sirao, T. (1959) On the Lipschitz's condition for Brownian motion, *J. Math. Soc. Japan* **11**, 263–274. *381*

[84] Chung, K. L. and Fuchs, W. H. (1951) On the distribution of values of sums of random variables, *Four Papers in Probability* (Mem. Amer. Math. Soc., No. 6, Providence, RI.), 1–12. *252,272*

[85] Çinlar, E. and Pinsky, M. (1971) A stochastic integral in storage theory, *Zeit. Wahrsch. Verw. Gebiete* **17**, 227–240. *117*

[86] Ciesielski, Z. and Taylor, S. J. (1962) First passage times and sojourn times for Brownian motion in space and the exact Hausdorff measure of the sample path, *Trans. Amer. Math. Soc.* **103**, 434–450. *379*

[87] Cramér, H. (1938) Sur un nouveau théorème-limite de la théorie des probabilités, *Colloque Consacré à la Théorie des Probabilités, III* (Actualités Scientifiques et Industrielles, No. 736, Hermann, Paris), 5–23. (Reprinted in *Harald Cramér Collected Works,* Vol. 2, Springer, Berlin, 1994.) *169,232*

[88] Csáki, E. (1978) On the lower limit of maxima and minima of Wiener process and partial sums, *Zeit. Wahrsch. Verw. Gebiete* **43**, 205–221. *368*

[89] Csáki, E. (1980) A relation between Chung's and Strassen's laws of the iterated logarithm, *Zeit. Wahrsch. Verw. Gebiete* **54**, 287–301. *368*

[90] Csörgő, S. and Mason, D. M. (1991) A probabilistic approach to the tails of infinitely divisible laws, *Sums, Trimmed Sums and Extremes* (ed. M. G. Hahn et al., Birkhäuser, Boston, Mass.), 317–335. *196*

[91] Cuppens, R. (1975) *Decomposition of Multivariate Probabilities,* Academic Press, New York. *9,42,68$_2$,460*

[92] Darling, D. A. (1956) The maximum of sums of stable random variables, *Trans. Amer. Math. Soc.* **83**, 164–169. *384*

[93] Davis, B. (1983) On Brownian slow points, *Zeit. Wahrsch. Verw. Gebiete* **64**, 359–367. *381*

[94] Davis, B. (1984) On the paths of symmetric stable processes, *Trans. Amer. Math. Soc.* **281**, 785–794. *382*

[95] DeBlassie, R. D. (1990) The first exit time of a two-dimensional symmetric stable process from a wedge, *Ann. Probab.* **18**, 1034–1070. *332*

[96] *de Haan, L. and Karandikar, R. L. (1989) Embedding a stochastic difference equation into a continuous-time process, *Stoch. Proc. Appl.* **32**, 225–235. *456*

[97] Dharmadhikari, S. and Joag-dev, K. (1988) *Unimodality, Convexity, and Applications,* Academic Press, San Diego. *426$_2$*

[98] Diédhiou, A. (1998) On the self-decomposability of the half-Cauchy distribution, *J. Math. Anal. Appl.* **220**, 42–64. *98*

[99] Dobrushin, R. L. and Pechersky, E. A. (1996) Large deviations for random processes with independent increments on infinite intervals, *Probability Theory and Mathematical Statistics* (ed. I. A. Ibragimov and A. Yu. Zaitsev, Gordon and Breach, Amsterdam), 41–74. *249*

[100] Dœblin, W. (1939) Sur les sommes d'un grand nombre des variables aléatoires indépendantes, *Bull. Sci. Math.* **63**, 23–32 and 35–64. *196*

[101] Doney, R. A. (1987) On Wiener–Hopf factorization and the distribution of extrema for certain stable processes, *Ann. Probab.* **15**, 1352–1362. *384*

[102] Doney, R. A. (1996) Increase of Lévy processes, *Ann. Probab.* **24**, 961–970. *378*

[103] *Doney, R. A. (2007) *Fluctuation Theory for Lévy Processes. Ecole d'Eté de Probabilités de Saint-Flour XXXV–2005*, Lect. Notes in Math. No. 1897, Springer, Berlin Heidelberg. *ix*

Doney, R. A. *see also* [**28, 29**]

Donsker, M. D. *see* [**18**]

[104] Doob, J. L. (1937) Stochastic processes depending on a continuous parameter, *Trans. Amer. Math. Soc.* **42**, 107–140. *30,68*

[105] *Doob, J. L. (1942) The Brownian movement and stochastic equations, *Ann. of Math.* **43**, 351–369. *435*

[106] Doob, J. L. (1953) *Stochastic Processes,* Wiley, New York. *28,68,144,167,245,272*

[107] Doob, J. L. (1954) Semimartingales and subharmonic functions, *Trans. Amer. Math. Soc.* **77**, 86–121. *331*

[108] Doob, J. L. (1955) A probability approach to the heat equation, *Trans. Amer. Math. Soc.* **80**, 216–280. *331*

[109] Doob, J. L. (1984) *Classical Potential Theory and Its Probabilistic Counterpart,* Springer, New York. *331*

[110] Dudley, R. M. (1989) *Real Analysis and Probability,* Wadsworth, Pacific Grove, Calif. *9,42*

[111] Dunford, N. and Schwartz, J. T. (1958) *Linear Operators, Part 1, General Theory,* Interscience, New York. *206,431,469*

[112] Dupuis, C. (1974) Mesure de Hausdorff de la trajectoire de certains processus à accroissements indépendants et stationnaires, *Séminaire de Probabilités VIII, Université de Strasbourg* (Lect. Notes in Math. No. 381, Springer, Berlin), 37–77. *361*

[113] Durrett, R. (1984) *Brownian Motion and Martingales in Analysis,* Wadsworth, Belmont, Calif. *30*

[114] Dvoretzky, A. (1963) On the oscillation of the Brownian motion process, *Israel J. Math.* **1**, 212–214. *381*

[115] Dvoretzky, A. and Erdős, P. (1951) Some problems on random walk in space, *Proc. Second Berkeley Symp. Math. Statist. Probab.* (ed. J. Neyman, Univ. California Press, Berkeley), 353–367. *360*

[116] Dvoretzky, A., Erdős, P. and Kakutani, S. (1958) Points of multiplicity $\mathfrak{c}$ of plane Brownian paths, *Bull. Res. Council Israel* **7** F, 175–180. (Reprinted in *Shizuo Kakutani: Selected Papers,* Vol. 2, Birkhäuser, Boston, Mass., 1986.) *380*

[117] Dvoretzky, A., Erdős, P. and Kakutani, S. (1961) Nonincrease everywhere of the Brownian motion process, *Proc. Fourth Berkeley Symp. Math. Statist. Probab.* (ed. J. Neyman, Univ. California Press, Berkeley), Vol. 2, 103–116. (Reprinted in *Shizuo Kakutani: Selected Papers,* Vol. 2, Birkhäuser, Boston, Mass., 1986.) *27,28*

[118] *Dvoretzky, A., Kiefer, J. and Wolfowitz, J. (1953) Sequential decision problems for processes with continuous time parameter. Testing hypotheses, *Ann. Math. Statist.* **24**, 254–264. *445*

[119] Dwass, M. and Teicher, H. (1957) On infinitely divisible random vectors, *Ann. Math. Statist.* **28**, 461–470. *68,458*

[120] Dynkin, E. B. (1952) Criteria of continuity and absence of discontinuity of the second kind for trajectories of a Markov process, *Izv. Akad. Nauk SSSR Ser. Mat.* **16**, 563–572 (in Russian). *59*

[121] Dynkin, E. B. (1965) *Markov Processes,* I, II, Springer, New York. [Russian original 1963] *6,108*

[122] Dynkin, E. B. (1984) Gaussian and non-Gaussian random fields associated with Markov processes, *J. Func. Anal.* **55**, 344–376. *328*

[123] Embrechts, P., Goldie, C. M. and Veraverbeke, N. (1979) Subexponentiality and infinite divisibility, *Zeit. Wahrsch. Verw. Gebiete* **49**, 335–347. *163,164,450*

[124] Embrechts, P., Klüppelberg, C. and Mikosch, T. (1997) *Modelling Extremal Events for Insurance and Finance,* Springer, Berlin. *164*

[125] *Embrechts, P. and Maejima, M. (2002) *Selfsimilar Processes,* Princeton Univ. Press, Princeton and Oxford. *435*

[126] Erdoğan, M. B. and Ostrovskii, I. V. (1997) Non-symmetric Linnik distributions, *C. R. Acad. Sci. Paris* **325**, Sér. I, 511–516. *203*

[127] Erdős, P. (1939) On a family of symmetric Bernoulli convolutions, *Amer. J. Math.* **61**, 974–976. *184,455*

[128] Erdős, P. (1942) On the law of the iterated logarithm, *Ann. Math.* **43**, 419–436. *358*

[129] Erdős, P. and Révész, P. (1997) On the radius of the largest ball left empty by a Wiener process, *Stud. Sci. Math. Hungar.* **33**, 117–125. *368*

Erdős, P. *see also* [**82, 83**], [**115**]–[**117**]

[130] Erickson, K. B. (1973) The strong law of large numbers when the mean is undefined, *Trans. Amer. Math. Soc.* **185**, 371–381. *250,256,257*

[131] *Erickson, K. B., and Maller, R. A. (2005) Generalised Ornstein-Uhlenbeck processes and the convergence of Lévy integrals, *Séminaire de Probabilités XXXVIII* (Lect. Notes in Math. No. 1857, Springer, Berlin), 70–94. *456*

[132] *Esscher, F. (1932) On the probability function in the collective theory of risk, *Skand. Aktuarietidskr.* **15**, 175–195. *445*

[133] Ethier, S. N. and Kurtz, T. G, (1986) *Markov Processes. Characterization and Convergence,* Wiley, New York. *108,206,236*

[134] Evans, S. N. (1987) Multiple points in the sample paths of a Lévy process, *Probab. Theory Related Fields* **76**, 359–367. *380*

[135] Feller, W. (1943) The general form of the so-called law of the iterated logarithm, *Trans. Amer. Math. Soc.* **54**, 373–402. *357,358*

[136] Feller, W. (1946) The law of the iterated logarithm for identically distributed random variables, *Ann. Math.* **47**, 631–638. *358*

[137] Feller, W. (1966) Infinitely divisible distributions and Bessel functions associated with random walks, *J. Soc. Indust. Appl. Math.* **14**, 864–875. *234*

[138] Feller, W. (1968) *An Introduction to Probability Theory and Its Applications,* Vol. 1, 3rd ed., Wiley, New York. *196,243,387,465*

[139] Feller, W. (1971) *An Introduction to Probability Theory and Its Applications,* Vol. 2, 2nd ed., Wiley, New York. *10,30,34,42,47,68,90,98,117,128,174,233,234, 236,245,246,383,384,388,*410_2*,426,454,*466_2*,*472_2

[140] Feller, W. and Orey, S. (1961) A renewal theorem, *J. Math. Mech.* **10**, 619–624. *472*

[141] Ferguson, T. S. and Klass, M. J. (1972) A representation of independent increment process without Gaussian component, *Ann. Math. Statist.* **43**, 1634–1643. *142*

[142] Fisz, M. and Varadarajan, V. S. (1963) A condition for absolute continuity of infinitely divisible distribution functions, *Zeit. Wahrsch. Verw. Gebiete* **1**, 335–339. *196*

[143] Fitzsimmons, P. J. and Getoor, R. K. (1992) On the distribution of the Hilbert transform of the local time of a symmetric Lévy processes, *Ann. Probab.* **20**, 1487–1497. *328*

[144] Fitzsimmons, P. J. and Kanda, M. (1992) On Choquet's dichotomy of capacity for Markov processes, *Ann. Probab.* **20**, 342–349. *332*

[145] Fitzsimmons, P. J. and Salisbury, T. S. (1989) Capacity and energy for multiparameter Markov processes, *Ann. Inst. Henri Poincaré* **25**, 325–350. *380*

[146] Forst, G. (1979) A characterization of self-decomposable probabilities on the half-line, *Zeit. Wahrsch. Verw. Gebiete* **49**, 349–352. *426*

Forst, G. *see also* [**21**]

Foss, S. *see* [**7**]

[147] Freedman, D. (1983) *Brownian Motion and Diffusion,* 2nd ed., Springer, New York. *28,30,327,368*

[148] Fristedt, B. E. (1964) The behavior of increasing stable processes for both small and large times, *J. Math. Mech.* **13**, 849–856. *359*

[149] Fristedt, B. E. (1967) Sample function behavior of increasing processes with stationary, independent increments, *Pacific J. Math.* **21**, 21–33. *359*

[150] Fristedt, B. (1974) Sample functions of stochastic processes with stationary, independent increments, *Advances in Probability,* Vol. 3 (ed. P. Ney and S. Port, Marcel Dekker, New York), 241–396. *117,319,359,379,381,384,475,478*

[151] Fristedt, B. and Gray, L. (1997) *A Modern Approach to Probability Theory,* Birkhäuser, Boston, Mass. *5,9,174,179,182,228,243,*245_2*,387,388*

[152] Fristedt, B. E. and Pruitt, W. E. (1971) Lower functions for increasing random walks and subordinators, *Zeit. Wahrsch. Verw. Gebiete* **18**, 167–182. *360,379*

[153] Fristedt, B. and Taylor, S. J. (1973) Strong variation for the sample functions of a stable process, *Duke Math. J.* **40**, 259–278. *381*

Fuchs, W. H. *see* [**84**]

[154] Fukushima, M. (1972) On transition probabilities of symmetric strong Markov processes, *J. Math. Kyoto Univ.,* **12**, 431–450. *288*

[155] Fukushima, M. (1976) Potential theory of symmetric Markov processes and its applications, *Proc. Third Japan–USSR Symp. Probab. Theory* (ed. G. Maruyama and J. V. Prokhorov, Lect. Notes in Math. No. 550, Springer, Berlin), 119–133. *288*

[156] Fukushima, M., Oshima, Y. and Takeda, M. (1994) *Dirichlet Forms and Symmetric Markov Processes,* de Gruyter, Berlin. *232*

[157] Gawronski, W. and Wießner, M. (1992) Asymptotics and inequalities for the mode of stable laws, *Statist. and Decisions* **10**, 183–197. *416*

[158] Getoor, R. K. (1965) Some asymptotic formulas involving capacity, *Zeit. Wahrsch. Verw. Gebiete* **4**, 248–252. *263*

[159] Getoor, R. K. (1966) Continuous additive functionals of a Markov process with applications to processes with independent increments, *J. Math. Anal. Appl.* **13**, 132–153. *332*

[160] Getoor, R. K. (1979) The Brownian escape process, *Ann. Probab.* **7**, 864–867. *101*

[161] Getoor, R. K. and Kesten, H. (1972) Continuity of local times for Markov processes, *Compositio Math.* **24**, 277–303. *328*

[162] Getoor, R. K. and Sharpe, M. J. (1973) Last exit times and additive functinals, *Ann. Probab.* **1**, 550–569. *303*

[163] Getoor, R. K. and Sharpe, M. J. (1994) On the arc-sine law for Lévy processes, *J. Appl. Probab.* **31**, 76–89. *373*

Getoor, R.K. *see also* [**41**]–[**47**], [**143**]

[164] Gihman, I. I. and Skorohod, A. V. (1974) *The Theory of Stochastic Processes,* Vol. 1, Springer, Berlin. [Russian original 1971] *68*

[165] Gihman, I. I. and Skorohod, A. V. (1975) *The Theory of Stochastic Processes,* Vol. 2, Springer, Berlin. [Russian original 1973] *29*

[166] Gnedenko, B. V. (1943) Sur la croissance des processus stochastiques homogènes à accroissements indépendants, *Izv. Akad. Nauk SSSR Ser. Mat.* **7**, 89–110 (in Russian with French summary). *366*

[167] Gnedenko, B. V. and Kolmogorov, A. N. (1968) *Limit Distributions for Sums of Independent Random Variables,* 2nd ed., Addison-Wesley, Reading, Mass. [Russian original 1949] *66,68$_2$,117*

[168] Goldie, C. (1967) A class of infinitely divisible random variables, *Proc. Cambridge Phil. Soc.* **63**, 1141–1143. *389,426$_2$*

Goldie, C. *see also* [**38, 123, 406**]

[169] Gradshteyn, I. S. and Ryzhik, I. M. (1980) *Table of Integrals, Series, and Products,* Corrected and enlarged ed., Academic Press, San Diego. *204,462,471*

Gray, L. *see* [**151**]

[170] Greenwood, P. and Perkins, E. A. (1983) A conditioned limit theorem for random walk and Brownian local time on square root boundaries, *Ann. Probab.* **11**, 227–261. *381*

[171] Greenwood, P. and Pitman, J. (1980) Fluctuation identities for Lévy processes and splitting at the maximum, *Adv. Appl. Probab.* **12**, 893–902. *377,384*

[172] *Grigelionis, B. (1999) Processes of Meixner type, *Liet. Mat. Rink.* **39**, 40-51. *445*

[173] *Grintsevichyus, A. K. (1974) On the continuity of the distribution of a sum of dependent variables connected with independent walks on lines, *Theory Probab. Appl.* **19**, 163–168. *453*

[174] Grosswald, E. (1976) The Student t-distribution of any degree of freedom is infinitely divisible, *Zeit. Wahrsch. Verw. Gebiete* **36**, 103–109. *46*

[175] Grübel, R. (1983) Über unbegrenzt teilbare Verteilungen, *Arch. Math.* **41**, 80–88. *164*

[176] Gruet, J.-C. and Shi, Z. (1996) The occupation time of Brownian motion in a ball, *J. Theoretic. Probab.* **9**, 429–445. *379*

[177] Gusak, D. V. and Korolyuk, V. S. (1969) On the joint distribution of a process with stationary independent increments and its maximum, *Theory Probab. Appl.* **14**, 400–409. *383*

Gusak, D. V. *see also* [**62**]

[178] Halgreen, C. (1979) Self-decomposability of the generalized inverse Gaussian and hyperbolic disributions, *Zeit. Wahrsch. Verw. Gebiete* **47**, 13–17. *98,234, 235,236$_2$,470$_2$*

Halgreen, C. *see also* [**13, 14**]

[179] Hall, P. (1984) On unimodality and rates of convergence for stable laws, *J. London Math. Soc.* **30**, 371–384. *416*

[180] Halmos, P. R. (1950) *Measure Theory,* Van Nostrand, Princeton, NJ. *5*

[181] Hamel, G. (1905) Eine Basis aller Zahlen und die unstetige Lösungen der Funktionalgleichungen: $f(x+y) = f(x) + f(y)$, *Math. Annalen* **60**, 459–462. *37*

Harn, K. van *see* [**502, 503**]

[182] Hartman, P. (1976) Completely monotone families of solutions of n-th order linear differential equations and infinitely divisible distributions, *Ann. Scuola Norm. Sup. Pisa* (4) **3**, 267–287; Errata, **3**, 725. *47*

[183] Hartman, P. and Wintner, A. (1941) On the law of the iterated logarithm, *Amer. J. Math.* **63**, 169–176. *357*

[184] Hartman, P. and Wintner, A. (1942) On the infinitesimal generators of integral convolutions, *Amer. J. Math.* **64**, 273–298. *196_3*

[185] Hawkes, J. (1970) Polar sets, regular points and recurrent sets for the symmetric and increasing stable processes, *Bull. London Math. Soc.* **2**, 53–59. *310*

[186] Hawkes, J. (1971) Some dimension theorems for the sample functions of stable processes, *Indiana Univ. Math. J.* **20**, 733–738. *380*

[187] Hawkes, J. (1971) A lower Lipschitz condition for the stable subordinator, *Zeit. Wahrsch. Verw. Gebiete* **17**, 23–32. *382*

[188] Hawkes, J. (1974) Local times and zero sets for processes with infinitely divisible distributions, *J. London Math. Soc.* (2) **8**, 517–525. *380*

[189] Hawkes, J. (1975) On the potential theory of subordinators, *Zeit. Wahrsch. Verw. Gebiete* **33**, 113–132. *475*

[190] Hawkes, J. (1979) Potential theory for Lévy processes, *Proc. London Math. Soc.* **38**, 335–352. *310,328_2,331,473*

[191] Hawkes, J. (1986) Local times as stationary processes, *From Local Times to Global Geometry, Control and Physics* (ed. K. D. Elworthy, Pitman Research Notes in Math. Ser. No. 150, Longman, Harlow, Essex), 111–120. *328*

[192] Hawkes, J. (1998) Exact capacity results for stable processes, *Probab. Theory Related Fields* **112**, 1–11. *380*

[193] Hawkes, J. and Pruitt, W. E. (1974) Uniform dimension results for processes with independent increments, *Zeit. Wahrsch. Verw. Gebiete* **28**, 277–288. *380_2*

Hawkes, J. *see also* [**10**]

[194] Hendricks, W. J. (1972) Hausdorff dimension in a process with stable components – an interesting counterexample, *Ann. Math. Statist.* **43**, 690–694. *382*

[195] Hendricks, W. J. (1973) A dimension theorem for sample functions of processes with stable components, *Ann. Probab.* **1**, 849–853. *382*

[196] Hendricks, W. J. (1974]) Multiple points for a process in R^2 with stable components, *Zeit. Wahrsch. Verw. Gebiete* **28**, 113–128. *382*

[197] Hengartner, W. and Theodorescu, R. (1973) *Concentration Functions,* Academic Press, New York. *114,384*

[198] Herz, C. S. (1964) *Théorie Elémentaire des Distributions de Beurling,* Publication du Séminaire de Mathématique d'Orsay, France. *332*

[199] Heyde, C. C. (1969) On the maximum of sums of random variables and the supremum functional for stable processes, *J. Appl. Probab.* **6**, 419–429. *384*

[200] Hida, T. (1980) *Brownian Motion,* Springer, New York. [Japanese original 1975] *22,30*

[201] Hille, E. (1948) *Functional Analysis and Semi-Groups,* Amer. Math. Soc., Providence, RI. *206*

[202] *Hiraba, S. (1994) Asymptotic behaviour for densities of multi-dimensional stable distributions, *Tsukuba J. Math.* **18**, 223–246. *449*

[203] *Hiraba, S. (2003) Asymptotic estimates for densities of multi-dimensional stable distributions, *Tsukuba J. Math.* **27**, 261–287. *449*

[204] Hirsch, W. M. (1965) A strong law for the maximum cumulative sum of independent random variables, *Comm. Pure Appl. Math.* **18**, 109–127. *368*

[205] Hoeffding, W. (1961) On sequences of sums of independent random vectors, *Proc. Fourth Berkeley Symp. Math. Statist. Probab.* (ed. J. Neyman, Univ. California Press, Berkeley), Vol. 2, 213–226. *196*

[206] Holgate, P. (1970) The modality of some compound Poisson distributions, *Biometrika* **57**, 666–667. *426*

Hopf, E. *see* [**574**]

[207] Horowitz, J. (1968) The Hausdorff dimension of the sample path of a subordinator, *Israel J. Math.* **6**, 176–182. *362,379*

[208] Hunt, G. A. (1956) Semigroups of measures on Lie groups, *Trans. Amer. Math. Soc.*. **81**, 264–293. *236*

[209] Hunt, G. A. (1957,1958) Markoff processes and potentials, I, II, and III, *Illinois J. Math.* **1**, 44–93, **1**, 316–369, and **2**, 151–213. *330,331*$_5$*,474*

[210] Ibragimov, I. A. (1956) On the composition of unimodal distributions, *Theory Probab. Appl.* **1**, 255–260. *426*

[211] Ibragimov, I. A. (1972) On a problem of C. R. Rao on i.d. laws, *Sankhyā* A **34**, 447–448. *458*

[212] Ibragimov, I. A. and Linnik, Yu. V. (1972) *Independent and Stationary Sequences of Random Variables,* Wolters-Noordhoff, Groningen, Netherlands. [Russian original 1965] *117*

[213] Ikeda, N. and Watanabe, S. (1962) On some relations between the harmonic measure and the Lévy measure for a certain class of Markov processes, *J. Math. Kyoto Univ.* **2**, 79–95. *236,384*

[214] Ikeda, N. and Watanabe, S. (1973) The local structure of a class of diffusions and related problems, *Proc. Second Japan–USSR Symp. Probab. Theory* (ed. G. Maruyama and Yu. V. Prokhorov, Lect. Notes in Math. No. 330, Springer, Berlin), 124–169. *378*$_2$

[215] Ikeda, N. and Watanabe, S. (1989) *Stochastic Differential Equations and Diffusion Processes,* 2nd ed., North-Holland/Kodansha, Amsterdam/Tokyo. *232*

[216] Inoue, K. (1996) Admissible perturbations of processes with independent increments, *Probab. Math. Statist.* **16**, 45–63. *236*

[217] Ismail, M. E. H. (1977) Bessel functions and the infinite divisibility of the Student t-distribution, *Ann. Probab.* **5**, 582–585. *46*

[218] Ismail, M. E. H. and Kelker, D. H. (1979) Special functions, Stieltjes transforms and infinite divisibility, *SIAM J. Math. Anal.* **10**, 884–901. *46,47,234*

[219] Isozaki, Y. (1996) Asymptotic estimates for the distribution of additive functionals of Brownian motion by the Wiener–Hopf factorization method, *J. Math. Kyoto Univ.* **36**, 211–227. *384*

[220] Itô, K. (1942) On stochastic processes, I (Infinitely divisible laws of probability), *Japan. J. Math.* **18**, 261–301. (Reprinted in *Kiyosi Itô Selected Papers*, Springer, New York, 1987.) *39,119,144*

[221] Itô, K. (1951) *On Stochastic Differential Equations,* Memoirs Amer. Math. Soc., No. 4, Providence, RI. (Reprinted in *Kiyosi Itô Selected Papers*, Springer, New York, 1987.) *236*

[222] Itô, K. (1953) *Kakuritsuron* (Probability theory), Gendai Sûgaku 14, Iwanami, Tokyo (in Japanese). *29,68,144*

[223] Itô, K. (2006) *Essentials of Stochastic Processes,* Amer. Math. Soc., Providence, RI. [Japanese original 1957] *68,236*

[224] Itô, K. (1961) *Lectures on Stochastic Processes,* Tata Institute of Fundamental Research, Bombay. *30*

[225] Ito, K. (2004) *Stochastic Processes. Lectures Given at Aarhus University* (ed. O. Barndorff-Nielsen and K. Sato), Springer, Berlin. [Original lecture notes 1969] *30,68*$_2$*,196*$_2$

[226] Itô, K. (1972) Poisson point processes attached to Markov processes, *Proc. Sixth Berkeley Symp. Math. Statist. Probab.* (ed. L. M. Le Cam et al., Univ. California Press, Berkeley), Vol. 3, 225–239. (Reprinted in *Kiyosi Itô Selected Papers*, Springer, New York, 1987.) *144,384*

[227] Itô, K. (1991) *Kakuritsuron* (Probability theory), Iwanami Kiso Sûgaku Sensho, Iwanami, Tokyo (in Japanese). *29*

[228] Itô, K. and McKean, H. P., Jr. (1965) *Diffusion Processes and Their Sample Paths,* Springer, Berlin. *30_2,327,426*

[229] Itô, M. (1983,1986) Transient Markov convolution semi-groups and the associated negative definite functions, *Nagoya Math. J.* **92**, 153–161; Remarks, **102**, 181–184. *254*

[230] Jacod, J. and Shiryaev, A. N. (1987) *Limit Theorems for Stochastic Processes,* Springer, Berlin. *144,236*

[231] Jain, N. (1998) Large time asymptotics of Lévy processes and random walks, *J. Korean Math. Soc.* **35**, 583–611. *249*

[232] Jain, N. and Pruitt, W. E. (1973) Maxima of partial sums of independent random variables, *Zeit. Wahrsch. Verw. Gebiete* **27**, 141–151. *361,368*

[233] Jain, N. and Pruitt, W. E. (1975) The other law of the iterated logarithm, *Ann. Probab.* **3**, 1046–1049. *368*

[234] Jain, N. and Pruitt, W. E. (1987) Lower tail probability estimates for subordinators and nondecreasing random walks, *Ann. Probab.* **15**, 75–101. *360*

[235] Jain, N. and Taylor, S. J. (1973) Local asymptotic laws for Brownian motion, *Ann. Probab.* **1**, 527–549. *368,381*

[236] Jajte, R. (1977) Semi-stable probability measures on R^N, *Studia Math.* **61**, 29–39. *118*

[237] *James, L. F., Roynette, B. and Yor, M. (2008) Generalized gamma convolutions, Dirichlet means, Thorin measures, with explicit examples, *Probab. Surv.* **5**, 346–415. *438*

[238] Janicki, A. and Weron, A. (1994) *Simulation and Chaotic Behavior of α-Stable Stochastic Processes,* Marcel Dekker, New York. *79*

[239] *Jeanblanc, M., Pitman, J. and Yor, M. (2002) Self-similar processes with independent increments associated with Lévy and Bessel processes, *Stoch. Proc. Appl.* **100**, 223–231. *435*

Jesiak, B. *see* [**424**]

[240] Jessen, B. and Wintner, A. (1935) Distribution functions and Riemann zeta function, *Trans. Amer. Math. Soc.* **38**, 48–88. *196*

Joag-dev, K. *see* [**97**]

[241] Johnson, N. L. and Kotz, S. (1970) *Distributions in Statistics. Continuous Univariate Distributions – 1,* Wiley, New York. *46_2,194*

[242] Johnson, N. L. and Rogers, C. A. (1951) The moment problem for unimodal distributions, *Ann. Math. Statist.* **22**, 433–439. *426*

[243] Jurek, Z. J. (1982) An integral representation of operator-selfdecomposable random variables, *Bull. Acad. Polonaise Sci. Sér. Sci. Math.* **30**, 385–393. *117*

[244] *Jurek, Z. J. (1985) Relations between the *s*-selfdecomposable and selfdecomposable measures, *Ann. Probab.* **13**, 592–608. *439*

[245] Jurek, Z. J. and Mason, J. D. (1993) *Operator-Limit Distributions in Probability Theory,* Wiley, New York. *118*

[246] Jurek, Z. J. and Vervaat, W. (1983) An integral representation for self-decomposable Banach space valued random variables, *Zeit. Wahrsch. Verw. Gebiete* **62**, 247–262. *117*

[247] Kac, M. (1951) On some connections between probability theory and differential and integral equations, *Proc. Second Berkeley Symp. Math. Statist. Probab.* (ed. J. Neyman, Univ. California Press, Berkeley), 189–215. *372*

[248] Kagan, A. M., Linnik, Yu. V. and Rao, C. R. (1973) *Characterization Problems in Mathematical Statistics,* Wiley, New York. [Russian original 1972] *117*

[249] Kakutani, S. (1944) On Brownian motion in n-space, *Proc. Imp. Acad. Japan* **20**, 648–652. (Reprinted in *Shizuo Kakutani: Selected Papers*, Vol. 2, Birkhäuser, Boston, Mass., 1986.) *331,380*

[250] Kakutani, S. (1944) Two-dimensional Brownian motion and harmonic functions, *Proc. Imp. Acad. Japan* **20**, 706–714. (Reprinted in *Shizuo Kakutani: Selected Papers*, Vol. 2, Birkhäuser, Boston, Mass., 1986.) *331*

[251] Kakutani, S. (1945) Two-dimensional Brownian motion and the type problem of Riemann surfaces, *Proc. Imp. Acad. Japan* **21**, 138–140. (Reprinted in *Shizuo Kakutani: Selected Papers*, Vol. 1, Birkhäuser, Boston, Mass., 1986.) *331*

[252] Kakutani, S. (1948) On equivalence of infinite product measures, *Ann. Math.* **49**, 214–224. (Reprinted in *Shizuo Kakutani: Selected Papers*, Vol. 2, Birkhäuser, Boston, Mass., 1986.) *218,225,440*

Kakutani, S. *see also* [**116, 117**]

[253] Kanda, M. (1975) Some theorems on capacity for isotropic Markov processes with stationary independent increments, *Japan. J. Math.* **1**, 37–66. *332*

[254] Kanda, M. (1976) Two theorems on capacity for Markov processes with stationary independent increments, *Zeit. Wahrsch. Verw. Gebiete* **35**, 159–165. 331_4*,332*

[255] Kanda, M. (1978) Characterization of semipolar sets for processes with stationary independent increments, *Zeit. Wahrsch. Verw. Gebiete* **42**, 141–154. *332*

[256] Kanda, M. (1983) On the class of polar sets for a certain class of Lévy processes on the line, *J. Math. Soc. Japan* **35**, 221–242. *332*

[257] Kanda, M. and Uehara, M. (1981) On the class of polar sets for symmetric Lévy processes on the line, *Zeit. Wahrsch. Verw. Gebiete* **58**, 55–67. *332*

Kanda, M. *see also* [**144**]

Karandikar, R. L. *see* [**96**]

[258] Karatzas, I. and Shreve, S. E. (1991) *Brownian Motion and Stochastic Calculus,* 2nd ed., Springer, New York. *30,327*

[259] Karlin, S. (1968) *Total Positivity,* Vol. 1, Stanford Univ. Press., Stanford, Calif. *422,426*

[260] Karlin, S. and Taylor, H. M. (1975) *A First Course in Stochastic Processes,* 2nd ed., Academic Press, New York. *30*

[261] Karlin, S. and Taylor, H. M. (1981) *A Second Course in Stochastic Processes,* Academic Press, New York. *30*

[262] Kasahara, Y. (1978) Tauberian theorems of exponential type, *J. Math. Kyoto Univ.* **18**, 209–219. *466*

[263] Kasahara, Y. (1984) Limit theorems for Lévy processes and Poisson point processes and their applications to Brownian excursions, *J. Math. Kyoto Univ.* **24**, 521–538. *143,332*

[264] Kasahara, Y. and Kotani, S. (1979) On limit processes for a class of additive functionals of recurrent diffusion processes, *Zeit. Wahrsch. Verw. Gebiete* **49**, 133–153. *332*

[265] Kasahara, Y. and Watanabe, S. (1986) Limit theorems for point processes and their functionals, *J. Math. Soc. Japan* **38**, 543–574. *144*

[266] Katti, S.K. (1967) Infinite divisibility of integer-valued random variables, *Ann. Math. Statist.* **38**, 1306–1308. *426*

[267] Kaufman, R. (1969) Une propriété métrique du mouvement brownien, *C. R. Acad. Sci. Paris* **268**, 727–728. *380*

[268] Kawata, T. (1972) *Fourier Analysis in Probability Theory,* Academic Press, New York. *174_2,175*

[269] Keilson, J. (1963) The first passage time density for homogeneous skip-free walks on the continuum, *Ann. Math. Statist.* **34**, 1003–1011. *383*

[270] Keilson, J. and Steutel, F.W. (1972) Families of infinitely divisible distributions closed under mixing and convolution, *Ann. Math. Statist.* **43**, 242–250. *425, 426,480*

Kelker, D.H. *see* [**218**]

[271] Kent, J.T. (1982) The spectral decomposition of a diffusion hitting time, *Ann. Probab.* **10**, 207–219. *426*

[272] Kesten, H. (1969) *Hitting Probabilities of Single Points for Processes with Stationary Independent Increments,* Mem. Amer. Math. Soc., No. 93, Providence, RI. *317,326,329,332*

[273] Kesten, H. (1970) The limit points of a normalized random walk, *Ann. Math. Statist.* **41**, 1173–1205. *249,256,368*

[274] Kesten, H. (1972) Sums of independent random variables – without moment conditions, *Ann. Math. Statist.* **43**, 701–732. *368*

Kesten, H. *see also* [**161**]

[275] Khintchine, A. (1924) Über einen Satz der Wahrscheinlichkeitsrechnung, *Fund. Math.* **6**, 9–20. *357*

[276] Khintchine, A. (1933) *Asymptotische Gesetze der Wahrscheinlichkeitsrechnung,* Springer, Berlin. *357*

[277] Khintchine, A. Ya. (1937) A new derivation of a formula of Paul Lévy, *Bull. Moscow Gov. Univ.* **1**, No. 1, 1–5 (in Russian). *40*

[278] Khintchine, A. (1937) Zur Theorie der unbeschränkt teilbaren Verteilungsgesetze, *Mat. Sbornik* **44**, No.1, 79–119. *47,68,142,144*

[279] Khintchine, A. Ya. (1938) *Limit Laws for Sums of Independent Random Variables,* ONTI, Moscow–Leningrad (in Russian). *68_2,116,117*

[280] Khintchine, A. Ya. (1938) Zwei Sätze über stochastische Prozesse mit stabilen Verteilungen, *Mat. Sbornik* **3**, 577–584 (in Russian with German summary). *359*

[281] Khintchine, A. Ya. (1938) On unimodal distributions, *Trudy NIIMM Tomsk. Gos. Univ.* **2**, 1–6 (in Russian). *195*

[282] Khintchine, A. Ya. (1939) Sur la croissance locale des processus stochastiques homogènes à accroissements indépendants, *Izv. Akad. Nauk SSSR* **3**, 487–508 (in Russian with French summary). *358*

[283] Khoshnevisan, D. (1997) Escape rates for Lévy processes, *Stud. Sci. Math. Hungar.* **33**, 177–183. *382*

Kiefer, J. *see* [**118**]

[284] Kingman, J.F.C. (1964) Recurrence properties of processes with stationary independent increments, *J. Austral. Math. Soc.* **4**, 223–228. *240,272*

[285] Kingman, J.F.C. (1967) Completely random measures, *Pacific J. Math.* **21**, 59–78. *144*

[286] Kingman, J.F.C. (1993) *Poisson Processes,* Clarendon Press, Oxford. *30*

[287] Kinney, J. R. (1953) Continuity properties of sample functions of Markov processes, *Trans. Amer. Math. Soc.* **74**, 280–302. *59*

Klass, M. J. *see* [**141**]

Klüppelberg, C. *see* [**124**]

[288] Knight, F. B. (1963) Random walks and a sojourn density process of Brownian motion, *Trans. Amer. Math. Soc.* **109**, 56–86. *328*

[289] Knight, F. B. (1981) *Essentials of Brownian Motion and Diffusion*, Amer. Math. Soc., Providence, RI. *22,28$_2$,30,327*

[290] Knight, F. B. (1981) Characterization of Levy measures of inverse local times of gap diffusions, *Seminar on Stochastic Processes, 1981* (ed. E. Çinlar et al., Birkhäuser, Boston, Mass.), 53–78. *426*

[291] Kolmogoroff, A. (1929) Über das Gesetz des iterierten Logarithmus, *Math. Annalen* **101**, 126–135. [English translation: *Selected Works of A. N. Kolmogorov*, Vol. 2 (Kluwer Acad. Pub., Dordrecht, Netherlands, 1992), 32–42.] *357*

[292] Kolmogoroff, A. N. (1940) Wienersche Spiralen und einige andere interessante Kurven im Hilbertschen Raum, *C. R. (Doklady) Acad. Sci. URSS* **26**, 115–118. [English translation: *Selected Works of A. N. Kolmogorov*, Vol. 1 (Kluwer Acad. Pub., Dordrecht, Netherlands, 1991), 303–307.] *117*

[293] Kolmogorov, A. N. (1950) *Foundations of the Theory of Probability*, Chelsea Pub., New York. [German original 1933] *4*

Kolmogorov, A. N. (Kolmogoroff), *see also* [**167**]

[294] Kôno, N. (1977) The exact Hausdorff measure of irregularity points for a Brownian path, *Zeit. Wahrsch. Verw. Gebiete* **40**, 257–282. *381*

Korolyuk, V. S. *see* [**177**]

Korshunov, D. *see* [**7**]

Kotani, S. *see* [**264**]

Kotz, S. *see* [**241**]

[295] Kristiansen, G. K. (1994) A proof of Steutel's conjecture, *Ann. Probab.* **22**, 442–452. *390*

Kristiansen, G. K. *see also* [**56**]

[296] Kruglov, V. M. (1970) A note on infinitely divisible distributions, *Theory Probab. Appl.* **15**, 319–324. *196$_3$*

[297] Kruglov, V. M. (1972) On the extension of the class of stable distributions, *Theory Probab. Appl.* **17**, 685–694. *117*

[298] Kruglov, V. M. (1972) Integrals with respect to infinitely divisible distributions in a Hilbert space, *Math. Notes* **11**, 407–411. *196*

[299] *Kunita, H. (2004) Stochastic differential equations based on Lévy processes and stochastic flows of diffeomorphisms, *Real and Stochastic Analysis. New Perspectives* (ed. M. M. Rao, Trends in Mathematics, Birkhäuser, Boston), 305–373. *ix*

[300] Kunita, H. and Watanabe, S. (1967) On square integrable martingales, *Nagoya Math. J.* **30**, 209–245. *144,217,236$_2$,440*

Kurtz, T. G. *see* [**133**]

[301] *Kuznetsov, A., Kyprianou, A. E. and Rivero, V. (2012) The theory of scale functions for spectrally negative Lévy processes, *Lecture Notes in Math.* (Springer) **2061**, Lévy Matters II, 97–186. *ix*

[302] Kwapień, S. and Woyczyński, W. A. (1992) *Random Series and Integrals: Single and Multiple*, Birkhäuser, Boston, Mass. *144,196,430,431*

[303] *Kyprianou, A. E. (2006) *Introductory Lectures on Fluctuations of Lévy Processes with Applications*, Springer, Berlin Heidelberg. *ix*

[304] *Kyprianou, A. E., Pardo, J. C. and Watson, A. (2013) Hitting distributions of α-stable processes via path censoring and self-similarity, *Ann. Probab.*, to appear. *ix*

Kyprianou, A. E. *see also* [**301**]

[305] Lamperti, J. (1962) Semi-stable stochastic processes, *Trans. Amer. Math. Soc.* **104**, 62–78. *117,435*

[306] *Lamperti, J. (1972) Semi-stable Markov processes, I, *Zeit. Wahrsch. Verw. Gebiete* **22**, 205–225. *ix*

[307] Le Gall, J.-F. (1992) Some properties of planar Brownian motion, *Ecole d'Eté de Probabilités de Saint-Flour XX–1990* (ed. P. L. Hennequin, Lect. Notes in Math. No. 1527, Springer, Berlin), 111-235. *380*

[308] Le Gall, J.-F., Rosen, J. and Shieh, N.-R. (1989) Multiple points of Lévy processes, *Ann. Probab.* **17**, 503–515. *380*

[309] LePage, R. (1980) *Multidimensional Infinitely Divisible Variables and Processes. Part I: Stable Case*, Statistics Department, Stanford Univ., Technical Report No. 292, Calif. (Reprinted in *Probability Theory on Vector Spaces IV, Proc., Łańcut 1987* (ed. S. Cambanis and A. Weron, Lect. Notes in Math. No. 1391, Springer, Berlin, 1989), 153–163.) *143*

[310] LePage, R. (1981) Multidimensional infinitely divisible variables and processes. Part II, *Probability in Banach Spaces III, Proc., Medford 1980* (ed. A. Beck, Lect. Notes in Math. No. 860, Springer, Berlin), 279–284. *144*

[311] Lévy, P. (1925) *Calcul des Probabilités,* Gauthier-Villars, Paris. *116,117*

[312] Lévy, P. (1934) Sur les intégrales dont les éléments sont des variables aléatoires indépendantes, *Ann. Scuola Norm. Sup. Pisa* (2) **3**, 337–366; **4**, 217–218. (Reprinted in *Œuvre de Paul Lévy,* Vol. 4, Gauthier-Villars, Paris, 1980.) *39, 68,119*

[313] Lévy, P. (1939) Sur certains processus stochastiques homogènes, *Compositio Math.* **7**, 283–339. (Reprinted in *Œuvre de Paul Lévy,* Vol. 4, Gauthier-Villars, Paris, 1980.) *372*

[314] Lévy, P. (1948) The arithmetical character of the Wishart distribution, *Proc. Cambridge Phil. Soc.* **44**, 295–297. (Reprinted in *Œuvre de Paul Lévy,* Vol. 3, Gauthier-Villars, Paris, 1976.) *458*

[315] Lévy, P. (1951) Wiener's random function, and other Laplacian random functions, *Proc. Second Berkeley Symp. Math. Statist. Probab.* (ed. J. Neyman, Univ. California Press, Berkeley), 171–187. (Reprinted in *Œuvre de Paul Lévy,* Vol. 4, Gauthier-Villars, Paris, 1980.) *98*

[316] Lévy, P. (1953) La mesure de Hausdorff de la courbe du mouvement brownien, *Giorn. Istit. Ital. Attuari* **16**, 1–37. (Reprinted in *Œuvre de Paul Lévy,* Vol. 5, Gauthier-Villars, Paris, 1980.) *379*

[317] Lévy, P. (1954) *Théorie de l'Addition des Variables Aléatoires,* 2e éd., Gauthier-Villars, Paris. (1e éd. 1937) *29,68$_2$,116,117$_3$, 119,196,381*

[318] Lévy, P. (1965) *Processus Stochastiques et Mouvement Brownien,* 2e éd., Gauthier-Villars, Paris. (1e éd. 1948) *28,29,30$_2$,68,327,358*

[319] Linde, W. (1983) *Probability Measures in Banach Spaces – Stable and Infinitely Divisible Distributions,* Wiley, New York. *117*

[320] *Lindner, A. and Maller, R. (2005) Lévy integrals and the stationarity of generalised Ornstein–Uhlenbeck processes, *Stoch. Proc. Appl.* **115**, 1701–1722. *456*

[321] *Lindner, A. and Sato, K. (2009) Continuity properties and infinite divisibility of stationary distributions of some generalized Ornstein–Uhlenbeck processes, *Ann. Probab.* **37**, 250–274. *ix,455*

[322] *Lindner, A. and Sato, K. (2011) Properties of stationary distributions of a sequence of generalized Ornstein–Uhlenbeck processes, *Math. Nachr.* **284**, 2225–2248. *ix,455*

[323] Linnik Yu. V. (1954) On stable probability laws with exponent less than 1, *Dokl. Akad. Nauk SSSR* **94**, 619–621 (in Russian). *88*

[324] Linnik, Yu. V. (1964) *Decomposition of Probability Distributions,* Oliver and Boyd, Edinburgh. [Russian original 1960] *68*

[325] Linnik, J. V. and Ostrovskii, I. V. (1977) *Decomposition of Random Variables and Vectors,* Amer. Math. Soc., Providence, RI. [Russian original 1972] *9,68, 116,458,463*

Linnik, J. V. (Yu. V.) *see also* [**212, 248**]

[326] *Loève, M. (1945) Nouvelles classes de lois limites, *Bull. Soc. Math. France* **73**, 107–126. *453_2*

[327] Loève, M. (1977,1978) *Probability Theory,* Vol. 1 and 2, 4th ed., Springer, New York. (1st ed., Van Nostrand, Princeton, NJ, 1955) *30,68_2,117,119,128,182,212, 245,453,454,462*

[328] Łuczak, A. (1981,1987) Operator semi-stable probability measures on $\mathbf{R}^N$, *Colloq. Math.* **45**, 287–300; Corrigenda, **52**, 167–169. *118*

[329] Lukacs, E. (1970) *Characteristic Functions,* 2nd ed., Griffin, London. *175,457*

[330] Lynch, J. and Sethuraman, J. (1987) Large deviations for processes with independent increments, *Ann. Probab.* **15**, 610–627. *249*

[331] Maejima, M. (1989) Self-similar processes and limit theorems, *Sugaku Expositions* **2**, 103–123. *117*

[332] Maejima, M. and Naito, Y. (1998) Semi-selfdecomposable distributions and a new class of limit theorems, *Probab. Theory Related Fields,* **112**, 13–31. *91,117,118*

[333] Maejima, M. and Sato, K. (1999) Semi-selfsimilar processes, *J. Theoretic. Probab.,* **12**, 347–373. *75,117_2*

[334] *Maejima, M. and Sato, K. (2003) Semi-Lévy processes, semi-selfsimilar additive processes, and semi-stationary Ornstein–Uhlenbeck processes, *J. Math. Kyoto Univ.* **43**, 609–639. *435,436*

[335] *Maejima, M. and Sato, K. (2009) The limits of nested subclasses of several classes of infinitely divisible distributions are identical with the closure of the class of stable distributions, *Probab. Theory Related Fields* **145**, 119–142. *439*

Maejima, M. *see also* [**15, 125**]

[336] Maisonneuve, B. (1975) Exit systems, *Ann. Probab.* **3**, 399–411. *384*

Maller, R. A. *see* [**131, 320**]

[337] Marcus, M. B. and Rosen, J. (1992) Sample path properties of the local times of strongly symmetric Markov processes via Gaussian processes, *Ann. Probab.* **20**, 1603–1684. *328*

[338] Marcus, M. B. and Rosen, J. (1992) *p*-variation of the local times of symmetric stable processes and of Gaussian processes with stationary increments, *Ann. Probab.* **20**, 1685–1713. *328*

Martin, W. T. *see* [**72**]

[339] Maruyama, G. (1954) On the transition probability functions of the Markov process, *Natural Science Report, Ochanomizu Univ.* **5**, 10–20. (Reprinted in *Gisiro Maruyama Selected Papers*, Kaigai Pub., Tokyo, 1988.) *232*

[340] Maruyama, G. (1970) Infinitely divisible processes, *Theory Probab. Appl.* **15**, 1–22. (Reprinted in *Gisiro Maruyama Selected Papers*, Kaigai Pub., Tokyo, 1988.) *68,427,428*

Mason, D. M. *see* [**90**]

Mason, J. D. *see* [**245**]

[341] *Matsumoto, H. and Yor, M. (2005) Exponential functionals of Brownian motion, I: Probability laws at fixed time; II: Some related diffusion processes, *Probab. Surv.* **2**, 312–347 and 348–384. *ix*

[342] McKean, H. P., Jr. (1955) Hausdorff–Besicovitch dimension of Brownian motion paths, *Duke Math. J.* **22**, 229–234. *380*

McKean, H. P., Jr., *see also* [**228**]

[343] Medgyessy, P. (1967) On a new class of unimodal infinitely divisible distribution functions and related topics, *Stud. Sci. Math. Hungar.* **2**, 441–446. *417*

[344] Memin, J. and Shiryayev, A. N. (1985) Distance de Hellinger–Kakutani des lois correspondant à deux processus à accroissements indépendants, *Zeit. Wahrsch. Verw. Gebiete* **70**, 67–89. *236,440*

[345] Meyer, P.-A. (1966) *Probabilités et Potentiel,* Hermann, Paris. [English version: *Probability and Potentials,* Blaisdell, Waltham, Mass., 1966] *245,331*

[346] Mijnheer, J. L. (1975) *Sample Path Properties of Stable Processes,* Math. Centre Tracts, No. 59, Math. Centrum, Amsterdam. *358*

Mikosch, T. *see* [**124**]

[347] Millar, P. W. (1971) Path behavior of processes with stationary independent increments, *Zeit. Wahrsch. Verw. Gebiete* **17**, 53–73. *194,380,465*

[348] Millar, P. W. (1973) Exit properties of stochastic processes with stationary independent increments, *Trans. Amer. Math. Soc.* **178**, 459–479. *378*

[349] Millar, P. W. (1973) Radial processes, *Ann. Probaab.* **1**, 613–626. *330*

[350] Millar, P. W. (1977) Zero–one laws and the minimum of a Markov process, *Trans. Amer. Math. Soc.* **226**, 365–391. *384*

[351] Millar, P. W. and Tran, L. T. (1974) Unbounded local times, *Zeit. Wahrsch. Verw. Gebiete* **30**, 87–92. *328*

[352] *Mišeikis, F. F. (1975) Interrelationship between certain classes of probability distributions, *Lithuanian Math. J.* **15**, 243–246. *454*

[353] Mogulskii, A. A. (1993) Large deviations for processes with independent increments, *Ann. Probab.* **21**, 202–215. *249*

[354] Molchanov, S. A. and Ostrovskii, E. (1969) Symmetric stable processes as traces of degenerate diffusion processes, *Theory Probab. Appl.* **14**, 128–131. *236*

[355] Monrad, D. and Silverstein, M. L. (1979) Stable processes: sample function growth at a local minimum, *Zeit. Wahrsch. Verw. Gebiete* **49**, 177–210. *384*

[356] Mori, T. (1972) A note on fluctuations of random walks without the first moment, *Yokohama Math. J.* **20**, 51–55. *250*

[357] Motoo, M. (1958) Proof of the law of the iterated logarithm through diffusion equation, *Ann. Inst. Statist. Math.* **10**, 21–28. *358*

[358] Motoo, M. (1967) Application of additive functionals to the boundary problem of Markov provesses (Lévy's system of U-processes), *Proc. Fifth Berkeley Symp. Math. Statist. Probab.* (ed. L. M. Le Cam and J. Neyman, Univ. California Press, Berkeley), Vol. 2, Part 2, 75–110. *384*

[359] Nagasawa, M. (1964) Time reversions of Markov processes, *Nagoya Math. J.* **24**, 177–204. *331*

Naito, Y. *see* [**332**]

[360] Newman, C. M. (1972) The inner product of path space measures corresponding to random processes with independent increments, *Bull. Amer. Math. Soc.* **78**, 268–271. *217,236,440*$_2$

[361] Newman, C. M. (1973) On the orthogonality of independent increment processes, *Topics in Probability Theory* (ed. D. W. Stroock and S. R. S. Varadhan, Courant Inst. Math. Sci., New York Univ., New York), 93–111. *217,236*$_3$*,440*$_2$*,442*

[362] *Niedbalska-Rajba, T. (1981) On decomposability semigroups on the real line, *Colloq. Math.* **44**, 347–358. *454*

[363] Orey, S. (1967) Polar sets for processes with stationary independent increments, *Markov Processes and Potential Theory* (ed. J. Chover, Wiley, New York), 117–126. *331,332*$_2$

[364] Orey, S. (1968) On continuity properties of infinitely divisible distribution functions, *Ann. Math. Statist.* **39**, 936–937. *183,190,194,196,293*

[365] Orey, S. and Taylor, S. J. (1974) How often on a Brownian path does the law of iterated logarithm fail? *Proc. London Math. Soc.* **28**, 174–192. *381*

Orey, S. *see also* [**140**]

[366] Ornstein, D. (1969) Random walks, I and II, *Trans. Amer. Math. Soc.* **138**, 1–43 and 45–60. *254*

Oshima, Y. *see* [**156**]

Ostrovskii, E. *see* [**354**]

Ostrovskii, I. V. *see* [**126, 325**]

Pardo, J. C. *see* [**304**]

[367] Parthasarathy, K. R. (1967) *Probability Measures on Metric Space,* Academic Press, New York. *68*

[368] Pecherskii, E. A. and Rogozin, B. A. (1969) On joint distribution of random variables associated with fluctuations of a process with independent increments, *Theory Probab. Appl.* **14**, 410–423. *68,334,383*$_2$*,384*$_2$

Pecherskii, E. A. (Pechersky) *see also* [**99**]

Pemantle, R. *see* [**5**]

[369] Peres, Y. and Solomyak, B. (1998) Self-similar measures and intersections of Cantor sets, *Trans. Amer. Math. Soc.* **350**, 4065–4087. *184,194,455*

[370] Perkins, E. A. and Taylor, S. J. (1987) Uniform measure results for the image of subsets under Brownian motion, *Probab. Theory Related Fields* **76**, 257–289. *380*

Perkins, E. A. *see also* [**11, 170**]

[371] Petrov, V. V. (1975) *Sums of Independent Random Variables*, Springer, Berlin. [Russian original 1972] *196*

[372] Petrowsky, I. (1935) Zur ersten Randwertaufgabe der Wärmleitungsgleichung, *Compositio Math.* **1**, 383–419. *358*

[373] Phillips, R. S. (1952) On the generation of semigroups of linear operators, *Pacific J. Math.* **2**, 343–369. *212*

[374] Pillai, R. N. (1971) Semi stable laws as limit distributions, *Ann. Math. Statist.* **42**, 780–783. *117*

[375] Pillai, R. N. (1990) On Mittag–Leffler functions and related distributions, *Ann. Inst. Statist. Math.* **42**, 157–161. *234,469*

Pinsky, M. *see* [**85**]

[376] Pitman, J. W. (1975) One-dimensional Brownian motion and the three-dimensional Bessel process, *Adv. Appl. Probab.* **7**, 511–526. *101*

[377] Pitman, J. and Yor, M. (1981) Bessel processes and infinitely divisible laws, *Stochastic Integrals Proc. LMS Durham Symp. 1980* (ed. D. Williams, Lect. Notes in Math. No. 851, Springer, Berlin), 285–370. *47*

Pitman, J. *see also* [**171, 239**]

[378] Pólya, G. (1921) Über eine Aufgabe der Wahrscheinlichkeitsrechnung betreffend die Irrfahrt im Straßennetz, *Math. Annalen* **84**, 149–160. *243*

[379] Pólya, G. and Szegö, G. (1976) *Problems and Theorems in Analysis*, Vol. 2, Springer, Berlin. [German original, 4th ed., 1971] *422*

[380] Port, S. C. (1966) Limit theorems involving capacities, *J. Math. Mech.* **15**, 805–832. *263*

[381] Port, S. C. (1967) Hitting times and potentials for recurrent stable processes, *J. Anal. Math.* **20**, 371–395. *332*

[382] Port, S. C. (1967) Potentials associated with recurrent stable processes, *Markov Processes and Potential Theory* (ed. J. Chover, Wiley, New York), 135–163. *332*

[383] Port, S. C. (1989) Stable processes with drift on the line, *Trans. Amer. Math. Soc.* **313**, 805–841. *263,332*

[384] Port, S. C. (1990) Asymptotic expansions for the expected volume of a stable sausage, *Ann. Probab.* **18**, 492–523. *263*

[385] Port, S. C. and Stone, C. J. (1969) The asymmetric Cauchy processes on the line, *Ann. Math. Statist.* **40**, 137–143. *332*

[386] Port, S. C. and Stone, C. J. (1971) Infinitely divisible processes and their potential theory, I and II, *Ann. Inst. Fourier* **21**, Fasc. 2, 157–275 and Fasc. 4, 179–265. *254,263,272,331_5,332,472*

[387] Port, S. C. and Stone, C. J. (1978) *Brownian Motion and Classical Potential Theory,* Academic Press, New York. *331,474_3*

[388] Port, S. C. and Vitale, R. A. (1988) Positivity of stable densities, *Proc. Amer. Math. Soc.* **102**, 1018–1023. *193*

[389] Prabhu, N. U. (1972) Wiener–Hopf factorization for convolution semigroups, *Zeit. Wahrsch. Verw. Gebiete* **23**, 103–113. *383,478*

[390] Prabhu, N. U. (1980) *Stochastic Storage Processes. Queues, Insurance Risk, and Dams*, Springer, New York. *384*

[391] Prabhu, N. U. and Rubinovitch, M. (1973) Further results for ladder processes in continuous time, *Stoch. Proc. Appl.* **1**, 151–168. *384*

[392] Pruitt, W. E. (1969) The Hausdorff dimension of the range of a process with stationary independent increments, *J. Math. Mech.* **19**, 371–378. *379*

[393] Pruitt, W. E. (1975) Some dimension results for processes with independent increments, *Stochastic Processes and Related Topics* (ed. M. L. Puri, Academic Press, New York), 133–165. *117,381*

[394] Pruitt, W. E. (1981) General one-sided laws of iterated logarithm, *Ann. Probab.* **9**, 1–48. *368*

[395] Pruitt, W. E. (1981) The growth of random walks and Lévy processes, *Ann. Probab.* **9**, 948–956. *362_2,367*

[396] Pruitt, W. E. (1983) The class of limit laws for stochastically compact normed sums, *Ann. Probab.* **11**, 962–969. *426*

[397] Pruitt, W. E. (1990) The rate of escape of random walk, *Ann. Probab.* **18**, 1417–1461. *368*

[398] Pruitt, W. E. (1991) An integral test for subordinators, *Random Walks, Brownian Motion, and Interacting Particle Systems. A Festschrift in Honor of Frank Spitzer* (ed. R. Durrett and H. Kesten, Birkhäuser, Boston, Mass.), 389–398. *360*

[399] Pruitt, W. E. and Taylor, S. J. (1969) Sample path properties of processes with stable components, *Zeit. Wahrsch. Verw. Gebiete* **12**, 267–289. *361,379,382*

[400] Pruitt, W. E. and Taylor, S. J. (1969) The potential kernel and hitting probabilities for the general stable process in R^N, *Trans. Amer. Math. Soc.* **146**, 299–321. *332,361,449*

[401] Pruitt, W. E. and Taylor, S. J. (1977) Some sample path properties of the asymmetric Cauchy processes, *Proc. Symp. Pure Math.* **31**, 111–123. *382*

[402] Pruitt, W. E. and Taylor, S. J. (1977) Hausdorff measure properties of the asymmetric Cauchy processes, *Ann. Probab.* **5**, 608–615. *379*

[403] Pruitt, W. E. and Taylor, S. J, (1983) The behavior of asymmetric Cauchy processes for large time, *Ann. Probab.* **11**, 302–327. *382*

[404] Pruitt, W. E. and Taylor, S. J. (1985) The local structure of the sample paths of asymmetric Cauchy processes, *Zeit. Wahrsch. Verw. Gebiete* **70**, 535–561. *382*

[405] Pruitt, W. E. and Taylor, S. J, (1996) Packing and covering indices for a general Lévy process, *Ann. Probab.* **24**, 971–986. *379*

Pruitt, W. E. *see also* [**152**], [**193**], [**232**]–[**234**]

[406] Puri, P. S. and Goldie, C. M. (1979) Poisson mixtures and quasi-infinite divisibility of distributions, *J. Appl. Probab.* **16**, 138–153. *426*

[407] *Rajput, B. and Rosinski, J. (1989) Spectral representations of infinitely divisible processes, *Probab. Theory Related Fields* **82**, 451–487. *430,431*

[408] Ramachandran, B. (1969) On characteristic functions and moments, *Sankhyā,* A **31**, 1–12. *196*

Rao, C. R. *see* [**248**]

[409] Rao, M. (1987) On polar sets for Lévy processes, *J. London Math. Soc.* **35**, 569–576. *332*

[410] Ray, D. (1958) Stable processes with an absorbing barrier, *Trans. Amer. Math. Soc.* **89**, 16–24. *305*

[411] Ray, D. (1963) Sojourn times of diffusion processes, *Illinois J. Math.* **7**, 615–630. *328*

[412] Ray, D. (1967) Some local properties of Markov processes, *Proc. Fifth Berkeley Symp. Math. Statist. Probab.* (ed. L. M. Le Cam and J. Neyman, Univ. California Press, Berkeley), Vol. 2, Part 2, 201–212. *379*

Ray, D. *see also* [**47**]

[413] Resnick, S. I. (1992) *Adventures in Stochastic Processes,* Birkhäuser, Boston, Mass. *30,272,387*

[414] Révész, P. (1990) *Random Walk in Random and Non-Random Environments,* World Scientific, Singapore. *382*

Révész, P. *see also* [**129**]

[415] Revuz, D. and Yor, M. (1999) *Continuous Martingales and Brownian Motion,* 3rd ed., Springer, Berlin. *30,474*

Rivero, V. *see* [**301**]

[416] *Rocha-Arteaga, A. and Sato, K. (2003) *Topics in Infinitely Divisible Distributions and Lévy Processes,* Aportaciones Matemáticas, Investigación 17, Sociedad Matemática Mexicana, México. *437*

[417] Rogers, C. A. (1970) *Hausdorff Measures,* Cambridge Univ. Press, Cambridge. *379*

Rogers, C. A. *see also* [**242**]

[418] Rogozin, B. A. (1965) On some classes of processes with independent increments, *Theory Probab. Appl.* **10**, 479–483. *236*

[419] Rogozin, B. A. (1966) On the distribution of functionals related to boundary problems for processes with independent increments, *Theory Probab. Appl.* **11**, 580–591. *236,334,363,383$_2$,384*

[420] Rogozin, B. A. (1968) Local behavior of processes with independent increments, *Theory Probab. Appl.* **13**, 482–486. *383$_2$*

Rogozin, B. A. *see also* [**368**]

Rosen, J. *see* [**308, 337, 338**]

Rosenblatt, M. *see* [**40**]

[421] Rosinski, J. (1990) On series representations of infinitely divisible random vectors, *Ann. Probab.* **18**, 405–430. *143,144,464$_2$*

[422] *Rosiński, J. (2007) Tempering stable processes, *Stoch. Proc. Appl.* **117**, 677–707. *439*

[423] Rosinski, J. and Samorodnitsky, G. (1993) Distributions of subadditive functionals of sample paths of infinitely divisible processes, *Ann. Probab.* **21**, 996–1014. *167*

Rosiński, J. (Rosinski) *see also* [**16, 407**]

[424] Rossberg, H.-J., Jesiak, B. and Siegel, G. (1985) *Analytic Methods of Probability Theory,* Akademie-Verlag, Berlin. *67,68*

Roynette, B. *see* [**237**]

[425] Rubin, H. (1967) Supports of convolutions of identical distributions, *Proc. Fifth Berkeley Symp. Math. Statist. Probab.* (ed. L. M. Le Cam and J. Neyman, Univ. California Press, Berkeley), Vol. 2, Part 1, 415–422. *196*

Rubinovitch, M. *see* [**391**]

[426] Rvačeva, E. L. (1962) On domains of attraction of multi-dimensional distributions, *Selected Transl. Math. Statist. and Probab.*, Vol. 2 (AMS, Providence, RI), 183–205. *117*

Ryzhik, I. M. *see* [**169**]

Salisbury, T. S. *see* [**145**]

[427] Samorodnitsky, G. and Taqqu, M. S. (1994) *Stable Non-Gaussian Random Processes,* Chapman & Hall, New York. *117*

Samorodnitsky, G. *see also* [**64, 423**]

[428] Sato, K. (1972) Potential operators for Markov processes, *Proc. Sixth Berkeley Symp. Math. Statist. Probab.* (ed. L. M. Le Cam et al., Univ. California Press, Berkeley), Vol. 3, 193–211. *236$_2$,384*

[429] Sato, K. (1972) Cores of potential operators for processes with stationary independent increments, *Nagoya Math. J.* **48**, 129–145. *236*

[430] Sato, K. (1973) A note on infinitely divisible distributions and their Lévy measures, *Sci. Rep. Tokyo Kyoiku Daigaku,* Sec. A, **12**, 101–109. *196$_2$*

[431] Sato, K. (1980) Class L of multivariate distributions and its subclasses, *J. Multivar. Anal.* **10**, 207–232. *117,118,437,480*

[432] Sato, K. (1981) *Mugen-Bunkai-Kanou Bunpu* (Infinitely divisible distributions), Seminar on Probability, Vol. 52, published by Kakuritsuron Seminar, Japan (in Japanese). *68*

[433] Sato, K. (1982) Absolute continuity of multivariate distributions of class L, *J. Multivar. Anal.* **12**, 89–94. *196$_2$*

[434] Sato, K. (1985) *Lectures on Multivariate Infinitely Divisible Distributions and Operator-Stable Processes,* Technical Report Series, Lab. Res. Statist. Probab. Carleton Univ. and Univ. Ottawa, No. 54, Ottawa. *68*

[435] Sato, K. (1986) Bounds of modes and unimodal processes with independent increments, *Nagoya Math. J.* **104**, 29–42. *415,416*

[436] Sato, K. (1986) Behavior of modes of a class of processes with independent increments, *J. Math. Soc. Japan* **38**, 679–695. *415*

[437] Sato, K. (1987) Modes and moments of unimodal distributions, *Ann. Inst. Statist. Math.* **39**, Part A, 407–415. *426*

[438] Sato, K. (1987) Strictly operator-stable distributions, *J. Multivar. Anal.* **22**, 278–295. *118*

[439] Sato, K. (1990) Subordination depending on a parameter, *Probability Theory and Mathematical Statistics, Proc. Fifth Vilnius Conf.* (ed. B. Grigelionis et al., VSP/Mokslas, Utrecht/Vilnius) Vol. 2, 372–382. *212,231,233*

[440] Sato, K. (1991) Self-similar processes with independent increments, *Probab. Theory Related Fields* **89**, 285–300. *117_2*

[441] Sato, K. (1992) On unimodality and mode behavior of Lévy processes, *Probability Theory and Mathematical Statistics, Proc. Sixth USSR-Japan Symp.* (ed. A. N. Shiryaev et al., World Scientific, Singapore), 292–305. *426*

[442] Sato, K. (1993) Convolution of unimodal distributions can produce any number of modes, *Ann. Probab.* **21**, 1543–1549. *395*

[443] Sato, K. (1994) Multimodal convolutions of unimodal infinitely divisible distributions, *Theory Probab. Appl.* **39**, 336–347. *420*

[444] Sato, K. (1994) Time evolution of distributions of Lévy processes from continuous singular to absolutely continuous, *Research Bulletin, College of General Education, Nagoya Univ.*, Ser. B, **38**, 1–11. *196*

[445] Sato, K. (1995) Time evolution in distributions of Lévy processes, *Southeast Asian Bull. Math.* **19**, No. 2, 17–26. *196,418,420_3,422,426*

[446] Sato, K. (1995) *Lévy Processes on Euclidean Spaces*, Lecture Notes, Institute of Mathematics, University of Zurich. *xii*

[447] Sato, K. (1996) Criteria of weak and strong transience for Lévy processes, *Probability Theory and Mathematical Statistics, Proc. Seventh Japan–Russia Symp.* (ed. S. Watanabe et al., World Scientific, Singapore), 438–449. *263*

[448] Sato, K. (1997) Time evolution of Lévy processes, *Trends in Probability and Related Analysis, Proc. SAP '96* (ed. N. Kono and N.-R. Shieh, World Scientific, Singapore), 35–82. *263_2,272_2,426*

[449] Sato, K. (1999) Semi-stable processes and their extensions, *Trends in Probability and Related Analysis, Proc. SAP '98* (ed. N. Kono and N.-R. Shieh, World Scientific, Singapore), 129–145. *316,326,356,366*

[450] *Sato, K. (2000) *Density Transformation in Lévy Processes*, Lecture Notes, No. 7, MaPhySto, Centre for Math. Physics and Stochastics, Univ. Aarhus. *440,442,443_2*

[451] *Sato, K. (2001) Basic results on Lévy processes, *Lévy Processes, Theory and Applications* (ed. O. E. Barndorff-Nielsen, T. Mikosch and S. I. Resnick, Birkhäuser, Boston), 3–37. *x*

[452] *Sato, K. (2004) Stochastic integrals in additive processes and application to semi-Lévy processes, *Osaka J. Math.* **41**, 211–236. *431_2,434_2*

[453] *Sato, K. (2006) Additive processes and stochastic integrals, *Illinois J. Math.* **50**, 825–821. *431_2,434_2*

[454] *Sato, K. (2006) Two families of improper stochastic integrals with respect to Lévy processes, *ALEA Lat. Am. J. Probab. Math. Statist.* **1**, 47–87. *434,439*

[455] *Sato, K. (2006) Monotonicity and non-monotonicity of domains of stochastic integral operators, *Probab. Math. Stat.* **26**, 23–39. *434*

[456] *Sato, K. (2007) Transformations of infinitely divisible distributions via improper stochastic integrals, *ALEA Lat. Am. J. Probab. Math. Statist.* **3**, 67–110. *434*

[457] *Sato, K. (2010) Fractional integrals and extensions of selfdecomposability, *Lecture Notes in Math.* (Springer) **2001**, Lévy Matters I, 1–91. *432,434$_2$,437,439$_4$*

[458] *Sato, K. (2011) Stochastic integrals with respect to Lévy processes and infinitely divisible distributions, *Sūgaku* **63**, 161–181 (in Japanese). [English translation to appear in *Sugaku Expositions*] *455*

[459] *Sato, K. (2013) Inversions of infinitely divisible distributions and conjugates of stochastic integral mappings, *J. Theoretic. Probab.* DOI 10.1007/s10959-012-0420-9 *439*

[460] *Sato, K. and Ueda, Y. (2013) Weak drifts of infinitely divisible distributions and their applications, *J. Theoretic. Probab.* DOI 10.1007/s10959-012-0419-2 *430,439*

[461] Sato, K., Watanabe, Toshiro, Yamamuro, K. and Yamazato, M. (1996) Multidimensional process of Ornstein–Uhlenbeck type with nondiagonalizable matrix in linear drift terms, *Nagoya Math. J.* **141**, 45–78. *272*

[462] Sato, K., Watanabe, Toshiro and Yamazato, M. (1994) Recurrence conditions for multidimensional processes of Ornstein–Uhlenbeck type, *J. Math. Soc. Japan* **46**, 245–265. *272*

[463] *Sato, K. and Watanabe, Toshiro (2004) Moments of last exit times for Lévy processes, *Ann. Inst. H. Poincaré Probab. Statist.* **40**, 207–225. *446*

[464] *Sato, K. and Watanabe, Toshiro (2005) Last exit times for transient semistable processes, *Ann. Inst. H. Poincaré Probab. Statist.* **41**, 929–951. *446,447,456*

[465] Sato, K. and Yamamuro, K. (1998) On selfsimilar and semi-selfsimilar processes with independent increments, *J. Korean Math. Soc.* **35**, 207–224. *272*

[466] Sato, K. and Yamazato, M. (1978) On distribution functions of class L, *Zeit. Wahrsch. Verw. Gebiete* **43**, 273–308. *193,409$_2$,410$_2$,415*

[467] Sato, K. and Yamazato, M. (1981) On higher derivatives of distribution functions of class L, *J. Math. Kyoto Univ.* **21**, 575–591. *193,415*

[468] Sato, K. and Yamazato, M. (1983) Stationary processes of Ornstein–Uhlenbeck type, *Probability Theory and Mathematical Statistics, Fourth USSR–Japan Symp., Proc. 1982* (ed. K. Itô and J. V. Prokhorov, Lect. Notes in Math. No. 1021, Springer, Berlin), 541–551. *118$_2$*

[469] Sato, K. and Yamazato, M. (1984) Operator-selfdecomposable distributions as limit distributions of processes of Ornstein–Uhlenbeck type, *Stoch. Proc. Appl.* **17**, 73–100. *114,117,118$_2$,426*

[470] Sato, K. and Yamazato, M. (1985) Completely operator-selfdecomposable distributions and operator-stable distributions, *Nagoya J. Math.* **97**, 71–94. *118*

Sato, K. *see also* [**15, 77, 321, 322**], [**333**]–[**335**], [**416**]

[471] Schilder, M. (1966) Some asymptotic formulae for Wiener integrals, *Trans. Amer. Math. Soc.* **125**, 63–85. *249*

[472] *Schilling, R. L., Song, R. and Vondraček, Z. (2012) *Bernstein Functions. Theory and Applications*, 2nd ed., De Gruyter, Berlin. *ix*

[473] *Schoutens, W. and Teugels, J. L. (1998) Lévy processes, polynomials and martingales, *Comm. Statist. Stochastic Models* **14**, 335–349. *445*

Schwartz, J. T. *see* [**111**]

[474] Seshadri, V. (1993) *The Inverse Gaussian Distribution,* Oxford Univ. Press, Oxford. *233*

Sethuraman, J. *see* [**330**]

[475] Shanbhag, D. N. and Sreehari, M. (1977) On certain self-decomposable distributions, *Zeit. Wahrsch. Verw. Gebiete* **38**, 217–222. *163$_2$,195,463,466*

[476] Shanbhag, D. N. and Sreehari, M. (1979) An extension of Goldie's result and further results in infinite divisibility, *Zeit. Wahrsch. Verw. Gebiete* **47**, 19–25. *98*

Shapiro, J. M. *see* [**19**]

[477] Sharpe, M. (1969) Operator-stable probability distributions on vector groups, *Trans. Amer. Math. Soc.* **136**, 51–65. *118*

[478] Sharpe, M. (1969) Zeroes of infinitely divisible densities, *Ann. Math. Statist.* **40**, 1503–1505. *193*

[479] *Sharpe, M. J. (1995) Supports of convolution semigroups and densities, *Probability Measures on Groups and Related Structures, XI* (Oberwolfach, 1994) (World Scientific, River Edge, NJ), 364–369. *446_3*

Sharpe, M. J. *see also* [**162, 163**]

[480] Shepp, L. A. (1962) Symmetric random walk, *Trans. Amer. Math. Soc.* **104**, 144–153. *263,265,272_3,473*

[481] Shepp, L. A. (1964) Recurrent random walks with arbitrary large steps, *Bull. Amer. Math. Soc.* **70**, 540–542. *263,269,272*

Shi, Z. *see* [**176**]

[482] Shieh, N.-R. (1998) Multiple points of dilation-stable Lévy processes, *Ann. Probab.* **26**, 1341–1355. *382*

Shieh, N.-R. *see also* [**308**]

[483] Shiga, T. (1990) A recurrence criterion for Markov processes of Ornstein–Uhlenbeck type, *Probab. Theory Related Fields* **85**, 425–447. *272*

[484] Shimizu, R. (1970) On the domain of partial attraction of semi-stable distribution, *Ann. Inst. Statist. Math.* **22**, 245–255. *117*

[485] *Shimura, T. and Watanabe, Toshiro (2005) Infinite divisibility and generalized subexponentiality, *Bernoulli* **11**, 445–469. *453*

Shiryaev, A. N. (Shiryayev) *see* [**230, 344**]

Shreve, S. E. *see* [**258**]

[486] Shtatland, E. S. (1965) On local properties of processes with independent increments, *Theory Probab. Appl.* **10**, 317–322. *323,383*

Siegel, G. *see* [**424**].

[487] Silverstein, M. L. (1980) Classification of coharmonic and coinvariant functions for a Lévy process, *Ann. Probab.* **8**, 539–575. *384*

Silverstein, M. L. *see also* [**355**]

[488] Sirao, T. (1953) On some asymptotic properties concerning homogeneous differential processes, *Nagoya Math. J.* **6**, 95–107. *368*

Sirao, T. *see also* [**83**]

[489] Skorohod, A. V. (1954) Asymptotic formulas for stable distribution laws, *Dokl. Akad. Nauk SSSR* **98**, 731–734 (in Russian). [English translation: *Selected Transl. Math. Statist. and Probab.*, Vol. 1 (AMS, Providence, RI, 1961), 157–161.] *88*

[490] Skorokhod, A. V. (1957) On the differentiability of measures which correspond to stochastic processes, I. Processes with independent increments, *Theory Probab. Appl.* **2**, 407–432. *217,236,440*

[491] Skorokhod, A. V. (1964) *Random Processes with Independent Increments,* Nauka, Moscow (in Russian). [English translaton: *Theory of Random Processes*, National Lending Library for Science and Technology, Boston Spa, Yorkshire, England, 1971.] *29,68,440*

[492] Skorokhod, A. V. (1965) *Studies in the Theory of Random Processes,* Addison-Wesley, Reading, Mass. [Russian original 1961] *217,236,440*

[493] Skorohod, A. V. (1986) *Random Processes with Independent Increments,* 2nd ed., Nauka, Moscow (in Russian). [English translation: Kluwer Academic Pub., Dordrecht, Netherlands, 1991.] *29,68,144,342,383,464*

Skorohod, A. V. (Skorokhod) *see also* [**164, 165**]

Solomyak, B. *see* [**369**]

Song, R. *see* [**472**]

[494] Spitzer, F. (1956) A combinatorial lemma and its application to probability theory, *Trans. Amer. Math. Soc.* **82**, 323–339. *333,363,373*

[495] Spitzer, F. (1958) Some theorems concerning 2-dimensional Brownian motion, *Trans. Amer. Math. Soc.* **87**, 187–197. *236,305,360*

[496] Spitzer, F. (1964) *Principles of Random Walk,* Van Nostrand, Princeton, NJ. (2nd ed. Springer, New York, 1976.) *254,332,333,383,384*

Sreehari, M. *see* [**475, 476**]

Stegun, I. A. *see* [**1**]

[497] Stein, E. M. and Weiss, G. (1971) *Introduction to Fourier Analysis on Euclidean Spaces,* Princeton Univ. Press, Princeton, NJ. *174_2*

[498] Steutel, F. W. (1967) Note on the infinite divisibility of exponential mixtures, *Ann. Math. Statist.* **38**, 1303–1305. *426*

[499] Steutel, F. W. (1970) *Preservation of Infinite Divisibility under Mixing and Related Topics,* Math. Centre Tracts, No. 33, Math. Centrum, Amsterdam. *46_2,389,390,425,426_4,478*

[500] Steutel, F. W. (1973) Some recent results in infinite divisibility, *Stoch. Proc. Appl.* **1**, 125–143. *46*

[501] Steutel, F. W. (1979) Infinite divisibility in theory and practice, *Scand. J. Statist.* **6**, 57–64. *47*

[502] Steutel, F. W. and Harn, K. van (1979) Discrete analogues of self-decomposability and stability, *Ann. Probab.* **7**, 893–899. *426*

[503] *Steutel, F. W. and Harn, K. van (2004) *Infinite Divisibility of Probability Distributions on the Real Line*, Marcel Dekker, New York. *ix,452*

Steutel, F. W. *see also* [**56, 270**]

[504] Stone, C. J. (1969) The growth of a random walk, *Ann. Math. Statist.* **40**, 2203–2206. *368*

[505] Stone, C. J. (1969) On the potential operator for one-dimensional recurrent random walks, *Trans. Amer. Math. Soc.* **136**, 413–426. *254*

Stone, C. J. *see also* [**385**]–[**387**]

[506] Strassen, V. (1964) An invariance principle for the law of the iterated logarithm, *Zeit. Wahrsch. Verw. Gebiete* **3**, 211–226. *368*

[507] Stroock, D. W. (1993) *Probability Theory, an Analytic View,* Cambridge Univ. Press, Cambridge. *368*

[508] Sugitani, S. (1979) On the limit distributions of decomposable Galton–Watson processes, *Proc. Japan Acad.* **55**, Ser. A, 334–336. *426*

[509] Sugitani, S. (1979) On the smoothness of infinitely divisible distributions corresponding to some ordinary differential equations, *Proc. Japan Acad.* **55**, Ser. A, 371–374. *426*

Szegö, G. *see* [**379**]

[510] Takács, L. (1967) *Combinatorial Methods in the Theory of Stochastic Processes,* Wiley, New York. *384*

[511] Takada, T. (1974) On potential densities of one-dimensional Lévy processes, *J. Math. Kyoto Univ.* **14**, 371–390. *378*

[512] Takahashi, Y. (1990) Absolute continuity of Poisson random fields, *Publ. Res. Inst. Math. Sci. Kyoto Univ.* **26**, 629–647. *236*

[513] Takano, K. (1988) On the Lévy representation of the characteristic function of the probability distribution $Ce^{-|x|}dx$, *Bull. Fac. Sci. Ibaraki Univ.*, Ser. A **20**, 61–65. *47*

[514] Takano, K. (1989) The Lévy representation of the characteristic function of the probability density $\Gamma(m+\frac{d}{2})\{\pi^{d/2}\Gamma(m)\}^{-1}(1+|x|^2)^{-m-d/2}$, *Bull. Fac. Sci. Ibaraki Univ.*, Ser. A **21**, 21–27. *47,98*

[515] Takano, K. (1989,1990) On mixtures of the normal distribution by the generalized gamma convolutions, *Bull. Fac. Sci. Ibaraki Univ.*, Ser. A **21**, 29–41; Correction and addendum, **22**, 49–52. *236*

Takeda, M. *see* [**156**]

[516] Takeuchi, J. (1964) On the sample paths of the symmetric stable processes in spaces, *J. Math. Soc. Japan* **16**, 109–127. *380*

[517] Takeuchi, J. (1964) A local asymptotic law for the transient stable process, *Proc. Japan Acad.* **40**, 141–144. *361*

[518] Takeuchi, J. and Watanabe, S. (1964) Spitzer's test for the Cauchy process on the line, *Zeit. Wahrsch. Verw. Gebiete* **3**, 204–210. *361*

[519] Takeuchi, J., Yamada, T. and Watanabe, S. (1962) *Antei Katei* (Stable processes), Seminar on Probability, Vol. 13, published by Kakuritsuron Seminar, Japan (in Japanese). *117,271,330*

[520] Takeuchi, J., Yamada, T. and Watanabe, S. (1963) *Kahou Katei* (Additive processes), Kakuritsuron no Tebiki (Guide to Probability Theory), Vol. 3, published by Kakuritsuron Seminar, Japan (in Japanese). *117*

[521] Tanaka, H. (1989) Time reversal of random walks in one dimension, *Tokyo J. Math.* **12**, 159–174. *384*

[522] Tanaka, H. (1993) Green operators of absorbing Lévy processes on the half line, *Stochastic Processes. A Festschrift in Honour of Gopinath Kallianpur* (ed. S. Cambanis et al., Springer, New York), 313–319. *384*

[523] Tanaka, H. (1993) Superharmonic transform of absorbing Lévy processes, *Inst. Statist. Math. Cooperative Research Report,* No. 51, 13–25 (in Japanese). *384*

Taqqu, M. S. *see* [**427**]

Taylor, H. M. *see* [**260, 261**]

[524] Taylor, S. J. (1964) The exact Hausdorff measure of the sample path for planar Brownian motion, *Proc. Cambridge Phil. Soc.* **60**, 253–258. *379*

[525] Taylor, S. J. (1966) Multiple points for the sample paths of the symmetric stable process, *Zeit. Wahrsch. Verw. Gebiete* **5**, 247–264. *313,380*

[526] Taylor, S. J. (1967) Sample path properties of a transient stable process, *J. Math. Mech.* **16**, 1229–1246. *193,361*

[527] Taylor, S. J. (1972) Exact asymptotic estimates of Brownian path variation, *Duke Math. J.* **39**, 219–241. *381*

[528] Taylor, S. J. (1973) Sample path properties of processes with stationary independent increments, *Stochastic Analysis* (ed. D. G. Kendall and E. F. Harding, Wiley, New York), 387–414. *117,381*

[529] Taylor, S. J. (1986) The measure theory of random fractals, *Math. Proc. Cambridge Phil. Soc.* **100**, 383–406. *117,381*

[530] Taylor, S. J. and Tricot, C. (1985) Packing measure, and its evaluation for a Brownian path, *Trans. Amer. Math. Soc.* **288**, 679–699. *379*

[531] Taylor, S. J. and Wendel, J. G. (1966) The exact Hausdorff measure of the zero set of a stable process, *Zeit. Wahrsch. Verw. Gebiete* **6**, 170–180. *380*

Taylor, S. J. *see also* [**11, 86, 153, 235, 365, 370**], [**399**]–[**405**]

Teicher, H. *see* [**119**]

Teugels, J. L. *see* [**38, 473**]

Theodorescu, R. *see* [**197**]

Thorbjørnsen, S. *see* [**16**]

[532] Thorin, O. (1977) On the infinite divisibility of the Pareto distribution, *Scand. Actuarial J.* **1977**, 31–40. *46,438,480*

[533] Thorin, O. (1977) On the infinite divisibility of the lognormal distribution, *Scand. Actuarial J.* **1977**, 121–148. *47,438*

[534] *Thorin, O. (1978) An extension of the notion of a generalized Γ-convolution, *Scand. Actuarial J.* **1978**, 141–149. *438*

[535] *Thu, Nguyen Van (1986) An alternative approach to multiply self-decomposable probability measures on Banach spaces, *Probab. Theory Related Fields* **72**, 35–54. *439*

[536] *Tortrat, A. (1988) Le support des lois indéfiniment divisibles dans un groupe Abélien localement compact, *Math. Z.* **197**, 231–250. *446*

Tran, L. T. *see* [**351**]

Tricot, C. *see* [**530**]

[537] Trotter, H. F. (1958) A property of Brownian motion paths. *Illinois J. Math.* **2**, 425–433. *328*

[538] Tucker, H. G. (1962) Absolute continuity of infinitely divisible distributions, *Pacific J. Math.* **12**, 1125–1129. *196*

[539] Tucker, H. G. (1964) On continuous singular infinitely divisible distribution functions, *Ann. Math. Statist.* **35**, 330–335. *196*

[540] Tucker, H. G. (1965) On a necessary and sufficient condition that an infinitely divisible distribution be absolutely continuous, *Trans. Amer. Math. Soc.* **118**, 316–330. *184_2,196*

[541] Tucker, H. G. (1975) The supports of infinitely divisible distribution functions, *Proc. Amer. Math. Soc.* **49**, 436–440. *196*

Tucker, H. G. *see also* [**69**]

Ueda, Y. *see* [**460**]

Uehara, M. *see* [**257**]

[542] Urbanik, K. (1969) Self-decomposable probability distributions on R^m, *Zastos Mat.* **10**, 91–97. *118*

[543] Urbanik, K. (1972) Lévy's probability measures on Euclidean spaces, *Studia Math.* **44**, 119–148. *118*

[544] Urbanik, K. (1972) Slowly varying sequences of random variables, *Bull. Acad. Polonaise Sci. Sér. Sci. Math. Astronom. Phys.* **20**, 679–682. *118,437,439,480*

[545] *Urbanik, K. (1973) Limit laws for sequences of normed sums satisfying some stability conditions, *Multivariate Analysis–III* (ed. P. R. Krishnaiah, Academic Press, New York), 225–237. *437*

[546] *Urbanik, K. and Woyczyński, W. A. (1967) A random integral and Orlicz spaces, *Bull. Acad. Polon. Sci.* **15**, 161–169. *431*

Varadarajan, V. S. *see* [**142**]

Veraverbeke, N. *see* [**123**]

Vervaat, W. *see* [**246**]

Vitale, R. A. *see* [**388**]

Vondraček, Z. *see* [**472**]

[547] Watanabe, S. (1962) On stable processes with boundary conditions, *J. Math. Soc. Japan* **14**, 170–198. *384*

[548] Watanabe, S. (1968) A limit theorem of branching processes and continuous state branching processes, *J. Math. Kyoto Univ.* **8**, 141–167. *236*

Watanabe, S. *see also* [**213**]–[**215**], [**265, 300**], [**518**]–[**520**]

[549] Watanabe, Takesi (1972) Some potential theory of processes with stationary independent increments by means of the Schwartz distribution theory, *J. Math. Soc. Japan* **24**, 213–231. *332*

[550] Watanabe, Takesi (1979) Infinitely divisible distributions and ordinary differential equations, *Proc. Japan Acad.* **55**, Ser. A, 375–378. *426*

[551] Watanabe, Toshiro (1989) Non-symmetric unimodal Lévy processes that are not of class L, *Japan. J. Math.* **15**, 191–203. *417*

[552] Watanabe, Toshiro (1991) On the strong unimodality of Lévy processes, *Nagoya Math. J.* **121**, 195–199. *423*

[553] Watanabe, Toshiro (1992) On unimodal Lévy processes on the nonnegative integers, *J. Math. Soc. Japan* **44**, 239–250. 426_2

[554] Watanabe, Toshiro (1992) On Yamazato's property of unimodal one-sided Lévy processes, *Kodai Math. J.* **15**, 50–64. 418_2,*426*

[555] Watanabe, Toshiro (1992) Sufficient conditions for unimodality of non-symmetric Lévy processes, *Kodai Math. J.* **15**, 82–101. *417*

[556] Watanabe, Toshiro (1993) Oscillation of modes of some semi-stable Lévy processes, *Nagoya Math. J.* **132**, 141–153. *419,462*

[557] Watanabe, Toshiro (1994) Some examples on unimodality of Lévy processes, *Kodai Math. J.* **17**, 38–47. 418_2,*419*

[558] Watanabe, Toshiro (1996) Sample function behavior of increasing processes of class L, *Probab. Theory Related Fields* **104**, 349–374. *117*

[559] Watanabe, Toshiro (1998) Sato's conjecture on recurrence conditions for multidimensional processes of Ornstein–Uhlenbeck type, *J. Math. Soc. Japan* **50**, 155–168. *272*

[560] Watanabe, Toshiro (2000) Absolute continuity of some semi-selfdecomposable distributions and self-similar measures, *Probab. Theory Related Fields* **117**, 387–405. *184*,196_2,*455*

[561] Watanabe, Toshiro (1999) On Bessel transforms of multimodal increasing Lévy processes, *Japan. J. Math.* **25**, 227–256. *420,421*,422_3,*423,426*

[562] *Watanabe, Toshiro (2000) Continuity properties of distributions with some decomposability, *J. Theoretic. Probab.* **13**, 169–191. *455*

[563] *Watanabe, Toshiro (2007) Asymptotic estimates of multi-dimensional stable densities and their applications, *Trans. Amer. Math. Soc.* **359**, 2851–2879. *ix,448*,449_2

[564] *Watanabe, Toshiro (2012) Asymptotic properties of Fourier transforms of b-decomposable distributions, *J. Fourier Anal. Appl.* **18**, 803–827. 455_2

[565] *Watanabe, Toshiro and Yamamuro, K. (2009) Local subexponentiality of infinitely divisible distributions, *J. Math-for-Ind.* **1**, 81–90. *450,451,453*

[566] *Watanabe, Toshiro and Yamamuro, K. (2010) Local subexponentiality and selfdecomposability, *J. Theoretic. Probab.* **23**, 1039–1067. 450_2,*451*,452_2,*453*

[567] *Watanabe, Toshiro and Yamamuro, K. (2010) Ratio of the tail of an infinitely divisible distribution on the line to that of its Lévy measure, *Electr. J. Probab.* **15**, 44–74. *453*

[568] *Watanabe, Toshiro and Yamamuro, K. (2012) Tail behaviors of semi-stable distributions, *J. Math. Anal. Appl.* **393**, 108–121. *456*

Watanabe, Toshiro *see also* [**461**]–[**464**], [**485**]

Watson, A. *see* [**304**]

[569] Watson, G.N. (1944) *A Treatise on the Theory of Bessel Functions,* Cambridge Univ. Press, Cambridge. *204*

[570] Wee, I.-S. (1988) Lower functions for processes with stationary independent increments, *Probab. Theory Related Fields* **77**, 551–566. *361*

[571] Wee, I.-S. (1990) Lower functions for asymmetric Lévy processes, *Probab. Theory Related Fields* **85**, 469–488. *361*

Weiss, G. *see* [**497**]

Wendel, J.G. *see* [**531**]

Weron, A. *see* [**238**]

[572] Widder, D.V. (1946) *The Laplace Transform,* Princeton Univ. Press, Princeton, NJ. *388*

[573] Wiener, N. (1923) Differential-space, *J. Math. and Phys.* **2**, 131–174. (Reprinted in *Norbert Wiener: Collected Works*, Vol. 1, MIT Press, Cambridge, Mass., 1976.) *21*

[574] Wiener, N. and Hopf, E. (1931) Über eine Klasse singulärer Integralgleichungen, *Sitzber. Deutsch. Akad. Wiss. Berlin, Kl. Math. Phys. Tech.* **1931**, 696–706. (Reprinted in *Norbert Wiener: Collected Works*, Vol. 3, MIT Press, Cambridge, Mass., 1981.) *334*

Wießner, M. *see* [**157**]

[575] Wintner, A. (1936) On a class of Fourier transforms, *Amer. J. Math.* **58**, 45–90. *195*

[576] Wintner, A. (1956) Cauchy's stable distributions and an "explicit formula" of Mellin, *Amer. J. Math.* **78**, 819–861. *403*

Wintner, A. *see also* [**183, 184, 240**]

[577] Wolfe, S.J. (1971) On the unimodality of L functions, *Ann. Math. Statist.* **42**, 912–918. *147,404*

[578] Wolfe, S.J. (1971) On the continuity of L functions, *Ann. Math. Statist.* **42**, 2064–2073. *193*

[579] Wolfe, S.J. (1975) On moments of probability distribution functions, *Fractional Calculus and Its Applications* (ed. B. Ross, Lect. Notes in Math. No. 457, Springer, Berlin), 306–316. *163*

[580] Wolfe, S.J. (1978) On the unimodality of mutivariate symmetric distribution functions of class L, *J. Multivar. Anal.* **8**, 141–145. *117*

[581] Wolfe, S.J. (1978) On the unimodality of infinitely divisible distribution functions, *Zeit. Wahrsch. Verw. Gebiete* **45**, 329–335. *147,416,422,425,480*

[582] Wolfe, S.J. (1982) On a continuous analogue of the stochastic difference equation $X_n = \rho X_{n-1} + B_n$, *Stoch. Proc. Appl.* **12**, 301–312. *118*

[583] Wolfe, S.J. (1983) Continuity properties of decomposable probability measures on Euclidean spaces, *J. Multivar. Anal.* **13**, 534–538. *194,196*

Wolfowitz, J. *see* [**118**]

Woyczyński, W.A. *see* [**302, 546**]

[584] Yamada, T. (1985) On the fractional derivative of Brownian local times, *J. Math. Kyoto Univ.* **25**, 49–58. *328*

Yamada, T. *see also* [**519, 520**]

[585] Yamamuro, K. (1995) On modality of Lévy processes corresponding to mixtures of two exponential distributions, *Proc. Japan Acad.* Ser. A, **71**, 98–100. *423*

[586] Yamamuro, K. (1998) On transient Markov processes of Ornstein–Uhlenbeck type, *Nagoya Math. J.* **149**, 19–32. *263*

[587] Yamamuro, K. (2000) Transience conditions for self-similar additive processes, *J. Math. Soc. Japan* **52**, 343–362. *272*

Yamamuro, K. *see also* [**461, 465**], [**565**]–[**568**]

[588] Yamazato, M. (1978) Unimodality of infinitely divisible distribution functions of class L, *Ann. Probab.* **6**, 523–531. 400_2,404_2

[589] Yamazato, M. (1982) On strongly unimodal infinitely divisible distributions, *Ann. Probab.* **10**, 589–601. 424_2,*425,426*

[590] Yamazato, M. (1983) Absolute continuity of operator-self-decomposable distributions on R^d, *J. Multivar. Anal.* **13**, 550–560. *196*

[591] Yamazato, M. (1990) Hitting time distributions of single points for 1-dimensional generalized diffusion processes, *Nagoya Math. J.* **119**, 143–172. *426*

[592] Yamazato, M. (1992) Characterization of the class of hitting time distributions of 1-dimensional generalized diffusion processes, *Probability Theory and Mathematical Statistics, Proc. Sixth USSR–Japan Symp.* (ed. A. N. Shiryaev et al., World Scientific, Singapore), 422–428. *426*

[593] Yamazato, M. (1994) Absolute continuity of transition probabilities of multidimensional processes with stationary independent increments, *Theory Probab. Appl.* **39**, 347–354. *196*

[594] Yamazato, M. (1995) On strongly unimodal infinitely divisible distributions of class CME, *Theory Probab. Appl.* **40**, 518–532. *424*

Yamazato, M. *see also* [**461, 462**], [**466**]–[**470**]

[595] Yor, M. (1982) Sur la transformée de Hilbert des temps locaux browniens et une extension de la formule d'Itô, *Séminaire de Probabilités XVI 1980/81* (ed. J. Azéma and M. Yor, Lect. Notes in Math. No. 920, Springer, Berlin), 238–247. *328*

[596] Yor, M. (1992) *Some Aspects of Brownian Motion, Part I: Some Special Functionals*, Birkhäuser, Basel. *47*

Yor, M. *see also* [**32, 237, 239, 341, 377, 415**]

[597] Yosida, K. (1948) On the differentiability and the representation of one-parameter semi-group of linear operators, *J. Math. Soc. Japan* **1**, 15–21. *206*

[598] Yosida, K. (1968) The existence of the potential operator associated with an equicontinuous semi-group of class (C_0), *Studia Math.* **31**, 531–533. *212*

[599] Zabczyk, J. (1970) Sur la théorie semi-classique du potentiel pour les processus à accroissements indépendants, *Studia Math.* **35**, 227–247. *332*

[600] *Zakusilo, O. (1978) Some properties of the class L_c of limit distributions, *Theory Probab. Math. Statist.* **15**, 67–72. *455*

[601] Zolotarev, V. M. (1954) Expression of the density of a stable distribution with exponent α greater than one by means of a frequency with exponent $1/\alpha$, *Dokl. Akad. Nauk SSSR* **98**, 735–738 (in Russian). [English translation: *Selected Transl. Math. Statist. and Probab.*, Vol. 1 (AMS, Providence, RI, 1961), 163–167.] *90*

[602] Zolotarev, V. M. (1957) Mellin–Stieltjes transforms in probability theory, *Theory Probab. Appl.* **2**, 433–460. *194*

[603] Zolotarev, V. M. (1958) Distribution of the superposition of infinitely divisible processes, *Theory Probab. Appl.* **3**, 185–188. 236_2

[604] Zolotarev, V. M. (1963) The analytic structure of infinitely divisible laws of class L, *Litovsk. Mat. Sb.* **3**, 123–140 (in Russian). *193*

[605] Zolotarev, V. M. (1964) Analog of the iterated logarithm law for semi-continuous stable processes, *Theory Probab. Appl.* **9**, 512–513. *359*

[606] Zolotarev, V. M. (1964) The first-passage time of a level and the behavior at infinity for a class of processes with independent increments, *Theory Probab. Appl.* **9**, 653–662. *383*

[607] Zolotarev, V. M. (1965) Asymptotic behavior of the distributions of processes with independent increments, *Theory Probab. Appl.* **10**, 28–44. *196*

[608] Zolotarev, V. M. (1986) *One-Dimensional Stable Distributions,* Amer. Math. Soc., Providence, RI. [Russian original 1983] *88,89,117,194,*416_2*,462*

Subject index

Underlined numbers show the pages where definitions are given.